Lecture Notes in Artificial Intelligence   5753

Edited by R. Goebel, J. Siekmann, and W. Wahlster

Subseries of Lecture Notes in Computer Science

Esra Erdem   Fangzhen Lin
Torsten Schaub (Eds.)

Logic Programming and Nonmonotonic Reasoning

10th International Conference, LPNMR 2009
Potsdam, Germany, September 14-18, 2009
Proceedings

 Springer

Series Editors

Randy Goebel, University of Alberta, Edmonton, Canada
Jörg Siekmann, University of Saarland, Saarbrücken, Germany
Wolfgang Wahlster, DFKI and University of Saarland, Saarbrücken, Germany

Volume Editors

Esra Erdem
Sabanci University
Faculty of Engineering and Natural Sciences
Orhanli, Tuzla, 34956 Istanbul, Turkey
E-mail: esraerdem@sabanciuniv.edu

Fangzhen Lin
Hong Kong University of Science and Technology
Department of Computer Science and Engineering
Clear Water Bay, Kowloon, Hong Kong
E-mail: flin@cs.ust.hk

Torsten Schaub
Universität Potsdam
Institut für Informatik
August-Bebel-Str. 89
14482 Potsdam, Germany
E-mail: torsten@cs.uni-potsdam.de

Library of Congress Control Number: 2009933627

CR Subject Classification (1998): I.2.3, I.2.4, F.1.1, F.4.1, D.1.6, G.2

LNCS Sublibrary: SL 7 – Artificial Intelligence

ISSN   1867-8211
ISBN-10   3-642-04237-6 Springer Berlin Heidelberg New York
ISBN-13   978-3-642-04237-9 Springer Berlin Heidelberg New York

springer.com

© Springer-Verlag Berlin Heidelberg 2009
Printed in Germany

Typesetting: Camera-ready by author, data conversion by Scientific Publishing Services, Chennai, India
Printed on acid-free paper   SPIN: 12752607   06/3180   5 4 3 2 1 0

Preface

This volume contains the proceedings of the 10th International Conference on Logic Programming and Nonmonotonic Reasoning (LPNMR 2009), held during September 14–18, 2009 in Potsdam, Germany.

LPNMR is a forum for exchanging ideas on declarative logic programming, nonmonotonic reasoning and knowledge representation. The aim of the conference is to facilitate interaction between researchers interested in the design and implementation of logic-based programming languages and database systems, and researchers who work in the areas of knowledge representation and nonmonotonic reasoning. LPNMR strives to encompass theoretical and experimental studies that have led or will lead to the construction of practical systems for declarative programming and knowledge representation.

The special theme of LPNMR 2009 was "Applications of Logic Programming and Nonmonotonic Reasoning" in general and "Answer Set Programming (ASP)" in particular. LPNMR 2009 aimed at providing a comprehensive survey of the state of the art of ASP/LPNMR applications.

The special theme was reflected by dedicating an entire day of the conference to applications. Apart from special sessions devoted to original and significant ASP/LPNMR applications, we solicited contributions providing an overview of existing successful applications of ASP/LPNMR systems. The presentations on applications were accompanied by two panels, one on existing and another on future applications of ASP/LPNMR.

The theme of the conference was also reflected in our invited talks given by:

- Armin Biere (Johannes Kepler University, Austria)
- Alexander Bockmayr (Freie Universität Berlin, Germany)
- Ilkka Niemelä (Helsinki University of Technology, Finland)

The conference was accompanied by several workshops and hosted the award ceremony of the Second ASP Competition, run prior to the conference by Marc Denecker's research group at the University of Leuven, Belgium.

LPNMR 2009 received 75 submissions, of which 55 were technical, 8 original applications, 9 system descriptions and 3 short papers. Out of these, we accepted 25 technical, 4 original applications, 10 system descriptions, and 13 short papers. We additionally received 13 summaries of existing successful application papers, which were handled by the Application Chair.

Finally, we would like to thank all members of the Program and Organizing Committee as well as all further reviewers and people locally supporting the organization of LPNMR 2009 at the University of Potsdam. A special thanks to Yi Zhou for his help in checking the copyright forms.

July 2009

Fangzhen Lin
Torsten Schaub
Esra Erdem

Conference Organization

Program Chairs

Fangzhen Lin and Torsten Schaub

Application Chair

Esra Erdem

Competition Chair

Marc Denecker

Program Committee

José Júlio Alferes	Andrea Formisano	Enrico Pontelli
Chitta Baral	Michael Gelfond	Chiaki Sakama
Leopoldo Bertossi	Giovambattista Ianni	John Schlipf
Richard Booth	Tomi Janhunen	Tran Cao Son
Gerhard Brewka	Antonis Kakas	Evgenia Ternovska
Pedro Cabalar	Joohyung Lee	Hans Tompits
Stefania Costantini	Nicola Leone	Francesca Toni
Marina De Vos	Vladimir Lifschitz	Mirosław Truszczyński
James Delgrande	Jorge Lobo	Kewen Wang
Marc Denecker	Robert Mercer	Yisong Wang
Yannis Dimopoulos	Pascal Nicolas	Stefan Woltran
Jürgen Dix	Ilkka Niemelä	Jia-Huai You
Agostino Dovier	Mauricio Osorio	Yan Zhang
Thomas Eiter	Ramon Otero	Yi Zhou
Wolfgang Faber	David Pearce	

Additional Reviewers

Mario Alviano	Sandeep Chintabathina	Michael Fink
Peter Antonius	Raffaele Cipriano	Jorge Gonzalez
Federico Banti	Sylvie Coste-Marquis	Gianluigi Greco
Tristan Behrens	Carlos Damasio	Stijn Heymans
Annamaria Bria	Minh Dao-Tran	Alberto Illobre
Francesco Calimeri	Phan Minh Dung	Katsumi Inoue
Carlos Chesevar	Jorge Fandino	Wojtek Jamroga

Tommi Junttila
Michael Koester
Thomas Krennwallner
Yuliya Lierler
Guohua Liu
Marco Maratea
Alessandra Martello
Yunsong Meng
Loizos Michael
Hootan Nakhost
Juan Carlos Nieves
Peter Novak

Johannes Oetsch
Ravi Palla
Juan Antonio
   Navarro Perez
Hou Ping
Jörg Pührer
Gabriele Puppis
Francesco Ricca
Javier Romero
Stefan Rmmele
Alessandro Saetti
Pietro Sala

Vadim Savenkov
Juan Manuel Serrano
Yi-Dong Shen
Guillermo Simari
Marco Sirianni
Giorgio Terracina
Agustin Valverde
Hanne Vlaeminck
Zhe Wang
Schueller Weinzierl
Claudia Zepeda
Neng-Fa Zhou

Organizing Committee

Jürgen Brandt-Mihram
Martin Gebser
Sabine Hübner
Roland Kaminski
Benjamin Kaufmann

Max Ostrowski
Torsten Schaub
Wolfgang Severin
Sven Thiele
Peter-Uwe Zettiér

Table of Contents

Session 3. Original Application Papers

Session 4. Short Papers

Session 5. System Descriptions

Session 6. Summaries of Existing Successful Applications Papers

Session 7. Short Application Papers

Session 8 (Panel on Future Applications). Position Papers by the Panelists

System Competition. Summary of System Competition

SAT, SMT and Applications

Armin Biere

Johannes Kepler University, Linz, Austria

Abstract. SAT solving has gained tremendous interest. On the practical side there have been considerable performance improvements, due to new highly efficient algorithms, new heuristics, and optimized data structures. There are new applications and reformulations of important classical problems, mainly in the context of formal methods, where SAT solving is also applied successfully in an industrial setting. These applications range from equivalence checking, configuration, over model checking to test case generation. SAT is becoming one of the most important core technology in all these areas. Many applications actually use Satisfiability Modulo Theory (SMT), which can be seen as an extension of SAT solving. SMT has it roots in automated theorem proving. But it heavily relies on SAT technology. We discuss some key technologies in practical SAT solving, e.g. how to write a fast solver, some aspects in lifting propositional SAT technology to richer domains, how competitions can help to improve the state-of-the-art and finally touch on applications in model checking, hardware and software verification.

References

1. Biere, A.: PicoSAT essentials. Journal on Satisfiability, Boolean Modeling and Computation (JSAT) 4 (2008)
2. Biere, A.: Tutorial on model checking: Modelling and verification in computer science. In: Horimoto, K., et al. (eds.) AB 2008. LNCS, vol. 5147, pp. 16–21. Springer, Heidelberg (2008)
3. Biere, A., Heule, M., van Maaren, H., Walsh, T. (eds.): Handbook of Satisfiability. IOS Press, Amsterdam (2009)
4. Prasad, M., Biere, A., Gupta, A.: A survey on recent advances in SAT-based formal verification. Software Tools for Technology Transfer (STTT) 7(2) (2005)

E. Erdem, F. Lin, and T. Schaub (Eds.): LPNMR 2009, LNCS 5753, p. 1, 2009.
© Springer-Verlag Berlin Heidelberg 2009

Logic-Based Modeling in Systems Biology

Alexander Bockmayr

DFG-Research Center MATHEON, FB Mathematik und Informatik,
Freie Universität Berlin, Arnimallee 6, 14195 Berlin, Germany
Alexander.Bockmayr@fu-berlin.de

Abstract. Systems biology is a new interdisciplinary research field that has received considerable attention in recent years. While traditional molecular biology studies the various components of a biological system (e.g. genes, RNAs, proteins) in isolation, systems biology aims to understand how these components interact in order to perform complex biological functions.

A variety of mathematical and computational methods is currently being used to model and analyze biological systems, ranging from continuous, stochastic, and discrete to various hybrid approaches.

In this talk, we focus on logic-based methods for systems biology, which arise at two distinct levels. On the one hand, Boolean or multi-valued logics provide a natural way to represent the structure of a regulatory biological network, which is given by positive and negative interactions (i.e., activation and inhibition) between its different components. On the other hand, temporal logics (e.g. CTL or LTL) may be used to reason about the dynamics of a biological system, represented by a state transition graph or Kripke model.

E. Erdem, F. Lin, and T. Schaub (Eds.): LPNMR 2009, LNCS 5753, p. 2, 2009.

Integrating Answer Set Programming and Satisfiability Modulo Theories*

Ilkka Niemelä

Helsinki University of Technology TKK
Department of Information and Computer Science
P.O. Box 5400, FI-02015 TKK, Finland
Ilkka.Niemela@tkk.fi

Abstract. In this talk we consider the problem of integrating answer set programming (ASP) and satisfiability modulo theories (SMT). We discuss a characterization of stable models of logic programs based on Clark's completion and simple difference constraints. The characterization leads to a method of translating a ground logic program to a linear size theory in difference logic, i.e. propositional logic extended with difference constraints between two integer variables, such that stable models of the program correspond to satisfying assignments of the resulting theory in difference logic. Many of the state-of-the-art SMT solvers support directly difference logic. This opens up interesting possibilities. On one hand, any solver supporting difference logic can be used immediately without modifications as an ASP solver for computing stable models of a logic program by translating the program to a theory in difference logic. On the other hand, SMT solvers typically support also other extensions of propositional logic such as linear real and integer arithmetic, fixed-size bit vectors, arrays, and uninterpreted functions. This suggests interesting opportunities to extend ASP languages with such constraints and to provide effective solver support for the extensions. Using the translation an extended language including logic program rules and, for example, linear real arithmetic can be translated to an extension of propositional logic supported by current SMT solvers. We discuss the effectiveness of state-of-the-art SMT solvers as ASP solvers and the possibilities of developing extended ASP languages based on SMT solver technology.

---

* This work is financially supported by Academy of Finland under the project *Methods for Constructing and Solving Large Constraint Models* (grant #122399).

E. Erdem, F. Lin, and T. Schaub (Eds.): LPNMR 2009, LNCS 5753, p. 3, 2009.

How Flexible Is Answer Set Programming?
An Experiment in Formalizing Commonsense in ASP

Marcello Balduccini

Intelligent Systems, OCTO
Eastman Kodak Company
Rochester, NY 14650-2102 USA
marcello.balduccini@gmail.com

Abstract. This paper describes an exercise in the formalization of commonsense
with Answer Set Programming aimed at finding the answer to an interesting riddle, whose solution is not obvious to many people. Solving the riddle requires a
considerable amount of commonsense knowledge and sophisticated knowledge
representation and reasoning techniques, including planning and adversarial reasoning. Most importantly, the riddle is difficult enough to make it unclear, at first
analysis, whether and how Answer Set Programming or other formalisms can be
used to solve it.

1 Introduction

This paper describes an exercise in the formalization of commonsense [1,2] with Answer Set Programming (ASP) [3], aimed at solving the riddle:

*"A long, long time ago, two cowboys where fighting to marry the daughter of the OK
Corral rancher. The rancher, who liked neither of these two men to become his future
son-in-law, came up with a clever plan. A horse race would determine who would be
allowed his daughter's hand. Both cowboys had to travel from Kansas City to the OK
Corral, and the one whose horse arrived LAST would be proclaimed the winner.*

*The two cowboys, realizing that this could become a very lengthy expedition, finally
decided to consult the Wise Mountain Man. They explained to him the situation, upon
which the Wise Mountain Man raised his cane and spoke four wise words. Relieved, the
two cowboys left his cabin: They were ready for the contest!*

Which four wise words did the Wise Mountain Man speak?"

This riddle is interesting because it is easy to understand, but not trivial, and the solution is not obvious to many people. The story can be simplified in various ways without
losing the key points. The story is also entirely based upon commonsense knowledge.
The amount of knowledge that needs to be encoded is not large, which simplifies the
encoding; on the other hand, as we will see in the remainder of this paper, properly
dealing with the riddle requires various sophisticated capabilities, including modeling
direct and indirect effects of actions, encoding triggers, planning, dealing with defaults
and their exceptions, and concepts from multi-agent systems such as adversarial reasoning. The riddle is difficult enough to make it unclear, at first analysis, whether and how
ASP or other formalisms can be used to formalize the story and underlying reasoning.

E. Erdem, F. Lin, and T. Schaub (Eds.): LPNMR 2009, LNCS 5753, pp. 4–16, 2009.

In the course of this paper we will discuss how the effects of the actions involved in the story can be formalized and how to address the main issues of determining that "this could be a lengthy expedition" and of answering the final question.

We begin with a brief introduction on ASP. Next, we show how the knowledge about the riddle is encoded and how reasoning techniques can be used to solve the riddle. Finally, we draw conclusions.

2   Background

ASP [3] is a programming paradigm based on language A-Prolog [4] and its extensions [5,6,7]. In this paper we use the extension of A-Prolog called CR-Prolog [5], which allows, among other things, simplified handling of exceptions, rare events. To save space, we describe only the fragment of CR-Prolog that will be used in this paper.

Let Σ be a signature containing constant, function, and predicate symbols. Terms and atoms are formed as usual. A literal is either an atom a or its strong (also called classical or epistemic) negation $\neg a$.

A *regular rule* (rule, for short) is a statement of the form:

$$h_1 \vee \ldots \vee h_k \leftarrow l_1, \ldots, l_m, \text{not } l_{m+1}, \ldots, \text{not } l_n$$

where h_i's and l_i's are literals and *not* is the so-called *default negation*.[1] The intuitive meaning of a rule is that a reasoner, who believes $\{l_1, \ldots, l_m\}$ and has no reason to believe $\{l_{m+1}, \ldots, l_n\}$, must believe one of h_i's.

A *consistency restoring rule* (cr-rule) is a statement of the form:

$$h_1 \vee \ldots \vee h_k \xleftarrow{+} l_1, \ldots, l_m, \text{not } l_{m+1}, \ldots, \text{not } l_n$$

where h_i's and l_i's are as before. The informal meaning of a cr-rule is that a reasoner, who believes $\{l_1, \ldots, l_m\}$ and has no reason to believe $\{l_{m+1}, \ldots, l_n\}$, may believe one of h_i's, but only if strictly necessary, that is, only if no consistent set of beliefs can be formed otherwise.

A *program* is a pair $\langle \Sigma, \Pi \rangle$, where Σ is a signature and Π is a set of rules and cr-rules over Σ. Often we denote programs by just the second element of the pair, and let the signature be defined implicitly.

Given a CR-Prolog program Π, we denote the set of its regular rules by Π^r and the set of its cr-rules by Π^{cr}. By $\alpha(r)$ we denote the regular rule obtained from cr-rule r by replacing the symbol $\xleftarrow{+}$ with $\leftarrow$. Given a set of cr-rules R, $\alpha(R)$ denotes the set obtained by applying α to each cr-rule in R. The semantics of a CR-Prolog program is defined in two steps.

Definition 1. *Given a CR-Prolog program Π, a minimal (with respect to set-theoretic inclusion) set R of cr-rules of Π, such that $\Pi^r \cup \alpha(R)$ is consistent is called an* abductive support *of Π.*

Definition 2. *Given a CR-Prolog program Π, a set of literals A is an* answer set *of Π if it is an answer set of the program $\Pi^r \cup \alpha(R)$ for some abductive support R of Π.*

---

[1] We also allow the use of SMODELS style choice rules, but omit their formal definition to save space.

To represent knowledge and reason about dynamic domains, we use ASP to encode dynamic laws, state constraints and executability conditions [8]. The laws are written directly in ASP, rather than represented using an action language [9], to save space and to have a more uniform representation.

The key elements of the representation are as follows; we refer the readers to e.g. [9] for more details. The evolution of a dynamic domain is viewed as a *transition diagram*, which is encoded in a compact way by means of an *action description* consisting of dynamic laws (describing the direct effects of actions), state constraints (describing the indirect effects), and executability conditions (stating when the actions can be executed). Properties of interest, whose truth value changes over time, are represented by *fluents* (e.g., $on(block_1, block_2)$). A state of the transition diagram is encoded as a consistent and complete set of fluent literals (i.e., fluents and their negations). The truth value of a fluent f is encoded by a statement of the form $h(f, s)$, where s is an integer denoting the step in the evolution of the domain, intuitively saying that f holds at step s. The fact that f is false is denoted by $\neg h(f, s)$. Occurrences of actions are traditionally represented by expressions of the form $o(a, s)$, saying that a occurs at step s.

3   Formalizing the Riddle

The next step is to encode the knowledge about the domain of the story. To focus on the main issues, we abstract from several details and concentrate on the horse ride. The objects of interest are the two competitors (a, b), the two horses $(hrs(a), hrs(b))$, and locations $start$, $finish$, and en_route. Horse ownership is described by relation $owns$, defined by the rule $owns(C, hrs(C)) \leftarrow competitor(C)$.

The fluents of interest and their informal meanings are: $at(X, L)$, "competitor or horse X is at location L"; $riding(C, H)$, "competitor C is riding horse H"; $crossed(X)$, "competitor or horse X has crossed the finish line."

The actions of interest are $wait$, $move$ (the actor moves to the next location along the race track), and $cross$ (the actor crosses the finish line). Because this domain involves multiple actors, we represent the occurrence of actions by a relation $o(A, C, S)$, which intuitively says that action A occurred, performed by competitor C, at step S.[2]

The formalization of action $move$ deserves some discussion. Typically, it is difficult to predict who will complete a race first, as many variables influence the result of a race. To keep our formalization simple, we have chosen a rather coarse-grained model of the movements from one location to the other. Because often one horse will be faster than the other, we introduce a relation $faster(H)$, which informally says that H is the faster horse. This allows us to deal with both simple and more complex situations: when it is known which horse is faster, we encode the information as a fact. When the information is not available, we use the disjunction $faster(hrs(a)) \lor faster(hrs(b))$. Action $move$ is formalized so that, when executed, the slower horse moves from location $start$ to en_route and from en_route to $finish$. The faster horse, instead, moves

---

[2] This simple representation is justified because the domain does not include exogenous actions. Otherwise, we would have to use a more sophisticated representation, such as specifying the actor as an argument of the terms representing the actions.

from *start* directly to *finish*.[3] The direct effects of the actions can be formalized in ASP as follows:[4]

- Action *move*:

% If competitor C is at start and riding the faster horse,
% action *move* takes him to the finish line.
$h(at(C, finish), S + 1) \leftarrow$
$\qquad h(at(C, start), S),$
$\qquad h(riding(C, H), S),$
$\qquad faster(H),$
$\qquad o(move, C, S).$

% If competitor C is at start and riding the slower horse,
% action *move* takes him to location "en route."
$h(at(C, en\_route), S + 1) \leftarrow$
$\qquad h(at(C, start), S),$
$\qquad h(riding(C, H), S),$
$\qquad not\ faster(H),$
$\qquad o(move, C, S).$

% Performing *move* while "en route" takes the actor
% to the finish line.
$h(at(C, finish), S + 1) \leftarrow$
$\qquad h(at(C, en\_route), S),$
$\qquad o(move, C, S).$

% *move* cannot be executed while at the finish line.
$\leftarrow o(move, C, S), h(at(C, finish), S).$

- Action *cross*:

% Action *cross*, at the finish line, causes the actor to
% cross the finish line.
$h(crossed(C), S + 1) \leftarrow$
$\qquad o(cross, C, S),$
$\qquad h(at(C, finish), S).$

% *cross* can only be executed at the finish line.
$\leftarrow o(cross, C, S), h(at(C, L), S), L \neq finish.$
% *cross* can be executed only once by each competitor.
$\leftarrow o(cross, C, S), h(crossed(C), S).$

---

[3] More refined modeling is possible, but is out of the scope of the present discussion. However, we would like to mention the possibility of using the recent advances in integrating ASP and constraint satisfaction [7] to introduce numerical distances, speed, and to take into account parameters such as stamina in their computation.

[4] Depending upon the context, executability conditions might be needed stating that each competitor must be riding in order to perform the *move* or *cross* actions. Because the story assumes that the competitors are riding at all times, we omit such executability conditions to save space.

No rules are needed for action *wait*, as it has no direct effects. The state constraints are:

- "Each competitor or horse can only be at one location at a time."

$$\neg h(at(X, L_2), S) \leftarrow \\ h(at(X, L_1), S), \\ L_1 \neq L_2.$$

- "The competitor and the horse he is riding on are always at the same location."

$$h(at(H, L), S) \leftarrow \\ h(at(C, L), S), \\ h(riding(C, H), S).$$

$$h(at(C, L), S) \leftarrow \\ h(at(H, L), S), \\ h(riding(C, H), S).$$

It is worth noting that, in this formalization, horses do not perform actions on their own (that is, they are viewed as "vehicles"). Because of that, only the first of the two rules above is really needed. However, the second rule makes the formalization more general, as it allows one to apply it to cases when the horses can autonomously decide to perform actions (e.g., the horse suddenly moves to the next location and the rider is carried there as a side-effect).

- "Each competitor can only ride one horse at a time; each horse can only have one rider at a time."

$$\neg h(riding(X, H2), S) \leftarrow \\ h(riding(X, H1), S), \\ H1 \neq H2.$$

$$\neg h(riding(C2, H), S) \leftarrow \\ h(riding(C1, H), S), \\ C1 \neq C2.$$

- "The competitor and the horse he is riding on always cross the finish line together."

$$h(crossed(H), S) \leftarrow \\ h(crossed(C), S), \\ h(riding(C, H), S).$$

$$h(crossed(C), S) \leftarrow \\ h(crossed(H), S), \\ h(riding(C, H), S).$$

As noted for the previous group of state constraints, only the first of these two rules is strictly necessary, although the seconds increases the generality of the formalization.

The action description is completed by the law of inertia [10], in its usual ASP representation (e.g. [9]):

$$h(F, S + 1) \leftarrow h(F, S), \text{not } \neg h(F, S + 1).$$

$$\neg h(F, S + 1) \leftarrow \neg h(F, S), \text{not } h(F, S + 1).$$

4   Reasoning about the Riddle

Let us now see how action description $\mathcal{AD}$, consisting of all of the rules from the previous section, is used to reason about the riddle.

The first task that we want to be able to perform is determining the winner of the race, based upon the statement from the riddle "the one whose horse arrived LAST would be proclaimed the winner." In terms of the formalization developed so far, arriving last means being the last to cross the finish line. Encoding the basic idea behind this notion is not difficult, but attention must be paid to the special case of the two horses crossing the finish line together. Commonsense seems to entail that, if the two horses cross the line together, then they are both first. (One way to convince oneself about this is to observe that the other option is to say that both horses arrived last. But talking about "last" appears to imply that they have been preceded by some horse that arrived "first.") The corresponding definition of relations $first\_to\_cross$ and $last\_to\_cross$ is:[5]

```
% first_to_cross(H): horse H crossed the line first.
first_to_cross(H₁) ←
        h(crossed(H₁), S₂),
        ¬h(crossed(H₂), S₁),
        S₂ = S₁ + 1,
        horse(H₂), H₁ ≠ H₂.
```

```
% last_to_cross(H): horse H crossed the line last.
last_to_cross(H₁) ←
        h(crossed(H₁), S₂),
        ¬h(crossed(H₁), S₁),
        S₂ = S₁ + 1,
        h(crossed(H₂), S₁), horse(H₂), H₁ ≠ H₂.
```

Winners and losers can be determined from the previous relations and from horse ownership:

```
% C wins if his horse crosses the finish line last.
wins(C) ← owns(C, H), last_to_cross(H).
```

```
% C loses if his horse crosses the finish line first.
loses(C) ← owns(C, H), first_to_cross(H).
```

---

[5] To save space, the definitions of these relations are given for the special case of a 2-competitor race. Extending the definitions to the general case is not difficult, but requires some extra rules.

Let W be the set consisting of the definitions of $last\_to\_cross$, $first\_to\_cross$, $wins$, and $loses$. It is not difficult to check that, given suitable input about the initial state, $AD \cup W$ entails intuitively correct conclusions. For example, let σ denote the intended initial state of the riddle, where each competitor is at the start location, riding his horse:

$$h(at(a, start), 0). \quad h(at(b, start), 0).$$

$$h(riding(C, H), 0) \leftarrow$$
$$owns(C, H),$$
$$not \ \neg h(riding(C, H), 0).$$

$$\neg h(F, 0) \leftarrow not \ h(F, 0).$$

The rule about fluent $riding$ captures the intuition that normally one competitor rides his own horse, but there may be exceptions. Also notice that the last rule in σ encodes the Closed World Assumption, and provides a compact way to specify the fluents that are false in σ. Also, notice that it is not necessary to specify explicitly the location of the horses, as that will be derived from the locations of their riders by state constraints of AD. Assuming that a's horse is the faster, let $F^a = \{faster(hrs(a))\}$. Let also O^0 denote the set $\{o(a, move, 0), o(b, move, 0)\}$. It is not difficult to see that $\sigma \cup F^a \cup O^0 \cup AD \cup W$ entails:

$$\{h(at(a, finish), 1), h(at(b, en\_route), 1)\},$$

meaning that a is expected to arrive at the finish, and b at location "en route." Similarly, given

$$O^1 = \begin{cases} o(a, move, 0). & o(b, move, 0). \\ o(a, wait, 1). & o(b, move, 1). \\ o(a, wait, 2). & o(b, cross, 2). \\ o(a, cross, 3). \end{cases}$$

the theory $\sigma \cup F^a \cup O^1 \cup AD \cup W$ entails:

$$\{h(at(a, finish), 1), \ h(at(b, finish), 2),$$
$$h(crossed(a), 4), \ h(crossed(b), 3),$$
$$last\_to\_cross(hrs(a)), \ first\_to\_cross(hrs(b)),$$
$$wins(a), \ loses(b)\},$$

meaning that both competitors crossed the finish line, but b's horse crossed it first, and therefore b lost the race.

The next task of interest is to use the theory developed so far to determine that the race "could become a very lengthy expedition." Attention must be paid to the interpretation of this sentence. Intuitively, the sentence refers to the fact that none of the competitors might be able to end the race. However, this makes sense only if interpreted with commonsense. Of course sequences of actions exist that cause the race to terminate. For example, one competitor could ride his horse as fast as he can to the finish line and then cross, but that is likely to cause him to lose the race.

We believe the correct interpretation of the sentence is that we need to check if the two competitors *acting rationally* (i.e. selecting actions in order to achieve their own goal) will ever complete the race. In the remainder of the discussion, we call this the *completion problem*. Notice that, under the assumption of rational acting, no competitor will just run as fast as he can to the finish line and cross it, without paying attention to where the other competitor is.

In this paper, we will focus on addressing the completion problem from the point of view of one of the competitors. That is, we are interested in the reasoning that one competitor needs to perform to solve the problem. So, we will define a relation *me*, e.g. $me(a)$. In the remainder of the discussion, we refer to the competitor whose reasoning we are examining as "our competitor," while the other competitor is referred to as the "adversary."

The action selection performed by our competitor can be formalized using the well-known ASP planning technique (e.g., [9]) based upon a generate-and-test approach, encoded by the set $\mathcal{P}_{me}$ of rules:

$$me(a).$$

$$1\{\ o(A, C, S) : relevant(A)\ \}1 \leftarrow me(C).$$
$$\leftarrow not\ wins(C), me(C), selected\_goal(win).$$

$$relevant(wait).\ relevant(move).\ relevant(cross).$$

where the first rule informally states that the agent should consider performing any action relevant to the task (and exactly one at a time), while the second rule says that sequences of actions that do not lead our competitor to a win should be discarded (if our competitor's goal is indeed to win). Relation *relevant* allows one to specify which actions are relevant to the task at hand, thus reducing the number of combinations that the reasoner considers.

Our competitor also needs to reason about his adversary's actions. For that purpose, our competitor possesses a model of the adversary's behavior,[6] based upon the following heuristics:

- Reach the finish line;
- At the finish line, if crossing would cause you to lose, then wait; otherwise cross.
- In all other cases, wait.

This model of the adversary's behavior could be more sophisticated – for example, it could include some level of non-determinism – but even such a simple model is sufficient to solve the completion problem for this simple riddle. The heuristics are encoded by the set $\mathcal{P}_{adv}$ of triggers:[7]

---

[6] The model here is hard-coded, but could be learned, e.g. [11,12].

[7] A discussion on the use of triggers can be found in the Conclusions section.

$$my\_adversary(C_2) \leftarrow me(C_1),\ C_1 \neq C_2.$$

$$
\begin{aligned}
o(move, C, S) \leftarrow \\
\quad my\_adversary(C), \\
\quad \neg h(at(C, finish\_line), S).
\end{aligned}
$$

$$
\begin{aligned}
o(wait, C_1, S) \leftarrow \\
\quad my\_adversary(C_1), \\
\quad h(at(C_1, finish), S), \\
\quad owns(C_1, H_1),\ crossing\_causes\_first(C_1, H_1, S).
\end{aligned}
$$

$$
\begin{aligned}
o(cross, C_1, S) \leftarrow \\
\quad my\_adversary(C_1), \\
\quad h(at(C_1, finish), S), \\
\quad \neg h(crossed(C_1), S), \\
\quad owns(C_1, H_1),\ \text{not } crossing\_causes\_first(C_1, H_1, S).
\end{aligned}
$$

$$
\begin{aligned}
crossing\_causes\_first(C_1, H_1, S) \leftarrow \\
\quad h(at(C_1, finish), S), \\
\quad h(riding(C_1, H_1), S), \\
\quad \neg h(crossed(C_1), S),\ C_1 \neq C_2,\ \neg h(crossed(C_2), S).
\end{aligned}
$$

$$
\begin{aligned}
\neg o(A_2, C, S) \leftarrow \\
\quad my\_adversary(C), \\
\quad o(A_1, C, S), \\
\quad A_2 \neq A_1.
\end{aligned}
$$

$$
\begin{aligned}
o(wait, C, S) \leftarrow \\
\quad my\_adversary(C), \\
\quad \text{not } \neg o(wait, C, S).
\end{aligned}
$$

At the core of the above set of rules is the definition of relation $crossing\_causes\_first$ (C, H, S), which intuitively means that C's crossing the finish line at S would cause H to be the first horse to cross. Such a determination is made by ensuring that (1) C is at the finish line, and thus can cross it; (2) C is riding H, and thus by crossing would cause H to cross as well; and (3) no competitor has already crossed.

Now let us see how the theory developed so far can be used to reason about the completion problem. Let $\mathcal{P}$ denote the set $\mathcal{P}_{me} \cup \mathcal{P}_{adv}$. It is not difficult to see that the theory

$$\sigma \cup F^a \cup \mathcal{AD} \cup \mathcal{W} \cup \mathcal{P}$$

is inconsistent. That is, a has no way of winning if his horse is faster. Let us now show that the result does not depend upon the horse's speed. Let $F^\vee$ denote the rule

$$faster(hrs(a))\ \vee\ faster(hrs(b)).$$

which informally says that it is not known which horse is faster. The theory

$$\sigma \cup F^\vee \cup \mathcal{AD} \cup \mathcal{W} \cup \mathcal{P}$$

is still inconsistent. That is, a cannot win no matter whose horse is faster. Therefore, because our competitor is acting rationally, he is not going to take part in the race. Because the domain of the race is fully symmetrical, it is not difficult to see that b cannot win either, and therefore we will refuse to take part in the race as well.

However, that is not exactly what the statement of the completion problem talks about. The statement in fact seems to suggest that, were the competitors to take part in the race (for example, because they hope for a mistake by the opponent), they would not be able to complete the race. To model that, we allow our competitor to have two goals with a preference relation among them: the goal to win, and the goal to at least not lose, where the former is preferred to the second. The second goal formalizes the strategy of waiting for a mistake by the adversary. To introduce the second goal and the preference, we obtain $\mathcal{P}'$ from $\mathcal{P}$ by adding to it the rules:

$$selected\_goal(win) \leftarrow$$
$$not\ \neg selected\_goal(win).$$

$$\neg selected\_goal(win) \leftarrow$$
$$selected\_goal(not\_lose).$$

$$\leftarrow lose(C), me(C), selected\_goal(not\_lose).$$

$$selected\_goal(not\_lose) \overset{+}{\leftarrow} .$$

The first rule says that our competitor's goal is to win, unless otherwise stated. The second rule says that one exception to this is if the selected goal is to not lose. The constraint says that, if the competitor's goal is to not lose, all action selections causing a loss must be discarded. The last rule says that our competitor may possibly decide to select the goal to just not lose, but only if strictly necessary (i.e., if the goal of winning cannot be currently achieved).

Now, it can be shown that the theory

$$\sigma \cup F^{\vee} \cup \mathcal{AD} \cup \mathcal{W} \cup \mathcal{P}'$$

is consistent. One of its answer sets includes for example the atoms:

$$\{faster(hrs(a)),$$
$$o(wait, a, 0), \quad o(move, b, 0),$$
$$o(wait, a, 1), \quad o(move, b, 1),$$
$$o(move, a, 2), \quad o(wait, b, 2),$$
$$o(wait, a, 3), \quad o(wait, b, 3),$$
$$o(wait, a, 4), \quad o(wait, b, 4)\ \}$$

which represent the possibility that, if a's horse is faster, a and b will reach the finish line and then wait there indefinitely. To confirm that the race will not be completed, let us introduce a set of rules $\mathcal{C}$ containing the definition of completion, together with a constraint that requires the race to be completed in any model of the underlying theory:

$$completed \leftarrow h(crossed(X), S).$$
$$\leftarrow not\ completed.$$

The first rule states that the race has been completed when one competitor has crossed the finish line (the result of the race at that point is fully determined). Because the theory

$$\sigma \cup F^\vee \cup \mathcal{AD} \cup \mathcal{W} \cup \mathcal{P}' \cup \mathcal{C}$$

is inconsistent, we can conclude formally that, if the competitors act rationally, they will not complete the race.

The last problem left to solve is answering the question "Which four wise words did the Wise Mountain Man speak?" In terms of our formalization, we need to find a modification of the theory developed so far that yields the completion of the race. One possible approach is to revisit the conclusions that were taken for granted in the development of the theory. Particularly interesting are the defaults used in the encoding. Is it possible that solving the riddle lies in selecting appropriate exceptions to some defaults?

The simple formalization given so far contains only one default, the rule for fluent $riding$ in σ:

$$h(riding(C, H), 0) \leftarrow$$
$$owns(C, H),$$
$$\text{not } \neg h(riding(C, H), 0).$$

To allow the reasoner to consider the possible exceptions to this default, we add a cr-rule stating that a competitor may possibly ride the opponent's horse, although that should happen only if strictly necessary.

$$h(riding(C, H2), 0) \xleftarrow{+}$$
$$owns(C, H1),$$
$$horse(H2),$$
$$H1 \neq H2.$$

We use a cr-rule to capture the intuition that the competitors will not normally switch horses. Let σ' be obtained from σ by adding the new cr-rule. It can be shown that the theory[8]

$$\sigma' \cup F^\vee \cup \mathcal{AD} \cup \mathcal{W} \cup \mathcal{P}$$

is consistent and its unique answer set contains:

$$\{faster(hrs(b)),$$
$$h(riding(a, hrs(b)), 0), \quad h(riding(b, hrs(a)), 0),$$
$$o(move, a, 0), \quad o(move, b, 0),$$
$$o(cross, a, 1), \quad o(move, b, 1),$$
$$o(wait, a, 2), \quad o(cross, b, 2),$$
$$o(wait, a, 3), \quad o(wait, b, 3),$$
$$o(wait, a, 4), \quad o(wait, b, 4) \}$$

which encodes the answer that, if the competitors switch horses and the horse owned by b is faster, then a can win by immediately reaching the finish line and crossing it.

---

[8] The same answer is obtained by replacing $\mathcal{P}$ by $\mathcal{P}'$. However, doing that would require specifying preferences between the cr-rule just added and the cr-rule in $\mathcal{P}'$. To save space, we use $\mathcal{P}$ to answer the final question of the riddle.

In agreement with commonsense, a does not expect to win if the horse that b owns is slower. On the other hand, it is not difficult to see that b will win in that case. That is, the race will be completed no matter what.

The conclusion obtained here formally agrees with the accepted solution of the riddle: "Take each other's horse."

5   Conclusions

In this paper we have described an exercise in the use of ASP for commonsense knowledge representation and reasoning, aimed at formalizing and reasoning about an easy-to-understand but non-trivial riddle. One reason why we have selected this particular riddle, besides its high content of commonsense knowledge, is the fact that upon an initial analysis, it was unclear whether and how ASP or other formalisms could be used to solve it. Solving the riddle has required the combined use of some of the latest ASP techniques, including using consistency restoring rules to allow the reasoner to select alternative goals, and to consider exceptions to the defaults in the knowledge base as a last resort, and has shown how ASP can be used for adversarial reasoning by employing it to encode a model of the adversary's behavior.

Another possible way of solving the riddle, not shown here for lack of space, consists in introducing a *switch_horses* action, made not relevant by default, but with the possibility to use it if no solution can be found otherwise. Such action would be *cooperative*, in the sense that both competitors would have to perform it together. However, as with many actions of this type in a competitive environment, rationally acting competitors are not always expected to agree to perform the action. An interesting continuation of our exercise will consist of an accurate formalization of this solution to the riddle, which we think may yield useful results in the formalization of sophisticated adversarial reasoning. We think that this direction of research may benefit from the recent application of CR-Prolog to the formalization of negotiation described in [13].

One last note should be made regarding the use of triggers to model the adversary's behavior. We hope the present paper has shown the usefulness of this technique and the substantial simplicity of implementation using ASP. This technique has limits, however, due to the fact that an a-priori model is not always available. Intuitively, it is possible to use ASP to allow a competitor to "simulate" the opponent's line of reasoning (e.g., by using choice rules). However, an accurate execution of this idea involves solving a number of non-trivial technical issues. We plan to expand on this topic in a future paper.

References

1. McCarthy, J.: Programs with Common Sense. In: Proceedings of the Third Biannual World Automaton Congress, pp. 75–91 (1998)
2. Mueller, E.T.: Commonsense Reasoning. Morgan Kaufmann, San Francisco (2006)
3. Marek, V.W., Truszczynski, M.: Stable models and an alternative logic programming paradigm. In: The Logic Programming Paradigm: a 25-Year Perspective, pp. 375–398. Springer, Berlin (1999)
4. Gelfond, M., Lifschitz, V.: Classical negation in logic programs and disjunctive databases. New Generation Computing, 365–385 (1991)

5. Balduccini, M., Gelfond, M.: Logic Programs with Consistency-Restoring Rules. In: Doherty, P., McCarthy, J., Williams, M.A. (eds.) International Symposium on Logical Formalization of Commonsense Reasoning, March 2003. AAAI 2003 Spring Symposium Series, pp. 9–18 (2003)
6. Brewka, G., Niemela, I., Syrjanen, T.: Logic Programs with Ordered Disjunction 20(2), 335–357 (2004)
7. Mellarkod, V.S., Gelfond, M., Zhang, Y.: Integrating Answer Set Programming and Constraint Logic Programming. Annals of Mathematics and Artificial Intelligence (2008)
8. Gelfond, M., Lifschitz, V.: Action Languages. Electronic Transactions on AI 3(16) (1998)
9. Gelfond, M.: Representing knowledge in A-prolog. In: Kakas, A.C., Sadri, F. (eds.) Computational Logic: Logic Programming and Beyond, Part II. LNCS (LNAI), vol. 2408, pp. 413–451. Springer, Heidelberg (2002)
10. Hayes, P.J., McCarthy, J.: Some Philosophical Problems from the Standpoint of Artificial Intelligence. In: Meltzer, B., Michie, D. (eds.) Machine Intelligence 4, pp. 463–502. Edinburgh University Press (1969)
11. Sakama, C.: Induction from answer sets in nonmonotonic logic programs. ACM Transactions on Computational Logic 6(2), 203–231 (2005)
12. Balduccini, M.: Learning Action Descriptions with A-Prolog: Action Language C. In: Amir, E., Lifschitz, V., Miller, R. (eds.) Procs. of Logical Formalizations of Commonsense Reasoning, 2007 AAAI Spring Symposium (March 2007)
13. Son, T.C., Sakama, C.: Negotiation Using Logic Programming with Consistency Restoring Rules. In: 2009 International Joint Conferences on Artificial Intelligence, IJCAI (2009)

Splitting a CR-Prolog Program

Marcello Balduccini

Intelligent Systems Department
Eastman Kodak Company
Rochester, NY 14650-2102 USA
marcello.balduccini@gmail.com

Abstract. CR-Prolog is an extension of A-Prolog, the knowledge representation language at the core of the Answer Set Programming paradigm. CR-Prolog is based on the introduction in A-Prolog of *consistency-restoring rules* (cr-rules for short), and allows an elegant formalization of events or exceptions that are unlikely, unusual, or undesired. The flexibility of the language has been extensively demonstrated in the literature, with examples that include planning and diagnostic reasoning. In this paper we hope to provide the technical means to further stimulate the study and use of CR-Prolog, by extending to CR-Prolog the Splitting Set Theorem, one of the most useful theoretical results available for A-Prolog. The availability of the Splitting Set Theorem for CR-Prolog is expected to simplify significantly the proofs of the properties of CR-Prolog programs.

1 Introduction

In recent years, Answer Set Programming (ASP) [1,2,3], a declarative programming paradigm with roots in the research on non-monotonic logic and on the semantics of default negation of Prolog, has been shown to be a useful tool for knowledge representation and reasoning (e.g., [4,5]). The underlying language, often called A-Prolog, is expressive and has a well-understood methodology of representing defaults, causal properties of actions and fluents, various types of incompleteness, etc. Over time, several extensions of A-Prolog have been proposed, aimed at improving even further the expressive power of the language.

One of these extensions, called CR-Prolog [6], is built around the introduction of *consistency-restoring rules* (cr-rules for short). The intuitive idea behind cr-rules is that they are normally not applied, even when their body is satisfied. They are only applied if the regular program (i.e., the program consisting only of conventional A-Prolog rules) is inconsistent. The language also allows the specification of a partial preference order on cr-rules, intuitively regulating the application of cr-rules.

Among the most direct uses of cr-rules is an elegant encoding of events or exceptions that are unlikely, unusual, or undesired (and preferences can be used to formalize the relative likelihood of these events and exceptions).

The flexibility of CR-Prolog has been extensively demonstrated in the literature [6,7,8,9,10], with examples including planning and diagnostic reasoning. For example, in [6], cr-rules have been used to model exogenous actions that may occur unobserved and cause malfunctioning in a physical system. In [10], instead, CR-Prolog has been used to formalize negotiations.

E. Erdem, F. Lin, and T. Schaub (Eds.): LPNMR 2009, LNCS 5753, pp. 17–29, 2009.

To further stimulate the study and use of CR-Prolog, theoretical tools are needed that simplify the proofs of the properties of CR-Prolog programs. Arguably, one of the most important such tools for A-Prolog is the Splitting Set Theorem [11]. Our goal in this paper is to extend the Splitting Set Theorem to CR-Prolog programs.

This paper is organized as follows. In the next section, we introduce the syntax and semantics of CR-Prolog. Section 3 gives key definitions and states various lemmas as well as the main result of the paper. Section 4 discusses the importance of some conditions involved in the definition of splitting set, and gives examples of the use of the Splitting Set Theorem to split CR-Prolog programs. In Section 5 we talk about related work and draw conclusions. Finally, in Section 6, we give proofs for the main results of this paper.

2 Background

The syntax and semantics of ASP are defined as follows. Let Σ be a signature containing constant, function, and predicate symbols. Terms and atoms are formed as usual. A literal is either an atom a or its strong (also called classical or epistemic) negation $\neg a$. The *complement* of an atom a is literal $\neg a$, while the complement of $\neg a$ is a. The complement of literal l is denoted by $\bar{l}$. The sets of atoms and literals formed from Σ are denoted by $atoms(\Sigma)$ and $lit(\Sigma)$, respectively.

A *regular rule* is a statement of the form:

$$[r]\ h_1\ \text{OR}\ h_2\ \text{OR}\ \ldots\ \text{OR}\ h_k \leftarrow l_1, \ldots, l_m, \text{not}\ l_{m+1}, \ldots, \text{not}\ l_n \qquad (1)$$

where r, called *name*, is a *possibly compound* term uniquely denoting the regular rule, h_i's and l_i's are literals and *not* is the so-called *default negation*. The intuitive meaning of the regular rule is that a reasoner who believes $\{l_1, \ldots, l_m\}$ and has no reason to believe $\{l_{m+1}, \ldots, l_n\}$, must believe one of h_i's.

A *consistency-restoring rule* (or *cr-rule*) is a statement of the form:

$$[r]\ h_1\ \text{OR}\ h_2\ \text{OR}\ \ldots\ \text{OR}\ h_k \overset{+}{\leftarrow} l_1, \ldots l_m, \text{not}\ l_{m+1}, \ldots, \text{not}\ l_n \qquad (2)$$

where r, h_i's, and l_i's are as before. The intuitive reading of a cr-rule is that a reasoner who believes $\{l_1, \ldots, l_m\}$ and has no reason to believe $\{l_{m+1}, \ldots, l_n\}$, *may possibly* believe one of h_i's. The implicit assumption is that this possibility is used as little as possible, only when the reasoner cannot otherwise form a consistent set of beliefs. A preference order on the use of cr-rules is expressed by means of the atoms of the form $prefer(r_1, r_2)$. Such an atom informally says that r_2 should not be used unless there is no way to obtain a consistent set of beliefs with r_1. More details on preferences in CR-Prolog can be found in [6,12,13].

By *rule* we mean a regular rule or a cr-rule. Given a rule ρ of the form (1) or (2), we call $\{h_1, \ldots, h_k\}$ the *head* of the rule, denoted by $head(\rho)$, and $\{l_1, \ldots, l_m, \text{not}\ l_{m+1}, \ldots, \text{not}\ l_n\}$ its *body*, denoted by $body(\rho)$. Also, $pos(\rho)$ denotes $\{l_1, \ldots, l_m\}$, $neg(\rho)$ denotes $\{l_{m+1}, \ldots, l_n\}$, $name(\rho)$ denotes name r, and $lit(\rho)$ denotes the set of all literals from ρ. When $l \in lit(\rho)$, we say that l *occurs* in ρ.

A *program* is a pair $\langle \Sigma, \Pi \rangle$, where Σ is a signature and Π is a set of rules over Σ. Often we denote programs by just the second element of the pair, and let the signature be defined implicitly. In that case, the signature of Π is denoted by $\Sigma(\Pi)$.

In practice, variables are often allowed to occur in ASP programs. A rule containing variables (called a *non-ground* rule) is then viewed as a shorthand for the set of its *ground instances*, obtained by replacing the variables in it by all of the possible ground terms. Similarly, a non-ground program is viewed as a shorthand for the program consisting of the ground instances of its rules.

Given a program Π, $\mu(\Pi)$ denotes the set of names of the rules from Π. In the rest of the discussion, letter r (resp. ρ), possibly indexed, denotes the name of a rule (resp., a rule). Given a set of rule names R, $\rho(R, \Pi)$ denotes the set of rules from Π whose name is in R. $\rho(r, \Pi)$ is shorthand for $\rho(\{r\}, \Pi)$. To simplify notation, we allow writing $r \in \Pi$ to mean that a rule with name r is in Π. We extend the use of the other set operations in a similar way. Also, given a program Π, $head(r)$ and $body(r)$ denote[1] the corresponding parts of $\rho(r, \Pi)$. Given a CR-Prolog program, Π, the *regular part* of Π is the set of its regular rules, and is denoted by $reg(\Pi)$. The set of cr-rules of Π is denoted by $cr(\Pi)$. Programs that do not contain cr-rules are legal ASP programs, and their semantics is defined as usual. Next, we define the semantics of arbitrary CR-Prolog programs. Let us begin by introducing some notation.

For every $\mathcal{R} \subseteq cr(\Pi)$, $\theta(\mathcal{R})$ denotes the set of regular rules obtained from $\mathcal{R}$ by replacing every connective $\xleftarrow{+}$ with $\leftarrow$. Given a program Π and a set R of rule names, $\theta(R)$ denotes the application of θ to the rules of Π whose name is in R.

A literal l is *satisfied* by a set of literals S ($S \models l$) if $l \in S$. An expression not l is satisfied by S if $l \notin S$. The body of a rule is satisfied by S if each element of the set is satisfied by S. A set of literals S entails $prefer^*(r_1, r_2)$ ($S \models prefer^*(r_1, r_2)$) if:

- $S \models prefer(r_1, r_2)$, or
- $S \models prefer(r_1, r_3)$ and $S \models prefer^*(r_3, r_2)$.

The semantics of CR-Prolog is given in three steps.

Definition 1. Let S be a set of literals and R be a set of names of cr-rules from Π. The pair $\mathcal{V} = \langle S, R \rangle$ is a *view* of Π if:

1. S is an answer set[2] of $reg(\Pi) \cup \theta(R)$, and
2. for every r_1, r_2, if $S \models prefer^*(r_1, r_2)$, then $\{r_1, r_2\} \not\subseteq R$, and
3. for every r in R, $body(r)$ is satisfied by S.

We denote the elements of $\mathcal{V}$ by $\mathcal{V}^S$ and $\mathcal{V}^R$ respectively. The cr-rules in $\mathcal{V}^R$ are said to be *applied*. This definition of view differs from the one given in previous papers (e.g., [6,13]) in that set R here is a set of names of cr-rules rather than a set of cr-rules. The change allows one to simplify the proofs of the theorems given later. Because of the one-to-one correspondence between cr-rules and their names, the two definitions are equivalent.

---

[1] The notation can be made more precise by specifying Π as an argument, but in the present paper Π will always be clear from the context.

[2] We only consider *consistent* answer sets.

For every pair of views of Π, $\mathcal{V}_1$ and $\mathcal{V}_2$, $\mathcal{V}_1$ *dominates* $\mathcal{V}_2$ if there exist $r_1 \in \mathcal{V}_1^R$, $r_2 \in \mathcal{V}_2^R$ such that $(\mathcal{V}_1^S \cap \mathcal{V}_2^S) \models prefer^*(r_1, r_2)$.

Definition 2. A view, $\mathcal{V}$, is a *candidate answer set* of Π if, for every view $\mathcal{V}'$ of Π, $\mathcal{V}'$ does not dominate $\mathcal{V}$.

Definition 3. A set of literals, S, is an *answer set* of Π if:

1. there exists a set R of names of cr-rules from Π such that $\langle S, R \rangle$ is a candidate answer set of Π, and
2. for every candidate answer set $\langle S', R' \rangle$ of Π, $R' \not\subset R$.

3   Splitting Set Theorem

Proceeding along the lines of [11], we begin by introducing the notion of splitting set for a CR-Prolog program, and then use this notion to state the main theorems.

A *preference set for cr-rule r with respect to a set of literals S* is the set

$$\pi(r, S) = \{r' \mid S \models prefer^*(r, r') \text{ or } S \models prefer^*(r', r)\}.$$

Given a program Π, the preference set of r with respect to the literals from the signature of Π is denoted by $\pi(r)$.

Definition 4. *Literal l is relevant to cr-rule r (for short, l is r-relevant) if:*

1. *l occurs in r, or*
2. *l occurs in some rule where a literal relevant to r occurs, or*
3. *$\bar{l}$ is relevant to r, or*
4. *$l = prefer(r, r')$ or $l = prefer(r', r)$.*

Definition 5. *A* splitting set *for a program Π is a set U of literals from $\Sigma(\Pi)$ such that:*

- *for every rule $r \in \Pi$, if $head(r) \cap U \neq \emptyset$, then $lit(r) \subseteq U$;*
- *for every cr-rule $r \in \Pi$, if some $l \in U$ is relevant to r, then every r-relevant literal belongs to U.*

Observation 1. *For programs that do not contain cr-rules, this definition of splitting set coincides with the one given in [11].*

Observation 2. *For every program Π and splitting set U for Π, if $l \in U$ is r-relevant and $r' \in \pi(r)$, then every r'-relevant literal from $\Sigma(\Pi)$ belongs to U.*

We define the notions of bottom and partial evaluation of a program similarly to [11]. The bottom of a CR-Prolog program Π relative to splitting set U is denoted by $b_U(\Pi)$ and consists of every rule $\rho \in \Pi$ such that $lit(\rho) \subseteq U$. Given a program Π and a set R of names of rules, $b_U(R)$ denotes the set of rule names in $b_U(\rho(R, \Pi))$.

The *partial evaluation* of a CR-Prolog program Π w.r.t. splitting set U and set of literals X, denoted by $e_U(\Pi, X)$, is obtained as follows:

- For every rule $\rho \in \Pi$ such that $pos(\rho) \cap U$ is part of X and $neg(\rho) \cap U$ is disjoint from X, $e_U(\Pi, X)$ contains the rule ρ' such that:

$$name(\rho') = name(\rho), \quad head(\rho') = head(\rho),$$
$$pos(\rho') = pos(\rho) \setminus U, \quad neg(\rho') = neg(\rho) \setminus U.$$

- For every other rule $\rho \in \Pi$, $e_U(\Pi, X)$ contains the rule ρ' such that:

$$name(\rho') = name(\rho), \quad head(\rho') = head(\rho),$$
$$pos(\rho') = \{\bot\} \cup pos(\rho) \setminus U, \quad neg(\rho') = neg(\rho) \setminus U.$$

Given Π, U, and X as above, and a set R of names of rules from Π, $e_U(R, X)$ denotes the set of rule names in $e_U(\rho(R, \Pi), X)$.

Observation 3. *For every program Π, splitting set U and set of literals X, $e_U(\Pi, X)$ is equivalent to the similarly denoted set of rules defined in [11].*

Observation 4. *For every program Π, set R of names of rules from Π, splitting set U, and set of literals X, $R = e_U(R, X)$.*

From Observations 1 and 3, and from the original Splitting Set Theorem [11], one can easily prove the following statement.

Theorem 1 (Splitting Set Theorem from [11]). *Let U be a splitting set for a program Π that does not contain any cr-rule. A set S of literals is an answer set of Π if and only if: (i) X is an answer set of $b_U(\Pi)$; (ii) Y is an answer set of $e_U(\Pi \setminus b_U(\Pi), X)$; (iii) $S = X \cup Y$ is consistent.*

We are now ready to state the main results of this paper. Complete proofs can be found in Section 6.

Lemma 1 (Splitting Set Lemma for Views). *Let U be a splitting set for a program Π, S a set of literals, and R a set of names of cr-rules from Π. The pair $\langle S, R \rangle$ is a view of Π if and only if:*

- $\langle X, b_U(R) \rangle$ *is a view of $b_U(\Pi)$;*
- $\langle Y, R \setminus b_U(R) \rangle$ *is a view of $e_U(\Pi \setminus b_U(\Pi), X)$;*
- $S = X \cup Y$ *is consistent.*

Lemma 2 (Splitting Set Lemma for Candidate Answer Sets). *Let U be a splitting set for a program Π. A pair $\langle S, R \rangle$ is a candidate answer set of Π if and only if:*

- $\langle X, b_U(R) \rangle$ *is a candidate answer set of $b_U(\Pi)$;*
- $\langle Y, R \setminus b_U(R) \rangle$ *is a candidate answer set of $e_U(\Pi \setminus b_U(\Pi), X)$;*
- $S = X \cup Y$ *is consistent.*

Theorem 2 (Splitting Set Theorem for CR-Prolog). *Let U be a splitting set for a program Π. A consistent set of literals S is an answer set of Π if and only if:*

- X *is an answer set of $b_U(\Pi)$;*
- Y *is an answer set of $e_U(\Pi \setminus b_U(\Pi), X)$;*
- $S = X \cup Y$ *is consistent.*

4 Discussion

Now that the Splitting Set Theorem for CR-Prolog has been stated, in this section we give examples of the application of the theorem and discuss the importance of the conditions of Definition 4 upon which the definition of splitting set and the corresponding theorem rely.

Let us begin by examining the role of the conditions of Definition 4.

Consider condition (3) of Definition 4. To see why the condition is needed, consider the following program, P_1 (as usual, rule names are omitted whenever possible to simplify the notation):

$$[r_1]\ q \xleftarrow{+}\ \text{not}\ p.$$

$$s \leftarrow \text{not}\ q.$$

$$\neg s.$$

It is not difficult to see that P_1 has the unique answer set $\{q, \neg s\}$, intuitively obtained from the application of r_1. Let us now consider set $U_1 = \{q, p, s\}$. Notice that U_1 satisfies the definition of splitting set, as long as the condition under discussion is dropped from Definition 4. The corresponding $b_{U_1}(P_1)$ is:

$$[r_1]\ q \xleftarrow{+}\ \text{not}\ p.$$

$$s \leftarrow \text{not}\ q.$$

$b_{U_1}(P_1)$ has a unique answer set, $X_1^a = \{s\}$, obtained without applying r_1. $e_{U_1}(P_1 \setminus b_{U_1}(P_1), \{s\})$ is:

$$\neg s.$$

which has a unique answer set, $Y_1^a = \{\neg s\}$. Notice that $X_1^a \cup Y_1^a$ is inconsistent. Because X_1^a and Y_1^a are unique answer sets of the corresponding programs, it follows that the answer set of P_1 cannot be obtained from the any of the answer sets of $b_{U_1}(P_1)$ and of the corresponding partial evaluation of P_1. Hence, dropping condition (3) of Definition 4 causes the splitting set theorem to no longer hold.

Very similar reasoning shows the importance of condition (2): just obtain P_2 from P_1 by (i) replacing $\neg s$ in P_1 by t and (ii) adding a constraint $\leftarrow t, s$, and consider the set $U_2 = \{q, p\}$. Observe that, if the condition is dropped, then U_2 is a splitting set for P_2, but the splitting set theorem does not hold.

Let us now focus on condition (4). Consider program P_3:

$$[r_1]\ q \xleftarrow{+}\ \text{not}\ p.$$
$$[r_2]\ s \xleftarrow{+}\ \text{not}\ t.$$

$$prefer(r_2, r_1).$$

$$\leftarrow \text{not}\ q.$$
$$\leftarrow \text{not}\ s.$$

Observe that P_3 is inconsistent, the intuitive explanation being that r_1 can only be used if there is no way to use r_2 to form a consistent set of beliefs, but the only way to form such a consistent set would be to use r_1 and r_2 together. Now consider set $U_3 = \{q, p, prefer(r_2, r_1)\}$. The corresponding $b_{U_3}(P_3)$ is:

$$[r_1]\; q \xleftarrow{+} not\; p.$$

$$prefer(r_2, r_1).$$

$$\leftarrow not\; q.$$

which has a unique answer set $X_3 = \{q, prefer(r_2, r_1)\}$. The partial evaluation $e_{U_3}(P_3 \setminus b_{U_3}(P_3), X_3)$ is:

$$[r_2]\; s \xleftarrow{+} not\; t.$$
$$\leftarrow not\; s.$$

whose unique answer set is $Y_3 = \{s\}$. If condition (4) is dropped from Definition 4, then U_3 is a splitting set for P_3. However, the splitting set theorem does not hold, as P_3 is inconsistent while $X_3 \cup Y_3$ is consistent.

Let us now give a few examples of the use of the Splitting Set Theorem to finding the answer sets of CR-Prolog programs. Consider program P_4:

$$[r_1]\; q \xleftarrow{+} not\; a.$$
$$[r_2]\; p \xleftarrow{+} not\; t.$$
$$a\; \text{OR}\; b.$$
$$s \leftarrow not\; b.$$
$$\leftarrow not\; q, b.$$
$$c\; \text{OR}\; d.$$
$$u \leftarrow z, not\; p.$$
$$z \leftarrow not\; u.$$

and the set $U_4 = \{q, a, b, s\}$. It is not difficult to check the conditions and verify that U_4 is a splitting set for P_4. In particular, observe that U_4 includes the r_1-relevant literals, and does not include the r_2-relevant literals. $b_{U_4}(P_4)$ is:

$$[r_1]\; q \xleftarrow{+} not\; a.$$
$$a\; \text{OR}\; b.$$
$$s \leftarrow not\; b.$$
$$\leftarrow not\; q, b.$$

Because P_4 contains a single cr-rule, from the semantics of CR-Prolog it follows that its answer sets are those of $reg(b_{U_4}(P_4))$, if the program is consistent, and those of $reg(b_{U_4}(P_4)) \cup \theta(\{r_1\})$ otherwise. $reg(b_{U_4}(P_4))$ has a unique answer set, $X_4 = \{a, s\}$, which is, then, also the answer set of $b_{U_4}(P_4)$. The partial evaluation $e_{U_4}(P_4 \setminus b_{U_4}(P_4), X_4)$ is:

$$[r_2]\; p \xleftarrow{+} not\; t.$$
$$c\; \text{OR}\; d.$$
$$u \leftarrow z, not\; p.$$
$$z \leftarrow not\; u.$$

Again, the program contains a single cr-rule. This time, $reg(e_{U_4}(P_4 \setminus b_{U_4}(P_4), X_4))$ is inconsistent. $reg(e_{U_4}(P_4 \setminus b_{U_4}(P_4), X_4)) \cup \theta(\{r_2\})$, on the other hand, has an answer set $Y_4 = \{p, c, z\}$, which is, then, also an answer set of $e_{U_4}(P_4 \setminus b_{U_4}(P_4), X_4)$. Therefore, an answer set of P_4 is

$$X_4 \cup Y_4 = \{a, s, p, c, z\}.$$

Now the following modification of P_4, P_5:

$$[r_1]\ q \xleftarrow{+} \text{not } a.$$
$$[r_2]\ p \xleftarrow{+} \text{not } t.$$
$$a\ \text{OR}\ b.$$
$$s \leftarrow \text{not } b.$$
$$\leftarrow \text{not } q, b.$$
$$c\ \text{OR}\ d \leftarrow v.$$
$$\neg c \leftarrow \text{not } v.$$
$$u \leftarrow z, \text{not } p.$$
$$z \leftarrow \text{not } u.$$
$$v \leftarrow \text{not } w.$$

The goal of this modification is to show how rules, whose literals are not relevant to any cr-rule, can be split. Let U_5 be $\{q, a, b, s, v, w\}$. Notice that U_5 is a splitting set for P_5 even though $v \in U_5$ and P_5 contains the rule $c\ \text{OR}\ d \leftarrow v$. In fact, v is not relevant to any cr-rule from P_5, and thus c and d are not required to belong to U_5. $b_{U_5}(P_5)$ is:

$$[r_1]\ q \xleftarrow{+} \text{not } a.$$
$$a\ \text{OR}\ b.$$
$$s \leftarrow \text{not } b.$$
$$\leftarrow \text{not } q, b.$$
$$v \leftarrow \text{not } w.$$

which has an answer set $X_5 = \{a, s, v\}$. $e_{U_5}(P_5 \setminus b_{U_5}(P_5), X_5)$ is:

$$[r_2]\ p \xleftarrow{+} \text{not } t.$$
$$c\ \text{OR}\ d.$$
$$\neg c \leftarrow \bot.$$
$$u \leftarrow z, \text{not } p.$$
$$z \leftarrow \text{not } u.$$

which has an answer set $Y_5 = \{p, c, z\}$. Hence, an answer set of P_5 is

$$X_5 \cup Y_5 = \{a, s, v, p, c, z\}.$$

5   Related Work and Conclusions

Several papers have addressed the notion of splitting set and stated various versions of splitting set theorems throughout the years. Notable examples are [11], with the original formulation of the Splitting Set Theorem for A-Prolog, [14], with a Splitting Set Theorem for default theories, and [15] with a Splitting Set Theorem for epistemic specifications.

In this paper we have defined a notion of splitting set for CR-Prolog programs, and stated the corresponding Splitting Set Theorem. We hope that the availability of this theoretical result will further stimulate the study and use of CR-Prolog, by making it easier to prove the properties of the programs written in this language. As the reader may have noticed, to hold for CR-Prolog programs (that include at least one cr-rule), the Splitting Set Theorem requires substantially stronger conditions than the Splitting Set Theorem for A-Prolog. We hope that future research will allow weakening the conditions of the theorem given here, but we suspect that the need for stronger conditions is strictly tied to the nature of cr-rules.

6 Proofs

Proof of Lemma 1. To be a view of a program, a pair $\langle S, R \rangle$ must satisfy all of the requirements of Definition 1. Let us begin from item (1) of the definition. We must show that S is an answer set of $reg(\Pi) \cup \theta(R)$ if and only if:

- X is an answer set of $reg(b_U(\Pi)) \cup \theta(b_U(R))$;
- Y is an answer set of $reg(e_U(\Pi \setminus b_U(\Pi), X)) \cup \theta(R \setminus b_U(R))$.

From Theorem 1, S is an answer set of $reg(\Pi) \cup \theta(R)$ iff:

- X is an answer set of

$$
b_U(reg(\Pi) \cup \theta(R)) =
$$
$$
b_U(reg(\Pi)) \cup b_U(\theta(R)) =
$$
$$
b_U(reg(\Pi)) \cup \theta(b_U(R)) =
$$
$$
reg(b_U(\Pi)) \cup \theta(b_U(R)).
$$

- Y is an answer set of

$$
e_U((reg(\Pi) \cup \theta(R)) \setminus b_U(reg(\Pi) \cup \theta(R)), X) =
$$
$$
e_U((reg(\Pi) \cup \theta(R)) \setminus (reg(b_U(\Pi)) \cup \theta(b_U(R))), X) =
$$
$$
e_U((reg(\Pi) \setminus reg(b_U(\Pi))) \cup (\theta(R) \setminus \theta(b_U(R))), X) =
$$
$$
e_U(reg(\Pi \setminus b_U(\Pi)) \cup \theta(R \setminus b_U(R)), X) =
$$
$$
e_U(reg(\Pi \setminus b_U(\Pi)), X) \cup e_U(\theta(R \setminus b_U(R)), X) =
$$
$$
reg(e_U(\Pi \setminus b_U(\Pi), X)) \cup \theta(e_U(R \setminus b_U(R), X)) =
$$
$$
reg(e_U(\Pi \setminus b_U(\Pi), X)) \cup \theta(R \setminus b_U(R)),
$$

where the last transformation follows from Observation 4.
- $S = X \cup Y$ is consistent.

This completes the proof for item (1) of the definition of view, and furthermore concludes that $S = X \cup Y$ is consistent. Let us now consider item (2) of the definition of view. We must show that

$$\forall r_1, r_2 \in R, \text{ if } S \models prefer^*(r_1, r_2), \text{ then } \{r_1, r_2\} \not\subseteq R$$

iff

$$\forall r_1, r_2 \in b_U(R), \text{ if } X \models prefer^*(r_1, r_2), \text{ then } \{r_1, r_2\} \not\subseteq b_U(R), \text{ and}$$
$$\forall r_1, r_2 \in R \setminus b_U(R), \text{ if } Y \models prefer^*(r_1, r_2), \text{ then } \{r_1, r_2\} \not\subseteq R \setminus b_U(R).$$

Left-to-right. The statement follows from the fact that $X \subseteq S$, $Y \subseteq S$, and $b_U(R) \subseteq R$.

Right-to-left. Proceeding by contradiction, suppose that, for some $r_1, r_2 \in R$, $S \models prefer^*(r_1, r_2)$, but $\{r_1, r_2\} \subseteq R$. By definition of preference set, $r_2 \in \pi(r_1)$. Let l be some literal from $head(r_1)$. Obviously, either $l \in U$ or $l \notin U$.

Suppose $l \in U$. By the definition of splitting set, $prefer(r_i, r_j) \in U$ for every $r_i, r_j \in \{r_1\} \cup \pi(r_1)$. Moreover, for every $r' \in \pi(r_1)$ and every $l' \in head(r')$, l' belongs to U. But $r_2 \in \pi(r_1)$. Hence, $X \models prefer^*(r_1, r_2)$ and $\{r_1, r_2\} \subseteq b_U(R)$. *Contradiction.*

Now suppose $l \notin U$. With reasoning similar to the previous case, we can conclude that $prefer(r_i, r_j) \in \overline{U}$ for every $r_i, r_j \in \{r_1\} \cup \pi(r_1)$. Moreover, for every $r' \in \pi(r_1)$ and every $l' \in head(r')$, l' belongs to $\overline{U}$. Because $r_2 \in \pi(r_1)$, $Y \models prefer^*(r_1, r_2)$ and $\{r_1, r_2\} \subseteq R \setminus b_U(R)$. *Contradiction.*

This completes the proof for item (2) of the definition of view. Let us now consider item (3). We must prove that

$$\forall r \in R, body(\rho(r, \Pi)) \text{ is satisfied by } S$$

iff

$$\forall r \in b_U(R), body(\rho(r, b_U(\Pi))) \text{ is satisfied by } X, \text{ and}$$
$$\forall r \in R \setminus b_U(R), body(\rho(r, e_U(\Pi \setminus b_U(\Pi), X))) \text{ is satisfied by } Y.$$

Left-to-right. The claim follows from the following observations: (i) for every $r \in R$, either $r \in b_U(R)$ or $r \in R \setminus b_U(R)$; (ii) $body(\rho(r, e_U(\Pi \setminus b_U(\Pi), X)))$ is satisfied by Y iff $body(\rho(r, \Pi))$ is satisfied by $X \cup Y = S$.

Right-to-left. Again, observe that, for every $r \in R$, either $r \in b_U(R)$ or $r \in R \setminus b_U(R)$.

Suppose $r \in b_U(R)$. Because $X \subseteq S$, $body(\rho(r, b_U(\Pi)))$ is satisfied by S. Because $b_U(\Pi) \subseteq \Pi$, $\rho(r, b_U(\Pi)) = \rho(r, \Pi)$.

Suppose $r \in R \setminus b_U(R)$. The notion of partial evaluation is defined in such a way that, if $body(\rho(r, e_U(\Pi \setminus b_U(\Pi), X)))$ is satisfied by Y, then $body(\rho(r, \Pi))$ is satisfied by $X \cup Y = S$.

Proof of Lemma 2. From Lemma 1, it follows that $\langle S, R \rangle$ is a view of Π iff (i) $\langle X, b_U(R) \rangle$ is a view of $b_U(\Pi)$, (ii) $\langle Y, R \setminus b_U(R) \rangle$ is a view of $e_U(\Pi \setminus b_U(\Pi), X)$, and (iii) $S = X \cup Y$ is consistent. Therefore, we only need to prove that:

$$\text{no view of } \Pi \text{ dominates } \langle S, R \rangle \tag{3}$$

if and only if

$$\text{no view of } b_U(\Pi) \text{ dominates } \langle X, b_U(R) \rangle, \text{ and} \qquad (4)$$

$$\text{no view of } e_U(\Pi \setminus b_U(\Pi), X) \text{ dominates } \langle Y, R \setminus b_U(R) \rangle. \qquad (5)$$

Left-to-right. Let us prove that (3) implies (4). By contradiction, suppose that:

$$\text{there exists a view } \mathcal{V}'_X = \langle X', R'_X \rangle \text{ of } b_U(\Pi) \text{ dominates } \mathcal{V}_X = \langle X, b_U(R) \rangle. \qquad (6)$$

Let (X'_D, X'_I) be the partition of X' such that X'_D is the set of the literals from X' that are relevant to the cr-rules in $b_U(R) \cup R'_X$. Let (X_D, X_U) be a similar partition of X. From Definition 4, it is not difficult to see that $\langle X_I \cup X'_D, R'_X \rangle$ is a view of $b_U(\Pi)$. Moreover, given $R' = (R \setminus b_U(R)) \cup R'_X$, $\langle Y, R' \setminus b_U(R') \rangle$ is a view of $e_U(\Pi \setminus b_U(\Pi), X_I \cup X'_D)$. By Lemma 1, $\langle X' \cup Y, R' \rangle$ is a view of Π. From hypothesis (6), it follows that there exist $r \in b_U(R) \subseteq R$ and $r' \in b_U(R') \subseteq R'$ such that $(X \cap X') \models prefer^*(r', r)$. But then $\langle X' \cup Y, R' \rangle$ dominates $\langle S, R \rangle$. *Contradiction.*

Let us prove that (3) implies (5). By contradiction, suppose that there exists a view $\mathcal{V}'_Y = \langle Y', R'_Y \rangle$ of $e_U(\Pi \setminus b_U(\Pi), X)$ that dominates $\mathcal{V}_Y = \langle Y, R \setminus b_U(R) \rangle$. That is,

$$\text{there exist } r \in R \setminus b_U(R), r' \in R'_Y \text{ such that } (Y \cap Y') \models prefer^*(r', r). \qquad (7)$$

Let R' be $R'_Y \cup b_U(R)$. By Lemma 1, $\langle X \cup Y', R' \rangle$ is a view of Π. From (7), it follows that there exist $r \in R, r' \in R'$ such that $(Y \cap Y') \models prefer^*(r', r)$. Therefore, $\langle X \cup Y', R' \rangle$ dominates $\langle S, R \rangle$. *Contradiction.*

Next, from (4) and (5), we prove (3). By contradiction, suppose that there exists a view $\mathcal{V}' = \langle S', R' \rangle$ of Π that dominates $\mathcal{V} = \langle S, R \rangle$. That is, there exist $r \in R, r' \in R'$ such that $(S \cap S') \models prefer^*(r', r)$. There are two cases: $head(r') \subseteq U$ and $head(r') \subseteq \overline{U}$.

Case 1: $head(r') \subseteq U$. From Lemma 1 it follows that $\langle S \cap U, b_U(R) \rangle$ is a view of $b_U(\Pi)$. Similarly, $\langle S' \cap U, b_U(R') \rangle$ is a view of $b_U(\Pi)$. Because $head(r') \subseteq U$, from the definition of splitting set it follows that: (i) $head(r) \subseteq U$; (ii) because $(S \cap S') \models prefer^*(r', r)$, $(S \cap S' \cap U) \models prefer^*(r', r)$ also holds. Therefore, $\langle S' \cap U, b_U(R') \rangle$ dominates $\langle S \cap U, b_U(R) \rangle$. *Contradiction.*

Case 2: $head(r') \subseteq \overline{U}$. From Lemma 1 it follows that $\langle S \setminus U, R \setminus b_U(R) \rangle$ is a view of $e_U(\Pi \setminus b_U(\Pi), S \cap U)$. Similarly, $\langle S' \setminus U, R' \setminus b_U(R') \rangle$ is a view of $e_U(\Pi \setminus b_U(\Pi), S' \cap U)$. Because $head(r') \subseteq \overline{U}$, from the definition of splitting set it follows that: (i) $head(r) \subseteq \overline{U}$; (ii) because $(S \cap S') \models prefer^*(r', r)$, $(S \setminus U) \cap (S' \setminus U) \models prefer^*(r', r)$ also holds.

Consider now set $Q' \subseteq S' \setminus U$, consisting of all of the literals of $S' \setminus U$ that are relevant to the cr-rules of $R \cup R' \setminus b_U(R \cup R')$. Also, let $Q \subseteq S \setminus U$ be the set of all of the literals of $S \setminus U$ that are relevant to the cr-rules of $R \cup R' \setminus b_U(R \cup R')$, and $\overline{Q} = S \setminus U \setminus Q$. That is, $\overline{Q}$ is the set of literals from $S \setminus U$ that are *not* relevant to any cr-rule of $R \cup R' \setminus b_U(R \cup R')$.

From Definition 4, it is not difficult to conclude that $\langle \overline{Q} \cup Q', R' \setminus b_U(R') \rangle$ is a view of $e_U(\Pi \setminus b_U(\Pi), S \cap U)$. Furthermore, $(S \setminus U) \cap (Q \cup Q') \models prefer^*(r', r)$. Therefore, $\langle \overline{Q} \cup Q', R' \setminus b_U(R') \rangle$ dominates $\langle S \setminus U, R \setminus b_U(R) \rangle$. *Contradiction.*

Proof of Theorem 2. From the definition of answer set and Lemma 2, it follows that there exists a set R of (names of) cr-rules from Π such that $\langle S, R \rangle$ is a candidate answer set of Π if and only if:

- $\langle X, b_U(R) \rangle$ is a candidate answer set of $b_U(\Pi)$;
- $\langle Y, R \setminus b_U(R) \rangle$ is a candidate answer set of $e_U(\Pi \setminus b_U(\Pi), X)$;
- $S = X \cup Y$ is consistent.

Therefore, we only need to prove that:

$$\text{for every candidate answer set } \langle S', R' \rangle \text{ of } \Pi, R' \not\subset R \tag{8}$$

if and only if

$$\text{for every candidate answer set } \langle S'_X, R'_X \rangle \text{ of } b_U(\Pi), R'_X \not\subset b_U(R), \text{ and} \tag{9}$$

$$\text{for every candidate answer set } \langle S'_Y, R'_Y \rangle \text{ of } e_U(\Pi \setminus b_U(\Pi), X),$$
$$R'_Y \not\subset R \setminus b_U(R). \tag{10}$$

Let us prove that (8) implies (9). By contradiction, suppose that, for every candidate answer set $\langle S', R' \rangle$ of Π, $R' \not\subset R$, but that there exists a candidate answer set $\langle X', R'_X \rangle$ of $b_U(\Pi)$ such that $R'_X \subset b_U(R)$. Let (X'_D, X'_I) be the partition of X' such that X'_D is the set of the literals from X' that are relevant to the cr-rules in $R'_X \cup b_U(R)$. Let (X_D, X_U) be a similar partition of X and $X^\sim = X_I \cup X'_D$. From Definition 4, it is not difficult to prove that $\langle X^\sim, R'_X \rangle$ is a candidate answer set of $b_U(\Pi)$. Furthermore, $e_U(\Pi \setminus b_U(\Pi), X) = e_U(\Pi \setminus b_U(\Pi), X^\sim)$. Hence, $\langle Y, R \setminus b_U(R) \rangle$ is a candidate answer set of $e_U(\Pi \setminus b_U(\Pi), X^\sim)$.

Notice that $R \setminus b_U(R) = (R'_X \cup (R \setminus b_U(R))) \setminus b_U(R'_X)$, and that $b_U(R'_X) = b_U(R'_X \cup (R \setminus b_U(R)))$. Therefore, $\langle Y, R \setminus b_U(R) \rangle = \langle Y, (R'_X \cup (R \setminus b_U(R))) \setminus b_U(R'_X \cup (R \setminus b_U(R))) \rangle$, which allows us to conclude that $\langle Y, (R'_X \cup (R \setminus b_U(R))) \setminus b_U(R'_X \cup (R \setminus b_U(R))) \rangle$ is a candidate answer set of $e_U(\Pi \setminus b_U(\Pi), X^\sim)$. By Lemma 2, $\langle X^\sim \cup Y, R'_X \cup (R \setminus b_U(R)) \rangle$ is a candidate answer set of Π. Because $R'_X \subset b_U(R)$ by hypothesis, $R'_X \cup (R \setminus b_U(R)) \subset R$, which contradicts the assumption that, for every candidate answer set $\langle S', R' \rangle$ of Π, $R' \not\subset R$.

Let us now prove that (8) implies (10). By contradiction, suppose that, for every candidate answer set $\langle S', R' \rangle$ of Π, $R' \not\subset R$, but that there exists a candidate answer set $\langle Y', R'_Y \rangle$ of $e_U(\Pi \setminus b_U(\Pi), X)$ such that $R'_Y \subset R \setminus b_U(R)$. Because $R'_Y \subset R \setminus b_U(R)$, from Lemma 2 we conclude that $\langle X \cup Y', b_U(R) \cup R'_Y \rangle$ is a candidate answer set of Π, and that $b_U(R) \cup R'_Y \subset R$. But the hypothesis was that, for every candidate answer set $\langle S', R' \rangle$ of Π, $R' \not\subset R$. Contradiction.

Let us now prove that (9) and (10) imply (8). By contradiction, suppose that, for every candidate answer set $\langle S'_X, R'_X \rangle$ of $b_U(\Pi)$, $R'_X \not\subset b_U(R)$, and, for every candidate answer set $\langle S'_Y, R'_Y \rangle$ of $e_U(\Pi \setminus b_U(\Pi), X)$, $R'_Y \not\subset R \setminus b_U(R)$, but that there exists a candidate answer set $\langle S', R' \rangle$ of Π such that $R' \subset R$. By Lemma 2, (i) $\langle S' \cap U, b_U(R') \rangle$ is a candidate answer set of $b_U(\Pi)$, and $\langle S' \setminus U, R' \setminus b_U(R') \rangle$ is a candidate answer set of $e_U(\Pi \setminus b_U(\Pi), S' \cap U)$. Notice that, because $R' \subset R$, either $b_U(R') \subset b_U(R)$ or $R' \setminus b_U(R') \subset R \setminus b_U(R)$.

Case 1: $b_U(R') \subset b_U(R)$. It follows that $\langle S' \cap U, b_U(R') \rangle$ is a candidate answer set of $b_U(\Pi)$ such that $b_U(R') \subset b_U(R)$. *Contradiction.*

Case 2: $R' \setminus b_U(R') \subset R \setminus b_U(R)$. Let $X' = S' \cap U$, and (X'_I, X'_D) be the partition of X' such that X'_D consists of all of the literals of X that are relevant to the cr-rules in $b_U(R') \cup b_U(R)$. Let (X_I, X_D) be a similar partition of X. From Definition 4, it is not difficult to prove that $\langle S' \setminus U, R' \setminus b_U(R') \rangle$ is a candidate answer set of $e_U(\Pi \setminus b_U(\Pi), X_I \cup X'_D)$. Moreover, $e_U(\Pi \setminus b_U(\Pi), X_I \cup X'_D) = e_U(\Pi \setminus b_U(\Pi), X)$. Therefore, $\langle S' \setminus U, R' \setminus b_U(R') \rangle$ is a candidate answer set of $e_U(\Pi \setminus b_U(\Pi), X)$, and $R' \setminus b_U(R') \subset R \setminus b_U(R)$. This violates the assumption that, for every candidate answer set $\langle S'_Y, R'_Y \rangle$ of $e_U(\Pi \setminus b_U(\Pi), X)$, $R'_Y \not\subset R \setminus b_U(R)$. *Contradiction.*

References

1. Gelfond, M., Lifschitz, V.: The stable model semantics for logic programming. In: Proceedings of ICLP 1988, pp. 1070–1080 (1988)
2. Gelfond, M., Lifschitz, V.: Classical negation in logic programs and disjunctive databases. New Generation Computing, 365–385 (1991)
3. Marek, V.W., Truszczynski, M.: Stable models and an alternative logic programming paradigm. In: The Logic Programming Paradigm: a 25-Year Perspective, pp. 375–398. Springer, Berlin (1999)
4. Gelfond, M.: Representing Knowledge in A-Prolog. In: Kakas, A.C., Sadri, F. (eds.) Computational Logic: Logic Programming and Beyond, Part II. LNCS (LNAI), vol. 2408, pp. 413–451. Springer, Heidelberg (2002)
5. Baral, C.: Knowledge Representation, Reasoning, and Declarative Problem Solving. Cambridge University Press, Cambridge (2003)
6. Balduccini, M., Gelfond, M.: Logic Programs with Consistency-Restoring Rules. In: Doherty, P., McCarthy, J., Williams, M.A. (eds.) International Symposium on Logical Formalization of Commonsense Reasoning. AAAI 2003 Spring Symposium Series, March 2003, pp. 9–18 (2003)
7. Balduccini, M.: USA-Smart: Improving the Quality of Plans in Answer Set Planning. In: Jayaraman, B. (ed.) PADL 2004. LNCS (LNAI), vol. 3057, pp. 135–147. Springer, Heidelberg (2004)
8. Baral, C., Gelfond, M., Rushton, N.: Probabilistic reasoning with answer sets. Journal of Theory and Practice of Logic Programming (TPLP) (2005)
9. Balduccini, M., Gelfond, M., Nogueira, M.: Answer Set Based Design of Knowledge Systems. Annals of Mathematics and Artificial Intelligence (2006)
10. Son, T.C., Sakama, C.: Negotiation Using Logic Programming with Consistency Restoring Rules. In: 2009 International Joint Conferences on Artificial Intelligence, IJCAI (2009)
11. Lifschitz, V., Turner, H.: Splitting a logic program. In: Proceedings of the 11th International Conference on Logic Programming (ICLP 1994), pp. 23–38 (1994)
12. Balduccini, M., Mellarkod, V.S.: CR-Prolog with Ordered Disjunction. In: International Workshop on Non-Monotonic Reasoning, NMR 2004 (June 2004)
13. Balduccini, M.: CR-MODELS: An Inference Engine for CR-Prolog. In: Baral, C., Brewka, G., Schlipf, J. (eds.) LPNMR 2007. LNCS (LNAI), vol. 4483, pp. 18–30. Springer, Heidelberg (2007)
14. Turner, H.: Splitting a Default Theory. In: Proceedings of AAAI 1996, pp. 645–651 (1996)
15. Watson, R.: A Splitting Set Theorem for Epistemic Specifications. In: Proceedings of the 8th International Workshop on Non-Monotonic Reasoning (NMR 2000) (April 2000)

Contextual Argumentation in Ambient Intelligence

Antonis Bikakis and Grigoris Antoniou

Institute of Computer Science, FO.R.T.H., Vassilika Voutwn
P.O. Box 1385, GR 71110, Heraklion, Greece
{bikakis,antoniou}@ics.forth.gr

Abstract. The imperfect nature of context in Ambient Intelligence environments and the special characteristics of the entities that possess and share the available context information render contextual reasoning a very challenging task. Most current Ambient Intelligence systems have not successfully addressed these challenges, as they rely on simplifying assumptions, such as perfect knowledge of context, centralized context, and unbounded computational and communicating capabilities. This paper presents a knowledge representation model based on the Multi-Context Systems paradigm, which represents ambient agents as autonomous logic-based entities that exchange context information through mappings, and uses preference information to express their confidence in the imported knowledge. On top of this model, we have developed an argumentation framework that exploits context and preference information to resolve conflicts caused by the interaction of ambient agents through mappings, and a distributed algorithm for query evaluation.

1 Introduction

The study of Ambient Intelligence environments has introduced new research challenges in the field of Distributed Artificial Intelligence. These are mainly caused by the imperfect nature of context and the special characteristics of the entities that possess and share the available context information. [1] characterizes four types of imperfect context: *unknown*, *ambiguous*, *imprecise*, and *erroneous*. The agents that operate in such environments are expected to have different goals, experiences and perceptive capabilities, limited computation capabilities, and use distinct vocabularies to describe their context. Due to the highly dynamic and open nature of the environment and the unreliable wireless communications that are restricted by the range of transmitters, ambient agents do not typically know a priori all other entities that are present at a specific time instance nor can they communicate directly with all of them.

So far, Ambient Intelligence systems have not managed to efficiently handle these challenges. As it has been already surveyed in [2], most of them follow classical reasoning approaches that assume perfect knowledge of context, failing to deal with cases of missing, inaccurate or inconsistent context information. Regarding the distribution of reasoning tasks, a common approach followed in most systems assumes the existence of a central entity, which is responsible for collecting and reasoning with all the available context information. However, Ambient Intelligence environments have much more demanding requirements. The dynamics of the network and the unreliable and restricted wireless communications inevitably lead to the decentralization of reasoning tasks.

E. Erdem, F. Lin, and T. Schaub (Eds.): LPNMR 2009, LNCS 5753, pp. 30–43, 2009.

In this paper, we propose a totally distributed approach for contextual reasoning in Ambient Intelligence. We model an ambient environment as a *Multi-Context System*, and ambient agents as autonomous logic-based entities that exchange information through *mapping* rules and use preference information to evaluate imported knowledge. Then we provide a semantic characterization of our approach using *arguments*. The use of arguments is a natural choice in multi-agent systems and aims at a more formal and abstract description of our approach. Conflicts that arise from the interaction of mutually inconsistent sources are captured through *attacking* arguments, and conflict resolution is achieved by *ranking* arguments according to a preference ordering. Finally, we provide an operational model in the form of a distributed algorithm for query evaluation. The algorithm has been implemented in Java and evaluated in a simulated P2P system, and the results are available in [3]. Here, we focus more on its formal properties.

The rest of the paper is structured as follows. Section 2 presents background information and related work on contextual reasoning and preference-based argumentation systems. Section 3 presents the representation model, while section 4 describes its argumentation semantics. Section 5 presents the distributed algorithm and studies its properties. The last section summarizes the main results and discusses future work.

2 Background and Related Work

2.1 Multi-Context Systems

Since the seminal work of McCarthy [4] on *context* and *contextual reasoning*, two main formalizations have been proposed to formalize context: the *Propositional Logic of Context (PLC* [5]), and the *Multi-Context Systems (MCS)* introduced in [6], which later became associated with the *Local Model Semantics* [7]. MCS have been argued to be most adequate with respect to the three dimensions of contextual reasoning (*partiality*, *approximation*, *proximity*) and shown to be technically more general than PLC [8]. A MCS consists of a set of *contexts* and a set of inference rules (known as *mapping* rules) that enable information flow between different contexts. A context can be thought of as a logical theory - a set of axioms and inference rules - that models local knowledge. Different contexts are expected to use different languages, and although each context may be locally consistent, global consistency cannot be required or guaranteed.

The MCS paradigm has been the basis of two recent studies that were the first to deploy non-monotonic features in contextual reasoning: (a) the non-monotonic rule-based MCS framework [9], which supports default negation in the mapping rules allowing to reason based on the absence of context information; and (b) the multi-context variant of Default Logic (*ConDL* [10]), which additionally handles the problem of mutually inconsistent information provided by two or more different sources using default mapping rules. However, *ConDL* does not provide ways to model the quality of imported context information, nor preference between different information sources, leaving the conflicts that arise in such cases unresolved. The use of Multi-Context Systems as a means of specifying and implementing agent architectures has been recently proposed in [11], which proposes *breaking* the logical description of an agent into a set of contexts, each of which represents a different component of the architecture, and the interactions between these components are specified by means of bridge rules between the contexts.

Here, we follow a different approach; a context does not actually represent a logical component of an agent, but rather the viewpoint of each different agent in the system.

Peer data managements systems can be viewed as special cases of MCS, as they consist of autonomous logic-based entities (*peers*) that exchange local information using *bridge rules*. Two prominent recent works that handle the problem of peers providing mutually inconsistent information are: (a) the approach of [12], which is based on non-monotonic epistemic logic; and (b) the propositional P2P Inference System of [13]. A major limitation of both approaches is that conflicts are not actually resolved using some external preference information; they are rather isolated. Our approach enables resolving such conflicts using a preference ordering on the information sources. Building on the work of [13], [14] proposed an argumentation framework and algorithms for inconsistency resolution in P2P systems using a preference relation on system peers. However, their assumptions of a single global language and a global preference relation are in contrast with the dimension of perspective in MCS. In our approach, each agent uses its own vocabulary to describe its context and defines its own preference ordering.

2.2 Preference-Based Argumentation Systems

Argumentation systems constitute a way to formalize non-monotonic reasoning, viz. as the construction and comparison of arguments for and against certain conclusions. A central notion in such systems is that of *acceptability* of arguments. In general, to determine whether an argument is acceptable, it must be compared to its counter-arguments; namely, those arguments that support opposite or conflicting conclusions. In preference-based argumentation systems, this comparison is enabled by a preference relation, which is either implicitly derived from elements of the underlying theory, or is explicitly defined on the set of arguments. Such systems can be classified into four categories. In the first category, which includes the works of [15] and [16], the preference relation takes into account the internal structure of arguments, and arguments are compared in terms of *specifity*. The second category includes systems in which preferences among arguments are derived from a priority relation on the rules in the underlying theory (e.g. [17,18]). In *Value Based Argumentation Frameworks*, the preference ordering on the set of arguments is derived from a preference ordering over the values that they promote (e.g. [19,20]). Finally, the abstract argumentation frameworks proposed by Amgoud and her colleagues ([21,22]) assume that preferences among arguments are induced by a preference relation defined on the underlying belief base.

Our argumentation framework is an extension of the framework of Governatori *et al.* [18], which is based on the grounded semantics of Dung's abstract argumentation framework [23] to provide an argumentative characterization of Defeasible Logic. In our framework, preferences are derived both from the structure of arguments - arguments that use local rules are considered stronger than those that use mapping rules - and from a preference ordering on the information sources (contexts). Our approach also shares common ideas with [21], which first introduced the notion of *contextual* preferences (in the form of several pre-orderings on the belief base), to take into account preferences that depend upon a particular context. The main differences are that in our case, these orderings are applied to the contexts themselves rather than directly to a set of arguments, and that we use a distributed underlying knowledge base.

3   Representation Model

We model a Multi-Context System C as a collection of distributed context theories C_i: A context is defined as a tuple of the form (V_i, R_i, T_i), where V_i is the vocabulary used by C_i, R_i is a set of rules, and T_i is a preference ordering on C.

V_i is a set of positive and negative literals. If q_i is a literal in V_i, $\sim q_i$ denotes the complementary literal, which is also in V_i. If q_i is a positive literal p then $\sim q_i$ is $\neg p$; and if q_i is $\neg p$, then $\sim q_i$ is p. We assume that each context uses a distinct vocabulary.

R_i consists of two sets of rules: the set of local rules and the set of mapping rules. The body of a local rule is a conjunction of *local* literals (literals that are contained in V_i), while its head contains a local literal:

$$r_i^l : a_i^1, a_i^2, ... a_i^{n-1} \to a_i^n$$

Local rules express local knowledge and are interpreted in the classical sense: whenever the literals in the body of the rule are consequences of the local theory, then so is the literal in the head of the rule. Local rules with empty body denote factual knowledge.

Mapping rules associate local literals with literals from the vocabularies of other contexts (*foreign literals*). The body of each such rule is a conjunction of local and foreign literals, while its head contains a single local literal:

$$r_i^m : a_i^1, a_j^2, ... a_k^{n-1} \Rightarrow a_i^n$$

r_i^m associates local literals of C_i (e.g. a_i^1) with local literals of C_j (a_j^2), C_k (a_k^{n-1}) and possibly other contexts. a_i^n is a local literal of the theory that has defined r_i^m (C_i).

Finally, each context C_i defines a total preference ordering T_i on C to express its confidence in the knowledge it imports from other contexts. This is of the form:

$$T_i = [C_k, C_l, ..., C_n]$$

According to T_i, C_k is preferred by C_i to C_l if C_k precedes C_l in T_i. The total preference ordering enables resolving all potential conflicts that may arise from the interaction of contexts through their mapping rules.

Example. Consider the following scenario. Dr. Amber has configured his mobile phone to decide whether it should ring based on his preferences and context. He has the following preferences: His mobile phone should ring in case of an incoming call (*in_call*) if it is in normal mode (*normal*) and he is not giving a course lecture (*lecture*). Dr. Amber is currently located in 'RA201' university classroom. It is class time, but he has just finished with a lecture and remains in the classroom reading his emails on his laptop. The mobile phone receives an incoming call, and it is in normal mode. The local knowledge of the mobile phone (C_1), which includes information about the mode of the phone and incoming calls, is encoded in the following local rules.

$r_{11}^l : in\_call_1, normal_1, \neg lecture_1 \to ring_1$
$r_{12}^l :\to in\_call_1$
$r_{13}^l :\to normal_1$

In case the mobile phone cannot reach a decision based on its local knowledge, it imports knowledge from other ambient agents. In this case, to determine whether Dr. Amber is giving a lecture, it connects through the university wireless network with Dr. Amber's laptop (C_2), a localization service (C_3) and the classroom manager (C_4, a stationary computer installed in 'RA201'), and imports information about Dr. Amber's scheduled events, location, and the classroom state through mapping rules r_{14}^m and r_{15}^m.

$$r_{14}^m : classtime_2, location\_RA201_3 \Rightarrow lecture_1$$
$$r_{15}^m : \neg class\_activity_4 \Rightarrow \neg lecture_1$$

The local context knowledge of the laptop, the localization service, and the classroom manager is expressed in rules r_{21}^l, r_{31}^l and $r_{41}^l - r_{42}^l$ respectively. The classroom manager infers whether there is active class activity, based on the number of people detected in the classroom (*detected*) by a person detection service.

$$r_{21}^l :\rightarrow classtime_2$$
$$r_{31}^l :\rightarrow location\_RA201_3$$
$$r_{41}^l :\rightarrow detected(1)_4$$
$$r_{42}^l : detected(1)_4 \rightarrow \neg class\_activity_4$$

The mobile phone is configured to give highest priority to information imported by the classroom manager and lowest priority to information imported by the laptop. This is encoded in preference ordering $T_1 = [C_4, C_3, C_2]$.

This example characterizes the type of applications, in which each ambient agent is aware of the type of knowledge that each of the other agents that it communicates with possesses, and has predefined how part of this knowledge relates to its local knowledge.

4   Argumentation Semantics

The argumentation framework that we propose uses arguments of local range, in the sense that each one is made of rules derived from a single context. Arguments made by different contexts are interrelated in the *Support Relation* through mapping rules. The Support Relation contains triples that represent proof trees for literals in the system. Each proof tree is made of rules of the context that the literal in its root is defined by. In case a proof tree contains mapping rules, for the respective triple to be contained in the Support Relation, similar triples for the foreign literals in the proof tree must have already been obtained. We should also note that, for sake of simplicity, we assume that there are no loops in the local context theories, and thus proof trees are finite. Loops in the local knowledge bases can be easily detected and removed without needing to interact with other agents. However, even if there are no loops in the local theories, the global knowledge base may contain loops caused by mapping rules.

Let $C = \{C_i\}$ be a MCS. The *Support Relation* of C (SR_C) is the set of all triples of the form (C_i, PT_{p_i}, p_i), where $C_i \in C$, $p_i \in V_i$, and PT_{p_i} is the proof tree for p_i based on the set of local and mapping rules of C_i. PT_{p_i} is a tree with nodes labeled by literals such that the root is labeled by p_i, and for every node with label q:

1. If $q \in V_i$ and $a_1, ..., a_n$ label the children of q then
   - If $\forall a_i \in \{a_1, ..., a_n\}$: $a_i \in V_i$ then there is a local rule $r_i \in C_i$ with body $a_1, ..., a_n$ and head q

- If $\exists a_j \in \{a_1, ..., a_n\}$ such that $a_j \notin V_i$ then there is a mapping rule $r_i \in C_i$ with body $a_1, ..., a_n$ and head q

2. If $q \in V_j \neq V_i$, then this is a leaf node of the tree and there is a triple of the form (C_j, PT_q, q) in SR_C

3. The arcs in a proof tree are labeled by the rules used to obtain them.

An *argument* A for a literal p_i is a triple (C_i, PT_{p_i}, p_i) in SR_C. Any literal labeling a node of PT_{p_i} is called *a conclusion* of A. However, when we refer to *the conclusion* of A, we refer to the literal labeling the root of PT_{p_i} (p_i). We write $r \in A$ to denote that rule r is used in the proof tree of A. A (proper) subargument of A is every argument with a proof tree that is (proper) subtree of the proof tree of A.

Based on the literals used in their proof trees, arguments are classified to *local* and *mapping* arguments. An argument A with conclusion $p_i \in V_i$ is a local argument of C_i if its proof tree contains only local literals of C_i (literals that are contained in V_i). Otherwise, A is a mapping argument of C_i. We denote as $Args_{C_i}$ the set of all arguments of C_i, while $Args_C$ is the set of all arguments in C: $Args_C = \bigcup Args_{C_i}$.

The derivation of local logical consequences in C_i is based on its local arguments. Actually, the conclusions of all local arguments in $Args_{C_i}$ are logical consequences of C_i. Distributed logical consequences are derived from a combination of local and mapping arguments in $Args_C$. In this case, we should also consider conflicts between competing rules, which are modeled as *attacks* between arguments, and preference orderings, which are used in our framework to *rank* mapping arguments.

The *rank of a literal* p in context C_i (denoted as $R(p, C_i)$) equals 0 if $p \in V_i$. If $p \in V_j \neq V_i$, then $R(p, C_i)$ equals the rank of C_j in T_i. The *rank of an argument* A in C_i (denoted as $R(A, C_i)$) equals the maximum between the ranks in C_i of the literals contained in A. It is obvious that for any three arguments A_1, A_2, A_3: If $R(A_1, C_i) \leq R(A_2, C_i)$ and $R(A_2, C_i) \leq R(A_3, C_i)$, then $R(A_1, C_i) \leq R(A_3, C_i)$; namely the preference relation $<$ on $Args^C$, which is build according to ordering T_i, is transitive.

The definitions of *attack* and *defeat* apply only for mapping arguments. An argument A *attacks* a mapping argument B at p_i, if p_i is a conclusion of B, $\sim p_i$ is a conclusion of A, and the subargument of B with conclusion p_i is not a local argument. An argument A *defeats* an argument B at p_i, if A attacks B at p_i, and for the subarguments of A, A' with conclusion $\sim p_i$, and of B, B' with conclusion p_i: $R(A', C_i) \leq R(B', C_i)$.

To link arguments through the mapping rules that they contain, we introduce in our framework the notion of *argumentation line*. An *argumentation line* A_L for a literal p_i is a sequence of arguments in $Args_C$, constructed in steps as follows:

- In the first step add in A_L one argument for p_i.
- In each next step, for each distinct literal q_j labeling a leaf node of the proof trees of the arguments added in the previous step, add one argument with conclusion q_j; the addition should not violate the following restriction.
- An argument B with conclusion q_j can be added in A_L only if A_L does not already contain a different argument D with conclusion q_j.

The argument for p_i added in the first step is called the *head argument* of A_L. If the number of steps required to build A_L is finite, then A_L is a finite argumentation line.

Infinite argumentation lines imply loops in the global knowledge base. Arguments contained in infinite lines participate in *attacks* against counter-arguments but may not be used to support the conclusion of their argumentation lines.

The notion of *supported* argument is meant to indicate when an argument may have an active role in proving or preventing the derivation of a conclusion. An argument A is *supported* by a set of arguments S if: (a) every proper subargument of A is in S; and (b) there is a finite argumentation line A_L with head A, such that every argument in $A_L - \{A\}$ is in S.

A mapping argument A is *undercut* by a set of arguments S if for every argumentation line A_L with head A, there is an argument B, such that B is supported by S, and B defeats a proper subargument of A or an argument in $A_L - \{A\}$. That an argument A is *undercut* by a set of arguments S means that we can show that some premises of A cannot be proved if we accept the arguments in S.

An argument A is *acceptable* w.r.t a set of arguments S if:

1. A is a local argument; or
2. A is supported by S and every argument defeating A is undercut by S

Intuitively, that an argument A is *acceptable* w.r.t. S means that if we accept the arguments in S as valid arguments, then we feel compelled to accept A as valid. Based on the concept of acceptable arguments, we define *justified arguments* and *justified literals*. J_i^C is defined as follows:

- $J_0^C = \emptyset$;
- $J_{i+1}^C = \{A \in Args_C \mid A \text{ is acceptable w.r.t. } J_i^C\}$

The set of *justified arguments* in a MCS C is $JArgs^C = \bigcup_{i=1}^{\infty} J_i^C$. A literal p_i is *justified* if it is the conclusion of an argument in $JArgs^C$. That an argument A is justified means that it resists every reasonable refutation. That a literal p_i is justified, it actually means that it is a logical consequence of C.

Finally, we also introduce the notion of *rejected arguments* and *rejected literals* for the characterization of conclusions that do not derive from C. An argument A is *rejected* by sets of arguments S, T when:

1. A is not a local argument, and either
2. (a) a proper subargument of A is in S; or
   (b) A is defeated by an argument supported by T; or
   (c) for every argumentation line A_L with head A there exists an argument $A' \in A_L - \{A\}$, such that either a subargument of A' is in S; or A' is defeated by an argument supported by T

That an argument is rejected by sets of arguments S and T means that either it is supported by arguments in S, which can be thought of as the set of already rejected arguments, or it cannot overcome an attack from an argument supported by T, which can be thought of as the set of justified arguments. Based on the definition of rejected arguments, we define R_i^C as follows:

- $R_0^C = \emptyset$;
- $R_{i+1}^C = \{A \in Args_C \mid A \text{ is rejected by } R_i^C, JArgs^C\}$

The set of *rejected arguments* in a MCS C is $RArgs^C = \bigcup_{i=1}^{\infty} R_i^C$. A literal p_i is *rejected* if there is no argument in $Args^C - RArgs^C$ with conclusion p_i. That a literal is rejected means that we are able to prove that it is not a logical consequence of C.

Example (continued). Given the MCS C of the example, in_call_1, $normal_1$, $classtime_2$, $location\_RA201_3$, $detected(1)_4$, and $\neg class\_activity_4$ are justified in C, since they are supported by local arguments; all these arguments are in J_i^C for every i. The argument $A=\{\neg class\_activity_4 \Rightarrow \neg lecture_1\}$ is supported by J_1^C, as there is an argument for $\neg class\_activity_4$ in J_1^C. Moreover, it is not defeated by attacking argument $B=\{classtime_2, location\_RA201_3 \Rightarrow lecture_1\}$, as B has higher rank than A in C_1. Hence, A is in J_2, and $\neg lecture_1$ is justified in C. Argument D for $ring_1$, which is derived from the arguments for in_call_1, $normal_1$ and $\neg lecture_1$ (A) and rule r_{11}^l, is supported by J_2^C and not defeated by attacking argument B. Therefore, D is justified, and $ring_1$ is justified in C.

Lemmata 1-3 describe some formal properties of the framework. Their proofs are available at: www.csd.uoc.gr/~bikakis/thesis.pdf. Lemma 1 refers to the monotonicity in J_i^C and $R_i^C(T)$, while Lemma 2 represents the fact that no argument is both "believed" and "disbelieved".

Lemma 1. *The sequences J_i^C and $R_i^C(T)$ are monotonically increasing.*

Lemma 2. *In a Multi-Context System C, no literal is both justified and rejected.*

If consistency is assumed in the local rules of a context theory (two complementary conclusions may not be derived as local consequences of a context theory), then using Lemma 2, it is easy to prove that the entire framework is consistent (Lemma 3).

Lemma 3. *If the set of justified arguments in C, $JArgs^C$, contains two arguments with complementary conclusions, then both are local arguments of the same context.*

5   Distributed Query Evaluation

$P2P\_DR$ is a distributed algorithm for query evaluation that implements the proposed argumentation framework. The specific problem that it deals with is: *Given a MCS C, and a query about literal p_i issued to context C_i, compute the truth value of p_i.* For an arbitrary literal p_i, $P2P\_DR$ returns one of the following values: (a) *true*; indicating that p_i is justified in C; (b) *false*; indicating that p_i is rejected in C; or (c) *undefined*; indicating that p_i is neither justified nor rejected in C.

5.1   Algorithm Description

$P2P\_DR$ proceeds in four main steps. In the first step (lines 1-8), $P2P\_DR$ determines whether p_i or its negation $\sim p_i$, are consequences of the local rules of C_i, using $local\_alg$ (described later in this section). If $local\_alg$ computes *true* as an answer for p_i or $\sim p_i$, $P2P\_DR$ returns *true* / *false* respectively as an answer for p_i and terminates.

In step 2 (lines 9-12), $P2P\_DR$ calls $Support$ (described later in this section) to determine whether there are *applicable* and *unblocked* rules with head p_i. We call *applicable* those rules that for all literals in their body $P2P\_DR$ has computed *true* as their truth value, while *unblocked* are the rules that for all literals in their body $P2P\_DR$ has computed either *true* or *undefined* as their truth values. *Support* returns two data structures for p_i: (a) the Supportive Set of p_i (SS_{p_i}), which is the set of foreign literals used in the most preferred (according to T_i) chain of applicable rules for p_i ; and (b) the Blocking Set of p_i (BS_{p_i}), which is the set of foreign literals used in the most preferred chain of unblocked rules for p_i. If there is no unblocked rule for p_i ($BS_{p_i} = \emptyset$), $P2P\_DR$ returns *false* as an answer and terminates. Similarly, in step 3 (lines 13-14), $P2P\_DR$ calls *Support* to compute the respective constructs for $\sim p_i$ ($SS_{\sim p_i}, BS_{\sim p_i}$).

In the last step (lines 15-24), $P2P\_DR$ uses the constructs computed in the previous steps and preference ordering T_i to compute the answer for p_i. In case there is no unblocked rule for $\sim p_i$ ($BS_{\sim p_i} = \emptyset$), or SS_{p_i} is computed by *Stronger* (described later in this section) to be *stronger* than $BS_{\sim p_i}$, $P2P\_DR$ returns Ans_{p_i}=*true*. That SS_{p_i} is *stronger* than BS_{p_i} means that the chains of applicable rules for p_i involve information from more preferred contexts to those that are involved in the chains of unblocked rules for $\sim p_i$. If there is at least one applicable rule for $\sim p_i$, and BS_{p_i} is *not stronger* than $SS_{\sim p_i}$, $P2P\_DR$ returns *false*. In any other case, it returns *undefined*.

The context that is called to evaluate the query for p_i (C_i) returns through Ans_{p_i} the truth value for p_i. SS_{p_i} and BS_{p_i} are returned to the querying context (C_0) only if the two contexts (C_0 and C_i) are actually the same context. Otherwise, the empty set is assigned to both SS_{p_i} and BS_{p_i} and returned to C_0. In this way, the size of the messages exchanged between different contexts is kept small. $Hist_{p_i}$ is a structure used by *Support* to detect loops in the global knowledge base. The algorithm parameters are:

- p_i: the queried literal (input)
- C_0: the context that issues the query (input)
- C_i: the context that defines p_i (input)
- $Hist_{p_i}$: the list of pending queries ($[p_1, ..., p_i]$) (input)
- T_i: the preference ordering of C_i (input)
- SS_{p_i}: a set of foreign literals of C_i denoting the Supportive Set of p_i (output)
- BS_{p_i}: a set of foreign literals of C_i denoting the Blocking Set of p_i (output)
- Ans_{p_i}: the answer returned for p_i (output)

P2P_DR$(p_i, C_0, C_i, Hist_{p_i}, T_i, SS_{p_i}, BS_{p_i}, Ans_{p_i})$

```
1:  call local_alg(p_i, localAns_{p_i})
2:  if localAns_{p_i} = true then
3:      Ans_{p_i} ← true, SS_{p_i} ← ∅, BS_{p_i} ← ∅
4:      terminate
5:  call local_alg(∼ p_i, localAns_{∼p_i})
6:  if localAns_{∼p_i} = true then
7:      Ans_{p_i} ← false, SS_{p_i} ← ∅, BS_{p_i} ← ∅
8:      terminate
9:  call Support(p_i, Hist_{p_i}, T_i, SS_{p_i}, BS_{p_i})
10: if BS_{p_i} = ∅ then
11:     Ans_{p_i} ← false, SS_{p_i} ← ∅, BS_{p_i} ← ∅
12:     terminate
13: Hist_{∼p_i} ← (Hist_{p_i} − {p_i}) ∪ {∼ p_i}
14: call Support(∼ p_i, Hist_{∼p_i}, T_i, SS_{∼p_i}, BS_{∼p_i})
15: if SS_{p_i} ≠ ∅ and (BS∼ p_i = ∅ or Stronger(SS_{p_i}, BS_{∼p_i}, T_i) = SS_{p_i}) then
```

16:     $Ans_{p_i} \leftarrow true$
17:     **if** $C_0 \neq C_i$ **then**
18:         $SS_{p_i} \leftarrow \emptyset, BS_{p_i} \leftarrow \emptyset$
19: **else if** $SS_{\sim p_i} \neq \emptyset$ and $Stronger(BS_{p_i}, SS_{\sim p_i}, T_i) \neq BS_{p_i}$ **then**
20:     $Ans_{p_i} \leftarrow false, SS_{p_i} \leftarrow \emptyset, BS_{p_i} \leftarrow \emptyset$
21: **else**
22:     $Ans_{p_i} \leftarrow undefined$
23:     **if** $C_0 \neq C_i$ **then**
24:         $SS_{p_i} \leftarrow \emptyset, BS_{p_i} \leftarrow \emptyset$

local_alg is called by $P2P\_DR$ to determine whether the truth value of the queried literal can be derived from the local rules of a context.

local_alg$(p_i, localAns_{p_i})$
1: **for all** $r_i \in R^s[p_i]$ **do**
2:     **for all** $b_i \in body(r_i)$ **do**
3:         call $local\_alg(b_i, localAns_{b_i})$
4:     **if** for all b_i: $localAns_{b_i} = true$ **then**
5:         $localAns_{p_i} \leftarrow true$
6:         terminate
7: $localAns_{p_i} \leftarrow false$

Support is called by $P2P\_DR$ to compute SS_{p_i} and BS_{p_i}. To compute these structures, it checks the applicability of the rules with head p_i, using the truth values of the literals in their body, as these are evaluated by $P2P\_DR$. To avoid loops, before calling $P2P\_DR$, it checks if the same query has been issued before during the running call of $P2P\_DR$. For each applicable rule r_i, *Support* builds its Supportive Set, SS_{r_i}; this is the union of the set of *foreign literals* contained in the body of r_i with the Supportive Sets of the local literals contained in the body of the rule. Similarly, for each unblocked rule r_i, it computes its Blocking Set BS_{r_i} using the Blocking Sets of its body literals. *Support* computes the Supportive Set of p_i, SS_{p_i}, as the *strongest* rule Supportive Set SS_{r_i}; and its Blocking Set, BS_{p_i}, as the *strongest* rule Blocking Set BS_{r_i}, using the *Stronger* function. The parameters of *Support* are:

- p_i: the queried literal (input)
- $Hist_{p_i}$: the list of pending queries ($[p_1, ..., p_i]$) (input)
- T_i: the preference ordering of C_i (input)
- SS_{p_i}: the Supportive Set of p_i (output)
- BS_{p_i}: the Blocking Set of p_i (output)

Support$(p_i, Hist_{p_i}, T_i, SS_{p_i}, BS_{p_i})$
1: **for all** $r_i \in R[p_i]$ **do**
2:     $cycle(r_i) \leftarrow false$
3:     $SS_{r_i} \leftarrow \emptyset, BS_{r_i} \leftarrow \emptyset$
4:     **for all** $b_t \in body(r_i)$ **do**
5:         **if** $b_t \in Hist_{p_i}$ **then**
6:             $cycle(r_i) \leftarrow true$
7:             $BS_{r_i} \leftarrow BS_{r_i} \cup \{d_t\}$ $\{d_t \equiv b_t$ if $b_t \notin V_i$; otherwise d_t is the first foreign literal of C_i added in $Hist_{p_i}$ after $b_t\}$

```
 8:        else
 9:            Hist_{b_t} ← Hist_{p_i} ∪ {b_t}
10:            call P2P_DR(b_t, C_i, C_t, Hist_{b_t}, T_t, SS_{b_t}, BS_{b_t}, Ans_{b_t})
11:            if Ans_{b_t} = false then
12:                stop and check the next rule
13:            else if Ans_{b_t} = undefined or cycle(r_i) = true then
14:                cycle(r_i) ← true
15:                if b_t ∉ V_i then
16:                    BS_{r_i} ← BS_{r_i} ∪ {b_t}
17:                else
18:                    BS_{r_i} ← BS_{r_i} ∪ BS_{b_t}
19:            else
20:                if b_t ∉ V_i then
21:                    BS_{r_i} ← BS_{r_i} ∪ {b_t}
22:                    SS_{r_i} ← SS_{r_i} ∪ {b_t}
23:                else
24:                    BS_{r_i} ← BS_{r_i} ∪ BS_{b_t}
25:                    SS_{r_i} ← SS_{r_i} ∪ SS_{b_t}
26:    if BS_{p_i} = ∅ or Stronger(BS_{r_i}, BS_{p_i}, T_i) = BS_{r_i} then
27:        BS_{p_i} ← BS_{r_i}
28:    if cycle(r_i) = false then
29:        if SS_{p_i} = ∅ or Stronger(SS_{r_i}, SS_{p_i}, T_i) = SS_{r_i} then
30:            SS_{p_i} ← SS_{r_i}
```

$Stronger(A, B, T_i)$ returns the *strongest* between two sets of literals, A and B, according to preference ordering T_i. A literal a_k is *preferred* to literal b_j, if C_k precedes C_l in T_i. The strength of a set is determined by the the least preferred literal in this set.

Stronger(A, B, T_i)

```
1:  if ∃b_j ∈ B: ∀a_k ∈ A: C_k has lower rank than C_j in T_i then
2:      Stronger = A
3:  else if ∃a_k ∈ A: ∀b_j ∈ B: C_j has lower rank than C_k in T_i then
4:      Stronger = B
5:  else
6:      Stronger = None
```

Example (continued). Given a query about $ring_1$, $P2P\_DR$ proceeds as follows. It fails to compute an answer based on C_1's local theory, and uses rules r_{14}^m and r_{15}^m to compute an answer for $\neg lecture_1$. Using the local rules of C_2, C_3 and C_4, it computes positive answers for $classtime_2$, $location\_RA201_3$ and $\neg class\_activity_4$ respectively, determines that both r_{14}^m and r_{15}^m are applicable, and computes their Supportive Sets: $SS_{r_{14}^m} = \{class_2, location\_RA201_3\}$ and $SS_{r_{15}^m} = \{\neg class\_activity_4\}$. As C_4 precedes C_2 in T_1, $P2P\_DR$ determines that $SS_{r_{15}^m}$ is stronger, computes a positive answer for $\neg lecture_1$, and eventually (using rule r_{11}^l) returns a positive answer (*true*) for $ring_1$.

5.2 Properties of the Algorithm

Below, we describe formal properties of $P2P\_DR$ regarding its termination, soundness and completeness w.r.t. the argumentation framework and complexity. The proofs for the following propositions are available at www.csd.uoc.gr/~bikakis/thesis.pdf. Proposition 1 is a consequence of the cycle detection process within the algorithm.

Proposition 1. *The algorithm is guaranteed to terminate returning one of the values true, false and undefined as an answer for the queried literal.*

Proposition 2 associates the answers produced by $P2P\_DR$ with the concepts of justified and rejected literals.

Proposition 2. *For a Multi-Context System C and a literal p_i in C, $P2P\_DR$ returns:*

1. *$Ans_{p_i} = true$ iff p_i is justified in C*
2. *$Ans_{p_i} = false$ iff p_i is rejected in C*
3. *$Ans_{p_i} = undefined$ iff p_i is neither justified nor rejected in C*

Propositions 3 and 4 are consequences of two states that we retain for each context, which keep track of the results for all incoming and outgoing queries. The worst case that both propositions refer to is when all rules of C_i contain either p_i (the queried literal) or $\sim p_i$ in their head and all system literals in their bodies.

Proposition 3. *The total number of messages exchanged between the system contexts for the evaluation of a query is, in the worst case, $O(n \times \sum P(n,k))$, where n stands for the total number of literals in the system, $\sum$ expresses the sum over $k = 0, 1, ..., n$, and $P(n,k)$ stands for the number of permutations with length k of n elements. In case, there are no loops in the global knowledge base, the number of messages is polynomial to the size of the global knowledge base.*

Proposition 4. *The number of operations imposed by one call of $P2P\_DR$ for the evaluation of a query for literal p_i is, in the worst case, proportional to the number of rules in C_i, and to the total number of literals in the system.*

6   Conclusion

This paper proposes a totally distributed approach for contextual reasoning in Ambient Intelligence, based on representing context knowledge of ambient agents as context theories in a Multi-Context System, and reasoning with the available knowledge using arguments. Using a total preference ordering on the system contexts, our approach enables resolving all conflicts that arise from the interaction of contexts through their mappings. The paper also presents a distributed algorithm for query evaluation that implements the proposed argumentation framework, and studies its formal properties.

Our ongoing work involves: (a) studying alternative methods for conflict resolution, which differ in the way that agents evaluate the imported context information; (b) adding non-monotonic features to the local context theories to support uncertainty in the local context knowledge; (c) extending our approach to support overlapping vocabularies, which will enable different contexts to use elements of common vocabularies (e.g. URIs); and (d) implementing real-world applications of our approach in Ambient Intelligence environments.

References

1. Henricksen, K., Indulska, J.: Modelling and Using Imperfect Context Information. In: Proceedings of PERCOMW 2004, Washington, DC, USA, pp. 33–37. IEEE Computer Society, Los Alamitos (2004)
2. Bikakis, A., Patkos, T., Antoniou, G., Plexousakis, D.: A Survey of Semantics-based Approaches for Context Reasoning in Ambient Intelligence. In: Constructing Ambient Intelligence. Communications in Computer and Information Science, pp. 14–23. Springer, Heidelberg (2008)
3. Bikakis, A., Antoniou, G., Hassapis, P.: Alternative Strategies for Conflict Resolution in Multi-Context Systems. In: Proceedings of the 5th IFIP Conference on Artificial Intelligence Applications and Innovations (AIAI 2009). Springer, Heidelberg (2009)
4. McCarthy, J.: Generality in Artificial Intelligence. Communications of the ACM 30(12), 1030–1035 (1987)
5. Buvac, S., Mason, I.A.: Propositional Logic of Context. In: AAAI, pp. 412–419 (1993)
6. Giunchiglia, F., Serafini, L.: Multilanguage hierarchical logics, or: how we can do without modal logics. Artificial Intelligence 65(1) (1994)
7. Ghidini, C., Giunchiglia, F.: Local Models Semantics, or contextual reasoning=locality+compatibility. Artificial Intelligence 127(2), 221–259 (2001)
8. Serafini, L., Bouquet, P.: Comparing formal theories of context in AI. Artificial Intelligence 155(1-2), 41–67 (2004)
9. Roelofsen, F., Serafini, L.: Minimal and Absent Information in Contexts. In: IJCAI, pp. 558–563 (2005)
10. Brewka, G., Roelofsen, F., Serafini, L.: Contextual Default Reasoning. In: IJCAI, pp. 268–273 (2007)
11. Sabater, J., Sierra, C., Parsons, S., Jennings, N.R.: Engineering Executable Agents using Multi-context Systems. Journal of Logic and Computation 12(3), 413–442 (2002)
12. Calvanese, D., De Giacomo, G., Lembo, D., Lenzerini, M., Rosati, R.: Inconsistency tolerance in P2P data integration: An epistemic logic approach. In: Bierman, G., Koch, C. (eds.) DBPL 2005. LNCS, vol. 3774, pp. 90–105. Springer, Heidelberg (2005)
13. Chatalic, P., Nguyen, G.H., Rousset, M.C.: Reasoning with Inconsistencies in Propositional Peer-to-Peer Inference Systems. In: ECAI, pp. 352–356 (2006)
14. Binas, A., Sheila, A.: Peer-to-Peer Query Answering with Inconsistent Knowledge. In: KR, pp. 329–339 (2008)
15. Simari, G.R., Loui, R.P.: A Mathematical Treatment of Defeasible Reasoning and its Implementation. Artificial Intelligence 53(2-3), 125–157 (1992)
16. Stolzenburg, F., García, A.J., Chesñevar, C.I., Simari, G.R.: Computing Generalized Specificity. Journal of Applied Non-Classical Logics 13(1), 87–113 (2003)
17. Prakken, H., Sartor, G.: Argument-Based Extended Logic Programming with Defeasible Priorities. Journal of Applied Non-Classical Logics 7(1) (1997)
18. Governatori, G., Maher, M.J., Billington, D., Antoniou, G.: Argumentation Semantics for Defeasible Logics. Journal of Logic and Computation 14(5), 675–702 (2004)
19. Bench-Capon, T.: Persuasion in Practical Argument Using Value-based Argumentation Frameworks. Journal of Logic and Computation 13, 429–448 (2003)
20. Kaci, S., van der Torre, L.: Preference-based argumentation: Arguments supporting multiple values. Internation Journal of Approximate Reasoning 48(3), 730–751 (2008)

21. Amgoud, L., Parsons, S., Perrussel, L.: An Argumentation Framework based on contextual Preferences. In: International Conference on Formal and Applied and Practical Reasoning (FAPR 2000), pp. 59–67 (2000)
22. Amgoud, L., Cayrol, C.: A Reasoning Model Based on the Production of Acceptable Arguments. Annals of Mathematic and Artificial Intelligence 34(1-3), 197–215 (2002)
23. Dung, P.M.: On the acceptability of arguments and its fundamental role in nonmonotonic reasoning, logic programming and n-person games. Artif. Intell. 77, 321–357 (1995)

Argumentation Context Systems: A Framework for Abstract Group Argumentation

Gerhard Brewka[1] and Thomas Eiter[2]

[1] Universität Leipzig, Augustusplatz 10-11, 04109 Leipzig, Germany
brewka@informatik.uni-leipzig.de
[2] Vienna University of Technology, Favoritenstraße 9-11, A-1040 Vienna, Austria
eiter@kr.tuwien.ac.at

Abstract. We introduce a modular framework for distributed abstract argumentation where the argumentation context, that is information about preferences among arguments, values, validity, reasoning mode (skeptical vs. credulous) and even the chosen semantics can be explicitly represented. The framework consists of a collection of abstract argument systems connected via mediators. Each mediator integrates information coming from connected argument systems (thereby handling conflicts within this information) and provides the context used in a particular argumentation module. The framework can be used in different directions; e.g., for hierarchic argumentation as typically found in legal reasoning, or to model group argumentation processes.

1 Introduction

In his seminal paper, Dung [10] introduced an abstract framework for argumentation (sometimes referred to as calculus of opposition) which proved to be extremely useful for analyzing various kinds of argumentation processes. His approach gives a convincing account of how to select a set of "acceptable" arguments out of a set of arguments which may attack each other.

Dung's approach is monolithic in the sense that there are no means to structure a set of arguments any further. This is at odds with real world argumentation scenarios, be they informal as in everyday conversation, or institutionalized as in the legal domain. In such scenarios, one typically finds *meta-arguments*, i.e., arguments about other arguments, which can be clearly distinguished from arguments about the domain at hand. Moreover, in multi-agent scenarios it is often important to keep track of where certain arguments came from, who put them forward, who opposed and the like.

For these reasons, our interest in this paper is on adding more structure to formal models of argumentation. In doing so, we want to stick as much as possible to the idea of abstract argumentation. However, instead of a single, unstructured set of arguments we consider clearly distinguishable, distributed argumentation modules and formalize ways in which they can possibly interact. This has several benefits.

E. Erdem, F. Lin, and T. Schaub (Eds.): LPNMR 2009, LNCS 5753, pp. 44–57, 2009.

1. Even in the single-agent case, the additional structure provided by argumentation modules gives a handle on complexity and diversity (e.g., in our framework
   it will be possible for an agent to be skeptical in critical parts, credulous in
   less critical ones);
2. a distributed framework provides a natural account of multi-agent argumentation scenarios including information flow and knowledge integration methods;
3. modules provide explicit means to model meta-argumentation where the arguments in one module are about the arguments in another module; this leads
   to more realistic accounts of, say, legal argumentation processes.

In the simplest multi-module situation, one module determines the context for
another. By a *context* we mean the available meta-information including, for
instance, arguments which should not be taken into account, preferences among
arguments, values, reasoning modes (skeptical vs. credulous), or even the semantics to be used in a particular situation. However, we also want to be able to
capture more general situations where modules form complex hierarchies, and
even cycles where modules mutually influence each other. For these reasons, we
consider arbitrary directed graphs of modules.

As different "parent" modules in the graph can contribute to the context of a
single module, we face the difficulty that the context information may become
inconsistent. Each module is equipped with a *mediator*[1] to deal with this issue.

To model the flow of information among argumentation modules, we use techniques developed in the area of multi-context systems ($\mathcal{MCS}$s), in particular
the systems devised by Giunchiglia and Serafini [14] and their extensions to
nonmonotonic $\mathcal{MCS}$s [17,6]. An $\mathcal{MCS}$ describes the information available in a
number of contexts and specifies the information flow between them.[2] So-called
bridge rules play a key role: they are used to provide input for a context based
on the beliefs held in other relevant contexts. Since different contexts may use
different representation languages, this may include "translating" information
from one language to another. In our case, the bridge rules are necessary to
transform abstract arguments accepted in one module into context statements
for a child module.

Our approach substantially generalizes a framework recently introduced by
Modgil [15]. His framework consists of a *linear* hierarchy of argument systems
$(\mathcal{A}_1, \ldots, \mathcal{A}_n)$ and allows *preferences* among arguments in $\mathcal{A}_i$ to be established
in $\mathcal{A}_{i+1}$. Our approach is more general in at least two respects: (1) we provide
means for argumentation not only about preferences, but also about values, acceptability of arguments and attacks, and even reasoning mode and semantics,
and (2) we consider arbitrary directed graphs. Since our modules may have multiple parents, methods for information integration, as provided by our mediators,
become essential. In summary, our contribution in this paper is twofold:

- We introduce *context-based argumentation* (Sect. 4). Here statements in
  a context language, which formally specify a context, allow us to control

---

[1] Cf. Wiederhold's [19] classic notion in information systems.

[2] A context in an $\mathcal{MCS}$ is a local inference system. This is in contrast with the more
   specific meaning of the term in this paper.

argumentation processes by determining their semantics, reasoning modality, and arguments.
• Based on it, we develop *argumentation context systems (ACS)*, which are composed of *argumentation modules* and *mediators* (Sect. 5) in a graph structure. A mediator collects context information for a module based on the arguments accepted by its parent modules. It then applies a consistency handling method, possibly based on preference information regarding its parents, to select a consistent subset of the context statements.

Both context-based argumentation and ACSs are generic and can be instantiated in various ways (we consider several of them).

2 Background

Abstract argumentation. We assume some familiarity with Dung-style abstract argumentation [10] and just recall the essential definitions. An argumentation framework is a pair $\mathcal{A} = (AR, attacks)$ where AR is a set of arguments, and *attacks* is a binary relation on AR (used in infix in prose). An argument $a \in AR$ is *acceptable with respect to a set S of arguments*, if each argument $b \in AR$ that attacks a is attacked by some $b' \in S$. A set S of arguments is *conflict-free*, if there are no arguments $a, b \in S$ such that a attacks b, and S is *admissible*, if in addition each argument in S is acceptable wrt. S.

Dung defined three different semantics for $\mathcal{A} = (AR, attacks)$:

– A *preferred extension* of $\mathcal{A}$ is a maximal (wrt. $\subseteq$) admissible set of $\mathcal{A}$.
– A *stable extension of $\mathcal{A}$* is a conflict-free set of arguments S which attacks each argument not belonging to S.
– The *grounded extension of $\mathcal{A}$* is the least fixpoint of the operator $F_{\mathcal{A}} : 2^{AR} \to 2^{AR}$ where $F_{\mathcal{A}}(S) = \{a \in AR \mid a$ is acceptable wrt. $S\}$.

The unique grounded extension is a subset of the intersection of all preferred extensions, and each stable extension is a preferred extension, but not vice versa. While the grounded and some preferred extension are guaranteed to exist, $\mathcal{A}$ may have no stable extension.[3]

A *preference based argumentation framework (PAF)* $\mathcal{P} = (AR, C, \geq)$ [1,8] is based on a relation C representing logical conflicts between arguments AR and a reflexive, transitive preference relation $\geq$ on AR for expressing that arguments are stronger than others; as usual, the associated strict preference $>$ is given by $a > b$ iff $a \geq b$ and $b \not\geq a$. The PAF $\mathcal{P}$ induces an ordinary argumentation framework $\mathcal{A}_P = (AR, attacks)$ where $attacks = \{(a, b) \in C \mid b \not> a\}$. The grounded, preferred, and stable extensions of $\mathcal{P}$ are then the respective extensions of $\mathcal{A}_P$.

Value based argumentation frameworks (VAFs) [2,3] derive preferences among arguments from the values they promote. An audience specific[4] VAF

---

[3] We confine the discussion here to Dung's original semantics. Recent proposals like semi-stable [7] and ideal [11] semantics can be easily integrated in our framework.
[4] We omit discussing audiences; in our framework, they are best modeled by modules.

$$\mathcal{V} = (AR, \textit{attacks}, V, \textit{val}, \textit{valprefs})$$

extends an ordinary argumentation framework $(AR, \textit{attacks})$ with a non-empty set of values V, a function $val : AR \rightarrow V$, and a strict preference relation $\textit{valprefs}$ on V; argument a is preferred over b whenever $(val(a), val(b)) \in \textit{valprefs}$. The preferences are then treated as in Amgoud & Cayrol's approach.

Inconsistency handling. Let $F = (F_1, \dots, F_n)$ be a sequence of sets of formulas. We will use 4 different methods to generate a consistent subset of $F_1 \cup \dots \cup F_n$ from F: a skeptical method $sub_\succ$ which goes back to [5] and uses a partial preference order $\succ$ on the sets F_i, together with its skeptical variant $sub_{sk,\succ'}$, and a majority-based method maj, together with its skeptical variant maj_{sk}. The former methods are used to integrate information in cases where some parent modules are more important than others, the latter in peer-to-peer situations where voting is more appropriate. As here the technical details are of minor relevance, we leave them for a longer version of the paper.

Multi-context systems. Such systems model, in the tradition of [14] and their nonmonotonic extensions [17,6], the information flow between different reasoning modules (called *contexts*) using so-called *bridge rules*, which may refer to other modules in their bodies. For our purposes, we only need to consider rules referring to a single other module that is implicitly given. Therefore, our bridge rules will look like ordinary logic programming rules of the form:

$$s \leftarrow p_1, \dots, p_j, \textbf{not } p_{j+1}, \dots, \textbf{not } p_m \tag{1}$$

where the head s is a context expression (defined in the Sect. 4) and the body literals are arguments p_i (possibly negated with **not**) from a parent argumentation framework.

3   Motivating Examples

Legal reasoning. Different proof standards determine which arguments are acceptable depending on the type of a trial: *beyond reasonable doubt* for criminal trials; *preponderance of evidence* for civil trials (we omit *scintilla of evidence* for simplicity). Consider a situation where a judge J has to decide which of the arguments put forward by a prosecutor P are acceptable. Assume also that criminal trials require grounded reasoning, whereas less skeptical reasoning methods may be acceptable in civil trials.[5]

The arguments and attacks put forward by P form an argumentation framework $\mathcal{A} = (\{a_1, \dots, a_n\}, \textit{attacks})$. The judge has information about which of P's arguments are doubtful and the type of a trial. Using arguments dt_i, bd_i and dr_i for a_i is doubtful, beyond reasonable doubt and disregarded, respectively, the judge's argumentation framework may look as in Fig. 1, where *crl* means criminal trial and *civ* civil trial (arcs are according to *attacks*).[6] Optional arguments represent information the judge may or may not have are in blue.

---

[5] A similar example was discussed in [21].

[6] For simplicity, names of abstract arguments in $\mathcal{M}_2$ reflect their intended meaning.

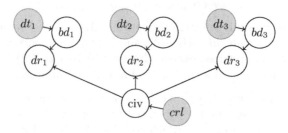

Fig. 1. The judge's argumentation framework for $n = 3$

The outcome of J's argumentation determines how P can argue, and sets a *context* for P; a language for representing contexts will be defined in the next section. As for now, let $\overline{\mathbf{arg}}(a_i)$ state that a_i is invalid for P, and that $\mathbf{sem}(grnd)$ fixes the semantics for P's argumentation framework to grounded semantics.

However, to build this context, we need to transform the relevant abstract arguments accepted by J into adequate context expressions for P that invalidate unusable arguments, pick the right semantics, etc. This is achieved by a *mediator*, which uses the following bridge rules $i \in \{1, \ldots, n\}$:

$$\{ \ \overline{\mathbf{arg}}(a_i) \leftarrow dr_i, \quad \mathbf{sem}(grnd) \leftarrow \mathbf{not} \ civ \ \}.$$

Not all arguments of J need to be "visible" for P. Privacy issues (e.g. hiding or summarizing arguments) can be modeled by choosing adequate bridge rules.

The example involving judge, prosecutor and mediator is clearly simplistic and one might ask why the two argumentation frameworks are not simply merged into one. There are the following aspects, though: (1) It is non-obvious how, for instance, a statement fixing a particular semantics could be eliminated this way. (2) Even if we could compile a single argumentation framework out of the example, the distinction between arguments and meta-arguments and their origin would be blurred. (3) Most importantly, the framework we develop aims at capturing much more complex scenarios where such compilations appear neither doable nor fruitful. We now briefly discuss such a scenario.

Conference reviewing. Consider the paper review process for a modern AI conference. This process can typically be characterized as follows

- There is a hierarchy consisting of a PC chair, several area chairs, many reviewers, and even more authors.
- The PC chair determines the review criteria, acceptance rates etc.
- Area chairs make sure reviewers make fair judgements and eliminate unjustified arguments from their reviews.
- Authors give feedback on the preliminary reviews of their papers. Information flow is thus cyclic.
- Reviewers exchange arguments in a peer-to-peer discussion.
- Area chairs generate a consistent recommendation out of the final reviews.
- PC chair takes recommendations as input for final decision.

What we see here is a complex argumentation scenario including hierarchic (the PC chair setting criteria) as well as peer-to-peer (the reviewers discussing the same paper) forms of argumentation in groups. It is also evident that information flow is cyclic: even the authors nowadays are able to feed arguments back into higher levels of the reviewing hierarchy.

Examples like this call for a flexible framework allowing for cyclic structures encompassing a variety of information integration methods. Exactly this kind of framework is what we are going to develop in the rest of the paper.

4   Context-Based Argumentation

We now give a simple language for representing context and define what it means for a set of arguments to be acceptable for an argumentation framework $\mathcal{A}$ given a context C. In the context language, we want to specify various aspects, including: a) preferences among arguments; b) values and value orderings; c) validity/invalidity of specific arguments; d) addition/deletion of attack relationships among arguments; e) a reasoning mode (sceptical vs. credulous); and f) an argumentation semantics (stable, preferred, grounded).

Definition 1. *A* context expression *for a set of arguments AR and a set of values V has one of the following forms ($a, b \in AR$; $v, v' \in V$):*

$\mathbf{arg}(a) \ / \ \overline{\mathbf{arg}}(a)$	*a is a valid / invalid argument*
$\mathbf{att}(a, b) \ / \ \overline{\mathbf{att}}(a, b)$	*(a, b) is a valid / invalid attack*
$a > b$	*a is strictly preferred to b*
$\mathbf{val}(a, v)$	*the value of a is v*
$v > v'$	*value v is strictly better than v'*
$\mathbf{mode}(r)$	*the reasoning mode is $r \in \{skep, cred\}$*
$\mathbf{sem}(s)$	*the chosen semantics is $s \in \{grnd, pref, stab\}$*

A context C is a set of context expressions (for given AR and V).

The preference and value expressions together define a preference order on arguments as follows:

Definition 2. *For a context C, the preference order $>_C$ induced by C is the smallest transitive relation such that $a >_C b$ if either (i) $a > b \in C$ or (ii) $\mathbf{val}(a, v_1) \in C$, $\mathbf{val}(b, v_2) \in C$, and (v_1, v_2) is in the transitive closure of $\{(v, v') \mid v > v' \in C\}$.*

A context C is *consistent*, if the following conditions hold:

1. $>_C$ is a strict partial order,
2. for no a both $\mathbf{arg}(a) \in C$ and $\overline{\mathbf{arg}}(a) \in C$,
3. for no (a, b) both $\mathbf{att}(a, b) \in C$ and $\overline{\mathbf{att}}(a, b) \in C$,
4. C contains at most one expression of the form $\mathbf{mode}(r)$; the same holds for $\mathbf{sem}(s)$ and for $\mathbf{val}(a, v)$, for each a.

Now we define the semantics of a consistent context. It acts as a modifier for an argumentation framework $\mathcal{A}$, which is evaluated under the argumentation semantics and reasoning mode specified in the context.

Definition 3. *Let $\mathcal{A} = (AR, attacks)$ be an argumentation framework, let V be a set of values, and let C be a consistent context for AR and V. The C-modification of $\mathcal{A}$ is the argumentation framework $\mathcal{A}^C = (AR^C, attacks^C)$, where*

- $AR^C = AR \cup \{\boldsymbol{def}\}$, *where* $\boldsymbol{def} \notin AR$.
- $attacks^C$ *is the smallest relation satisfying the following conditions:*
  1. *if* $\boldsymbol{att}(a, b) \in C$, *then* $(a, b) \in attacks^C$,
  2. *if* $(a, b) \in attacks$, $\overline{\boldsymbol{att}}(a, b) \notin C$ *and* $b \not>_C a$, *then* $(a, b) \in attacks^C$,
  3. *if* $\overline{\boldsymbol{arg}}(a) \in C$ *or* $(\boldsymbol{arg}(b) \in C \wedge (a, b) \in attacks^C)$ *then* $(\boldsymbol{def}, a) \in attacks^C$.

The basic idea is that the new, non-attackable argument **def** defeats invalid arguments as well as attackers of valid arguments. In this way, it is guaranteed that (in)valid arguments will indeed be (in)valid, independently of the chosen semantics. Moreover, the definition of $attacks^C$ guarantees that valid attacks are taken into account while invalid ones are left out. It also ensures that preferences among arguments are handled correctly, by disregarding any original attack (a, b) where b is more preferred than a. The preferences among arguments may be stated directly, or indirectly by the argument values and their preferences.

Example 1. Let $\mathcal{A} = (\{a, b, c, d\}, attacks)$ with $attacks = \{(a, b), (b, d), (d, b), (b, c)\}$. Moreover, let $C = \{\overline{\boldsymbol{arg}}(a), \boldsymbol{val}(b, v_1), \boldsymbol{val}(d, v_2), v_1 > v_2, c > b\}$. We obtain $\mathcal{A}^C = (\{a, b, c, d, \boldsymbol{def}\}, \{(\boldsymbol{def}, a), (a, b), (b, d)\})$.

Based on the C-modification, we define the sets of acceptable arguments. We adopt credulous reasoning and preferred semantics by default, i.e., whenever a context C contains no expression of the form $\boldsymbol{mode}(m)$ (resp., $\boldsymbol{sem}(s)$), we implicitly assume $\boldsymbol{mode}(cred) \in C$ (resp., $\boldsymbol{sem}(pref) \in C$).

Definition 4. *Let $\mathcal{A} = (AR, attacks)$ be an argumentation framework, let V be a value set, and let C be a consistent context for AR and V such that $\boldsymbol{sem}(s) \in C$, $\boldsymbol{mode}(m) \in C$. A set $S \subseteq AR$ is an* acceptable C-extension *for $\mathcal{A}$, if either*

- $m = cred$ *and* $S \cup \{\boldsymbol{def}\}$ *is an s-extension[7] of $\mathcal{A}^C$, or*
- $m = skep$ *and* $S \cup \{\boldsymbol{def}\}$ *is the intersection of all s-extensions of $\mathcal{A}^C$.*

We call a context *purely preferential* (resp., *purely value-based*), if it contains besides **mode**- and **sem**-statements only expressions of form $a > b$ (resp., $\boldsymbol{val}(a, v)$ or $v > v'$). The next proposition shows that our definition "does the right thing:"

Proposition 1. *Suppose C, $\mathcal{A}$ and V are as in Definition 4. Then,*

1. *if C is purely preferential, the acceptable C-extensions of $\mathcal{A}$ coincide with the s-extensions of the PAF $\mathcal{P} = (AR, attacks, \geq_{\mathcal{P}})$, if the strict partial order induced by $\geq_{\mathcal{P}}$ coincides with $>_C$.*

---

[7] s-extension means grounded, preferred, stable extension if $s = grnd, pref, stab$, resp.

2. *if C is purely value-based and $AR = \{a \mid \boldsymbol{val}(a,v) \in C\}$, the acceptable C-extensions of $\mathcal{A}$ coincide with the s-extensions of the VAF $\mathcal{V} = (AR, attacks, V, val, valprefs)$ where $val(a) = v$ iff $\boldsymbol{val}(a,v) \in C$ and $(v,v') \in valprefs$ iff (v,v') is in the transitive closure of $\{(v,v') \mid v > v' \in C\}$.*
3. *if $\boldsymbol{arg}(a) \in C$ (resp., $\overline{\boldsymbol{arg}}(a) \in C$) and S is an acceptable C-extension, then $a \in S$ (resp., $a \notin S$).*

5   Argumentation Context Systems

We now develop *argumentation context systems* ($\mathcal{ACS}$s) as a flexible framework for distributed argumentation, encompassing arbitrary directed graphs of *argumentation modules*. Such modules $\mathcal{M} = (\mathcal{A}, Med)$ consist of a Dung-style argument framework $\mathcal{A}$ and a *mediator Med*, which determines a context C for $\mathcal{A}$. To this end, it translates arguments accepted by $\mathcal{M}$'s parents (i.e., its direct ancestor modules) into context expressions using bridge rules and combines them with local context information. To ensure consistency of C, the mediator uses a *consistency handling method* that resolves any inconsistency between the local and the parent information (multiple parents are possible). Figure 2 shows the structure of an example $\mathcal{ACS}$. The module $\mathcal{M}_1$ has the single parent $\mathcal{M}_3$, and Med_1 receives input from $\mathcal{A}_3$; we say Med_1 is *based on* $\mathcal{A}_3$. The module $\mathcal{M}_2$ has the parents $\mathcal{M}_3$ and $\mathcal{M}_4$, and Med_2 is based on $\mathcal{A}_3$ and $\mathcal{A}_4$.

5.1   Mediators

We now define precisely what we mean by a mediator.

Definition 5. *Let $\mathcal{A}_1, \mathcal{A}_2, \ldots, \mathcal{A}_k$, $k \geq 1$, be argumentation frameworks. A mediator for $\mathcal{A} = \mathcal{A}_1$ based on $\mathcal{A}_2, \ldots, \mathcal{A}_k$ is a tuple*

$$Med = (E_1, R_2, \ldots, R_k, choice)$$

where

- *E_1 is a set of context expressions for $\mathcal{A}$;*
- *R_i, $2 \leq i \leq k$, is a set of rules of form (1) where s is a context expression for $\mathcal{A}$ and $p_1, \ldots, p_m$ are arguments in $\mathcal{A}_i$ (bridge rules for $\mathcal{A}$ based on $\mathcal{A}_i$);*

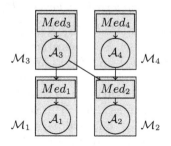

Fig. 2. Example argumentation context system $\mathcal{F} = \{(\mathcal{A}_i, Med_i) \mid 1 \leq i \leq 4\}$

– *choice* $\in \{ sub_\succ, sub_{sk,\succ}, maj, maj_{sk} \}$, *where* $\succ$ *is a strict partial order on* $\{1, \ldots, k\}$.

Intuitively, the $\mathcal{A}_i$, are from the parent modules $\mathcal{M}_i$ of $\mathcal{M}$. The set E_1 contains context statements for $\mathcal{A}$ which the mediator itself considers adequate; it is also used to provide the context for modules without parents. The bridge rules determine the relevant context expressions for $\mathcal{A}$ based on arguments accepted in $\mathcal{A}_i$ (i.e., in $\mathcal{M}_i$). This overcomes the following problem: argumentation in Dung-style systems is *abstract*, i.e., the content of arguments is not analyzed. In order to use arguments of a parent module in determining the context for a child module, this abstract view must be given up to some extent, as context expressions must be associated with arguments of the parent modules.

The parameter *choice* fixes a inconsistency handling strategy. The order $\succ$ is used to prioritize context expressions from different modules. Index 1 is included in the ordering; this makes it possible to specify whether statements in E_1 can be overridden by statements coming from parent modules or not. For the majority-based inconsistency methods maj and maj_{sk}, the order is not needed since the number of supporting modules is implicitly used as the preference criterion. Other consistency methods than those considered can be easily integrated.

Given sets of accepted arguments for all parent modules of $\mathcal{M}$, the mediator defines the consistent acceptable contexts for its argument system $\mathcal{A}$. There may be more than one acceptable context.

Let R_i be a set of bridge rules for $\mathcal{A}$ based on $\mathcal{A}_i$, and S_i a set of arguments of $\mathcal{A}_i$. $R_i(S_i)$ is the set of context expressions

$$R_i(S_i) = \{h \mid h \leftarrow a_1, \ldots, a_j, \mathbf{not}\ b_1, \ldots, \mathbf{not}\ b_n \in R_i,$$
$$\{a_1, \ldots, a_j\} \subseteq S_i, \{b_1, \ldots, b_n\} \cap S_i = \emptyset \},$$

Intuitively, $R_i(S_i)$ contains the context statements for $\mathcal{A}$ derivable through rules in R_i given the arguments S_i are accepted by $\mathcal{A}_i$.

The sets $E_1, R_2(S_2), \ldots, R_k(S_k)$ provide context information for $\mathcal{A}$ coming from the mediator for $\mathcal{A}$, respectively from the argument frameworks $\mathcal{A}_2, \ldots, \mathcal{A}_k$ in the parent modules. This information is integrated into a consistent context using the chosen consistency handling method.

Definition 6. *Let* $Med = (E_1, R_2, \ldots, R_k, choice)$ *be a mediator for* $\mathcal{A}$ *based on* $\mathcal{A}_2, \ldots, \mathcal{A}_k$. *A context* C *for* $\mathcal{A}$ *is* acceptable *wrt. sets of arguments* $S_2, \ldots, S_k$ *of* $\mathcal{A}_2, \ldots, \mathcal{A}_k$, *if* C *is a* choice-preferred set *for* $(E_1, R_2(S_2), \ldots, R_k(S_k))$.

Example 2. Consider a mediator $Med = (E_1, R_2, R_3, sub_\succ)$ based on argumentation frameworks $\mathcal{A}_2$ and $\mathcal{A}_3$. Let $E_1 = \{\mathbf{sem}(grnd), v > v'\}$ and $1 \succ 2 \succ 3$. Let $R_2 = \{\mathbf{arg}(a_1) \leftarrow b_1, a_2 > a_3 \leftarrow b_2\}$ and $R_3 = \{\overline{\mathbf{arg}}(a_1) \leftarrow c_1, \mathbf{val}(a_2, v') \leftarrow c_2, \mathbf{val}(a_3, v) \leftarrow c_3\}$. Suppose $S_2 = \{b_1, b_2\}$ is accepted in $\mathcal{A}_2$ and $S_3 = \{c_1, c_2, c_3\}$ in $\mathcal{A}_3$. We obtain $R_2(S_2) = \{\mathbf{arg}(a_1), a_2 > a_3\}$ and $R_3(S_3) = \{\overline{\mathbf{arg}}(a_1), \mathbf{val}(a_2, v'), \mathbf{val}(a_3, v)\}$. Note that $E_1 \cup R_2(S_2)$ is consistent. From the least preferred context information $R_3(S_3)$, only $\mathbf{val}(a_2, v')$ or $\mathbf{val}(a_3, v)$ can be consistently added to the more preferred context information. The acceptable contexts are $C_1 = E_1 \cup R_2(S_2) \cup \{\mathbf{val}(a_2, v')\}$ and $C_2 = E_1 \cup R_2(S_2) \cup \{\mathbf{val}(a_3, v)\}$.

5.2   The Framework

We are now in a position to introduce our framework.

Definition 7. *An* module $\mathcal{M} = (\mathcal{A}, Med)$ *consists of an argument framework* $\mathcal{A}$ *and a mediator Med for $\mathcal{A}$ (based on some argumentation frameworks).*

An argument context system is a collection of modules that fulfills certain structural conditions.

Definition 8. *An* argumentation context system $(\mathcal{ACS})$ *is a set*[8]

$$\mathcal{F} = \{\mathcal{M}_1, \ldots, \mathcal{M}_n\}$$

of modules $\mathcal{M}_i = (\mathcal{A}_i, Med_i)$, $1 \leq i \leq n$, *where Med_i is based on argumentation frameworks* $\mathcal{A}_{i_1}, \ldots, \mathcal{A}_{i_{k_i}}$, $\{i_1, \ldots, i_{k_i}\} \subseteq \{1, \ldots, n\}$ *(self-containedness).*

Self-containedness naturally induces the following graph and structural notion.

Definition 9. *The* module graph *of an $\mathcal{ACS}$ $\mathcal{F}$ as in Definition 8 is the directed graph $G(\mathcal{F}) = (\mathcal{F}, E)$ where $\mathcal{M}_j \to \mathcal{M}_i$ is in E iff $\mathcal{A}_j$ is among the $\mathcal{A}_{i_1}, \ldots, \mathcal{A}_{i_{k_i}}$ on which Med_i is based. We call $\mathcal{F}$ hierarchic, if $G(\mathcal{F})$ is acyclic.*

We next define the acceptable states for our framework. Intuitively, such a state consists of a context and a set of arguments for all modules $\mathcal{M}_i$ such that in each case the chosen arguments form an acceptable set for $\mathcal{A}_i$ given the respective context, and this context (determined by Med_i) is acceptable with respect to the argument sets chosen for the parent modules of $\mathcal{A}_i$. More formally,

Definition 10. *Let $\mathcal{F} = \{\mathcal{M}_1, \ldots, \mathcal{M}_n\}$ be an $\mathcal{ACS}$. A state of $\mathcal{F}$ is a function S that assigns each $\mathcal{M}_i = (\mathcal{A}_i, Med_i)$ a pair $S(\mathcal{M}_i) = (Acc_i, C_i)$ of a subset Acc_i of the arguments of $\mathcal{A}_i$ and a set C_i of context expressions for $\mathcal{A}_i$.*

*   *A state S is* acceptable, *if (i) each Acc_i is an acceptable C_i-extension for $\mathcal{A}_i$, and (ii) each C_i is an acceptable context for Med_i wrt. all Acc_j such that $\mathcal{M}_j \to \mathcal{M}_i$.*

As stable semantics may be chosen for argumentation frameworks in modules, clearly an acceptable state is not guaranteed to exist. However, non-existence can arise even in absence of stable semantics and negation in bridge rules.

Example 3. Let $\mathcal{F} = (\mathcal{M}_1, \mathcal{M}_2)$ with modules $\mathcal{M}_i = (\mathcal{A}_i, Med_i)$, $i = 1, 2$, where

* $\mathcal{A}_1 = (\{a\}, \emptyset)$, $Med_1 = (\{\mathbf{sem}(s_1)\}, \{\overline{\mathbf{arg}}(a) \leftarrow b_2\}, c_1)$;
* $\mathcal{A}_2 = (\{b_1, b_2\}, \{(b_1, b_2)\})$, $Med_2 = (\{\mathbf{sem}(s_2)\}, \{\overline{\mathbf{arg}}(b_1) \leftarrow a\}, c_2)$.

Note that $G(\mathcal{F})$ is cyclic, as $\mathcal{M}_1 \to \mathcal{M}_2$ and $\mathcal{M}_2 \to \mathcal{M}_1$. Now for arbitrary s_1, c_1 and s_2, c_2, no acceptable state exists: assume $Acc_1 = \{a\}$. Using its single bridge rule, Med_2 will derive $\overline{\mathbf{arg}}(b_1)$, which is consistent with $\mathbf{sem}(s_2)$ and will belong to the context C_2, regardless of the chosen consistency method c_2. This,

---

[8] For multiple occurrences of the same module, this is viewed as an indexed set.

however, will lead to acceptance of b_2 independently of the semantics s_2, as the only potential attacker of b_2, namely b_1, is invalid. Now, if b_2 is in Acc_2, then the single bridge rule of Med_1 invalidates argument a; hence, there is no acceptable state with $Acc_1 = \{a\}$.

The other option, namely $Acc_1 = \emptyset$, also fails: now b_1 is not invalidated in $\mathcal{M}_2$ and thus defeats b_2. As a consequence, Med_1's bridge rule does not fire and there is no justification not to accept argument a in $\mathcal{A}_1$.

To guarantee existence of acceptable states, we need more conditions; for example, excluding stable semantics and cycles between modules.

Proposition 2. *Suppose $\mathcal{F}$ is a hierarchic $\mathcal{ACS}$ and that $sem(stab)$ does not occur in any mediator of $\mathcal{F}$. Then $\mathcal{F}$ possesses at least one acceptable state.*

Computing some acceptable state is still intractable in this setting, due to the intractability of recognizing preferred extensions [9]. However,

Proposition 3. *If $\mathcal{F}$ is a hierarchic $\mathcal{ACS}$ and all modules use grounded semantics and either $sub_\succ$ or maj for choice, then some acceptable state of $\mathcal{F}$ is computable in polynomial time.*

Regarding the complexity in the general case, we note the following result.

Proposition 4. *Deciding whether an arbitrary given $\mathcal{ACS}$ $\mathcal{F}$ has some acceptable state is Σ_3^p-complete.*

Intuitively, an acceptable state can be guessed and verified in polynomial time with the help of an Σ_2^p oracle; note that the reasoning tasks in Definition 4 are all feasible polynomial time with an Σ_2^p oracle. On the other hand, skeptical inference from all preferred extensions is Π_2^p-complete [12], which in combination with the module framework generates the Σ_3^p-hardness. This can be shown by a reduction from suitable quantified Booleans formulas, which uses no negation in bridge rules and and an arbitrary inconsistency handling method *choice*.

Depending on the various parameters and the graph structure, the complexity decreases. The complexity of C-extensions is dominated by the underlying argumentation framework, and in several situations, the mediator framework does not increase complexity. A more detailed analysis is left for further work.

5.3   Relationship to EAFs

Modgil [16] recently introduced an interesting extension of argumentation frameworks where arguments may not only attack other arguments, but also attacks. An *extended argumentation framework* (EAF) $\mathcal{E} = (AR, attacks, D)$ adds to a Dung-style argumentation framework $(AR, attacks)$ a set $D \subseteq AR \times attacks$ of *attacks against attacks*. Whenever $(a_1, (b, c)) \in D$ and $(a_2, (c, b)) \in D$ it is required that *attacks* contains both (a_1, a_2) and (a_2, a_1). $S \subseteq AR$ is conflict free iff for all $a, b \in S$: if $(a, b) \in attacks$, then $(b, a) \notin attacks$ and there is $c \in S$ such that $(c, (a, b)) \in D$. Argument a S-defeats b iff $(a, b) \in attacks$ and, for no

$c \in S$, $(c,(a,b)) \in D$. S is a stable extension of $\mathcal{E}$ iff S is conflict free and each $c \in AR \setminus S$ is S-defeated by some element of S. We refer to [16] for numerous interesting applications of EAFs.

We can show that under stable semantics, any EAF $\mathcal{E}$ can be modeled as an $\mathcal{ACS}$ $\mathcal{F}(\mathcal{E})$ that consists of a single module with self-feedback. In detail, we construct $\mathcal{F}(\mathcal{E}) = \{\mathcal{M}^*\}$, where $\mathcal{M}^* = (\mathcal{A}^*, Med^*)$ such that $\mathcal{A}^* = (AR, attacks)$ and $Med^* = (\{\mathbf{sem}(stab), \mathbf{mode}(cred)\}, R, choice)$ is based on $\mathcal{A}^*$, where

$$R = \{\,\overline{\mathbf{att}}(a,b) \leftarrow c, b \mid (c,(a,b)) \in D, (b,a) \notin attacks\,\} \cup$$
$$\{\,\overline{\mathbf{att}}(a,b) \leftarrow c, \mathbf{not}\ b \mid (c,(a,b)) \in D\,\}.$$

and *choice* is arbitrary. The first type of rules in R handles conflict freeness, while the second ensures that an argument can only be defeated by an attack which is not successfully attacked.

Proposition 5. *Let $\mathcal{E}$ be an EAF. Then, (i) for each acceptable state S of $\mathcal{F}(\mathcal{E})$ with $S(\mathcal{M}^*) = (T,C)$, T is a stable extension of $\mathcal{E}$, and (ii) for each stable extension T of $\mathcal{E}$, $\mathcal{F}(\mathcal{E})$ has some acceptable state S such that $S(\mathcal{M}^*) = (T,C)$.*

The relationship under the other semantics is trickier; this is basically due to the fact that Modgil and Dung have very different notions of argument acceptability. A thorough investigation is an interesting topic for future work.

6   Related Work and Conclusion

We presented a flexible, modular framework for abstract argumentation. It builds on existing proposals extending them in various respects: argumentation is based on contexts described in a native language, integrating preference- and value-based argumentation, direct (in)validation of arguments and attacks, and specification of reasoning mode and semantics. Context information is integrated by a mediator. Arbitrary directed module graphs cover a wide range of applications involving multi-agent meta-argumentation.

Models of meta- and hierarchic argumentation are not new. The approaches by Modgil [15,16] were already discussed. Wooldridge, McBurney and Parsons [20] develop a meta-level approach which shares motivation with our work, but has different focus: a meta-level defines arguments, attacks, extensions etc. for the lower level based on formal provability in its logic. In contrast, we take the basic notions for granted and determine, at a generic level, how a module in an arbitrary module graph influences argumentation in others by determining its context. The many parameter combinations yield then a range of concrete systems with different properties.

A framework for distributed argumentation was presented in [18], based on defeasible argumentation as in [13], where a moderator integrates the argumentation structures of a group of agents. In contrast, we stick to abstract argumentation and allow for more general relationships between modules.

Our approach also differs from recent work by Binas and McIlraith [4] on distributed query answering: while our mediators use "classical" preference based inconsistency methods to establish argumentation contexts, they use techniques based on prioritized argumentation to define distributed entailment for graphs of heterogeneous reasoners. Moreover, they focus on information integration rather than meta-reasoning as we do.

Our framework specializes the multi-context systems of Brewka and Eiter [6] by fixing Dung style argument systems as reasoning components. The use of mediators to integrate meta-information and the context language to control argumentation clearly goes beyond these systems.

Our future work includes an investigation of more expressive mediator languages (both in terms of constructs and bridge rules used), and a detailed study of computational aspects, comprising complexity and implementation.

Acknowledgement. This work has been supported by the Austrian Science Fund (FWF) project P20841 and the Vienna Science and Technology Fund (WWTF) project ICT08-020.

References

1. Amgoud, L., Cayrol, C.: On the acceptability of arguments in preference-based argumentation. In: Proc. Fourteenth Conference on Uncertainty in Artificial Intelligence, UAI 1998, pp. 1–7 (1998)
2. Bench-Capon, T.J.M.: Value-based argumentation frameworks. In: Proc. 9th International Workshop on Non-Monotonic Reasoning, NMR 2002, Toulouse, France, pp. 443–454 (2002)
3. Bench-Capon, T.J.M.: Persuasion in practical argument using value-based argumentation frameworks. J. Log. Comput. 13(3), 429–448 (2003)
4. Binas, A., McIlraith, S.: Peer-to-peer query answering with inconsistent knowledge. In: Proc. 11th International Conference on Principles of Knowledge Representation and Reasoning, KR 2008, Sydney, Australia, pp. 329–339 (2008)
5. Brewka, G.: Preferred subtheories: An extended logical framework for default reasoning. In: Proc. IJCAI 1989, pp. 1043–1048 (1989)
6. Brewka, G., Eiter, T.: Equilibria in heterogeneous nonmonotonic multi-context systems. In: Proc. AAAI 2007, pp. 385–390 (2007)
7. Caminada, M.: Semi-stable semantics. In: Proc. Computational Models of Argument, COMMA 2006, pp. 121–130 (2006)
8. Dimopoulos, Y., Moraitis, P., Amgoud, L.: Theoretical and computational properties of preference-based argumentation. In: Proc. ECAI 2008, Patras, Greece, pp. 463–467 (2008)
9. Dimopoulos, Y., Torres, A.: Graph theoretical structures in logic programs and default theories. Theor. Comput. Sci. 170(1-2), 209–244 (1996)
10. Dung, P.M.: On the acceptability of arguments and its fundamental role in nonmonotonic reasoning, logic programming and n-person games. Artif. Intell. 77(2), 321–358 (1995)
11. Dung, P.M., Mancarella, P., Toni, F.: Computing ideal sceptical argumentation. Artif. Intell. 171(10-15), 642–674 (2007)
12. Dunne, P.E., Bench-Capon, T.J.M.: Coherence in finite argument systems. Artif. Intell. 141(1/2), 187–203 (2002)

13. García, A.J., Simari, G.R.: Defeasible logic programming: An argumentative approach. TPLP 4(1-2), 95–138 (2004)
14. Giunchiglia, F., Serafini, L.: Multilanguage hierarchical logics, or: how we can do without modal logics. Artificial Intelligence 65(1), 29–70 (1994)
15. Modgil, S.: Hierarchical argumentation. In: Fisher, M., van der Hoek, W., Konev, B., Lisitsa, A. (eds.) JELIA 2006. LNCS (LNAI), vol. 4160, pp. 319–332. Springer, Heidelberg (2006)
16. Modgil, S.: Reasoning about preferences in argumentation frameworks. Artif. Intell. (2007) (to appear)
17. Roelofsen, F., Serafini, L.: Minimal and absent information in contexts. In: Proc. IJCAI 2005, pp. 558–563 (2005)
18. Thimm, M., Kern-Isberner, G.: A distributed argumentation framework using defeasible logic programming. In: Proc. Computational Models of Argument, COMMA 2008, pp. 381–392 (2008)
19. Wiederhold, G.: Mediators in the architecture of future information systems. IEEE Computer 25(3), 38–49 (1992)
20. Wooldridge, M., McBurney, P., Parsons, S.: On the meta-logic of arguments. In: Proc. AAMAS 2005, pp. 560–567 (2005)
21. Wyner, A.Z., Bench-Capon, T.J.M.: Modelling judicial context in argumentation frameworks. In: Proceedings COMMA 2008, pp. 417–428 (2008)

A Revised Concept of Safety for General Answer Set Programs

Pedro Cabalar[1], David Pearce[2,*], and Agustín Valverde[3,*]

[1] Universidade da Coruña
cabalar@udc.es
[2] Universidad Politécnica de Madrid, Spain
david.pearce@upm.es
[3] Universidad de Málaga, Málaga, Spain
a_valverde@ctima.uma.es

Abstract. To ensure a close relation between the answer sets of a program and those of its ground version, some answer set solvers deal with variables by requiring a safety condition on program rules. If we go beyond the syntax of disjunctive programs, for instance by allowing rules with nested expressions, or perhaps even arbitrary first-order formulas, new definitions of safety are required. In this paper we consider a new concept of safety for formulas in quantified equilibrium logic where answer sets can be defined for arbitrary first-order formulas. The new concept captures and generalises two recently proposed safety definitions: that of Lee, Lifschitz and Palla (2008) as well as that of Bria, Faber and Leone (2008). We study the main metalogical properties of safe formulas.

1 Introduction

In answer set programming (ASP) more and more tools are being created to enable a fuller use of first-order languages and logic. Recent developments include efforts to extend the syntax of programs as well as to deal more directly with variables in a full, first-order context. In several cases, assumptions such as standard names (SNA) are being relaxed and issues involving programming in open domains are being addressed.

A stable model semantics for first-order structures and languages was defined in the framework of equilibrium logic in [17,18,19]. As a logical basis the non-classical logic of quantified here-and-there, **QHT**, is used (see also [15]). By expanding the language to include new predicates, this logic can be embedded in classical first-order logic, [20], and this permits an alternative but equivalent formulation of the concept of stable model for first-order formulas, expressed in terms of classical, second-order logic [7]. The latter definition of answer set has been further studied in [11,12] where the basis of a first-order programming language, RASPL-1, is described. An alternative approach to a first-order ASP language is developed in [8].

Several implementations of ASP deal with variables by requiring a safety condition on program rules. In the language of disjunctive LPs, this condition is usually expressed

---

* Partially supported by the MEC (now MICINN) projects TIN2006-15455-(C01,C02,C03) and CSD2007-00022, and the Junta de Andalucia project P6-FQM-02049.

E. Erdem, F. Lin, and T. Schaub (Eds.): LPNMR 2009, LNCS 5753, pp. 58–70, 2009.

by saying that a rule is safe if any variable in the rule also appears in its positive body – this condition will be referred here as *DLP safety*. Programs are safe if all their rules are safe. The safety of a program ensures that its answer sets coincide with the answer sets of its ground version and thus allows ASP systems to be based on computations at the level of propositional logic which may include for example the use of SAT-solvers.

What if we go beyond the syntax of disjunctive programs? Adding negation in the heads of program rules will not require a change in the definition of safety. But for more far reaching language extensions, such as allowing rules with nested expressions, or perhaps even arbitrary first-order formulas, new definitions of safety are required. Again the desideratum is that safe programs should be in a certain sense reducible to their ground versions, whether or not these can be compiled into an existing ASP-solver.

Recently, there have been two new extensions of the notion of safety to accommodate different syntactic extensions of the ASP language in the first-order case. In [2], Bria, Faber and Leone introduce an extension to a restricted variant of the language of programs with nested expressions (or nested programs, for short) [16]. The rules are called *normal form nested* or NFN rules. The safety of such rules which will be referred here as *BFL-safety* is designed to guarantee domain independence and permit their efficient translation into the language of disjunctive programs. In this manner, NFN programs can be compiled into the standard DLV system. A second example is the definition of safety introduced by Lee, Lifschitz and Palla in [12] (we will call it here *LLP-safety*) defined essentially for arbitrary first-order formulas in such a way that the stable model of such a formula can be obtained by considering its ground version. It is also shown that the stable models of safe formulas form a (first-order) definable class. Here the motivation is not so much to compile safe programs into existing ASP languages, but rather to find tractable, first-order extensions of the customary languages.[1]

On the face of it, the approach followed in [12] is much more general than that of [2] and so we might expect that the safety of general formulas from [12] embraces also the safe NFN rules of [2]. But this is not the case: safe NFN rules need not be safe formulas in the sense of [12]. In fact, there are simple examples of formulas that would normally be regarded as safe that fail to be so under the [12] definition. Consider the following example.

$$p \leftarrow q(X) \lor r \tag{1}$$

This rule belongs to the syntactic class of programs with nested expressions as introduced in [16], although (1) contains a variable, something not considered in that work. Still, this rule is strongly equivalent to the conjunction of rules

$$p \leftarrow q(X) \; ; \; p \leftarrow r$$

and both are DLP safe (note that this is now a normal logic program) and so, they are BFL-safe and LLP-safe as well. Therefore, it is reasonable to consider (1) as safe. According to [12], however, for (1) to be safe, X must occur in some implication $G \rightarrow H$ such that X belongs to what are called the restricted variables of G. In this case, the only option for G is the rule body $q(X) \lor r$ whose set of restricted variables is empty,

---

[1] *Open* answer set programming studies decidable language extensions via guarded fragments rather than safety conditions,[8].

implying that the rule is not safe. [2] also uses a similar concept of restricted variable but only imposes restrictions on variables occurring in the head of the rule or within the scope of negation; so according to this account, (1) is safe.

In this paper we propose a new concept of safety that, while defined for arbitrary first order formulas, is strictly weaker than LLP-safety, while it captures BFL-safety when restricted to the same syntactic class of NFN-programs.

Desiderata for safety. For our purposes there are three key desiderata associated with safety that we should keep distinct. The first property is very simple and does not refer to grounding. It says merely that a stable model should not contain unnamed individuals. This condition is formally expressed by Theorem 1 below and is fulfilled by formulas that we call *semi-safe*. Secondly we have the property usually most associated with safety. This is called *domain independence* and it does refer to grounding. It says that grounding a program with respect to any superset of the program's constants will not change the class of stable models. This is satisfied by formulas that we call *safe* and is expressed by Proposition 4 and Theorem 3. The third property, also satisfied by safe formulas, is expressed in Theorem 4: it says that the class of stable models should be first-order definable. This may be relevant for establishing properties such as interpolation and for computational purposes, being exploited for instance by the method of loop formulas.[2]

Let us consider some inherent limitations of a safety definition. Arbitrary theories are, in general, not domain independent. Even simple normal rules like, for instance:

$$p \leftarrow not \; q(X) \tag{2}$$

are not domain independent. To see why, take (2) plus the fact $q(1)$. If we ground this program on the domain $\{1\}$ we get the stable model $\{q(1)\}$, but if we use the extended domain $\{1, 2\}$, the only stable model becomes now $\{q(1), p\}$. Unfortunately, directly checking domain independence of an arbitrary formula does not seem a trivial task. It seems much more practical to find a (computationally) simple syntactic condition, like safety, that suffices to guarantee domain independence. However, due to its syntactic nature, a condition of this kind will just be *sufficient* but, most probably, not *necessary*. For instance, while as we saw above, (2) cannot be considered safe under any definition, if we consider the conjunction of (2) with the (variable-free) constraint:

$$\bot \leftarrow q(1) \tag{3}$$

the resulting program is *domain independent*. The reason is that (2) is strongly equivalent to the formula $(\exists x \; \neg q(x)) \rightarrow p$ whereas (3) is equivalent to $\neg q(1)$ that, in its turn, implies $(\exists x \; \neg q(x))$ in **QHT**. This example shows that checking domain independence may require considering semantic interrelations of the formula or program as a whole. On the other hand, the fact that safety is sensitive to redundancies or that the safety of a formula may vary under a strongly equivalent transformation is unavoidable. For instance, the program consisting of (2) and the fact p is clearly domain independent, as in the presence of p, (2) becomes redundant and can be removed.

---

[2] Of course there are other definable classes of stable models, such as those determined by formulas with a finite complete set of *loops* [13].

As regards our methodology, it is important to note that in order to handle explicit quantifiers and possible relaxations of the SNA it is indispensable to use a genuine first-order (or higher-order) semantics. Our approach here will be to analyse safety from the standpoint of (quantified) equilibrium logic in which a first-order definition of stable or equilibrium model can be given in the logic **QHT**. One specific aim is to generalise the safety concept from [12] so that rules such as (1), and indeed all the safe rules from [2], can be correctly categorised as safe. At least as regards simplicity and intuitiveness, we feel that our approach has some advantages over the methodology of [12] that uses a mixture of **QHT** and classical second-order logic.[3]

2   Review of Quantified Equilibrium Logic and Answer Sets

In this paper we restrict attention to function-free languages with a single negation symbol, '$\neg$', working with a quantified version of the logic *here-and-there*. In other respects we follow the treatment of [19]. We consider first-order languages $\mathcal{L} = \langle C, P \rangle$ built over a set of *constant* symbols, C, and a set of *predicate* symbols, P. The sets of $\mathcal{L}$-formulas, $\mathcal{L}$-sentences and atomic $\mathcal{L}$-sentences are defined in the usual way. We work here mainly with *sentences*. If D is a non-empty set, we denote by $At(D, P)$ the set of ground atomic sentences of the language $\langle D, P \rangle$. By an $\mathcal{L}$-interpretation I over a set D we mean a subset of $At(D, P)$. A classical $\mathcal{L}$-structure can be regarded as a tuple $\mathcal{M} = \langle (D, \sigma), I \rangle$ where I is an $\mathcal{L}$-interpretation over D and $\sigma \colon C \cup D \to D$ is a mapping, called the *assignment*, such that $\sigma(d) = d$ for all $d \in D$. If $D = C$ and $\sigma = id$, $\mathcal{M}$ is an *Herbrand structure*.

A *here-and-there* $\mathcal{L}$-structure with static domains, or **QHT**$(\mathcal{L})$-*structure*, is a tuple $\mathcal{M} = \langle (D, \sigma), I_h, I_t \rangle$ where $\langle (D, \sigma), I_h \rangle$ and $\langle (D, \sigma), I_t \rangle$ are classical $\mathcal{L}$-structures such that $I_h \subseteq I_t$. We can think of a here-and-there structure $\mathcal{M}$ as similar to a first-order classical model, but having two parts, or components, h and t, that correspond to two different points or "worlds", 'here' and 'there', in the sense of Kripke semantics for intuitionistic logic [6], where the worlds are ordered by $h \leq t$. At each world $w \in \{h, t\}$ one verifies a set of atoms I_w in the expanded language for the domain D. We call the model static, since, in contrast to, say, intuitionistic logic, the same domain serves each of the worlds. Since $h \leq t$, whatever is verified at h remains true at t. The satisfaction relation for $\mathcal{M}$ is defined so as to reflect the two different components, so we write $\mathcal{M}, w \models \varphi$ to denote that φ is true in $\mathcal{M}$ with respect to the w component. The recursive definition of the satisfaction relation forces us to consider formulas from $\langle C \cup D, P \rangle$. Evidently we should require that an atomic sentence is true at w just in case it belongs to the w-interpretation. Formally, if $p(t_1, \ldots, t_n) \in At(C \cup D, P)$ and $w \in \{h, t\}$ then

$$\mathcal{M}, w \models p(t_1, \ldots, t_n) \quad \text{iff} \quad p(\sigma(t_1), \ldots, \sigma(t_n)) \in I_w.$$
$$\mathcal{M}, w \models t = s \quad \text{iff} \quad \sigma(t) = \sigma(s)$$

Then $\models$ is extended recursively as follows, conforming to the usual Kripke semantics for intuitionistic logic given our assumptions about the two worlds h and t and the

---

[3] For reasons of space, in the sequel some proofs have been shortened or omitted. For a more detailed version, see http://www.ia.urjc.es/~dpearce.

single domain D, see eg. [6]. We shall assume that $\mathcal{L}$ contains the constants $\top$ and $\bot$ and regard $\neg\varphi$ as an abbreviation for $\varphi \rightarrow \bot$.

- $\mathcal{M}, w \models \top$, $\mathcal{M}, w \not\models \bot$
- $\mathcal{M}, w \models \varphi \wedge \psi$ iff $\mathcal{M}, w \models \varphi$ and $\mathcal{M}, w \models \psi$.
- $\mathcal{M}, w \models \varphi \vee \psi$ iff $\mathcal{M}, w \models \varphi$ or $\mathcal{M}, w \models \psi$.
- $\mathcal{M}, t \models \varphi \rightarrow \psi$ iff $\mathcal{M}, t \not\models \varphi$ or $\mathcal{M}, t \models \psi$.
- $\mathcal{M}, h \models \varphi \rightarrow \psi$ iff $\mathcal{M}, t \models \varphi \rightarrow \psi$ and $\mathcal{M}, h \not\models \varphi$ or $\mathcal{M}, h \models \psi$.
- $\mathcal{M}, t \models \forall x\varphi(x)$ iff $\mathcal{M}, t \models \varphi(d)$ for all $d \in D$.
- $\mathcal{M}, h \models \forall x\varphi(x)$ iff $\mathcal{M}, t \models \forall x\varphi(x)$ and $\mathcal{M}, h \models \varphi(d)$ for all $d \in D$.
- $\mathcal{M}, w \models \exists x\varphi(x)$ iff $\mathcal{M}, w \models \varphi(d)$ for some $d \in D$.

Truth of a sentence in a model is defined as follows: $\mathcal{M} \models \varphi$ iff $\mathcal{M}, w \models \varphi$ for each $w \in \{h, t\}$. In a model $\mathcal{M}$ we also use the symbols H and T, possibly with subscripts, to denote the interpretations I_h and I_t respectively; so, an $\mathcal{L}$-structure may be written in the form $\langle U, H, T \rangle$, where $U = (D, \sigma)$. A structure $\langle U, H, T \rangle$ is called *total* if $H = T$, whence it is equivalent to a classical structure. We shall also make use of an equivalent semantics based on many-valued logic; the reader is referred to [18].

The resulting logic is called *Quantified Here-and-There Logic with static domains and decidable equality*, and denoted in [15] by $\mathbf{SQHT}^=$, where a complete axiomaisation can be found. In terms of satisfiability and validity this logic is equivalent to the logic previously introduced in [18]. To simplify notation we drop the labels for static domains and equality and refer to this logic simply as quantified here-and-there, $\mathbf{QHT}$. In the context of logic programs, the following often play a role. Let $\sigma|_C$ denote the restriction of the assignment σ to constants in C. In the case of both classical and $\mathbf{QHT}$ models, we say that the *parameter names assumption* (PNA) applies in case $\sigma|_C$ is surjective, i.e., there are no unnamed individuals in D; the *unique names assumption* (UNA) applies in case $\sigma|_C$ is injective; in case both the PNA and UNA apply, the *standard names assumption* (SNA) applies, i.e. $\sigma|_C$ is a bijection. We will speak about PNA-, UNA-, or SNA-models, respectively, depending on σ.

As in the propositional case, quantified equilibrium logic, or QEL, is based on a suitable notion of minimal model.

Definition 1. *Let φ be an $\mathcal{L}$-sentence. An* equilibrium *model of φ is a total model $\mathcal{M} = \langle (D, \sigma), T, T \rangle$ of φ such that there is no model of φ of the form $\langle (D, \sigma), H, T \rangle$ where H is a proper subset of T.*

2.1   Relation to Answer Sets

We assume the reader is familiar with the usual definitions of answer set based on *Herbrand* models and ground programs, eg. [1]. Two variations of this semantics, the open [8] and generalised open answer set [9] semantics, consider non-ground programs and open domains, thereby relaxing the PNA.

For the present version of QEL the correspondence to answer sets can be summarised as follows (see [18,19,3]). If φ is a universal sentence in $\mathcal{L} = \langle C, P \rangle$ (see §3 below), a total $\mathbf{QHT}$ model $\langle U, T, T \rangle$ of φ is an equilibrium model of φ iff $\langle T, T \rangle$ is a propositional equilibrium model of the grounding of φ with respect to the universe U.

By the usual convention, when $\varPi$ is a logic program with variables we consider the models of its universal closure expressed as a set of logical formulas. It follows that if $\varPi$ is a logic program (of any form), a total **QHT** model $\langle U, T, T \rangle$ of $\varPi$ is an equilibrium model of $\varPi$ iff it is a generalised open answer set of $\varPi$ in the sense of [9]. If we assume all models are UNA-models, we obtain the version of QEL found in [18]. There, the following relation of QEL to (ordinary) answer sets for logic programs with variables was established. If $\varPi$ is a logic program, a total UNA-**QHT** model $\langle U, T, T \rangle$ of $\varPi$ is an equilibrium model of $\varPi$ iff it is an open answer set of $\varPi$.

[7] provides a new definition of stable model for arbitrary first-order formulas, defining the property of being a stable model syntactically via a second-order condition. However [7] also shows that the new notion of stable model is equivalent to that of equilibrium model defined here. In a sequel to this paper, [11] applies the new definition and makes the following refinements. The *stable models* of a formula are defined as in [7] while the *answer sets* of a formula are those Herbrand models of the formula that are stable in the sense of [7]. Using this new terminology, it follows that in general stable models and equilibrium models coincide, while answer sets are equivalent to SNA-**QHT** models that are equilibrium models.

3 Safety

We work with the same concept of restricted variable as used in [11,12]. To every quantifier-free formula φ the set $\mathrm{RV}(\varphi)$ of its *restricted variables* is defined as follows:

- For atomic φ, if φ is an equality between two variables then $\mathrm{RV}(\varphi) = \varnothing$; otherwise, $\mathrm{RV}(\varphi)$ is the set of all variables occurring in φ;
- $\mathrm{RV}(\bot) = \varnothing$;
- $\mathrm{RV}(\varphi_1 \wedge \varphi_2) = \mathrm{RV}(\varphi_1) \cup \mathrm{RV}(\varphi_2)$;
- $\mathrm{RV}(\varphi_1 \vee \varphi_2) = \mathrm{RV}(\varphi_1) \cap \mathrm{RV}(\varphi_2)$;
- $\mathrm{RV}(\varphi_1 \rightarrow \varphi_2) = \varnothing$.

A sentence is said to be in *prenex* form if it has the following shape, for some $n \geq 0$:

$$Q_1 x_1 \ldots Q_n x_n \psi \tag{4}$$

where Q_i is $\forall$ or $\exists$ and ψ is quantifier-free. A sentence is said to be *universal* if it is in prenex form and all quantifiers are universal. A universal theory is a set of universal sentences. For **QHT**, normal forms such as prenex and Skolem forms were studied in [18]. In particular it is shown there that in quantified here-and-there logic every sentence is logically equivalent to a sentence in prenex form. A similar observation regarding first-order formulas under the new stable model semantics of [7] was made in [14]. Thus from a logical point of view there is no loss of generality in defining safety for prenex formulas.

3.1 Semi-safety

A variable assignment ξ in a universe (D, σ) is a mapping from the set of variables to D. If $\varphi \in \mathcal{L}$ has free-variables, φ^ξ is the closed formula obtained by replacing every free variable x by $\xi(x)$. On the other hand, in the following, if $T \subset At(D, P)$, we denote by $T|_C$ the subset of T whose atoms contain terms only from $\sigma(C)$.

Lemma 1. *Let φ be a quantifier-free formula and ξ a variable assignment in a universe (D, σ). If $\langle (D, \sigma), T|_C, T \rangle \models \varphi^\xi$, then $\xi(x) \in \sigma(C)$ for all $x \in \mathrm{RV}(\varphi)$.*

As in [12], we define a concept of *semi-safety* of a prenex form sentence φ in terms of the *semi-safety* of all its variable occurrences. Formally, this is done by defining an operator NSS that collects the variables that have *non-semi-safe* occurrences in a formula φ.

Definition 2 (NSS and semi-safety)

1. *If φ is an atom, NSS(φ) is the set of variables in φ.*
2. $\mathrm{NSS}(\varphi_1 \wedge \varphi_2) = \mathrm{NSS}(\varphi_1) \cup \mathrm{NSS}(\varphi_2)$
3. $\mathrm{NSS}(\varphi_1 \vee \varphi_2) = \mathrm{NSS}(\varphi_1) \cup \mathrm{NSS}(\varphi_2)$
4. $\mathrm{NSS}(\varphi_1 \to \varphi_2) = \mathrm{NSS}(\varphi_2) \smallsetminus \mathrm{RV}(\varphi_1)$.

A sentence φ is said to be semi-safe *if* $\mathrm{NSS}(\varphi) = \varnothing$. □

In other words, a variable x is semi-safe in φ if every occurrence is inside some subformula $\alpha \to \beta$ such that, either x is restricted in α or x is semi-safe in β. This condition of semi-safety is a relaxation of that of [12], where the implication $\alpha \to \beta$ should always satisfy that x is restricted in α. As a result, (1) is semi-safe, but it is *not* considered so under the definition in [12]. Similarly, our definition implies, for instance, that any occurrence of a variable x in a negated subformula, $\neg\alpha(x)$, will be semi-safe – it corresponds to an implication $\alpha(x) \to \bot$ with no variables in the consequent. Other examples of semi-safe formulas are, for instance:

$$\neg p(x) \to (q(x) \to r(x)) \tag{5}$$
$$p(x) \vee q \to \neg r(x) \tag{6}$$

Note how in (6), x is not restricted in $p(x) \vee q$ but the consequent $\neg r(x)$ is semi-safe and thus the formula itself. On the contrary, the following formulas are not semi-safe:

$$p(x) \vee q \to r(x) \tag{7}$$
$$\neg\neg p(x) \wedge \neg r(x) \to q(x) \tag{8}$$

Lemma 2. *Let φ be a quantifier-free formula and ξ a variable assignment in (D, σ) s.t. $\xi(x) \in \sigma(C)$ for all $x \in \mathrm{NSS}(\varphi)$. If $\langle (D, \sigma), T, T \rangle \models \varphi^\xi$, then $\langle (D, \sigma), T|_C, T \rangle \models \varphi^\xi$.*

Proof. By induction over $\psi = \varphi^\xi$. If ψ is atomic the result is trivial and the induction step is also trivial for conjunctive and disjunctive formulas. So, let us assume that $\varphi = \varphi_1 \to \varphi_2$ where, by the induction hypothesis, φ_2 is such that, if $\langle (D, \sigma), T, T \rangle \models \varphi_2^\xi$ and $\xi(x) \in \sigma(C)$ for all $x \in \mathrm{NSS}(\varphi_2)$, then $\langle (D, \sigma), T|_C, T \rangle \models \varphi_2^\xi$. Assume

$$\langle (D, \sigma), T, T \rangle \models \varphi_1^\xi \to \varphi_2^\xi \tag{9}$$

and $\xi(x) \in \sigma(C)$ for all $x \in \mathrm{NSS}(\varphi_1 \to \varphi_2)$. We must prove that

$$\langle (D, \sigma), T|_C, T \rangle \models \varphi_1^\xi \to \varphi_2^\xi. \tag{10}$$

First, suppose that $\langle (D,\sigma), T|_C, T \rangle \not\models \varphi_1^\xi$. If $\langle (D,\sigma), T|_C, T \rangle \models \neg\varphi_1^\xi$ then clearly $\langle (D,\sigma), T|_C, T \rangle \models \varphi_1^\xi \rightarrow \varphi_2^\xi$ and we are done. Otherwise, $\langle (D,\sigma), T|_C, T \rangle \models \neg\neg\varphi_1^\xi$. Then, by (9) $\langle (D,\sigma), T|_C, T \rangle, t \models \varphi_1^\xi \rightarrow \varphi_2^\xi$, and since $\langle (D,\sigma), T|_C, T \rangle, h \not\models \varphi_1^\xi$, (10) follows. Suppose therefore that $\langle (D,\sigma), T|_C, T \rangle \models \varphi_1^\xi$, then $\langle (D,\sigma), T, T \rangle \models \varphi_1^\xi$ and thus

$$\langle (D,\sigma), T, T \rangle \models \varphi_2^\xi \tag{11}$$

On the other hand, if $\langle (D,\sigma), T|_C, T \rangle \models \varphi_1^\xi$, by Lemma 1, $\xi(x) \in \sigma(C)$ for all $x \in \mathrm{RV}(\varphi_1)$ and by hypothesis, $\xi(x) \in \sigma(C)$ for all $x \in \mathrm{NSS}(\varphi_1 \rightarrow \varphi_2)$; in particular, for $\mathrm{NSS}(\varphi_1 \rightarrow \varphi_2) \cup \mathrm{RV}(\varphi_1) \supseteq \mathrm{NSS}(\varphi_2)$ we have:

$$\xi(x) \in \sigma(C) \text{ for all } x \in \mathrm{NSS}(\varphi_2) \tag{12}$$

By the ind. hyp., (11) and (12), we conclude that $\langle (D,\sigma), T|_C, T \rangle \models \varphi_2^\xi$. $\qquad\square$

This lemma allows us to conclude the main property of semi-safe formulas: their equilibrium models only refer to objects from the language.

Proposition 1. *If φ is semi-safe, and $\langle (D,\sigma), T, T \rangle \models \varphi$, then $\langle (D,\sigma), T|_C, T \rangle \models \varphi$.*

Proof. (sketch) Assume the formula is in prenex form and proceed by induction over the length of the prefix. $\qquad\square$

Theorem 1. *If φ is semi-safe, and $\langle (D,\sigma), T, T \rangle$ is an equilibrium model of φ, then $T|_C = T$.*

3.2   Safe Formulas

The concept of safety relies on semi-safety plus an additional condition on variable occurrences that can be defined in terms of Kleene's three-valued logic [10]. Given a three-valued interpretation $\nu\colon At \rightarrow \{0, 1/2, 1\}$, we extend it to evaluate arbitrary formulas $\nu(\varphi)$ as follows:

$$\nu(\varphi \wedge \psi) = \min(\nu(\varphi), \nu(\psi)) \qquad\qquad \nu(\bot) = 0$$
$$\nu(\varphi \vee \psi) = \max(\nu(\varphi), \nu(\psi)) \qquad \nu(\varphi \rightarrow \psi) = \max(1 - \nu(\varphi), \nu(\psi))$$

from which we can derive $\nu(\neg\varphi) = \nu(\varphi \rightarrow \bot) = 1 - \nu(\varphi)$ and $\nu(\top) = \nu(\neg\bot) = 1$.

Definition 3 (ν_x **operator**). *Given any quantifier-free formula φ and any variable x, we define the three-valued interpretation so that for any atom α, $\nu_x(\alpha) = 0$ if x occurs in α and $\nu_x(\alpha) = 1/2$ otherwise.*

Intuitively, $\nu_x(\varphi)$ fixes all atoms containing the variable x to 0 (falsity) leaving all the rest undefined and then evaluates φ using Kleene's three-valued operators, that is nothing else but exploiting the defined values 1 (true) and 0 (false) as much as possible. For instance, $\nu_x(p(x) \rightarrow q(x))$ would informally correspond to $\nu_x(0 \rightarrow 0) = \max(1 - 0, 0) = 1$ whereas $\nu_x(p(x) \vee r(y) \rightarrow q(x)) = \nu_x(0 \vee 1/2 \rightarrow 0) = \max(1 - \max(0, 1/2), 0) = 1/2$.

For the following definition, we need to recall the following terminology: a subexpression of a formula is said to be positive in it if the number of implications that contain that subexpression in the antecedent is even, and negative if it is odd.

Definition 4 (Weakly-restricted variable). *An occurrence of a variable x in $Qx\ \varphi$ is* weakly-restricted *if it occurs in a subformula ψ of φ such that:*

- $Q = \forall$, ψ *is positive and* $\nu_x(\psi) = 1$
- $Q = \forall$, ψ *is negative and* $\nu_x(\psi) = 0$
- $Q = \exists$, ψ *is positive and* $\nu_x(\psi) = 0$
- $Q = \exists$, ψ *is negative and* $\nu_x(\psi) = 1$

In all cases, we further say that ψ makes the ocurrence weakly restricted in φ.

Definition 5. *A semi-safe sentence is said to be* safe *if all its positive occurrences of universally quantified variables, and all its negative occurrences of existentially quantified variables are weakly restricted.*

For instance, the formula $\varphi = \forall x(\neg q(x) \rightarrow (r \vee \neg p(x)))$ is safe: the occurrence of x in $p(x)$ is negative, whereas the occurrence in $q(x)$ is inside a positive subformula, φ itself, for which x is weakly-restricted, since $\nu_x(\varphi) = \neg 0 \rightarrow (1/2 \vee \neg 0) = 1$. The occurrence of x in $\neg q(x)$ is not restricted and as a consequence the formula is not LLP-safe. The other aspect where Definition 5 is more general than LLP-safety results from the fact that to be safe x can occur negatively in φ given $Q_i = \forall$ or positively given $Q_i = \exists$. Another example of a safe formula is $\forall x((\neg\neg p(x) \wedge q(x)) \rightarrow r)$.

Lemma 3. *Let $\varphi(x)$ a prenex formula that has no free variables other than x. Let $\langle(D,\sigma), H, T\rangle$ be a model such that $T \subset \mathrm{At}(\sigma(C), P)$. Then:*

1. *If $\forall x \varphi(x)$ is safe:* $\langle(D,\sigma), H, T\rangle \models \forall x \varphi(x)$ *iff* $\langle(D,\sigma), H, T\rangle \models \bigwedge_{c \in C} \varphi(c)$.

2. *If $\exists x \varphi(x)$ is safe:* $\langle(D,\sigma), H, T\rangle \models \exists x \varphi(x)$ *iff* $\langle(D,\sigma), H, T\rangle \models \bigvee_{c \in C} \varphi(c)$.

Proof. (sketch) The proof makes use of the monotonic properties of the many-valued assignments with respect to the positive and negative ocurrences of subformulas. □

4   Grounding

Let (D, σ) be a domain and $D' \subseteq D$ a finite subset; the grounding over D' of a sentence φ, denoted by $\mathrm{Gr}_{D'}(\varphi)$, is defined recursively: the operator does not modify ground formulas, it propagates through propositional connectives and:

$$\mathrm{Gr}_{D'}(\forall x \varphi(x)) = \bigwedge_{d \in D'} \mathrm{Gr}_{D'}\varphi(d) \qquad \mathrm{Gr}_{D'}(\exists x \varphi(x)) = \bigvee_{d \in D'} \mathrm{Gr}_{D'}\varphi(d)$$

The following property of grounding holds (cf [18] for a version for universal theories.)

Proposition 2. $\langle(D,\sigma), H, T\rangle \models \varphi$ *if and only if* $\langle(D,\sigma), H, T\rangle \models \mathrm{Gr}_D(\varphi)$

In particular, if we work with PNA-models ($D = \sigma(C)$), we can ground using the initial constants, but we need the first-order semantics because with the mapping σ two constants may denote the same element of the domain; then we would obtain:

$$\langle(D,\sigma), H, T\rangle \models \varphi \text{ iff } \langle(D,\sigma), H, T\rangle \models \mathrm{Gr}_C(\varphi) \qquad (13)$$

Lemma 4. *If $T \subset At(\sigma(C), P)$ and there exists $a \in D \setminus \sigma(C)$ then,*

1. $\langle (D, \sigma), H, T \rangle \models \forall x \varphi(x)$ *if and only if* $\langle (D, \sigma), H, T \rangle \models \bigwedge_{d \in \sigma(C) \cup \{a\}} \varphi(d)$
2. $\langle (D, \sigma), H, T \rangle \models \exists x \varphi(x)$ *if and only if* $\langle (D, \sigma), H, T \rangle \models \bigvee_{d \in \sigma(C) \cup \{a\}} \varphi(d)$

Proof. (sketch) If $T \subset At(\sigma(C), P)$, then the role of any object outside of $\sigma(C)$ is the same. So, we can choose just one element to represent all of them.   □

Proposition 3. *If $T \subset At(\sigma(C), P)$ and $a \in D \setminus \sigma(C)$, then $\langle (D, \sigma), H, T \rangle \models \varphi$ if and only if $\langle (D, \sigma), H, T \rangle \models \mathrm{Gr}_{C \cup \{a\}}(\varphi)$.*

Theorem 2. *Let φ be a safe prenex formula. Let $\langle (D, \sigma), H, T \rangle$ be a model such that $T \subset \mathrm{At}(\sigma(C), P)$. Then:*

$$\langle (D, \sigma), H, T \rangle \models \varphi \text{ if and only if } \langle (D, \sigma), H, T \rangle \models \mathrm{Gr}_C(\varphi)$$

Proof. By induction on the length of the prefix.   □

We can now show as promised the property that safe formulas satisfy domain independence. The following is immediate from Theorems 1 and 2 and the definition of equilibrium model.

Proposition 4. *Let φ be a safe prenex formula, then: $\langle (D, \sigma), T, T \rangle$ is an equilibrium model of φ if and only if it is an equilibrium model of $\mathrm{Gr}_C(\varphi)$.*

Here is an alternative formulation using language expansions.

Theorem 3. *Let φ be a safe prenex formula. Suppose we expand the language $\mathcal{L}$ by considering a set of constants $C' \supset C$. A total **QHT**-model $\langle (D, \sigma), T, T \rangle$ is an equilibrium model of $\mathrm{Gr}_{C'}(\varphi)$ if an only if it is an equilibrium model of $\mathrm{Gr}_C(\varphi)$.*

The concept of domain independence from [2] is obtained as a special case if we assume that the unique name assumption applies.

5   Definability

An important property of safe formulas is that they form a definable class in first-order logic.[4] Let φ be a ground sentence and $(\alpha_1, \ldots, \alpha_n)$ the sequence of atoms in φ. If $\beta = (\beta_1, \ldots, \beta_n)$ is another sequence of ground atoms, the formula $\varphi[\beta]$ is built recursively:

- $\alpha_i[\beta] = \beta_i$
- $\bot[\beta] = \bot$
- $(\psi_1 \wedge \psi_2)[\beta] = \psi_1[\beta] \wedge \psi_2[\beta]$
- $(\psi_1 \vee \psi_2)[\beta] = \psi_1[\beta] \vee \psi_2[\beta]$
- $(\psi_1 \rightarrow \psi_2)[\beta] = (\psi_1[\beta] \rightarrow \psi_2[\beta]) \wedge (\varphi \rightarrow \psi)$
- $(\neg \psi)[\beta] = \neg \psi[\beta] \wedge \neg \psi$

---

[4] Shown for LLP-safety in [12]. An alternative approach to exhibiting definable classes of stable models, using loop formulas, can be found in [13].

Lemma 5. *(i) Let φ be a ground sentence and $\alpha = (\alpha_1, \ldots, \alpha_n)$ the sequence of atoms in φ. Let $\mathbf{u}$ be a sequence $(u_1, \ldots, u_n)$ such that, for every i, either $u_i = \bot$ or $u_i = \alpha_i$ and there is some i such that $u_i = \bot$. (ii) Then $\langle (D, \sigma), T, T \rangle$ is an equilibrium model of φ if and only if T is a classical model of $\varphi \wedge \neg \bigvee_{\mathbf{u} \in U} \varphi[\mathbf{u}]$.*

Satisfaction in classical logic can be encoded by satisfaction by total models in here-and-there logic [20]; so we can characterise the classical models of the formula in the previous lemma by the total models of a formula in **QHT**, as follows:

Corollary 1. *Given statement (i) of Lemma 5, $\langle (D, \sigma), T, T \rangle$ is an equilibrium model of φ if and only if it is a **QHT**-model of*

$$\bigwedge_{1 \leq i \leq n} (\neg\neg\alpha_i \rightarrow \alpha_i) \wedge \varphi[\neg\neg\alpha] \wedge \neg \bigvee_{\mathbf{u} \in U} \varphi[\neg\neg\mathbf{u}]$$

As a consequence of this and Theorem 2 we obtain:

Theorem 4. *For every safe sentence φ, there exists a ground formula ψ in the same language such that $\langle (D, \sigma), T, T \rangle$ is an equilibrium model of φ if and only if it is a **QHT**-model of ψ.*

6   Discussion and Relation to Other Concepts of Safety

As remarked in the introduction, one of our aims is to generalise the safety concept of [11,12] and at the same time capture logic programs that are safe according to [2]. Although for reasons of space we cannot present here the full definition of LLP-safety from [12], we explained after Definition 5 the main features of our concept that generalise that of [12] (see also the examples discussed below). To verify the second objective we need to consider the class of logic programs treated by Bria, Faber and Leone in [2]. These programs are called *normal form nested*, or NFN for short. They comprise rules of a special form; in logical notation they are formulas that have the shape:

$$\bigwedge_{1 \leq i \leq m} \Delta_i \rightarrow \bigvee_{\leq i \leq n} \Gamma_j \tag{14}$$

where each Δ_i is a disjunction of literals and each Γ_j is a conjunction of atoms. A Δ_i is called a *basic disjunction* and is said to be *positive* if it contains only positive literals or atoms. In a rule r of form (14) the antecedent is called the *body* and the consequent the *head* of the rule. According to [2] r is safe if each variable in its head literals and its negative body literals is safe, where a variable x is said to be safe in r if if there exists a positive basic disjunction Δ_i in the body of r such that x is a variable of α for every atom $\alpha \in \Delta_i$. It is easy to check that the BFL-safe variables in r are weakly restricted in r, therefore:

Proposition 5. *All rules of an NFN program that are safe according to [2] are safe formulas in the sense of Definition 5.*

Let us consider some examples showing that the safety concept introduced in Definition 5 is quite general and, in many cases, is better behaved with respect to strongly equivalent transformations than other definitions presented before. There are more examples where our definition detects safe sentences that are not LLP-safe. Eg, while

$$\neg\neg p(x) \rightarrow q \tag{15}$$

is LLP-safe, once we add $\neg r(x)$ to the body

$$\neg\neg p(x) \wedge \neg r(x) \rightarrow q \tag{16}$$

it becomes an LLP-unsafe sentence, whereas it still preserves safety under the current definition. Notice that the new x occurs negatively in the body but it is "made safe" by $\neg\neg p(x)$ and the fact that x does not occur in the head. In fact, if the head becomes $q(x)$ like in (8), the formula is not domain independent and so is not LLP-safe nor safe under our definition (in fact, as we saw, it is not even semi-safe).

The idea of modifying the definition of semi-safe formulas to check, as in [2], that a variable is restricted only if occurs in the head, can also be applied to existentially quantified formulas. For instance, this formula

$$\exists x \left(\neg\neg(p(x) \vee q) \right) \tag{17}$$

should be safe. It is strongly equivalent to $\neg\exists x\, p(x) \rightarrow q$ and this can be easily captured by introducing an auxiliary predicate in a safe way as follows:

$$aux_1 \leftarrow p(X)$$
$$q \leftarrow \neg aux_1$$

In fact, (17) is safe. However, it is not even semi-safe according to the definition in [11]: there is no implication containing x restricted in the antecedent.

7 Conclusion

We have presented a new concept of safety for general first-order formulas under stable model semantics and hence for very general kinds of answer set programs. As a logical formalism we have used quantified equilibrium logic based on the non-classical logic of quantified here-and-there, **QHT**, in which equilibrium or stable models can be easily defined as minimal models and their logical properties readily studied. We showed that the new concept of safety extends and generalises recent definitions of safety from [12] and [2]. Unlike in [2], in our approach we do not have to make the unique name assumption. We showed that safe formulas satisfy three main properties or criteria of adequacy: their stable models do not contain unnamed individuals, they satisfy the condition of domain independence, and they form a first-order definable class of structures.

A topic of ongoing work concerns program transformations. In a sequel to this paper [5] we have begun to study classes of formulas whose safety is preserved under strong-equivalence preserving program transformations (in particular from [4]). This work may help to guide future improvements in the definition of a general safety concept, both from theoretical and from computational standpoints.

References

1. Baral, C.: Knowledge Representation, Reasoning and Declarative Problem Solving. CUP, Cambridge (2002)
2. Bria, A., Faber, W., Leone, N.: Normal Form Nested Programs. In: Hölldobler, S., Lutz, C., Wansing, H. (eds.) JELIA 2008. LNCS (LNAI), vol. 5293, pp. 76–88. Springer, Heidelberg (2008)
3. de Bruijn, J., Pearce, D., Polleres, A., Valverde, A.: Quantified Equilibrium Logic and Hybrid Rules. In: Marchiori, M., Pan, J.Z., Marie, C.d.S. (eds.) RR 2007. LNCS, vol. 4524, pp. 58–72. Springer, Heidelberg (2007)
4. Cabalar, P., Pearce, D., Valverde, A.: Reducing Propositional Theories in Equilibrium Logic to Logic Programs. In: Bento, C., Cardoso, A., Dias, G. (eds.) EPIA 2005. LNCS (LNAI), vol. 3808, pp. 4–17. Springer, Heidelberg (2005)
5. Cabalar, P., Pearce, D., Valverde, A.: Safety Preserving Transformations for General Answer Set Programs. In: Logic-Based program Synthesis and Transformation. Proc. LOPSTR 2009 (2009)
6. van Dalen, D.: Logic and Structure. Springer, Heidelberg (2004)
7. Ferraris, P., Lee, J., Lifschitz, V.: A New Perspective on Stable Models. In: Veloso, M. (ed.) 20th International Joint Conference on Artificial Intelligence, IJCAI 2007, pp. 372–379 (2007)
8. Heymans, S., van Nieuwenborgh, D., Vermeir, D.: Open Answer Set Programming with Guarded Programs. ACM Trans. Comp. Logic 9, Article 26 (2008)
9. Heymans, S., Predoiu, L., Feier, C., de Bruijn, J., van Nieuwenborgh, D.: G-hybrid Knowledge Bases. In: ALPSWS 2006, CEUR Workshop Proceedings, vol. 196 (2006)
10. Kleene, S.C.: On Notation for Ordinal Numbers. The J. of Symbolic Logic 3(4) (1938)
11. Lee, J., Lifschitz, V., Palla, R.: A Reductive Semantics for Counting and Choice in Answer Set Programming. In: Proceedings AAAI 2008, pp. 472–479 (2008)
12. Lee, J., Lifschitz, V., Palla, R.: Safe formulas in the general theory of stable models (Preliminary report). In: Garcia de la Banda, M., Pontelli, E. (eds.) ICLP 2008. LNCS, vol. 5366, pp. 672–676. Springer, Heidelberg (2008); longer version by personal communication (June 2008)
13. Lee, J., Meng, V.: On Loop Formulas with Variables. In: Proc. of AAAI 2008. AAAI, Menlo Park (2008)
14. Lee, J., Palla, R.: Yet Another Proof of the Strong Equivalence between Propositional Theories and Logic Programs. In: CENT 2007, Tempe, AZ, May 2007. CEUR Workshop Proceedings, vol. 265, CEUR-WS.org (2007)
15. Lifschitz, V., Pearce, D., Valverde, A.: A Characterization of Strong Equivalence for Logic Programs with Variables. In: Baral, C., Brewka, G., Schlipf, J. (eds.) LPNMR 2007. LNCS (LNAI), vol. 4483, pp. 188–200. Springer, Heidelberg (2007)
16. Lifschitz, V., Tang, L., Turner, H.: Nested Expressions in Logic Programs. Annals of Mathematics and Artificial Intelligence 25(3-4), 369–389 (1999)
17. Pearce, D., Valverde, A.: Towards a first order equilibrium logic for nonmonotonic reasoning. In: Alferes, J.J., Leite, J. (eds.) JELIA 2004. LNCS (LNAI), vol. 3229, pp. 147–160. Springer, Heidelberg (2004)
18. Pearce, D., Valverde, A.: A First-Order Nonmonotonic Extension of Constructive Logic. Studia Logica 80, 321–346 (2005)
19. Pearce, D., Valverde, A.: Quantified Equilibrium Logic. Tech. report, Univ. Rey Juan Carlos (2006), http://www.matap.uma.es/investigacion/tr/ma06_02.pdf
20. Pearce, D., Valverde, A.: Quantified Equilibrium Logic and Foundations for Answer Set Programs. In: Garcia de la Banda, M., Pontelli, E. (eds.) ICLP 2008. LNCS, vol. 5366, pp. 546–560. Springer, Heidelberg (2008)

Magic Sets for the Bottom-Up Evaluation
of Finitely Recursive Programs

Francesco Calimeri, Susanna Cozza, Giovambattista Ianni, and Nicola Leone

Department of Mathematics, University of Calabria, I-87036 Rende (CS), Italy
{calimeri,cozza,ianni,leone}@mat.unical.it

Abstract. The support for function symbols in logic programming under answer set semantics allows to overcome some modeling limitations of traditional Answer Set Programming (ASP) systems, such as the inability of handling infinite domains. On the other hand, admitting function symbols in ASP makes inference undecidable in the general case. Lately, the research community is focusing on finding proper subclasses for which decidability of inference is guaranteed. The two major proposals, so far, are finitary programs and finitely-ground programs. These two proposals are somehow complementary: indeed, the former is conceived to allow decidable querying (by means of a top-down evaluation strategy), while the latter supports the computability of answer-sets (by means of a bottom-up evaluation strategy). One of the main advantages of finitely-ground programs is that they can be "directly" evaluated by current ASP systems, which are based on a bottom-up computational model. However, there are also some interesting programs which are suitable for top-down query evaluation; but do not fall in the class of finitely-ground programs.

In this paper, we focus on disjunctive finitely-recursive positive (DFRP) programs. We present a proper adaptation of the magic-sets technique for DFRP programs, which ensures query equivalence under both brave and cautious reasoning. We show that, if the input program is DFRP, then its magic-set rewriting is guaranteed to be finitely ground. Thus, reasoning on DFRP programs turns out to be decidable, and we provide an effective method for its computation on the ASP system DLV.

1   Introduction

Disjunctive Logic Programming (DLP) under the answer set semantics, often referred to as Answer Set Programming (ASP) [1,2,3,4,5], evolved significantly during the last decade, and has been recognized as a convenient and powerful method for declarative knowledge representation and reasoning. Lately, the ASP community has clearly perceived the strong need to extend ASP by functions, and many relevant contributions have been done in this direction [6,7,8,9,10,11,12,13]. Supporting function symbols allows to overcome one of the major limitation of traditional ASP systems, i.e. the ability of handling finite sets of constants only. On the other hand, admitting function symbols in ASP makes the common inference tasks undecidable in the general case. The identification of expressive decidable classes of ASP programs with functions is therefore an important task. Two relevant decidable classes, resulting from alternative approaches (top-down vs bottom-up), are finitary programs [8] and finitely-ground programs [12].

E. Erdem, F. Lin, and T. Schaub (Eds.): LPNMR 2009, LNCS 5753, pp. 71–86, 2009.
© Springer-Verlag Berlin Heidelberg 2009

Finitary programs [8] is a class of logic programs that allows function symbols yet preserving decidability of ground querying by imposing restrictions both on recursion and on the number of potential sources of inconsistency. Recursion is restricted by requiring each ground atom to depend on finitely many ground atoms; such programs are called *finitely recursive* [13]. Moreover, potential sources of inconsistency are limited by requiring that the number of odd-cycles (cycles of recursive calls involving an odd number of negative subgoals) is finite. Thanks to these two restrictions, consistency checking and ground queries are decidable, provided that the atoms involved in odd-cycles are supplied [14]; while non-ground queries are semi-decidable.

The class of *finitely-ground* ($\mathcal{FG}$) programs [12], more recently proposed, can be seen as a "dual" class of finitary programs. Indeed, while the latter is suitable for a top-down evaluation, the former allows a bottom-up computation. Basically, for each program P in this class, there exists a finite subset P' of its instantiation, called *intelligent instantiation*, having precisely the same answer sets as P. Importantly, such a subset P' is computable for $\mathcal{FG}$ programs. Both finitary programs and $\mathcal{FG}$ programs can express any computable function, and preserve decidability for ground queries.[1] However, answer sets and non-ground queries are computable on $\mathcal{FG}$ programs, while they are not computable on finitary programs. Furthermore, the bottom-up nature of the notion of $\mathcal{FG}$ programs allows an immediate implementation in ASP systems (as ASP instantiators are based on a bottom-up computational model). Indeed, the DLV system [15], for instance, has already been adapted to deal with $\mathcal{FG}$ program by extending its instantiator [16].

Though membership to both the above mentioned classes is not decidable (semidecidable for $\mathcal{FG}$ programs), they are not comparable. In particular, the class of $\mathcal{FG}$ programs does not include some programs for which ground querying can be computed in a top-down fashion, like, in particular, or-free finitely-recursive positive programs. Despite of its simplicity, this latter class includes many significative programs, such as most of the standard predicates for lists manipulation. For instance, the following program, performing the check for membership of an element in a list, is finitely recursive and positive, yet not finitely ground.

$$member(X, [X|Y]). \qquad member(X, [Y|Z]) :\!\!- member(X, Z).$$

In this paper, we shed some light on the relationships between finitely recursive and $\mathcal{FG}$ programs, evidentiating a sort of "dual" behaviour of the two classes. We show that a suitable magic-set rewriting transforms finitely recursive positive programs into $\mathcal{FG}$ programs. In this way, we devise a strategy for the bottom-up evaluation of finitely-recursive positive programs. Importantly, we effectively deal also with *disjunctive* finitely-recursive positive programs, which were unknown to be decidable so far. In summary, the paper focuses on disjunctive finitely-recursive positive programs (DFRP programs) and queries, providing the following main contribution:

- We design a suitable adaptation of the *magic-sets* rewriting technique for disjunctive programs with functions, which exploits the peculiarities of finitely recursive programs.

---

[1] For finitary programs, decidability requires that the atoms involved in odd-cycles are additionally supplied in the input [14].

- We show that our magic-sets rewriting $RW(Q, P)$ of a (ground) query Q on a DFRP program P enjoys the following properties:
  - for both brave and cautious reasoning, we have that $P \models Q$ iff RW $(Q, P) \models Q$;
  - if Q belongs to the herein defined class of finitely recursive queries on P, then $RW(Q, P)$ is finitely ground.
  - the size of $RW(Q, P)$ is linear in the size of the input program;
- We then show that both brave and cautious reasoning on DFRP programs are decidable.

Importantly, we provide not only a theoretical decidability result for reasoning on DFRP programs, but we also supply a concrete implementation method. Indeed, by applying a light-weight magic-sets rewriting on the input program,[2] any query on a DFRP program can be evaluated by the ASP system DLV, or by any other system supporting $\mathcal{FG}$ programs.

The remainder of the paper is structured as follows. Section 2 motivates our work by means of few significative examples; for the sake of completeness, in Section 3 we report some needed preliminaries; Section 4 illustrates our adaptation of the magic-sets rewriting technique to the class of positive finitary programs; in Section 5 we present a number of theoretical results about the rewritten programs; Section 6 analyzes some related works, and, eventually, Section 7 draws our conclusions and depicts the future work.

2   Motivation

In the or-free case, positive finitely-recursive programs might be seen as the simplest subclass of finitary programs. As finitary programs, they enjoy all nice properties of this class. In particular, consistency checking is decidable as well as reasoning with ground queries (while reasoning is semi-decidable in case of non ground queries). Unfortunately, even if an or-free program P is finitely recursive, it is not suited for the bottom-up evaluation for two main reasons:

1. A bottom-up evaluation of a finitely-recursive program would generate some new terms at each iteration, thus iterating for ever.

   Example 1. Consider the following program, defining the natural numbers:

   $nat(0).$ $nat(s(X)) :\text{-} \ nat(X).$

   The program is positive and finitely recursive, so every ground query (such as for instance $nat(s(s(s(0))))$?) can be answered in a top-down fashion; but its bottom-up evaluation would iterate for ever, as, for any positive integer n, the n-th iteration would derive the new atom $nat(s^n(0))$.

2. Finitely-recursive programs do not enforce the range of a head variable to be restricted by a body occurrence (i.e., "bottom-up" safety is not required). A bottom-up evaluation of these "unsafe" rules would cause the derivation of non-ground

---

[2] Note that a working translation module, actually not considering the peculiar optimization techniques herein presented for finitely-recursive queries, is described in [17].

facts, standing for infinite instances, which are not admissible by present grounding algorithms.

Actually, in this paper we deal also with *disjunctive* finitely recursive programs, which were not even known to be decidable so far, also in the positive case.

Example 2. Consider the following program, computing all the possible 2-coloring for an *infinite* chain of nodes and defines coupled nodes as pairs of nodes which are successive and share the same color. Simple programs like this could be easily adapted for dealing with real problems.

$$color(X,b) \vee color(X,g). \quad coupled(X,next(X),C) :- color(X,C), color(next(X),C). \tag{1}$$

The program is positive and finitely recursive; nevertheless, a bottom-up evaluation is unfortunately unfeasible: the first rule, indeed, represents an infinite set of atoms.

Example 3. The following program P_2 defines the comparison operator 'less than' between two natural numbers (the function symbol s represents the successor of a natural number):

$$lessThan(X, s(X)). \qquad lessThan(X, s(Y)) :- lessThan(X, Y). \tag{2}$$

In this case, bottom-up evaluation is unfeasible, both because of the first rule, and because of the infinite recursion in the second rule.

3   Preliminaries

This section reports the formal specification of the ASP language with function symbols. The subclass of ASP programs herein considered are positive disjunctive logic programs (i.e., negation is not allowed).

3.1   Syntax

A *term* is either a *simple term* or a *functional term*.[3] A *simple term* is either a constant or a variable. If $t_1 \ldots t_n$ are terms and f is a function symbol (*functor*) of arity n, then $f(t_1, \ldots, t_n)$ is a *functional term*.

Each predicate p has a fixed arity $k \geq 0$. If $t_1, \ldots, t_k$ are terms and p is a *predicate* of arity k, then $p(t_1, \ldots, t_k)$ is an *atom*. An atom having p as predicate name is usually referred as $p(\bar{t})$.

A (positive) *disjunctive rule* r is of the form: $\alpha_1 \vee \cdots \vee \alpha_k :- \beta_1, \cdots, \beta_n.$, where $k > 0$; $\alpha_1, \ldots, \alpha_k$ and $\beta_1, \ldots, \beta_n$ are atoms. The disjunction $\alpha_1 \vee \cdots \vee \alpha_k$ is called *head* of r, while the conjunction $\beta_1, \cdots, \beta_n.$ is the *body* of r. We denote by $H(r)$ the set of the head atoms, by $B(r)$ the set of body atoms; we refer to all atoms occurring in a rule with $Atoms(r) = H(r) \cup B(r)$. A rule having precisely one head atom (i.e., $k = 1$ and then $|H(r)| = 1$) is called a *normal rule*. If r is a normal rule

---

[3] We will use traditional square-bracketed list constructors as shortcut for the representation of lists by means of nested functional terms (see, for instance, [12]). The usage "à la prolog", or any different, is only a matter of syntactic sugar.

with an empty body (i.e., $n = 0$ and then $B(r) = \emptyset$) we usually omit the " :– " sign; and if it contains no variables, then it is referred to as a *fact*. An ASP program P is a finite set of rules. A v-free program P is a program consisting of normal rules only.

Given a predicate p, a *defining rule* for p is a rule r such that some $p(\bar{t})$ occurs in $H(r)$. If all defining rules of a predicate p are facts, then p is an *EDB predicate*; otherwise p is an *IDB predicate*.[4] The set of all facts of P is denoted by $Facts(P)$; the set of instances of all EDB predicates is denoted by $EDB(P)$. A program (a rule, an atom, a term) is *ground* if it contains no variables. A *query* Q is a ground atom.[5]

3.2 Semantics

The most widely accepted semantics for ASP programs is based on the notion of answer-set, proposed in [3] as a generalization of the concept of stable model [2].

Given a program P, the *Herbrand universe* of P, denoted by U_P, consists of all (ground) terms that can be built combining constants and functors appearing in P. The *Herbrand base* of P, denoted by B_P, is the set of all ground atoms obtainable from the atoms of P by replacing variables with elements from U_P.[6] A *substitution* for a rule $r \in P$ is a mapping from the set of variables of r to the set U_P of ground terms. A *ground instance* of a rule r is obtained applying a substitution to r. Given a program P the *instantiation (grounding) $grnd(P)$* of P is defined as the set of all ground instances of its rules. Given a ground program P, an *interpretation* I for P is a subset of B_P. An atom a is true w.r.t. I if $a \in I$; it is false otherwise. Given a ground rule r, we say that r is satisfied w.r.t. I if some atom appearing in $H(r)$ is true w.r.t. I or some atom appearing in $B(r)$ is false w.r.t. I. Given a ground program P, we say that I is a *model* of P, iff all rules in $grnd(P)$ are satisfied w.r.t. I.

A model M of P is an *answer set* of P if it is *minimal*, i.e., there is no model N for P such that $N \subset M$. The set of all answer sets for P is denoted by $AS(P)$. A program P *bravely entails* (resp., *cautiously entails*) a query Q, denoted by $P \models_b Q$ (resp., $P \models_c Q$) if Q is true in some (resp., all) $M \in AS(P)$.

4 Magic Sets Rewriting of Finitely-Recursive Queries

In this Section we first give some basics on the *magic-sets* technique and provide the definition of finitely-recursive queries; then, a suitable adaptation of the magic-sets rewriting technique for programs with functions is presented, that allows a bottom-up evaluation of such queries over positive programs.

4.1 The Magic-Sets Technique

The magic-sets method is a strategy for simulating the top-down evaluation of a query by modifying the original program by means of additional rules, which narrow the

---

[4] EDB and IDB stand for Extensional Database and Intensional Database, respectively.

[5] Note that this definition of a query is not as restrictive as it may seem, as one can include appropriate rules in the program for expressing unions of conjunctive queries (and more).

[6] With no loss of generality, we assume that constants appearing in any query Q always appear in P. Since we focus on query answering, this allows us to restrict to Herbrand universe/base.

computation to what is relevant for answering the query. Intuitively, the goal of the magic-sets method is to use the constants appearing in the query to reduce the size of the instantiation by eliminating 'a priori' a number of ground instances of the rules which cannot contribute to the (possible) derivation of the query goal.

We next provide a brief and informal description of the magic-sets rewriting technique, which has originally been defined in [18] for non-disjunctive Datalog (i.e., with no function symbols) queries only, and afterwards many generalizations have been proposed. The reader is referred to [19] for a detailed presentation.

The method is structured in four main phases which are informally illustrated below by example, considering the query $path(1,5)$ on the following program:

$$path(X,Y) :- edge(X,Y). \qquad path(X,Y) :- edge(X,Z), path(Z,Y).$$

1. Adornment Step: The key idea is to materialize, by suitable adornments, binding information for IDB predicates which would be propagated during a top-down computation. Adornments are strings consisting of the letters b and f, denoting 'bound' and 'free' respectively, for each argument of an IDB predicate. First, adornments are created for query predicates so that an argument occurring in the query is adorned with the letter b if it is a constant, or with the letter f if it is a variable. The query adornments are then used to propagate their information into the body of the rules defining it, simulating a top-down evaluation. It is worth noting that adorning a rule may generate new adorned predicates. Thus, the adornment step is repeated until all adorned predicates have been processed, yielding the *adorned program*. Although a so-called "generalized" magic-sets method has been defined in [20], for simplicity of presentation, we next adopt the "basic" magic-sets method as defined in [18], in which binding information within a rule comes only from the adornment of the head predicate, from EDB predicates in the (positive) rule body, and from constants.

Example 4. Adorning the query $path(1,5)$ generates the adorned predicate $path^{bb}$ since both arguments are bound, and the adorned program is:

$$path^{bb}(X,Y) :- edge(X,Y). \qquad path^{bb}(X,Y) :- edge(X,Z), path^{bb}(Z,Y).$$

2. Generation Step: The adorned program is used to generate *magic rules*, which simulate the top-down evaluation scheme and single out the atoms which are relevant for deriving the input query. Let the *magic version* $\text{magic}(p^{\alpha}(\bar{t}))$ for an adorned atom $p^{\alpha}(\bar{t})$ be defined as the atom $magic\_p^{\alpha}(\bar{t}')$, where $\bar{t}'$ is obtained from $\bar{t}$ by eliminating all arguments corresponding to an f label in α, and where $magic\_p^{\alpha}$ is a new predicate symbol obtained by attaching the prefix '$magic\_$' to the predicate symbol p^{α}. Then, for each adorned atom A in the body of an adorned rule r_a, a magic rule r_m is generated such that (i) the head of r_m consists of magic(A), and (ii) the body of r_m consists of the magic version of the head atom of r_a, followed by all the (EDB) atoms of r_a which can propagate the binding on A.

3. Modification Step: The adorned rules are subsequently modified by including magic atoms generated in Step 2 in the rule bodies, which limit the range of the head variables avoiding the inference of facts which cannot contribute to deriving the query. The resulting rules are called *modified rules*. Each adorned rule r_a is modified as follows. Let

H be the head atom of r_a. Then, atom $magic(H)$ is inserted in the body of the rule, and the adornments of all non-magic predicates are stripped off.

4. Processing of the Query: Let the query goal be the adorned IDB atom g^α. Then, the magic version (also called *magic seed*) is produced as $magic(g^\alpha)$ (see step 2 above). For instance, in our example we generate $magic\_path^{bb}(1,5)$.

The complete rewritten program consists of the magic, modified, and query rules.

Example 5. The complete rewriting of our example program is:

$magic\_path^{bb}(1,5)$. $magic\_path^{bb}(Z,Y) :- magic\_path^{bb}(X,Y), edge(X,Z)$.
$path(X,Y) :- magic\_path^{bb}(X,Y), edge(X,Y)$.
$path(X,Y) :- magic\_path^{bb}(X,Y), edge(X,Z), path(Z,Y)$.

The adorned rule $path^{bb}(X,Y) :- edge(X,Y)$ does not produce any magic rule, since it does not contain any adorned predicate in its body. Hence, we only generate $magic\_path^{bb}(Z,Y) :- magic\_path^{bb}(X,Y), edge(X,Z)$. Moreover, the last two rules above results from the modification of the adorned program in Example 4. In this rewriting, $magic\_path^{bb}(X,Y)$ represents the start- and end-nodes of all potential sub-paths of paths from 1 to 5. Therefore, when answering the query, only these sub-paths will be actually considered in bottom-up computations.

4.2 Disjunctive Magic Sets

While dealing with disjunctive programs, bindings must be propagated also head-to-head, in order to preserve soundness: if a predicate p is relevant for the query, and a disjunctive rule r contains $p(X)$ in the head, the binding from $p(X)$ must be propagated to the body atoms, and to the other head atoms of r as well; thus, in disjunctive rules also head atoms must be all adorned. From now on, we refer to the proposal of [21]. The technique therein presented take also into account the possible generation of multiple rules from a disjunctive rule, due to the different adornments for the head predicates, and that may affect the semantics of the program. The main differences w.r.t. the non-disjunctive case concern the generation and modification steps. While generating the magic rule from an adorned version of a disjunctive rule and a selected head atom a, first a non-disjunctive intermediate rule is produced, by moving all head atoms different from a into the body; then, the standard generation step described in Section 4.1 is applied. Furthermore, while modifying a disjunctive rule a magic atom has to be added to the body for each head atom.

Example 6. Let us consider the query $Q = coupled(1, next(1), g)$? on the program P_1 of Example 2. After the adornment phase, all predicates, in this case, have a completely bound adornment. We show now what happens for the disjunctive rule. The generation phase first produces two intermediate rules: $color^{bb}(X,b) :- color^{bb}(X,g)$. and $color^{bb}(X,g) :- color^{bb}(X,b)$, that cause the two magic rules: $magic\_color^{bb}(X,g) :- magic\_color^{bb}(X,b)$. and $magic\_color^{bb}(X,b) :- magic\_color^{bb}(X,g)$., respectively. Then, the modification phase causes the rule $color(X,b) \lor color(X,g) :- magic\_color^{bb}(X,b), magic\_color^{bb}(X,g)$.[7] to be produced.

---

[7] The adornments from non-magic predicates in modified rules are stripped-off in order to preserve the semantics. See [21].

The complete rewritten program consists of the generated and modified rules above, plus what comes out from the query and the second rule (with no difference w.r.t. what previously described for the normal case).

4.3 Finitely-Recursive Queries

We next provide the definition of finitely-recursive queries and programs.

Definition 1. Let P be a program. A ground atom a *depends* on ground atom b with a 1 degree (denoted by $a \geqslant_1 b$) if there is some rule $r \in grnd(P)$ with $a \in H(r)$ and $b \in Atoms(r)$. The "depends on" relation is reflexive and transitive. For any ground atom a, $a \geqslant_0 a$. If $a \geqslant_1 b$, and $b \geqslant_i c$, then $a \geqslant_{i+1} c$ (a depends on c with a $i+1$ degree). The degree of the dependencies is often omitted, hence simply saying that a *depends on* b, if $a \geqslant_k b$ holds for some $k \geq 0$.

Definition 2. Given a query Q on P, define
$relAtoms(Q, P) = \{a \in B_P : Q \text{ depends on } a\}$
$relRules(Q, P) = \{r \in grnd(P) : r \text{ defines } a \text{ for some } a \in relAtoms(Q, P)\}$
Then: (i) a query Q is *finitely recursive* on P if $relAtoms(Q, P)$ is finite; (ii) a program P is *finitely recursive* if every query on P is finitely recursive.

The sets $relAtoms(Q, P)$ and $relRules(Q, P)$ are called, respectively, the *relevant atoms* and the *relevant subprogram* of P w.r.t. Q. It is worth noting that, if Q is finitely recursive, then also its relevant subprogram is finite.

Example 7. Consider the following program:

$lessThan(X, s(X))$. $lessThan(X, s(Y)) :- lessThan(X, Y)$.
$q(f(f(0)))$. $q(X) :- q(f(X))$.
$r(X) \lor t(X) :- lessThan(X, Y), q(X)$.

The program is not finitely recursive (because of rule $q(X) :- q(f(X))$. Nevertheless, one may have many finitely-recursive queries on it. All atoms like $lessThan(c_1, c_2)$, for instance, with c_1 and c_2 constant values, are examples of finitely-recursive queries.

4.4 Rewriting Algorithm

If we restrict our attention to finitely-recursive queries, some steps of the magic-sets technique reported in Section 4.1 can be significantly simplified. In particular, the adornment phase is no longer needed, given that the IDB predicates involved in the query evaluation would have a completely bound adornment. Indeed, for a program P and a finitely-recursive query Q:

- Q is ground, so it would have a completely bound adornment;
- since the relevant subprogram $relRules(Q, P)$ is supposed to be finitely recursive, we can state that all rules involved in a top-down evaluation of the query:[8]

---

[8] Indeed, function symbols make the universe infinite, and local variables in a rule would make its head depend on an infinite number of other ground atoms. Local variables could obviously appear in a function-free program, but this could be easily bottom-up evaluated.

- cannot have local variables (i.e. variables appearing only in the body of the rule);
- cannot have unbound head variables.

Thus, starting from a ground query, the only adorned predicates introduced are bounded on all their variable, and adorned rules contain bound variables only.

In the generation step, it is no longer necessary to include any other atom in the body of the generated magic rule, apart from the magic version of the considered head atom. Again, this is due to the absence of local variables, so that all the needed bindings are provided through the magic version of the head atom.

The algorithm MS_{FR} in Figure 1 implements the magic-sets method for finitely-recursive queries. Starting from a positive finitely-recursive program P and a ground query Q, the algorithm outputs a program $RW(Q, P)$ consisting of a set of *modified*

Input: a program P and a finitely-recursive query $Q = \mathtt{g}(\bar{c})$? on P
Output: the rewritten program $RW(Q, P)$.
Main Vars: S: **stack** of predicates to rewrite;   *Done*: **set** of predicates;
                $modifiedRules(Q, P)$, $magicRules(Q, P)$: **set** of rules;
begin
  1. $modifiedRules(Q, P) := \emptyset$; *Done* := $\emptyset$
  2. $magicRules(Q, P) := \{\mathtt{magic\_g}(\bar{c}).\}$;
  3. $S.\mathbf{push}(\mathtt{g})$;
  4. **while** $S \neq \emptyset$ **do**
  5. $u := S.\mathbf{pop}()$;
  6.   **if** $u \notin Done$ **then**
  7. $Done := Done \cup \{u\}$;
  8.   **for each** $r \in P$ and **for each** $u(\bar{t}) \in H(r)$ defining u **do**
  9.   **if** $B(r) \neq \emptyset$ **or** $|H(r)| > 1$ **or** $Vars(r) \neq \emptyset$ **then**
                // let r be $\mathtt{u}(\bar{t}) \vee \mathtt{v_1}(\bar{t_1}) \vee \ldots \vee \mathtt{v_m}(\bar{t_m}) :- \mathtt{v_{m+1}}(\bar{t_{m+1}}), \ldots, \mathtt{v_n}(\bar{t_n})$.
  10. $mod_r = \{\mathtt{u}(\bar{t}) \vee \mathtt{v_1}(\bar{t_1}) \vee \ldots \vee \mathtt{v_m}(\bar{t_m}) :-$
                  $\mathtt{magic\_u}(\bar{t}), \mathtt{magic\_v_1}(\bar{t_1}), \ldots, \mathtt{magic\_v_m}(\bar{t_m}), \mathtt{v_{m+1}}(\bar{t_{m+1}}), \ldots, \mathtt{v_n}(\bar{t_n}).\}$;
  11.   **if** $mod_r \notin modifiedRules(Q, P)$ **then**
  12. $modifiedRules(Q, P) := modifiedRules(Q, P) \cup \{mod_r\}$;
  13.   **for each** $v_i : 1 <= i <= n$ **and** $v_i \in IDB(P)$ **do**
  14. $magicRules(Q, P) := magicRules(Q, P) \cup$
                  $\{\mathtt{magic\_v_i}(\bar{t_i}) :- \mathtt{magic\_u}(\bar{t}).\}$;
  15. $S.\mathbf{push}(v_i)$;
  16.   **end for**
  17.   **else**
  18. $modifiedRules(Q, P) := modifiedRules(Q, P) \cup r$
  19.   **end if**
  20.   **end for**
  21.   **end if**
  22. **end while**
  23. $RW(Q, P) := magicRules(Q, P) \cup modifiedRules(Q, P)$;
  24. **return** $RW(Q, P)$;
end.

Fig. 1. Magic Sets rewriting algorithm for finitely-recursive queries

and *magic* rules (denoted by *modifiedRules* and *magicRules*, respectively), which are generated on a rule-by-rule basis.

A stack S stores all predicates that have still to be used for propagating the query binding. At first, the set of magic rules is initialized with the magic version of the query, and the query predicate is pushed on S. At each step, an element u is removed from S; if u has not been already considered (the auxiliary variable $Done$ checks this), all the rules defining u are processed one-at-a-time. For each such rule r, let $u(\bar{t})$ be the occurrence of predicate u in $H(r)$; if r features some variable, or if it has a non-empty body, first a modified version is created; moreover, all other atoms in $H(r) - \{u\}$, if some, and all IDB atoms in $B(r)$, are pushed on the stack S, once a proper set of magic rules, one per each such atoms, is generated. In case r is a fact, i.e., its body is empty and there are no variables, it is added to the *modifiedRules* set as it is. Finally, once all the predicates involved in the query evaluation have been processed (thus, S is empty), the algorithm outputs the program $RW(Q, P)$ as the union of all modified rules and generated magic rules.[9] Some rewriting examples are reported next.

Example 8. Let us consider the finitely-recursive query $Q = p(f(g(1)))$ on the following program: $p(1).p(f(X)) \lor p(g(X)) :\text{-} p(X).$

We will depict, step by step, the execution performed by the MS_{FR} algorithm. After the initialization of variables, the first magic rule deriving from the query is generated (*lines* $1-2$): $magic\_p(f(g(1)))$. The predicate p is then pushed onto the stack S (*line* 3) and the first iteration of the main loop (*line* 4) starts. The predicate p is extracted from S and marked as done (*lines* $5 - 7$). In this case, it is the only predicate to be considered. All defining rules for p are then processed (*lines* $8 - 20$). The first rule defining p is a fact $(p(1).)$: both conditions of line 9 are false ($Vars(r)$ denotes the set of variables occurring in r), so the rule is added to $ModifiedRules$ (*line* 18) unchanged. The second rule defining p is a recursive rule. It is worth noting that this rule will be processed twice, because its head contains two occurrences of predicate p (namely, $p(f(X))$ and $p(g(X))$). Thus, the first iteration (because of $p(f(X))$) causes the modified rule:

$$p(f(X)) \lor p(g(X)) :\text{-} magic\_p(f(X)), magic\_p(g(X)), p(X).$$

to be added to the $ModifiedRules$ set (*lines* $10 - 12$). Then the magic rules below are generated, one for $p(g(X))$ in the head, and one for $p(X)$ in the body, and the p predicate is pushed onto the stack S (*lines* $13 - 16$):

$$magic\_p(g(X)) :\text{-} magic\_p(f(X)). \qquad\qquad magic\_p(X) :\text{-} magic\_p(f(X)).$$

Then, the second iteration (because of $p(g(X))$) would cause the same modified rule as above, which thus is not added twice to $ModifiedRules$. The following magic rules are generated, one for $p(f(X))$ in the head, and one for $p(X)$ in the body, and the p predicate is pushed onto the stack S (*lines* $13 - 16$):

$$magic\_p(f(X)) :\text{-} magic\_p(g(X)). \qquad\qquad magic\_p(X) :\text{-} magic\_p(g(X)).$$

---

[9] Note that duplicate rules could be generated. Some further cares might be taken in order to prevent this, but this is out of the scope of this work; some examples of optimization methods can be found in [21].

Then, a second and a third iteration of the main loop start; both immediately ends, as the predicate extracted from the stack S is the already considered predicate p.

Finally, S is found empty; no further iterations are needed, and the algorithm outputs the following complete rewritten program $P_8 = RW(Q, P)$:

$p(1).$ $p(f(X)) \vee p(g(X)) \text{:-} \ magic\_p(f(X)), magic\_p(g(X)), p(X).$
$magic\_p(f(g(1))).$
$magic\_p(g(X)) \text{:-} \ magic\_p(f(X)).$ $magic\_p(X) \text{:-} \ magic\_p(f(X)).$
$magic\_p(f(X)) \text{:-} \ magic\_p(g(X)).$ $magic\_p(X) \text{:-} \ magic\_p(g(X)).$

Example 9. Considering the query $Q_9 = lessThan(s(s(0)), s(0))$? on the program P_2 of Example 3, the algorithm outputs the following rewritten program $P_3 = RW(Q, P)$:

$$
\begin{aligned}
&magic\_lessThan(s(s(0)), s(0)). \\
&magic\_lessThan(X, Y) \text{:-} \ magic\_lessThan(X, s(Y)). \\
&lessThan(X, s(X)) \text{:-} \ magic\_lessThan(X, s(X)). \\
&lessThan(X, s(Y)) \text{:-} \ magic\_lessThan(X, s(Y)), lessThan(X, Y).
\end{aligned}
\tag{3}
$$

5   Properties of Rewritten Programs

In this Section we report some relevant properties of the rewritten programs produced by the algorithm described in Section 4.

Definition 3. Given a program P and a set M of ground atoms of the form $magic\_a$ (where $a \in B_P$), we denote by $eval(P, M)$ the set of rules obtained from $grnd(P)$ as follows: remove from $grnd(P)$ any rule r such that some atom $magic\_a \in B(r)$ and $magic\_a \notin M$; remove any atom of the form $magic\_a$ from the body of the remaining rules.

Lemma 1. Let Q be a query on a program P; if Q is finitely recursive, then

  a. $magicRules(Q, P)$ has a unique answer set M_{mr};
  b. $\{a(\bar{t}) \mid magic\_a(\bar{t}) \in M_{mr}\} = relAtoms(Q, P)$;
  c. M_{mr} is finite;
  d. $magicRules(Q, P)$ is finitely ground;
  e. $eval(modifiedRules(Q, P), M_{mr}) = relRules(Q, P)$.

Proof. (*Sketch*)

  a. $magicRules(Q, P)$ is positive and normal ($\vee$-free), thus the statement follows.
  b. Q is finitely recursive, thus $relAtoms(Q, P)$ can be seen as consisting of the finite union of finite subsets $RA_i(Q, P)$, $0 \leq i \leq k$, with $RA_i(Q, P)$ containing each atom a such that Q depends on a at degree i. Furthermore, since $magicRules$ (Q, P) is positive and normal, any bottom-up application of the immediate consequence operator T_P[22] produces a subset of its unique answer set. The statement can hence be proved by showing that the i-th application of $T_P(\emptyset)$ on $magicRules$ (Q, P) (denoted by $T_{MP}^i(\emptyset)$) derives *all* and *only* the atoms of the form $magic\_a_i$ s.t. $a_i \in RA_i(Q, P)$.

We prove this by induction. *(Basis.)* For $i = 0$, the only relevant atom is Q itself, and $T_{MP}^0(\emptyset) = \{magic\_Q\}$. *(Induction.)* Assume that the statement holds for $i-1$. Then, we first prove that $T_{MP}^i(\emptyset)$ derives *all* $magic\_a_i$ s.t. $a_i \in RA_i(Q, P)$. Indeed, if an atom $a_i \in RA_i(Q, P)$, then there is some rule $r \in grnd(P)$ s.t. $a_i \in Atoms(r)$ and there is at least an atom $a_{i-1} \in H(r)$ s.t. $a_{i-1} \in RA_{i-1}(Q, P)$. This implies that in $grnd(magicRules(Q, P))$ there exists a rule of the form $magic\_a_i :\!- magic\_a_{i-1}$. By induction hypothesis, $magic\_a_{i-1}$ has been derived by $T_{MP}^{i-1}(\emptyset)$, thus $magic\_a_i$ will be derived by $T_{MP}^i(\emptyset)$. Analogous considerations allow us to prove that $T_{MP}^i(\emptyset)$ derives *only* atoms of the form $magic\_a_i$ s.t. $a_i \in RA_i(Q, P)$.

c. Since Q is finitely recursive on P, then $relAtoms(Q, P)$ is finite, hence the statement follows from point (b.).

d. As shown in the proof of (b.), the i-th application of $T_{MP}(\emptyset)$ derives the atoms of the form $magic\_a_i$ s.t. Q depends on a_i at degree i. Since Q is finitely ground on P, the number of these atoms is finite, and $T_{MP}(\emptyset)$ converges finitely (hence, $magicRules(Q, P)$ is finitely ground).

e. The statement easily follows from Definition 3 and from (b.).   □

The next theorem points out a relationship between finitely-recursive queries and $\mathcal{FG}$ programs [12]. For space constraints, we remind the reader to the aforementioned paper for the definition of $\mathcal{FG}$ programs and their properties.

Theorem 1. Given a ground query Q on a program P, if Q is finitely recursive on P, then $RW(Q, P)$ is finitely ground.

Proof. (Sketch) The rewritten program $RW(Q, P)$ can be partitioned in two modules, namely $magicRules(Q, P)$ and $modifiedRules(Q, P)$. Note that $grnd(magicRules$ $(Q, P))$ is a bottom for $grnd(RW(Q, P))$ in the sense of [23]. It follows that $eval(modifiedRules(Q, P), M_{mr}) \subseteq grnd(RW(Q, P))$, and it is finite from Lemma 1, point (b.). The intelligent instantiation [12] finitely generates this program. Hence, the statement follows.   □

The next theorem provides query equivalence results for the rewritten programs.

Theorem 2. Given a ground query Q on a program P, if Q is finitely recursive on P, then $P \models_b Q$ (resp., $P \models_c Q$) if and only if $RW(Q, P) \models_b Q$ (resp., $RW(Q, P) \models_c Q$).

Proof. (Sketch) From Lemma 1, point (e.), we know that $eval(modifiedRules(Q, P),$ $M_{mr}) = relRules(Q, P)$. Thus, we have that $AS(RW(Q, P)) = AS(relRules(Q, P))$. On the other hand, $relRules(Q, P)$ is a bottom for P in the sense of [23]. P is positive, has always at least one answer set, and each answer set of $relRules(Q, P)$ can be extended to an answer set of P. Thus, if we take brave (resp., cautious) reasoning in account, Q can appear in some (resp., all) $M \in AS(P)$ iff Q appears in some (resp., all) answer sets in $AS(relRules(Q, P))$, which, in turn, coincides with $AS(RW(Q, P))$.

□

These results are quite relevant; they also imply that all nice properties of $\mathcal{FG}$ programs hold for rewritten finitely-recursive queries too. This includes, in particular, bottom-up computability of answer sets, and hence full decidability of reasoning, as stated from the next Corollary.

Corollary 1. Both cautious and brave reasoning on a DFRP program P are decidable.

Proof. Any query on a finitely-recursive program P is finitely recursive. Hence, the result immediately follows from Theorem 1 above and Theorem 3 of [12].   □

Example 10. The intelligent instantiation of the program P_3 of Example 9 is the following:

$$magic\_lessThan(s(s(0)), s(0)).$$
$$magic\_lessThan(s(s(0)), 0) :\text{-} \ magic\_lessThan(s(s(0)), s(0)).$$

It is finite; hence, as expected, the originating rewritten program is finitely ground. It has the unique finite answer set $\{magic\_lessThan(s(s(0)), s(0)), magic\_lessThan (s(s(0)), 0)\}$, which is easily computable, thus allowing to answer to the query Q_9 of Example 9 with 'no'.

Next, we are going to prove a result about the efficiency of the rewriting algorithm. To this aim, we need to introduce the definition of what we mean for size of a program.

Definition 4. Let P be a (non-ground) logic program. The *size* $\|t\|$ of a term t is 1, if the term is a constant or a variable; the size of a functional term $f(t_1, \ldots, t_n)$ is defined as $1 + \|t_1\| + \ldots, \|t_n\|$. The *size of an atom* is given by the sum of the size of its terms; if the atom has arity 0, size is 1. The *size of the program* P, denoted by $\|P\|$, is the sum of the sizes of all atoms occurring in P. Note that, if the same atom occurs in two different rules of P, it accounts for twice its size.

Example 11. The program P_2 in Example 3 has size $\|P\| = 8$.

Theorem 3. Given a finitely-recursive query Q on a program P, the size of $RW(Q, P)$ is linear in the size of the input P and Q; that is $\|RW(Q, P)\| = O(\|P\| + \|Q\|)$.

Proof. (*Sketch*) The program $RW(Q, P)$ is obtained as the union of the two sets of rules $modifiedRules(Q, P)$ and $magicRules(Q, P)$.

In the worst case, the number of atoms in the first set is given by the number of atoms in P plus as many atoms as the number of rules in P (at most one magic atom is added for each rule in P). So, $\|modifiedRules(Q, P)\|$ is definitely equal to $O(\|P\|)$.

Let us consider now the $magicRules(Q, P)$ program. For each IDB atom occurring in the body of a rule in P, at most one magic rule with exactly two atoms is generated. Then, in the worst case, the number of atoms in $magicRules(Q, P)$ is not greater than $2 \cdot \|P\|$.

It is worth noting that, as far as the algorithm is defined, arities of all new predicates are equal or less than the arities of original ones; furthermore, terms are left unchanged. Thus, from the considerations above, the statement $\|RW(Q, P)\| = O(\|P\| + \|Q\|)$ immediately follows.   □

6 Related Works

There are many proposals for treating functional terms in ASP: ω-*restricted programs* [6], Finitary Programs [7,8], $\mathbb{FDNC}$ programs [9], $\mathcal{FG}$ programs [12], disjunctive finitely-recursive programs [13], and the works in [10,11]. However, only the works in [7,9,12,13] deal with disjunctive programs. In particular, the most related works to our paper are [7,13], since they try to single-out classes of computable queries.

The work in [7] extends finitary programs to preserve decidability in case of disjunction. To this end, a condition on disjunctive heads is added to the original definition of finitary programs [8]; while the concept of atoms dependencies remains unchanged as in the or-free case. Conversely, in [13], the concept of finitely-recursive programs has been redefined, taking into account head-to-head dependencies which are due to disjunction (as we do in the present work).

Interestingly, the class of DFRP programs herein defined, which enjoys the decidability of reasoning (as proved in Theorem 1), enlarges the positive subclass of disjunctive finitary programs of [7]. Indeed, while all positive finitary programs trivially belong to the class of DFRP programs, the above mentioned third condition is not guaranteed to be fulfilled when negation is forbidden, as witnessed by the following program:

$$p(X) \vee q(X) :\!- s(X). \qquad p(f(X)) :\!- q(X).$$
$$q(X) :\!- p(X). \qquad p(1).$$

Finitely-recursive programs, initially introduced in [8] as a super-class of finitary programs allowing function symbols and negation, has been redefined in [13], in order to deal with disjunction. The authors provide a compactness property result for such programs and some interesting semi-decidability results for cautious ground querying, but no decidability results about positive disjunctive programs. On the contrary, we focus on decidability results for disjunctive finitely-recursive positive programs, also providing an *effective* strategy for the actual computation of all ground reasoning tasks.

The introduction of functional terms (or similar constructs) have been studied in several other fields, besides Logic Programming, such as deductive databases (see $\mathcal{LDL}$ [24]); furthermore, studies on computable fragments of logic programs with functions are also related to termination studies of SLD-resolution for Prolog programs (see e.g. [25,26,27]).

Are also related to this work some papers about the magic-set technique [18,19,20], for which different extensions and refinements have been proposed. Among the more recent works, an adaptation for soft-stratifiable programs [28], the generalization to the disjunctive case [21] and to Datalog with (possibly unstratified) negation [29] are worth remembering.

7 Conclusions

We presented an adaptation of the magic-sets technique that allows query answering over disjunctive finitely recursive positive programs also by means of standard bottom-up techniques. This allows us to enrich the collection of logic programs with function symbols for which ground query answering can be performed by the current ASP solvers.

Future work will focus on overcoming the limitation of considering only ground queries, by identifying the minimal set of variables required to be bound in order to preserve decidability. Next step will then deal with negation.

References

1. Baral, C.: Knowledge Representation, Reasoning and Declarative Problem Solving. CUP (2003)
2. Gelfond, M., Lifschitz, V.: The Stable Model Semantics for Logic Programming. In: ICLP/SLP 1988, pp. 1070–1080. MIT Press, Cambridge (1988)
3. Gelfond, M., Lifschitz, V.: Classical Negation in Logic Programs and Disjunctive Databases. NGC 9, 365–385 (1991)
4. Lifschitz, V.: Answer Set Planning. In: ICLP 1999, pp. 23–37 (1999)
5. Marek, V.W., Truszczyński, M.: Stable Models and an Alternative Logic Programming Paradigm. In: The Logic Programming Paradigm – A 25-Year Perspective, pp. 375–398 (1999)
6. Syrjänen, T.: Omega-restricted logic programs. In: Eiter, T., Faber, W., Truszczyński, M. (eds.) LPNMR 2001. LNCS (LNAI), vol. 2173, p. 267. Springer, Heidelberg (2001)
7. Bonatti, P.A.: Reasoning with infinite stable models II: Disjunctive programs. In: Stuckey, P.J. (ed.) ICLP 2002. LNCS, vol. 2401, pp. 333–346. Springer, Heidelberg (2002)
8. Bonatti, P.A.: Reasoning with infinite stable models. AI 156(1), 75–111 (2004)
9. Simkus, M., Eiter, T.: FDNC: Decidable Non-monotonic Disjunctive Logic Programs with Function Symbols. In: Dershowitz, N., Voronkov, A. (eds.) LPAR 2007. LNCS (LNAI), vol. 4790, pp. 514–530. Springer, Heidelberg (2007)
10. Lin, F., Wang, Y.: Answer Set Programming with Functions. In: KR 2008, Sydney, Australia, pp. 454–465. AAAI Press, Menlo Park (2008)
11. Cabalar, P.: Partial Functions and Equality in Answer Set Programming. In: Garcia de la Banda, M., Pontelli, E. (eds.) ICLP 2008. LNCS, vol. 5366, pp. 392–406. Springer, Heidelberg (2008)
12. Calimeri, F., Cozza, S., Ianni, G., Leone, N.: Computable Functions in ASP: Theory and Implementation. In: Garcia de la Banda, M., Pontelli, E. (eds.) ICLP 2008. LNCS, vol. 5366, pp. 407–424. Springer, Heidelberg (2008)
13. Baselice, S., Bonatti, P.A., Criscuolo, G.: On Finitely Recursive Programs. Tech. Report 0901.2850v1, arXiv.org (2009); To appear in TPLP (TPLP)
14. Bonatti, P.A.: Erratum to: Reasoning with infinite stable models. AI 172(15), 1833–1835 (2008)
15. Leone, N., Pfeifer, G., Faber, W., Eiter, T., Gottlob, G., Perri, S., Scarcello, F.: The DLV System for Knowledge Representation and Reasoning. ACM TOCL 7(3), 499–562 (2006)
16. Calimeri, F., Cozza, S., Ianni, G., Leone, N.: DLV-Complex, homepage (since 2008), http://www.mat.unical.it/dlv-complex
17. Marano, M., Ianni, G., Ricca, F.: A Magic Set Implementation for Disjunctive Logic Programming with Function Symbols. Submitted to CILC 2009 (2009)
18. Bancilhon, F., Maier, D., Sagiv, Y., Ullman, J.D.: Magic Sets and Other Strange Ways to Implement Logic Programs. In: PODS 1986, Cambridge, Massachusetts, pp. 1–15 (1986)
19. Ullman, J.D.: Principles of Database and Knowledge Base Systems, vol. 2. Computer Science Press (1989)
20. Beeri, C., Ramakrishnan, R.: On the Power of Magic. In: PODS 1987, pp. 269–284. ACM, New York (1987)

21. Cumbo, C., Faber, W., Greco, G., Leone, N.: Enhancing the magic-set method for disjunctive datalog programs. In: Demoen, B., Lifschitz, V. (eds.) ICLP 2004. LNCS, vol. 3132, pp. 371–385. Springer, Heidelberg (2004)
22. Van Gelder, A., Ross, K.A., Schlipf, J.S.: The Well-Founded Semantics for General Logic Programs. J. ACM 38(3), 620–650 (1991)
23. Lifschitz, V., Turner, H.: Splitting a Logic Program. In: ICLP 1994, pp. 23–37. MIT Press, Cambridge (1994)
24. Naqvi, S., Tsur, S.: A logical language for data and knowledge bases. CS Press, NY (1989)
25. Schreye, D.D., Decorte, S.: Termination of Logic Programs: The Never-Ending Story. JLP 19/20, 199–260 (1994)
26. Bossi, A., Cocco, N., Fabris, M.: Norms on Terms and their use in Proving Universal Termination of a Logic Program. Theoretical Computer Science 124(2), 297–328 (1994)
27. Bruynooghe, M., Codish, M., Gallagher, J.P., Genaim, S., Vanhoof, W.: Termination analysis of logic programs through combination of type-based norms. ACM TOPLAS 29(2), 10 (2007)
28. Behrend, A.: Soft stratification for magic set based query evaluation in deductive databases. In: PODS 2003, San Diego, CA, USA, pp. 102–110 (2003)
29. Faber, W., Greco, G., Leone, N.: Magic Sets and their Application to Data Integration. JCSS 73(4), 584–609 (2007)

Relevance-Driven Evaluation of Modular Nonmonotonic Logic Programs*

Minh Dao-Tran, Thomas Eiter, Michael Fink, and Thomas Krennwallner

Institut für Informationssysteme, Technische Universität Wien
Favoritenstraße 9-11, A-1040 Vienna, Austria
{dao,eiter,fink,tkren}@kr.tuwien.ac.at

Abstract. Modular nonmonotonic logic programs (MLPs) under the answer-set semantics have been recently introduced as an ASP formalism in which modules can receive context-dependent input from other modules, while allowing (mutually) recursive module calls. This can be used for more succinct and natural problem representation at the price of an exponential increase of evaluation time. In this paper, we aim at an efficient top-down evaluation of MLPs, considering only calls to relevant module instances. To this end, we generalize the well-known Splitting Theorem to the MLP setting and present notions of call stratification, for which we determine sufficient conditions. Call-stratified MLPs allow to split module instantiations into two parts, one for computing input of module calls, and one for evaluating the calls themselves with subsequent computations. Based on these results, we develop a top-down evaluation procedure that expands only relevant module instantiations. Finally, we discuss syntactic conditions for its exploitation.

1 Introduction

Modularity is an important element of high-level programming languages that has beneficial effects on problem decomposition, which allows one to structure a program into parts that solve subproblems appropriately. Its importance has also been recognized in the area of logic programming (see [1] for a historic account), and in particular in Answer Set Programming (ASP), as witnessed by the early conception of Splitting Sets [2], a generalization of the notion of stratification for program decomposition.

Since then, modularity aspects have been considered in several works, cf. [1, 3, 4, 5, 6, 7, 8], that aim at practicable formalisms for modular logic programming. The approaches are classified into Programming-in-the-small, building on abstraction and scoping mechanisms (e.g., generalized quantifiers [1], macros [5], and templates [6] have been developed in ASP), and Programming-in-the-large, where compositional operators serve the combination of separate and independent modules based on standard semantics. A prominent representative of the latter in ASP is DLP-functions [3, 4].

---
* This research has been supported by the Austrian Science Fund (FWF) project P20841, the Vienna Science and Technology Fund (WWTF) project ICT08-020, and the EC ICT Integrated Project Ontorule (FP7 231875).

E. Erdem, F. Lin, and T. Schaub (Eds.): LPNMR 2009, LNCS 5753, pp. 87–100, 2009.

Recently, modular logic programs (MLPs) have been introduced in [9], which can be viewed as a generalization of DLP-functions. They overcome a restriction of a preliminary approach in [1], in which module calls must be acyclic (which prohibits the use of recursion through modules), as well as anomalies of the semantics due to the Gelfond-Lifschitz reduct, which is replaced by the FLP reduct [10]. The latter was also used for the semantics of HEX-programs [11], a generalization of [1] to the HiLog setting. However, both [1] and [11] defined models of a single module resp. program, and no global semantics for a collection of modules resp. programs is evident.

As the semantics of MLPs is based on module instantiations (which takes possible input values into account), a naive evaluation following the definition is—similar as with grounding of ordinary ASP programs—infeasible in practice; in general, a module may have double exponentially many instances. Towards implementation, efficient evaluation strategies are thus essential, which are sensitive to (sub-)program classes that do not require a simple guess-and-check procedure on the instantiation, but allow for a guided model building process. Starting from the main module, instances of modules may be created on demand as needed by module calls, focusing on relevant module instances.

Restrictions on programs, like stratification of normal MLPs in [9], may be helpful in this regard. However, the notion of stratification is very strict. It requires that *all* module instances are stratified. Moreover, the fix-point semantics for stratified programs given there is inherently bottom-up and only applies to normal programs, excluding a large class of programs which exploit recursion in a common and natural way and are evaluable top-down, even if they are not normal or unstratified in the sense of [9].

For illustration, consider the following example with an MLP consisting of three modules, one is main and the other two are libraries. Each consists of a module name, with (optional) formal input parameters, and a set of rules. One can inquiry a library module for the extensions of its predicates, with input fed into the module via the input parameters. This example exploits the mutual recursive calls between two library modules to determine whether a set has even cardinality.

Example 1. Let $\mathbf{P}$ be an MLP consisting of three modules $m_1 = (P_1, R_1)$, $m_2 = (P_2[q_2], R_2)$, and $m_3 = (P_3[q_3], R_3)$, where $R_1 = \{q(a).\ q(b).\ ok \leftarrow P_2[q].even.\}$,

$$
R_2 = \left\{
\begin{array}{c}
q_2'(X) \vee q_2'(Y) \leftarrow q_2(X), q_2(Y), \\
X \neq Y. \\
skip_2 \leftarrow q_2(X), \text{not } q_2'(X). \\
even \leftarrow \text{not } skip_2. \\
even \leftarrow skip_2, P_3[q_2'].odd.
\end{array}
\right\}, \quad
R_3 = \left\{
\begin{array}{c}
q_3'(X) \vee q_3'(Y) \leftarrow q_3(X), q_3(Y), \\
X \neq Y. \\
skip_3 \leftarrow q_3(X), \text{not } q_3'(X). \\
odd \leftarrow skip_3, P_2[q_3'].even.
\end{array}
\right\}
$$

Intuitively, m_1 calls m_2 to check if the number of facts for predicate q is even. The call to m_2 'returns' *even*, if either the input q_2 to m_2 is empty (as then $skip_2$ is false), or the call of m_3 with q_2' resulting from q_2 by randomly removing one element (then $skip_2$ is true) returns *odd*. Module m_3 returns *odd* for input q_3, if a call to R_2 with q_3' analogously constructed from q_3 returns *even*. In any answer set of $\mathbf{P}$, *ok* is true.

This program is not normal, and shifting head disjunctions yields a program which is not stratified as per [9]. However, along the mutual recursive chain of calls $P_3[q_2'].odd$, $P_2[q_3'].even$ the inputs q_2' and q_3' gradually decrease until the base, i.e., the empty input, is reached. Taking such decreasing inputs of the relevant module calls into account, we can evaluate MLPs efficiently along the relevant call graph using a finer grained notion of stratification, tolerating also disjunctive or unstratified rules in modules.

Capturing this intuition formally and developing a suitable evaluation algorithm for respective MLPs are the main contributions of this work, which are as follows:

- We develop appropriate notions of *call stratification* and *input stratification* for MLPs, and generalize the well-known Splitting Theorem [2] to this setting. Moreover, we establish a sufficient condition to determine call (and input) stratification at the schematic level, i.e., without the requirement to consider all module instantiations (Section 3).
- By the previous results, module instances calling other modules can be locally split into an input preparation part and a calling part. Based on this, a top-down evaluation procedure is developed, expanding only relevant module instances (Section 4).
- Finally, we discuss syntactic conditions for its exploitation, outline a module rewriting technique for self-recursive modules, and consider related work (Section 5).

The envisaged programming style of call-stratified MLPs is to exploit the natural way of specifying recursive problems with decreasing input. Applications emerge, e.g., in temporal reasoning with an ontology of (partially ordered) time points, reasoning about recurrent properties of sets, or games with 'decreasing input'.

Modular ASP in which modules can be used in an unrestricted and natural way for problem solving, including recursion, is an important requirement for the further development of the ASP paradigm. In this paper, we contribute to this goal, providing efficient evaluation techniques, which are essential for its realization.

2   Preliminaries

Modular logic programs (MLPs) [9] consist of modules as a way to structure nonmonotonic logic programs under answer set semantics [12]. Moreover, such modules allow for input provided by other modules, and may call each other in a (mutually) recursive way.

Syntax. We consider programs in a function-free first-order (Datalog) setting. Let V be a vocabulary C, P, X, and M of mutually disjoint sets of *constants*, *predicate*, *variable*, and *module names*, respectively, where each $p \in P$ has a fixed arity $n \geq 0$, and each module name in M has a fixed associated list $q = q_1, \ldots, q_k$ ($k \geq 0$) of predicate names $q_i \in P$ (the formal input parameters). Unless stated otherwise, elements from X (resp., $C \cup P$) are denoted with first letter in upper case (resp., lower case).

Each $t \in C \cup X$ is a *term*. An ordinary atom (simply atom) has the form $p(t_1, \ldots, t_n)$, where $p \in P$ and $t_1, \ldots, t_n$ are terms; $n \geq 0$ is its *arity*. A *module atom* has the form $P[p_1, \ldots, p_k].o(t_1, \ldots, t_l)$, where $P \in M$ is a module name with associated q, $p_1, \ldots, p_k$ is a list of predicate names $p_i \in P$, called *module input list*, such that p_i has the arity of q_i in q, and $o \in P$ is a predicate name such that $o(t_1, \ldots, t_l)$ is an ordinary atom. Intuitively, a module atom provides a way for deciding the truth value of a ground atom $o(c)$ in a program P depending on the extension of a set of input predicates.

A *rule* r is of the form

$$\alpha_1 \vee \cdots \vee \alpha_k \leftarrow \beta_1, \ldots, \beta_m, \text{not } \beta_{m+1}, \ldots, \text{not } \beta_n \quad (k \geq 1, m, n \geq 0), \quad (1)$$

where all α_i are atoms and each β_j is an ordinary or a module atom. We define $H(r) = \{\alpha_1, \ldots, \alpha_k\}$ and $B(r) = B^+(r) \cup B^-(r)$, where $B^+(r) = \{\beta_1, \ldots, \beta_m\}$ and $B^-(r) = \{\beta_{m+1}, \ldots, \beta_n\}$. If $B(r) = \emptyset$ and $H(r) \neq \emptyset$, then r is a *fact*; r is *ordinary*, if it contains only ordinary atoms. We denote by $at(r)$ the set $H(r) \cup B(r)$.

We now formally define the syntax of modules and MLPs. A *module* is a pair $m = (P, R)$, where $P \in \mathcal{M}$ with associated input q, and R is a finite set of rules. It is either a *main module* (then $|q| = 0$) or a *library module*, and is *ordinary* (resp., *ground*), iff all rules in R are ordinary (ground). We refer with $R(m)$ to the rule set of m, and omit empty $[]$ and $()$ from (main) modules and module atoms if unambiguous.

A *modular logic program (MLP)* is a tuple $\mathbf{P} = (m_1, \ldots, m_n)$, $n \geq 1$, where all m_i are modules and at least one is a main module, where $\mathcal{M} = \{P_1, \ldots, P_n\}$. $\mathbf{P}$ is *ground*, iff each module is ground.

Example 2. For instance, the MLP in Example 1 consists of three modules $\mathbf{P} = (m_1, m_2, m_3)$ where m_2 and m_3 are library modules, and m_1 is a (ground) main module.

Semantics. The semantics of MLPs is defined in terms of Herbrand interpretations and grounding as customary in traditional logic programming and ASP.

The *Herbrand base* w.r.t. vocabulary $\mathcal{V}$, $HB_{\mathcal{V}}$, is the set of all ground ordinary and module atoms that can be built using $\mathcal{C}$, $\mathcal{P}$ and $\mathcal{M}$; if $\mathcal{V}$ is implicit from an MLP $\mathbf{P}$, it is the *Herbrand base of* $\mathbf{P}$ and denoted by $HB_{\mathbf{P}}$. The grounding of a rule r is the set $gr(r)$ of all ground instances of r w.r.t. $\mathcal{C}$; the grounding of rule set R is $gr(R) = \bigcup_{r \in R} gr(r)$, and the one of a module m, $gr(m)$, is defined by replacing the rules in $R(m)$ by $gr(R(m))$; the grounding of an MLP $\mathbf{P}$ is $gr(\mathbf{P})$, which is formed by grounding each module m_i of $\mathbf{P}$. The semantics of an arbitrary MLP $\mathbf{P}$ is given in terms of $gr(\mathbf{P})$.

Let $S \subseteq HB_{\mathbf{P}}$ be any set of atoms. For any list of predicate names $\boldsymbol{p} = p_1, \ldots, p_k$ and $\boldsymbol{q} = q_1, \ldots, q_k$, we use the notation $S|_{\boldsymbol{p}} = \{p_i(\boldsymbol{c}) \in S \mid i \in \{1, \ldots, k\}\}$ and $S|_{\boldsymbol{p}}^{\boldsymbol{q}} = \{q_i(\boldsymbol{c}) \mid p_i(\boldsymbol{c}) \in S, i \in \{1, \ldots, k\}\}$.

For a $P \in \mathcal{M}$ with associated formal input q we say that $P[S]$ is a *value call with input* S, where $S \subseteq HB_{\mathbf{P}}|_q$. Let $VC(\mathbf{P})$ denote the set of all value calls $P[S]$ with some S (note that $VC(\mathbf{P})$ is also used as index set). A *rule base* is an (indexed) tuple $\mathbf{R} = (R_{P[S]} \mid P[S] \in VC(\mathbf{P}))$ of sets of rules $R_{P[S]}$. For a module $m_i = (P_i[\boldsymbol{q_i}], R_i)$ from $\mathbf{P}$, its *instantiation with* $S \subseteq HB_{\mathbf{P}}|_{q_i}$, is $I_{\mathbf{P}}(P_i[S]) = R_i \cup S$. For an MLP $\mathbf{P}$, its *instantiation* is the rule base $I(\mathbf{P}) = (I_{\mathbf{P}}(P_i[S]) \mid P_i[S] \in VC(\mathbf{P}))$.

We next define (Herbrand) interpretations and models of MLPs.

Definition 1 (model). *An interpretation* $\mathbf{M}$ *of an MLP* $\mathbf{P}$ *is an (indexed) tuple* $(M_i/S \mid P_i[S] \in VC(\mathbf{P}))$, *where all* $M_i/S \subseteq HB_{\mathbf{P}}$ *contain only ordinary atoms. An interpretation* $\mathbf{M}$ *of an MLP* $\mathbf{P}$ *is a model of*

- *a ground atom* $\alpha \in HB_{\mathbf{P}}$ *at* $P_i[S]$, *denoted* $\mathbf{M}, P_i[S] \models \alpha$, *iff (i)* $\alpha \in M_i/S$ *when* α *is ordinary, and (ii)* $o(\boldsymbol{c}) \in M_k/((M_i/S)|_{\boldsymbol{p}}^{\boldsymbol{q_k}})$, *when* $\alpha = P_k[\boldsymbol{p}].o(\boldsymbol{c})$ *is a module atom;*
- *a ground rule* r *at* $P_i[S]$ *(* $\mathbf{M}, P_i[S] \models r$*), iff* $\mathbf{M}, P_i[S] \models H(r)$ *or* $\mathbf{M}, P_i[S] \not\models B(r)$, *where (i)* $\mathbf{M}, P_i[S] \models H(r)$, *iff* $\mathbf{M}, P_i[S] \models \alpha$ *for some* $\alpha \in H(r)$, *and (ii)* $\mathbf{M}, P_i[S] \models B(r)$, *iff* $\mathbf{M}, P_i[S] \models \alpha$ *for all* $\alpha \in B^+(r)$ *and* $\mathbf{M}, P_i[S] \not\models \alpha$ *for all* $\alpha \in B^-(r)$;

- *a set of ground rules R at $P_i[S]$ ($\mathbf{M}, P_i[S] \models R$) iff $\mathbf{M}, P_i[S] \models r$ for all $r \in R$;*
- *a ground rule base $\mathbf{R}$ ($\mathbf{M} \models \mathbf{R}$) iff $\mathbf{M}, P_i[S] \models R_{P_i[S]}$ for all $P_i[S] \in VC(\mathbf{P})$.*

Finally, $\mathbf{M}$ is a model of an MLP $\mathbf{P}$, denoted $\mathbf{M} \models \mathbf{P}$, iff $\mathbf{M} \models I(\mathbf{P})$ in case $\mathbf{P}$ is ground resp. $\mathbf{M} \models gr(\mathbf{P})$, if $\mathbf{P}$ is nonground. An MLP $\mathbf{P}$ is satisfiable, *iff it has a model.*

For any interpretations $\mathbf{M}$ and $\mathbf{M}'$ of $\mathbf{P}$, we define $\mathbf{M} \leq \mathbf{M}'$, iff $M_i/S \subseteq M_i'/S$ for every $P_i[S] \in VC(\mathbf{P})$, and $\mathbf{M} < \mathbf{M}'$, iff $\mathbf{M} \neq \mathbf{M}'$ and $\mathbf{M} \leq \mathbf{M}'$. A model $\mathbf{M}$ of $\mathbf{P}$ (resp., a rule base $\mathbf{R}$) is *minimal*, if $\mathbf{P}$ (resp., $\mathbf{R}$) has no model $\mathbf{M}'$ such that $\mathbf{M}' < \mathbf{M}$.

We next proceed to define answer sets for MLPs. In order to focus on relevant modules, we introduce the formal notion of a call graph. Intuitively, a call graph represents the relationship between module instantiations and potential module calls: nodes correspond to instantiations and an edge indicates that there is a presumptive call from one module instantiation to another. Labels on the edges distinguish different syntactical calls. Given an interpretation, one can determine the actual calls as edges with labels such that the respective predicates match in the interpretation of the corresponding module instantiations. Edges satisfying this condition, their incident nodes, and the nodes representing main modules constitute the relevant call graph.

Definition 2 (call graph). *The* call graph *of an MLP $\mathbf{P}$ is a labeled digraph $CG_\mathbf{P} = (V, E, l)$ with vertex set $V = VC(\mathbf{P})$ and an edge e from $P_i[S]$ to $P_k[T]$ in E iff $P_k[\boldsymbol{p}].o(\boldsymbol{t})$ occurs in $R(m_i)$; furthermore, e is labeled with an input list $\boldsymbol{p}$, denoted $l(e)$. Given an interpretation $\mathbf{M}$, the* relevant call graph *$CG_\mathbf{P}(\mathbf{M}) = (V', E')$ of $\mathbf{P}$ w.r.t. $\mathbf{M}$ is the subgraph of $CG_\mathbf{P}$ where E' contains all edges from $P_i[S]$ to $P_k[T]$ of $CG_\mathbf{P}$ such that $(M_i/S)|_{l(e)}^{q_k} = T$, and V' contains all $P_i[S]$ that are main module instantiations or induced by E'; any such $P_i[S]$ is called* relevant *w.r.t. $\mathbf{M}$.*

Example 3. Let in Example 1 $S_\emptyset^i = \emptyset$, $S_a^i = \{q_i(a)\}$, $S_b^i = \{q_i(b)\}$, and $S_{ab}^i = \{q_i(a), q_i(b)\}$. Then $VC(\mathbf{P}) = \{P_1[\emptyset], P_2[S_v^2], P_3[S_w^3]\}$, where $v, w \in \{\emptyset, a, b, ab\}$, and $CG_\mathbf{P}$ has edges $P_1[\emptyset] \xrightarrow{q} P_2[S_v^2]$, $P_2[S_v^2] \xrightarrow{q_2'} P_3[S_w^3]$, and $P_3[S_w^3] \xrightarrow{q_3'} P_2[S_v^2]$. For the interpretation $\mathbf{M}$ such that $M_1/\emptyset = \{q(a), q(b), ok\}$, $M_2/S_{ab}^2 = \{q_2(a), q_2(b), q_2'(a), skip_2, even\}$, $M_2/\emptyset = \{even\}$, and $M_3/S_a^3 = \{q_3(a), skip_3, odd\}$, the nodes of $CG_\mathbf{P}(\mathbf{M})$ are $P_1[\emptyset]$, $P_2[S_{ab}^2]$, $P_2[\emptyset]$, and $P_3[S_a^3]$.

For answer sets of an MLP $\mathbf{P}$, we use a reduct of the instantiated program as customary in ASP. As $\mathbf{P}$ might have inconsistent module instantiations, compromising the existence of an answer set of $\mathbf{P}$, we contextualize reducts and answer sets. We denote the vertex and edge set of a graph G by $V(G)$ and $E(G)$, respectively.

Definition 3 (context-based reduct). *A* context *for an interpretation $\mathbf{M}$ of an MLP $\mathbf{P}$ is any set $C \subseteq VC(\mathbf{P})$ such that $V(CG_\mathbf{P}(\mathbf{M})) \subseteq C$. The* reduct *of $\mathbf{P}$ at $P[S]$ w.r.t. $\mathbf{M}$ and C, denoted $f\mathbf{P}(P[S])^{\mathbf{M},C}$, is the rule set $I_{gr(\mathbf{P})}(P[S])$ from which, if $P[S] \in C$, all rules r such that $\mathbf{M}, P[S] \not\models B(r)$ are removed. The* reduct *of $\mathbf{P}$ w.r.t. $\mathbf{M}$ and C is $f\mathbf{P}^{\mathbf{M},C} = (f\mathbf{P}(P[S])^{\mathbf{M},C} \mid P[S] \in VC(\mathbf{P}))$.*

That is, outside C the module instantiations of $\mathbf{P}$ resp. $gr(\mathbf{P})$ remain untouched, while inside C the FLP-reduct [10] is applied.

Definition 4 (answer set). *Let* $\mathbf{M}$ *be an interpretation of a ground MLP* $\mathbf{P}$. *Then* $\mathbf{M}$ *is an* answer set *of* $\mathbf{P}$ *w.r.t. a context* C *for* $\mathbf{M}$, *iff* $\mathbf{M}$ *is a minimal model of* $f\mathbf{P}^{\mathbf{M},C}$.

Note that C is a parameter that allows to select a degree of overall-stability for answer sets of $\mathbf{P}$. The minimal context $C = V(CG_{\mathbf{P}}(\mathbf{M}))$ is the relevant call graph of $\mathbf{P}$. From now on we consider this as the default context and omit C from notation.

Example 4. Recall $\mathbf{M}$ from Example 3. For every $P_i[S] \in V(CG_{\mathbf{P}}(\mathbf{M}))$, M_i/S is a $\subseteq$-minimal set that satisfies $f\mathbf{P}(P_i[S])^{\mathbf{M}}$. Thus, any such $\mathbf{M}$ is an answer set of $\mathbf{P}$ iff for every $P_k[T] \in VC(\mathbf{P}) \setminus V(CG_{\mathbf{P}}(\mathbf{M}))$, M_k/T is a $\subseteq$-minimal set satisfying $I_{\mathbf{P}}(P_i[S])$.

3 Splitting for Modular Nonmonotonic Logic Programs

We investigate splitting for MLPs at two different levels: the global (module instantiation) level along the relevant call graph, and the local level ('inside' module instantiations) w.r.t. the (instance) dependency graph. These two notions reveal a class of MLPs, for which an efficient top-down algorithm can be developed for answer set computation.

3.1 Global Splitting for Call-Stratified MLPs

We start by introducing *call stratified* MLPs, whose module instantiations can be split into different layers and evaluated in a stratified way.

Definition 5. *Let* $\mathbf{M}$ *be an interpretation of an MLP* $\mathbf{P}$. *We say that* $\mathbf{P}$ *is* c-stratified (call stratified) *w.r.t.* $\mathbf{M}$ *iff cycles in* $CG_{\mathbf{P}}(\mathbf{M})$ *contain only nodes of the form* $P_i[\emptyset]$.

The intuition is to evaluate module instantiations of c-stratified MLPs in a particular order along the call chain, such that potential 'self-stabilizing' effects of cycles have to be taken into account only at the base, i.e., for module instantiations with empty input.

Example 5. Consider the MLP $\mathbf{P}$ and the interpretation $\mathbf{M}$ from Example 3. It is easily verified that $\mathbf{P}$ is c-stratified w.r.t. $\mathbf{M}$. One possible call chain for evaluation is

$$P_1[\emptyset] \xrightarrow{q} P_2[\{q(a), q(b)\}] \xrightarrow{q_2'} P_3[\{q_2'(a)\}] \xrightarrow{q_3'} P_2[\emptyset] \ .$$

Definition 6. *Let* $\mathbf{M}$ *be an interpretation of an MLP* $\mathbf{P}$ *and* R *be a rule set. We say that* M_i/S *is an* answer set *of* R *relative to* $\mathbf{M}$, *iff* $\mathbf{M}$ *is an answer set of the rule base* $(R_{P[S]} \mid P[S] \in VC(\mathbf{P}))$, *where* $R_{P_i[S]} = R$ *and* $R_{P_j[T]} = M_j/T$ *for* $i \neq j$ *or* $S \neq T$.

In particular, M_i/S is an answer set of $R = I_{\mathbf{P}}(P_i[S])$ relative to $\mathbf{M}$, if it is an answer set of R while other instances are fixed by corresponding elements in $\mathbf{M}$, i.e., all module calls in R are fixed.

Example 6. Consider $\mathbf{P}$ from Example 1 and $\mathbf{M}$ from Example 3, then M_2/S_{ab}^2 is an answer set of $I_{\mathbf{P}}(P_2[S_{ab}^2])$ relative to $\mathbf{M}$.

Proposition 1. *Let* $\mathbf{M}$ *be an interpretation of a c-stratified MLP* $\mathbf{P}$. *Suppose that along* $CG_{\mathbf{P}}(\mathbf{M})$, M_i/S *is an answer set of* $I_{\mathbf{P}}(P_i[S])$ *relative to* $\mathbf{M}$ *for each* $P_i[S] \in V$ $(CG_{\mathbf{P}}(\mathbf{M}))$. *If there is an answer set of* $\mathbf{P}$ *that coincides with* $\mathbf{M}$ *for every* $P_i[\emptyset]$ *on a cycle in* $CG_{\mathbf{P}}(\mathbf{M})$, *then* $\mathbf{P}$ *has an answer set that coincides with* $\mathbf{M}$ *on* $CG_{\mathbf{P}}(\mathbf{M})$.

Proposition 1 already indicates a top-down way to evaluate c-stratified MLPs. For a concrete procedure, we need a notion of "local" splitting inside module instances, introduced in the next section.

3.2 Local Splitting for Input and Call Stratified MLPs

Towards local splitting, we will first extend the notion of Splitting Sets [2] to MLPs. Then, for pratical purposes, we are interested in splitting a module instance w.r.t. module calls. To this end, we introduce a general and another specific notion of *input splitting sets* in Definition 7. Given a set R of ground rules and a list of predicate names $\boldsymbol{p} = \{p_1, \ldots, p_k\}$, let $def(\boldsymbol{p}, R) = \{p_\ell(\boldsymbol{d}) \mid \exists r \in R, p_\ell(\boldsymbol{d}) \in H(r), p_\ell \in \boldsymbol{p}\}$.

Definition 7 (splitting set). *Let* $\mathbf{P}$ *be an MLP,* R *be a set of ground rules and* α *be a ground module atom of form* $P_k[\boldsymbol{p}].o(\boldsymbol{c})$.

(a) A splitting set of R *is a set* $U \subseteq HB_{\mathbf{P}}$ *s.t. (i) for any rule* $r \in R$, *if* $H(r) \cap U \neq \emptyset$ *then* $at(r) \subset U$; *and (ii) if* $\alpha \in U$ *then* $def(\boldsymbol{p}, R) \subseteq U$.
(b) Let U *be a splitting set of* R. *We say that* U *is an* input splitting set *of* R *for* α, *iff* $\alpha \notin U$ *and* $def(\boldsymbol{p}, R) \subseteq U$.

As usual, the bottom of a set of ground rules R w.r.t. a set of atoms $A \subseteq HB_{\mathbf{P}}$ is $b_A(R) = \{r \in R \mid H(r) \cap A \neq \emptyset\}$.

Example 7. Consider $\mathbf{P}$ from Example 1 and $P_2[S_{ab}^2]$ from Example 3. Let R be the instantiation $gr(I_{\mathbf{P}}(P_2[S_{ab}^2]))$. A possible splitting set for R is $U = \{q_2(a), q_2(b), q_2'(a), q_2'(b), skip_2\}$. Then the bottom $b_U(R)$ is

$$\{q_2(\alpha).\ \ skip_2 \leftarrow q_2(\alpha), \text{not } q_2'(\alpha).\ \ q_2'(\alpha) \lor q_2'(\beta) \leftarrow q_2(\alpha), q_2(\beta) \mid \alpha \neq \beta \in \{a, b\}\}.$$

Based on the extended notion of a Splitting Set, the Splitting Theorem [2] straightforwardly applies to c-stratified MLPs.

Theorem 1. *Let* $\mathbf{M}$ *be an interpretation of a c-stratified MLP* $\mathbf{P}$, R *be the instantiation* $gr(I_{\mathbf{P}}(P_i[S]))$ *for* $P_i[S] \in VC(\mathbf{P})$, *and let* U *be a splitting set for* R. *Then* M_i/S *is an answer set of* R *relative to* $\mathbf{M}$ *iff it is an answer set of* $\{R \setminus b_U(R)\} \cup N$, *where* N *is an answer set of* $b_U(R)$ *relative to* $\mathbf{M}$.

Example 8. Consider $\mathbf{P}$ from Example 1, $\mathbf{M}$ from Example 3, and R from Example 7. An answer set of R is $N = \{q_2(a), q_2(b), q_2'(a), skip_2\}$. By updating R to $\{R \setminus b_U(R)\} \cup N$, we obtain $R' = \{q_2'(a).\ \ q_2(a).\ \ q_2(b).\ \ skip_2.\ \ even \leftarrow skip_2, P_3[q_2'].odd.\ \ even \leftarrow \text{not } skip_2.\}$. Then M_2/S_{ab}^2 is an answer set of R' relative to $\mathbf{M}$.

In the sequel, we single out a subclass of c-stratified MLPs, namely *input and call stratified (ic-stratified)* MLPs, which guarantee that input splitting sets exist for their local splitting. We define the property of *input stratification* at two different levels of the dependency graph: the schematic level and the instance level. Comparing these two options, checking the property at the schematic level is easier, but is often too strong and misses input stratification at the instance level.

In the remainder of the paper, we assume without loss of generality that each predicate occurs in ordinary atoms of at most one module.

Let $\mathbf{P}$ be an MLP. The dependency graph of $\mathbf{P}$ is the digraph $G_{\mathbf{P}} = (V, E)$. The vertex set V contains all $p \in \mathcal{P} \cup \mathcal{E}$, with p appearing somewhere in $\mathbf{P}$, and $\mathcal{E}$ is the set of module atoms in $\mathbf{P}$. The edge set E is as follows:

Let $r \in R(m_i)$. There is a $\star$-edge $p \rightarrow^\star q$ in $G_{\mathbf{P}}, \star \in \{+, -, \vee\}$, if either

(i) $p(\boldsymbol{t_1}) \in H(r)$ and $q(\boldsymbol{t_2}) \in B^\star(r)$,
(ii) $p(\boldsymbol{t_1}), q(\boldsymbol{t_2}) \in H(r)$ and $\star = \vee$, or
(iii) $p(\boldsymbol{t_1}) \in H(r)$ and q is a module atom in $B^\star(r)$.

Moreover, for $\alpha = P_j[\boldsymbol{p}].o(\boldsymbol{t}) \in B(r)$, the set E contains all edges

(iv) $\alpha \rightarrow^{in} q_\ell$, for every $q_\ell \in \boldsymbol{q_j}$ of $P_j[\boldsymbol{q_j}]$,
(v) $\alpha \rightarrow^m o$, and
(vi) $q_\ell \rightarrow^b p_\ell$, where $q_\ell \in \boldsymbol{q_j}$ of $P_j[\boldsymbol{q_j}]$ and $p_\ell \in \boldsymbol{p}$ of α.

This notion of dependency graph refines the one in [9] concerning the labels of arcs (types of dependencies) and allows us to capture input stratification as follows:

Definition 8. *An MLP $\mathbf{P}$ is* si-stratified *(input stratified at the schematic level), iff no cycle in $G_{\mathbf{P}}$ has in-edges.*

For example, one can easily verify that the MLP in Example 1 is si-stratified.

For any module atoms $\alpha_1, \alpha_2 \in \mathcal{E}$, we say that α_1 locally depends on α_2, if $\alpha_1 \rightsquigarrow \alpha_2$, where $\rightsquigarrow = \rightarrow^+ \cup \rightarrow^- \cup \rightarrow^\vee \cup \rightarrow^{in}$. For each module m_i of a si-stratified MLP $\mathbf{P}$, we define a local labeling function $ll_i \colon V \rightarrow \mathbb{N}$ s.t. $ll_i(\alpha_1) > ll_i(\alpha_2)$ if $\alpha_1 \rightsquigarrow \alpha_2$.

Instance stratification. Proceeding to finer grained level of instances, we define the instance dependency graph $G_{\mathbf{P}}^{\mathbf{M}} = (IV, IE)$ of $\mathbf{P}$ w.r.t. an interpretation $\mathbf{M}$. The idea is to distinguish different predicate names and module atoms in different module instances by associating them with the corresponding value call. Hence, a node in IV is a pair $(p, P_i[S])$ or $(\alpha, P_i[S])$, where p (resp., α) is a predicate name (resp., module atom) appearing in module m_i, and S is the input for a value call $P_i[S] \in VC(\mathbf{P})$.

$G_{\mathbf{P}}^{\mathbf{M}}$ has edges (i')–(iv') similar to (i)–(iv) in $G_{\mathbf{P}}$, except that appropriate value calls are added to predicate names/module atoms; the real difference is made by edges (v') and (vi'); for a module atom of the form $\alpha = P_j[\boldsymbol{p}].o(\boldsymbol{t})$ in $R(m_i)$, $G_{\mathbf{P}}^{\mathbf{M}}$ has edges

(v') $(\alpha, P_i[S]) \rightarrow^m (o, P_i[(M_i/S)|_{\boldsymbol{p}}^{\boldsymbol{q_j}}])$; and
(vi') $(q_\ell, P_j[(M_i/S)|_{\boldsymbol{p}}^{\boldsymbol{q_j}}]) \rightarrow^b (p_\ell, P_i[S])$, where $q_\ell \in \boldsymbol{q_j}$ of $P_j[\boldsymbol{q_j}]$ and $p_\ell \in \boldsymbol{p}$ of α.

Intuitively, these edges capture the relationship between a module atom and the corresponding (ordinary) output atom, respectively between formal input parameters and actual input provided. Restricting to concrete applicable instances $(P_j[(M_i/S)|_{\boldsymbol{p}}^{\boldsymbol{q_j}}])$, they do not just schematically extend (v) and (vi).

Definition 9. *Let* $\mathbf{M}$ *be an interpretation of an MLP* $\mathbf{P}$. *We say that* $\mathbf{P}$ *is* i-stratified w.r.t. $\mathbf{M}$, *iff cycles with in-edges in* $G_{\mathbf{P}}^{\mathbf{M}}$ *contain only nodes of the form* $(X, P_i[\emptyset])$. *Moreover,* $\mathbf{P}$ *is* ic-stratified w.r.t. $\mathbf{M}$ *iff it is both i-stratified and c-stratified w.r.t.* $\mathbf{M}$.

The following theorem shows that ic-stratification is sufficient for the existence of input splitting sets for module atoms in relevant instances.

Theorem 2. *Let* $\mathbf{M}$ *be an interpretation of an ic-stratified MLP* $\mathbf{P}$, $P_i[S]$ *be a value call in* $V(CG_{\mathbf{P}}(\mathbf{M}))$, *and let* $R = gr(I_{\mathbf{P}}(P_i[S]))$. *Then, for every ground module atom* α *occurring in* R, *there exists an input splitting set* U *of* R *for* α.

Example 9. In Example 7, U is an input splitting set for $P_3[S_a^3].odd$. As $\mathbf{P}$ is c-stratified w.r.t. $\mathbf{M}$ (cf. Example 5) and si-stratified, $\mathbf{P}$ is ic-stratified w.r.t. $\mathbf{M}$. Thus, by Theorem 2, all module atoms in grounded instances from $V(CG_{\mathbf{P}}(\mathbf{M}))$ have input splitting sets.

Naturally, si-stratification implies i-stratification, but not vice versa. However, the following condition identifies a case in which i-stratification holds at the instance level while si-unstratification holds at the schematic level.

Definition 10. *Consider a si-unstratified MLP* $\mathbf{P}$. *If all cycles in* $G_{\mathbf{P}}$ *which include an in-edge also contain an m-edge, then we say that* $\mathbf{P}$ *is* psi-unstratified.

Proposition 2. *If an MLP* $\mathbf{P}$ *is c-stratified and psi-unstratified, then* $\mathbf{P}$ *is i-stratified w.r.t. all interpretations* $\mathbf{M}$, *hence ic-stratified.*

Since ic-stratification (of an MLP $\mathbf{P}$ w.r.t. $\mathbf{M}$) ensures that no cycle in $G_{\mathbf{P}}^{\mathbf{M}}$ has in-edges, it yields intended local splits, where the input for any module atom is *fully prepared* before this module atom is called. Extending the notion of local labeling to an instance local labeling function $ill_i \colon IV \to \mathbb{N}$ s.t. $ill_i(\alpha_1, P_i[S]) > ill_i(\alpha_2, P_i[S])$ if $(\alpha_1, P_i[S]) \rightsquigarrow (\alpha_2, P_i[S])$, one can exploit input splitting sets, starting with a module atom α where $ill_i(\alpha, P_i[S])$ is smallest. For ic-stratified MLPs, such input splitting sets consist of ordinary atoms only, hence respective answer sets can be computed in the usual way. By iteration, this inspires an evaluation algorithm presented next.

4   Top-Down Evaluation Algorithm

A top-down evaluation procedure *comp* for building the answer sets of ic-stratified MLPs along the call graph is shown in Algorithm 1. Intuitively, *comp* traverses the relevant call graph from top to the base and back. In forward direction, it gradually prepares input to each module call in a set R of rules, in the order given by the instance local labeling function for R. When all calls are solved, R is rewritable to a set of ordinary rules, and standard methods can be used to find the answer sets, which are fed back to a calling instance, or returned as the result if we are at the top level.

The algorithm has several parameters: a current set of value calls C, a list of sets of value calls *path* storing the recursion chain of value calls up to C, a partial interpretation $\mathbf{M}$ for assembling a (partial stored) answer set, an indexed set $\mathbf{A}$ of split module atoms (initially, all M_i/S and A_i/S are *nil*), and a set $\mathcal{AS}$ for collecting answer sets. It uses the following subroutines:

Algorithm 1. $comp$(in: $\mathbf{P}, C, path, \mathbf{M}, \mathbf{A}$, in/out: $\mathcal{AS}$)

Input: MLP $\mathbf{P}$, set of value calls C, list of sets of value calls $path$, partial model $\mathbf{M}$,
indexed set of sets of module atoms $\mathbf{A}$, set of answer sets $\mathcal{AS}$

(a) **if** $\exists P_i[S] \in C$ s.t. $P_i[S] \in C_{prev}$ *for some* $C_{prev} \in path$ **then**
     if $S \neq \emptyset$ *for some* $P_i[S] \in C$ **then return**
     repeat
         $C' := tail(path)$ and remove the last element of $path$
         if $\exists P_j[T] \in C'$ s.t. $T \neq \emptyset$ **then return else** $C := C \cup C'$
     until $C' = C_{prev}$

     $R := rewrite(C, \mathbf{M}, \mathbf{A})$
     if R *is ordinary* **then**
         if $path$ *is empty* **then**
(b)   **forall** $N \in ans(R)$ **do** $\mathcal{AS} := \mathcal{AS} \cup \{\mathbf{M} \uplus mlpize(N, C)\}$
         else
             $C' := tail(path)$ and remove the last element of $path$
             forall $P_i[S] \in C$ **do** $A_i/S := fin$
             forall $N \in ans(R)$ **do** $comp(\mathbf{P}, C', path, \mathbf{M} \uplus mlpize(N, C), \mathbf{A}, \mathcal{AS})$

     else
(c)   pick an $\alpha := P_j[\mathbf{p}].o(\mathbf{c})$ in R with smallest $ill_R(\alpha)$ and find splitting set U of R for α
         forall $P_i[S] \in C$ **do if** $A_i/S = nil$ **then** $A_i/S := \{\alpha\}$ **else** $A_i/S := A_i/S \cup \{\alpha\}$
         forall $N \in ans(b_U(R))$ **do**
             $T := N|_{\mathbf{p}}^{q_j}$
             if $(M_j/T \neq nil) \wedge (A_j/T = fin)$ **then** $C' := C$ and $path' := path$
             else $C' := \{P_j[T]\}$ and $path' := append(path, C)$
(d) $comp(\mathbf{P}, C', path', \mathbf{M} \uplus mlpize(N, C), \mathbf{A}, \mathcal{AS})$

$mlpize(N, C)$: Convert a set of ordinary atoms N to a *partial* interpretation $\mathbf{N}$ (having undefined components nil), by projecting atoms in N to module instances $P_i[S] \in C$, removing module prefixes, and putting the result at position N_i/S in $\mathbf{N}$.

$ans(R)$: Find the answer sets of a set of ordinary rules R.

$rewrite(C, \mathbf{M}, \mathbf{A})$: For all $P_i[S] \in C$, put into a set R all rules in $I_\mathbf{P}(P_i[S])$, and M_i/S as facts if not nil, prefixing every ordinary atom (appearing in a rule or fact) with $P_i[S]$. Futhermore, replace each module atom $\alpha = P_j[\mathbf{p}].o(\mathbf{t})$ in R, such that $\alpha \in A_i/S$, by o prefixed with $P_j[T]$, where $T = (M_i/S)|_{\mathbf{p}_i}^{q_j}$, and $\mathbf{p}_i$ is $\mathbf{p}$ without prefixes; moreover add any atoms from $(M_j/T)|_o$ prefixed by $P_j[T]$ to R.

The algorithm first checks if a value call $P_i[S] \in C$ appears somewhere in $path$ (Step (a)). If yes, a cycle is present and all value calls along $path$ until the first appearance of $P_i[S]$ are joined into C. If a value call in this cycle has non-empty input, then $\mathbf{P}$ is not ic-stratified for any completion of $\mathbf{M}$, and $comp$ simply returns. After checking for (and processing) cycles, all instances in C are merged into R by the function $rewrite$.

If R is ordinary, meaning that all module atoms (if any) are solved, ans can be applied to find answer sets of R. Now, if $path$ is empty, then a main module is reached and $\mathbf{M}$ can be completed by the answer sets of R and put into $\mathcal{AS}$ (Step (b)). Otherwise, i.e., $path$ is nonempty, $comp$ marks all instances in C as finished, and goes back to the tail of *path* where a call to C was issued. In both cases, the algorithm uses an operator $\uplus$ for combining two partial interpretations as follows: $\mathbf{M} \uplus \mathbf{N} = \{M_i/S \uplus N_i/S \mid P_i[S] \in VC(\mathbf{P})\}$, where $x \uplus y = x \cup y$ if $x, y \neq nil$ and $x \uplus nil = x$, $nil \uplus x = x$.

When R is not ordinary, *comp* splits R according to a module atom α with smallest $ill_R(\alpha)$ in Step (c). If $C = \{P_i[S]\}$, then $ill_R = ill_i$, otherwise it is a function compliant with every ill_i s.t. $P_i[S] \in C$. Then, *comp* adds α to $\mathbf{A}$ for all value calls in C, and computes all answer sets of the bottom of R, which fully determine the input for α. If the called instance $P_j[T]$ has already been fully evaluated, then a recursive call with the current C and *path* yields a proper rewriting of α. Otherwise, the next, deeper level of recursion is entered, keeping the chain of calls in *path* for coming back (Step (d)).

Example 10. Consider Algorithm 1 on $\mathbf{P}$ from Example 1 and 3. The call chain $P_1[\emptyset] \xrightarrow{q} P_2[\{q(a), q(b)\}] \xrightarrow{q_2'} P_3[\{q_2'(a)\}] \xrightarrow{q_3'} P_2[\emptyset] \xrightarrow{q_2'} P_3[\emptyset] \xrightarrow{q_3'} P_2[\emptyset]$ will be reflected by the list $\{P_1[S_\emptyset^1]\}, \{P_2[S_{a,b}^2]\}, \{P_3[S_a^3]\}, \{P_2[S_\emptyset^2]\}, \{P_3[S_\emptyset^3]\}$ in *path*, and a current set of value calls $C = \{P_2[S_\emptyset^2]\}$. At this point, the last two elements of the path will be removed and joined with C yielding $C = \{P_2[S_\emptyset^2], P_3[S_\emptyset^3]\}$. The rewriting R w.r.t. C is[1]

$$\left\{ \begin{array}{llll} q_\emptyset^{i'}(X) \vee q_\emptyset^{i'}(Y) \leftarrow q_\emptyset^i(X), q_\emptyset^i(Y), X \neq Y. & skip_\emptyset^i \leftarrow q_\emptyset^i(X), \text{not } q_\emptyset^i(X). \mid i = 1, 2 \\ even_\emptyset^2 \leftarrow skip_\emptyset^2, odd_\emptyset^3. & odd_\emptyset^3 \leftarrow skip_\emptyset^3, even_\emptyset^2. & even_\emptyset^2 \leftarrow \text{not } skip_\emptyset^2. \end{array} \right\}$$

The only answer set of R is $\{even_\emptyset^2\}$. On the way back, $even_v^2$ is toggled with odd_w^3, and at P_1 the answer set $\{q_\emptyset^1(a), q_\emptyset^1(b), ok_\emptyset^1\}$ is built; *comp* adds a respective (partial) interpretation $\mathbf{M}$ to $\mathcal{AS}$, i.e., where $M_2/\emptyset = \{even\}$, $M_3/\emptyset = \emptyset$, etc., and $M_1/\emptyset = \{q(a), q(b), ok\}$. Following the chain $P_1[\emptyset] \xrightarrow{q} P_2[\{q(a), q(b)\}] \xrightarrow{q_2'} P_3[\{q_2'(b)\}] \xrightarrow{q_3'} P_2[\emptyset] \rightarrow \cdots$, *comp* finds another answer set of $\mathbf{P}$.

The following proposition shows that *comp* works for ic-stratified answer sets.

Proposition 3. *Suppose $\mathbf{P}$ is an MLP with single main module $m_1 = (P_1[], R_1)$. Set $\mathcal{AS} = \emptyset$, path $= \epsilon$, $\mathbf{M}$ and $\mathbf{A}$ to have nil at all components. Then, comp$(\mathbf{P}, \{P_1[]\}, path, \mathbf{M}, \mathbf{A}, \mathcal{AS})$ computes in $\mathcal{AS}$ all answer sets $\mathbf{N}$ of $\mathbf{P}$ s.t. $\mathbf{P}$ is ic-stratified w.r.t. $\mathbf{N}$ (disregarding irrelevant module instances, i.e., $N_i/S = nil$ iff $P_i[S] \notin V(CG_\mathbf{P}(\mathbf{N}))$).*

This can be extended to $\mathbf{P}$ with multiple main modules. Compared to a simple guess-and-check approach, *comp* can save a lot of effort as it just looks into the relevant part of the call graph. Allowing non-ic-stratified answer sets, e.g., loops with non-empty S, is a subject for further work.

5 Discussion

Determining c-stratification of an MLP $\mathbf{P}$ requires checking for cycles in the call graph, which is rather expensive. In practice, it seems useful to perform a syntactic analysis of the rules as a sound yet incomplete test that ic-stratification is given, and to exploit information provided by the programmer.

In a simple form, the programmer makes an *assertion* for specific module calls that when processing these calls recursively, inputs to module calls will always be fully

---

[1] Rather than prefixes, we use superscripts and subscripts like for instances (cf. Example 3).

prepared and no call with the same input (except at the base level) is issued. Ideally, the assertion is made for all calls, as possible e.g. in the odd/even example or the Cyclic Hanoi Tower example. While this may sound to put a burden on programmer, in fact one tends quite often, especially for recursive applications, to drive the chain of calls to a base case (e.g., instances with empty input). In such cases, the programmer can confidently provide this information, which can tremendously improve performance.

If no assertions are provided by the programmer, a syntactic analysis might be helpful to compare the inputs of a module call and the module specification. We discuss one such case here. Consider a module atom $P_j[\boldsymbol{p}].o(\boldsymbol{t})$ in module $m_i = (P_i[\boldsymbol{q_i}], R_i)$. Let q_ℓ and p_ℓ be corresponding predicate names in $\boldsymbol{q_i}$ and $\boldsymbol{p}$, respectively. Assume p_ℓ is concluded from q_ℓ in one step, i.e., by a rule r where $p_\ell(\boldsymbol{X}) \in H(r)$ and $q_\ell(\boldsymbol{X}) \in B(r)$. Suppose that for all such rules, (i) $q_\ell(\boldsymbol{X}) \in B^+(r)$, (ii) $\nexists q_\ell(\boldsymbol{Y}) \in B(r) \mid \boldsymbol{Y} \neq \boldsymbol{X}$, and (iii) all variables not in $\boldsymbol{X}$ in r are safe. Then, all module atoms have the same or smaller input compared to $\boldsymbol{q_i}$. If p_ℓ is concluded from q_ℓ through a chain of rules $r_1, \ldots, r_m$ where $p_\ell(\boldsymbol{X}) \in H(r_1)$, $q_\ell(\boldsymbol{Y}) \in B(r_m)$, then conditions similar to (i)-(iii) must be respected by each r_i, $1 \leq i \leq m$, taking shared atoms between $B^+(r_i)$ and $H(r_{i+1})$ into account.

Now when compared to a module instance, the input to calls in it is either (a) the same or (b) smaller, the evaluation process can branch into handling these cases, and program rewriting is applicable. For simplicity, we discuss this here for self-recursive MLPs, i.e., module calls within module P_i (different from main) are always to P_i.

Rewrite self-recursive MLPs. For a call atom $\alpha = P_i[\boldsymbol{p}].o(\boldsymbol{t})$ in $P_i[\boldsymbol{q_i}]$, we can guess whether case (a) applies; if so, we can replace α by $o(\boldsymbol{t})$. The resulting rules contain fewer module atoms (if none is left, they are ordinary and can be evaluated as usual). In case (b), if $\mathbf{P}$ is psi-unstratified, we can apply Algorithm 1 (with ill_i replaced by ll_i) to a rewritten program in which additional constraints ensure the decrease of the input.

More formally, let $\mathbf{P} = (m_1, \ldots, m_n)$ where $m_i = (P_i[\boldsymbol{q_i}], R_i)$, $i \in \{1, \ldots, n\}$. Let α be as above and let noc and $\overline{noc}$ be two fresh predicate names. We define two types of rewriting functions, ν and μ, as follows:

- Let $\nu_i(p(\boldsymbol{t})) = p(\boldsymbol{t})$, for each predicate $p \in \mathcal{P}$, and $\nu_i(\alpha) = o(\boldsymbol{t})$. For a rule r of form (1), let $\nu_i(r) = \alpha_1 \vee \cdots \vee \alpha_k \leftarrow noc, \nu_i(\beta_1), \ldots, \nu_i(\beta_m), \text{not } \nu_i(\beta_{m+1}), \ldots,$ $\text{not } \nu_i(\beta_n)$. Then, $\nu(R_i) = \{\nu_i(r) \mid r \in R_i\}$.
- For a rule r, $\mu(r)$ adds $\overline{noc}$ to $B(r)$, and $\mu(R_i) = \{\mu(r) \mid r \in R_i\}$.

Let r_g be $noc \vee \overline{noc} \leftarrow$ and let $Eq_i(\alpha)$ and $Con_i(\alpha)$ be the following sets of rules, where q_ℓ and p_ℓ are corresponding predicate names in the formal input list $\boldsymbol{q_i}$ of $P_i[\boldsymbol{q_i}]$ and the actual input list $\boldsymbol{p}$ of α:

$$Eq_i(\alpha) = \left\{ \begin{array}{l} fail \leftarrow p_\ell(\boldsymbol{X}), \text{not } q_\ell(\boldsymbol{X}), noc, \text{not } fail. \\ fail \leftarrow q_\ell(\boldsymbol{X}), \text{not } p_\ell(\boldsymbol{X}), noc, \text{not } fail. \end{array} \mid 1 \leq \ell \leq |\boldsymbol{q_i}| \right\} ,$$

$$Con_i(\alpha) = \left\{ \begin{array}{l} fail \leftarrow p_\ell(\boldsymbol{X}), \text{not } q_\ell(\boldsymbol{X}), \overline{noc}, \text{not } fail. \\ ok \leftarrow q_\ell(\boldsymbol{X}), \text{not } p_\ell(\boldsymbol{X}), \overline{noc}. \\ fail \leftarrow \overline{noc}, \text{not } ok, \text{not } fail. \end{array} \mid 1 \leq \ell \leq |\boldsymbol{q_i}| \right\} .$$

Let Eq_i and Con_i stand for the union of $Eq_i(\alpha)$ and $Con_i(\alpha)$ for all module atoms α appearing in R_i, resp. For each module m_i, let $\tau(m_i) = \nu(R_i) \cup \mu(R_i) \cup \{r_g\} \cup Eq_i \cup Con_i$. Finally, let $\tau(\mathbf{P}) = (\tau(m_1), \ldots, \tau(m_n))$.

Proposition 4. *Let* **P** *be a psi-unstratified MLP* **P** *where the input to any module call is either equal or strictly smaller compared to the input of the module instance issuing the call. Then the answer sets of* $\tau(\mathbf{P})$ *correspond 1-1 to those of* **P**.

The method can be extended to non-self recursive MLPs, where calls to different modules are allowed. Here, one needs to keep track of the module call chain, or assume an ordering on the module names to determine input decrease; we omit the details.

Concerning complexity, using our algorithm suitably, answer-set existence of ic-stratified MLPs is decidable in EXPSPACE, i.e., more efficiently than arbitrary MLPs (2NEXP^{NP}-complete [9]). Since already best practical algorithms for ordinary, call-free programs (NEXP^{NP}-complete) require exponential space, the algorithm has reasonable resource bounds. A detailed complexity analysis is planned for the extended paper.

6 Related Work and Conclusion

In the ASP context, several modular logic programming formalisms have been proposed (cf. Introduction). We already mentioned the modular logic programs of [1] and DLP-functions [3]. For the former, a rich taxonomy of notions of stratification was given in [1]; however, they essentially address merely the module schema level, and no specific algorithms were described. For DLP-functions, Janhunen et al. [3,4] developed a Module Theorem which allows to compose the answer sets of multiple modules; however, no specific account of stratification was given in [3,4]. As MLPs can be viewed as a generalization of DLP-Functions, our results may be transferred to the DLP context.

In an upcoming paper, Ferraris et al. present Symmetric Splitting [13] as a generalization of the Module Theorem [3,4] allowing to decompose also nonground programs like MLPs do. Similar to [3,4], this technique is only applicable to programs with no positive cycles in the dependency graph. Studying the relationship between Symmetric Splitting and our notions of stratification is an interesting subject for future work.

Several other issues remain for further work, including extensions and refinements of the stratification approach. For example, while we have focused here on decreasing inputs in terms of set inclusion, the extension of the method to other partial orderings of inputs that have bounded decreasing chains is suggestive. This and investigating complexity issues as well as implementation are on our agenda.

References

1. Eiter, T., Gottlob, G., Veith, H.: Modular Logic Programming and Generalized Quantifiers. In: Fuhrbach, U., Dix, J., Nerode, A. (eds.) LPNMR 1997. LNCS, vol. 1265, pp. 290–309. Springer, Heidelberg (1997)
2. Lifschitz, V., Turner, H.: Splitting a logic program. In: ICLP 1994, pp. 23–37. MIT Press, Cambridge (1994)
3. Janhunen, T., Oikarinen, E., Tompits, H., Woltran, S.: Modularity Aspects of Disjunctive Stable Models. In: Baral, C., Brewka, G., Schlipf, J. (eds.) LPNMR 2007. LNCS (LNAI), vol. 4483, pp. 175–187. Springer, Heidelberg (2007)

4. Oikarinen, E., Janhunen, T.: Achieving compositionality of the stable model semantics for Smodels programs. Theory Pract. Log. Program. 8(5-6), 717–761 (2008)
5. Baral, C., Dzifcak, J., Takahashi, H.: Macros, Macro calls and Use of Ensembles in Modular Answer Set Programming. In: Etalle, S., Truszczyński, M. (eds.) ICLP 2006. LNCS, vol. 4079, pp. 376–390. Springer, Heidelberg (2006)
6. Calimeri, F., Ianni, G.: Template programs for Disjunctive Logic Programming: An operational semantics. AI Commun. 19(3), 193–206 (2006)
7. Bugliesi, M., Lamma, E., Mello, P.: Modularity in Logic Programming. J. Log. Program. 19/20, 443–502 (1994)
8. Brogi, A., Mancarella, P., Pedreschi, D., Turini, F.: Modular logic programming. ACM Trans. Program. Lang. Syst. 16(4), 1361–1398 (1994)
9. Dao-Tran, M., Eiter, T., Fink, M., Krennwallner, T.: Modular Nonmonotonic Logic Programming Revisited. In: ICLP 2009, pp. 145–159. Springer, Heidelberg (2009)
10. Faber, W., Leone, N., Pfeifer, G.: Recursive Aggregates in Disjunctive Logic Programs: Semantics and Complexity. In: Alferes, J.J., Leite, J. (eds.) JELIA 2004. LNCS (LNAI), vol. 3229, pp. 200–212. Springer, Heidelberg (2004)
11. Eiter, T., Ianni, G., Schindlauer, R., Tompits, H.: Effective integration of declarative rules with external evaluations for semantic web reasoning. In: Sure, Y., Domingue, J. (eds.) ESWC 2006. LNCS, vol. 4011, pp. 273–287. Springer, Heidelberg (2006)
12. Gelfond, M., Lifschitz, V.: Classical negation in logic programs and deductive databases. New Gener. Comput. 9, 365–385 (1991)
13. Ferraris, P., Lee, J., Lifschitz, V., Palla, R.: Symmetric Splitting in the General Theory of Stable Models. In: IJCAI 2009, pp. 797–803. AAAI Press, Menlo Park (2009)

Complexity of the Stable Model Semantics for Queries on Incomplete Databases

Jos de Bruijn[1] and Stijn Heymans[2]

[1] KRDB Research Center, Free University of Bozen-Bolzano, Italy
debruijn@inf.unibz.it
[2] Institute of Information Systems, Vienna University of Technology, Austria
heymans@kr.tuwien.ac.at

Abstract. We study the complexity of consistency checking and query answering on incomplete databases for languages ranging from non-recursive Datalog to disjunctive Datalog with negation under the stable model semantics. We consider both possible and certain answers and both closed- and open-world interpretation of *C-databases* with and without conditions. By reduction to stable models of logic programs we find that, under closed-world interpretation, adding negation to (disjunctive) Datalog does not increase the complexity of the considered problems for C-databases, but certain answers for databases without conditions are easier for Datalog without than with negation. Under open-world interpretation, adding negation to non-recursive Datalog already leads to undecidability, but the complexity of certain answers for negation-free queries is the same as under closed-world interpretation.

1 Introduction

In applications of relational databases a need often arises for representing incomplete information [5], typically in the form of *null values*. For example, in data exchange [8,17] anomalies in the semantics for solutions may arise if nulls are not treated with care. In data integration [1,12,16] incomplete information arises when integrating different complete data sources using a global schema; a materialized view may be incomplete with respect to local sources and local sources may be incomplete with respect to the global schema or constraints of other sources.

Null values in incomplete databases are represented using variables. A single database represents several possible instances, called *representations*. In our treatment we follow the landmark paper by Imieliński and Lipski [14], which considers both *closed-world* and *open-world* interpretation of incomplete databases. In the former, representations are in direct correspondence with valuations of the variables; each tuple in a representation is the valuation of a tuple in the database. In the latter, representations may include additional tuples not originating from the database. In addition, *local* conditions may be attached to tuples and a *global* condition to the database. Such incomplete databases with conditions are called *C-databases*.

So far, most research on query answering has been concerned with first-order and Datalog queries [25,3], and has focused mainly on data complexity. However, since those landmark papers, the formal properties of more expressive query languages such

E. Erdem, F. Lin, and T. Schaub (Eds.): LPNMR 2009, LNCS 5753, pp. 101–114, 2009.
© Springer-Verlag Berlin Heidelberg 2009

as Datalog with disjunction [7] and (unstratified) negation with the accompanying Stable Model Semantics [9] (Datalog$^{\neg,\vee}$) established themselves firmly as well-accepted expressive Knowledge Representation languages. Sufficient reason for having a closer look again at those query languages for incomplete databases and thus going beyond PTime queries – queries that can be answered in polynomial time on complete databases (e.g., stratified Datalog [2]).

We study the data and combined complexity of consistency, and possible and certain answers for languages ranging from nonrecursive Datalog to Datalog$^{\neg,\vee}$. We consider both the open- and closed-world interpretation of C-databases with and without conditions. Our main contributions are summarized as follows:

– We show that answering Datalog$^{\neg,\vee}$ queries on incomplete databases under closed-world interpretation can be reduced to common reasoning tasks in logic programming, by an encoding of incomplete databases into logic programs.

– We present complete pictures of the data and combined complexity of consistency, and possible and certain answers for languages ranging from non-recursive Datalog to Datalog$^{\neg,\vee}$, complementing earlier results [3,25] with combined complexity results for (fragments of) stratified Datalog$^{\neg}$ queries and novel data and combined complexity results for queries beyond PTime. The results for closed-world interpretation are summarized in Table 1 on page 110, and for open-world interpretation in Table 2 on page 111.

– Finally, we show that results about checking uniform and strong equivalence of queries from the areas of complete databases and logic programming apply immediately to the case of incomplete databases.

To the best of our knowledge, ours are the first results about answering Datalog queries with disjunction and/or stable model negation on incomplete databases. Related to Datalog$^{\neg}$ queries on incomplete databases are the techniques for consistent query answering in data integration based under the local-as-view using Datalog$^{\neg}$ programs, by Bertossi and Bravo [4]. The precise relationship with query answering on incomplete databases is an open question.

Under open-world interpretation, adding negation leads to undecidability of consistency and query answering, by undecidability of finite satisfiability in first-order logic [22]. Under closed-world interpretation, for all query languages ranging from non-recursive Datalog to Datalog$^{\neg}$, the data complexity of possible answers on databases without conditions is NP-complete and of certain answers on C-databases it is coNP-complete; for certain answers coNP-completeness holds additionally for Datalog$^{\vee}$ queries. For positive queries these results also apply under open-world interpretation. This shows that, for possible answers and for certain answers on general C-databases, there is no computational justification for restricting oneself to PTime languages such as Datalog and stratified Datalog$^{\neg}$.

In Section 2 we review incomplete databases and define Datalog$^{\neg,\vee}$ queries. In Section 3 we reduce query answering under closed-world interpretation to standard logic programming reasoning tasks. We present our complexity analysis in Section 4. Finally, we discuss related and future work in Sections 5 and 6.

An extended version of this paper, which includes the proofs of Propositions 4, 7, 10, and 11, can be found online at:
`http://www.debruijn.net/publications/sm-incomplete-db.pdf`

2 Incomplete Databases and Queries

We consider C-databases, as defined by Imieliński and Lipski [14], and the Stable Model Semantics for logic programs, as defined by Gelfond and Lifschitz [9].

Incomplete Databases. Let $\mathcal{D}$ be a countably infinite set of *constants*, called the *domain*, and let $\mathcal{V}$ be a finite set of variables, disjoint from $\mathcal{D}$. A *condition* ψ is a formula of the form $\varphi_1 \vee \cdots \vee \varphi_m$, where φ_j are conjunctions of equality atoms $x = y$ and inequality atoms $x \neq y$, with $x, y \in \mathcal{D} \cup \mathcal{V}$. A *C-table* (Conditional table) of arity n is a finite subset of $(\mathcal{D} \cup \mathcal{V})^n$ such that a *local* condition ϕ_t is associated with each tuple t in the relation. We sometimes omit ϕ_t if it is $x = x$.

A *schema* is a list $\mathcal{T} = R_1, \ldots, R_k$ of predicate symbols R_i each with an arity $n_i \geq 1$. We assume a constant bound l on the arities. A *C-database* over $\mathcal{T}$ is a tuple $\mathbf{T} = (T_1, \ldots, T_k)$ with associated condition $\Phi_{\mathbf{T}}$, such that each T_i is a C-table with arity n_i. We write individual tuples $(a_1, \ldots, a_{n_i}) \in T_i$ as $R_i(a_1, \ldots, a_{n_i})$; if $R_i(a_1, \ldots, a_{n_i})$ contains no variables, it is a *fact*. With $preds(\mathbf{T})$ we denote the set $\{R_1, \ldots, R_k\}$. We call a C-database *condition-free* if every condition is $x = x$. A *complete database* or *instance I* is a variable- and condition-free C-database.

Validity of variable-free conditions is defined as follows: $c_1 = c_1$ is valid; $c_1 \neq c_2$ is valid, for c_1, c_2 distinct constants; this extends to conditions in the natural way. A *valuation* is a mapping $\sigma : \mathcal{V} \cup \mathcal{D} \to \mathcal{D}$ such that $\sigma(c) = c$, for every $c \in \mathcal{D}$. This extends to tuples and conditions in the natural way. For C-tables T and C-databases $\mathbf{T}$ we define $\sigma(T) = \{\sigma(t) \mid t \in T \ \& \ \sigma(\phi_t) \text{ is valid}\}$ and $\sigma(\mathbf{T}) = (\sigma(T_1), \ldots, \sigma(T_k))$.

The *closed-world interpretation* (CWI) of a C-database $\mathbf{T}$ with arity $(n_1, \ldots, n_k)$ is defined as:

$$rep(\mathbf{T}) = \{\sigma(\mathbf{T}) \mid \sigma \text{ is a valuation such that } \sigma(\Phi_{\mathbf{T}}) \text{ is valid}\} \tag{1}$$

The *open-world interpretation* (OWI) of $\mathbf{T}$ is defined as:

$$Rep(\mathbf{T}) = \{\mathbf{R} \subseteq \mathcal{D}^{n_1} \times \cdots \times \mathcal{D}^{n_k} \mid \exists \sigma. \sigma(\Phi_{\mathbf{T}}) \text{ is valid,}$$
$$\sigma(\mathbf{T}) \subseteq \mathbf{R}, \text{ and } \mathbf{R} \text{ is finite}\} \tag{2}$$

Lemma 1 (Implicit in [14]). *Let $\mathbf{T}$ be a C-database. Then, $rep(\mathbf{T}) \subseteq Rep(\mathbf{T})$ and for every $I \in Rep(\mathbf{T})$ there is an $I' \in rep(\mathbf{T})$ such that $I' \subseteq I$.*

Datalog$^{\neg,\vee}$ *Queries. Atoms* are of the form $p(a_1, \ldots, a_n)$, where the a_i's are terms and p is an n-ary predicate symbol, $n \geq 1$. *Positive literals* are atoms α and *negative literals* are negated atoms *not* α. A Datalog$^{\neg,\vee}$ rule r is of the form:

$$h_1 \vee \cdots \vee h_l \leftarrow b_1, \ldots, b_k \tag{3}$$

where the h_i's are atoms and the b_j's are literals, such that every variable in r occurs in some positive b_j. We call $H(r) = \{h_1, \ldots, h_l\}$ the *head* and $B(r) = \{b_1, \ldots, b_k\}$ the *body* of r. If r contains no negation, then it is a Datalog$^\vee$ rule. If $l = 1$, then r is a Datalog$^\neg$ rule. If r is both a Datalog$^\vee$ and Datalog$^\neg$ rule, then it is a Datalog rule. A Datalog$^{\neg,\vee}$ *program* P is a countable set of Datalog$^{\neg,\vee}$ rules. Datalog$^\neg$, Datalog$^\vee$, and Datalog programs are defined analogously.

The set of predicate symbols of P, denoted $preds(P)$, is partitioned into sets of *intentional* $(int(P))$ and *extensional* $(ext(P))$ predicates such that there is no $p \in ext(P)$ in the head of any $r \in P$. We assume that each variable occurs in at most one $r \in P$.

The dependency graph of P is a directed graph $G(P) = \langle N, E \rangle$: $N = preds(P)$ and E is the smallest set that includes an edge $(p, q) \in preds(P)^2$ if there is an $r \in P$ such that p in some $h \in H(r)$ and q in some $b \in B(r)$; (p, q) is labeled *negative* if b is a a negative literal. P is *non-recursive* if $G(P)$ contains no cycles and *stratified* if $G(P)$ contains no cycles involving a negative edge. We use the prefixes nr- and st- for class of non-recursive and stratified programs.

Given a set $\Delta \subseteq \mathcal{D}$, the *grounding* of P with respect to Δ, denoted $gr_\Delta(P)$, is defined as the union of all substitutions of variables in P with elements of Δ.

Definition 1 (Queries). *If* X *is a class of programs, then an* X *query* Q *with signature* $(e_1, \ldots, e_n) \to (o_1, \ldots, o_m)$ *is a finite* X *program without constants and without empty rule heads such that* $\{e_1, \ldots, e_n\} = ext(Q)$ *are the input and* $\{o_1, \ldots, o_m\} \subseteq int(Q)$ *are the output predicates.* Q *is* well-defined *with respect to a database* $\mathbf{T}$ *if* $ext(Q) \subseteq preds(\mathbf{T})$.

We assume in the remainder that all queries are well-defined with respect to the database under consideration; further, let Δ be a set of constants, P a program, I an instance, and $\mathbf{I}$ a set of instances.

An *interpretation* M is a set of facts formed using predicate symbols in $preds(P)$ and constants Δ. Given a set of predicate symbols or constants Υ, with $M|_\Upsilon$ we denote the restriction of M to Υ.

If P is negation- and variable-free, M is a *model* of P if, for every $r \in P$, whenever $B(r) \subseteq M$, $H(r) \cap M \neq \emptyset$. The *reduct* $P^{M,\Delta}$ is obtained from $gr_\Delta(P)$ by (a) removing every rule $r \in gr_\Delta(P)$ such that *not* $b \in B(r)$ for some $b \in M$ and (b) removing all negative literals from the remaining rules.

M is a *stable Δ-model* of P with respect to I if $M|_{ext(P)} = I|_{ext(P)}$, M is a model of $P^{M,\Delta}$ and there is no model M' of $P^{M,\Delta}$ such that $M'|_{ext(P)} = M|_{ext(P)}$ and $M' \subsetneq M$. We leave out I if $I = \emptyset$ and Δ if $\Delta = \mathcal{D}$. We note that if Δ includes the constants in I and P, then the stable Δ-models of P with respect to I are the same as the stable models of $P \cup I$.

Example 1. Let $\Delta = \{a\}$. Consider the instance $I = \{p(a)\}$ and the program $P = \{q(x) \vee r(x) \leftarrow p(x), not\ s(x)\}$. $M = \{p(a), q(a)\}$ is a model of the reduct $P^{M,\Delta} = \{q(a) \vee r(a) \leftarrow p(a)\}$ and a stable Δ-model of P with respect to I; the other stable Δ-model is $\{p(a), r(a)\}$.

Definition 2 (Query Answers). *Let* I *be an instance,* $\mathbf{I}$ *a set of instances, and* Q *a* Datalog$^{\neg,\vee}$ *query with signature* $(e_1, \ldots, e_n) \to (o_1, \ldots, o_m)$.

$$Q(I) = \{(M|_{\{o_1\}}, \ldots, M|_{\{o_m\}}) \mid M \text{ is a stable model of } Q \text{ with respect to } I\}$$
$$Q(\mathbf{I}) = \bigcup\{Q(I) \mid I \in \mathbf{I}\}$$

The closed-world interpretation of a query Q on a C-database $\mathbf{T}$ is written $Q(rep(\mathbf{T}))$ and the open-world interpretation is written $Q(Rep(\mathbf{T}))$.

3   Logic Programming Characterization of Queries under CWI

We reduce queries on incomplete databases under closed-world interpretation to logic programs with negation. Specifically, we show that there is a polynomial embedding of C-databases $\mathbf{T}$ into Datalog$^{\neg}$ programs $P_{\mathbf{T}}$ such that the answers to a query Q on $\mathbf{T}$ correspond with the stable models of $Q \cup P_{\mathbf{T}}$ with respect to the output predicates $(o_1, \ldots, o_m)$.

Recall that the domain $\mathcal{D}$ is infinite, and thus there may be infinitely many valuations for a given variable in $\mathbf{T}$. The following lemma shows we need to consider only a finite subset.

Lemma 2 (Implicit in [3]). *Let Q be a* Datalog$^{\neg,\vee}$ *query, $\mathbf{T}$ a C-database, $\Delta \subset \mathcal{D}$ include the constants in $\mathbf{T}$, and V the set of variables in $\mathbf{T}$. Then there is a set of constants $\Delta' \subset \mathcal{D}$ with cardinality $|V|$ such that $\Delta \cap \Delta' = \emptyset$ and.*

$$Q(rep(\mathbf{T}))|_\Delta = \{I|_\Delta \mid \sigma : V \rightarrow \Delta \cup \Delta', I \in Q(\sigma(\mathbf{T})), \text{ and } \sigma(\Phi_{\mathbf{T}}) \text{ is valid}\}$$

Note that for given Q, $\mathbf{T}$, Δ, and V, such a Δ' is finite, since $|V|$ is finite.

Definition 3. *Let $\mathbf{T}$ be a C-database and $\Delta \subset \mathcal{D}$ include constants in $\mathbf{T}$. For each tuple $t = R(\boldsymbol{a})$ in $\mathbf{T}$, with $\phi_t = \varphi_{t,1} \vee \cdots \vee \varphi_{t,m}$, the program $P_{\mathbf{T},\Delta}$ contains the rules*

$$R(\boldsymbol{a}) \leftarrow \varphi'_{t,i}, v_{x_1}(x_1), \ldots, v_{x_k}(x_k) \tag{4}$$

for $1 \leq i \leq m$, where $\varphi'_{t,i}$ is obtained from $\varphi_{t,i}$ by replacing '$\wedge$' with ',', and $x_1, \ldots, x_k$ are the variables occurring in t or $\varphi_{t,i}$.

$P_{\mathbf{T},\Delta}$ contains $D(c) \leftarrow$, for every $c \in \Delta \cup \Delta'$, with Δ' as in Lemma 2,

$$
\begin{aligned}
v_x(z) &\leftarrow not\ v'_x(z), D(z) & &\leftarrow v_x(z), v_x(y), z \neq y \\
v'_x(z) &\leftarrow not\ v_x(z), D(z) & e_x &\leftarrow v_x(z) \\
& & &\leftarrow not\ e_x
\end{aligned}
\tag{5}
$$

for every variable x in $\mathbf{T}$. Finally, for $\Phi_{\mathbf{T}} = \varphi_{\mathbf{T},1} \vee \cdots \vee \varphi_{\mathbf{T},l}$, $P_{\mathbf{T},\Delta}$ contains

$$g \leftarrow \varphi'_{\mathbf{T},i}, v_{x_1}(x_1), \ldots, v_{x_k}(x_k) \qquad \leftarrow not\ g \tag{6}$$

for $1 \leq i \leq l$, where $\varphi'_{\mathbf{T},i}$ is obtained from $\varphi_{\mathbf{T},i}$ as before and $x_1, \ldots, x_k$ are the variables in $\varphi_{\mathbf{T},i}$. $P_{\mathbf{T},\Delta}$ contains no other rules.

Note that equality and inequality can be straightforwardly axiomatized using Datalog¬ rules, such that $P_{\mathbf{T},\Delta}$ is indeed a Datalog¬ program.

Intuitively, the rules (5) ensure the presence of an atom $v_x(c)$ in every stable model, indicating that the variable x is assigned to c. The constraints ensure that there is such a guess for each variable and this guess is unique. The rules (4) subsequently ensure evaluating the conditions.

The following proposition establishes correspondence between the answers to Q on $rep(\mathbf{T})$ and the stable models of $Q \cup P_{\mathbf{T},\Delta}$.

Proposition 1. *Let Q be a query with signature $(e_1, \ldots, e_n) \to (o_1, \ldots, o_m)$ on a C-database $\mathbf{T}$ and let Δ be a superset of the set of constants in $\mathbf{T}$. Then,*

$$Q(rep(\mathbf{T}))|_\Delta = \{(M|_{\{o_1\}}, \ldots, M|_{\{o_m\}}) \mid M \text{ is a stable model of } Q \cup P_{\mathbf{T},\Delta}\}|_\Delta$$

Proof. One can verify that M is a stable model of $P_{\mathbf{T},\Delta}$ iff $M = \sigma(\mathbf{T})$ for a $\sigma : V \to \Delta \cup \Delta'$ such that $\sigma(\Phi_{\mathbf{T}})$ is valid. The proposition then straightforwardly follows from the definition and Lemma 2. □

Observe that the grounding of the program $P_{\mathbf{T},\Delta}$ is in general exponential in the size of $\mathbf{T}, \Delta$, since the size of the non-ground rules (4) depends on the size of $\mathbf{T}$. However, we will see in Proposition 5 that using an intelligent polynomial grounding, the stable models of $P_{\mathbf{T},\Delta}$ can be computed in time NP.

Example 2. Consider a C-database $\mathbf{T}$ with ternary table T describing the flights of a plane on a particular day. $\mathbf{T}$ contains the tuples $t_1 = T(v, x_1, y_1)$ and $t_2 = T(i, x_2, y_2)$, with variables x_1, x_2, y_1, y_2, indicating that the plane flies from v to a destination x_1 with a pilot y_1 and from i to x_2 with y_2. As mg and mc are the only pilots certified to fly on i, t_1 has associated condition $x_1 \neq i \lor y_1 = mg \lor y_1 = mc$ and t_2 has condition $y_2 = mg \lor y_2 = mc$. Additionally, a pilot may not fly two stretches, hence the global condition $y_1 \neq y_2 \land x_1 \neq v \land x_2 \neq i$.

Let Δ be the set of constants in $\mathbf{T}$. Besides the guess rules (5), $P_{\mathbf{T},\Delta}$ contains

$$T(v, x_1, y_1) \leftarrow x_1 \neq i, v_{x_1}(x_1), v_{y_1}(y_1) \quad T(i, x_2, y_2) \leftarrow y_2 = mg, v_{x_2}(x_2), v_{y_2}(y_2)$$
$$T(v, x_1, y_1) \leftarrow y_1 = mg, v_{x_1}(x_1), v_{y_1}(y_1) \quad T(i, x_2, y_2) \leftarrow y_2 = mc, v_{x_2}(x_2), v_{y_2}(y_2)$$
$$T(v, x_1, y_1) \leftarrow y_1 = mc, v_{x_1}(x_1), v_{y_1}(y_1)$$
$$g \leftarrow y_1 \neq y_2, x_1 \neq v, x_2 \neq i, v_{y_1}(y_1), v_{y_2}(y_2), v_{x_1}(x_1), v_{x_2}(x_2) \qquad \leftarrow not\ g$$

Among the stable models of $P_{\mathbf{T},\Delta}$ (restricted to $preds(\mathbf{T})$) are $M_1 = \{T(v, i, mg), T(i, v, mc)\}$ and $M_2 = \{T(v, i, mc), T(i, v, mg)\}$. One can verify that these indeed correspond to elements of $rep(\mathbf{T})$. Consider the Datalog¬ query Q

$$flying(x, y, z) \leftarrow T(x, y, z)$$
$$flying(x, y, z) \leftarrow T(x, u, z), flying(u, y, z)$$
$$roundtrip(z) \leftarrow flying(x, x, z)$$
$$stranded(x) \leftarrow not\ roundtrip(z), flying(x, y, z)$$

where *flying* is the transitive closure of the trips in T and the pilot is *stranded* if the departure and final destination do not coincide. The output predicate is *stranded*.

Consider the stable models M_1' and M_2' of $Q \cup P_{\mathbf{T},\Delta}$, which are extensions of M_1 and M_2, respectively. Both M_1' and M_2' contain $stranded(mg)$ and $stranded(mc)$. However, $Q \cup P_{\mathbf{T},\Delta}$ also has the stable models $\{T(v, c_{x_1}, c_{y_1}), T(i, v, mc), stranded(mc), stranded(c_{y_1})\}$ and $\{T(v, c_{x_1}, c_{y_1}), T(i, v, mg), stranded(mg), stranded(c_{y_1})\}$, and so neither $stranded(mg)$ nor $stranded(mc)$ is included in every stable model.

4   Complexity Analysis

In this section we study the complexity of checking consistency (cons) and of query answering, under the possible (poss) and certain (cert) answer semantics.

We consider two notions of complexity (cf. [23]): *combined complexity* is measured in the combined size of the database and the query and *data complexity* is measured in the size of the database – the query is considered *fixed*. We consider the following decision problems. As inputs (in parentheses) we consider a set of facts A, a C-database $\mathbf{T}$, and a query Q.

cons$(\mathbf{T}, Q)$ *question:* is there an $I \in Q(rep(\mathbf{T}))$ such that $I \neq \emptyset$?
poss$(A, \mathbf{T}, Q)$ *question:* is there an $I \in Q(rep(\mathbf{T}))$ such that $A \subseteq I$?
cert$(A, \mathbf{T}, Q)$ *question:* for all $I \in Q(rep(\mathbf{T}))$, $A \subseteq I$?

cons^Q, poss^Q, and cert^Q are like the above except that Q is not part of the input.

We denote the consistency and certain answer problems under open-world interpretation with the symbols Cons and Cert, respectively. Their definitions are obtained from the above by replacing $rep(\cdot)$ with $Rep(\cdot)$. We do not consider possible answers in the open-world case, since representations may include facts not justified by tuples in the database.

With a problem Y (resp., Y^Q) for a class X of queries, we mean the restriction of the problem Y (resp., Y^Q) such the queries Q in the input (resp., parameter) are in the class X. We use the following notation for complexity classes: LSpace (logarithmic space), PTime, NP, coNP, $\Sigma_2^p = \mathrm{NP}^{\mathrm{NP}}$, $\Pi_2^p = \mathrm{coNP}^{\mathrm{NP}}$, PSpace, Exp (exponential time), NExp, coNExp, NExp^{NP}, and coNExp^{NP}. See, e.g., [6, Section 3] for definitions.

We consider the closed-world interpretation in Section 4.1 and the open-world interpretation in Section 4.2.

4.1   Complexity of Closed-World Interpretation

In order to give a full picture of the complexity, we repeat some results from literature of query answering over incomplete databases in Proposition 2 and 3. Queries in these propositions have no negation or only stratified negation such that a stable model semantics coincides with the usual minimal model semantics as defined in the respective literature. The following result is due to Abiteboul et al. [3].

Proposition 2 ([3]). *The problem* poss^Q *is NP-complete and* cert^Q *is coNP-complete for* nr-Datalog, nr-Datalog$^\neg$, Datalog, *and* st-Datalog$^\neg$ *queries.*

In addition, Grahne [11] showed that when restricting local conditions ϕ_t to conjunctions of equalities and the global condition $\Phi_{\mathbf{T}}$ to a conjunction of Horn clauses, the problem cert^Q can be solved in PTime for Datalog queries.

The hardness results in the following proposition follow from the hardness results for the case of complete databases; see [6,7,26]. Observe that the complement of poss for Datalog$^\vee$ queries is easily reduced to cert for stratified Datalog$^{\neg,\vee}$ queries; Π_2^p (resp., coNExp^{NP})-hardness of cert^Q (resp., cert) for st-Datalog$^{\neg,\vee}$ follows immediately from Σ_2^p (resp., NExp^{NP})-hardness of poss^Q (resp., poss) for Datalog$^\vee$ queries [7]. Observe also that the problems poss and cert correspond for nr-Datalog queries on complete databases; PSpace-hardness of cert was established by Vorobyov and Voronkov [26].

Proposition 3. *The problem*
- *poss^Q is Σ_2^p-hard for* Datalog$^\vee$ *queries,*
- *cert^Q is Π_2^p-hard for* st-Datalog$^{\neg,\vee}$ *queries,*
- *poss and cert are PSpace-hard for* nr-Datalog *queries,*
- *poss is NExp^{NP}-hard for* Datalog$^\vee$ *queries, and*
- *cert is coNExp^{NP}-hard for* st-Datalog$^{\neg,\vee}$ *queries.*

We state our novel hardness result of combined complexity for Datalog in Proposition 4. Our novel membership results for queries beyond Datalog are in Propositions 5 and 6. The results are summarized in Table 1 on page 110.

Proposition 4. *The problem poss is NExp-hard and cert is coNExp-hard for* Datalog *queries.*

Proof (Sketch). The proof is by an encoding of nondeterministic Turing machines that run in exponential time into Datalog queries Q on C-databases $\mathbf{T}$. Q, which has one output predicate *accept*, encodes all possible transitions of the machine, using binary coding for time points and positions on the tape. $\mathbf{T}$ encodes a guess for each time point j. Conditions ensure that in a given representation, exactly one guess is made for each time point. We have that $accept(1) \in I$ for some (resp., all) $I \in Q(rep(\mathbf{T}))$ iff some (resp., all) run(s) of T are accepting. Consequently, poss is NExp-hard and cert is coNExp-hard for Datalog queries.

The complete encoding can be found in the extended version.   □

We obtain the following membership results with the help of the reduction to logic programs in Section 3 (see Proposition 1).

Proposition 5. *The problem*
- *poss^Q is in NP and cert^Q is in coNP for* Datalog$^\neg$ *queries,*
- *poss^Q is in Σ_2^p and cert^Q is in Π_2^p for* Datalog$^{\neg,\vee}$ *queries,*
- *poss and cert are in PSpace for* nr-Datalog$^\neg$ *queries,*
- *poss is in NExp and cert is in coNExp for* Datalog$^\neg$ *queries, and*
- *poss is in NExp^{NP} and cert is in coNExp^{NP} for* Datalog$^{\neg,\vee}$ *queries.*

Proof. Let $\mathbf{T}$ be a C-database, A a set of facts, Q a Datalog$^{\neg,\vee}$ query, Δ the set of constants occurring in $\mathbf{T}$ or A, V the set of variables in $\mathbf{T}$, and $P_{\mathbf{T},\Delta}$ the logic program that encodes $\mathbf{T}$ (see Definition 3). Without loss of generality we assume that the facts in A all involve output predicates of Q. By Proposition 1, poss$(A, \mathbf{T}, Q)$ (resp., cert$(A, \mathbf{T}, Q)$) iff for some (resp, all) stable model(s) M of $P_{\mathbf{T},\Delta} \cup Q$, $A \subseteq M$.

Consider the following algorithm for computing the stable models of $P_{\mathbf{T},\Delta} \cup Q$. Observe that for each stable model M of $P_{\mathbf{T},\Delta} \cup Q$ it must hold, by the rules (5), that (†) for each v_{x_i}, with $x_i \in V$, there is exactly one $v_{x_i}(t_{x_i}) \in M$.

1. Guess an interpretation M for $P_{\mathbf{T},\Delta} \cup Q$ such that (†) holds.
2. Check whether M is a minimal model of $(gr_{\Delta \cup \Delta'}(P'_{\mathbf{T},\Delta} \cup Q))^M$, where $P'_{\mathbf{T},\Delta}$ is obtained from $P_{\mathbf{T},\Delta}$ by replacing every $v_{x_i}(x_i)$ with $v_{x_i}(t_{x_i})$.

The size of the guess M is clearly polynomial in $\mathbf{T}$. The reduct $(gr_{\Delta \cup \Delta'}(P'_{\mathbf{T},\Delta} \cup Q))^M$ can be computed in time polynomial in the size of $\mathbf{T}$ (since every predicate has bounded arity) and exponential in the combined size of $\mathbf{T}$ and Q. Then, checking whether M is a minimal model of the reduct can be done in PTime if $Q \in$ Datalog$^\neg$ and with an NP oracle if $Q \in$ Datalog$^{\neg,\vee}$ (cf. [6]). The first, second, fourth, and fifth bullet follow immediately.

Finally, if Q does not contain recursion, it is not necessary to consider the complete grounding; the algorithm can consider the possible variable substitutions one at a time. This requires polynomial space; the third bullet follows from the fact that nondeterministic PSpace =PSpace.   □

For determining the complexity of the certain answer semantics for Datalog$^\vee$ queries we exploit the fact that entailment from Datalog$^\vee$ programs corresponds to propositional consequence from its ground instantiation.

Proposition 6. *The problem* cert^Q *is in coNP and the problem* cert *is in coNExp for* Datalog$^\vee$ *queries.*

Proof. We have that cert$^Q(A, \mathbf{T})$ iff $A \subseteq M$ for every $I \in rep(\mathbf{T})$ and stable model M of $Q \cup I$, which is in turn equivalent to $gr_\Delta(Q \cup I) \models A$, where $\models$ is propositional consequence and Δ is the set of constants in I. The problem *there is an $I \in rep(\mathbf{T})$ such that $gr_\Delta(Q \cup I) \not\models A$* can be decided as follows: (1) guess a valuation σ for the variables in $\mathbf{T}$ and a propositional valuation γ for the atoms in $gr_\Delta(Q \cup \sigma(\mathbf{T}))$ and (2) check $\sigma(\mathbf{T}) \in rep(\mathbf{T})$, $\gamma \models gr_\Delta(Q \cup \sigma(\mathbf{T}))$, and $\gamma \not\models A$. Clearly, the algorithm runs in NP in the size of $\mathbf{T}$ and in NExp in the combined size. It follows that cert^Q can be decided in coNP and cert in coNExp.   □

We observe that poss can be straightforwardly reduced to cons, and vice versa.

Proposition 7. *There exists an LSpace reduction from* cons *(resp.,* cons^Q*) for a class of queries X to* poss *(resp.,* poss$^{Q'}$*) for X, and vice versa.*

Therefore, our results for consistency correspond with those for possible answers.

When considering C-databases without conditions (called *V-databases* in [14]), Abiteboul et al. [3] showed that cert^Q is in PTime for Datalog, while poss^Q is NP-complete for nr-Datalog queries and cert^Q is coNP-complete for nr-Datalog$^\neg$ queries. We complement these results as follows.

Proposition 8. *When considering C-databases without conditions,* cert^Q *is in LSpace for nr-Datalog queries,* cert *is Exp-complete for Datalog queries, and* poss *is NExp-complete for Datalog queries.*

Table 1. Complexity results for C-databases with/without conditions under closed-world interpretation

	cons^Q	poss^Q	cert^Q	cons	poss	cert
nr-Datalog	**NP**	NP	coNP/LSpace	**PSpace**	**PSpace**	**PSpace**
nr-Datalog$^{\neg}$	**NP**	NP	coNP	**PSpace**	**PSpace**	**PSpace**
Datalog	**NP**	NP	coNP/P	**NExp**	**NExp**	coNExp/Exp
st-Datalog$^{\neg}$	**NP**	NP	coNP	**NExp**	**NExp**	**coNExp**
Datalog$^{\neg}$	**NP**	NP	**coNP**	**NExp**	**NExp**	**coNExp**
Datalog$^{\vee}$	Σ_2^P	Σ_2^P	**coNP**	$\mathbf{NExp^{NP}}$	$\mathbf{NExp^{NP}}$	**coNExp**
st-Datalog$^{\neg,\vee}$	Σ_2^P	Σ_2^P	Π_2^P	$\mathbf{NExp^{NP}}$	$\mathbf{NExp^{NP}}$	$\mathbf{coNExp^{NP}}$
Datalog$^{\neg,\vee}$	Σ_2^P	Σ_2^P	Π_2^P	$\mathbf{NExp^{NP}}$	$\mathbf{NExp^{NP}}$	$\mathbf{coNExp^{NP}}$

Proof. For deciding cert, variables in C-databases without conditions can be treated as constants (Skolemization), and so the database can be treated as if it were a complete database (implicit in [14,24]). Exp-completeness of cert for Datalog and membership in LSpace of cert^Q for nr-Datalog queries follows from the results for complete databases.

By Proposition 6, poss is in NExp. Hardness is proved by a slight modification of the proof of Proposition 4: the guess of the next computation step i at time point j is performed using a single variable x_j, which may or may not be valuated with a valid i. This means that not all $I \in Q(rep(\mathbf{T}))$ correspond to runs, but still T has an accepting run iff there is an $I \in Q(rep(\mathbf{T}))$ such that $accept(1) \in I$.   □

The further complexity results for V-databases are the same as for C-databases. We note that the stated complexity results about V-databases apply even if variables may not occur twice in the database. Such V-databases are called *Codd databases* in [14].

Table 1 summarizes the complexity results for consistency and query answering under closed-world interpretation (CWI), both for databases with and without conditions (separated by the '/' symbol). Where the two cases correspond, only one complexity class is written. The results in boldface are novel. Note that all results in the table, save the LSpace result, are completeness results.

We can observe from the table that problems for query languages that are complete for (the complement of) a nondeterministic complexity class when considering complete databases (e.g., Datalog$^{\neg}$) do not increase in complexity when considering incomplete databases. So, for Datalog with disjunction and/or negation, answering queries on incomplete databases is not harder than answering queries on complete databases.

All considered PTime query languages jump to NP (resp., coNP) when considering data complexity and queries on C-databases. However, differences arise when considering the size of the query: for example, the combined complexity of the Datalog NExp-complete, whereas it is PSpace-complete for nr-Datalog.

Finally, we can observe that problems for queries on databases without conditions are only easier than those with conditions when considering PTime queries without negation and even then only certain answers are easier; possible answers and consistency are just as hard.

Table 2. Complexity results for C-databases with/without conditions under open-world interpretation

	Cons^Q	Cert^Q	Cons	Cert
nr-Datalog	**NP**/constant	coNP/LSpace	**NP/LSpace**	PSpace
nr-Datalog$^\neg$	Undec.	Undec.	Undec.	Undec.
Datalog	**NP**/constant	coNP/P	**NP/P**	**coNExp**/Exp
st-Datalog$^\neg$	Undec.	Undec.	Undec.	Undec.
Datalog$^\neg$	Undec.	Undec.	Undec.	Undec.
Datalog$^\vee$	**NP**/constant	**coNP**	Σ_2^p	**coNExp**
st-Datalog$^{\neg,\vee}$	Undec.	Undec.	Undec.	Undec.
Datalog$^{\neg,\vee}$	Undec.	Undec.	Undec.	Undec.

4.2 Complexity of Open-World Interpretation

For positive queries, certain answers under open-world interpretation (OWI) correspond to certain answers under CWI, which is a straightforward consequence of Lemma 1.

Proposition 9. *Let* $\mathbf{T}$ *be C-database,* Q *a* Datalog$^\vee$ *query, and* A *a set of facts. Then,* $\mathsf{Cert}(A, \mathbf{T}, Q)$ *iff* $\mathsf{cert}(A, \mathbf{T}, Q)$.

Checking $\mathsf{Cons}(\mathbf{T}, Q)$ for consistent $\mathbf{T}$ (i.e., $Rep(\mathbf{T}) \neq \emptyset$) corresponds to checking satisfiability of Q, which is known to be decidable for $Q \in$ Datalog [2, Theorem 12.5.2]. Observe that databases without conditions are trivially consistent. We establish the complexity of Cons^Q and Cons in the following two propositions.

Proposition 10. *Satisfiability of* Datalog *queries is PTime-hard and satisfiability of* Datalog$^\vee$ *queries is* Σ_2^p*-hard.*

Proposition 11. *The problems* Cons^Q *and* Cons *are NP-complete for* nr-Datalog *and* Datalog; Cons^Q *is NP-complete and* Cons *is in* Σ_2^p *for* Datalog$^\vee$ *queries.*
When considering C-databases without conditions, Cons *is in LSpace for* nr-Datalog *queries and in PTime for* Datalog *queries.*

Adding negation to any of the considered query languages results in undecidability, by the undecidability of finite satisfiability of nr-Datalog$^\neg$ queries [22]. Table 2 summarizes the complexity results under OWI where "Undec." is short for "Undecidable" and "constant" means "decidable in constant time". All results, save the two LSpace results, are completeness results.

As can be seen from the table, checking consistency under OWI is often easier than checking consistency under CWI. Intuitively, this is the case because under CWI one needs to take the absence of tuples in the database into account.

5 Related Work

Variations on Query Languages. Reiter [18] devised an algorithm for evaluating certain answers to queries on logical databases, which are essentially condition-free

C-databases under CWI. The algorithm, based on relational algebra, is complete for positive first-order queries (i.e., nr-Datalog) and for conjunctive queries extended with negation in front of atomic formulas (i.e., a subsef of st-Datalog$^\neg$). We obtain sound and complete reasoning for free by our translation of queries on C-databases to calculating the stable models of a logic program.

Rosati [19] considers condition- and variable-free databases under OWI and certain answers for conjunctive queries and unions of conjunctive queries, as well as extensions with inequality and negation. The data complexity of such queries is polynomial as long as the queries are safe, but becomes undecidable when considering unions of conjunctive queries extended with negation involving universally quantified variables.

We considered nr-Datalog, which generalize (unions of) conjunctive queries, but did not (yet) consider extensions with inequality and restricted forms of negation. A topic for future work is query answering on C-databases for such languages, both under CWI and OWI.

Logic Programming with Open Domains. One traditionally assumes in Logic Programming information regarding individuals is complete. Hence, the grounding of logic programs with the constants in the program. Approaches that allow for *incomplete information* in that sense, e.g., where one does not need all relevant constants in the program to deduce correct satisfiability results, are, the finite k-belief sets of [10,20,21] and its generalization[1], *open answer sets* [13]. Both deal with incomplete information by not a priori assuming that all relevant constants are present in the program under consideration. It is not clear what the exact relation with C-databases is; this is part of future work.

Or-sets. An alternative way of representing incomplete information is through objects with *or-sets* [15]. For example, a tuple $(John, \{30, 31\})$ indicates that John has age 30 or 31. This notion of incompleteness (which assumes closed-world interpretation) is somewhat simpler than C-databases and could be simulated using disjunctions of equality atoms. In [15], one shows that certain answers for existentially quantified conjunctive first-order formulas are data complete for coNP. A result that conforms with the coNP result for cert^Q with nr-Datalog queries in Table 1.

6 Outlook

We studied query languages ranging from nr-Datalog to Datalog$^{\neg,\vee}$. Besides extensions of positive query languages (including conjunctive queries) with inequality and limited forms of negation (e.g., only in front of extensional predicates), in future work we plan to consider integrity constraints, both as part of the query language, as is common in logic programming, and as part of the database. While under OWI adding integrity constraints to the database leads to undecidability already for very simple query languages [19], query answering under integrity constraints for databases under CWI is largely uncharted territory. We suspect that there are cases that are undecidable under OWI, but solvable under CWI, because by Lemma 2 we need to consider only a

---

[1] Both finite and infinite open answer sets are allowed.

finite subset of $rep(\mathbf{T})$. We note that Vardi [24] showed that checking integrity of an incomplete database is often harder under CWI than under OWI.

Abiteboul and Duschka [1] argue that a materialized view (e.g., the result of data integration) should be seen as an incomplete database, where the source predicates are seen as incomplete. Indeed, viewing a global schema as a sound view – essentially a condition- and variable-free incomplete database – is common in data integration [12,16]. Considering variables and, possibly, also conditions in (materialized) global views is a natural extension in this scenario; for example, local relations may have fewer columns than global relations, requiring view definitions of the form $\exists Y.v(X, Y, Z) \leftarrow s(X, Z)$. In future work we intend to consider query answering using such views.

Acknowledgements. We thank the anonymous reviewers for useful comments and feedback. The work in this paper was partially supported by the European Commission under the project ONTORULE (IST-2009-231875).

References

1. Abiteboul, S., Duschka, O.M.: Complexity of answering queries using materialized views. In: Proc. PODS, pp. 254–263 (1998)
2. Abiteboul, S., Hull, R., Vianu, V.: Foundations of Databases. Addison-Wesley, Reading (1995)
3. Abiteboul, S., Kanellakis, P.C., Grahne, G.: On the representation and querying of sets of possible worlds. Theoretical Computer Science 78(1), 158–187 (1991)
4. Bertossi, L.E., Bravo, L.: Consistent query answers in virtual data integration systems. In: Bertossi, L., Hunter, A., Schaub, T. (eds.) Inconsistency Tolerance. LNCS, vol. 3300, pp. 42–83. Springer, Heidelberg (2005)
5. Codd, E.F.: Extending the database relational model to capture more meaning. ACM ToDS 4(4), 397–434 (1979)
6. Dantsin, E., Eiter, T., Gottlob, G., Voronkov, A.: Complexity and expressive power of logic programming. ACM Computing Surveys 33(3), 374–425 (2001)
7. Eiter, T., Gottlob, G., Mannila, H.: Disjunctive Datalog. ACM ToDS 22(3), 364–418 (1997)
8. Fagin, R., Kolaitis, P.G., Miller, R.J., Popa, L.: Data exchange: semantics and query answering. Theor. Comput. Sci. 336(1), 89–124 (2005)
9. Gelfond, M., Lifschitz, V.: Classical negation in logic programs and disjunctive databases. New Generation Computing 9(3-4), 365–386 (1991)
10. Gelfond, M., Przymusinska, H.: Reasoning on open domains. In: Proc. LPNMR, pp. 397–413 (1993)
11. Grahne, G.: Horn tables - an efficient tool for handling incomplete information in databases. In: Proc. PODS, pp. 75–82 (1989)
12. Halevy, A.Y.: Answering queries using views: A survey. VLDB J. 10(4), 270–294 (2001)
13. Heymans, S., Nieuwenborgh, D.V., Vermeir, D.: Open answer set programming with guarded programs. ACM ToCL 9(4) (2008)
14. Imieliński, T., Lipski, W.: Incomplete information in relational databases. Journal of the ACM 31(4), 761–791 (1984)
15. Imieliński, T., Naqvi, S.A., Vadaparty, K.V.: Incomplete objects - a data model for design and planning applications. In: Proc. SIGMOD, pp. 288–297 (1991)
16. Lenzerini, M.: Data integration: A theoretical perspective. In: PODS, pp. 233–246 (2002)
17. Libkin, L.: Data exchange and incomplete information. In: Proc. PODS, pp. 60–69 (2006)

18. Reiter, R.: A sound and sometimes complete query evaluation algorithm for relational databases with null values. Journal of the ACM 33(2), 349–370 (1986)
19. Rosati, R.: On the decidability and finite controllability of query processing in databases with incomplete information. In: Proc. PODS, pp. 356–365 (2006)
20. Schlipf, J.: Some Remarks on Computability and Open Domain Semantics. In: Proc. WS on Structural Complexity and Recursion-Theoretic Methods in Logic Programming (1993)
21. Schlipf, J.: Complexity and Undecidability Results for Logic Programming. Annals of Mathematics and Artificial Intelligence 15(3-4), 257–288 (1995)
22. Trakhtenbrot, B.: Impossibility of an algorithm for the decision problem for finite models. Dokl. Akad. Nauk SSSR 70, 569–572 (1950)
23. Vardi, M.Y.: The complexity of relational query languages (extended abstract). In: ACM Symposium on Theory of Computing, pp. 137–146 (1982)
24. Vardi, M.Y.: On the integrity of databases with incomplete information. In: Proc. PODS, pp. 252–266 (1986)
25. Vardi, M.Y.: Querying logical databases. Journal of Computer and System Sciences 33(2), 142–160 (1986)
26. Vorobyov, S.G., Voronkov, A.: Complexity of nonrecursive logic programs with complex values. In: PODS, pp. 244–253 (1998)

Manifold Answer-Set Programs for Meta-reasoning[*]

Wolfgang Faber[1] and Stefan Woltran[2]

[1] University of Calabria, Italy
wf@wfaber.com
[2] Vienna University of Technology, Austria
woltran@dbai.tuwien.ac.at

Abstract. In answer-set programming (ASP), the main focus usually is on computing answer sets which correspond to solutions to the problem represented by a logic program. Simple reasoning over answer sets is sometimes supported by ASP systems (usually in the form of computing brave or cautious consequences), but slightly more involved reasoning problems require external postprocessing. Generally speaking, it is often desirable to use (a subset of) brave or cautious consequences of a program P_1 as input to another program P_2 in order to provide the desired solutions to the problem to be solved. In practice, the evaluation of the program P_1 currently has to be decoupled from the evaluation of P_2 using an intermediate step which collects the desired consequences of P_1 and provides them as input to P_2. In this work, we present a novel method for representing such a procedure within a *single* program, and thus within the realm of ASP itself. Our technique relies on rewriting P_1 into a so-called *manifold program*, which allows for accessing all desired consequences of P_1 within a single answer set. Then, this manifold program can be evaluated jointly with P_2 avoiding any intermediate computation step. For determining the consequences within the manifold program we use *weak constraints*, which is strongly motivated by complexity considerations. As an application, we present an encoding for computing the ideal extension of an abstract argumentation framework.

1   Introduction

In the last decade, *Answer Set Programming* (ASP) [1,2], also known as A-Prolog [3,4], has emerged as a declarative programming paradigm. ASP is well suited for modelling and solving problems which involve common-sense reasoning, and has been fruitfully applied to a wide variety of applications including diagnosis, data integration, configuration, and many others. Moreover, the efficiency of the latest tools for processing ASP programs (so-called ASP solvers) reached a state that makes them applicable for problems of practical importance [5]. The basic idea of ASP is to compute answer sets (usually stable models) of a logic program from which the solutions of the problem encoded by the program can be obtained.

---

[*] This work was supported by the Vienna Science and Technology Fund (WWTF), grant ICT08-028, and by M.I.U.R. within the Italia-Austria internazionalization project "Sistemi basati sulla logica per la rappresentazione di conoscenza: estensioni e tecniche di ottimizzazione.".

E. Erdem, F. Lin, and T. Schaub (Eds.): LPNMR 2009, LNCS 5753, pp. 115–128, 2009.
© Springer-Verlag Berlin Heidelberg 2009

However, frequently one is interested not only in the solutions per se, but rather in reasoning tasks that have to take some or even all solutions into account. As an example, consider the problem of database repair, in which a given database instance does not satisfy some of the constraints imposed in the database. One can attempt to modify the data in order to obtain a consistent database by changing as little as possible. This will in general yield multiple possibilities and can be encoded conveniently using ASP (see, e.g., [6]). However, usually one is not interested in the repairs themselves, but in the data which is present in *all* repairs. For the ASP encoding, this means that one is interested in the elements which occur in all answer sets; these are also known as *cautious consequences*. Indeed, ASP systems provide special interfaces for computing cautious consequences by means of query answering. But sometimes one has to do more, such as answering a complex query over the cautious consequences (not to be confused with complex queries over answer sets). So far, ASP solvers provide no support for such tasks. Instead, computations like this have to be done outside ASP systems, which hampers usability and limits the potential of ASP.

In this work, we tackle this limitation by providing a technique, which transforms an ASP program P into a *manifold program* M_P which we use to identify all consequences of a certain type[1] within a *single* answer set. The main advantage of the manifold approach is that the resulting program can be extended by additional rules representing a query over the brave (or cautious, definite) consequences of the original program P, thereby using ASP itself for this additional reasoning. In order to identify the consequences, we use *weak constraints* [8], which are supported by the ASP-solver DLV [9]. Weak constraints have been introduced to prefer a certain subset of answer sets via penalization. Their use for computing consequences is justified by a complexity-theoretic argument: One can show that computing consequences is complete for the complexity classes $FP_{||}^{NP}$ or $FP_{||}^{\Sigma_2^P}$ (depending on the presence of disjunction), for which also computing answer sets for programs with weak constraints is complete[2], which means that an equivalent compact ASP program without these extra constructs does not exist, unless the polynomial hierarchy collapses. In principle, other preferential constructs similar to weak constraints could be used as well for our purposes, as long as they meet these complexity requirements.

We discuss two particular applications of the manifold approach. First, we specify an encoding which decides the SAT-related *unique minimal model problem*, which is closely related to closed-world reasoning [10]. The second problem stems from the area of argumentation (cf. [11] for an overview) and concerns the computation of the ideal extension [12] of an argumentation framework. For both problems we make use of manifold programs of well-known encodings (computing all models of a CNF-formula

---

[1] We consider here the well-known concepts of brave and cautious consequence, but also definite consequence [7].

[2] The first of these results is fairly easy to see, for the second, it was shown [8] that the related decision problem is complete for the class Θ_2^P or Θ_3^P, from which the $FP_{||}^{NP}$ and $FP_{||}^{\Sigma_2^P}$ results can be obtained. Also note that frequently cited NP, Σ_2^P, and co-NP, Π_2^P completeness results hold for brave and cautious query answering, respectively, but not for computing brave and cautious consequences.

for the former application, computing all admissible extensions of an argumentation framework for the latter) in order to compute consequences. Extensions by a few more rules then directly provide the desired solutions, requiring little effort in total.

Organization and Main Results. After introducing the necessary background in the next section, we

- introduce in Section 3 the concept of a manifold program for rewriting propositional programs in such a way that all brave (resp. cautious, definite) consequences of the original program are collected into a single answer set;
- lift the results to the non-ground case (Section 4); and
- present applications for our technique in Section 5. In particular, we provide an ASP encoding for computing the ideal extension of an argumentation framework.

The paper concludes with a brief discussion of related and further work.

2 Preliminaries

In this section, we review the basic syntax and semantics of ASP with weak constraints, following [9], to which we refer for a more detailed definition.

An *atom* is an expression $p(t_1, \ldots, t_n)$, where p is a *predicate* of arity $\alpha(p) = n \geq 0$ and each t_i is either a variable or a constant. A *literal* is either an atom a or its negation not a. A *(disjunctive) rule* r is of the form

$$a_1 \vee \cdots \vee a_n :\!\!- b_1, \ldots, b_k, \text{ not } b_{k+1}, \ldots, \text{ not } b_m$$

with $n \geq 0, m \geq k \geq 0, n + m > 0$, and where $a_1, \ldots, a_n, b_1, \ldots, b_m$ are atoms.

The *head* of r is the set $H(r) = \{a_1, \ldots, a_n\}$, and the *body* of r is the set $B(r) = \{b_1, \ldots, b_k, \text{ not } b_{k+1}, \ldots, \text{ not } b_m\}$. Furthermore, $B^+(r) = \{b_1, \ldots, b_k\}$ and $B^-(r) = \{b_{k+1}, \ldots, b_m\}$. We will sometimes denote a rule r as $H(r) :\!\!- B(r)$.

A *weak constraint* [8] is an expression wc of the form

$$:\!\sim b_1, \ldots, b_k, \text{ not } b_{k+1}, \ldots, \text{ not } b_m. \ [w : l]$$

where $m \geq k \geq 0$ and $b_1, \ldots, b_m$ are literals, while $weight(wc) = w$ (the *weight*) and l (the *level*) are positive integer constants or variables. For convenience, w and/or l may be omitted and are set to 1 in this case. The sets $B(wc), B^+(wc)$, and $B^-(wc)$ are defined as for rules. We will sometimes denote a weak constraint wc as $:\!\sim B(wc)$.

A *program* P is a finite set of rules and weak constraints. $Rules(P)$ denotes the set of rules and $WC(P)$ the set of weak constraints in P. w_{max}^P and l_{max}^P denote the maximum weight and maximum level over $WC(P)$, respectively. A program (rule, atom) is *propositional* or *ground* if it does not contain variables. A program is called *strong* if $WC(P) = \emptyset$, and *weak* otherwise.

For any program P, let U_P be the set of all constants appearing in P (if no constant appears in P, an arbitrary constant is added to U_P); let B_P be the set of all ground literals constructible from the predicate symbols appearing in P and the constants of U_P; and let $Ground(P)$ be the set of rules and weak constraints obtained by applying, to each rule and weak constraint in P all possible substitutions from the variables in P to elements of U_P. U_P is usually called the *Herbrand Universe* of P and B_P the *Herbrand Base* of P.

A ground rule r is *satisfied* by a set I of ground atoms iff $H(r) \cap I \neq \emptyset$ whenever $B^+(r) \subseteq I$ and $B^-(r) \cap I = \emptyset$. I satisfies a ground program P, if each $r \in P$ is satisfied by I. For non-ground P, I satisfies P iff I satisfies $Rules(Ground(P))$. A ground weak constraint wc is *violated* by I, iff $B^+(wc) \subseteq I$ and $B^-(wc) \cap I = \emptyset$; it is satisfied otherwise.

Following [13], a set $I \subseteq B_P$ of atoms is an *answer set* for a strong program P iff it is a subset-minimal set that satisfies the *reduct*

$$P^I = \{H(r) :\text{-}\, B^+(r) \mid I \cap B^-(r) = \emptyset, r \in Ground(P)\}.$$

A set of atoms $I \subseteq B_P$ is an *answer set* for a weak program P iff I is an answer set of $Rules(P)$ and $H^{Ground(P)}(I)$ is minimal among all the answer sets of $Rules(P)$, where the penalization function $H^P(I)$ for weak constraint violation of a ground program P is defined as follows:

$$H^P(I) = \sum_{i=1}^{l_{max}^P} \left(f_P(i) \cdot \sum_{w \in N_i^P(I)} weight(w) \right)$$
$$f_P(1) = 1, \text{ and}$$
$$f_P(n) = f_P(n-1) \cdot |WC(P)| \cdot w_{max}^P + 1 \text{ for } n > 1.$$

where $N_i^P(I)$ denotes the set of weak constraints of P in level i violated by I. For any program P, we denote the set of its answer sets by $AS(P)$. In this paper, we use only weak constraints with weight and level 1, for which $H^{Ground(P)}(I)$ amounts to the number of weak constraints violated in I.

A ground atom a is a *brave* (sometimes also called credulous or possible) consequence of a program P, denoted $P \models_b a$, if $a \in A$ holds for at least one $A \in AS(P)$. A ground atom a is a *cautious* (sometimes also called skeptical or certain) consequence of a program P, denoted $P \models_c a$, if $a \in A$ holds for all $A \in AS(P)$. A ground atom a is a *definite* consequence [7] of a program P, denoted $P \models_d a$, if $AS(P) \neq \emptyset$ and $a \in A$ holds for all $A \in AS(P)$. The sets of all brave, cautious, definite consequences of a program P are denoted as $BC(P)$, $CC(P)$, $DC(P)$, respectively.

3 Propositional Manifold Programs

In this section, we present a translation which essentially creates a copy of a given strong propositional program for each of (resp. for a subset of) its atoms. Thus, we require several copies of the alphabet used by the given program.

Definition 1. *Given a set I of literals, a collection $\mathcal{I}$ of sets of literals, and an atom a, define $I^a = \{p^a \mid atom\ p \in I\} \cup \{not\ p^a \mid not\ p \in I\}$ and $\mathcal{I}^a = \{I^a \mid I \in \mathcal{I}\}$.*

The actual transformation to a manifold is given in the next definition. We copy a given program P for each atom a in a given set S, whereby the transformation guarantees the existence of an answer set by enabling the copies conditionally.

Definition 2. *For a strong propositional program P and $S \subseteq B_P$, define its* manifold *as*

$$P_S^{tr} = \bigcup_{r \in P} \{H(r)^a :\text{-}\, \{c\} \cup B(r)^a \mid a \in S\} \cup \{c :\text{-} not\ i \ ; \ i :\text{-} not\ c\}.$$

We assume $B_P \cap B_{P_S^{tr}} = \emptyset$, that is, all symbols in P_S^{tr} are assumed to be fresh.

Example 1. Consider $\Phi = \{p \vee q :\text{-} \ ; \ r :\text{-} p \ ; \ r :\text{-} q\}$ for which $AS(\Phi) = \{\{p, r\},$ $\{q, r\}\}, BC(\Phi) = \{p, q, r\}$ and $CC(\Phi) = DC(\Phi) = \{r\}$. When forming the manifold for $B_\Phi = \{p, q, r\}$, we obtain

$$\Phi_{B_\Phi}^{tr} = \left\{ \begin{array}{l} p^p \vee q^p :\text{-} c \ ; \ r^p :\text{-} c, p^p \ ; \ r^p :\text{-} c, q^p \ ; \ c :\text{-} \text{not } i \ ; \\ p^q \vee q^q :\text{-} c \ ; \ r^q :\text{-} c, p^q \ ; \ r^q :\text{-} c, q^q \ ; \ i :\text{-} \text{not } c \ ; \\ p^r \vee q^r :\text{-} c \ ; \ r^r :\text{-} c, p^r \ ; \ r^r :\text{-} c, q^r \end{array} \right\}$$

Note that given a strong program P and $S \subseteq B_P$, the construction of P_S^{tr} can be done in polynomial time (w.r.t. the size of P). The answer sets of the transformed program consist of all combinations (of size $|S|$) of answer sets of the original program (augmented by c) plus the special answer set $\{i\}$ which we shall use to indicate inconsistency of P.

Proposition 1. *For a strong propositional program P and a set $S \subseteq B_P$, $AS(P_S^{tr}) = A \cup \{\{i\}\}$, where*

$$A = \{\bigcup_{i=1}^{|S|} A_i \cup \{c\} \mid \langle A_1, \ldots, A_{|S|} \rangle \in \prod_{a \in S} AS(P)^a\}.$$

Note that $\prod$ denotes the Cartesian product in Proposition 1.

Example 2. For Φ of Example 1, we obtain that $AS(\Phi_{B_\Phi}^{tr})$ consists of $\{i\}$ plus (copies of $\{q, r\}$ are underlined for readability)

$$\{c, p^p, r^p, p^q, r^q, p^r, r^r\}, \{c, \underline{q^p}, r^p, p^q, r^q, p^r, r^r\}, \{c, p^p, r^p, \underline{q^q}, r^q, p^r, r^r\},$$
$$\{c, p^p, r^p, p^q, r^q, \underline{q^r}, r^r\}, \{c, \underline{q^p}, r^p, \underline{q^q}, r^q, p^r, r^r\}, \{c, \underline{q^p}, r^p, p^q, r^q, \underline{q^r}, r^r\},$$
$$\{c, p^p, r^p, \underline{q^q}, r^q, \underline{q^r}, r^r\}, \{c, \underline{q^p}, r^p, \underline{q^q}, r^q, \underline{q^r}, r^r\}.$$

Using this transformation, each answer set encodes an association of an atom with some answer set of the original program. If an atom a is a brave consequence of the original program, then a witnessing answer set exists, which contains the atom a^a. The idea is now to prefer those atom-answer set associations where the answer set is a witness. We do this by means of weak constraints and penalize each association where the atom is not in the associated answer set, that is, where a^a is not in the answer set of the transformed program. Doing this for each atom means that an optimal answer set will not contain a^a only if there is no answer set of the original program that contains a, so each a^a contained in an optimal answer set is a brave consequence of the original program.

Definition 3. *Given a strong propositional program P and $S \subseteq B_P$, let*

$$P_S^{bc} = P_S^{tr} \cup \{:\sim \text{not } a^a \mid a \in S\} \cup \{:\sim i\}$$

Observe that all weak constraints are violated in the special answer set $\{i\}$, while in the answer set $\{c\}$ (which occurs if the original program has an empty answer set) all but $:\sim i$ are violated. The following result would also hold without $:\sim i$ being included.

Proposition 2. *Given a strong propositional program P and $S \subseteq B_P$, for any $A \in AS(P_S^{bc})$, $\{a \mid a^a \in A\} = BC(P) \cap S$.*

Example 3. For the program Φ as given Example 1, $\Phi^{bc}_{B_\Phi}$ is given by $\Phi^{tr}_{B_\Phi} \cup \{:\sim$ not p^p ; $:\sim$ not q^q ; $:\sim$ not r^r ; $:\sim i\}$. We obtain that $AS(\Phi^{bc}_{B_\Phi}) = \{A_1, A_2\}$, where $A_1 = \{c, p^p, r^p, q^q, r^q, p^r, r^r\}$ and $A_2 = \{c, p^p, r^p, q^q, r^q, q^r, r^r\}$, as these two answer sets are the only ones that violate no weak constraint. We can observe that $\{a \mid a^a \in A_1\} = \{a \mid a^a \in A_2\} = \{p, q, r\} = BC(\Phi)$.

Concerning cautious consequences, we first observe that if a program is inconsistent (in the sense that it does not have any answer set), each atom is a cautious consequence. But if P is inconsistent, then P^{tr}_S will have only $\{i\}$ as an answer set, so we will need to find a suitable modification in order to deal with this in the correct way. In fact, we can use a similar approach as for brave consequences, but penalize those associations where an atom is contained in its associated answer set. Any optimal answer set will thus contain a^a for an atom only if a is contained in each answer set. If an answer set containing i exists, it is augmented by all atoms a^a, which also causes all weak constraints to be violated.

Definition 4. *Given a strong propositional program P and $S \subseteq B_P$, let*

$$P^{cc}_S = P^{tr}_S \cup \{:\sim a^a \mid a \in S\} \cup \{a^a :\text{-} i \mid a \in S\} \cup \{:\sim i\}$$

As for P^{bc}_S, the following result also holds without including $:\sim i$.

Proposition 3. *Given a strong propositional program P and $S \subseteq B_P$, for any $A \in AS(P^{cc}_S)$, $\{a \mid a^a \in A\} = CC(P) \cap S$.*

Example 4. Recall program Φ from Example 1. We have $\Phi^{cc}_{B_\Phi} = \Phi^{tr}_{B_\Phi} \cup \{:\sim p^p$; $:\sim q^q$; $:\sim r^r$; $p^p :\text{-} i$; $q^q :\text{-} i$; $r^r :\text{-} i$; $:\sim i\}$. We obtain that $AS(\Phi^{cc}_{B_\Phi}) = \{A_3, A_4\}$, where $A_3 = \{c, q^p, r^p, p^q, r^q, p^r, r^r\}$ and $A_4 = \{c, q^p, r^p, p^q, r^q, q^r, r^r\}$, as these two answer sets are the only ones that violate only one weak constraint, namely $:\sim r^r$. We observe that $\{a \mid a^a \in A_3\} = \{a \mid a^a \in A_4\} = \{r\} = CC(\Phi)$.

We next consider the notion of definite consequences. Different to cautious consequences, we do not add the annotated atoms to the answer set containing i. However, this answer set should never be among the optimal ones unless it is the only one. Therefore we inflate it by new atoms i^a, all of which incur a penalty. This guarantees that this answer set will incur a higher penalty ($|B_P| + 1$) than any other ($\leq |B_P|$).

Definition 5. *Given a strong propositional program P and $S \subseteq B_P$, let*

$$P^{dc}_S = P^{tr}_S \cup \{:\sim a^a;\ i^a :\text{-} i;\ :\sim i^a \mid a \in S\} \cup \{:\sim i\}$$

Proposition 4. *Given a strong propositional program P and $S \subseteq B_P$, for any $A \in AS(P^{dc}_S)$, $\{a \mid a^a \in A\} = DC(P) \cap S$.*

Example 5. Recall program Φ from Example 1. We have $\Phi^{dc}_{B_\Phi} = \Phi^{tr}_{B_\Phi} \cup \{:\sim p^p$; $:\sim q^q$; $:\sim r^r$; $i^p :\text{-} i$; $i^q :\text{-} i$; $i^r :\text{-} i :\sim i^p$; $:\sim i^q$; $:\sim i^r$; $:\sim i\}$. As in Example 4, A_3 and A_4 are the only ones that violate only one weak constraint, namely $:\sim r^r$, and thus are the answer sets of $\Phi^{dc}_{B_\Phi}$.

Obviously, one can compute all brave, cautious, or definite consequences of a program by choosing $S = B_P$. We also note that the programs from Definitions 3, 4 and 5 yield multiple answer sets. However each of these yields the same atoms a^a, so it is sufficient to compute one of these. The programs could be extended in order to admit only one answer set by suitably penalizing all atoms a^b ($a \neq b$). To avoid interference with the weak constraints already used, these additional weak constraints would have to pertain to a different level.

4 Non-ground Manifold Programs

We now generalize the techniques introduced in Section 3 to non-ground strong programs. In principle, one could annotate each predicate (rather than atom as in Section 3) with ground atoms of a subset of the Herbrand Base. However, one can also move the annotations to the non-ground level: For example, instead of annotating a rule $p(X,Y) \coloneq q(X,Y)$ by the set $\{r(a), r(b)\}$ yielding $p^{r(a)}(X,Y) \coloneq q^{r(a)}(X,Y)$ and $p^{r(b)}(X,Y) \coloneq q^{r(b)}(X,Y)$ we will annotate using only the predicate r and extend the arguments of p, yielding the compact rule $d_p^r(X,Y,Z) \coloneq d_q^r(X,Y,Z)$ (we use predicate symbols d_p^r and d_q^r rather than p^r and q^r just for pointing out the difference between annotation by predicates versus annotation by ground atoms). In this particular example we have assumed that the program is to be annotated by all ground instances of $r(Z)$; we will use this assumption also in the following for simplifying the presentation. In practice, one can clearly add atoms to the rule body for restricting the instances of the predicate by which we annotate, in the example this would yield $p^r(X,Y,Z) \coloneq q^r(X,Y,Z), dom(Z)$ where the predicate dom should be defined appropriately. In the following, recall that $\alpha(p)$ denotes the arity of a predicate p.

Definition 6. *Given an atom $a = p(t_1, \ldots, t_n)$ and a predicate q, let a_q^{tr} be the atom $d_p^q(t_1, \ldots, t_n, X_1, \ldots, X_{\alpha(q)})$ where $X_1, \ldots, X_{\alpha(q)}$ are fresh variables and d_p^q is a new predicate symbol with $\alpha(d_p^q) = \alpha(p) + \alpha(q)$. Furthermore, given a set $\mathcal{L}$ of literals, and a predicate q, let $\mathcal{L}_q^{tr}$ be $\{a_q^{tr} \mid atom\ a \in \mathcal{L}\} \cup \{not\ a_q^{tr} \mid not\ a \in \mathcal{L}\}$.*

Note that we assume that even though the variables $X_1, \ldots, X_{\alpha(q)}$ are fresh, they will be the same for each a_q^{tr}. One could define similar notions also for partially ground atoms or for sets of atoms characterized by a collection of defining rules, from which we refrain here for the ease of presentation. We define the manifold program in analogy to Definition 2, the only difference being the different way of annotating.

Definition 7. *Given a strong program P and a set S of predicates, define its* manifold *as*

$$P_S^{tr} = \bigcup_{r \in P} \{H(r)_q^{tr} \coloneq \{c\} \cup B(r)_q^{tr} \mid q \in S\} \cup \{c \coloneq not\ i\ ;\ i \coloneq not\ c\}.$$

Example 6. Consider program $\Psi = \{p(X) \vee q(X) \coloneq r(X);\ ;\ r(a) \coloneq\ ;\ r(b) \coloneq\ \}$ for which $AS(\Psi) = \{\{p(a), p(b), r(a), r(b)\}, \{p(a), q(b), r(a), r(b)\}, \{q(a), p(b), r(a), r(b)\}, \{q(a), q(b), r(a), r(b)\}\}$. Hence, $BC(\Psi) = \{p(a), p(b), q(a), q(b), r(a), r(b)\}$ and $CC(\Psi) = DC(\Psi) = \{r(a), r(b)\}$. Forming the manifold for $S = \{p\}$, we obtain

$$\Psi_S^{tr} = \left\{ \begin{array}{l} \mathrm{d}_p^p(X, X_1) \vee \mathrm{d}_q^p(X, X_1) :\text{-} \mathrm{d}_r^p(X, X_1), c \ ; \\ \mathrm{d}_r^p(a, X_1) :\text{-} c \ ; \quad \mathrm{d}_r^p(b, X_1) :\text{-} c \ ; \quad c :\text{-} \mathrm{not} \ i \ ; \quad i :\text{-} \mathrm{not} \ c \end{array} \right\}$$

$AS(\Psi_S^{tr})$ consists of $\{i\}$ plus 16 answer sets, corresponding to all combinations of the 4 answer sets in $AS(\Psi)$.

Now we are able to generalize the encodings for brave, cautious, and definite consequences. These definitions are direct extensions of Definitions 3, 4, and 5, the differences are only due to the non-ground annotations. In particular, the diagonalization atoms a^a should now be written as $\mathrm{d}_p^p(X_1, \ldots, X_{\alpha(p)}, X_1, \ldots, X_{\alpha(p)})$ which represent the set of ground instances of $p(X_1, \ldots, X_{\alpha(p)})$, each annotated by itself. So, a weak constraint $:\sim \mathrm{d}_p^p(X_1, \ldots, X_{\alpha(p)}, X_1, \ldots, X_{\alpha(p)})$ gives rise to $\{:\sim \mathrm{d}_p^p(c_1, \ldots, c_{\alpha(p)}, c_1, \ldots, c_{\alpha(p)}) \mid c_1, \ldots, c_{\alpha(p)} \in U\}$ where U is the Herbrand base of the program in question, that is one weak constraint for each ground instance annotated by itself.

Definition 8. *Given a strong program P and a set S of predicate symbols, let*

$$P_S^{bc} = P_S^{tr} \cup \{:\sim \mathrm{not} \ \Delta_q \mid q \in S\} \cup \{:\sim i\}$$
$$P_S^{cc} = P_S^{tr} \cup \{:\sim \Delta_q; \ \Delta_q :\text{-} i \mid q \in S\} \cup \{:\sim i\}$$
$$P_S^{dc} = P_S^{tr} \cup \{:\sim \Delta_q; \ I_q :\text{-} i; \ :\sim I_q \mid q \in S\} \cup \{:\sim i\}$$

where $\Delta_q = \mathrm{d}_q^q(X_1, \ldots, X_{\alpha(q)}, X_1, \ldots, X_{\alpha(q)})$ and $I_q = i_q(X_1, \ldots, X_{\alpha(q)})$.

Proposition 5. *Given a strong program P and a set S of predicates, for an arbitrary $A \in AS(P_S^{bc})$, (resp., $A \in AS(P_S^{cc})$, $A \in AS(P_S^{dc})$), the set $\{p(c_1, \ldots, c_{\alpha(p)}) \mid \mathrm{d}_p^p(c_1, \ldots, c_{\alpha(p)}, c_1, \ldots, c_{\alpha(p)}) \in A\}$ is the set of brave (resp., cautious, definite) consequences of P with a predicate in S.*

Example 7. Consider again Ψ and $S = \{p\}$ from Example 6. We obtain $\Psi_S^{bc} = \Psi_S^{tr} \cup \{:\sim \mathrm{not} \ \mathrm{d}_p^p(X_1, X_1) \ ; \ :\sim i\}$ and we can check that $AS(\Psi_S^{bc})$ consists of the sets

$R \cup \{\mathrm{d}_p^p(a, a), \mathrm{d}_p^p(b, b), \mathrm{d}_q^p(a, b), \mathrm{d}_q^p(b, a)\}, R \cup \{\mathrm{d}_p^p(a, a), \mathrm{d}_p^p(b, b), \mathrm{d}_p^p(a, b), \mathrm{d}_q^p(b, a)\},$
$R \cup \{\mathrm{d}_p^p(a, a), \mathrm{d}_p^p(b, b), \mathrm{d}_q^p(a, b), \mathrm{d}_p^p(b, a)\}, R \cup \{\mathrm{d}_p^p(a, a), \mathrm{d}_p^p(b, b), \mathrm{d}_p^p(b, a), \mathrm{d}_p^p(b, a)\};$

where $R = \{\mathrm{d}_r^p(a, a), \mathrm{d}_r^p(a, b), \mathrm{d}_r^p(b, a), \mathrm{d}_r^p(b, b)\}$. For each A of these answer sets we obtain $\{p(t) \mid \mathrm{d}_p^p(t, t) \in A\} = \{p(a), p(b)\}$ which corresponds exactly to the brave consequences of Ψ with a predicate of $S = \{p\}$.

For cautious consequences, $\Psi_S^{cc} = \Psi_S^{tr} \cup \{:\sim \mathrm{d}_p^p(X_1, X_1) \ ; \ \mathrm{d}_p^p(X_1, X_1) :\text{-} i \ ; \ :\sim i\}$ and we can check that $AS(\Psi_S^{cc})$ consists of the sets

$R \cup \{\mathrm{d}_q^p(a, a), \mathrm{d}_q^p(b, b), \mathrm{d}_q^p(a, b), \mathrm{d}_q^p(b, a)\}, R \cup \{\mathrm{d}_q^p(a, a), \mathrm{d}_q^p(b, b), \mathrm{d}_p^p(a, b), \mathrm{d}_q^p(b, a)\},$
$R \cup \{\mathrm{d}_q^p(a, a), \mathrm{d}_q^p(b, b), \mathrm{d}_q^p(a, b), \mathrm{d}_p^p(b, a)\}, R \cup \{\mathrm{d}_q^p(a, a), \mathrm{d}_q^p(b, b), \mathrm{d}_p^p(b, a), \mathrm{d}_p^p(b, a)\};$

where $R = \{\mathrm{d}_r^p(a, a), \mathrm{d}_r^p(a, b), \mathrm{d}_r^p(b, a), \mathrm{d}_r^p(b, b)\}$. For each A of these answer sets we obtain $\{p(t) \mid \mathrm{d}_p^p(t, t) \in A\} = \emptyset$ and indeed there are no cautious consequences of Ψ with a predicate of $S = \{p\}$.

Finally, for definite consequences, $\Psi_S^{dc} = \Psi_S^{tr} \cup \{:\sim \mathrm{d}_p^p(X_1, X_1) \ ; \ i_p(X_1) :\text{-} i \ ; \ :\sim i_p(X_1) \ ; \ :\sim i\}$. It is easy to see that $AS(\Psi_S^{dc}) = AS(\Psi_S^{cc})$ and so $\{p(t) \mid \mathrm{d}_p^p(t, t) \in A\} = \emptyset$ for each answer set A of Ψ_S^{dc}, and indeed there is also no definite consequence of Ψ with a predicate of $S = \{p\}$.

These definitions exploit the fact that the semantics of non-ground programs is defined via their grounding with respect to their Herbrand Universe. So the fresh variables introduced in the manifold will give rise to one copy of a rule for each ground atom.

In practice, ASP systems usually require rules to be safe, that is, that each variable occurs (also) in the positive body. The manifold for a set of predicates may therefore contain unsafe rules (because of the fresh variables). But this can be repaired by adding a *domain atom* $dom_q(X_1, \ldots, X_m)$ to a rule which is to be annotated with q. This predicate can in turn be defined by a rule $dom_q(X_1, \ldots, X_m) :- u(X_1), \ldots, u(X_m)$ where u is defined using $\{u(c) \mid c \in U_P\}$. One can also provide smarter definitions for dom_q by using a relaxation of the definition for q.

We also observe that ground atoms that are contained in all answer sets of a program need not be annotated in the manifold. Note that these are essentially the cautious consequences of a program and therefore determining all of those automatically before rewriting does not make sense. But for some atoms this property can be determined only by the structure of the program. For instance, facts will be in all answer sets. In the sequel we will not annotate extensional atoms (those defined only by facts) in order to obtain more concise programs. One could also go further and omit the annotation of atoms which are defined using nondisjunctive stratified programs.

As an example, we present an ASP encoding for boolean satisfiability and then create its manifold program for resolving the following problem: Given a propositional formula in CNF φ, compute all atoms which are true in all models of φ. We provide a fixed program which takes a representation of φ as facts as input. To apply our method we first require a program whose answer sets are in a one-to-one correspondence to the models of φ. To start with, we fix the representation of CNFs. Let φ (over atoms A) be of the form $\bigwedge_{i=1}^{n} c_i$. Then, $D_\varphi = \{at(a) \mid a \in A\} \cup \{cl(i) \mid 1 \leq i \leq n\} \cup \{pos(a, i) \mid$ atom a occurs positively in $c_i\} \cup \{neg(a, i) \mid$ atom a occurs negatively in $c_i\}$. We construct program SAT as the set of the following rules.

$$true(X) :- \text{not } false(X), at(X); \quad false(X) :- \text{not } true(X), at(X);$$
$$ok(C) :- true(X), pos(C, X); ok(C) :- false(X), neg(C, X); \quad :- \text{not } ok(C), cl(C).$$

It can be checked that the answer sets of $\text{SAT} \cup D_\varphi$ are in a one-to-one correspondence to the models (over A) of φ. In particular, for any model $I \subseteq A$ of φ there exists an answer set M of $\text{SAT} \cup D_\varphi$ such that $I = \{a \mid true(a) \in M\}$. We now consider $\text{SAT}^{cc}_{\{true\}}$ which consists of the following rules.

$$d_{true}^{true}(X, Y) :- c, \text{not } d_{false}^{true}(X, Y), at(X); \quad c :- \text{not } i; \qquad\qquad i :- \text{not } c;$$
$$d_{false}^{true}(X, Y) :- c, \text{not } d_{true}^{true}(X, Y), at(X); \quad :- c, \text{not } d_{ok}^{true}(C, Y), cl(C);$$
$$d_{ok}^{true}(C, Y) :- c, d_{true}^{true}(X, Y), pos(C, X); \quad :\sim d_{true}^{true}(X, X); \qquad\qquad :\sim i;$$
$$d_{ok}^{true}(C, Y) :- c, d_{false}^{true}(X, Y), neg(C, X); \qquad\qquad d_{true}^{true}(X, X) :- i.$$

Given Proposition 5, it is easy to see that, given some answer set A of $\text{SAT}^{cc}_{\{true\}} \cup D_\varphi$, $\{a \mid d_{true}^{true}(a, a) \in A\}$ is precisely the set of atoms which are true in all models of φ.

5 Applications

In this section, we put our technique to work and show how to use meta-reasoning over answer sets for two application scenarios. The first one is a well-known problem from propositional logic, and we will reuse the example from above. The second example takes a bit more background, but presents a novel method to compute ideal extensions for argumentation frameworks.

5.1 The Unique Minimal Model Problem

As a first example, we show how to encode the problem of deciding whether a given propositional formula φ has a unique minimal model. This problem is known to be in Θ_2^P and to be co-NP-hard (the exact complexity is an open problem). Let I be the intersection of all models of φ. Then φ has a unique minimal model iff I is also a model of φ. We thus use our example from the previous section, and define the program UNIQUE as $\mathrm{SAT}^{cc}_{\{true\}}$ augmented by rules $ok(C) :\text{-} \, \mathrm{d}^{true}_{true}(X, X), \mathrm{pos}(C, X);$ $ok(C) :\text{-} \, \mathrm{not} \, \mathrm{d}^{true}_{true}(X, X), \mathrm{neg}(C, X); :\text{-} \, \mathrm{not} \, ok(C), \mathrm{cl}(C).$

Theorem 1. *For any CNF formula φ, it holds that φ has a unique minimal model, if and only if program UNIQUE $\cup \, D_\varphi$ has at least one answer set.*

A slight adaption of this encoding allows us to formalize CWA-reasoning [10] over a propositional knowledge base φ, since the atoms a in φ, for which the corresponding atoms $\mathrm{d}^{true}_{true}(a, a)$ are not contained in an answer set of $\mathrm{SAT}^{cc}_{\{true\}} \cup D_\varphi$, are exactly those which are added negated to φ for CWA-reasoning.

5.2 Computing the Ideal Extension

Our second example is from the area of argumentation, where the problem of computing the ideal extension [12] of an abstract argumentation framework was recently shown to be complete for $\mathrm{FP}^{\mathrm{NP}}_{||}$ in [14]. Thus, this task cannot be compactly encoded via normal programs (under usual complexity theoretic assumptions). On the other hand, the complexity shows that employing disjunction is not necessary, if one instead uses weak constraints. We first give the basic definitions following [15].

Definition 9. *An* argumentation framework *(AF) is a pair $F = (A, R)$ where $A \subseteq U$ is a set of arguments and $R \subseteq A \times A$. $(a, b) \in R$ means that a attacks b. An argument $a \in A$ is* defended *by $S \subseteq A$ (in F) if, for each $b \in A$ such that $(b, a) \in R$, there exists a $c \in S$, such that $(c, b) \in R$. An argument a is* admissible *(in F) w.r.t. a set $S \subseteq A$ if each $b \in A$ which attacks a is defended by S.*

Semantics for argumentation frameworks are given in terms of so-called extensions. The next definitions introduce two such notions which also underly the concept of an ideal extension.

Definition 10. *Let $F = (A, R)$ be an AF. A set $S \subseteq A$ is said to be* conflict-free *(in F), if there are no $a, b \in S$, such that $(a, b) \in R$. A set S is an* admissible extension *of*

F, *if S is conflict-free in F and each a $\in$ S is admissible in F w.r.t. S. The collection of admissible extensions is denoted by adm(F). An admissible extension S of F is a preferred extension of F, if for each T $\in$ adm(F), S $\not\subseteq$ T. The collection of preferred extensions of F is denoted by pref(F).*

Definition 11. *Let F be an AF. A set S is called* ideal *for F, if S $\in$ adm(F) and S $\subseteq \bigcap_{T \in pref(F)} T$. A maximal (w.r.t. set-inclusion) ideal set of F is called an* ideal extension *of F.*

It was shown that for each AF F, a unique ideal extension exists. In [14], the following algorithm to compute the ideal extension of an AF $F = (A, R)$ is proposed. Let $X_F^- = A \setminus \bigcup_{S \in adm(F)} S$ and $X_F^+ = \{a \in A \mid \forall b, c : (b, a), (a, c) \in R \Rightarrow b, c \in X_F^-\} \setminus X_F^-$, and define an AF $F^* = (X_F^+ \cup X_F^-, R^*)$ where $R^* = R \cap \{(a, b), (b, a) \mid a \in X_F^+, b \in X_F^-\}$. F^* is a bipartite AF in the sense that R^* is a bipartite graph.

Proposition 6 ([14]). *The ideal extension of AF F is given by $\bigcup_{S \in adm(F^*)} (S \cap X_F^+)$.*

The set of all admissible atoms for a bipartite AF F can be computed in polynomial time using Algorithm 1 of [16]. This is basically a fixpoint iteration identifying arguments in X_F^+ that cannot be in an admissible extension: First, arguments in $X_0 = X_F^+$ are excluded, which are attacked by unattacked arguments (which are necessarily in X_F^-), yielding X_1. Now, arguments in X_F^- may be unattacked by X_1, and all arguments in X_1 attacked by such newly unattacked arguments should be excluded. This process is iterated until either no arguments are left or no more argument can be excluded. There may be at most $|X_F^+|$ iterations in this process.

We exploit this technique to formulate an ASP-encoding IDEAL. We first report a program the answer sets of which characterize admissible extensions. Then, we use the brave manifold of this program in order to determine all arguments contained in some admissible extension. Finally, we extend this manifold program in order to identify F^* and to simulate Algorithm 1 of [16].

The argumentation frameworks will be given to IDEAL as sets of input facts. Given an AF $F = (A, R)$, let $D_F = \{a(x) \mid x \in A\} \cup \{r(x, y) \mid (x, y) \in R\}$. Program ADM, given by the rules below, computes admissible extensions (cf. [17,18]):

$$\text{in}(X) :\!- \text{not out}(X), a(X); \quad \text{out}(X) :\!- \text{not in}(X), a(X); \quad \text{def}(X) :\!- \text{in}(Y), r(Y, X);$$
$$:\!- \text{in}(X), \text{in}(Y), r(X, Y); \quad :\!- \text{in}(X), r(Y, X), \text{not def}(Y).$$

Indeed one can show that, given an AF F, the answer sets of $\text{ADM} \cup D_F$ are in a one-to-one correspondence to the admissible extensions of F via the $\text{in}(\cdot)$ predicate. In order to determine the brave consequences of ADM for predicate in, we form $\text{ADM}_{\{\text{in}\}}^{bc}$, and extend it by collecting all brave consequences of $\text{ADM} \cup D_F$ in predicate $\text{in}(\cdot)$, from which we can determine X_F^- (represented by $\text{in}^-(\cdot)$), X_F^+ (represented by $\text{in}^+(\cdot)$, using auxiliary predicate $\text{not\_in}^+(\cdot)$), and R^* (represented by $q(\cdot, \cdot)$).

$$\text{in}(X) :\!- d_{\text{in}}^{\text{in}}(X, X); \quad \text{in}^-(X) :\!- a(X), \text{not in}(X); \quad \text{in}^+(X) :\!- \text{in}(X), \text{not not\_in}^+(X);$$
$$\text{not\_in}^+(X) :\!- \text{in}(Y), r(X, Y); \qquad \text{not\_in}^+(X) :\!- \text{in}(Y), r(Y, X);$$
$$q(X, Y) :\!- r(X, Y), \text{in}^+(X), \text{in}^-(Y); \qquad q(X, Y) :\!- r(X, Y), \text{in}^-(X), \text{in}^+(Y).$$

In order to simulate Algorithm 1 of [16], we use the elements in X_F^+ for marking the iteration steps. To this end, we use an arbitrary order $<$ on ASP constants (all ASP systems provide such a predefined order) and define successor, infimum and supremum among the constants representing X_F^+ w.r.t. the order $<$.

$$\text{nsucc}(X,Z) :- \text{in}^+(X), \text{in}^+(Y), \text{in}^+(Z), X{<}Y, Y{<}Z;$$
$$\text{succ}(X,Y) :- \text{in}^+(X), \text{in}^+(Y), X{<}Y, \text{not nsucc}(X,Y);$$
$$\text{ninf}(Y) :- \text{in}^+(X), \text{in}^+(Y), X{<}Y; \quad \text{nsup}(X) :- \text{in}^+(X), \text{in}^+(Y), X{<}Y;$$
$$\text{inf}(X) :- \text{in}^+(X), \text{not ninf}(X); \quad \text{sup}(X) :- \text{in}^+(X), \text{not nsup}(X).$$

We now use this to iteratively determine arguments that are not in the ideal extension, using $\text{nid}(\cdot,\cdot)$, where the first argument is the iteration step. In the first iteration (identified by the infimum) all arguments in X_F^+ which are attacked by an unattacked argument are collected. In subsequent iterations, all arguments from the previous steps are included and augmented by arguments that are attacked by an argument not attacked by arguments in X_F^+ that were not yet excluded in the previous iteration. Finally, arguments in the ideal extension are those that are not excluded from X_F^+ in the final iteration (identified by the supremum).

$$\text{att}_0(X) :- q(Y,X); \quad \text{att}_i(J,Z) :- q(Y,Z), \text{in}^+(Y), \text{not nid}(J,Y), \text{in}^+(J);$$
$$\text{ideal}(X) :- \text{in}^+(X), \text{sup}(I), \text{not nid}(I,X); \quad \text{nid}(I,Y) :- \text{succ}(J,I), \text{nid}(J,Y);$$
$$\text{nid}(I,Y) :- \text{inf}(I), q(Z,Y), \text{in}^+(Y), \text{not att}_0(Z);$$
$$\text{nid}(I,Y) :- \text{succ}(J,I), q(Z,Y), \text{in}^+(Y), \text{not att}_i(J,Z).$$

If we put $\text{ADM}_{\{\text{in}\}}^{bc}$ and all of these additional rules together to form the program IDEAL, we obtain the following result:

Theorem 2. *Let F be an AF and $A \in AS(\text{IDEAL} \cup D_F)$. Then, the ideal extension of F is given by $\{a \mid \text{ideal}(a) \in A\}$.*

6   Conclusion

In this paper, we provided a novel method to rewrite ASP-programs in such a way that reasoning over all answer sets of the original program can be formulated within the same program. Our method exploits the well-known concept of weak constraints. We illustrated the impact of our method by encoding the problems of (i) deciding whether a propositional formula in CNF has a unique minimal model, and (ii) computing the ideal extension of an argumentation framework. Known complexity results witness that our encodings are adequate in the sense that efficient ASP encodings without weak constraints or similar constructs are assumed to be infeasible.

The manifold program for cautious consequences is also closely related to the concept of data disjunctions [19] (this paper also contains a detailed discussion about the complexity class Θ_2^P and related classes for functional problems). Related work has also been done in the area of default logic, where a method for reasoning within a single extension has been proposed [20]. That method uses set-variables which characterize the set of generating defaults of the original extensions. Such an approach differs considerably from ours as it encodes certain aspects of the semantics (which ours does not), which puts it closer to meta-programming (cf. [21]).

As future work, we are interested in developing a suitable language for expressing reasoning with brave, cautious and definite consequences, allowing also for mixing different reasoning modes. This language should serve as a platform for natural encodings of problems in complexity classes Θ_2^P, Θ_3^P, $\mathrm{FP}_{||}^{\mathrm{NP}}$, and $\mathrm{FP}_{||}^{\Sigma_2^P}$. Moreover, we intend studying the use of alternative preferential constructs in place of weak constraints.

References

1. Marek, V.W., Truszczyński, M.: Stable models and an alternative logic programming paradigm. In: The Logic Programming Paradigm – A 25-Year Perspective, pp. 375–398 (1999)
2. Niemelä, I.: Logic programming with stable model semantics as a constraint programming paradigm. Ann. Math. Artif. Intell. 25(3-4), 241–273 (1999)
3. Baral, C.: Knowledge Representation, Reasoning and Declarative Problem Solving. CUP (2002)
4. Gelfond, M.: Representing knowledge in A-Prolog. In: Kakas, A.C., Sadri, F. (eds.) Computational Logic: Logic Programming and Beyond. LNCS (LNAI), vol. 2408, pp. 413–451. Springer, Heidelberg (2002)
5. Gebser, M., Liu, L., Namasivayam, G., Neumann, A., Schaub, T., Truszczyński, M.: The first answer set programming system competition. In: Baral, C., Brewka, G., Schlipf, J. (eds.) LPNMR 2007. LNCS (LNAI), vol. 4483, pp. 3–17. Springer, Heidelberg (2007)
6. Bravo, L., Bertossi, L.E.: Logic programs for consistently querying data integration systems. In: IJCAI 2003, pp. 10–15 (2003)
7. Saccà, D.: Multiple total stable models are definitely needed to solve unique solution problems. Inf. Process. Lett. 58(5), 249–254 (1996)
8. Buccafurri, F., Leone, N., Rullo, P.: Enhancing disjunctive datalog by constraints. IEEE Trans. Knowl. Data Eng. 12(5), 845–860 (2000)
9. Leone, N., Pfeifer, G., Faber, W., Eiter, T., Gottlob, G., Perri, S., Scarcello, F.: The dlv system for knowledge representation and reasoning. ACM Trans. Comput. Log. 7(3), 499–562 (2006)
10. Reiter, R.: On closed world data bases. In: Logic and Databases, pp. 55–76. Plenum Press (1978)
11. Bench-Capon, T.J.M., Dunne, P.E.: Argumentation in artificial intelligence. Artif. Intell. 171(10-15), 619–641 (2007)
12. Dung, P.M., Mancarella, P., Toni, F.: Computing ideal sceptical argumentation. Artif. Intell. 171(10-15), 642–674 (2007)
13. Gelfond, M., Lifschitz, V.: Classical negation in logic programs and disjunctive databases. New Generation Comput. 9(3/4), 365–386 (1991)
14. Dunne, P.E.: The computational complexity of ideal semantics I: Abstract argumentation frameworks. In: COMMA 2008, pp. 147–158. IOS Press, Amsterdam (2008)
15. Dung, P.M.: On the acceptability of arguments and its fundamental role in nonmonotonic reasoning, logic programming and n-person games. Artif. Intell. 77(2), 321–358 (1995)
16. Dunne, P.E.: Computational properties of argument systems satisfying graph-theoretic constraints. Artif. Intell. 171(10-15), 701–729 (2007)

17. Osorio, M., Zepeda, C., Nieves, J.C., Cortés, U.: Inferring acceptable arguments with answer set programming. In: ENC 2005, pp. 198–205 (2005)
18. Egly, U., Gaggl, S., Woltran, S.: Answer-set programming encodings for argumentation frameworks. In: Proceedings ASPOCP 2008 (2008)
19. Eiter, T., Veith, H.: On the complexity of data disjunctions. Theor. Comput. Sci. 288(1), 101–128 (2002)
20. Delgrande, J.P., Schaub, T.: Reasoning credulously and skeptically within a single extension. Journal of Applied Non-Classical Logics 12(2), 259–285 (2002)
21. Eiter, T., Faber, W., Leone, N., Pfeifer, G.: Computing preferred answer sets by meta-interpretation in answer set programming. TPLP 3(4-5), 463–498 (2003)

A Deductive System for FO(ID) Based on Least Fixpoint Logic*

Ping Hou and Marc Denecker

Department of Computer Science, Katholieke Universiteit Leuven, Belgium
{Ping.Hou,Marc.Denecker}@cs.kuleuven.be

Abstract. The logic FO(ID) uses ideas from the field of logic programming to extend first order logic with non-monotone inductive definitions. The goal of this paper is to extend Gentzen's sequent calculus to obtain a deductive inference method for FO(ID). The main difficulty in building such a proof system is the representation and inference of unfounded sets. It turns out that we can represent unfounded sets by least fixpoint expressions borrowed from stratified least fixpoint logic (SLFP), which is a logic with a least fixpoint operator and characterizes the expressibility of stratified logic programs. Therefore, in this paper, we integrate least fixpoint expressions into FO(ID) and define the logic FO(ID,SLFP). We investigate a sequent calculus for FO(ID,SLFP), which extends the sequent calculus for SLFP with inference rules for the inductive definitions of FO(ID). We show that this proof system is sound with respect to a slightly restricted fragment of FO(ID) and complete for a more restricted fragment of FO(ID).

1 Introduction

Inductive definitions are common in mathematical practice. For instance, the non-monotone inductive definition of the satisfaction relation $\models$ can be found in most textbooks on first order logic (FO). This prevalence of inductive definitions indicates that these offer a natural and well-understood way of representing knowledge. It is well-known that, in general, inductive definitions cannot be expressed in first order logic.

It turns out, however, that certain knowledge representation logics do allow a natural and uniform formalization of the most common forms of inductive definitions. The authors of [6,7] pointed out that semantical studies in the area of logic programming might contribute to a better understanding of such generalized forms of induction. In particular, it was argued that the well-founded semantics of logic programming [19] extends monotone induction and formalizes induction over well-founded sets and iterated induction. The language of FO(ID) uses the well-founded semantics to extend classical first order logic with a new "inductive definition" primitive. In the resulting formalism, all kinds of definitions regularly found in mathematical practice – e.g., monotone inductive

---

* This work is supported by FWO-Vlaanderen and by GOA/2003/08.

E. Erdem, F. Lin, and T. Schaub (Eds.): LPNMR 2009, LNCS 5753, pp. 129–141, 2009.

definitions, non-monotone inductive definitions over a well-ordered set, and iterated inductive definitions – can be represented in a uniform way. Moreover, this representation neatly coincides with the form such definitions would take in a mathematical text. For instance, in FO(ID) the transitive closure of a graph can be defined as:

$$\left\{ \begin{array}{c} \forall x, y \ TransCl(x, y) \leftarrow Edge(x, y) \\ \forall x, y \ TransCl(x, y) \leftarrow (\exists z TransCl(x, z) \wedge TransCl(z, y)) \end{array} \right\}.$$

However, FO(ID) is able to handle more than only mathematical concepts. Indeed, inductive definitions are also crucial in declarative Knowledge Representation. Not only non-inductive definitions are frequent in common-sense reasoning as argued in [2], also inductive definitions are. For instance, in [8], it was shown that situation calculus can be given a natural representation as an iterated inductive definition. The resulting theory is able to correctly handle tricky issues such as recursive ramifications, and is in fact, to the best of our knowledge, the most general representation of this calculus to date. It thus appears that FO(ID) has very strong links to several KR-paradigms.

As for every formal logical system, the development of deductive inference methods is an important research topic. For instance, it is well-known that deductive reasoning is a distinguished feature of Description Logics. However, because FO(ID) is not even semi-decidable, it is clear that a sound and complete proof system for FO(ID) cannot exist. As such, we will have to investigate deductive systems for FO(ID) and the subclasses of FO(ID) for which these systems are complete.

The initial motivation for the research of this paper is to extend Gentzen's sequent calculus **LK** [12,18] to obtain a proof system for FO(ID). We intended to build a sequent calculus for FO(ID) which is sound in general and complete for a useful subclass of FO(ID). The main challenge in building such a calculus is the representation and inference of unfounded sets [10]. In our approach to this problem, we represent unfounded sets by least fixpoint expressions borrowed from SLFP [5]. Compton described stratified least fixpoint logic (SLFP) in [5], which is a logic with a least fixpoint operator and characterizes the expressibility of stratified logic programs. He used sequent calculus to investigate deductive inference method for SLFP and proved the soundness and completeness of such a proof system. This work is an extension of the previous work in [14], which presents a sequent calculus for the propositional fragment of FO(ID), to the first order case.

In our deductive system, we use Compton's inference system for SLFP. Our strategy consists of two steps. The first is to integrate least fixpoint expressions into FO(ID), leading to the logic FO(ID,SLFP). In the logic FO(ID,SLFP), we then build a proof calculus, which extends Compton's sequent calculus for SLFP with inference rules for the inductive definitions of FO(ID). By restricting input to FO(ID), this proof calculus yields a proof system for FO(ID) then. However, in proofs of FO(ID) formulae, some inference rules may introduce least fixpoint expressions, and thus, proofs may contain expressions of FO(ID,SLFP).

The contributions of this paper can be summarized as follows: (a) We study a deductive inference method for FO(ID) based on stratified least fixpoint logic by defining the logic FO(ID,SLFP) and introducing a sequent calculus for FO(ID,SLFP). (b) We show that the deductive system is sound for a slightly restricted fragment of FO(ID,SLFP), where all definitions have to be total. Also, we investigate a more restricted fragment of FO(ID) and show the completeness result of the sequent calculus for FO(ID,SLFP) for this fragment.

This work is a step forward in the development of domain independent deductive reasoning for FO(ID). This form of inference is similar as that in Description Logics or classical logic but differs from Answer Set Programming which imposes Domain Closure. Our deductive calculus and the completeness result for a fragment of FO(ID), can be viewed as a step in developing tools similar as in the field of description logics. In a similar aim, a decidable guarded fragment of FO(ID) was described recently in [20].

The structure of this paper is as follows. We introduce FO(ID,SLFP) in Section 2. We present a deductive system for FO(ID,SLFP) in Section 3. The main results of the soundness and completeness of the deductive system are presented in Section 4. We finish with conclusions, related and future work.

2   Preliminaries

We start by defining the logic FO(ID,SLFP), which is the extension of FO with both inductive definitions and stratified least fixpoint expressions.

2.1   Syntax of FO(ID,SLFP)

In this subsection, we present the syntax of FO(ID,SLFP), which is an integration of FO(ID) and SLFP.

We assume familiarity with classical logic. A vocabulary Σ consists of Σ^c and Σ^v where Σ^c consists of (countable) predicate constants and function symbols while Σ^v is a (countable) set of predicate variables. Object symbols and propositional symbols are 0-ary function symbols, respectively predicate symbols. Terms and first order formulae (FO formulae) of Σ are defined as usual, and are built inductively from object symbols, function symbols, predicate symbols and logical connectives and quantifiers. A predicate symbol P has a negative (positive) occurrence in a formula F if P has an occurrence in the scope of an odd (even) number of occurrences of the negation symbol $\neg$.

First, we introduce the notation of a definition. A *definition* over Σ is a set of rules of the form $\forall \bar{x}(P(\bar{x}) \leftarrow \varphi)$, where $\bar{x}$ is a tuple of object variables over Σ^c, P is a predicate constant over Σ^c and φ is an FO formula over Σ^c such that all the free variables of φ occur in $\bar{x}$. We call $P(\bar{x})$ the *head* of the rule and φ the *body*. The connective $\leftarrow$ is called *definitional implication* and is to be distinguished from material implication $\supset$. A predicate appearing in the head of a rule of a definition D is called a *defined predicate* of D, any other symbol is called an *open symbol* of D. The set of defined predicates, respectively open symbols of D are denoted by Σ_D^d, respectively, $\Sigma_D^o = \Sigma \setminus \Sigma_D^d$. Notice that there are no predicate variables occurring in the definition D.

Now we are ready to define the formulae of FO(ID,SLFP).

Definition 1. *An FO(ID,SLFP) formula over Σ is defined by the following induction:*

- *If X is an n-ary predicate symbol (predicate constant or predicate variable) and $t_1, \ldots, t_n$ are terms then $X(t_1, \ldots, t_n)$ is a formula.*
- *If D is a definition then D is a formula.*
- *If ψ is a formula containing no free predicate variables, then $\neg \psi$ is a formula.*
- *If φ, ψ are formulae, then so are $\varphi \wedge \psi$ and $\varphi \vee \psi$.*
- *If ψ is a formula, then $\exists x \psi$ is a formula.*
- *If ψ is a formula containing no free predicate variables, then $\forall x \psi$ is a formula.*
- *If $\psi, \theta_1, \ldots, \theta_n$ are formulae containing no definitions, $X_1, \ldots, X_n$ are predicate variables, then $[LFP_{X_1(\bar{x}_1), \ldots, X_n(\bar{x}_n)}(\theta_1, \ldots, \theta_n)] \psi$ is a formula and called a* stratified least fixpoint expression.

Note that the subformulae $\psi, \theta_1, \ldots, \theta_n$ of a stratified least fixpoint expression $[LFP_{X_1(\bar{x}_1), \ldots, X_n(\bar{x}_n)}(\theta_1, \ldots, \theta_n)] \psi$ may not contain definitions, but may contain stratified least fixpoint expressions. Indeed, nesting of stratified least fixpoint expressions is allowed in FO(ID,SLFP), but nesting of definitions is not. All subformulae $\psi, \theta_1, \ldots, \theta_n$ of an unnested stratified least fixpoint expression contain only positive occurrences of predicate variables.

Definition 2. *Each occurrence of an object variable x in formulae $\exists x \varphi$ and $\forall x \varphi$ is* bound. *All occurrences of variables $\bar{x}_1, \ldots, \bar{x}_n$ in the subformulae $\theta_1, \ldots, \theta_n$ of a stratified least fixpoint expression $[LFP_{X_1(\bar{x}_1), \ldots, X_n(\bar{x}_n)}(\theta_1, \ldots, \theta_n)] \psi$ are bound. Each occurrence of a predicate variable X_i inside subformulae $\psi, \theta_1, \ldots, \theta_n$ of a stratified least fixpoint expression $[LFP_{X_1(\bar{x}_1), \ldots, X_n(\bar{x}_n)}(\theta_1, \ldots, \theta_n)] \psi$ is bound. All other occurrences of an object or predicate variable is* free. *Let us denote $free(\varphi)$ the set of object and predicate variables with a free occurrence in φ.*

An FO(ID,SLFP) sentence is a formula without free variables. An FO formula is an FO(ID,SLFP) formula without definitions and stratified least fixpoint expressions. An FO(ID) formula is an FO(ID,SLFP) formula without stratified least fixpoint expressions, and an SLFP formula is one without definitions. An FO(ID,SLFP) theory is a set of FO(ID,SLFP) sentences.

2.2   Semantics of FO(ID,SLFP)

The semantics of FO(ID,SLFP) is an integration of the semantics of FO(ID) and the semantics of SLFP.

The semantics of the FO(ID) is an integration of standard two-valued FO semantics with the well-founded semantics of definitions. For technical reasons, we introduce some concepts from three-valued logic. Consider the set of truth values $\{\mathbf{t}, \mathbf{f}, \mathbf{u}\}$. The *truth order* $\leq$ on this set is induced by $\mathbf{f} \leq \mathbf{u} \leq \mathbf{t}$; the *precision order* $\leq_p$ is induced by $\mathbf{u} \leq_p \mathbf{f}$ and $\mathbf{u} \leq_p \mathbf{t}$. Define $\mathbf{f}^{-1} = \mathbf{t}$, $\mathbf{u}^{-1} = \mathbf{u}$ and $\mathbf{t}^{-1} = \mathbf{f}$.

Given a domain $\mathcal{D}$, a *value* for an n-ary function symbol is a function from $\mathcal{D}^n$ to $\mathcal{D}$. A value for an n-ary predicate symbol is a function from $\mathcal{D}^n$ to $\{\mathbf{t}, \mathbf{f}, \mathbf{u}\}$. A Σ-interpretation I consists of a domain $\mathcal{D}^I$, and a value σ^I for each symbol $\sigma \in \Sigma$. A two-valued interpretation is one in which predicates have range $\{\mathbf{t}, \mathbf{f}\}$. For each interpretation F for the function symbols of Σ, both truth and precision order have a pointwize extension to an order on all Σ-interpretations extending F. A domain atom of I is a tuple of a predicate $P \in \Sigma$ and a tuple $(a_1, \ldots, a_n) \in \mathcal{D}^n$; it will be denoted by $P(a_1, \ldots, a_n)$, or more compactly, $P(\bar{a})$.

For a given Σ-interpretation I, symbol σ and a value v for σ, we denote by $I[\sigma/v]$ the $\Sigma \cup \{\sigma\}$-interpretation, that assigns to all symbols the same value as I, expect that $\sigma^{I[\sigma/v]} = v$. Likewise, for a domain atom $P(\bar{a})$ and a truth value $v \in \{\mathbf{t}, \mathbf{f}, \mathbf{u}\}$, we define $I[P(\bar{a})/v]$ as the interpretation I' identical to I except that $P(\bar{a})^{I'} = P^{I'}(\bar{a}) = v$. Similarly, for any set U of domain atoms, $I[U/v]$ is identical to I except that all atoms in U have value v. When $\Sigma' \subseteq \Sigma$, we denote the restriction of a Σ-interpretation I to the symbols of Σ' by $I|_{\Sigma'}$.

When all symbols of term t are interpreted in I, we define its value t^I using the standard induction. The truth value φ^I of an FO sentence φ in I is defined by the standard induction on the subformula order.

We now define the semantics of definitions. Firstly, we generalize the well-known concept of an unfounded set [19].

Definition 3. *Given a definition D and a three-valued Σ-interpretation I, an* unfounded set *of D in I is a non-empty set U of defined domain atoms such that each $P(\bar{a}) \in U$ is unknown in I and for each rule $\forall \bar{x}(P(\bar{x}) \leftarrow \varphi(\bar{x})) \in D$, $\varphi(\bar{a})^{I[U/\mathbf{f}]} = \mathbf{f}$.*

The set of all unfounded sets of D in I is denoted by $\mathcal{U}_D(I)$. The *maximal unfounded set* of D in I is defined by $\bigcup_{X \in \mathcal{U}_D(I)} X$.

Definition 4. *[10] We define a* well-founded induction *of a definition D extending a Σ_D^o-interpretation I_O as a sequence $(I^\xi)_{\xi \leq \alpha}$ of three-valued Σ-interpretations extending I_O such that:*

- *for every defined predicate symbol P, P^{I^0} is the constant function $\mathbf{u}$,*
- *for each limit ordinal $\lambda \leq \alpha$, $I^\lambda = lub_{\leq_p}(\{I^\xi \mid \xi < \lambda\})$, and*
- *for each ordinal ξ, $I^{\xi+1}$ relates to I^ξ in one of the following ways:*

  1. $I^{\xi+1} := I^\xi[P(\bar{a})/\mathbf{t}]$, *for some domain atom $P(\bar{a})$, unknown in I^ξ, such that for some rule $\forall \bar{x}(P(\bar{x}) \leftarrow \varphi(\bar{x})) \in D$, $\varphi(\bar{a})^{I^\xi} = \mathbf{t}$;*
  2. $I^{\xi+1} := I^\xi[U/\mathbf{f}]$, *where U is an unfounded set of D in I^ξ.*

A well-founded induction is terminal *if it can not be extended anymore.*

It can be shown that each well-founded induction of D extending I_O is strictly increasing in precision and the limit of every terminal well-founded induction of D extending I_O is the *well-founded model* of D [10].

For a given three-valued interpretation I and definition D, we define $D^I = \mathbf{t}$ if I is the well-founded model of D extending $I|_{\Sigma_D^o}$, and $D^I = \mathbf{f}$ otherwise. We are now ready to define the semantics of FO(ID). An interpretation I satisfies an FO(ID) formula φ (is a model of φ) if I is two-valued and $\varphi^I = \mathbf{t}$. As usual, this is denoted by $I \models \varphi$. I satisfies an FO(ID) theory T if I satisfies every $\varphi \in T$. Note that the semantics is two-valued. The restriction to consider only two-valued well-founded models boils down to the requirement that a definition D should be *total*, i.e., should define the truth of all defined domain atoms (see [9]).

Definition 5. *A definition D is total if for each Σ_D^o-interpretation I_O, the well-founded model of D extending I_O is two-valued.*

We now give the semantics of stratified least fixpoint expressions (see [5]). Let φ be the form of $[LFP_{X_1(\bar{x}_1),\ldots,X_n(\bar{x}_n)}(\theta_1,\ldots,\theta_n)]\psi$. Let α be an assignment to all variables in this expression other than $X_1,\ldots,X_n,\bar{x}_1,\ldots,\bar{x}_n$. A structure $\mathfrak{A}$ satisfies φ in the assignment α if $\mathfrak{A}$ satisfies ψ in α', where α' is identical to α except it assigns the least relations to every X_i ($1 \leq i \leq n$) satisfying the implications $\forall \bar{x}_1(X_1(\bar{x}_1) \supset \theta_1),\ldots,\forall \bar{x}_n(X_n(\bar{x}_n) \supset \theta_n)$.

3   The Deductive System for FO(ID,SLFP)

In this section, we present **LFO(ID,SLFP)**, a proof system for FO(ID,SLFP) based on Gentzen's sequent calculus **LK** for first order logic [12,18].

First, we introduce some basic definitions and notations. Let capital Greek letters $\Gamma, \Delta, \ldots$ denote (possibly empty) multisets of FO(ID,SLFP) formulae. Γ, Δ denotes $\Gamma \cup \Delta$ and Γ, φ denotes $\Gamma \cup \{\varphi\}$. By $\bigwedge \Gamma$, respectively $\bigvee \Gamma$, we denote the conjunction, respectively disjunction of all formulae in Γ. By $\neg \Gamma$, we denote the multiset obtained by taking the negation of each formula in Γ. By $\Gamma \setminus \Delta$, we denote the multiset obtained by deleting from Γ all occurrences of formulae that occur in Δ. Given an FO(ID,SLFP) formula φ, by $\varphi(x/t)$, we denote the formula obtained by substituting all free occurrences of the object variable x in φ by term t.

A *sequent* is an expression of the form $\Gamma \rightarrow \Delta$. Γ and Δ are respectively called the *antecedent* and *succedent* of the sequent and each formula in Γ and Δ is called a *sequent formula*. We will denote sequents by $S, S_1, \ldots$. A sequent $\Gamma \rightarrow \Delta$ is *valid*, denote by $\models \Gamma \rightarrow \Delta$, if every model of $\bigwedge \Gamma$ satisfies some formula in Δ. The sequent $\Gamma \rightarrow$ is equivalent to $\Gamma \rightarrow \bot$ and $\rightarrow \Delta$ is equivalent to $\top \rightarrow \Delta$, where $\bot, \top$ are logical constants denoting *false* and *true*, respectively.

An *inference rule* is an expression of the form $\dfrac{S_1; \ldots; S_n}{S}$ ($n \geq 0$). Each S_i is called a *premise* of the inference rule, S is called *consequence*. Intuitively, an inference rule means that S can be inferred, given that all $S_1,\ldots,S_n$ are already inferred.

The sequent calculus for SLFP, mentioned in [5], contains an infinitary inference rule for stratified least fixpoint expression with countably infinite number of premises. In [5], it is shown that SLFP is not compact. It follows that we must have some sort of infinitary rule in any complete sequent calculus for SLFP. Such infinitary rules are expressed in the form: $\dfrac{\Gamma_m \to \Delta_m \;\; (m \in \omega)}{\Gamma \to \Delta}$.

The *initial sequents*, or *axioms*, of **LFO(ID,SLFP)** are the sequents of the form

$$\varphi, \Gamma \to \Delta, \varphi \;\; \text{or} \;\; \bot \to \Delta \;\; \text{or} \;\; \Gamma \to \top \;\; \text{or} \;\; \Gamma \to \Delta, t = t$$

The inference rules for **LFO(ID,SLFP)** consists of the structural rules, logical rules, equality rules, the stratified least fixpoint rules and definition rules. We use the standard structural rules, logical rules and equality rules as given in many sources (see e.g. [4]).

The inference rules for the stratified least fixpoint expressions follow those in [5], which are of the form:

Stratified least fixpoint rules

left: $\dfrac{\Gamma, [LFP_{P_1(\bar{x}_1),\ldots,P_n(\bar{x}_n)}(\theta_1,\ldots,\theta_n)]_m \psi \to \Delta \;\; (m \in \omega)}{\Gamma, [LFP_{P_1(\bar{x}_1),\ldots,P_n(\bar{x}_n)}(\theta_1,\ldots,\theta_n)]\psi \to \Delta}$;

right: $\dfrac{\Gamma \to \Delta, [LFP_{P_1(\bar{x}_1),\ldots,P_n(\bar{x}_n)}(\theta_1,\ldots,\theta_n)]_m \psi}{\Gamma \to \Delta, [LFP_{P_1(\bar{x}_1),\ldots,P_n(\bar{x}_n)}(\theta_1,\ldots,\theta_n)]\psi}$

Notice that the left stratified least fixpoint rule is infinitary: it has countably many premises while in the right stratified least fixpoint rule there is just one premise: m is a fixed nonnegative integer.

To these rules we add the *definition rules* of **LFO(ID,SLFP)**. The definition rules consist of the *right definition rule*, the *left definition rule* and the *definition introduction rule*. The left definition rule, respectively, the right definition rule, can introduce the defined predicates in the antecedents, respectively, in the succedents, of the sequents. The definition introduction rule can introduce the total definitions in the succedents of the sequents. Without loss of generality, we assume from now on that there is only one rule with the head $P(\bar{x})$ in a definition D for every $P \in \Sigma_D^d$. We refer to this rule as *the rule for P in D* and denote it by $\forall \bar{x}(P(\bar{x}) \leftarrow \varphi_P(\bar{x}))$. Indeed, any set of rules $\{\forall \bar{x}(P(\bar{x}) \leftarrow \varphi_1),\ldots,\forall \bar{x}(P(\bar{x}) \leftarrow \varphi_n)\}$ can be transformed into a single rule $\forall \bar{x}(P(\bar{x}) \leftarrow \varphi_1 \vee \ldots \vee \varphi_n)$.

The right definition rule allows inferring the truth of a defined atomic formula from a definition D. It is closely related to the step (1) of Definition 4. Let D be a definition and P a defined predicate of D. Then the right definition rule for P is given as follows.

Right definition rule for P $\dfrac{\Gamma \to \Delta, \varphi_P(\bar{t})}{D, \Gamma \to \Delta, P(\bar{t})}$.

Example 1. Consider the definition D of the transitive closure of a directed graph G as $\{ \forall x, y(T(x,y) \leftarrow G(x,y) \vee \exists z(T(x,z) \wedge T(z,y))) \}$, $\Gamma = \{G(a,b), G(b,c)\}$ and Δ an empty set. The instance of the right definition rule for $T(a,c)$ is

$$\frac{G(a,b), G(b,c) \rightarrow G(a,c) \vee \exists z(T(a,z) \wedge T(z,c))}{D, G(a,b), G(b,c) \rightarrow T(a,c)}.$$

The left definition rule allows inferring the falsity of defined atoms from a definition D and is therefore related to step (2) of Definition 4. We first introduce some notations. Given a set U of predicate symbols, let U' be a set consisting of new predicate symbol P' for every $P \in U$. The vocabulary Σ augmented with these new symbols is denoted by Σ'. Given an FO formula φ, φ' denotes the formula obtained by replacing all positive occurrences of every $P \in U$ in φ by P'. We call φ' the *renaming* of φ with respect to U. By $\neg\varphi'$, we mean $\neg(\varphi')$.

Let D be a definition and $U = \{P_1, \ldots, P_n\}$ a non-empty set of defined predicates of D. Let $\psi_1(\bar{x^1}), \ldots, \psi_n(\bar{x^n})$ be n arbitrary FO formulae over Σ or stratified least fixpoint expressions over Σ with $\bar{x^i}$ as many fresh variables as the arity of $P_i(\bar{x^i})$. Then the left definition rule for $(\{P_1(\bar{x^1}) \mid \psi_1(\bar{x^1})\}, \ldots, \{P_n(\bar{x^n}) \mid \psi_n(\bar{x^n})\})$ is given as follows.

Left definition rule for $\{P_i(\bar{x^i}) \mid \psi_i(\bar{x^i})\}_{i \in [1,n]}$

$$\Gamma, \forall \bar{x^1}(\psi_1(\bar{x^1}) \supset \neg P_1'(\bar{x^1})), \ldots, \forall \bar{x^n}(\psi_n(\bar{x^n}) \supset \neg P_n'(\bar{x^n})) \rightarrow \Delta, \forall \bar{x^1}(\psi_1(\bar{x^1}) \supset \neg\varphi_{P_1}'(\bar{x^1}))$$

$$\vdots$$

$$\frac{\Gamma, \forall \bar{x^1}(\psi_1(\bar{x^1}) \supset \neg P_1'(\bar{x^1})), \ldots, \forall \bar{x^n}(\psi_n(\bar{x^n}) \supset \neg P_n'(\bar{x^n})) \rightarrow \Delta, \forall \bar{x^n}(\psi_n(\bar{x^n}) \supset \neg\varphi_{P_n}'(\bar{x^n}))}{D, \Gamma, \exists \bar{x^i}(\psi_i(\bar{x^i}) \wedge P_i(\bar{x^i})) \rightarrow \Delta}$$

where Γ and Δ are multisets of FO(ID,SLFP) formulae over Σ and $\varphi_{P_i}'(\bar{x^i})$ is the renaming of $\varphi_{P_i}(\bar{x^i})$ with respect to U.

Actually, in the left definition rule, the set of atoms $\{P_1(\bar{x^1}) \mid \psi_1(\bar{x^1})\} \cup \ldots \cup \{P_n(\bar{x^n}) \mid \psi_n(\bar{x^n})\}$ is a symbolic representation of a candidate unfounded set of D.

Example 2. Given the definition D of the transitive closure of a directed graph G as $\{ \forall x, y(T(x,y) \leftarrow G(x,y) \vee \exists z(T(x,z) \wedge T(z,y))) \}$, $\Gamma = \{\neg \exists x G(x,a)\}$ and Δ an empty set. The instance of the left definition rule for $(\{T(x,y) \mid y = a\})$ is as follows:

$$\frac{\neg \exists x G(x,a), \forall x, y(y = a \supset \neg T'(x,y)) \rightarrow \quad}{D, \neg \exists x G(x,a), \exists x, y(y = a \wedge T(x,y)) \rightarrow}.$$
$$\forall x, y(y = a \supset \neg G(x,y) \wedge \neg \exists z(T'(x,z) \wedge T'(z,y)))$$

Notice that the set $\{T(x,y) \mid y = a\}$ is the symbolic representation of a unfounded set of D here.

The definition introduction rule allows inferring the truth of a total definition from FO(ID,SLFP) formulae. We introduce some notations. Let D be a total definition. Denote by P^R a new defined predicate for each $P \in \Sigma_D^d$. Denote by Σ^R the vocabulary $\Sigma \cup \{P^R \mid P \in \Sigma_D^d\}$. Denote by D^R the definition over the new vocabulary Σ^R obtained by replacing each occurrence of each defined predicate P in D by P^R. Let Γ and Δ be multisets of FO(ID,SLFP) formulae over Σ. The definition introduction rule for D is given as follows.

Definition introduction rule

$$\frac{\Gamma, D^R \to \forall \bar{x^1}(P_1^R(\bar{x^1}) \equiv P_1(\bar{x^1})), \Delta; \ldots; \Gamma, D^R \to \forall \bar{x^n}(P_n^R(\bar{x^n}) \equiv P_n(\bar{x^n})), \Delta}{\Gamma \to \Delta, D},$$

where $P_1, \ldots, P_n$ are all defined predicates of D.

Example 3. Given a definition $D = \{ \forall x(P(x) \leftarrow P(x) \land O(x)) \}$, $\Gamma = \{\forall x O(x),$ $\forall x \neg P(x)\}$ and Δ an empty set. The instance of the definition introduction rule for D is as follows:

$$\frac{D^R, \forall x O(x), \forall \neg P(x) \to \forall x(P^R(x) \equiv P(x))}{\forall x O(x), \forall x \neg P(x) \to D},$$

where $D^R = \{ \forall x(P^R(x) \leftarrow P^R(x) \land O(x)) \}$.

It is necessary to emphasize that the definition D in the definition introduction rule is required to be total. We will give an example to show that this inference rule is not sound if D is not total. Actually, there are practically important syntactic classes of definitions which are known to be total, including positive definitions, stratified definitions and definitions over a well-founded order (see [9]).

We now come to the notion of an **LFO(ID,SLFP)**-*proof* for a sequent.

Definition 6. *A* proof *in* **LFO(ID,SLFP)** *or* **LFO(ID,SLFP)**-proof *for a sequent S, is a tree T of sequents with root S. Moreover, each leaf of T must be an axiom and for each interior node S' there exists an inference rule such that S' is the consequence of that inference rule while the children of S' are precisely the premises of that inference rule. T is called a* proof tree *for S. A sequent S is called* provable *in* **LFO(ID,SLFP)**, *or* **LFO(ID,SLFP)**-provable, *if there is an* **LFO(ID,SLFP)**-*proof for it.*

4   Main Results

In this section, we will present that the deductive system **LFO(ID,SLFP)** is sound for a slightly restricted fragment of FO(ID,SLFP) and it is complete with respect to a more restricted fragment of FO(ID,SLFP).

It is trivial to verify that all axioms of **LFO(ID,SLFP)** are valid and that the structural and logical rules and the rules for equality are sound. The soundness of the stratified least fixpoint rules can be shown analogously to that in [5]. Hence, only the soundness of the definition rules of **LFO(ID,SLFP)** must be presented.

Lemma 1 (Soundness of the right definition rule). *If $\models \Gamma \to \Delta, \varphi_P(\bar{t})$ then $\models D, \Gamma \to \Delta, P(\bar{t})$.*

Lemma 2 (Soundness of the left definition rule). *If* $\models \Gamma, \forall \bar{x}^1(\psi(\bar{x}^1) \supset \neg P'_1(\bar{x}^1)), \ldots, \forall \bar{x}^n(\psi_n(\bar{x}^n) \supset \neg P'_n(\bar{x}^n)) \to \Delta, \forall \bar{x}^i(\psi_i(\bar{x}^i) \supset \neg \varphi'_{P_i}(\bar{x}^i))$ *for every* $i \in [1, n]$, *then* $\models D, \Gamma, \exists \bar{x}^i(\psi_i(\bar{x}^i) \wedge P_i(\bar{x}^i)) \to \Delta$ *for all* $i \in [1, n]$.

Lemma 3 (Soundness of the definition introduction rule). *Let* D *be a total definition. If* $\models D^R, \Gamma \to \Delta, \forall \bar{x}(P(\bar{x}) \equiv P^R(\bar{x}))$ *for every* $P \in \Sigma_D^d$, *then* $\models \Gamma \to \Delta, D$.

Note that the definition introduction rule is not sound if D is not total. We illustrate it with an example.

Example 4. Given a definition $D = \{ \forall x(P(x) \leftarrow \neg P(x)) \}$ and Γ an empty set. It is obvious that D^R is non-total. Thus, $\models D^R \to \forall x(P^R(x) \equiv P(x))$ but $\not\models \to D$, which shows that the definition introduction rule is not sound when D is non-total.

By the fact that all inference rules in **LFO(ID,SLFP)** are sound if all definitions occurring in them are total and a straightforward induction, the soundness of **LFO(ID,SLFP)** can now be obtained.

Theorem 1 (Soundness of LFO(ID,SLFP)). *If a sequent of* $\Gamma \to \Delta$ *is provable in* **LFO(ID,SLFP)** *and all definitions in* Γ *and* Δ *are total, then* $\models \Gamma \to \Delta$.

In the following, we present the completeness property of **LFO(ID,SLFP)** for a fragment of FO(ID,SLFP). First, we define a special class of definitions.

Definition 7. *A definition* D *is an* S-*definition if the body of each rule in* D *satisfies the following conditions:*

- *it is in negation normal form (NNF).*
- *all occurrences of defined predicates in its subformula* $\exists x \varphi$ *are positive.*
- *all occurrences of defined predicates in its subformula* $\forall x \varphi$ *are negative.*
- *all occurrences of open predicates are in arbitrary ways.*

Example 5. $D = \{ \forall x, y(T(x, y) \leftarrow G(x, y) \vee \exists z(T(x, z) \wedge T(z, y))) \}$ is an S-definition. $D = \left\{ \begin{array}{l} \forall x, y(T(x, y) \leftarrow G(x, y) \vee \exists z(Q(x, z) \wedge \neg Q(z, y))) \\ \forall x, y(T(x, y) \leftarrow \neg O(x, y) \vee \forall z(\neg T(x, z) \vee T(z, y))) \end{array} \right\}$ is not an S-definition because there is a positive occurrence of T in $\forall z(\neg T(x, z) \vee T(z, y))$ while there is a negative occurrence of Q in $\exists z(Q(x, z) \wedge \neg Q(z, y))$.

It is obvious that some familiar types of definitions such as non-recursive definitions and positive definitions [9] are S-definitions. However, not all S-definitions are total. For example, $\forall x(P(x) \leftarrow \neg P(x))$ is an S-definition but it is not total.

An *SFO(ID,SLFP) formula* is an FO(ID,SLFP) formula with the restriction that all definitions occurring in it are S-definitions. SFO(ID,SLFP) is the complete fragment of FO(ID,SLFP) with respect to **LFO(ID,SLFP)**, as shown later.

The maximal unfounded set of an S-definition can be represented by the negation of stratified least fixpoint expressions, which is an important property of $\mathcal{S}$-definitions and will be applied in the completeness proof. It is demonstrated by the following proposition, where the notation of φ'_{P_i} is same as that in the left definition rule and $\varphi'_{P_i}(P'_1/X_1, \ldots, P'_m/X_m)$ denotes the formula obtained by replacing every occurrence of every P'_j ($j \in [1, n]$) in φ'_{P_i} by X_j.

Proposition 1. *Let D be an $\mathcal{S}$-definition and $(I^\xi)_{\xi \leq \alpha}$ a well-founded induction of D. Let $I^{\xi+1} = I^\xi[U/\mathbf{f}]$ where U is the maximal unfounded set of D in I^ξ. Suppose that $\{P_1, \ldots, P_m\}$ is the set of all defined predicates occurring in U. Then $\{P_1(\bar{x}^1) \mid$*

$\neg([LFP_{X_1(\bar{x}^1), \ldots, X_m(\bar{x}^m)}(\varphi'_{P_1}(P'_1/X_1, \ldots, P'_m/X_m), \ldots, \varphi'_{P_m}(P'_1/X_1, \ldots, P'_m/X_m))]X_1(\bar{x}^1))\}$
$\cup \ldots \cup \{P_m(\bar{x}^1) \mid$
$\neg([LFP_{X_1(\bar{x}^1), \ldots, X_m(\bar{x}^m)}(\varphi'_{P_1}(P'_1/X_1, \ldots, P'_m/X_m), \ldots, \varphi'_{P_m}(P'_1/X_1, \ldots, P'_m/X_m))]X_m(\bar{x}^m))\}$

is a symbolic representation of U.

We now turn to show the completeness of **LFO(ID,SLFP)** for $\mathcal{S}$FO(ID,SLFP). The proof is an extension of the direct style of completeness proof for Gentzen's **LK** as given in e.g. [4]. The structure of the completeness proof is then roughly as follows:

1. Assume that $\Gamma \to \Delta$ is not provable, construct from $\Gamma \to \Delta$ a *limit sequent* $\Gamma_\omega \to \Delta_\omega$, where Γ_ω and Δ_ω are infinite sets, such that no finite subsequent of $\Gamma_\omega \to \Delta_\omega$ is provable by using a uniform proof-search procedure, namely *schedule*, on which every formula of FO(ID,SLFP) appears infinitely often.
2. Define an equivalence relation $\sim$ on the terms of Σ that essentially factors out the equality formulae appearing in Γ_ω and use $\Gamma_\omega \to \Delta_\omega$ to construct a partial Σ-interpretation I_ω that interprets all symbols in Σ except all defined predicates.
3. Prove that $\Gamma_\omega \setminus \Pi_\omega \to \Delta_\omega \setminus \Upsilon_\omega$ is false in I_ω, where Π_ω is the multiset of all definitions appearing in Γ_ω and Υ_ω is the multiset of all definitions appearing in Δ_ω.
4. Use I_ω to construct a model M_ω and prove that M_ω is the well-founded model for every definition appearing in Γ_ω and M_ω is not the well-founded model for every definition appearing in Δ_ω.
5. It now follows from step 3 and step 4 that every finite subsequent of $\Gamma_\omega \to \Delta_\omega$ is false in M_ω, including $\Gamma \to \Delta$, so $\Gamma \to \Delta$ is not valid.

Steps 1,2,3 and the analogous version of step 5 also appears in standard completeness proof for first order logic. However, in our proof, the construction of the limit sequent in step 1 must take into account of the stratified least fixpoint rules. Applications of stratified least fixpoint rules must also be accounted for in step 3. The new work in our proof goes into establishing that M_ω defined in step 4 is indeed a well-founded model for every definition appearing in Γ_ω but M_ω is not a well-founded model for every definition appearing in Δ_ω, where the Proposition 1 is applied.

Theorem 2 (Completeness of $\mathcal{S}$FO(ID,SLFP)). *Let Γ, Δ be multisets of $\mathcal{S}FO(ID,SLFP)$ formulae. If $\Gamma \to \Delta$ is valid and all definitions occurring in Γ and Δ are total, then $\Gamma \to \Delta$ is provable in **LFO(ID,SLFP)**.*

5 Conclusions, Related and Future Work

We present a deductive system for FO(ID) based on least fixpoint logic by introducing the logic FO(ID,SLFP) and extending Gentzen's sequent calculus for first order logic to the proof system **LFO(ID,SLFP)** for FO(ID,SLFP). The main technical results are the soundness theorem of **LFO(ID,SLFP)** for a slightly restricted fragment of FO(ID) and the completeness theorem of **LFO(ID,SLFP)** for a more restricted fragment of FO(ID).

Related work is provided by Brotherston in [3]. He introduced the language FOL_{ID} of first order logic with the schema for inductive definitions, which is based upon Martin-Löf's "ordinary production" [17] and developed a proof system which is sound and complete with respect to a standard model in FOL_{ID}. A similar work to Brotherston's is studied by Hagiya and Sakurai in [13]. They proposed to interpret a (stratified) logic program as iterated inductive definitions of Martin-Löf and developed a proof theory which is sound with respect to the perfect model, and hence, the well-founded semantics of logic programming. Actually, both the FOL_{ID} and the stratified logic programs as iterative definitions can be generalized in FO(ID). A formal proof system based on tableau methods for analyzing computation for Answer Set Programming (ASP) was given as well by Gebser and Schaub [11]. As shown in [15], ASP is closely related to FO(ID). Their approach furnishes declarative and fine-grained instruments for characterizing operations as well as strategies of ASP-solvers and provides a uniform proof-theoretic framework for analyzing and comparing different algorithms.

The first topic for future work is the development of tools to check the correctness of the outputs generated by FO(ID) model generators such as MiniSat(ID) [16]. Given an FO(ID) theory T and a finite domain D as input, a model generator outputs a model for T with domain D or concludes that T is unsatisfiable in D. In the former case, an independent *model checker* can be used to check whether the output is indeed a model of T. However, when the model generator concludes that T is unsatisfiable in D, it is less obvious how to check the correctness of this answer. A similar problem arises in SAT, where certain solvers can output a trace of a failed computation in the form of a resolution proof [21]. An independent *proof checker* can then be used to check this formal proof. For the future, we aim to study how to transform a trace of the computation of a model generator such as MiniSat(ID) into a proof of (a propositional version of) our proof system. Model and proof checkers can be a great help to detect bugs in model generators. Also we intend to develop semi-automated and automated theorem proving systems for FO(ID) from the current deductive system for FO(ID) based on the least fixpoint logic, which may have more applications and implementations.

References

1. Baral, C., Brewka, G., Schlipf, J. (eds.): LPNMR 2007. LNCS (LNAI), vol. 4483. Springer, Heidelberg (2007)
2. Brachman, R.J., Levesque, H.J.: Competence in knowledge representation. In: National Conference on Artificial Intelligence (AAAI 1982), pp. 189–192 (1982)

3. Brotherston, J.: Sequent Calculus Proof System for Inductive Definitions. PhD thesis, University of Edinburgh (2006)
4. Buss, S.R.: Handbook of Proof Theory. Elsevier, Amsterdam (1998)
5. Compton, K.J.: Stratified least fixpoint logic. Theoretical Computer Science 131(1), 95–120 (1994)
6. Denecker, M.: The well-founded semantics is the principle of inductive definition. In: Dix, J., Fariñas del Cerro, L., Furbach, U. (eds.) JELIA 1998. LNCS (LNAI), vol. 1489, pp. 1–16. Springer, Heidelberg (1998)
7. Denecker, M., Bruynooghe, M., Marek, V.: Logic programming revisited: Logic programs as inductive definitions. ACM Transactions on Computational Logic (TOCL) 2(4), 623–654 (2001)
8. Denecker, M., Ternovska, E.: Inductive Situation Calculus. In: Ninth International Conference on Principles of Knowledge Representation and Reasoning (KR 2004), pp. 545–553 (2004)
9. Denecker, M., Ternovska, E.: A logic of non-monotone inductive definitions. ACM Transactions On Computational Logic (TOCL) 9(2) (March 2008)
10. Denecker, M., Vennekens, J.: Well-founded semantics and the algebraic theory of non-monotone inductive definitions. In: Baral et al [1], pp. 84–96
11. Gebser, M., Schaub, T.: Tableau calculi for answer set programming. In: Etalle, S., Truszczyński, M. (eds.) ICLP 2006. LNCS, vol. 4079, pp. 11–25. Springer, Heidelberg (2006)
12. Gentzen, G.: Untersuchungen über das logische schließen. Mathematische Zeitschrift 39, 176–210, 405–431 (1935)
13. Hagiya, M., Sakurai, T.: Foundation of Logic Programming Based on Inductive Definition. New Generation Computing 2, 59–77 (1984)
14. Hou, P., Wittocx, J., Denecker, M.: A deductive system for PC(ID). In: Baral et al [1], pp. 162–174
15. Mariën, M., Gilis, D., Denecker, M.: On the relation between ID-Logic and Answer Set Programming. In: Alferes, J.J., Leite, J. (eds.) JELIA 2004. LNCS (LNAI), vol. 3229, pp. 108–120. Springer, Heidelberg (2004)
16. Mariën, M., Wittocx, J., Denecker, M., Bruynooghe, M.: SAT(ID): Satisfiability of propositional logic extended with inductive definitions. In: Kleine Büning, H., Zhao, X. (eds.) SAT 2008. LNCS, vol. 4996, pp. 211–224. Springer, Heidelberg (2008)
17. Martin-Löf, P.: Hauptsatz for the intuitionistic theory of iterated inductive definitions. In: Fenstad, J.e. (ed.) Second Scandinavian Logic Symposium, pp. 179–216 (1971)
18. Szabo, M.E. (ed.): The Collected Papers of Gerhard Gentzen. North-Holland Publishing Co., Amsterdam (1969)
19. Van Gelder, A., Ross, K.A., Schlipf, J.S.: The well-founded semantics for general logic programs. Journal of the ACM 38(3), 620–650 (1991)
20. Vennekens, J., Denecker, M.: FO(ID) as an extension of DL with rules. In: The 6th Annual European Semantic Web Conference (accepted, 2009)
21. Zhang, L., Malik, S.: Validating sat solvers using an independent resolution-based checker: Practical implementations and other applications. In: 2003 Design, Automation and Test in Europe Conference and Exposition (DATE 2003), Munich, Germany, March 3-7, pp. 10880–10885. IEEE Computer Society, Los Alamitos (2003)

Computing Stable Models via Reductions to Difference Logic⋆

Tomi Janhunen, Ilkka Niemelä, and Mark Sevalnev

Helsinki University of Technology TKK
Department of Information and Computer Science
P.O. Box 5400, FI-02015 TKK, Finland
{Tomi.Janhunen,Ilkka.Niemela}@tkk.fi, msevalne@tcs.hut.fi

Abstract. Propositional satisfiability (SAT) solvers provide a promising computational platform for logic programs under the stable model semantics. However, computing stable models of a logic program using a SAT solver presumes translating the program into a set of clauses which is the input form accepted by most SAT solvers. This leads to fairly complex super-linear translations. There are, however, interesting extensions to plain clausal propositional representations such as difference logic. A number of solvers have been developed for difference logic, in particular in the context of the satisfiability modulo theories (SMT) framework, and the goal of the paper is to study whether such engines could be harnessed to the computation of stable models for logic programs in an effective way. To this end, we provide succinct translations from logic programs to theories of difference logic and evaluate the potential of SMT solvers in the computation of stable models using these translations and a selection of benchmarks.

1 Introduction

Normal logic programs, or just *normal programs* for short, form a simple yet attractive formalism for knowledge representation under the *stable model* semantics [1] and the basis for solving search problems according to the *answer set programming* (ASP) paradigm [2,3,4]. The fragment of normal logic programs is comprehensively supported by the current ASP systems, such as SMODELS [5], DLV [6], and CLASP [7] which are otherwise different when it comes to language extensions.

In addition to the native ASP systems listed above, a number of other systems have been developed to exploit SAT solvers in the computation of stable models. The idea is, of course, to benefit from the rapid development of SAT solvers. The basic approach comprises of (i) computing Clark's completion $\mathrm{Comp}(P)$ [8] for an input program P, (ii) transforming $\mathrm{Comp}(P)$ into conjunctive normal form, and (iii) invoking a SAT solver to find a satisfying assignment for $\mathrm{Comp}(P)$. However, such a procedure is correct for only certain subclasses of normal programs, out of which *tight* [9] logic programs are perhaps most well-known. For a tight program P, the sets of stable models $\mathrm{SM}(P)$, *supported models* $\mathrm{SuppM}(P)$, and *classical models* of $\mathrm{Comp}(P)$ coincide.[1]

⋆ A preliminary version of this paper was presented at the Second International Workshop on Logic and Search (LaSh 2008). This research has been partially funded by the Academy of Finland under project #122399.

[1] To this end, the completion $\mathrm{Comp}(P)$ must be formed without new atoms.

E. Erdem, F. Lin, and T. Schaub (Eds.): LPNMR 2009, LNCS 5753, pp. 142–154, 2009.

To deal with non-tight programs further constraints become necessary in addition to Comp(P). One possibility is to introduce *loop formulas* [10] on the fly to exclude models of the completion which are not stable. As witnessed by the ASSAT system [10] this strategy can be highly efficient. A drawback of the approach is that, at worst, the number of loop formulas introduced by the solver can become exponential in program length (denoted $\|P\|$). Although the number of loop formulas often stays reasonably low, e.g., when computing just one stable model, they can become a bottleneck when finding all models—or proving their non-existence—is of interest. There is a follow-up system CMODELS [11] which implements an improved *ASP-SAT algorithm* detailed in [12]. As demonstrated therein, there are logic programs for which the number of supported models is exponentially higher than that of stable models. Therefore, extra bookkeeping is required in order to avoid repetitive disqualification of supported but unstable models. An analogous problem is encountered when the task is to compute all stable models for a logic program having an exponential number of stable models and the recomputation of models should be avoided. To reduce space consumption when searching for stable models, the implementation of ASP-SAT algorithm in CMODELS attempts to identify and record small reasons for disqualifying models.

The translations from normal programs into propositional theories [13,14,15] provide a way to circumvent the worst-case behavior of the strategy based on loop formulas. However, the translation from [13] does not yield a one-to-one correspondence of models and the one from [14] is quadratic which makes it infeasible for larger program instances. The latest proposal [15] exploits a characterization of stable models in terms of *level numberings*—which lead to a translation with the size of the order of $\|P\| \times \log_2 |\mathrm{At}(P)|$ where $\mathrm{At}(P)$ is the set of atoms appearing in P. The implementation, the LP2SAT system, compared favorably with CMODELS especially when used with RELSAT to compute all stable models for logic programs [15].

Recently, it has been shown that stable models of normal programs can be captured in an extension of propositional logic, called difference logic, using an approach [16] which is quite close to the level numbering technique in [15]. Difference logic [17] can be seen as an instance of the *satisfiability modulo theories* (SMT) framework [18,19] where propositional logic is extended by allowing simple linear constraints of the form $x_i + k \geq x_j$ where x_i, x_j are integer variables and k is an integer constant. Such linear constraints enable to express differences $x_j - x_i \leq k$ which are useful in various applications. Moreover, efficient implementation techniques for difference logic are emerging in the SMT framework [17,20].

The idea of the paper is to develop effective compact translations from logic programs to difference logic based on results in [15,16]. The translations reduce stable model computation to finding satisfying interpretations of difference logic theories. Hence, SAT solvers for difference logic can be used for computing stable models without any modifications. This provides a novel way of implementing stable model computation where it is possible to exploit directly the rapidly improving SMT solver technology. Moreover, the translations offer an interesting method for generating challenging benchmarks for these solvers and can help to boost their development even further. We provide some experimental results on the effectiveness of the approach using off-the-shelf solvers for difference logic. The results suggest that this approach can potentially lead to a very competitive technique for stable model computation.

The rest of the paper is structured as follows. In the next section we explain briefly the necessary basics of normal logic programs and difference logic. In Section 3 the translations from logic programs to difference logic are explained. Section 4 provides an experimental evaluation of the translations compared to state-of-the-art techniques for computing stable models. The paper is concluded in Section 5.

2   Preliminaries

A propositional *normal program* P is a set of *rules* of the form

$$a \leftarrow b_1, \dots, b_n, \sim c_1, \dots, \sim c_m \qquad (1)$$

where a, $b_1, \dots, b_n$, and $c_1, \dots, c_m$ are propositional atoms, and $\sim$ denotes *default negation*. In the sequel, propositional atoms are called just *atoms* for short. Intuitively speaking, a rule r of the form (1) is applied as follows: the *head atom* $\mathrm{H}(r) = a$ can be inferred by r if the *positive body atoms* in $\mathrm{B}^+(r) = \{b_1, \dots, b_n\}$ are inferable by the rules of the program, but not the *negative body atoms* in $\mathrm{B}^-(r) = \{c_1, \dots, c_m\}$. We write $\mathrm{B}(r)$ for the entire body $\mathrm{B}^+(r) \cup \{\sim c \mid c \in \mathrm{B}^-(r)\}$ of r. In addition, the positive part r^+ of a rule r is defined as $\mathrm{H}(r) \leftarrow \mathrm{B}^+(r)$ hence omitting negative body conditions. A normal program P is called *positive*, if $r = r^+$ holds for every rule $r \in P$.

To define the semantics of normal programs, we write $\mathrm{At}(P)$ for the set of atoms that appear in a program P. An *interpretation* $I \subseteq \mathrm{At}(P)$ of P determines which atoms $a \in \mathrm{At}(P)$ are *true* ($a \in I$) and which atoms are *false* ($a \in \mathrm{At}(P) \setminus I$). A rule r is satisfied in I, denoted by $I \models r$, iff $I \models \mathrm{H}(r)$ is implied by $I \models \mathrm{B}(r)$ where $\sim$ is treated classically, i.e., $I \models \sim c_i$ iff $I \not\models c_i$. An interpretation I is a (*classical*) *model* of P, denoted $I \models P$, iff $I \models r$ for each $r \in P$. A model $M \models P$ is a *minimal model* of P iff there is no model $M' \models P$ such that $M' \subset M$. In particular, every positive normal program P has a unique minimal model, i.e., the *least model* $\mathrm{LM}(P)$ of P.

The least model semantics can be extended to cover an arbitrary normal program P [1] by reducing P into a positive program $P^M = \{r^+ \mid r \in P \text{ and } M \cap \mathrm{B}^-(r) = \emptyset\}$ with respect to any *model candidate* $M \subseteq \mathrm{At}(P)$. In particular, an interpretation $M \subseteq \mathrm{At}(P)$ is a *stable model* of P iff $M = \mathrm{LM}(P^M)$. The number of stable models can vary in general and we write $\mathrm{SM}(P)$ for the set of stable models of P.

The stable model semantics was preceded by an alternative semantics, namely the one based on *supported models* [21]. A classical model M of a normal program P is a supported model of P iff for every atom $a \in M$ there is a rule $r \in P$ such that $\mathrm{H}(r) = a$ and $M \models \mathrm{B}(r)$. Inspired by this idea, we define for any program P and $I \subseteq \mathrm{At}(P)$, the set of *supporting rules* $\mathrm{SuppR}(P, I) = \{r \in P \mid I \models \mathrm{B}(r)\}$. As shown in [22], stable models are also supported models, but not necessarily vice versa.

The *positive dependency graph* $\mathrm{DG}^+(P)$ of P is a pair $\langle \mathrm{At}(P), \leq \rangle$ where $b \leq a$ iff there is a rule $r \in P$ such that $\mathrm{H}(r) = a$ and $b \in \mathrm{B}^+(r)$. A *strongly connected component* (SCC) of $\mathrm{DG}^+(P)$ is a non-empty and maximal subset $S \subseteq \mathrm{At}(P)$ such that $a \leq^* b$ and $b \leq^* a$ hold for each $a, b \in S$ and the *reflexive* and *transitive* closure $\leq^*$ of $\leq$. The set of SCCs of $\mathrm{DG}^+(P)$ is denoted by $\mathrm{SCC}^+(P)$. The set of rules associated with a component $S \in \mathrm{SCC}^+(P)$ is $\mathrm{Def}_P(S) = \{r \in P \mid \mathrm{H}(r) \in S\}$. The same notation is also overloaded for single atoms by setting $\mathrm{Def}_P(a) = \mathrm{Def}_P(\{a\})$. The component $S \in \mathrm{SCC}^+(P)$ associated with an atom $a \in \mathrm{At}(P)$ is denoted by $\mathrm{SCC}(a)$.

The *completion* of a normal program P, denoted by $\mathrm{Comp}(P)$, contains a formula $a \leftrightarrow \bigvee_{r \in \mathrm{Def}_P(a)} (\bigwedge_{b \in \mathrm{B}^+(r)} b \wedge \bigwedge_{c \in \mathrm{B}^-(r)} \neg c)$ for each atom $a \in \mathrm{At}(P)^2$. It is well-known that the set of supported models $\mathrm{SuppM}(P)$ coincides with the set of *classical models* of $\mathrm{Comp}(P)$, denoted by $\mathrm{CM}(\mathrm{Comp}(P))$.

Formulas of difference logic (see, e.g., [17]) combine atomic propositions and linear constraints of the form $x_i + k \geq x_j$ where k is an arbitrary integer constant and x_i, x_j are integer variables with propositional connectives $(\neg, \vee, \wedge, \rightarrow, \leftrightarrow)$. Hence, for instance, $(x_1 + 2 \geq x_2) \leftrightarrow (p_1 \rightarrow \neg(x_2 + 2 \geq x_1))$ is a formula in difference logic. An interpretation in difference logic assigns truth values to atomic propositions and integers to integer variables. For example, an interpretation where p_1 is false and both integer variables x_1 and x_2 are given the value 1 satisfies the formula shown above.

As difference logic contains classical propositional logic as a special case, the satisfiability problem of deciding whether there is a satisfying interpretation for a formula in difference logic is NP-hard. The problem is also NP-complete, and, for example, [17,20] present SMT-based techniques for solving the satisfiability problem.

3   Translations

As suggested in Section 1, Clark's completion [8] forms a basic step when a normal logic program P is translated into a set of formulas in difference logic. For the completion we introduce a new atom for each rule body to make the rest of the translation more compact. The resulting representation of $\mathrm{Comp}(P)$ is denoted by $\mathrm{CompN}(P)$ for clarity. In order to define the contribution of each atom $a \in \mathrm{At}(P)$ in $\mathrm{CompN}(P)$, we refer to the set of rules $\mathrm{Def}_P(a) = \{r_1, \ldots, r_k\}$ and introduce a new name bd_a^i for each rule r_i involved. Each individual rule $r_i \in \mathrm{Def}_P(a)$ of the form (1) is encoded by an equivalence (2) and the entire definition $\mathrm{Def}_P(a)$ is captured by (3).

$$\mathrm{bd}_a^i \leftrightarrow b_1 \wedge \cdots \wedge b_n \wedge \neg c_1 \wedge \cdots \wedge \neg c_m \tag{2}$$

$$a \leftrightarrow \mathrm{bd}_a^1 \vee \cdots \vee \mathrm{bd}_a^k \tag{3}$$

Example 1. Consider a logic program P consisting of the following six rules:

1. $a \leftarrow b, c.$ 3. $c \leftarrow \sim d.$ 5. $a \leftarrow d.$
2. $b \leftarrow a, \sim d.$ 4. $d \leftarrow \sim c.$ 6. $b \leftarrow a, \sim c.$

Corresponding to the bodies of these rules, the theory $\mathrm{CompN}(P)$ contains

$$\mathrm{bd}_1 \leftrightarrow b \wedge c, \qquad \mathrm{bd}_3 \leftrightarrow \neg d, \qquad \mathrm{bd}_5 \leftrightarrow d,$$
$$\mathrm{bd}_2 \leftrightarrow a \wedge \neg d, \qquad \mathrm{bd}_4 \leftrightarrow \neg c, \qquad \mathrm{bd}_6 \leftrightarrow a \wedge \neg c$$

in addition to equivalences (3) for the definitions of individual atoms:

$$a \leftrightarrow \mathrm{bd}_1 \vee \mathrm{bd}_5, \quad b \leftrightarrow \mathrm{bd}_2 \vee \mathrm{bd}_6, \quad c \leftrightarrow \mathrm{bd}_3, \quad \text{and } d \leftrightarrow \mathrm{bd}_4.$$

---

[2] As usual we treat a disjunction over the empty set as $\bot$ and a conjunction as $\top$.

Due to new atoms $bd_1, \ldots, bd_6$, the models of the completion $\text{Comp}(P)$ are obtained as projections of models of $\text{CompN}(P)$ over the original signature $\text{At}(P) = \{a, b, c, d\}$. For the program P given above, the classical models of $\text{CompN}(P)$ are $N_1 = \{a, b, c, bd_1, bd_2, bd_3\}$, $N_2 = \{c, bd_3\}$, and $N_3 = \{a, b, d, bd_4, bd_5, bd_6\}$. The respective members of $\text{CM}(\text{Comp}(P))$ are $M_1 = \{a, b, c\}$, $M_2 = \{c\}$, and $M_3 = \{a, b, d\}$. However, only the last two models are stable models of P.   ∎

In addition to the completion $\text{CompN}(P)$ defined above, the translation involves *ranking constraints* which characterize stable models using simple linear constraints of the form allowed in difference logic. The approach captures stable models using mappings from the atoms in a model to integers: a supported model of a program is stable exactly when the model has such a mapping satisfying a set of simple linear constraints [16]. Such a mapping is called a *level ranking* in [16] and in this paper we call the corresponding linear constrains *weak ranking constraints* enforcing the stability of the models obtained for the completion. In [16] it is shown that each stable model has a unique level ranking if a further set of linear constraints, here called *strong ranking constraints*, is added.

First we describe the translation $\text{Tr}_{\text{DIFF}}^{\text{w}}(P)$ which extends $\text{CompN}(P)$ with weak ranking constraints. Our encoding utilizes two new atoms ext_a and int_a for each atom $a \in \text{At}(P)$ having a *non-trivial* component $\text{SCC}(a)$ in the graph $\text{SCC}^+(P)$ satisfying $|\text{SCC}(a)| > 1$. The idea is to classify rules in $\text{SuppR}(\text{Def}_P(a), M)$ given a supported model $M \subseteq \text{At}(P)$ as a candidate for a stable model. Now $\text{Tr}_{\text{DIFF}}^{\text{w}}(P)$ includes for each such atom a the *external* and *internal* support formulas

$$\text{ext}_a \leftrightarrow \bigvee_{r_i \in \text{Def}_P(a),\ B^+(r_i) \cap \text{SCC}(a) = \emptyset} bd_a^i \qquad (4)$$

$$\text{int}_a \leftrightarrow \bigvee_{r_i \in \text{Def}_P(a),\ B^+(r_i) \cap \text{SCC}(a) \neq \emptyset} [bd_a^i \wedge \bigwedge_{b \in B^+(r_i) \cap \text{SCC}(a)} (x_a - 1 \geq x_b)] \qquad (5)$$

where x_a and x_b are integer variables introduced for all atoms in $\text{SCC}(a)$. The relationship of external and internal support is formalized by including, for each atom a residing in a non-trivial component, formulas $a \rightarrow \text{ext}_a \vee \text{int}_a$ and $\neg\text{ext}_a \vee \neg\text{int}_a$. In addition, we fix the value of the variable x_a when the atom a has external support by introducing an implication $\text{ext}_a \rightarrow (x_a = z)^3$ if there is at least one defining rule $r \in \text{Def}_P(a)$ such that $B^+(r) \cap \text{SCC}(a) = \emptyset$. The translation above is a slight extension of that in [16] as it explicates the internal and external support and their relationship.

The weak ranking constraints require that there is an assignment of integers to the variables x_a for atoms a residing in non-trivial SCCs such that each such atom a true in the model candidate M but lacking external support, has internal support. This means that the variable x_a is assigned an integer so that there is a rule $r \in \text{SuppR}(\text{Def}_P(a), M)$ with $B^+(r) \cap \text{SCC}(a) \neq \emptyset$ for which $x_a > x_b$ holds for every $b \in B^+(r) \cap \text{SCC}(a)$. Such an assignment of integers to variables x_a gives a weak level ranking for the atoms a in the model candidate M and it can be shown that a supported model with such

---

3 Since $x_a = 0$ is not directly expressible in pure difference logic, we use a special variable z assumed to hold the value 0 and the shorthand $(x_a = z)$ for the formula $(x_a \geq z) \wedge (z \geq x_a)$.

a ranking is guaranteed to be stable [16]. The net effect is the elimination of models of the completion, i.e., supported models, which are not stable. Hence, weak ranking constraints are sufficient whenever the goal is to compute only one stable model, or to check the existence of stable models.

Example 2. The program P of Example 1 has only one non-trivial SCC, i.e., the component $\text{SCC}(a) = \text{SCC}(b) = \{a, b\}$. The external and internal supports of a and b, respectively, are captured by the following equivalences:

$$\text{ext}_a \leftrightarrow \text{bd}_5, \qquad \text{int}_a \leftrightarrow \text{bd}_1 \wedge (x_a - 1 \geq x_b),$$
$$\text{ext}_b \leftrightarrow \bot, \qquad \text{int}_b \leftrightarrow [\text{bd}_2 \wedge (x_b - 1 \geq x_a)] \vee [\text{bd}_6 \wedge (x_b - 1 \geq x_a)].$$

Moreover, external and internal support are related in terms of formulas $a \rightarrow \text{ext}_a \vee \text{int}_a$, $\neg\text{ext}_a \vee \neg\text{int}_a$, $\text{ext}_a \rightarrow (x_a = z)$, $b \rightarrow \text{ext}_b \vee \text{int}_b$, $\neg\text{ext}_b \vee \neg\text{int}_b$. There are two kinds of models for the translation $\text{Tr}^{\text{w}}_{\text{DIFF}}(P)$ introduced so far:

1. $N_2 = \{c, \text{bd}_3\}$ together with arbitrary assignments to x_a, x_b, and z; and
2. $N_3 \cup \{\text{ext}_a, \text{int}_b\} = \{a, b, d, \text{bd}_4, \text{bd}_5, \text{bd}_6, \text{ext}_a, \text{int}_b\}$ augmented by any assignments satisfying $x_a = z$ and $x_b > z$.

In particular, the interpretation $N_1 = \{a, b, c, \text{bd}_1, \text{bd}_2, \text{bd}_3\}$ given in Example 1 cannot be extended to a model, as ext_a and ext_b must be false and, consequently, int_a and int_b must be true. Hence, inequalities $x_a - 1 \geq x_b$ and $x_b - 1 \geq x_a$ have to be simultaneously satisfied. But this is impossible—witnessing the instability of M_1. ∎

As suggested by Example 2, weak ranking constraints do not guarantee the uniqueness of models obtained, i.e., the same stable model can in principle be encountered a number of times—which may effectively pre-empt the computation of *all* stable models, or *counting* the number of stable models. In the sequel, we explore two variants of strong ranking constraints. Roughly, the idea is to strengthen the constraint $x_a - 1 \geq x_b$, i.e., $x_a > x_b$ used above, to $x_b + 1 = x_a$ for some b. For an atom a in a non-trivial SCC, the *local* strong ranking constraints include for every rule $r_i \in \text{Def}_P(a)$ with $\text{B}^+(r_i) \cap \text{SCC}(a) \neq \emptyset$ the formula (6) and the *global* strong ranking constraint is (7).

$$\text{bd}_a^i \rightarrow \bigvee_{b \in \text{B}^+(r_i) \cap \text{SCC}(a)} (x_b + 1 \geq x_a) \tag{6}$$

$$\text{int}_a \rightarrow \bigvee_{r_i \in \text{Def}_P(a),\, \text{B}^+(r_i) \cap \text{SCC}(a) \neq \emptyset} [\text{bd}_a^i \wedge \bigvee_{b \in \text{B}^+(r_i) \cap \text{SCC}(a)} (x_b + 1 = x_a)] \tag{7}$$

In [16] the local version of the strong ranking constraints has been discussed but the global version inspired by the level numbering characterization in [15] is novel.

Consider the case that we extend the translation $\text{Tr}^{\text{w}}_{\text{DIFF}}(P)$ forcing weak ranking constraints with local strong ranking constraints for each atom a residing in a non-trivial SCC. Then the resulting assignment of numbers to the integer variables x_a for atoms a true in the supported model candidate satisfies an extra condition stating that each rule for a whose body is satisfied by the model candidate has a positive body literal whose ranking is at least that of a minus one. This is enough to force a unique ranking for the atoms in the model candidate provided that the starting point of the ranking is fixed [16]. The global strong ranking constraints require that for such an atom a there is a rule for a whose body is satisfied and it has a positive body literal whose ranking is exactly that of a minus one. This also guarantees a unique ranking for the atoms in the model candidate assuming that the lowest number used in the ranking is fixed. Constraints arising from (6) and (7) are mutually compatible and give rise to three further translations:

- $\mathrm{Tr}^{\mathrm{wl}}_{\mathrm{DIFF}}(P)$ extends $\mathrm{Tr}^{\mathrm{w}}_{\mathrm{DIFF}}(P)$ with local strong ranking constraints,
- $\mathrm{Tr}^{\mathrm{wg}}_{\mathrm{DIFF}}(P)$ extends it with global strong ranking constraints, and
- $\mathrm{Tr}^{\mathrm{wlg}}_{\mathrm{DIFF}}(P)$ extends it with both local and global strong ranking constraints.

All these variants include $\neg a \to (x_a = z)$ for atoms a involved in non-trivial SCCs to prune extra models. Note that each translation $\mathrm{Tr}^{*}_{\mathrm{DIFF}}(P)$ is linear in $\|P\|$.

Example 3. For the program P of Example 1, the translation $\mathrm{Tr}^{\mathrm{w}}_{\mathrm{DIFF}}(P)$ was already presented in Example 2. The local strong ranking constraints (6) are

$$\mathrm{bd}_1 \to (x_b + 1 \geq x_a) \qquad \mathrm{bd}_2 \to (x_a + 1 \geq x_b) \qquad \mathrm{bd}_6 \to (x_a + 1 \geq x_b)$$

whereas the global ones (7) become $\mathrm{int}_a \to [\mathrm{bd}_1 \wedge (x_b + 1 = x_a)]$ and $\mathrm{int}_b \to [\mathrm{bd}_2 \wedge (x_a + 1 = x_b)] \vee [\mathrm{bd}_6 \wedge (x_a + 1 = x_b)]$. In addition to these formulas, the translations $\mathrm{Tr}^{\mathrm{wl}}_{\mathrm{DIFF}}(P)$, $\mathrm{Tr}^{\mathrm{wg}}_{\mathrm{DIFF}}(P)$, and $\mathrm{Tr}^{\mathrm{wlg}}_{\mathrm{DIFF}}(P)$ include $\neg a \to (x_a = z)$ and $\neg b \to (x_b = z)$. Regardless of the translation, the values of integer variables become unique up to the value assigned to z. The first model, corresponding to $M_2 = \{c\}$, is $N_2 = \{c, \mathrm{bd}_3\}$ augmented by the values of x_a, x_b, and z so that $x_a = z = x_b$. The second model, an incarnation of $M_3 = \{a, b, d\}$, is $N_3 \cup \{\mathrm{ext}_a, \mathrm{int}_b\} = \{a, b, d, \mathrm{bd}_4, \mathrm{bd}_5, \mathrm{bd}_6, \mathrm{ext}_a, \mathrm{int}_b\}$ completed by the requirements $x_a = z$ and $x_b = z + 1$ on variable values. ∎

4   Implementation and Experiments

In the following, we report our first experiments with a prototype implementation of the translations $\mathrm{Tr}^{*}_{\mathrm{DIFF}}(\cdot)$ described in Section 3. The main component is translator called LP2DIFF (v. 1.19)[4] which expects its input in the internal file format of the SMOD-ELSsystem. In this paper, we assume that only *basic rules* of the format are used. The output of LP2DIFF is in the QF_IDL format of the SMT library. The translations given in Section 3 have been implemented in LP2DIFF with the following modifications. (i) If an atom a has no defining rule, then $\neg a$ is used instead of formula (3). (ii) If an atom has a unique defining rule, then formulas (3) and (2) are replaced by a single equivalence where the atom is used as the left hand side. (iii) In formulas (5) and (7) a new atom is introduced for each disjunct in the main disjunction. (iv) Instead of the variable z employed in Section 3 constant 0 is used directly because this is supported by the QF_IDL format. Models of the resulting difference logic theory are searched for using BCLT (v. 1.3),[5] YICES (v. 1.0.21),[6] and Z3 (v. 1.3)[7] as back-ends. The last is only available as a Windows binary and hence it is run under Linux using WINE[HQ] (v. 1.1.15).

---

[4] See asptools, smodels, and lp2sat at http://www.tcs.hut.fi/Software/

[5] http://www.lsi.upc.edu/~oliveras/bclt-main.html

[6] http://yices.csl.sri.com/

[7] http://research.microsoft.com/projects/Z3/

Table 1. Unsatisfiable 3-coloring problems: median / minimum times (s)

Nodes	2001	4002	6003	8004
CLASP 1.2.1	0.2/0.2	0.6/0.5	0.9/0.9	1.3/1.3
CMODELS 3.79	0.7/0.7	1.5/1.5	2.4/2.4	3.3/3.3
LP2ATOMIC 1.14	1.5/1.3	2.9/2.8	4.5/4.5	6.3/6.2
SMODELS 2.34	0.6/0.6	1.3/1.3	2.1/2.0	2.8/2.8
LP2DIFF+BCLT	75.7/75.4	298.6/298.4	670.2/668.3	-/-
LP2DIFF+YICES	3.1/3.0	**mem**	**mem**	**mem**
LP2DIFF+Z3	6.9/6.8	18.2/17.8	33.0/32.3	52.5/51.5
Rules	67 837	135 887	203 881	271 931
Atoms	25 967	51 984	77 987	104 004

The goal of our experiments was to compare the performance of the resulting systems with a number of reference systems, namely CLASP (v. 1.2.1)[8] [7], CMODELS (v. 3.79)[9] [11] with ZCHAFF (v. 2007.3.12)[10] as a back-end, LP2ATOMIC (v. 1.14) with LP2SAT (v. 1.13)[15] and MINISAT (v. 1.14)[11] as back-ends, and SMODELS (v. 2.34) [5]. We were obliged to exclude ASSAT [10] which turned out to be outdated for present-day compilers. For fairness reasons, we use SMODELS (command line options -internal and -nolookahead) with LP2ATOMIC and the three SMT solvers to simplify the input program using the well-founded semantics. The symbol table is also compressed using LPCAT (v. 1.14)before the actual translation.

Our first two benchmarks involve randomly generated planar graphs which were generated using PLANARand Stanford graph base. The first benchmark[12] is about coloring the nodes of a graph with n colors so that any two nodes connected by an edge are colored with different colors. Finding such an n-coloring is known to be impossible for planar graphs in general when $n < 4$. To formalize this problem, we used an encoding from [4] based on a *tight* and *normal* logic program. For such programs, the translation produced by LP2DIFF is essentially the completion of the program and involving no ranking constraints. Hence, this benchmark basically tests how effectively the plain completion is handled by different solvers. Starting from a graph with 2001 nodes, we increased the number of nodes by increments of 2001. Timings for $n = 3$ and graphs of 2001–8004 nodes are collected in Table 1. For each number of nodes, nine random planar graphs were generated and the respective ground logic programs produced by LPARSEwere then randomly shuffled. We report the median and minimum CPU (user) times in seconds as measured by the /usr/bin/time Linux system utility. As hardware we used Intel Core2 1.86 Ghz CPUs with 2 GB of main memory. The measured running times cover all processing steps after grounding each input program. Because the instances are relatively big, simplification using SMODELS accounts for 0.9–3.8

---

[8] http://www.cs.uni-potsdam.de/clasp/

[9] http://www.cs.utexas.edu/users/tag/cmodels.html

[10] http://www.princeton.edu/~chaff/zchaff.html

[11] http://minisat.se/

[12] All available at http://www.tcs.tkk.fi/Software/benchmarks/lpnmr09/

Table 2. Satisfiable 4-coloring problems: median / minimum times (s) / timeouts

Nodes	2501	3002	3503	4004
CLASP 1.2.1	544.8 / 84.1 /1	871.5 /270.3 /4	- / - /9	- / - /9
CMODELS 3.79	96.5 / 56.9 /0	168.2 / 36.8 /0	414.2 /264.2 /0	682.8 /421.2 /0
LP2ATOMIC 1.14	29.7 / 12.1 /0	78.3 / 62.6 /0	119.1 / 97.6 /0	238.7 /212.4 /0
SMODELS 2.34	156.8 /153.7 /0	232.8 /223.0 /0	324.7 /313.7 /0	380.6 /366.7 /0
LP2DIFF+BCLT	234.8 /234.3 /0	338.4 /337.5 /0	460.9 /458.1 /0	602.3 /601.3 /0
LP2DIFF+YICES	41.9 / 28.1 /0	**mem**	**mem**	**mem**
LP2DIFF+Z3	490.5 /464.4 /1	766.5 /702.5 /0	- / - /9	- / - /9
Rules	117 321	140 828	164 425	187 942
Atoms	37 474	44 981	52 506	60 015

Table 3. Hamiltonian cycle problems: median / minimum times (s) / timeouts

Nodes	40	60	80	100
CLASP 1.2.1	0.0/ 0.0/0	0.1/ 0.0/0	2.1/ 0.0/0	99.4/ 4.2/0
CMODELS 3.79	0.2/ 0.1/0	3.3/ 0.6/0	11.1/ 2.4/0	99.5/ 8.0/0
LP2ATOMIC 1.14	1.7/ 1.0/0	37.8/ 8.3/1	551.5/ 39.7/1	-/726.2/8
SMODELS 2.34	0.6/ 0.1/0	-/409.0/7	-/133.3/8	-/ -/9
LP2DIFF^w+BCLT	0.7/ 0.4/0	16.8/ 2.8/0	541.3/ 5.7/3	-/ 20.0/6
LP2DIFF^{wl}+BCLT	0.8/ 0.4/0	12.6/ 0.7/0	56.8/ 4.6/0	167.2/ 23.1/1
LP2DIFF^{wg}+BCLT	0.9/ 0.4/0	9.0/ 1.3/0	78.5/ 1.5/0	895.2/ 73.6/4
LP2DIFF^{wlg}+BCLT	0.6/ 0.4/0	5.9/ 1.3/0	74.0/ 6.0/0	-/128.5/5
LP2DIFF^w+YICES	0.5/ 0.1/0	14.3/ 2.6/1	68.8/ 11.8/2	-/211.1/7
LP2DIFF^{wl}+YICES	0.8/ 0.2/0	11.3/ 3.2/0	242.2/ 12.2/2	-/627.6/8
LP2DIFF^{wg}+YICES	0.9/ 0.1/0	32.7/ 3.3/1	-/ 52.1/6	-/195.5/5
LP2DIFF^{wlg}+YICES	0.5/ 0.2/0	5.8/ 1.0/0	68.6/ 19.6/1	-/177.4/5
LP2DIFF^w+Z3	0.6/ 0.4/0	11.6/ 1.9/0	105.6/ 23.0/0	-/164.5/6
LP2DIFF^{wl}+Z3	0.5/ 0.4/0	8.3/ 2.8/0	97.3/ 17.4/0	-/114.9/5
LP2DIFF^{wg}+Z3	0.5/ 0.4/0	7.8/ 1.8/0	44.3/ 4.6/0	-/687.4/8
LP2DIFF^{wlg}+Z3	0.6/ 0.4/0	6.8/ 2.4/0	34.8/ 7.8/0	-/ 89.4/5
Rules	2 747	4 377	5 901	7 647
Atoms	724	1 130	1 506	1 912

seconds of the run times reported for LP2ATOMIC and LP2DIFF. Due to timeouts, i.e., runs exceeding 900 seconds, certain entries remain undefined as indicated by dashes. The "**mem**" keywords denote memory overflows. The results of Table 1 indicate that our reference systems handle tight programs (resp. plain completion) very effectively in this benchmark. The performance of SMT-based approaches, however, degrades much faster and solvers require increasing amounts of memory.

We tried out also satisfiable problems instances generated by the number of colors $n = 4$. The results are collected in Table 2 for graphs having 2501–4004 nodes. In

Table 4. Hand-crafted Hamiltonian cycle problems: median / minimum times (s) / timeouts

Graph	2×p30.3	2×p30.4	4×p20.2	4×p20.3
CLASP 1.2.1	0.0 / 0.0 /0	13.3 / 10.6 /0	0.0 / 0.0 /0	0.0 / 0.0 /0
CMODELS 3.79	10.6 / 2.2 /0	307.6 / 68.4 /0	0.6 / 0.1 /0	1.3 / 1.0 /0
LP2ATOMIC 1.14	221.9 / 4.7 /0	- / - /9	84.7 / 15.3 /0	132.7 / 58.4 /0
SMODELS 2.34	- / - /9	- / - /9	0.4 / 0.2 /4	0.0 / 0.0 /0
LP2DIFF^w+BCLT	2.8 / 1.4 /0	302.1 /148.7 /0	2.6 / 1.0 /0	6.3 / 5.3 /0
LP2DIFF^{wl}+BCLT	3.0 / 0.6 /0	- /529.3 /6	2.8 / 1.1 /0	13.3 / 12.3 /0
LP2DIFF^{wg}+BCLT	3.3 / 0.8 /0	- /663.2 /8	3.2 / 1.8 /0	15.5 / 14.2 /0
LP2DIFF^{wlg}+BCLT	3.1 / 2.0 /0	- / - /9	3.6 / 1.4 /0	15.0 / 13.0 /0
LP2DIFF^w+YICES	16.7 / 1.0 /0	49.5 / 29.9 /0	4.7 / 1.5 /0	2.8 / 1.2 /0
LP2DIFF^{wl}+YICES	41.1 / 2.0 /0	53.5 / 36.0 /0	6.7 / 3.9 /0	2.5 / 1.3 /0
LP2DIFF^{wg}+YICES	37.1 / 9.0 /0	52.4 / 35.4 /0	5.8 / 2.9 /0	3.5 / 1.9 /0
LP2DIFF^{wlg}+YICES	48.4 / 4.6 /0	51.4 / 29.1 /0	5.4 / 2.3 /0	2.3 / 1.9 /0
LP2DIFF^w+Z3	9.8 / 2.5 /0	82.2 / 38.6 /0	2.9 / 1.1 /0	3.4 / 1.7 /0
LP2DIFF^{wl}+Z3	5.5 / 0.6 /0	252.4 /184.0 /0	16.2 / 1.3 /0	6.6 / 5.2 /0
LP2DIFF^{wg}+Z3	6.5 / 0.7 /0	411.2 /278.2 /0	8.0 / 2.5 /0	6.9 / 5.8 /0
LP2DIFF^{wlg}+Z3	2.7 / 1.0 /0	304.4 / 88.1 /0	12.2 / 2.1 /0	7.2 / 5.5 /0
Rules	3 981	4 027	4 665	4 749
Atoms	1 070	1 076	1 338	1 350

addition to median and minimum times spent on finding a coloring for 9 different graphs of the size in question, we also report the number of timeouts. It is interesting that the performance of native answer set solvers degrades remarkably when we shift from unsatisfiable to satisfiable instances. Another observation from Tables 1 and 2 is that SAT-based approaches, viz. CMODELS and LP2ATOMIC, are more efficient and robust to handle plain completion than the three SMT solvers evaluated here.

For our second benchmark reported in Table 3 we chose the problem of finding Hamiltonian circuits and a *non-tight* encoding taken from [4] to study the effects of ranking constraints in particular. We tested the four variants of $\mathrm{Tr}^*_{\mathrm{DIFF}}(P)$ from Section 3 referred to by the superscripts w, wl, wg, and wlg, respectively, and implemented in LP2DIFF (command line options -l and -g). Table 3 reports, similarly as in the first benchmark, run times for finding one stable model for a randomly shuffled ground normal program obtained using the non-tight encoding for nine random planar graphs of 40–100 nodes, respectively. The results indicate that the performance obtained with LP2DIFF and SMT solvers is not too far from other ASP solvers. We note that all SMT solvers benefit from both local and global strong ranking constraints. Table 4 presents the data for both satisfiable (2×p30.3 and 4×p20.2) and unsatisfiable (2×p30.4 and 4×p20.3) instances based on hand-crafted graphs [10, Table 5]. The same non-tight encoding is used but each ground instance is randomly shuffled for nine distinct runs. For unsatisfiable instances, strong ranking constraints seem to slow down computations systematically. On the other hand, the combination LP2DIFF+Z3 performs best on the satisfiable instance 2×p30.3 in the presence of both local and global constraints.

Table 5. Random non-tight normal logic programs: median times (s)

Atoms Instance	50					60				
	u-b11	s-b14	u-b15	u-b16	u-b17	u-b9	u-b11	u-b12	u-b13	s-b19
CLASP 1.2.1	6.4	1.1	13.3	5.0	1.1	135.1	69.8	126.7	76.8	14.1
CMODELS 3.79	44.5	6.3	130.2	32.8	3.6	-	-	-	-	22.2
LP2ATOMIC 1.14	156.1	28.6	290.0	117.5	23.5	-	-	-	-	818.4
SMODELS 2.34	37.1	11.7	67.4	31.6	10.2	-	803.2	-	714.6	133.1
LP2DIFF^w+BCLT	22.8	4.8	64.2	16.5	2.6	-	-	-	-	174.3
LP2DIFF^{wl}+BLCT	34.9	4.9	94.2	25.3	3.3	-	-	-	-	196.2
LP2DIFF^{wg}+BCLT	41.5	12.6	107.5	32.4	4.4	-	-	-	-	193.8
LP2DIFF^{wlg}+BCLT	48.1	13.2	121.1	34.0	5.0	-	-	-	-	85.5
LP2DIFF^w+YICES	8.3	1.5	16.0	6.4	1.3	265.9	97.1	227.1	110.0	25.4
LP2DIFF^{wl}+YICES	11.3	3.9	21.7	8.5	1.7	396.3	142.9	275.0	171.0	34.3
LP2DIFF^{wg}+YICES	13.4	2.5	25.0	9.7	1.9	380.0	171.5	491.8	318.3	58.4
LP2DIFF^{wlg}+YICES	25.5	4.4	46.7	17.2	3.4	624.3	284.7	581.5	278.5	53.5
LP2DIFF^w+Z3	61.8	25.0	99.9	44.1	10.2	-	453.0	-	492.2	121.2
LP2DIFF^{wl}+Z3	62.7	15.4	115.3	44.1	11.4	-	513.4	-	597.1	237.5
LP2DIFF^{wg}+Z3	80.8	38.3	134.2	56.1	13.4	-	668.7	-	678.1	221.2
LP2DIFF^{wlg}+Z3	86.8	27.7	131.4	56.6	14.6	-	678.1	-	729.5	311.7

The third benchmark involved a number of randomly generated *non-tight* logic program instances taken from the Asparagus collection[13]. Instances with 50 and 60 atoms involve roughly 760 and 990 rules—creating only one strongly connected component which encompasses all atoms. The results are collected in Table 5. Each instance was randomly shuffled nine times to obtain nine distinct runs on each solver. The results indicate that the satisfiable instances (s-b14 and s-b19) are the easiest for each number of atoms. Overall, the systems CLASP and LP2DIFF+YICES show the best performance for these benchmark instances, CLASP being circa 1.2–2.0 times faster. Moreover, all SMT-based approaches seem to be at least as robust as the SAT-based ones.

Table 6 reports the results of a benchmark on Yoshio Murase's hand-made instances of the Sokoban game[14]. We translated the instances as sets of facts and used an encoding of Sokoban by Tommi Syrjänen which employs recursive rules and requires an upper bound on the number of steps in the solution. Table 6 shows results for two instances (levels 2 and 9) with 16 as the minimal number of steps for a solution. Hence, instances lev2-s16 and lev9-s16 have stable models but lev2-s15 and lev9-s15 not. The instance lev2-s100 is a big one allowing for 100 steps. Since the problem encoding involves also other SMODELS rule types, yet another translator LP2NORMALwas used to transform each ground program into a normal one. The results indicate that our approach based on difference logic (particularly when using Z3) compares favorably to the SAT based approaches and scales very similarly to the native ASP solver CLASP.

---

[13] See http://asparagus.cs.uni-potsdam.de/ for details.

[14] http://www.sourcecode.se/sokoban/levtext.php?file=Handmade.slc

Table 6. Sokoban instances: median / minimum times (s)

Instance	lev2-s15	lev2-s16	lev2-s100	lev9-s15	lev9-s16
CLASP 1.2.1	0.4 / 0.3	0.5 / 0.3	93.4 / 4.6	149.6 /108.8	22.5 / 4.9
CMODELS 3.79	82.6 / 26.6	53.4 / 25.0	- / -	- / -	452.0 /103.9
LP2ATOMIC 1.14	32.3 / 19.8	27.2 / 13.9	- / -	- / -	- / -
SMODELS 2.34	- / -	- /491.1	- / -	- / -	- / -
LP2DIFF^w+BCLT	5.0 / 4.6	6.3 / 5.5	777.2 /682.0	- /783.0	896.8 /302.7
LP2DIFF^{wl}+BCLT	12.4 / 11.3	14.4 / 13.6	- / -	392.0 /346.0	457.9 /280.0
LP2DIFF^{wg}+BCLT	14.0 / 12.3	16.5 / 15.9	- / -	464.1 /395.0	505.5 /366.1
LP2DIFF^{wlg}+BCLT	13.7 / 12.2	17.9 / 17.1	- / -	447.5 /332.7	471.7 /262.7
LP2DIFF^w+YICES	1.4 / 1.1	1.6 / 1.0	- / -	439.9 /335.8	356.4 / 26.4
LP2DIFF^{wl}+YICES	1.6 / 1.3	1.9 / 1.5	573.4 /353.1	499.9 /438.6	420.7 /136.9
LP2DIFF^{wg}+YICES	1.7 / 1.4	1.9 / 1.4	- / -	605.1 /551.0	143.5 / 65.9
LP2DIFF^{wlg}+YICES	1.8 / 1.5	2.0 / 1.5	595.1 /422.3	520.0 /427.0	187.5 / 52.6
LP2DIFF^w+Z3	3.0 / 2.1	2.7 / 2.2	157.4 /103.8	493.3 /381.6	371.3 /101.8
LP2DIFF^{wl}+Z3	3.6 / 3.1	4.1 / 1.9	194.1 /128.5	590.4 /422.6	340.9 / 67.4
LP2DIFF^{wg}+Z3	3.8 / 3.2	4.1 / 3.5	202.5 /120.3	524.3 /494.1	316.5 / 70.1
LP2DIFF^{wlg}+Z3	4.0 / 3.1	4.3 / 2.6	161.2 /118.0	545.0 /461.0	239.8 /113.3
Rules	7 110	7 629	51 225	18 047	19 396
Atoms	5 137	5 475	33 867	11 450	12 207

5   Conclusions

SAT solver technology has been advancing significantly and provides a very promising computational platform for implementing also ASP systems. Recently, increasingly efficient extensions to the basic clausal SAT solvers have been developed, in particular in the SMT framework. The goal of the paper is to develop compact translations from logic programs to difference logic, where SMT-based solver technology is advancing steadily. Based on the results in [15,16] we present a collection of translations providing a tight connection between logic programs and difference logic. The translations enable one to map any normal program to a theory in difference logic without substantial blow-up in size so that any solver for difference logic can be used for computing stable models *without modifications*. This is different from translations to SAT where either the solver has to be modified and extended to allow a dynamic translation where clauses are added on the fly or where the translations are of super-linear size. On the other hand, the translations offer an interesting method for generating challenging benchmarks for these solvers and can help to boost their development. Our experimental results indicate that with the translations even the current SMT solver technology shows performance which is close to that of state-of-the-art ASP solvers. Given the large amount effort put into developing SMT solvers[15] there is potential that the translations devised herein provide a basis of a cutting-edge technology for implementing ASP systems.

---

[15] See, e.g., the SMT competition website at http://www.smtcomp.org/

References

1. Gelfond, M., Lifschitz, V.: The stable model semantics for logic programming. In: ICLP, pp. 1070–1080. MIT Press, Cambridge (1988)
2. Lifschitz, V.: Answer set planning. In: ICLP, pp. 23–37. MIT Press, Cambridge (1999)
3. Marek, V., Truszczyński, M.: Stable models and an alternative logic programming paradigm. In: The Logic Programming Paradigm: a 25-Year Perspective, pp. 375–398. Springer, Heidelberg (1999)
4. Niemelä, I.: Logic programming with stable model semantics as a constraint programming paradigm. Ann. Math. Artif. Intell. 25(3-4), 241–273 (1999)
5. Simons, P., Niemelä, I., Soininen, T.: Extending and implementing the stable model semantics. Artificial Intelligence 138(1-2), 181–234 (2002)
6. Leone, N., Pfeifer, G., Faber, W., Eiter, T., Gottlob, G., Scarcello, F.: The DLV System for Knowledge Representation and Reasoning. ACM TOCL 7(3), 499–562 (2006)
7. Gebser, M., Kaufmann, B., Neumann, A., Schaub, T.: Clasp: A conflict-driven answer set solver. In: Baral, C., Brewka, G., Schlipf, J. (eds.) LPNMR 2007. LNCS (LNAI), vol. 4483, pp. 260–265. Springer, Heidelberg (2007)
8. Clark, K.: Negation as failure. In: Logic and Data Bases, pp. 293–322. Plenum Press (1978)
9. Erdem, E., Lifschitz, V.: Tight logic programs. TPLP 3(4-5), 499–518 (2003)
10. Lin, F., Zhao, Y.: ASSAT: computing answer sets of a logic program by SAT solvers. AIJ 157(1-2), 115–137 (2004)
11. Lierler, Y.: Cmodels – SAT-based disjunctive answer set solver. In: Baral, C., Greco, G., Leone, N., Terracina, G. (eds.) LPNMR 2005. LNCS (LNAI), vol. 3662, pp. 447–451. Springer, Heidelberg (2005)
12. Giunchiglia, E., Lierler, Y., Maratea, M.: Answer set programming based on propositional satisfiability. Journal of Automated Reasoning 36(4), 345–377 (2006)
13. Ben-Eliyahu, R., Dechter, R.: Propositional Semantics for Disjunctive Logic Programs. Ann. Math. Artif. Intell. 12(1-2), 53–87 (1994)
14. Lin, F., Zhao, J.: On tight logic programs and yet another translation from normal logic programs to propositional logic. In: IJCAI, pp. 853–858 (2003)
15. Janhunen, T.: Representing normal programs with clauses. In: ECAI, pp. 358–362. IOS Press, Amsterdam (2004)
16. Niemelä, I.: Stable models and difference logic. Ann. Math. Artif. Intell. 53(1-4), 313–329 (2008)
17. Nieuwenhuis, R., Oliveras, A.: DPLL(T) with exhaustive theory propagation and its application to difference logic. In: Etessami, K., Rajamani, S.K. (eds.) CAV 2005. LNCS, vol. 3576, pp. 321–334. Springer, Heidelberg (2005)
18. Bozzano, M., Bruttomesso, R., Cimatti, A., Junttila, T., van Rossum, P., Schulz, S., Sebastiani, R.: MathSAT: Tight integration of sat and mathematical decision procedures. Journal of Automated Reasoning 35(1-3), 265–293 (2005)
19. Nieuwenhuis, R., Oliveras, A., Tinelli, C.: Solving SAT and SAT Modulo Theories: From an abstract Davis–Putnam–Logemann–Loveland procedure to DPLL(T). J. ACM 53(6), 937–977 (2006)
20. Cotton, S., Maler, O.: Fast and flexible difference constraint propagation for DPLL(T). In: Biere, A., Gomes, C.P. (eds.) SAT 2006. LNCS, vol. 4121, pp. 170–183. Springer, Heidelberg (2006)
21. Apt, K., Blair, H., Walker, A.: Towards a theory of declarative knowledge. In: Foundations of Deductive Databases and Logic Programming, pp. 89–148. Morgan Kaufmann, San Francisco (1988)
22. Marek, V.W., Subrahmanian, V.S.: The relationship between stable, supported, default and autoepistemic semantics for general logic programs. TCS 103, 365–386 (1992)

A Module-Based Framework for Multi-language Constraint Modeling*

Matti Järvisalo, Emilia Oikarinen, Tomi Janhunen, and Ilkka Niemelä

Helsinki University of Technology TKK
Department of Information and Computer Science
P.O. Box 5400, FI-02015 TKK, Finland
{matti.jarvisalo,emilia.oikarinen,tomi.janhunen,
ilkka.niemela}@tkk.fi

Abstract. We develop a module-based framework for constraint modeling where
it is possible to combine different constraint modeling languages and exploit their
strengths in a flexible way. In the framework a constraint model consists of modules with clear input/output interfaces. When combining modules, apart from
the interface, a module is a black box whose internals are invisible to the outside world. Inside a module a chosen constraint language (approaches such as
CP, ASP, SAT, and MIP) can be used. This leads to a clear modular semantics
where the overall semantics of the whole constraint model is obtained from the
semantics of individual modules. The framework supports multi-language modeling without the need to develop a complicated joint semantics and enables the
use of alternative semantical underpinnings such as default negation and classical
negation in the same model. Furthermore, computational aspects of the framework are considered and, in particular, possibilities of benefiting from the known
module structure in solving constraint models are studied.

1 Introduction

There are several constraint-based approaches to solving combinatorial search and optimization problems: constraint programming (CP), answer set programming (ASP),
mixed integer programming (MIP), linear programming (LP), propositional satisfiability checking (SAT) and its extension to satisfiability modulo theories (SMT). Each has
its particular strengths: for example, CP systems support global constraints, ASP recursive definitions and default negation, MIP constraints on real-valued variables, and
SAT efficient solver technology. In larger applications it is often necessary to exploit
the strengths of several languages and to reuse and combine available components. For
example, in scheduling problems involving a large amount data and constraints, multi-
language modeling can be very useful (as also exemplified in this paper in Sect. 5).

In this work we develop a module-based framework for modeling complex problems with constraints using a combination of different modeling languages. Rather than
taking one language as a basis and extending it, we develop a framework for multi-
language modeling where different languages are treated on equal terms. The starting

---

* This work is financially supported by Academy of Finland under the project *Methods for
Constructing and Solving Large Constraint Models* (grant #122399).

E. Erdem, F. Lin, and T. Schaub (Eds.): LPNMR 2009, LNCS 5753, pp. 155–168, 2009.

point is to use modules with clear input/output (I/O) interfaces. When combining modules, apart from the interface, a module is a black box whose internals are invisible to the outside world. Inside a module a chosen constraint language (for example, CP, ASP, MIP) with its normal semantics can be used. In this way a clear modular semantics is obtained: the overall semantics of the whole constraint model (consisting of modules) is obtained by "composing" the semantics of individual modules.

We see substantial advantages of this approach for modeling. The clear module interfaces enable support for multi-language modeling without the need to develop a complicated joint semantics capturing arbitrary combinations of special constraints available in different languages. It is also possible to use alternative semantical underpinnings such as default negation and classical negation in the same model. The module-based approach brings the benefits of modular programming to developing constraints models and enables to create libraries to enhance module reuse. It also improves elaboration tolerance and facilitates maintaining and updating a constraint model. Moreover, extending the approach with further languages is conceptually straightforward.

Computational aspects of the framework are also promising. Module interfaces and separation of inputs and outputs can be exploited in decision methods, for example, with more top-down solution techniques where the overall output of the constraint model can be used to identify the relevant parts of the model. The module-based approach allows optimizing the computational efficiency of a model in a structured way: a module can be replaced by another (more optimized) version without altering the solutions of the model as long as the I/O relation of the module is not changed. Similarly, the framework supports modular testing, validation, and debugging of constraint models.

This module-based framework for multi-language modeling seems to be a novel approach. Several approaches to adding modularity to ASP languages [1,2,3,4] have been proposed. However, in these approaches modular multi-language modeling is not directly supported (although the combination of propositional ASP and SAT modules is studied in [5]). A large number of extended modeling languages have also been previously proposed. On one hand, ASP languages have been extended with constraints or other externally defined relations (see, e.g., [6,7,8,9,10,11]). On the other hand, Prolog systems have been extended with ASP features [12,13,14]. Extended modeling languages have been developed also for constraint programming, including ESRA [15], ESSENCE [16], and Zinc [17]. However, none of the approaches supports modular multi-language modeling where different languages are treated on equal terms. Instead, they can all be seen as extensions of a given basic language with features from other languages.

The rest of this paper is organized as follows. As preliminaries, we first give a generic definition of constraints and related notation (Sect. 2). Then constraint modules, the basic building blocks of module systems, are introduced (Sect. 3). The language of module systems, based on composing constraint modules, is discussed in Sect. 4. Then, in Sect. 5 we discuss how the framework can be instantiated in practice: A larger application is considered in order to illustrate the issues arising in using a multi-language modeling approach, and the required language interface for constructing a multi-language module system is sketched. Before conclusions (Sect. 7), computational aspects, and especially, possibilities of benefiting from the explicit modular constraint model description when solving such a model are highlighted (Sect. 6).

2 Constraints

In this section we give necessary definitions and notation related to the generic concept of constraints applied in this work. These serve as basic building blocks for constraint modules which are then combined to form complex constraint models.

Let $\mathcal{X}$ be a set of variables. For each variable $x \in \mathcal{X}$, we associate a set of *values* $D(x)$, called the *domain* of x. Given a set $X \subseteq \mathcal{X}$ of variables, an assignment over X is a function

$$\tau : X \to \bigcup_{x \in X} D(x),$$

which maps variables in X to values in their domains. A *constraint* $\mathcal{C}$ over a set of variables X is characterized by a set $\mathsf{Solutions}(\mathcal{C})$ of assignments over X, called the *satisfying assignments* of $\mathcal{C}$. We denote by $\mathsf{Vars}(\mathcal{C})$ the set X of variables.

It is important to notice that, since the satisfying assignments solely characterize the constraint, this generic way of describing constraints does not specify how a constraint should be implemented, i.e., the modeling language and semantics used for realizing the constraint declaratively remain unspecified.

Example 1. Let $\mathcal{C}$ be a constraint over a set of Boolean variables $\{a, b\}$, i.e., $D(a) = D(b) = \{\mathbf{t}, \mathbf{f}\}$, characterized by $\mathsf{Solutions}(\mathcal{C}) = \{\tau_1, \tau_2\}$, where $\tau_1 = \{a \mapsto \mathbf{t}, b \mapsto \mathbf{f}\}$ and $\tau_2 = \{a \mapsto \mathbf{f}, b \mapsto \mathbf{t}\}$. Now, $\mathcal{C}$ can be implemented, for example, as a *normal logic program* $\{a \leftarrow \sim b.\ b \leftarrow \sim a\}$ or as a *disjunctive logic program* $\{a \lor b \leftarrow\}$ in ASP, or as a *conjunctive normal form (CNF) formula* $\{a \lor \neg b, \neg a \lor b\}$ in SAT.

Given an assignment τ and a set of variables X, the *projection* $\pi_X(\tau)$ of τ on X is the assignment that maps each variable $x \in X$ for which $\tau(x)$ is defined to $\tau(x)$. For instance, the projection $\pi_{\{a\}}(\tau_1)$ for τ_1 from Example 1 is the assignment $\pi_{\{a\}}(\tau_1) = \{a \mapsto \mathbf{t}\}$ over the set $\{a\}$.

Given a constraint $\mathcal{C}$, and an assignment τ over a set X of variables, the *restriction* $\mathcal{C}[\tau]$ of $\mathcal{C}$ to τ is characterized by

$$\mathsf{Solutions}(\mathcal{C}[\tau]) = \{\tau' \in \mathsf{Solutions}(\mathcal{C}) \mid \pi_{\mathsf{Vars}(\mathcal{C}) \cap X}(\tau') = \pi_{\mathsf{Vars}(\mathcal{C}) \cap X}(\tau)\}.$$

For instance, let $\tau_3 = \{b \mapsto \mathbf{f}\}$ be an assignment over $\{b\}$. Now, the restriction $\mathcal{C}[\tau_3]$ of $\mathcal{C}$ from Example 1 is a constraint characterized by $\{\tau_1\} \subseteq \mathsf{Solutions}(\mathcal{C})$, i.e., $\mathsf{Solutions}(\mathcal{C}[\tau_3]) = \{\tau_1\}$.

Given two constraints $\mathcal{C}$ and $\mathcal{C}'$, an assignment τ over $\mathsf{Vars}(\mathcal{C})$ is *compatible* with an assignment τ' over $\mathsf{Vars}(\mathcal{C}')$ if $\pi_{\mathsf{Vars}(\mathcal{C}) \cap \mathsf{Vars}(\mathcal{C}')}(\tau) = \pi_{\mathsf{Vars}(\mathcal{C}) \cap \mathsf{Vars}(\mathcal{C}')}(\tau')$. The union $\tau \cup \tau'$ of two compatible assignments, τ and τ' over X and X', respectively, is the assignment over $X \cup X'$ mapping each $x \in X$ to $\tau(x)$ and each $x \in X' \setminus X$ to $\tau'(x)$.

Example 2. Let $\mathcal{C}'$ be a constraint over a set of Boolean variables $\{b, c\}$ characterized by $\mathsf{Solutions}(\mathcal{C}') = \{\tau'\}$ such that $\tau' = \{b \mapsto \mathbf{f}, c \mapsto \mathbf{f}\}$. Consider $\mathcal{C}$ from Example 1. The assignment τ_1 is compatible with τ', because $\{a, b\} \cap \{b, c\} = \{b\}$ and $\tau_1(b) = \mathbf{f} = \tau'(b)$. On the other hand, τ_2 is not compatible with τ', because $\tau_2(b) = \mathbf{t} \neq \tau'(b)$. The union $\tau_1 \cup \tau' = \{a \mapsto \mathbf{t}, b \mapsto \mathbf{f}, c \mapsto \mathbf{f}\}$ is an assignment over the set $\{a, b, c\}$.

3   Constraint Modules

The view to constructing complex constraint models proposed in this work is based on expressing such models as *module systems*. Module systems are built from *constraint modules* which are combined together in a controlled fashion. In this section we introduce the generic concept of a constraint module. Constraint modules are based on a chosen constraint, with the addition of an explicit I/O interface. Our definition for a constraint module is generic in the sense that it does not insist on a specific implementation of the constraint on the declarative level. The aim here is to allow implementing the constraint using different declarative languages, offering the implementer of a module the possibility to choose the constraint language and the semantics.

Definition 1. *A constraint module $\mathcal{M}$ is a triple $\langle \mathcal{C}, \mathcal{I}, \mathcal{O} \rangle$, where $\mathcal{C}$ is a constraint, and $\mathcal{I}$ and $\mathcal{O}$, with $\mathcal{I} \cap \mathcal{O} = \emptyset$, define the I/O interface of $\mathcal{M}$:*

- *$\mathcal{I} \subseteq \mathsf{Vars}(\mathcal{C})$ is the input specification of $\mathcal{M}$, and*
- *$\mathcal{O} \subseteq \mathsf{Vars}(\mathcal{C})$ is the output specification of $\mathcal{M}$.*

A module $\mathcal{M}$ is thus a constraint with a fixed I/O interface. In analogy to the characterization of a constraint, a module $\mathcal{M} = \langle \mathcal{C}, \mathcal{I}, \mathcal{O} \rangle$ is characterized by a set $\mathsf{Solutions}(\mathcal{M})$ of assignments over $\mathcal{I} \cup \mathcal{O}$ called the *satisfying assignments of the module*. Given a constraint module $\mathcal{M} = \langle \mathcal{C}, \mathcal{I}, \mathcal{O} \rangle$ and an assignment τ_I over $\mathcal{I}$, the set of consistent outputs of $\mathcal{M}$ w.r.t. τ_I is

$$\mathsf{SolutionOut}(\mathcal{M}, \tau_I) := \{\pi_{\mathcal{O}}(\tau) \mid \tau \in \mathsf{Solutions}(\mathcal{C}[\tau_I])\}.$$

The satisfying assignments of a module are obtained by considering all possible input assignments.

Definition 2. *Given a constraint module $\mathcal{M} = \langle \mathcal{C}, \mathcal{I}, \mathcal{O} \rangle$, the set $\mathsf{Solutions}(\mathcal{M})$ of satisfying assignments of $\mathcal{M}$ is the union of the sets $\{\tau_I \cup \tau_O \mid \tau_O \in \mathsf{SolutionOut}(\mathcal{M}, \tau_I)\}$ for all assignments τ_I over $\mathcal{I}$.*

Those variables in $\mathsf{Vars}(\mathcal{C})$ which are not in $\mathcal{I} \cup \mathcal{O}$ are *local to* $\mathcal{M}$; the assignments in $\mathsf{Solutions}(\mathcal{M})$ do not assign values to them. Notice that the possibility of local variables enables *encapsulation* and *information hiding*. A module offers through its I/O interface to the user a black-box implementation of a specific constraint. The idea behind this abstract way of defining a module is that, looking from the outside of a module when using the module as a part of a constraint model, the user is interested in the input-output relationship, i.e., the functionality of the module. This can be highlighted by making explicit the conditions under which two modules are considered equivalent.

Definition 3. *Two constraint modules, $\mathcal{M}_1 = \langle \mathcal{C}_1, \mathcal{I}_1, \mathcal{O}_1 \rangle$ and $\mathcal{M}_2 = \langle \mathcal{C}_2, \mathcal{I}_2, \mathcal{O}_2 \rangle$, are* equivalent, *denoted by $\mathcal{M}_1 \equiv \mathcal{M}_2$, if and only if $\mathcal{I}_1 = \mathcal{I}_2$, $\mathcal{O}_1 = \mathcal{O}_2$, and $\mathsf{Solutions}(\mathcal{M}_1) = \mathsf{Solutions}(\mathcal{M}_2)$.*

Example 3. Consider $\mathcal{M} = \langle \mathcal{C}, \{a\}, \{b\} \rangle$, where a and b are Boolean variables, and let $\mathsf{Solutions}(\mathcal{M}) = \{\tau_1, \tau_2\}$ where $\tau_1 = \{a \mapsto \mathbf{t}, b \mapsto \mathbf{f}\}$ and $\tau_2 = \{a \mapsto \mathbf{f}, b \mapsto \mathbf{t}\}$. Since τ_1 and τ_2 are the same as in Example 1, $\mathcal{M}$ can be implemented using any of the implementations of the constraint described in Example 1.

Moreover, the set of variables used in implementing $\mathcal{C}$ is not limited to $\{a, b\}$. For instance, a *logic program module* [4] $\langle P, I, O \rangle = \langle \{c \leftarrow \sim a.\ b \leftarrow c\}, \{a\}, \{b\} \rangle$ is an implementation of $\mathcal{C}$ such that $\mathsf{Solutions}(\mathcal{C}) = \{\tau_3, \tau_4\}$ where $\tau_3 = \{a \mapsto \mathbf{t}, b \mapsto \mathbf{f}, c \mapsto \mathbf{f}\}$ and $\tau_4 = \{a \mapsto \mathbf{f}, b \mapsto \mathbf{t}, c \mapsto \mathbf{t}\}$.[1] Now, there are two possible assignments over $\{a\}$. If $\tau_I = \{a \mapsto \mathbf{t}\}$ we obtain $\mathsf{SolutionOut}(\mathcal{M}, \tau_I) = \{\pi_{\{b\}}(\tau_3)\}$ since $\mathsf{Solutions}(\mathcal{C}[\tau_I]) = \{\tau_3\}$ as $\tau_3(a) = \tau_I(a) = \mathbf{t}$. For the other possible input assignment $\tau_I' = \{a \mapsto \mathbf{f}\}$, we obtain $\mathsf{SolutionOut}(\mathcal{M}, \tau_I') = \{\pi_{\{b\}}(\tau_4)\}$. Finally, notice that $\tau_I \cup \pi_{\{b\}}(\tau_3) = \tau_1$ and $\tau_I' \cup \pi_{\{b\}}(\tau_4) = \tau_2$. Thus, $\mathsf{Solutions}(\mathcal{M}) = \{\tau_1, \tau_2\}$.

4 Module Systems

In this section we discuss how larger module systems are built from individual constraint modules. The idea is that module systems are constructed by connecting smaller module systems through the I/O interfaces offered by such systems. In other words, in analogy to constraint modules, a module system has an I/O interface, and constraint modules are seen as primitive module systems. We will start by introducing a formal language for expressing such systems and then introduce the semantics for module systems which are *well-formed*.

Definition 4 (The language of module systems)
1. *A constraint module* $\mathcal{M} = \langle \mathcal{C}, \mathcal{I}, \mathcal{O} \rangle$ *is a module system with* $\mathsf{Input}(\mathcal{M}) = \mathcal{I}$ *and* $\mathsf{Output}(\mathcal{M}) = \mathcal{O}$.
2. *If* $\mathcal{M}$ *is a module system and* X *a set of variables, then* $\pi_X(\mathcal{M})$ *is a module system with* $\mathsf{Input}(\pi_X(\mathcal{M})) = \mathsf{Input}(\mathcal{M})$ *and* $\mathsf{Output}(\pi_X(\mathcal{M})) = \mathsf{Output}(\mathcal{M}) \cap X$.
3. *If* $\mathcal{M}$ *and* $\mathcal{M}'$ *are module systems, then* $(\mathcal{M} \rhd \mathcal{M}')$ *is a module system with* $\mathsf{Input}(\mathcal{M}_1 \rhd \mathcal{M}_2) = \mathsf{Input}(\mathcal{M}_1) \cup (\mathsf{Input}(\mathcal{M}_2) \setminus \mathsf{Output}(\mathcal{M}_1))$ *and* $\mathsf{Output}(\mathcal{M}_1 \rhd \mathcal{M}_2) = \mathsf{Output}(\mathcal{M}_1) \cup \mathsf{Output}(\mathcal{M}_2)$.

Notice that Definition 4 is purely syntactical. Our next goal is to define the semantics for more complex module systems as we have already defined the sets of satisfying assignments for individual constraint modules. This is achieved by formalizing the semantics of the two operators: intuitively, projection π_X offers a way of filtering the output of a module system, whereas composition $\rhd$ is used for merging two module systems into one. We start by defining the conditions under which two module systems are *composable* and *independent*.

Definition 5 (Composable and independent module systems). *Two module systems* $\mathcal{M}_1$ *and* $\mathcal{M}_2$ *are* composable *if* $\mathsf{Output}(\mathcal{M}_1) \cap \mathsf{Output}(\mathcal{M}_2) = \emptyset$. *Module system* $\mathcal{M}_1$ *is* independent *from module system* $\mathcal{M}_2$ *if* $\mathsf{Input}(\mathcal{M}_1) \cap \mathsf{Output}(\mathcal{M}_2) = \emptyset$.

Composability is used to ensure that if two module systems interfere with each others' output, they cannot be put together. Independence allows us to ensure that two modules are not in cyclic dependency. Notice that the independence of $\mathcal{M}_1$ from $\mathcal{M}_2$ does not imply that $\mathcal{M}_2$ is independent from $\mathcal{M}_1$.

---

[1] Notice that unlike other formalisms mentioned so far, the logic program modules in [4] already facilitate I/O interfaces, and their semantics differs from the standard stable model semantics since input variables have a classical interpretation.

Definition 6 (Module composition). *Given two composable module systems $\mathcal{M}_1$ and $\mathcal{M}_2$, their composition $\mathcal{M}_1 \triangleright \mathcal{M}_2$ is defined if and only if $\mathcal{M}_1$ is independent from $\mathcal{M}_2$. The set of satisfying assignments of $\mathcal{M}_1 \triangleright \mathcal{M}_2$, Solutions$(\mathcal{M}_1 \triangleright \mathcal{M}_2)$, is*

$$\{(\tau_1 \cup \tau_2) \mid \tau_1 \in \text{Solutions}(\mathcal{M}_1), \tau_2 \in \text{Solutions}(\mathcal{M}_2), \text{ and } \tau_2 \text{ is compatible with } \tau_1\}.$$

Example 4. Let $\mathcal{M} = \langle \mathcal{C}, \{a\}, \{n\}\rangle$ and $\mathcal{M}' = \langle \mathcal{C}', \{n\}, \{m\}\rangle$ be constraint modules where a is a Boolean variable, $D(n) = D(m) = \{1, 2, 3\}$, Solutions$(\mathcal{M}) = \{\tau_1\}$, and Solutions$(\mathcal{M}') = \{\tau_2, \tau_3\}$ where $\tau_1 = \{a \mapsto \mathbf{f}, n \mapsto 3\}$, $\tau_2 = \{n \mapsto 1, m \mapsto 1\}$, and $\tau_3 = \{n \mapsto 3, m \mapsto 2\}$. Since $\mathcal{M}$ is independent from $\mathcal{M}'$, their composition $\mathcal{M} \triangleright \mathcal{M}'$ is defined. Notice that $n \in$ Output$(\mathcal{M}) \cap$ Input$(\mathcal{M}')$ provides the connection between $\mathcal{M}$ and $\mathcal{M}'$, i.e., $n \in$ Output$(\mathcal{M})$ is the input for $\mathcal{M}'$ because $n \in$ Input$(\mathcal{M}')$. Furthermore, Input$(\mathcal{M} \triangleright \mathcal{M}') = \{a\}$, Output$(\mathcal{M} \triangleright \mathcal{M}') = \{n, m\}$, and Solutions$(\mathcal{M} \triangleright \mathcal{M}') = \{\tau_1 \cup \tau_3\}$, because τ_1 is not compatible with τ_2 and τ_1 is compatible with τ_3.

As a special case, the *empty module* $\mathcal{E}$ is a constraint module such that Input$(\mathcal{E}) =$ Output$(\mathcal{E}) = \emptyset$ and Solutions$(\mathcal{E}) = \{\tau_e\}$, where τ_e is the *empty assignment*. Given any module system $\mathcal{M}$, both $\mathcal{E} \triangleright \mathcal{M}$ and $\mathcal{M} \triangleright \mathcal{E}$ are defined, and $\mathcal{E} \triangleright \mathcal{M} \equiv \mathcal{M} \triangleright \mathcal{E} \equiv \mathcal{M}$.

Definition 7 (Projecting output of a module system). *Given a module system $\mathcal{M}$ and set of variables $\mathcal{O}$, the module system $\pi_{\mathcal{O}}(\mathcal{M})$ is defined if and only if $\mathcal{O} \subseteq$ Output$(\mathcal{M})$. The set of satisfying assignments of $\pi_{\mathcal{O}}(\mathcal{M})$, Solutions$(\pi_{\mathcal{O}}(\mathcal{M}))$, is*

$$\{\pi_{\mathcal{O} \cup \text{Input}(\mathcal{M})}(\tau) \mid \tau \in \text{Solutions}(\mathcal{M})\}.$$

Example 5. Consider the module system $\mathcal{M} \triangleright \mathcal{M}'$ from Example 4 and assume that we are not interested in the values assigned to n. Thus, we consider the projection $\mathcal{M}_\pi = \pi_{\{m\}}(\mathcal{M} \triangleright \mathcal{M}')$. Now Input$(\mathcal{M}_\pi) = \{a\}$, Output$(\mathcal{M}_\pi) = \{m\}$, and Solutions$(\mathcal{M}_\pi) = \{\tau_\pi\}$ where $\tau_\pi = \pi_{\{a,m\}}(\tau_1 \cup \tau_3) = \{a \mapsto \mathbf{f}, m \mapsto 2\}$.

We are interested in so called *well-formed* module systems that respect the conditions for applying $\triangleright$ (independence) and π_X (projection is focused on output).

Definition 8 (Well-formed module system). *A module system is well-formed if each composition and projection operation is defined in the sense of Definitions 6 and 7.*

Determining whether an arbitrary module system is well-formed consists of a syntactic check on the compositionality and compatibility of the I/O interfaces ($\triangleright$) and subset relation (π). From now on we use the term *module system* to refer to a well-formed module system. The graph formed by taking into account the input-output dependencies of parts of a module system is *directed* and *acyclic*, and is referred to as the *module dependency graph*. More precisely, the module dependency graph of a given module system $\mathcal{M}$ has the set of constraint modules appearing in $\mathcal{M}$ as the set of vertices. There is a edge from a constraint module $\mathcal{M}_1$ to module $\mathcal{M}_2$ if and only if at least one output variable of $\mathcal{M}_1$ is an input variable of $\mathcal{M}_2$. Notice that the acyclicity comes from that fact that recursive definitions can be stated only inside individual modules.

By definition, the semantics of a well-formed module system is *compositional*: compatible solutions for individual parts form a solution for the whole system and a solution for the module system gives solutions for the individual parts.

Remark 1. Operators for ▷ and π_X provide flexible ways for building complex module systems. Additional operators useful in practice can be defined as combinations of these basic operators. For instance, by combining composition with projection we obtain $\mathcal{M}_1 \blacktriangleright \mathcal{M}_2$ defined as $\pi_{\mathsf{Output}(\mathcal{M}_2)}(\mathcal{M}_1 \triangleright \mathcal{M}_2)$. One could also be interested in a non-deterministic choice of solutions for $\mathcal{M}_1$ and $\mathcal{M}_2$ (denoted $\mathcal{M}_1 \cup \mathcal{M}_2$) or common solutions for $\mathcal{M}_1$ and $\mathcal{M}_2$ (denoted $\mathcal{M}_1 \cap \mathcal{M}_2$). In order to define $\mathcal{M}_1 \cup \mathcal{M}_2$ and $\mathcal{M}_1 \cap \mathcal{M}_2$, we cannot assume that $\mathcal{M}_1$ and $\mathcal{M}_2$ are composable. However, even these operators can be expressed in terms of composition and projection using an additional renaming scheme for variables.

5   Module Systems in Practice

We now outline how the framework for module systems developed in the previous section can be instantiated in practice. First, we consider a demanding application to illustrate the issues arising in using a multi-language modeling approach. Then we sketch the required language interface for constructing a multi-language module system.

5.1   Modular Representation for the Timetabling Domain

For illustrating multi-language modeling, we describe components involved in a modular constraint model for *university timetabling*, variants of which have previously been formalized in SAT, CP, and ASP [18,19]. Designing a feasible weekly schedule for events related to courses in a university curriculum is a challenging task. The problem is not just about allocation time and space resources; the interdependencies of courses and the respective events give rise to a rich body of constraints. For modeling, one needs to express the mutual exclusion of events as regards, e.g., placing any two events in the same lecture hall at the same time. A straightforward representation of such a constraint with clauses or rules may require quadratic space. In contrast, a concise encoding can be obtained with global constraints such as *all-different* or *cumulative* constraints typically supported by constraint programming systems. On the other hand, there are features which are cumbersome to describe in CP. For example, exceptions like the temporary unavailability of a particular lecture hall in a timetable are easy to represent with non-monotonic rules such as those used in ASP. Moreover, rules provide a flexible way of defining new relations on the basis of existing ones.

The structure of a modular constraint model for the university timetabling domain is given in Fig. 1. The two ASP modules at the bottom define relations specific to a particular problem instance. The first module, *eventData*, defines which events are involved in the problem. The second, *resourcesData*, formalizes the time and space resources available for scheduling. An individual *resource* is conceptualized as a pair $\langle r, s \rangle$ where r is a room and s is a session. The ASP module on top of these two modules, *dataViews*, defines a number of subsidiary relations, such as ROOM(r) (available rooms) and LECTURER(l) (involved lecturers), on the basis of the relations provided by modules *eventData* and *resourcesData*. The relations MAXEVENT(n) and MAXRESOURCE(m) hold (only) for the numbers of events n and resources m, respectively. After suitable type conversions (represented by the circles in Fig. 1), these two size parameters serve as input for the CP module *allDifferent* whose purpose is to assign different resources (represented by integers in the range $1 \ldots m$) to all events (represented by an array of integers indexed by $1 \ldots n$). Through such a conversion, a constraint library implementation of *allDifferent* which works only on integer-valued variables can be directly used. The resulting array of assignments of integers, RESOURCEOF, is then converted to a relation for events e and resources r and the ASP module *allocation* is used to restore the representation of resources as integers back to pairs of rooms and sessions. The outcome relation OCCURS(e, r, s) denotes the fact that an event e takes place in room r during session s. The topmost module *testAllocation* ensures that the given allocation of resources to events, i.e., the relation OCCURS(e, r, s) meets further criteria of interest. For instance, one could insist on the property that sessions related with a particular lecture hall are always reserved in a contiguous manner, i.e., no gaps are allowed between reservations in the respective schedule.

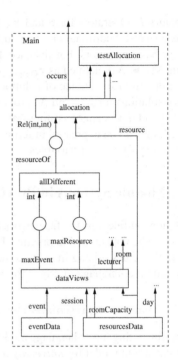

Fig. 1. Example of a Module System

5.2 Language Interface for Combining Constraint Modules

Referring to the theory developed in Sect. 3 and 4, we distinguish two types of module declarations. An individual constraint module is written in a particular constraint language accompanied by an appropriate I/O interface specification. The language of each constraint module is declared using an identifier "SAT", "ASP", "CP", etc. A module system is effectively a definition of the interconnections between submodules encapsulated by it. Since module systems are not confined to a particular constraint language the identifier "SYSTEM" is used. In addition, simple *type converters* are declared when needed, as outlined above.

In practice, a module system is not described as an expression (recall Definition 4) using explicitly composition and projection operators. Instead, it is very useful to give primitive constraint module descriptions as schemata which can be reused by instantiating them with appropriate input and output variables. To support this we follow an approach which handles module instantiation and composition simultaneously. Modules are instantiated using a declaration [outputlist] = modulename(inputlist); where modulename is the name of the module being instantiated, and inputlist and outputlist are the lists of input and output variables, respectively. This allows for writing a module composition $\mathcal{M}_1 \triangleright \mathcal{M}_2$ as suitable module instantiations: $[x_1, x_2, \ldots] = \mathcal{M}_1(\ldots); [\ldots] = \mathcal{M}_2(x_1, x_2, \ldots)$; where appropriate output variables of $\mathcal{M}_1$ are used as input variables of $\mathcal{M}_2$. A module system is described as a sequence $\mathcal{M}_1; \mathcal{M}_2; \ldots; \mathcal{M}_n$; of such instantiation declarations which is acyclic, i.e.,

```
#module ASP dataViews
  (Rel(int, string, string, int, string, int) event,
  Rel(int, string, int) resource,
  Rel(int) session,
  Rel(string,int) roomCapacity)
  [Rel(int) maxEvent,
  Rel(int) maxResource,
  Rel(string) room,
  Rel(string) lecturer]

% Determine problem dimensions
eventId(I) :- event(I,CC,T,D,L,C).
maxEvent(I) :- eventId(I), not eventId(I+1).
resourceId(I) :- resource(I,R,S).
maxResource(I) :- resourceId(I), not resourceId(I+1).

% Rooms and personnel
room(R) :- resource(I,R,S).
lecturer(L) :- event(I,CC,l,D,L,C), L!=noname.

...

#endmodule
```

```
#module SYSTEM main()

% Data (problem instance)
[event] = eventData();
[day,session,resource,roomCapacity] =
    resourcesData();

% Different views of data
[maxEvent, maxResource, room, lecturer] =
    dataViews(event,resource,session,roomCapacity);

% Allocating resources
[resourceOf] =
    allDifferent(indexOfTrueElement(maxEvent),
                 indexOfTrueElement(maxResource));

% Recover rooms and sessions from resources
[occurs] = allocation(resource,
                      arrayToRel(resourceOf));

% Checking the feasibility allocation
[] = testAllocation(occurs);

#solve[occurs]

#endmodule
```

Fig. 2. Examples of a constraint module and a module system as illustrated in Fig. 1

output variables of $\mathcal{M}_i$ cannot be used as input variables for any $\mathcal{M}_j$, $j \leq i$. This guarantees that the set of declarations can be seen as a well-formed composition $\mathcal{M}_1' \triangleright (\mathcal{M}_2' \triangleright (\ldots \triangleright \mathcal{M}_n') \ldots)$ where $\mathcal{M}_i'$s are the corresponding instantiated constraint modules. The projection operator is handled implicitly in the instantiation of modules. For the top level of a module system we provide an explicit projection operator as the #solve[·] directive for defining the actual output variables of the whole module system.

A simplified example of a constraint module and a module system is given in Fig. 2. Each module description begins with a header line. The keyword "#module" is followed by (i) the language identifier, e.g., SAT, ASP, CP, or SYSTEM, (ii) the name of the module, and (iii) the specification of input and output variables enclosed in parentheses "(. . .)" and brackets "[. . .]", respectively. The types of variables are declared using elementary types (int, string, . . .) and type constructors such as Rel.[2] Local variables (if any) and their types are declared with lines that begin with the keyword #type. A module description ends with a line designated by a keyword #endmodule. The module instantiation declarations need to be well-typed, i.e., the given input and output variables must conform to the module interfaces. The top-level module is distinguished by the reserved name main and the #solve directive for defining the output variables of the whole module system can be used only there.

6   Computational Aspects and Benefits of the Modular Approach

In this section we consider computational aspects related to module systems. First we analyze how certain computational properties of individual constraint modules are related to those of more complex module systems. Then we show how the structure of a

---

[2] Description of a complete typing mechanism is beyond the scope of this paper. For now, we aim at type specifications which allow for static type checking.

module system can be exploited when one is interested in finding a satisfying assignment for a subset of the output variables of the module system.

We describe computational properties of a constraint module under the terms *checkable*, *solvable*, and *finite output for fixed input*, defined as follows.

Definition 9. *A constraint module* $\mathcal{M} = \langle \mathcal{C}, \mathcal{I}, \mathcal{O} \rangle$

- *is checkable if and only if given any assignment τ over the variables in $\mathcal{I} \cup \mathcal{O}$, it can be decided whether $\tau \in$ Solutions($\mathcal{M}$);*
- *is solvable if and only if there is a computable function that, given any assignment τ over the variables in $\mathcal{I}$, returns an assignment in* SolutionOut($\mathcal{M}, \tau$) *if one exists, and reports unsatisfiability otherwise; and*
- *has FOFI (finite output for fixed input) if and only if (i) the set* SolutionOut($\mathcal{M}, \tau$) *is finite for any assignment τ over the variables in $\mathcal{I}$, and (ii) there is a computable function that, given any assignment τ over the variables in $\mathcal{I}$, outputs* SolutionOut($\mathcal{M}, \tau$).

In general, a constraint module which has FOFI is both checkable and solvable. However, a solvable (checkable, respectively) module is not necessarily checkable (solvable, respectively).

The knowledge about a specific property for $\mathcal{M}$ and $\mathcal{M}'$ is not necessarily enough to guarantee that the property holds for a module system obtained using $\mathcal{M}$ and $\mathcal{M}'$. Clearly, if $\mathcal{M}$ and $\mathcal{M}'$ are checkable, then $\mathcal{M} \triangleright \mathcal{M}'$ is checkable, too. Solvability of $\mathcal{M}$ and $\mathcal{M}'$ does not, however, imply that $\mathcal{M} \triangleright \mathcal{M}'$ is solvable. For instance, let $\mathcal{M} = \langle \mathcal{C}, \emptyset, \{a\} \rangle$ and $\mathcal{M}' = \langle \mathcal{C}', \{a\}, \{b\} \rangle$ be solvable constraint modules such that Solutions($\mathcal{M}$) $= \{\{a \mapsto 1\}, \{a \mapsto 2\}\}$, Solutions($\mathcal{M}'$) $= \{\{a \mapsto 2, b \mapsto 2\}\}$, and $\mathcal{M} \triangleright \mathcal{M}'$ is defined. Assume that the computable function for $\mathcal{M}$ always returns $\tau = \{a \mapsto 1\}$. Now, SolutionOut($\mathcal{M}', \tau$) $= \emptyset$, and which leads us to think that Solutions($\mathcal{M} \triangleright \mathcal{M}'$) $= \emptyset$. But this is in contradiction with Solutions($\mathcal{M} \triangleright \mathcal{M}'$) $= \{\{a \mapsto 2, b \mapsto 2\}\}$. If we in addition assume that $\mathcal{M}$ and $\mathcal{M}'$ have the FOFI property, then $\mathcal{M} \triangleright \mathcal{M}'$ is solvable and, moreover, has the FOFI property.

For projection, the situation is slightly different. If $\mathcal{M}$ is a checkable constraint module, then $\pi_{\mathcal{O}}(\mathcal{M})$ is not necessarily checkable for $\mathcal{O} \subset$ Output($\mathcal{M}$). Given τ over Input($\mathcal{M}$) $\cup \mathcal{O} \subset$ Input($\mathcal{M}$) $\cup$ Output($\mathcal{M}$), we cannot decide whether $\tau \in$ Solutions($\pi_{\mathcal{O}}(\mathcal{M})$) as we do not know the assignment for variables in Output($\mathcal{M}$) $\setminus \mathcal{O}$. If, in addition, $\mathcal{M}$ is solvable, then using the projection τ' of τ to Input($\mathcal{M}$) we can compute $\tau'' \in$ SolutionOut($\mathcal{M}, \tau'$) and the projection of τ'' to $\mathcal{O}$. Thus $\pi_{\mathcal{O}}(\mathcal{M})$ is solvable.

Proposition 1. *Let $\mathcal{M}$ and $\mathcal{M}'$ be constraint modules s.t. $\mathcal{M} \triangleright \mathcal{M}'$ is defined, and $\mathcal{O} \subseteq$ Output($\mathcal{M}$). If $\mathcal{M}$ and $\mathcal{M}'$ are checkable, $\mathcal{M} \triangleright \mathcal{M}'$ is checkable. If $\mathcal{M}$ is solvable, $\pi_{\mathcal{O}}(\mathcal{M})$ is solvable. If $\mathcal{M}$ and $\mathcal{M}'$ have FOFI, $\mathcal{M} \triangleright \mathcal{M}'$ and $\pi_{\mathcal{O}}(\mathcal{M})$ have FOFI.*

Based on the concepts of *total module systems* and *don't care variables*, the *cone-of-influence* of a system is intuitively the part of the system that may influence the values of output variables of interest. We will define the *cone-of-influence reduction* for module systems which can be used in disregarding parts of a module system in the case we are only interested in the values assigned to a subset of the output of the system.

Definition 10. *A constraint module* $\mathcal{M}$ *is* total *if* $\mathsf{SolutionOut}(\mathcal{M}, \tau) \neq \emptyset$ *for all assignments* τ *over* $\mathsf{Input}(\mathcal{M})$.

If $\mathcal{M}_1$ and $\mathcal{M}_2$ are total module systems such that $\mathcal{M}_1 \triangleright \mathcal{M}_2$ is defined, then $\mathcal{M}_1 \triangleright \mathcal{M}_2$ is total. Furthermore $\pi_\mathcal{O}(\mathcal{M})$ is total for any total $\mathcal{M}$ and $\mathcal{O} \subseteq \mathsf{Output}(\mathcal{M})$.

Seen as a black-box entity, testing totality from the outside is hard even on the level of constraint modules. However, if the declarative implementation of the module is known, there are easy-to-test syntactic conditions guaranteeing totality. For example, in Boolean circuit satisfiability, we know that if no gate of a circuit is constrained to a specific truth value, any module implemented by such a Boolean circuit is total. In practice, when implementing reusable modules for inclusion in a module library, the totality of a module could be explicitly declared in the module interface specification.

Definition 11. *Given a constraint module* $\mathcal{M}$, $x \in \mathsf{Input}(\mathcal{M})$, *and* $y \in \mathsf{Output}(\mathcal{M})$, *we say that* x *is a* $\mathcal{M}$*-don't care w.r.t.* y, *if for any assignment* τ *over* $\mathsf{Input}(\mathcal{M}) \setminus \{x\}$,
$$\{\pi_{\{y\}}(\tau') \mid \tau' \in \mathsf{SolutionOut}(\mathcal{M}, \tau \cup \tau_1)\} = \{\pi_{\{y\}}(\tau') \mid \tau' \in \mathsf{SolutionOut}(\mathcal{M}, \tau \cup \tau_2)\}$$
for all pairs of assignments τ_1, τ_2 *for* x.

As in the case of totality, in general checking whether a given input variable is a don't care is hard when constraint modules are seen as black-box entities. But again, if the declarative implementation of the module is known, there are easy-to-test syntactic conditions which guarantee that a variable is a don't care. For example, if a CNF formula can be split into two disjoint components, i.e., sets of clauses which do not share variables. A similar check can be done, e.g., for ASP programs and CSP instances.

In addition to totality and don't cares, we use the concept of *relevant I/O variables*. Let $\mathsf{CM}(\mathcal{M})$ denote the set of constraint modules appearing in a module system $\mathcal{M}$. For instance, if $\mathcal{M} = \pi_\mathcal{O}(\mathcal{M}_1 \triangleright \mathcal{M}_2)$ then $\mathsf{CM}(\mathcal{M}) = \{\mathcal{M}_1, \mathcal{M}_2\}$.

Definition 12. *Given a module system* $\mathcal{M}$ *and* $\mathcal{O} \subseteq \mathsf{Output}(\mathcal{M})$, *the set of relevant I/O variables in* $\mathcal{M}$ *w.r.t.* $\mathcal{O}$, *denoted by* $\mathsf{Rel}(\mathcal{M}, \mathcal{O})$, *is the smallest set* $S \supseteq \mathcal{O}$ *of variables that fulfills the following conditions:*

- $\mathsf{Input}(\mathcal{M}') \subseteq S$ *for each non-total* $\mathcal{M}' \in \mathsf{CM}(\mathcal{M})$.
- *If* $y \in S$, *then for each total* $\mathcal{M}' \in \mathsf{CM}(\mathcal{M})$ *such that* $y \in \mathsf{Output}(\mathcal{M}')$,
  $\{x \in \mathsf{Input}(\mathcal{M}') \mid x \text{ is not } \mathcal{M}'\text{-don't care w.r.t. } y\} \subseteq S$.

The cone-of-influence reduction allows the parts not belonging to the cone-of-influence to be neglected when solving the constraint model.

Definition 13. *Given a module system* $\mathcal{M}$ *and a set* X *of variables, the* module system reduction $\mathcal{M}|_X$ *is defined as follows.*

- *If* $\mathcal{M}$ *is a constraint module, then*
  $$\mathcal{M}|_X = \begin{cases} \mathcal{E} \text{ (the empty module)}, & \text{if } \mathsf{Output}(\mathcal{M}) \cap X = \emptyset \text{ and } \mathcal{M} \text{ is total} \\ \mathcal{M} & , \text{ otherwise.} \end{cases}$$
- *If* $\mathcal{M}$ *is of the form* $\mathcal{M}_1 \triangleright \mathcal{M}_2$, *then* $\mathcal{M}|_X = (\mathcal{M}_1|_X \triangleright \mathcal{M}_2|_X)$.
- *If* $\mathcal{M}$ *is of the form* $\pi_O(\mathcal{M}')$, *then* $\mathcal{M}|_X = \pi_{\mathsf{Output}(\mathcal{M}'|_X) \cap O}(\mathcal{M}'|_X)$.

Given a module system $\mathcal{M}$ *and a set of variables* $\mathcal{O}$, *the* cone-of-influence reduction *of* $\mathcal{M}$ *w.r.t.* $\mathcal{O}$ *is the module system* $\mathcal{M}|_{\mathsf{Rel}(\mathcal{M}, \mathcal{O})}$.

For finding a satisfying assignment for $\mathcal{O} \subseteq \mathsf{Output}(\mathcal{M})$ of a module system $\mathcal{M}$, one needs to consider only the *subsystem* $\mathcal{M}|_{\mathsf{Rel}(\mathcal{M},\mathcal{O})}$ of $\mathcal{M}$.

Proposition 2. *Given a module system $\mathcal{M}$ and a set of variables $\mathcal{O} \subseteq \mathsf{Output}(\mathcal{M})$, then $\{\pi_{\mathcal{O}}(\tau) \mid \tau \in \mathsf{Solutions}(\mathcal{M})\} = \{\pi_{\mathcal{O}}(\tau) \mid \tau \in \mathsf{Solutions}(\mathcal{M}|_{\mathsf{Rel}(\mathcal{M},\mathcal{O})})\}$.*

Example 6. Consider the module system $\mathcal{M} = (\mathcal{M}_1 \triangleright \mathcal{M}_2) \triangleright (\mathcal{M}_3 \triangleright \mathcal{M}_4)$ illustrated in Fig. 3. Thus, $\mathsf{Input}(\mathcal{M}) = \{a, b, c\}$ and $\mathsf{Output}(\mathcal{M}) = \{d, e, f, g\}$. The constraint module $\mathcal{M}_2$ represented with gray in Fig. 3 is not total, while the other constraint modules in $\mathsf{CM}(\mathcal{M})$, i.e., $\mathcal{M}_1$, $\mathcal{M}_3$, and $\mathcal{M}_4$, are total. Assume that, in addition, it is known that e and f are $\mathcal{M}_4$-don't cares w.r.t. g. Assume that we are only interested in finding a satisfying assignment for $\mathcal{O} = \{g\}$. By Proposition 2 we can exploit the cone-of-influence reduction. The set of relevant I/O variables $\mathsf{Rel}(\mathcal{M}, \mathcal{O}) = X = \{a, b, c, d, g\}$ because $\mathcal{O} \subseteq \mathsf{Rel}(\mathcal{M}, \mathcal{O})$, $\mathsf{Input}(\mathcal{M}_2) \subseteq \mathsf{Rel}(\mathcal{M}, \mathcal{O})$, d is not $\mathcal{M}_4$-don't care w.r.t. $g \in \mathsf{Output}(\mathcal{M}_4)$, and a and b are not $\mathcal{M}_1$-don't cares w.r.t. $d \in \mathsf{Output}(\mathcal{M}_1)$. Using the set of relevant I/O variables, the cone-of-influence reduction of $\mathcal{M}$ w.r.t. $\mathcal{O}$ is

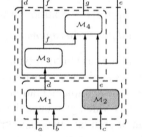

$$\begin{aligned}
\mathcal{M}|_{\mathsf{Rel}(\mathcal{M},\mathcal{O})} &= (\mathcal{M}_1 \triangleright \mathcal{M}_2)|_X \triangleright (\mathcal{M}_3 \triangleright \mathcal{M}_4)|_X \\
&= (\mathcal{M}_1|_X \triangleright \mathcal{M}_2|_X) \triangleright (\mathcal{M}_3|_X \triangleright \mathcal{M}_4|_X) \\
&= (\mathcal{M}_1 \triangleright \mathcal{M}_2) \triangleright (\mathcal{E} \triangleright \mathcal{M}_4) \\
&= (\mathcal{M}_1 \triangleright \mathcal{M}_2) \triangleright \mathcal{M}_4.
\end{aligned}$$

Fig. 3. A module system

7 Conclusions

We develop a generic framework for module-based constraint modeling using multiple modeling languages within the same model. In the framework, constraint models are constructed as module systems which are composed of constraint modules each having an explicit input/output interface specification. This approach has many interesting properties. First of all, individual constraint modules can be implemented using a constraint language most suitable for modeling the constraint in question. The approach paves the way for reusable constraint module libraries and also allows for multiple modelers to implement parts of a constraint model in parallel. Our framework supports modular multi-language modeling by treating different constraint languages on equal terms whereas previous approaches can be seen as extensions of a given basic language with features from other languages. The modular construction of constraint models as module systems yields in itself a structured view to the model which can be exploited when solving the model. We describe a system-level cone-of-influence reduction, which allows parts of the module system to be disregarded when solving a constraint model, without the need to consider properties specific to the constraint languages employed in implementing the individual constraint modules.

For further work, we see a number of possible approaches to solving constraint models expressed using the multi-language framework. In a hybrid system individual constraint modules (or parts of the module system modeled using the same constraint language) are solved using language-specific solvers which have to interact in order to compute solutions to the whole constraint model. In a translation-based approach all parts of the model are mapped into a single constraint language for which highly efficient off-the-shelf solvers are available. For example, there is interesting recent work on bit-blasting more general CP models into SAT [20]. Another interesting paradigm is the extension of SAT to Satisfiability Module Theories (SMT), into which e.g. stable model computation can be very compactly encoded [21]. Additionally, the modular structure of module systems poses interesting research topics such as harnessing the I/O interfaces in developing novel decision heuristics and devising techniques to instantiate and ground module schemata lazily.

References

1. Eiter, T., Gottlob, G., Veith, H.: Modular logic programming and generalized quantifiers. In: Fuhrbach, U., Dix, J., Nerode, A. (eds.) LPNMR 1997. LNCS, vol. 1265, pp. 290–309. Springer, Heidelberg (1997)
2. Baral, C., Dzifcak, J., Takahashi, H.: Macros, macro calls and use of ensembles in modular answer set programming. In: Etalle, S., Truszczyński, M. (eds.) ICLP 2006. LNCS, vol. 4079, pp. 376–390. Springer, Heidelberg (2006)
3. Balduccini, M.: Modules and signature declarations for A-Prolog: Progress report. In: SEA, pp. 41–55 (2007)
4. Oikarinen, E., Janhunen, T.: Achieving compositionality of the stable model semantics for smodels programs. Theory and Practice of Logic Programming 8(5-6), 717–761 (2008)
5. Janhunen, T.: Modular equivalence in general. In: ECAI, pp. 75–79. IOS Press, Amsterdam (2008)
6. Eiter, T., Ianni, G., Schindlauer, R., Tompits, H.: A uniform integration of higher-order reasoning and external evaluations in answer-set programming. In: IJCAI, pp. 90–96 (2005)
7. Elkabani, I., Pontelli, E., Son, T.: Smodels[a] - a system for computing answer sets of logic programs with aggregates. In: Baral, C., Greco, G., Leone, N., Terracina, G. (eds.) LPNMR 2005. LNCS (LNAI), vol. 3662, pp. 427–431. Springer, Heidelberg (2005)
8. Gebser, M., et al.: Clingcon (2009), http://www.cs.uni-potsdam.de/clingcon/
9. Tari, L., Baral, C., Anwar, S.: A language for modular answer set programming: Application to ACC tournament scheduling. In: ASP, pp. 277–292 (2005)
10. Baselice, S., Bonatti, P.A., Gelfond, M.: Towards an integration of answer set and constraint solving. In: Gabbrielli, M., Gupta, G. (eds.) ICLP 2005. LNCS, vol. 3668, pp. 52–66. Springer, Heidelberg (2005)
11. Mellarkod, V., Gelfond, M., Zhang, Y.: Integrating answer set programming and constraint logic programming. Ann. Math. Artif. Intell. 53(1-4), 251–287 (2008)
12. Castro, L., Swift, T., Warren, D.: Xasp (2009), http://xsb.sourceforge.net/
13. El-Khatib, O., Pontelli, E., Son, T.: Integrating an answer set solver into Prolog: ASP-PROLOG. In: Baral, C., Greco, G., Leone, N., Terracina, G. (eds.) LPNMR 2005. LNCS (LNAI), vol. 3662, pp. 399–404. Springer, Heidelberg (2005)
14. Pontelli, E., Son, T., Baral, C.: A logic programming based framework for intelligent web services composition. In: Managing Web Services Quality: Measuring Outcomes and Effectiveness. IDEA Group Publishing (2008)
15. Flener, P., Pearson, J., Ågren, M.: Introducing ESRA, a relational language for modelling combinatorial problems. In: Bruynooghe, M. (ed.) LOPSTR 2004. LNCS, vol. 3018, pp. 214–232. Springer, Heidelberg (2004)
16. Frisch, A., Harvey, W., Jefferson, C., Hernández, B.M., Miguel, I.: ESSENCE: A constraint language for specifying combinatorial problems. Constraints 13(3), 268–306 (2008)

17. Marriott, K., Nethercote, N., Rafeh, R., Stuckey, P., de la Banda, M.G., Wallace, M.: The design of the Zinc modelling language. Constraints 13(3), 229–267 (2008)
18. Goltz, H.J., Matzke, D.: University timetabling using constraint logic programming. In: Gupta, G. (ed.) PADL 1999. LNCS, vol. 1551, pp. 320–334. Springer, Heidelberg (1999)
19. Perri, S., Scarcello, F., Catalano, G., Leone, N.: Enhancing DLV instantiator by backjumping techniques. Ann. Math. Artif. Intell. 51(2-4), 195–228 (2007)
20. Huang, J.: Universal Booleanization of constraint models. In: Stuckey, P.J. (ed.) CP 2008. LNCS, vol. 5202, pp. 144–158. Springer, Heidelberg (2008)
21. Niemelä, I.: Stable models and difference logic. Ann. Math. Artif. Intell. 53(1-4), 313–329 (2008)

Induction on Failure:
Learning Connected Horn Theories

Tim Kimber, Krysia Broda, and Alessandra Russo

Imperial College London, UK
{timothy.kimber06,k.broda,a.russo}@imperial.ac.uk

Abstract. Several learning systems based on Inverse Entailment (IE) have been proposed, some that compute single clause hypotheses, exemplified by Progol, and others that produce multiple clauses in response to a single seed example. A common denominator of these systems is a restricted hypothesis search space, within which each clause must individually explain some example E, or some member of an abductive explanation for E. This paper proposes a new IE approach, called *Induction on Failure* (IoF), that generalises existing Horn clause learning systems by allowing the computation of hypotheses within a larger search space, namely that of *Connected Theories*. A proof procedure for IoF is proposed that generalises existing IE systems and also resolves Yamamoto's example. A prototype implementation is also described. Finally, a semantics is presented, called *Connected Theory Generalisation*, which is proved to extend Kernel Set Subsumption and to include hypotheses constructed within this new IoF approach.

Keywords: Inductive Logic Programming, Inverse Entailment, Abduction.

1 Introduction

Inductive Logic Programming (ILP) uses the expressive power of first-order logic and the sound theoretical foundations of logic programming, to create a branch of machine learning that seeks the construction of explanations for given examples relative to some background knowledge, in a form easily understood by the user. Among ILP systems that perform best on practical applications are those that use the *Inverse Entailment* (IE) approach to learn Horn theories [1,2,3,4]. These systems first construct a *most specific* hypothesis and then search through formulas that subsume it. The restricted search space that IE defines contributes to the efficiency of these systems. However, the assumptions made by current Horn theory IE systems, regarding the number of clauses necessary to explain a given example, mean that some hypotheses that correctly explain the observations are excluded.

The Progol systems [1,5] assume that every clause in a hypothesis H must individually explain at least one example E. This excludes hypotheses in which two

E. Erdem, F. Lin, and T. Schaub (Eds.): LPNMR 2009, LNCS 5753, pp. 169–181, 2009.
© Springer-Verlag Berlin Heidelberg 2009

or more clauses combine to explain some E, a situation particularly, though not exclusively, relevant to *non-observational learning*. Consider, for example, the observation $E = \{p(a, b)\}$, background knowledge $B = \{p(X, Y) \leftarrow q(X), r(Y)\}$ and hypothesis $H = \{q(X)\} \cup \{r(X)\}$. This H is outside the Progol search space. A single clause that does explain some E may still be excluded, because the assumption is applied at the ground level first. For example, if the background clause above was in fact $p(X, Y) \leftarrow q(X), q(Y)$, the clause $H = q(X)$ explains E. However, the ground explanation for $p(a, b)$ is $\{q(a)\} \cup \{q(b)\}$ which contains two clauses, meaning it and its subsuming hypothesis $q(X)$ are not returned. The HAIL [3] and ALECTO [4] systems allow a multiple clause explanation of a single *seed* example, and do compute the hypotheses described so far. However, the single clause assumption is only weakened in these systems, not removed completely. Thus, for the previous example HAIL computes the ground unit explanation $\Delta = \{q(a), q(b)\}$ by abduction, and generates one, and only one, clause for each member of Δ. The next example shows that this approach also excludes some correct explanations.

Example 1.

$$E_1 = vehicle(focus)$$
$$B_1 = \{doors(focus, 5)\} \cup \{car(X) \leftarrow hatchback(X)\}$$
$$H_1 = \{vehicle(X) \leftarrow car(X)\} \cup \{hatchback(X) \leftarrow doors(X, 5)\}$$

◆

The abductive explanation $\Delta_1 = \{vehicle(focus)\}$, will constrain HAIL's search space to be the same as Progol's for this example, including only single clauses subsuming $H_1' = vehicle(X) \leftarrow doors(X, 5)$. The two-clause theory H_1 is outside this search, and cannot be computed by any of the ILP systems described.

Intuitively, one can see that H_1 is a structured explanation for E_1, and if the observation set includes data on cars and hatchbacks then it will be preferred to H_1' since it will cover examples with those predicates. This paper proposes a new IE approach, within the setting of Horn clauses, that enables the automatic computation of hypotheses such as H_1, as well as those considered by existing Horn ILP systems. This approach, called *Induction on Failure* (IoF), is based on the notion of a *Connected Theory*. A Connected Theory contains clauses that depend on one another, either directly or via clauses in the background knowledge. IoF is designed to find hypotheses as generalisations of a most specific Connected Theory, denoted $T_\perp$. The semantics of *Connected Theory Generalisation* is presented in Sect. 3, and is shown to define a set of hypotheses that extends those of Progol's Bottom Generalisation (BG) and HAIL's Kernel Set Subsumption (KSS).

The IoF approach computes the theory $T_\perp$ via a recursive inductive procedure. Where current Horn IE systems use a purely deductive *saturation* step to generate the body of clauses, IoF considers body literals which initially lack explanation, starting a new induction process for them. This recursive induction has as base case the standard saturation process adopted by existing systems. A full description of the IoF framework is given in Sect. 4.

In Sect. 5 it is also shown, using a prototype implementation called Imparo, that the proposed IoF approach is general enough to resolve Yamamoto's counter example to the completeness of IE [6], demonstrating its suitability for learning mutually recursive programs. Section 2 gives the necessary background information on ILP, and Sect. 6 describes related and future work.

2   Background

This section presents notation and terminologies used throughout the paper and briefly reviews the ILP task, the IE principle and the methods of Bottom Generalisation and Kernel Set Subsumption.

Notation and Terminology. All formulas are assumed to be constructed from a first-order logic signature. A first-order *clause* is a finite disjunction of zero or more literals $A_1 \vee \ldots \vee A_m \vee \neg A_{m+1} \vee \ldots \vee \neg A_n$, which is assumed to be universally quantified and will be written as the implication $A_1, \ldots, A_m \leftarrow A_{m+1}, \ldots, A_n$. A Horn clause is either the *empty clause* $\Box$, or a *fact* A, or a *denial*, $\leftarrow A_1, \ldots, A_n$, or a *definite clause* $A_0 \leftarrow A_1, \ldots, A_n$, where A_0 is the *head* and $A_1, \ldots, A_n$ is the *body* of the clause. Capitalised identifiers denote meta variables. If C is a Horn clause, then C^+ is the set containing the head atom (if any) of C, and C^- is the set of body atoms of C. Moreover, given a set $S = \{C_1, \ldots, C_n\}$ of Horn clauses, S^+ denotes $C_1^+ \cup \ldots \cup C_n^+$ and S^- denotes $C_1^- \cup \ldots \cup C_n^-$. A *theory* is a set of clauses. The symbol $\models$ represents classical entailment. So, if S and T are theories and C is a clause, $S \models C$ means that C is satisfied in every model of S, and $S \models T$ means that $S \models D$ for every $D \in T$. The term *subsumption* and the symbol $\succeq$ refer to θ-subsumption. So, if C and D are clauses, $C \succeq D$ means that the set of literals in $C\theta$ is a (non-strict) subset of the set of literals in D for some substitution θ, and if S and T are theories, $S \succeq T$ means that every clause in T is subsumed by at least one clause in S. Throughout this paper the term *clause* will refer to a Horn clause, unless otherwise stated.

Inductive Logic Programming. Inductive Logic Programming (ILP) is concerned with the task of generalizing positive and negative examples with respect to a background theory. Formally, given background theory B, positive examples E_{pos} and negative examples E_{neg}, the task of ILP is to compute a theory H, called a *hypothesis*, that satisfies the conditions $B \cup H \models E_{pos}$ (*posterior sufficiency*) and $\forall x \in E_{neg}(B \cup H \not\models x)$ (*posterior satisfiability*). The ILP task of finding such an hypothesis H is called *inductive generalisation*. Standard preconditions of an ILP task are *prior necessity*, for which the given background theory must not already entail all the positive examples, $B \not\models E_{pos}$ and *prior satisfiability*, for which B must not entail any of the negative examples, $\forall x \in E_{neg}(B \not\models x)$.

Inverse Entailment. Various existing ILP systems build their approaches to the ILP task upon the principle of *Inverse Entailment* (IE), which states that the negation of every inductive hypothesis H for some example E may be deduced from B and $\neg E$. That is to say $B \cup \{\neg E\} \models \neg H$. This relationship is used to derive a *most specific* hypothesis ϕ, as defined below, which allows the search space for the inductive generalisation to be limited to those H satisfying $H \models \phi$.

Definition 1 (Most Specific Hypothesis). *Let B be a set of clauses and E be a clause such that $B \not\models E$. Then a set of clauses H is a* most specific hypothesis *for E given B if and only if $B \cup H \models E$ and there does not exist a set H' of clauses such that $B \cup H' \models E$ and $H \models H'$ and $H' \not\models H$.*

Systems such as Progol and Aleph [2] employ IE within the technique of *Bottom Generalisation*. Bottom Generalisation computes a most specific hypothesis $\perp(B, E)$ called the *Bottom Clause* of B and E, as defined below. Since $\perp(B, E)$ is a single clause, its negation is a set of (possibly skolemised) ground literals each of which can be derived from B and $\neg E$. Any clause H which implies $\perp(B, E)$ is said to be derivable by BG, as formalised in Definition 3.

Definition 2 (Bottom Clause [1]). *Let B be a Horn theory and E a Horn clause. Then $\perp(B, E)$, the* Bottom Clause *of B and E, is the disjunction of the ground literals in the set $\{ L \mid B \cup \{\neg E\} \models \neg L \}$.*

Definition 3 (Bottom Generalisation [1]). *Let B be a Horn theory and E a Horn clause. A Horn clause H is said to be* derivable from B and E by Bottom Generalisation *if and only if $H \models \perp(B, E)$.*

The Progol system [1], implements Bottom Generalisation by means of Mode Directed Inverse Entailment (MDIE), which computes a definite bottom clause $A_0 \leftarrow A_1, \ldots, A_n$ based on $\perp(B, E)$. Ray et al. [3] showed that such a definite bottom clause could be defined as a head atom A_0 that, together with the background knowledge, explains the head of the example clause (i.e. $B \cup \{A_0\} \models E^+$), and a set of body atoms $\{A_1, \ldots, A_n\}$ that are *directly* derivable from the background knowledge and the body of E (i.e. $B \cup E^- \models A_i$ for each $1 \le i \le n$). *Kernel Set Subsumption* (defined below) extends this notion by defining a most specific hypothesis K, the Kernel Set, that is a *set* of ground Horn clauses where the set of head atoms of K explains the head of the example ($B \cup K^+ \models E^+$) and the body atoms of K are directly derivable from the background knowledge and E^- ($B \cup E^- \models K^-$).

Definition 4 (Kernel Set [3]). *Let B be a Horn theory and E be a Horn clause such that $B \not\models E$. Then a ground Horn theory $K = \{C_1, \ldots, C_k\}$ $(k \ge 1)$ is a Kernel Set of B and E if and only if each clause $C_i, 1 \le i \le k$ is given by $A_0^i \leftarrow A_1^i, \ldots, A_{n_i}^i$, where $B \cup \{A_0^1, \ldots, A_0^k\} \models E^+$, and $B \cup E^- \models \{A_1^1, \ldots, A_{n_k}^k\}$.*

Definition 5 (Kernel Set Subsumption [3]). *Let B be a Horn theory and E be a Horn clause such that $B \not\models E$. A set of Horn clauses H is said to be* derivable from B and E by Kernel Set Subsumption, *denoted $B, E \vdash_{KSS} H$, if and only if there is a K such that K is a Kernel Set of B and E, and $H \succeq K$.*

Kernel Set Subsumption extends Bottom Generalisation since it has been shown that all clauses that can be derived by BG can also be derived by KSS, and that some single clauses not found by BG are found by KSS [3]. Not only this, but KSS allows theories to be derived from a single seed example, whereas BG is limited to clauses. However, the nature of the Kernel Set imposes limits on the possible theories that KSS can compute. The next section describes Connected Theory Generalisation (CTG), a new semantic approach which further generalises KSS, and computes a "more complete" set of hypotheses.

3   Connected Theory Generalisation

This section proposes a new semantic approach to ILP, called *Connected Theory Generalisation* (CTG). This approach is based on the notion of a *Connected Theory*, defined below, and is proved to extend the semantics of Kernel Set Subsumption, which has been shown [3] to extend Bottom Generalisation [1]. The motivation is to extend the class of hypotheses learnable by Horn theory ILP systems.

The Kernel Set extends the notion of a Bottom Clause by replacing the single atom explanation of the given example, $\{A_0\}$, with a set of atoms $\{A_0^1, \ldots, A_0^k\}$. A Connected Theory extends the basic notion even further, not only generalising the head of $\bot(B, E)$ to be a set of atoms, but also extending the set of body atoms beyond those that are direct consequences of the background knowledge. Intuitively, all body atoms in $\bot(B, E)$, a single clause, must be implied by B for $B \cup \{\bot(B, E)\}$ to entail E. In a hypothesis which is a set of clauses this need not be true for a given member of the set. The Kernel Set retains this restriction, but, as explained below, it is relaxed in a Connected Theory leading to a wider class of most specific hypotheses, and thus a larger search space.

In general, for a set T of ground Horn clauses to be a hypothesis for an example E given background knowledge B, T must include a set T_1 of *root clauses* whose head atoms form an (abductive) explanation of the example E:

$$B \cup T_1^+ \models E .$$

If the body atoms of clauses in T_1 are all implied by B ($B \models T_1^-$), then T_1 is a hypothesis for E ($B \cup T_1 \models E$), and also a Kernel Set for B and E. On the other hand, if some body atoms in T_1 are not directly derivable from the background knowledge, T_1 is not an explanation for E. In this case, those body atoms in T_1 that are not consequences of B must themselves be explained. Such atoms are referred to as *secondary examples*, and require that T contains a set T_2 of *auxiliary clauses*, in addition to T_1. In the ground hypothesis shown for Example 1, the atom $car(focus)$ is a secondary example, and is explained by the auxiliary clause $hatchback(focus) \leftarrow doors(focus, 5)$. By analogy with T_1, the heads of the T_2 clauses must abductively explain the secondary examples:

$$B \cup T_2^+ \models T_1^- ,$$

and either $T_1 \cup T_2$ is now a hypothesis for E, or T_2 contains secondary examples, and T must include a further subset T_3, etc. A ground hypothesis such as T is called a *Connected Theory* for B and E, as formalised below.

Definition 6 (Connected Theory). *Let B be a Horn theory and E be a ground Horn clause such that $B \not\models E$. Let $T_1, \ldots, T_n$ be n sets of ground Horn clauses $(n \geqslant 1)$, and let $T = T_1 \cup \ldots \cup T_n$. T is a* Connected Theory *for B and E if and only if (i) $B \cup T_1^+ \models E^+$, (ii) $B \cup E^- \cup T_{i+1}^+ \models T_i^-$ $(1 \leqslant i < n)$, (iii) $B \cup E^- \models T_n^-$, and (iv) $B \cup T \not\models \square$.*

A set of Horn clauses which entails a Connected Theory is said to be derivable by Connected Theory Generalisation, as formalised in Definition 7 below. Theorem 1 shows such a set of clauses to be a correct hypothesis.

Definition 7 (Connected Theory Generalisation). *Let B be a Horn theory and E be a ground Horn clause such that $B \not\models E$. A set H of Horn clauses is said to be* derivable from B and E by Connected Theory Generalisation, *denoted $B, E \vdash_{CTG} H$, if and only if there is a T such that T is a Connected Theory for B and E and $H \models T$.*

Theorem 1 (Soundness of Connected Theory Generalisation). *Let B and H be Horn theories and E be a ground Horn clause such that $B \not\models E$. If $B, E \vdash_{CTG} H$ then $B \cup H \models E$.*

Proof. By Definitions 6 and 7, H entails some Connected Theory T for B and E, and T comprises n subsets $(n > 0)$ $T_1 \ldots T_n$, such that $B \cup T_1^+ \models E^+$, and $B \cup E^- \cup T_{i+1}^+ \models T_i^-$ $(1 \leqslant i < n)$, and $B \cup E^- \models T_n^-$. Since $B \cup E^- \models T_n^-$, and $T \cup T_n^- \models T_n^+$, then $B \cup E^- \cup T \models T_n^+$ by transitivity of $\models$. Also, since $B \cup E^- \cup T_{i+1}^+ \models T_i^-$ and $T \cup T_i^- \models T_i^+$ $(1 \leqslant i < n)$, then $B \cup E^- \cup T \models T_i^+$ $(1 \leqslant i < n)$, and since $B \cup T_1^+ \models E^+$, then $B \cup E^- \cup T \models E^+$ or $B \cup T \models E$. Finally, since $H \models T$, $B \cup H \models E$. $\square$

Thus Connected Theory Generalisation is a sound inductive learning method for Horn theories. Theorem 2, below, shows that CTG extends Kernel Set Subsumption, which itself has been shown to extend Bottom Generalisation [3].

Theorem 2 (CTG extends Kernel Set Subsumption). *Let B be a Horn theory and E be a Horn clause such that $B \not\models E$. The set of hypotheses derivable from B and E by Connected Theory Generalisation strictly includes the set of hypotheses derivable from B and E by Kernel Set Subsumption.*

Proof. The proof is in two parts. First it is shown that all hypotheses derivable by KSS are derivable by CTG. Then it is shown, by means of a counter example, that some hypotheses derivable by CTG are not derivable by KSS. Part (i). Assume $B, E \vdash_{KSS} H$ where H is a set of Horn clauses. Then H subsumes some Kernel Set K of B and E. By Definition 4, K is a set of clauses $\{A_0^j \leftarrow A_1^j, \ldots, A_{n_j}^j, 1 \leq j \leq k\}$, for some k, where $B \cup \{A_0^1, \ldots, A_0^k\} \models E^+$ and $B \cup E^- \models A_i^j$, for $1 \leq i \leq n_j$, and $1 \leq j \leq k$. Therefore $B, E \vdash_{CTG} H$ with the Connected Theory $T = \{A_0^j \leftarrow A_1^j, \ldots, A_{n_j}^j, 1 \leq j \leq k\}$. Hence $B, E \vdash_{KSS} H$ only if $B, E \vdash_{CTG} H$. Part (ii). Let $p/1$ and $q/1$ be predicates, let a and b be constants, let $B = \{q(a) \leftarrow p(b)\} \cup \{q(b)\}$ and let $E = p(a)$. The hypothesis $H = \{p(X) \leftarrow q(X)\}$

is *not* derivable by KSS since it does not θ-subsume $K = \{p(a) \leftarrow q(b)\}$, but H *is* derivable by CTG as it subsumes $T = \{p(a) \leftarrow q(a), q(b)\} \cup \{p(b) \leftarrow q(b)\}$. Hence $B, E \vdash_{CTG} H$ does not imply $B, E \vdash_{KSS} H$.   □

As with both the Bottom Clause and Kernel Set, restricting a Connected Theory to Horn clause logic results in several alternative most specific hypotheses. If the Bottom Clause and the Kernel (the non-Horn formula underpinning the Kernel Set) are represented thus:

$$\bot(B, E) = A_0^1 \vee \ldots \vee A_0^m \leftarrow A_1 \wedge \ldots \wedge A_n \ ;$$
$$Ker = \Delta_1 \vee \ldots \vee \Delta_m \leftarrow A_1 \wedge \ldots \wedge A_n \ ;$$

then alternative definite Bottom Clauses $\bot_i$ (resp. Kernel Sets K) can be seen to comprise one of the atoms (resp. conjunctions of atoms) on the left of the implication, together with all the atoms to the right. All alternative most specific Connected Theories are captured by the equivalent formula

$$T_\bot = S_1 \vee \ldots \vee S_m \leftarrow A_1 \wedge \ldots \wedge A_n \ ,$$

where each S_i is an alternative *set of Horn clauses* that, together with the background knowledge, explains E^+ ($B \cup S_i \models E^+$, $1 \leq i \leq m$) and where each A_i is entailed by the background knowledge and E^- ($B \cup E^- \models A_i$, $1 \leq i \leq n$). This reflects the fact that the alternative most specific Connected Theories differ in their body atoms as well as their heads, as illustrated by Example 2.

Example 2. Find a Horn theory that explains E_2, given background knowledge B_2. Hypothesis clauses may have x, u, v or w in the head, and p, a or b in the body.

$$E_2 = x \qquad B_2 = \{a \leftarrow u, v\} \cup \{b \leftarrow v, w\} \cup \{p\} \qquad ◆$$

There are five alternative most specific hypotheses $T_\bot$ for B_2 and E_2 as shown in Table 1. The simplest, Theory 1, is the Kernel Set. Each of the other theories in the table is also a most specific hypothesis by virtue of the inclusion of different secondary examples, and auxiliary clauses to explain them.

Since none of the different $T_\bot$ imply one another, each is the bottom element of a separate implication lattice that, taken together, form the full search space for the inductive problem. Depending on the amount of underlying structure in the data (in the case of the example, whether u, v or w have also been observed), generalisation of any of the alternative $T_\bot$ could yield the preferred hypothesis.

Table 1. All possible most specific connected theories for B_2 and E_2

Theory	1	2	3	4	5
$T_\bot$	$x \leftarrow p$	$x \leftarrow p, a$ $u \leftarrow p$ $v \leftarrow p$	$x \leftarrow p, b$ $v \leftarrow p$ $w \leftarrow p$	$x \leftarrow p, a, b$ $u \leftarrow p, b$ $v \leftarrow p$ $w \leftarrow p$	$x \leftarrow p, a, b$ $u \leftarrow p$ $v \leftarrow p$ $w \leftarrow p, a$

4 Induction on Failure

This section describes the IoF proof procedure for computing Connected Theories. The key features of the procedure are the inclusion of secondary examples in the saturation phase, and their recursive inductive explanation.

The top level covering loop is shown in Fig. 1. The initial set of examples is divided into positive examples, E_{pos}, which should be provable from the theory at the end of *Cover*, and negative examples, E_{neg} which should not. The hypothesis language is defined by sets of head (M_h) and body (M_b) mode declarations as defined in [1] and briefly explained below. M_h, which is translated into a set A of abducible atoms, M_b and E_{neg} are constant for a given application, and so global to all procedures. The cover loop proceeds by selecting a seed example $E \in E_{pos}$, and generating a hypothesis H to explain it. H is added to B and all members of E_{pos} implied, or "covered", by the new program are removed from the set. The loop continues until E_{pos} is empty.

The goal $\leftarrow E$ is queried against the background program in *Abduce*, based on the Kakas and Mancarella (KM) abductive procedure [7], collecting any ground atoms that are needed to refute it into the set Δ, which is thus an abductive explanation for E. The members of Δ will form the heads of the root clauses. These head atoms are "saturated" by adding body atoms to them, forming $T_\perp$, a most specific Connected Theory for B and E. The *Saturate* algorithm is shown in Fig. 2. As explained in Sect. 3, in general there are multiple $T_\perp$, each alternative being computed using a restricted set of abducibles $A_{aux} \subseteq A$. The final step in computing H is to search the lattice of sets of clauses which subsume $T_\perp$. The *Search* procedure returns the preferred H according to criteria appropriate to the particular learning task. A common criterion is to select the most *compressive* hypothesis, the H with the highest ratio of positive examples covered to number of literals in the hypothesis, which does not cover negative examples. The final H chosen for a given seed example E is the one with the best measure across the entire search space defined by all $T_\perp$.

Begin Cover
     given B, E_{pos}
     let $A = Initialise(M_h)$
     while $E_{pos} \neq \emptyset$
          select a seed example $E \in E_{pos}$
          let $H = \{E\}$
          for each $\Delta = Abduce(E, B, A)$
               for each $T_\perp = Saturate(\Delta, \emptyset, \{E\}, E_{pos} \cup \Delta, B, A_{aux})$, where $A_{aux} \subseteq A$
                    $H = Search(T_\perp, B, E_{pos}, H)$
          $B = B \cup H$
          $E_{pos} = E_{pos} - \{ex \in E_{pos} \mid B \models ex\}$
End Cover

Fig. 1. The Cover loop algorithm

Begin Saturate

1   given $\Delta, Sat, E_{sec}, Abd, B, A_{aux}$

2   let $T_\perp = \emptyset$

3   for each $\alpha \in \Delta$ and $\alpha \notin Sat$:

4   let $R = \alpha$

5          $A'_{aux} = A_{aux} - \{\alpha\}$

6   for each $\beta \in M_b$:

7   replace all placeholders in β with input terms or fresh variables

8   for each $(\theta, \Delta_\beta) = C\text{-}Abduce(\beta, B, A'_{aux}, Abd)$ and $\beta\theta \notin E_{sec}$

10                  $R = R \vee \neg\beta\theta$

11   if $\Delta_\beta \neq \emptyset$

12                     $E'_{sec} = E_{sec} \cup \{\beta\theta\}$

13                     $Abd = Abd \cup \Delta_\beta$

14                     $T_\perp = T_\perp \cup Saturate(\Delta_\beta, Sat, E'_{sec}, Abd, B, A'_{aux})$

15          $T_\perp = T_\perp \cup \{R\}$

16          $Sat = Sat \cup \{\alpha\}$

17   return $T_\perp$

End Saturate

Fig. 2. The Saturate algorithm

The inputs to a call to *Saturate* are the set Δ of ground atoms to be saturated, the set Sat of ground atoms already saturated, the set E_{sec} of secondary examples in the current branch of the computation, the set Abd of atoms abduced so far, B and A_{aux}. The output is a most specific Connected Theory $T_\perp$.

Figure 3 shows a tree depiction of the *Saturate* computation which generates Theory 4 in Example 2:

$$T_\perp = \{x \leftarrow a, b, p\} \cup \{u \leftarrow b, p\} \cup \{v \leftarrow p\} \cup \{w \leftarrow p\} \ .$$

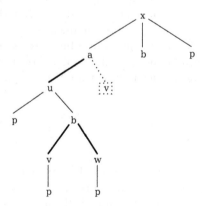

Fig. 3. A tree depiction of the saturation of $\Delta = \{x\}$ in Example 2. The branch containing the node v, and connected to a with a dotted edge, is pruned and does not form part of $T_\perp$.

Thin edges connect clause heads (parent nodes) to their body atoms (children). Bold edges connect secondary examples (parent) to clauses which explain them (subtrees). In this example the abductive explanation for the seed, x, is $\Delta = \{x\}$. Each atom $\alpha \in \Delta$ becomes the head of a new root clause R (line 4), unless it has already been saturated (line 3) in some previous iteration. *Saturate* proceeds to look for body atoms for each α that are compatible with M_b. In a first-order case each body mode declaration β is instantiated with relevant input terms (line 7) to produce a ground (if all placeholders are inputs), or partially ground atom. This atom is now queried against the background program by *C-Abduce*, which extends the KM procedure by checking "global consistency" with E_{neg} and *Abd*. If successful *C-Abduce* instantiates β with substitution θ, and returns a (possibly empty) set Δ_β of abduced atoms.

In Fig. 3 the first body atom for x to be proved by *C-Abduce* is a, with $\Delta_a = \{u, v\}$. Since Δ_a is not empty, a is a secondary example, requiring its own $T_\perp$ explanation. This new auxiliary hypothesis is induced via a recursive call to *Saturate* with input $\Delta = \{u, v\}$. This is the fundamental feature of the Induction on Failure procedure. Standard saturation procedures, using deduction rather than abduction to generate the body of clauses, would fail to prove atoms such as a and terminate. Induction on Failure instead induces new rules to explain a.

The tree also demonstrates how the procedure efficiently produces a theory which is consistent with the background knowledge and example. Two pruning steps are used. Firstly, the condition $\beta\theta \notin E_{sec}$ (line 8) prevents a secondary example appearing twice in the same branch of the tree. This, for instance allowing a to be a child of u, v or w in the figure, would correspond to the secondary example forming part of its own explanation, meaning $T_\perp$ would no longer be a correct hypothesis. The check eliminates such cyclic definitions. If the set of possible body atoms is finite, this check also guarantees that all branches of the tree terminate. Secondly, the condition $\alpha \notin Sat$ (line 3) prevents clauses being generated twice in different parts of the tree. So, even though v is part of the explanation for a in the example, the figure shows that the computation proceeds to generate the tree in a depth first manner, and a v clause is first added under b. The loop check prevents this clause being regenerated when completing the explanation of a and the second v node, shown with a dotted line, is pruned. Consequently, an abduced (head) atom can appear only once in the entire tree.

5   Imparo

Imparo, a prototype implementation of the IoF procedure, has been written in Sictus Prolog. A general-to-specific search of the hypothesis space bounded by the theory consisting of the empty clause, $\{\Box\}$ and the ground $T_\perp$ theories is carried out using a branch and bound algorithm, similar to that of Aleph [2]. Coverage of each theory searched is tested using OLDT (tabled) resolution [8] to ensure that queries with a finite set of solutions will terminate.

As an illustration of the execution of the system, the result of applying Imparo to Yamamoto's main example in [6] is presented. This example is used in [6] to highlight the incompleteness of Bottom Generalisation for finding single clause hypotheses, since the clause $odd(s(X)) \leftarrow even(X)$ cannot be derived by BG.

Example 3 (Yamamoto). [1]

$$B_3 = \left\{ \begin{array}{l} even(s(X)) :\text{-}\ odd(X) \\ even(0) \end{array} \right\} \cup \left\{ \begin{array}{l} nat(0) \\ nat(s(X)) :\text{-}\ nat(X) \end{array} \right\}$$

$$E_3 = odd(s^3(0)), E_{pos} = \{odd(s^5(0))\}, E_{neg} = \{odd(0), odd(s^2(0))\}$$

$$M_3 = \{modeh(*, odd(+nat)), modeb(*, even(+nat)), modeb(*, +nat = s(-nat)))\}$$

♦

The $s/1$ function is handled by adding the $+nat = s(-nat)$ declaration to M_b, which generates equality body literals. Imparo returns the following output:

```
The seed example is odd(s(s(s(0))))
New most specific explanation:
-----------------------------------
odd(A):-A=s(B),even(B),B=s(C),C=s(D),even(D)
odd(A):-A=s(B),even(B)
-----------------------------------

Searching generalisations...
...

The most successful hypothesis is:
-----------------------------------
odd(A):-A=s(B),even(B)
-----------------------------------

19 nodes searched. Time taken: 0.21 seconds.
```

The most specific hypothesis generated by Imparo contains two ground clauses:

$$\left\{ \begin{array}{l} odd(s^3(0)) \leftarrow s^3(0) = s(s^2(0)), even(s^2(0)), s^2(0) = s(s^1(0)), s^1(0) = s(0), even(0) \\ odd(s^1(0)) \leftarrow s^1(0) = s(0), even(0) \end{array} \right\}$$

computed from the seed example $odd(s^3(0))$. Saturation of the abductive explanation $\Delta_1 = \{odd(s^3(0))\}$ yields five body literals including $even(0)$ which can be proved directly from the background knowledge, and $even(s^2(0))$ which is a secondary example. The other possible $even/1$ literals, $even(s^3(0))$ and $even(s(0))$ are inconsistent with the negative examples. The explanation of $even(s^2(0))$ is $\Delta_2 = \{odd(s(0))\}$. Saturation of Δ_2 produces no secondary examples and so computation of $T_\perp$ terminates. Distinct variables replace each ground term in $T_\perp$ in the program output. The single clause theory $\{odd(A) \leftarrow A = s(B), even(B)\}$

---

[1] E_{pos} and E_{neg} do not form part of Yamamoto's formulation of the example. They are added here only to guide the search towards choosing $odd(s(X)) \leftarrow even(X)$ as the preferred hypothesis (the simplest clause with full coverage of the examples).

is returned by the search, which is equivalent to $H_3 = odd(s(X)) \leftarrow even(X)$, in which the equality has been "flattened".

H_3 cannot be derived by BG or KSS as it does not imply the bottom clause (and kernel set) $odd(s^3(0)) \leftarrow even(0)$. However, H_3 does imply both clauses in $T_\perp$, and can be learned by Imparo. This example demonstrates the extended search space of IoF, and its suitability to mutually recursive learning tasks.

6   Related Work and Conclusion

This section relates the IoF approach to other existing IE approaches, focusing mainly on those that generalise the search space beyond that of Bottom Generalization, and summarises the contribution of this paper.

The HAIL (Hybrid Abductive Inductive Learning) proof procedure [3] is a multiple clause IE learning approach that implements the KSS semantics described in Sects. 2 and 3. CTG extends KSS as shown by Theorem 2.

The method of *CF-Induction* described by Inoue in [9] is a full clausal consequence finding approach, with a semantics based on *characteristic clauses*. The characteristic clauses, $Carc(S, \mathcal{P})$, of a program S are those that are implied by S and not subsumed by any other such clause. $\mathcal{P}$ is a so-called *production field* and specifies language bias. A hypothesis H is derivable by CF-Induction if $Carc(B \wedge \neg E, \mathcal{P}) \models CC(B, E)$ and $H \models \neg CC(B, E)$. So, the bridge formula $CC(B, E)$ is equivalent to a (negated) most specific hypothesis, and is in fact a set of ground instances of clauses in $Carc(B \wedge \neg E, \mathcal{P})$. CF-Induction is complete for full clause hypotheses. Yamamoto and Fronhöfer also report a complete multiple clause IE method in full clausal logic [10]. This approach is based on *residue hypotheses*. A residue hypothesis $Res(T)$ for a ground theory T is obtained by deleting all tautological clauses from $\overline{T}$. The procedure described in [10] generates a most specific hypothesis by taking some subset S of the ground instances of $B \cup \{\neg E\}$ and computing $Res(S)$. Since CF-Induction and the Residue method are complete for full clausal theories they are also complete for Horn theories, and so are capable of learning Connected Theories, if appropriate selection of their bridge formulas is made. In IoF, abduction is used to guide selection of the connected clauses. It is also the authors' conjecture that the IoF procedure is complete for Horn theories, with respect to the CTG semantics. Further investigation is required to formally relate the properties and performance of IoF and these full clausal systems.

In summary, this paper has proposed the notion of a Connected Theory as a new type of bridge formula for inverse entailment in Horn programs. A proof procedure, Induction on Failure, has been described which computes most specific Connected Theories via a recursive abductive algorithm. The semantics of Connected Theory Generalisation has been shown to be more complete for single clause hypotheses than the Bottom Generalisation of Progol, and more complete for multiple clause hypotheses than the Kernel Set Subsumption of HAIL. Further work on Induction on Failure will include characterisation of the completeness of the procedure, and the extension of the approach to compute normal program hypotheses.

References

1. Muggleton, S.H.: Inverse Entailment and Progol. New Generat. Comput. 13, 245–286 (1995)
2. Srinivasan, A.: The Aleph Manual, version 4 (2003),
   http://web.comlab.ox.ac.uk/oucl/research/areas/machlearn/Aleph/index.html
3. Ray, O., Broda, K., Russo, A.: Hybrid Abductive Inductive Learning: A Generalisation of Progol. In: Horváth, T., Yamamoto, A. (eds.) ILP 2003. LNCS (LNAI), vol. 2835, pp. 311–328. Springer, Heidelberg (2003)
4. Moyle, S.A.: An Investigation into Theory Completion Techniques in Inductive Logic Programming. PhD thesis, University of Oxford (2000)
5. Muggleton, S.H., Bryant, C.H.: Theory Completion Using Inverse Entailment. In: Cussens, J., Frisch, A.M. (eds.) ILP 2000. LNCS (LNAI), vol. 1866, pp. 130–146. Springer, Heidelberg (2000)
6. Yamamoto, A.: Which Hypotheses Can Be Found with Inverse Entailment? In: Džeroski, S., Lavrač, N. (eds.) ILP 1997. LNCS, vol. 1297, pp. 296–308. Springer, Heidelberg (1997)
7. Kakas, A.C., Mancarella, P.: Database Updates through Abduction. In: 16th International Conference on Very Large Databases, pp. 650–661. Morgan Kaufmann, San Francisco (1990)
8. Tamaki, H., Sato, T.: OLD Resolution with Tabulation. In: Shapiro, E. (ed.) ICLP 1986. LNCS, vol. 225, pp. 84–98. Springer, Heidelberg (1986)
9. Inoue, K.: Induction as Consequence Finding. Mach. Learn. 55, 109–135 (2004)
10. Yamamoto, A., Fronhöfer, B.: Hypotheses Finding via Residue Hypotheses with the Resolution Principle. In: Arimura, H., Sharma, A.K., Jain, S. (eds.) ALT 2000. LNCS (LNAI), vol. 1968, pp. 156–165. Springer, Heidelberg (2000)

On Reductive Semantics of Aggregates
in Answer Set Programming

Joohyung Lee and Yunsong Meng

Computer Science and Engineering
Arizona State University, Tempe, AZ, USA
{joolee,Yunsong.Meng}@asu.edu

Abstract. Several proposals of the semantics of aggregates are based on
different extensions of the stable model semantics, which makes it difficult to compare them. In this note, building upon a reductive approach
to designing aggregates, we provide reformulations of some existing semantics in terms of propositional formulas, which help us compare the
semantics and understand their properties in terms of their propositional
formula representations. We also present a generalization of semantics of
aggregates without involving grounding, and define loop formulas for
programs with aggregates guided by the reductive approach.

1 Introduction

Defining a reasonable semantics of aggregates under the stable model semantics
has turned out to be a non-trivial task. An obvious "reductive" approach to
understand an aggregate as shorthand for a nested expression [1] in the form
of disjunctions over conjunctions leads to unintuitive results. For instance, one
would expect $\{p(0), p(1)\}$ to be the only answer set of the following program.

$$p(1) \qquad p(0) \leftarrow \text{SUM}\langle\{x : p(x)\}\rangle = 1.$$

Assuming that the domain is $\{0, 1\}$, one may try to identify $\text{SUM}\langle\{x : p(x)\}\rangle = 1$
with the disjunction over two "solutions,"—one in which only $p(1)$ is true, and
the other in which both $p(0)$ and $p(1)$ are true. However, the resulting program

$$p(1) \qquad p(0) \leftarrow (p(0), p(1)) \ ; \ (not\ p(0), p(1))$$

has no answer sets.[1]

The difficulty led to several interesting extensions of the stable model semantics to account for aggregates, such as an extended definition of the reduct [2],
an extension of T_P operator with "conditional satisfaction" [3], and an extension
of the standard approximating operator Φ_P to Φ_P^{aggr} [4]. On the other hand, a
few reasonable "reductive" semantics of aggregates were also developed. In [4]

---

[1] Dropping negative literals in forming each conjunct does not work either. For in-
stance, consider $p(1) \leftarrow \text{SUM}\langle\{x : p(x)\}\rangle \neq 1$, which intuitively has no answer sets,
but its translation, assuming that the domain is $\{1\}$, results in $p(1) \leftarrow \top$.

E. Erdem, F. Lin, and T. Schaub (Eds.): LPNMR 2009, LNCS 5753, pp. 182–195, 2009.

and [5], the authors defined translations of aggregates into nested expressions, which are somewhat complex than the naive approach above. In [6], instead of translating into nested expressions, Ferraris proposed to identify an aggregate with conjunctions of implications under his extension of the answer set semantics for arbitrary propositional formulas. The extended semantics is essentially a reformulation of the equilibrium logic [7], and was generalized to arbitrary first-order formulas in [8].

While most semantics agree on monotone and anti-monotone aggregates, they have subtle differences in understanding arbitrary aggregates. For example, the following program Π_1

$$p(2) \leftarrow not \ \text{SUM}\langle\{x : p(x)\}\rangle < 2$$
$$p(-1) \leftarrow \text{SUM}\langle\{x : p(x)\}\rangle \geq 0$$
$$p(1) \leftarrow p(-1) \ .$$

has no answer sets according to [3,4], one answer set $\{p(-1), p(1)\}$ according to [2], and two answer sets $\{p(-1), p(1)\}$ and $\{p(-1), p(1), p(2)\}$ according to [6].

In this paper we make further developments on the reductive approach. We note that the semantics by Pelov, Denecker and Bruynooghe [4] and the semantics by Ferraris [6] are closely related to each other in terms of propositional formula representations of aggregates, yielding a few interesting alternative characterizations. Furthermore we show that the semantics by Faber, Leone and Pfeifer [2] can also be reformulated in terms of propositional formulas. Such uniform characterization helps us compare the semantics and understand their properties by turning to their propositional formula representations. We define loop formulas for programs with aggregates guided by the reductive approach; such loop formulas contain aggregates, and when these aggregates are turned into corresponding propositional formulas, the resulting formulas are the same as the loop formulas as defined in [9] for the propositional theory corresponding to the program with aggregates.

2   Background

2.1   Answer Sets of First-Order Formulas

We review the definition of an answer set from [8]. Let $\mathbf{p}$ be a list of predicate constants $p_1, \ldots, p_n$, and let $\mathbf{u}$ be a list of predicate variables $u_1, \ldots, u_n$. By $\mathbf{u} \leq \mathbf{p}$ we denote the conjunction of the formulas $\forall \mathbf{x}(u_i(\mathbf{x}) \rightarrow p_i(\mathbf{x}))$ for all $i = 1, \ldots n$ where $\mathbf{x}$ is a list of distinct object variables of the same length as the arity of p_i, and by $\mathbf{u} < \mathbf{p}$ we denote $(\mathbf{u} \leq \mathbf{p}) \wedge \neg(\mathbf{p} \leq \mathbf{u})$.

For any first-order sentence F, $\text{SM}[F]$ stands for the second-order sentence

$$F \wedge \neg\exists \mathbf{u}((\mathbf{u} < \mathbf{p}) \wedge F^*(\mathbf{u})), \tag{1}$$

where $\mathbf{p}$ is the list $p_1, \ldots, p_n$ of all predicate constants occurring in F, $\mathbf{u}$ is a list $u_1, \ldots, u_n$ of distinct predicate variables, and $F^*(\mathbf{u})$ is defined recursively:

- $p_i(t_1, \ldots, t_m)^* = u_i(t_1, \ldots, t_m);$

- $(t_1 = t_2)^* = (t_1 = t_2);$ $- \perp^* = \perp;$
- $(F \odot G)^* = (F^* \odot G^*),$ where $\odot \in \{\wedge, \vee\};$
- $(F \to G)^* = (F^* \to G^*) \wedge (F \to G);$
- $(QxF)^* = QxF^*,$ where $Q \in \{\forall, \exists\}.$

(There is no clause for negation here, $\neg F$ is treated as shorthand for $F \to \perp$.)

Let $\sigma(F)$ be the signature consisting of the object, function and predicate constants occurring in F. According to [8], an interpretation of $\sigma(F)$ that satisfies $\mathrm{SM}[F]$ is called a *stable model* of F. If F contains at least one object constant, an Herbrand stable model of F is called an *answer set* of F. The answer sets of a logic program Π are defined as the answer sets of the FOL-representation of Π (i.e., the conjunction of the universal closure of implications corresponding to the rules).

Ferraris *et al.* [8] shows that this definition, if restricted to the syntax of traditional logic programs, is equivalent to the traditional definition of an answer set based on grounding and the reduct [10], and, if restricted to the syntax of arbitrary propositional formulas, is equivalent to the definition of an answer set given by Ferraris [6].

2.2   Syntax of a Program with Aggregates

An *aggregate function* is any function that maps multisets of objects into numbers, such as *count, sum, times, min* and *max*. For this paper we assume that all numbers are integers. The domain of an aggregate function is defined as usual. For instance, *sum, times, min* and *max* are defined for multisets of numbers; *min* and *max* do not allow the empty set in their domains.

An *aggregate expression* is of the form

$$\mathrm{OP}\langle\{\mathbf{x} : F(\mathbf{x})\}\rangle \succeq b \tag{2}$$

where

- OP is a symbol for an *aggregate function* op;
- $\mathbf{x}$ is a nonempty list of distinct object variables;
- $F(\mathbf{x})$ is an arbitrary quantifier-free formula;
- $\succeq$ is a symbol for a binary relation over integers, such as $\leq, \geq, <, >, =, \neq$;
- b is an integer constant.

A *rule (with aggregates)* is an expression of the form

$$A_1 ; \ldots ; A_l \leftarrow E_1, \ldots, E_m, not\ E_{m+1}, \ldots, not\ E_n \tag{3}$$

($l \geq 0;\ n \geq m \geq 0$), where each A_i is an atomic formula (possibly containing equality) and each E_i is an atomic formula or an aggregate expression. A *program (with aggregates)* is a finite set of rules.

Throughout this paper, unless otherwise noted (e.g., Section 4.1), we assume that the program contains no function constants of positive arity. We do not consider symbols OP and $\succeq$ as part of the signature.

We say that an occurrence of a variable v in a rule (3) is *bound* if the occurrence is in an aggregate expression (2) such that v is in $\mathbf{x}$; otherwise it is *free*. We say that v is *free* in the rule if some occurrence of v is free in it. Given a program Π, by $\sigma(\Pi)$ we mean the signature consisting of object and predicate constants that occur in Π. By $Ground(\Pi)$ we denote the program without free variables that is obtained from Π by replacing every free occurrence of variables with every object constant from $\sigma(\Pi)$ in all possible ways.

2.3   Review: FLP Semantics

The FLP semantics [2] is based on an alternative definition of the reduct and the notion of satisfaction extended to aggregate expressions. Let Π be a program such that $\sigma(\Pi)$ contains at least one object constant.[2] We consider Herbrand interpretations of $\sigma(\Pi)$ only. Consider any aggregate expression (2) occurring in $Ground(\Pi)$ and any Herbrand interpretation I of $\sigma(\Pi)$. Let S_I be the multiset consisting of all $\mathbf{c}[1]$ (i.e., the first element of $\mathbf{c}$) in the Herbrand universe where

- $\mathbf{c}$ is a list of object constants of $\sigma(\Pi)$ whose length is the same as the length of $\mathbf{x}$, and
- I satisfies $F(\mathbf{c})$.

A set I of ground atoms of $\sigma(\Pi)$ satisfies the aggregate expression if S_I is in the domain of op, and $op(S_I) \succeq b$. [3]

The *FLP reduct* of Π relative to I is obtained from $Ground(\Pi)$ by removing every rule whose body is not satisfied by I. Set I is an *FLP answer set* of Π if it is minimal among the sets of atoms that satisfy the FLP reduct of Π relative to I. For example, in program Π_1 (Section 1), the FLP reduct of Π_1 relative to $\{p(-1), p(1)\}$ contains the last two rules only. Set $\{p(-1), p(1)\}$ is minimal among the sets of atoms that satisfy the reduct, and thus is an FLP answer set of Π_1. One can check that this is the only FLP answer set.

2.4   Review: Ferraris Semantics

The Ferraris semantics [6] can be extended to allow variables as follows.

Notation. Given a multiset of object constants $\{\!\{c_1, \ldots, c_n\}\!\}$,

$$\mathrm{OP}\langle\{\!\{c_1, \ldots, c_n\}\!\}\rangle \succeq b$$

if

- b is an integer constant,
- multiset $\{\!\{c_1, \ldots, c_n\}\!\}$ is in the domain of op, and
- $op(\{\!\{c_1, \ldots, c_n\}\!\}) \succeq b$.

$\mathrm{OP}\langle\{\!\{c_1, \ldots, c_n\}\!\}\rangle \not\succeq b$ if it is not the case that $\mathrm{OP}\langle\{\!\{c_1, \ldots, c_n\}\!\}\rangle \succeq b$.

Let $E = \mathrm{OP}\langle\{\mathbf{x} : F(\mathbf{x})\}\rangle \succeq b$ be an aggregate expression occurring in $Ground(\Pi)$, let $\mathbf{O}_\Pi(E)$ be the set of all lists of object constants of $\sigma(\Pi)$ whose

---

[2] The syntax in [2] requires $F(\mathbf{x})$ to be a conjunction of atoms.
[3] By an *atom* we mean a non-equality atomic formula of the form $p(t_1, \ldots, t_n)$.

length is the same as the length of $\mathbf{x}$, and let $\overline{C}_{\varPi}(E)$ be the set of all subsets $\mathbf{C}$ of $\mathbf{O}_{\varPi}(E)$ such that $\text{OP}\langle\{\!\{\mathbf{c}[1] : \mathbf{c} \in \mathbf{C}\}\!\}\rangle \not\succeq b$. For instance, in program $\varPi_1$ (Section 1), for $E_1 = \text{SUM}\langle\{x : p(x)\}\rangle < 2$, set $\mathbf{O}_{\varPi_1}(E_1)$ is $\{-1, 1, 2\}$, and $\overline{C}_{\varPi_1}(E_1)$ is $\{\{2\}, \{1, 2\}, \{-1, 1, 2\}\}$. Similarly, for $E_2 = \text{SUM}\langle\{x : p(x)\}\rangle \geq 0$, set $\overline{C}_{\varPi_1}(E_2)$ is $\{\{-1\}\}$.

By $Fer_{\varPi}(E)$ we denote

$$\bigwedge_{\mathbf{C}\in\overline{C}_{\varPi}(E)} \left(\bigwedge_{\mathbf{c}\in\mathbf{C}} F(\mathbf{c}) \rightarrow \bigvee_{\mathbf{c}\in\mathbf{O}_{\varPi}(E)\backslash\mathbf{C}} F(\mathbf{c}) \right). \tag{4}$$

For instance, $Fer_{\varPi_1}(E_1)$ is

$$(p(2) \rightarrow p(-1) \vee p(1)) \wedge (p(1) \wedge p(2) \rightarrow p(-1)) \wedge (p(-1) \wedge p(1) \wedge p(2) \rightarrow \bot) .$$

By $Fer(\varPi)$ we denote the propositional formula obtained from $Ground(\varPi)$ by replacing every aggregate expression E in it by $Fer_{\varPi}(E)$. The *Ferraris answer sets* of $\varPi$ are defined as the answer sets of $Fer(\varPi)$ in the sense of Section 2.1. For example, the Ferraris answer sets of $\varPi_1$ are the answer sets of the following formula $Fer(\varPi_1)$:[4]

$$\begin{aligned} &(\neg[(p(2) \rightarrow p(-1) \vee p(1)) \wedge (p(1) \wedge p(2) \rightarrow p(-1)) \wedge (p(-1) \wedge p(1) \wedge p(2) \rightarrow \bot)] \rightarrow p(2)) \\ &\wedge ([p(-1) \rightarrow p(1) \vee p(2)] \rightarrow p(-1)) \\ &\wedge (p(-1) \rightarrow p(1)) . \end{aligned}$$
$$\tag{5}$$

This formula has two answer sets: $\{p(-1), p(1)\}$ and $\{p(-1), p(1), p(2)\}$.

2.5   Review: SPT-PDB Semantics

Son and Pontelli [5] presented two equivalent definitions of aggregates, one in terms of "unfolding" into nested expressions, and the other in terms of "conditional satisfaction." The latter notion was simplified by Son, Pontelli and Tu [3]. Lemma 6 from [5] shows that these definitions are equivalent to the definition by Pelov, Denecker and Bruynooghe [4], which is in terms of translation into nested expressions. Thus we group them together and review only the last one.[5]

Under the SPT-PDB semantics, an aggregate can be identified with a nested expression in the form of disjunctions over conjunctions, but unlike the naive attempt given in the introduction, it involves the notion of a "(maximal) local power set."

Given a set A of some sets, a pair $\langle B, T \rangle$ where $B, T \in A$ and $B \subseteq T$ is called a *local power set (LPS)* of A if every S such that $B \subseteq S \subseteq T$ belongs to A as well. A local power set is called *maximal* if there is no other local power set $\langle B', T' \rangle$ of A such that $B' \subseteq B$ and $T \subseteq T'$.

The SPT-PDB semantics eliminates the negation in front of an aggregate expression using an equivalent transformation. Let $Pos(\varPi)$ be a program obtained from $\varPi$ by replacing *not* E_i in each rule (3) where $E_i = \text{OP}\langle\{\mathbf{x} : F(\mathbf{x})\}\rangle \succeq b$ with $\text{OP}\langle\{\mathbf{x} : F(\mathbf{x})\}\rangle \prec b$ ($\prec$ is the symbol for the relation complementary to $\succeq$).

---

[4] We underline the parts of a formula that correspond to aggregates.

[5] We ignore some differences in the syntax, and allow disjunctions in the head.

Clearly, $Pos(\Pi)$ contains no negation in front of aggregate expressions. For instance, the first rule of $Pos(\Pi_1)$ is

$$p(2) \leftarrow \text{SUM}\langle\{x : p(x)\}\rangle \geq 2.$$

Let $E = \text{OP}\langle\{\mathbf{x} : F(\mathbf{x})\}\rangle \succeq b$ be an aggregate expression occurring in $Ground(Pos(\Pi))$, let HU_Π be the set of all ground atoms that can be constructed from $\sigma(\Pi)$. Let $\mathcal{I}_\Pi(E)$ be the set of all Herbrand interpretations I of $\sigma(\Pi)$ such that $I \models E$ (satisfaction as defined in Section 2.3). For instance, in Example Π_1, HU_{Π_1} is $\{p(-1), p(1), p(2)\}$, and, for $\overline{E_1} = \text{SUM}\langle\{x : p(x)\}\rangle \geq 2$, $\mathcal{I}_{\Pi_1}(\overline{E_1})$ is $\{\{p(2)\}, \{p(1), p(2)\}, \{p(-1), p(1), p(2)\}\}$, and $\mathcal{I}_{\Pi_1}(E_2)$ is

$$\{\emptyset, \{p(1)\}, \{p(2)\}, \{p(-1), p(1)\}, \{p(-1), p(2)\}, \{p(1), p(2)\}, \{p(-1), p(1), p(2)\}\}.$$

The maximal local power sets of $\mathcal{I}_{\Pi_1}(\overline{E_1})$ are

$$\langle\{p(2)\}, \{p(1), p(2)\}\rangle, \quad \langle\{p(1), p(2)\}, \{p(-1), p(1), p(2)\}\rangle,$$

and the maximal local power sets of $\mathcal{I}_{\Pi_1}(E_2)$ are

$$\langle\emptyset, \{p(1), p(2)\}\rangle, \quad \langle\{p(1)\}, \{p(-1), p(1), p(2)\}\rangle, \quad \langle\{p(2)\}, \{p(-1), p(1), p(2)\}\rangle.$$

For any aggregate expression E occurring in $Ground(Pos(\Pi))$, by $SPT\text{-}PDB_\Pi(E)$ we denote

$$\bigvee_{\langle B, T\rangle \text{ is a maximal LPS of } \mathcal{I}_\Pi(E)} \left(\bigwedge_{A \in B} A \wedge \bigwedge_{A \in HU_\Pi \setminus T} \neg A\right). \tag{6}$$

For instance, $SPT\text{-}PDB_{\Pi_1}(\overline{E_1})$ is $(p(2) \wedge \neg p(-1)) \vee (p(1) \wedge p(2))$.

By $SPT\text{-}PDB(\Pi)$ we denote the propositional formula obtained from $Ground(Pos(\Pi))$ by replacing all aggregate expressions E in it by $SPT\text{-}PDB_\Pi(E)$. The $SPT\text{-}PDB$ answer sets of Π are defined as the answer sets of $SPT\text{-}PDB(\Pi)$ in the sense of Section 2.1. For example, Π_1 has no SPT-PDB answer sets, and neither does the following formula $SPT\text{-}PDB(\Pi_1)$:

$$\begin{aligned}&([[(p(2) \wedge \neg p(-1)) \vee (p(1) \wedge p(2))] \rightarrow p(2)) \\ &\wedge ([\neg p(-1) \vee p(1) \vee p(2)] \rightarrow p(-1)) \\ &\wedge (p(-1) \rightarrow p(1)).\end{aligned} \tag{7}$$

3 Comparison of the Semantics of Aggregates

3.1 A Reformulation of Ferraris Semantics

The propositional formula representation of an aggregate according to the Ferraris semantics can be written in a more compact way by considering maximal local power sets as in the SPT-PDB semantics.

For an aggregate expression that contains no free variables, by $MLPS\text{-}Fer_\Pi(E)$ we denote

$$\bigwedge_{\langle B,T\rangle \text{ is a maximal LPS of } \overline{C}_\Pi(E)} \left(\bigwedge_{\mathbf{c}\in B} F(\mathbf{c}) \to \bigvee_{\mathbf{c}\in \mathbf{O}_\Pi(E)\setminus T} F(\mathbf{c}) \right). \qquad (8)$$

One can prove that formulas (4) and (8) are strongly equivalent [11] to each other, which provides another characterization of Ferraris answer sets. We define MLPS-Ferraris answer sets of Π same as the Ferraris answer sets of Π except that we refer to (8) in place of (4). The following proposition follows from the strong equivalence between (4) and (8).

Proposition 1. *The MLPS-Ferraris answer sets of Π are precisely the Ferraris answer sets of Π.*

For example, in program Π_1, the maximal local power sets of $\overline{C}_{\Pi_1}(E_1)$ are $\langle\{2\},\{1,2\}\rangle$, $\langle\{1,2\},\{-1,1,2\}\rangle$. The maximal local power set of $\overline{C}_{\Pi_1}(E_2)$ is $\langle\{-1\},\{-1\}\rangle$. Formula (5) is strongly equivalent to this shorter formula

$$(\neg[(p(2)\to p(-1)) \wedge (p(1)\wedge p(2)\to\bot)] \to p(2))$$
$$\wedge\ ([p(-1)\to p(1)\vee p(2)] \to p(-1))$$
$$\wedge\ (p(-1) \to p(1))\ .$$

3.2 A Reformulation of FLP Semantics

The FLP semantics can also be defined by reduction to propositional formulas. For an aggregate expression E that contains no free variables, let $\overline{\mathcal{I}}_\Pi(E)$ be the set of all Herbrand interpretations I of $\sigma(\Pi)$ such that $I \not\models E$ (as defined in Section 2.3). Clearly $\overline{\mathcal{I}}_\Pi(E)$ and $\mathcal{I}_\Pi(E)$ partition HU_Π. By $FLP_\Pi(E)$ we denote

$$\bigwedge_{I\in \overline{\mathcal{I}}_\Pi(E)} \left(\bigwedge_{A\in I} A \to \bigvee_{A\in HU_\Pi\setminus I} A \right). \qquad (9)$$

As with the SPT-PDB semantics, before turning a program to the propositional formula representation for the FLP semantics, we eliminate the negation in front of an aggregate expression using an equivalent transformation. By $FLP(\Pi)$ we denote the propositional formula obtained from $Ground(Pos(\Pi))$ by replacing all aggregate expressions E in it by $FLP_\Pi(E)$ (Recall the definition of $Pos(\Pi)$ in Section 2.5).

Proposition 2. *For any program Π, the FLP answer sets of Π (Section 2.3) are precisely the answer sets of $FLP(\Pi)$.*

For example, in program Π_1, $\overline{\mathcal{I}}_{\Pi_1}(E_1)$ is $\{\emptyset, \{p(-1)\}, \{p(1)\}, \{p(-1),p(1)\}, \{p(-1),p(2)\}\}$ and $\overline{\mathcal{I}}_{\Pi_1}(E_2)$ is $\{\{p(-1)\}\}$, so that $FLP(\Pi_1)$ is

$$([[(p(-1)\vee p(1)\vee p(2)) \wedge (p(-1)\to p(1)\vee p(2)) \wedge (p(1)\to p(-1)\vee p(2))$$
$$\wedge(p(-1)\wedge p(1)\to p(2)) \wedge (p(-1)\wedge p(2)\to p(1))] \to p(2))$$
$$\wedge\ ([p(-1)\to p(1)\vee p(2)] \to p(-1))$$
$$\wedge\ (p(-1) \to p(1))\ .$$

Similar to (8), formula $FLP(\Pi)$ can also be simplified using the notion of maximal local power sets, which provides yet another characterization of the FLP semantics. We call the resulting propositional formula representation $MLPS\text{-}FLP(\Pi)$.

Lemma 1. *Formula (9) is strongly equivalent to*

$$\bigwedge_{\langle B,T\rangle \text{ is a maximal LPS of } \overline{\mathcal{I}}_{\Pi}(E)} \left(\bigwedge_{A\in B} A \to \bigvee_{A\in HU_{\Pi}\setminus T} A \right). \tag{10}$$

For example, $FLP(\Pi_1)$ has the same answer sets as the following $MLPS\text{-}FLP$ (Π_1):

$$([p(2) \wedge (p(-1) \to p(1))] \to p(2)) \ \wedge \ ([p(-1) \to p(1) \vee p(2)] \to p(-1)) \ \wedge \ (p(-1) \to p(1)) . \tag{11}$$

3.3 A Reformulation of SPT-PDB Semantics

Consider the following formula modified from (9) by simply eliminating implications in favor of negations and disjunctions as in classical logic:

$$\bigwedge_{I\in\overline{\mathcal{I}}_{\Pi}(E)} \left(\bigvee_{A\in I} \neg A \ \vee \ \bigvee_{A\in HU_{\Pi}\setminus I} A \right) . \tag{12}$$

Formulas (12) and (9) are classically equivalent to each other, but not strongly equivalent. However, interestingly, (12) is strongly equivalent to (6), which in turn provides a simple reformulation of the SPT-PDB semantics, without involving the notion of local power sets. We define *modified FLP answer sets* of Π same as in Section 3.2 except that we refer to (12) in place of (9).

Proposition 3. *For any program Π, the modified FLP answer sets of Π are precisely the SPT-PDB answer sets of Π.*

For instance, the SPT-PDB answer sets of Π_1 are the same as the answer sets of the following formula:

$$\begin{aligned} ([(p(-1)\vee p(1)\vee p(2)) \wedge (\neg p(-1)\vee p(1)\vee p(2)) \wedge (\neg p(1)\vee p(-1)\vee p(2)) \\ \wedge(\neg p(-1)\vee \neg p(1)\vee p(2)) \wedge (\neg p(-1)\vee \neg p(2)\vee p(1))] \to p(2)) \\ \wedge \ ([\neg p(-1)\vee p(1)\vee p(2)] \to p(-1)) \\ \wedge \ (p(-1) \to p(1)) . \end{aligned} \tag{13}$$

Similar to the Ferraris and the FLP semantics, considering maximal local power sets can yield shorter propositional formula representation as the following lemma tells.

Lemma 2. *Formula (12) is strongly equivalent to*

$$\bigwedge_{\langle B,T\rangle \text{ is a maximal LPS of } \overline{\mathcal{I}}_{\Pi}(E)} \left(\bigvee_{A\in B} \neg A \ \vee \ \bigvee_{A\in HU_{\Pi}\setminus T} A \right) . \tag{14}$$

Again note the similarity between (14) and (10). They are classically equivalent to each other, but not strongly equivalent.

3.4   Relationship between the Semantics

The characterizations of each semantics in terms of the uniform framework of propositional formulas give new insights into their relationships. Note that for any aggregate expression E, formulas $SPT\text{-}PDB_\Pi(E)$, $FLP_\Pi(E)$, $Fer_\Pi(E)$ are classically equivalent to each other, but not strongly equivalent.

It is not difficult to check that for any aggregate expression E occurring in $Ground(Pos(\Pi))$, formula $SPT\text{-}PDB_\Pi(E)$ entails $FLP_\Pi(E)$ under the logic of Here-and-There, but not the other way around. Using this fact, we can prove the following.

Proposition 4. *[5, Theorem 2] Every SPT-PDB answer set of Π is an FLP answer set of Π.*

For program Π_1, its only FLP answer set is a Ferraris answer set. Indeed, such relationship holds if the program is "semi-positive." We call a program *semi-positive* if, for every aggregate expression (2) occurring in it, $F(\mathbf{x})$ is a quantifier-free formula that contains no implications (this, in particular, means that there are no negations since we treat $\neg G$ as shorthand for $G \rightarrow \bot$). For example, Π_1 is semi-positive.

Proposition 5. *For any semi-positive program Π, every FLP answer set of Π is a Ferraris answer set of Π.*

However, the relationship does not hold for arbitrary programs. For instance, the following non-semi-positive program

$$p(a) \leftarrow \text{COUNT}\langle \{x : \neg\neg p(x) \vee q(x)\}\rangle \neq 1$$
$$q(b) \leftarrow p(a)$$
$$p(a) \leftarrow q(b)$$

has no Ferraris answer sets while it has only one FLP answer set $\{p(a), q(b)\}$.

The following proposition is a slight extension of Theorem 3 from [6], which describes a class of programs whose FLP answer sets coincide with Ferraris answer sets.

Proposition 6. *For any semi-positive program Π, the FLP answer sets of Π are precisely the Ferraris answer sets of $Pos(\Pi)$.*

4   Generalized Definition of Aggregates

4.1   Syntax and Semantics of Aggregate Formulas

In this section we provide a general definition of a stable model that applies to arbitrary "aggregate formulas" in the style of the definition in Section 2.1, by extending the notion F^* to aggregate expressions in a way similar to other connectives and using the extended notion of satisfaction as in the FLP semantics (Section 2.3).

We allow the signature to contain any function constants of positive arity, and allow b in aggregate expression (2) to be any term. We define *aggregate formulas* as an extension of first-order formulas by treating aggregate expressions as a base case in addition to (standard) atomic formulas (including equality) and $\perp$ (falsity). In other words, aggregate formulas are constructed from atomic formulas and aggregate expressions using connectives and quantifiers as in first-order logic. For instance,

$$(\text{SUM}\langle\{x : p(x)\}\rangle \geq 1 \ \lor \ \exists y\, q(y)) \to r(x)$$

is an aggregate formula.

We say that an occurrence of a variable v in an aggregate formula H is *bound* if the occurrence is in a part of H of the form $\{\mathbf{x} : F(\mathbf{x})\}$ where v is in $\mathbf{x}$, or in a part of H of the form $Qv G$. Otherwise it is *free*. We say that v is *free* in H if H contains a free occurrence of v. An aggregate sentence is an aggregate formula with no free variables.

The definition of an interpretation is the same as in first-order logic. Consider an interpretation I of a first-order signature σ that may contain any function constants of positive arity. By $\sigma^{|I|}$ we mean the signature obtained from σ by adding distinct new object constants d^*, called *names*, for all d in the universe of I. We identify an interpretation I of σ with its extension to $\sigma^{|I|}$ defined by $I(d^*) = d$.

The notion of satisfaction in first-order logic is extended to aggregate sentences, similar to the definition given in Section 2.3. The integer constants and built-in symbols, such as $+, -, \leq, \geq$ are evaluated in the standard way, and we consider only those "standard" interpretations.[6] Let I be an interpretation of signature σ. Consider any aggregate expression (2) that has no free variables. Let S_I be the multiset consisting of all $\mathbf{d}[1]$ in the universe of I where

- $\mathbf{d}^*$ is a list of object names of $\sigma^{|I|}$ whose length is the same as the length of $\mathbf{x}$, and
- I satisfies $F(\mathbf{d}^*)$.

An interpretation I satisfies the aggregate expression if S_I is in the domain of *op*, and $op(S_I) \succeq b^I$.

For any aggregate sentence F, expression $SM[F]$ stands for (1) where $F^*(\mathbf{u})$ is extended to aggregate expressions as

- $(\text{OP}\langle\{\mathbf{x} : F(\mathbf{x})\}\rangle \succeq b)^* = (\text{OP}\langle\{\mathbf{x} : F^*(\mathbf{x})\}\rangle \succeq b) \land (\text{OP}\langle\{\mathbf{x} : F(\mathbf{x})\}\rangle \succeq b)$.

By a *stable model* of F we mean a model of $SM[F]$ (under the extended notion of satisfaction).

---

[6] For instance, we assume that, when x or y is not an integer, $x \leq y$ evaluates to false, and $x + y$ has an arbitrary value according to the interpretation.

4.2   Programs with Aggregates as a Special Case

The *AF-representation* ("Aggregate Formula representation") of (3) is the universal closure of the aggregate formula

$$E_1 \wedge \cdots \wedge E_m \wedge \neg E_{m+1} \wedge \cdots \wedge \neg E_n \rightarrow A_1 \vee \cdots \vee A_l. \tag{15}$$

The *AF-representation* of Π is the conjunction of the AF-representation of its rules.

   The *stable models* of Π are defined as the stable models of the AF-representation of Π. The following proposition shows that this definition is a proper generalization of the Ferraris semantics.

Proposition 7. *Let Π be a program that contains no function constants of positive arity and let F be its AF-representation. The Herbrand stable models of F whose signature is $\sigma(\Pi)$ are precisely the Ferraris answer sets of Π.*

5   Loop Formulas for Programs with Aggregates

Let us identify rule (3) with

$$A \leftarrow B, C, N \tag{16}$$

where $A = \{A_1, \ldots, A_l\}$, B is the set of all atoms (i.e., non-equality atomic formulas) from $\{E_1, \ldots, E_m\}$, C is the set of all aggregate expressions from $\{E_1, \ldots, E_m\}$ and N is the set of the remaining expressions in the body. We assume that the rules contain no free variables and no function constants of positive arity and that $F(\mathbf{x})$ in every aggregate expression (2) is a conjunction of atoms. Following [12], for any aggregate expression $E = \mathrm{OP}\langle\{\mathbf{x} : F(\mathbf{x})\}\rangle \succeq b$ and any finite set Y of ground atoms, formula $NFES_E(Y)$ is defined as the conjunction of

$$\mathrm{OP}\langle\{\mathbf{x} \ : \ F(\mathbf{x}) \wedge \bigwedge_{\substack{p_i(\mathbf{t}) \text{ occurs in } F(\mathbf{x}) \\ p_i(\mathbf{t}') \in Y}} \mathbf{t} \neq \mathbf{t}'\}\rangle \succeq b$$

and E. For instance, in Example Π_1, formula $NFES_{E_2}(\{p(-1), p(1)\})$ is

$$\mathrm{SUM}\langle\{x : p(x) \wedge x \neq -1 \wedge x \neq 1\}\rangle \geq 0 \ \wedge \ \mathrm{SUM}\langle\{x : p(x)\}\rangle \geq 0 \ .$$

For any finite set Y of expressions, by $Y^\wedge$ and $Y^\vee$ we denote the conjunction and, respectively, disjunction of the elements of Y. We define the *external support formula* of Y for Π, denoted by $ES_\Pi(Y)$, as the disjunction of

$$B^\wedge \wedge \bigwedge_{E \in C} NFES_E(Y) \wedge N^\wedge \wedge \neg(A \setminus Y)^\vee$$

for all rules (3) in Π such that $A \cap Y \neq \emptyset$ and $B \cap Y = \emptyset$. The *(conjunctive) aggregate loop formula* of Y for Π is the aggregate formula

$$Y^\wedge \rightarrow ES_\Pi(Y) \ . \tag{17}$$

This definition extends the definition of a loop formula given in [13], which is limited to programs with monotone aggregates.

For instance, if Y is $\{p(-1), p(1)\}$, the loop formula of Y for Π_1 is

$$p(-1) \wedge p(1) \rightarrow (\text{SUM}\langle\{x : p(x) \wedge x \neq -1 \wedge x \neq 1\}\rangle \geq 0) \wedge \text{SUM}\langle\{x : p(x)\}\rangle \geq 0 \ .$$

The *PL representation* of (17) is the propositional formula obtained from (17) by replacing all occurrences of aggregate expressions E in it with $MLPS\text{-}Fer_\Pi(E)$.

The *Ferraris dependency graph* of Π is the directed graph such that

- its vertices are the ground atoms of $\sigma(\Pi)$;
- for every rule (16) in $Ground(\Pi)$, it has edges from each element of A to $p(\mathbf{t})$
  - if $p(\mathbf{t})$ is an element of B, or
  - if there are an aggregate expression (2) in C, a maximal local power set $\langle B', T\rangle$ of $\overline{C}_\Pi((2))$ and an element $\mathbf{c}$ in $\mathbf{O}_\Pi((2)) \setminus T$ such that $p(\mathbf{t})$ belongs to $F(\mathbf{c})$.

It is not difficult to check that the Ferraris dependency graph of Π according to this definition is the same as the dependency graph of the propositional formula $MLPS\text{-}Fer(\Pi)$ according to [9]. A *loop* is a nonempty set L of ground atoms of $\sigma(\Pi)$ such that the subgraph of the dependency graph of Π induced by L is strongly connected. Again, L is a loop of Π according to this definition iff it is a loop of $MLPS\text{-}Fer(\Pi)$ according to [9].[7] For example, Π_1 has four loops: $\{p(-1)\}$, $\{p(1)\}$, $\{p(2)\}$, $\{p(-1), p(1)\}$.

Proposition 8. *For any set X of ground atoms of $\sigma(\Pi)$ that satisfies Π, the following conditions are equivalent to each other.*

(a) X is a Ferraris answer set of Π;
(b) for every loop Y of Π, X satisfies the aggregate loop formula of Y for Π;
(c) for every loop Y of Π, X satisfies the PL representation of the aggregate loop formula of Y for Π;
(d) for every loop Y of $MLPS\text{-}Fer(\Pi)$ according to [9], X satisfies the loop formula of Y for $MLPS\text{-}Fer(\Pi)$ according to [9].

This result can be extended to the general case when $F(\mathbf{x})$ in an aggregate expression (2) is an arbitrary quantifier-free formula, by using the notion of $NFES_F$ that is defined in [12]. The definition of external support formula above is closely related to the definition of unfounded sets under the FLP semantics given in [14]. Indeed, one can define loop formulas and loops under the FLP semantics in a similar way based on the reductive approach.

You and Liu [15] presented the definition of loop formulas under the SPT-PDB semantics. We note that a set of ground atoms is a loop of Π according to their definition iff it is a loop of $SPT\text{-}PDB(\Pi)$ according to [9]. The same can be said about loop formulas.

---

[7] Note that $Fer(\Pi)$ may contain redundant loops not present in $MLPS\text{-}Fer(\Pi)$. For example, consider $p(1) \leftarrow \text{SUM}\langle\{x : p(x)\}\rangle \geq 1 \quad p(-1) \leftarrow p(1)$.

6   Conclusion

The paper presented several reformulations of the semantics of aggregates in terms of propositional formulas. The resulting formulas are classically equivalent to each other but not strongly equivalent, which results in different semantics. The reformulations give us insights into each of the semantics in terms of the underlying general language. Guided by the reduction, we defined the loop formulas of a program with aggregates, which result in the same as loop formulas of the corresponding propositional formula representation.

The reductive approach led us to the general semantics of aggregates presented in Section 4, which extends the definition of a stable model of a first-order formula to an aggregate formula, using a notion of satisfaction extended from the one used in the FLP semantics. The new semantics is more general than that of RASPL-1 [16] in that it allows arbitrary aggregates and non-Herbrand stable models, along with built-in functions. On the other hand, it is not fully reductive; it requires the notion of satisfaction be extended to aggregate expressions, while a counting aggregate expression in RASPL-1 was defined as an abbreviation for a first-order formula.

Acknowledgements. We are grateful to Vladimir Lifschitz and Tran Cao Son for helpful discussions and to the anonymous referees for their useful comments on this paper. This work was partially supported by the National Science Foundation under Grant IIS-0839821.

References

1. Lifschitz, V., Tang, L.R., Turner, H.: Nested expressions in logic programs. Annals of Mathematics and Artificial Intelligence 25, 369–389 (1999)
2. Faber, W., Leone, N., Pfeifer, G.: Recursive aggregates in disjunctive logic programs: Semantics and complexity. In: Alferes, J.J., Leite, J. (eds.) JELIA 2004. LNCS (LNAI), vol. 3229, pp. 200–212. Springer, Heidelberg (2004)
3. Son, T.C., Pontelli, E., Tu, P.H.: Answer sets for logic programs with arbitrary abstract constraint atoms. J. Artif. Intell. Res. (JAIR) 29, 353–389 (2007)
4. Pelov, N., Denecker, M., Bruynooghe, M.: Translation of aggregate programs to normal logic programs. In: Proc. Answer Set Programming (2003)
5. Son, T.C., Pontelli, E.: A constructive semantic characterization of aggregates in answer set programming. TPLP 7(3), 355–375 (2007)
6. Ferraris, P.: Answer sets for propositional theories. In: Baral, C., Greco, G., Leone, N., Terracina, G. (eds.) LPNMR 2005. LNCS (LNAI), vol. 3662, pp. 119–131. Springer, Heidelberg (2005)
7. Pearce, D.: A new logical characterization of stable models and answer sets. In: Dix, J., Pereira, L., Przymusinski, T. (eds.) NMELP 1996. LNCS (LNAI), vol. 1216, pp. 57–70. Springer, Heidelberg (1997)
8. Ferraris, P., Lee, J., Lifschitz, V.: A new perspective on stable models. In: Proc. International Joint Conference on Artificial Intelligence (IJCAI), pp. 372–379 (2007)
9. Ferraris, P., Lee, J., Lifschitz, V.: A generalization of the Lin-Zhao theorem. Annals of Mathematics and Artificial Intelligence 47, 79–101 (2006)

10. Gelfond, M., Lifschitz, V.: The stable model semantics for logic programming. In: Kowalski, R., Bowen, K. (eds.) Proceedings of International Logic Programming Conference and Symposium, pp. 1070–1080. MIT Press, Cambridge (1988)
11. Lifschitz, V., Pearce, D., Valverde, A.: Strongly equivalent logic programs. ACM Transactions on Computational Logic 2, 526–541 (2001)
12. Lee, J., Meng, Y.: On loop formulas with variables. In: Proc. International Conference on Knowledge Representation and Reasoning (KR), pp. 444–453 (2008)
13. Liu, L., Truszczynski, M.: Properties and applications of programs with monotone and convex constraints. J. Artif. Intell. Res. (JAIR) 27, 299–334 (2006)
14. Faber, W.: Unfounded sets for disjunctive logic programs with arbitrary aggregates. In: Baral, C., Greco, G., Leone, N., Terracina, G. (eds.) LPNMR 2005. LNCS (LNAI), vol. 3662, pp. 40–52. Springer, Heidelberg (2005)
15. You, J.H., Liu, G.: Loop formulas for logic programs with arbitrary constraint atoms. In: Proc. AAAI Conference on Artificial Intelligence (AAAI), pp. 584–589 (2008)
16. Lee, J., Lifschitz, V., Palla, R.: A reductive semantics for counting and choice in answer set programming. In: Proc. AAAI Conference on Artificial Intelligence (AAAI), pp. 472–479 (2008)

A First Order Forward Chaining Approach for Answer Set Computing

Claire Lefèvre and Pascal Nicolas

LERIA, University of Angers, France
`claire.lefevre@univ-angers.fr`, `pascal.nicolas@univ-angers.fr`
2, bd Lavoisier, F-49045 Angers Cedex 01

Abstract. The natural way to use Answer Set Programming (ASP) to represent knowledge in Artificial Intelligence or to solve a Constraint Satisfaction Problem is to elaborate a first order logic program with default negation. In a preliminary step this program, with variables, is translated in an equivalent propositional one by a first tool: the grounder. Then, the propositional program is given to a second tool: the solver. This last one computes (if they exist) one or many answer sets (models) of the program, each answer set encoding one solution of the initial problem. Until today, almost all ASP systems apply this two steps computation.

In this work, our major contribution is to introduce a new approach of answer set computing that escapes the preliminary phase of rule instantiation by integrating it in the search process. Our methodology applies a forward chaining of first order rules that are grounded on the fly by means of previously produced constants. We have implemented this strategy in our new ASP solver `ASPeRiX`. The first benefit of our work is to avoid the bottleneck of instantiation phase arising for some problems because of the huge amount of memory needed to ground all rules of a program, even if these rules are not really useful in certain cases. The second benefit is to make the treatment of function symbols easier and without syntactic restriction provided that rules are safe.

1 Introduction

Answer Set Programming (ASP) is a very convenient paradigm to represent knowledge in Artificial Intelligence (AI) and to encode Constraint Satisfaction Problems (CSP). It has its roots in non monotonic reasoning and logic programming and has led to a lot of works since the seminal paper [9]. But, beyond its ability to formalize various problems from AI or CSP, ASP is also became a very interesting way to practically solve them since some efficient solvers are available. In few words, if someone wants to use ASP to solve an issue, and whatever is the variant of ASP that he uses, he has to write a logic program in a purely declarative manner in such a way that the stable models (also called answer sets) of the program represent the solutions of his original problem (see [1,18] for more details). Usually, the program contains different kind

E. Erdem, F. Lin, and T. Schaub (Eds.): LPNMR 2009, LNCS 5753, pp. 196–208, 2009.

of rules. The simplest ones are facts as $bird(tweety) \leftarrow .$, $edge(4, 10) \leftarrow .$, representing data of the particular problem. Some ones are about background knowledge as $path(X, Y) \leftarrow edge(X, Z), path(Z, Y)$. expressing a well-know property about path in a graph for instance. Some others can be non monotonic, as $fly(X) \leftarrow bird(X), not\ penguin(X)$., for reasoning with incomplete knowledge. In other cases, especially for CSP, default negation is also used to encode alternative potential solutions of a problem as $red(X) \leftarrow v(X), not\ blue(X)$. and $blue(X) \leftarrow v(X), not\ red(X)$., expressing the two exclusive possibilities to color a vertex in a graph. Last but not least, special headless rules are used to represent constraints of the problem to solve as $\leftarrow edge(X, Y),\ red(X), red(Y)$. in order to not color with red two vertices linked by an edge.

Depending which solver is used to compute the answer sets of the program, one can also use some particular atoms for (in)equalities and simple arithmetic calculus. Other constructions using aggregates functions, cardinality constraints, weights, ... are also possible but they are out of the scope of this work in which we restrict our attention to original stable model semantics [9]. In fact, with these above examples we want to point out that knowledge representation in ASP is done by means of first order rules. But, from a theoretical point of view, the models of such a first order program are those of its ground instantiation with respect to its Herbrand universe. So, from a practical point of view, every available ASP solver begins its work by an instantiation phase in order to obtain a propositional program. After this first phase, called *grounding*, the solver starts the real phase of answer set computation by dealing with a finite, but sometimes huge, propositional program.

Let us note that there exists some more or less recent works [11,4,5,15] dealing with first order non monotonic logic programs. These works establish some relations between stable model semantics and constraints systems or second order logic or circumscription but are not really concerned by the explicit computation of answer sets. On our side, the aim of our present work is to propose a new approach of answer set computation that escapes this preliminary grounding phase by integrating it in the search process. In Sect. 2 we recall the theoretical backgrounds about ASP necessary to the comprehension of our work. In Sect. 3 we present our new approach of answer set computation that is first order rule oriented. In Sect. 4 we point out some positive consequences of our methodology. Mainly, the huge amount of memory needed during the usual pregrounding phase for certain problems can now be avoided and function symbols can be used inside rule predicates very easily. We conclude in Sect. 5 by citing some new perspectives for ASP as a result of our innovative approach.

2   ASP Backgrounds

A *normal logic program* is a set of rules like

$$c \leftarrow a_1,\ \ldots,\ a_n,\ not\ b_1,\ \ldots,\ not\ b_m.\quad n \geq 0, m \geq 0 \tag{1}$$

where $c, a_1, \ldots, a_n, b_1, \ldots, b_m$ are ground atoms. For a rule r (or by extension for a rule set), we note $head(r) = c$ its *head*, $body^+(r) = \{a_1, \ldots, a_n\}$

its *positive body*, $body^-(r) = \{b_1, \ldots, b_m\}$ its *negative body* and $body(r) = body^+(r) \cup body^-(r)$. When the negative body of a rule is not empty we say that this rule is *non-monotonic*. The Gelfond-Lifschitz reduct of a program P by an atom set X is the program $P^X = \{head(r) \leftarrow body^+(r). \mid body^-(r) \cap X = \emptyset\}$. Since it has no default negation, such a program is definite and then it has a unique minimal Herbrand model denoted with $Cn(P)$. By definition, an *answer set* (originally called a *stable model* [9]) of P is an atom set S such that $S = Cn(P^S)$. For instance the program $\{a \leftarrow not\ b., \ b \leftarrow not\ a.\}$ has two answer sets $\{a\}$ and $\{b\}$. Special headless rules, called *constraints*, are admitted and considered equivalent to rules like $bug \leftarrow \ldots, \ not\ bug.$ where bug is a new symbol appearing nowhere else. For instance, the program $\{a \leftarrow not\ b., \ b \leftarrow not\ a., \leftarrow a.\}$ has one, and only one, answer set $\{b\}$.

As presented in our introduction, in many cases a problem is encoded in ASP with a logic program P containing rules with variables that we call *first order rules*. More formally, these rules are of type (1) where a_i's and b_j's are non ground atoms, like $p(X, 3, f(Y))$, built from an n-ary predicate, constants, variables and function symbols. Since answer set definition is given for propositional programs, P has to be seen as an intensional version of the propositional program $ground(P)$ defined as follows. Given a rule r, $ground(r)$ is the set of all fully instantiated rules that can be obtained by substituting every variable in r by every constant of the Herbrand universe of P and then, $ground(P) = \bigcup_{r \in P} ground(r)$.

Example 1. The program P_1 is a shorthand for the program $ground(P_1)$.

$$P_1 = \left\{ \begin{array}{l} n(1) \leftarrow ., \ n(2) \leftarrow ., \\ a(X) \leftarrow n(X), \ not\ b(X)., \\ b(X) \leftarrow n(X), \ not\ a(X). \end{array} \right\} \quad ground(P_1) = \left\{ \begin{array}{l} n(1) \leftarrow ., \ n(2) \leftarrow ., \\ a(1) \leftarrow n(1), \ not\ b(1)., \\ b(1) \leftarrow n(1), \ not\ a(1)., \\ a(2) \leftarrow n(2), \ not\ b(2)., \\ b(2) \leftarrow n(2), \ not\ a(2). \end{array} \right\}$$

$ground(P_1)$ has four answer sets $\{a(1), a(2), n(1), n(2)\}$, $\{a(1), b(2), n(1), n(2)\}$, $\{a(2), b(1), n(1), n(2)\}$ and $\{b(1), b(2), n(1), n(2)\}$ that are considered as the answer sets of P_1.

Until today, almost all systems available to compute the answer sets of a program follow the architecture described in Fig. 1. For the grounder box we can cite Lparse [21] and Gringo [8], and for the solver box Clasp [7] and Smodels [20]. A particular family of solvers are Assat [14], Cmodels [10] and Pbmodels [17], since they transform the answer set computation problem into a (pseudo) boolean model computation problem and use a (pseudo) SAT solver as an internal black box. In the system DLV [13], symbolized in Fig. 1 by the dash-line rectangle, the grounder ([3] describes a parallel version) is incorporated as an internal function.

The main goal of each grounding system is to generate all propositional rules that can be relevant for a solver and only these ones, while preserving answer sets of the original program. But, whatever the methodology is, the grounding phase is firstly and fully processed before computing the answer sets. In the next section, we present our alternative approach that integrates this grounding phase into the search of model.

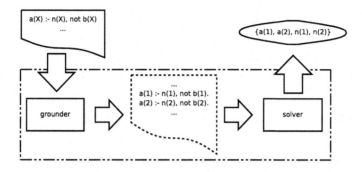

Fig. 1. Architecture of answer set computation

3 A First Order Rule Based Approach

We first present the characterization of answer sets for ground normal logic programs based on an abstract notion of *computation* proposed in [16]. A computation is a sequence of atom sets starting with the empty set. At each step, the heads of some applicable rules (see Definition 2 below) w.r.t. actual state are added. When no more atom can be added, one must check that the rules that have been fired are still applicable.

Definition 1. *(from [16]) Let P be a normal logic program. A computation for P is a sequence $\langle X_i \rangle_{i=0}^{\infty}$ of atom sets that satisfies the following conditions :*

- $X_0 = \emptyset$
- *(Revision)* $\forall i \geq 1, X_i \subseteq T_P(X_{i-1})$
- *(Persistence of beliefs)* $\forall i \geq 1, X_{i-1} \subseteq X_i$
- *(Convergence)* $X_\infty = \bigcup_{i=0}^{\infty} X_i = T_P(X_\infty)$
- *(Persistence of reasons)* $\forall i \geq 1, \forall a \in X_i \backslash X_{i-1}, \exists r_a \in P$ *s.t.* $head(r_a) = a$, *and* $\forall j \geq i - 1, body^+(r_a) \subseteq X_j, body^-(r_a) \cap X_j = \emptyset$

where $T_P(X) = \{a \mid \exists r \in P, head(r) = a, body^+(r) \subseteq X, body^-(r) \cap X = \emptyset\}$

Theorem 1. *(from [16]) Let P be a normal logic program and X be an atom set. Then, X is an answer set of P iff there is a computation $\langle X_i \rangle_{i=0}^{\infty}$ for P such that $X_\infty = X$.*

This computation is fundamentally based on a forward chaining of rules by means of operator T_P. It builds incrementally the answer set of the program and does not require the whole set of ground atoms since the beginning of the process. So, it is well suited to deal directly with first order rules by instantiating them during the computation and that is what we introduce in the following.

The only syntactic restriction required by our methodology is that every rule of a program must be *safe*. That is, all variables occurring in the head and all variables occurring in the negative body of a rule occur also in its positive body. Moreover, every constraint (ie : headless rule) is considered given with the particular head $\bot$ and is also safe. For the moment we do not consider function symbols but their use will be described in Sect. 4.

A *partial interpretation* for a program P is a pair $\langle IN, OUT \rangle$ of disjoint atom sets included in the Herbrand base of P. Intuitively, all atoms in IN belong to a seeking answer set and all atoms in OUT do not. If $I_1 = \langle A_1, B_1 \rangle$ and $I_2 = \langle A_2, B_2 \rangle$ are partial interpretations, $I_1 \subseteq I_2$ iff $A_1 \subseteq A_2 \wedge B_1 \subseteq B_2$.

Definition 2. *Let r be a ground rule and $I = \langle IN, OUT \rangle$ a partial interpretation. We say*

- r *is* supported *w.r.t. I when $body^+(r) \subseteq IN$,*
- r *is* unsupported *w.r.t. I when $body^+(r) \cap OUT \neq \emptyset$*
- r *is* blocked *w.r.t. I when $body^-(r) \cap IN \neq \emptyset$,*
- r *is* unblocked *w.r.t. I when $body^-(r) \subseteq OUT$,*
- r *is* applicable *w.r.t. I when r is supported and not blocked[1]*

The approach of answer set computation that we introduce here is a particular class of computations (Definition 1), obtained by restricting the principle of revision that originally enables to fire any subset of the supported and not blocked rules at each step. In our strategy, we follow a forward chaining that instantiates and applies one unique rule at each iteration. Our revision principle distinguishes two kinds of inference: a monotonic step of *propagation* and a non monotonic step of *choice*. The two functions defined below realize in the same time the grounding and the selection of the rule to apply at each step.

Definition 3. *Let P be a set of first order rules, $\langle IN, OUT \rangle$ be a partial interpretation and R be a set of ground rules.*

- γ_{pro} *is a non deterministic function selecting one supported monotonic rule or one supported and unblocked non monotonic rule $\gamma_{pro}(P, IN, OUT, R)$ in $ground(P) \setminus R$, or returns false if no such a rule exists.*
- γ_{cho} *is a non deterministic function selecting one supported and not blocked non monotonic rule $\gamma_{cho}(P, IN, OUT, R)$ in $ground(P) \setminus R$, or returns false if no such a rule exists.*

To avoid any confusion, we insist on the fact that the set $ground(P)$, mentioned in the above definition, is never explicitly given. It is in accordance with the principal aim of our work that is to avoid its extensive construction. The two functions γ_{pro} and γ_{cho} select and ground first order rules of P with propositional atoms occurring in IN and OUT in order to return a new (not already occurring in R) fully ground rule. Because of the safety constraint on rules this full grounding is always possible.

We give below the whole definition of an *ASPeRiX computation*[2] that we characterize as a sequence of partial interpretations instead of atom sets as in [16].

---

[1] The negation of blocked, *not blocked*, is different from *unblocked*.

[2] ASPeRiX is the name of the solver that we have developed following this principle (see Sect. 4).

Definition 4. *Let P be a first order normal logic program. An* `ASPeRiX` *compu-tation for P is a sequence $\langle X_i \rangle_{i=0}^{\infty}$ of partial interpretations $X_i = \langle IN_i, OUT_i \rangle$ that satisfies the following conditions :*

- *$X_0 = \langle \emptyset, \{\perp\} \rangle$,*
- *(Revision) $\forall i \geq 1, X_i = \langle IN_{i-1} \cup \{head(r_i)\}, OUT_{i-1} \rangle$ for some rule $r_i = \gamma_{pro}(P, IN_{i-1}, OUT_{i-1}, \bigcup_{k=1}^{i-1}\{r_k\})$ if it exists else, $X_i = \langle IN_{i-1} \cup \{head(r_i)\}, OUT_{i-1} \cup body^-(r_i) \rangle$ for some rule $r_i = \gamma_{cho}(P, IN_{i-1}, OUT_{i-1}, \bigcup_{k=1}^{i-1}\{r_k\})$ if it exists else, $X_i = X_{i-1}$,*
- *(Persistence of beliefs) $\forall i \geq 1, X_{i-1} \subseteq X_i$,*
- *(Convergence) $IN_{\infty} = \bigcup_{i=0}^{\infty} IN_i = T_P(IN_{\infty})$,*
- *(Persistence of reasons) $\forall i \geq 1, \forall a \in IN_i \setminus IN_{i-1}, \exists r_a \in ground(P)$ s.t. $head(r_a) = a$, and $\forall j \geq i - 1, body^+(r_a) \subseteq IN_j, body^-(r_a) \cap IN_j = \emptyset$.*

Theorem 2. *Let P be a normal logic program and A be an atom set. Then, A is an answer set of P iff there is an* `ASPeRiX` *computation $\langle X_i \rangle_{i=0}^{\infty}$ for P such that $IN_{\infty} = A$.*

Proof. (sketch) By restricting our attention to the sequence $\langle IN_i \rangle_{i=0}^{\infty}$, it is easy to verify that an `ASPeRiX` computation is a computation and thus, by Theorem 1, converges to an answer set. For the other direction, every answer set can be mapped into a computation by Theorem 1. On its turn, this computation can be mapped into an `ASPeRiX` computation because persistence of reasons allows us to build a set $\{r_a \mid a \in X_{\infty}\}$ that can be ordered in such a way that it corresponds to the successive application of rules in an `ASPeRiX` computation. $\square$

Note that in order to respect the revision principle of an `ASPeRiX` computa-tion every sequence of partial interpretations must be generated by using the

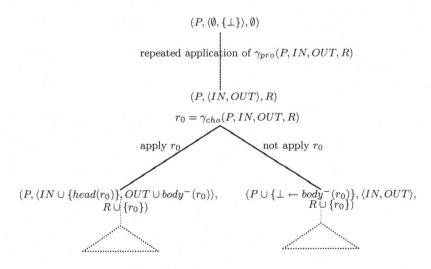

Fig. 2. Overview of the search procedure

propagation inference based on γ_{pro} as long as possible before to use the choice function γ_{cho} in order to apply a non monotonic rule. Then, because of the non determinism of γ_{cho}, the natural implementation of our approach leads to a usual search procedure that has to decide to apply or not every rule chosen by γ_{cho}. This strategy is illustrated in Fig. 2 in which we can see that persistence of reasons is ensured by adding to OUT all ground atoms from the negative body of the non monotonic rule chosen to be applied. On the other branch, where the rule is not applied, the translation of its negative body into a new constraint ensures that it becomes impossible to find later an answer set in which this rule is not blocked.

Algorithm 1. Algorithm for a first order rule-based answer set computing

```
Function Solve(P_R, P_K, IN, OUT, CR);
repeat // propagation phase
    r_0 ← γ_pro(P_R ∪ P_K, IN, OUT, CR);
    if r_0 then
        IN ← IN ∪ {head(r_0)};
        CR ← CR ∪ {r_0};
until ¬r_0;
if IN ∩ OUT ≠ ∅ then // contradiction detected
    return false;
else
    r_0 ← γ_cho(P_R, IN, OUT, CR);
    if ¬r_0 then
        if γ_cho(P_K, IN, OUT, ∅) then // constraint non satisfied
            return false;
        else// an answer set is found
            return IN;
    else//choice point
        stop ← solve(P_R, P_K, IN ∪ {head(r_0)}, OUT ∪ body⁻(r_0), CR ∪ {r_0});
        if ¬stop then
            stop ← solve(P_R, P_K ∪ {⊥ ← body⁻(r_0)}, IN, OUT, CR ∪ {r_0});
        return stop ;
```

Our search procedure of answer set computing for a program P is detailed in Algorithm 1 that must be called by $solve(P_R, P_K, \emptyset, \{\bot\}, \emptyset)$, knowing that $P_K = \{r \in P \mid head(r) = \bot\}$ (the constraint set) and $P_R = P \setminus P_K$. CR (for chosen rules) is the set of ground rules built by γ_{pro} and γ_{cho} all along the search. In the propagation phase we treat equally the constraints and the rules in P_R. By this way, if a constraint is selected by γ_{pro}, then it is applied and $\bot$ is added in IN. Since $\bot$ is also in OUT a contradiction is detected and the algorithm returns false. On its side γ_{cho} restricts its attention to rules in P_R firstly. But, if no new non monotonic rule is found, then it tries to find an applicable constraint (again, it is the grounding of a first order constraint). If it is the case, then the convergence principle is not respected ($\bot$ can never be added to IN) and the current partial interpretation can not be an ASPeRiX computation. Thus the algorithm returns false. Otherwise, no constraint is violated and then an answer set is returned. Even if our algorithm describes the computation of one answer set (or no one if the program is inconsistent) it can easily be extended to the computation of an arbitrary number of (or all) answer sets of P.

4 Benefits of a First Order Rule Based Computation

Following Algorithm 1, we have implemented in C++ a new solver called ASPeRiX available at http://www.info.univ-angers.fr/pub/claire/asperix. The name has been chosen to draw attention to rules (R) and variables (X) since our system is able to deal with rules with variables by grounding them just as it applies them. We have to mention that there exists another ASP system named GASP [19], and up to our knowledge it is the only one, that realizes the grounding of the program during the search of an answer set[3]. GASP is an implementation in Prolog and Constraint Logic Programming over finite domains of the notion of computation (see Definition 1). Its strategy is formally presented as a computation of well founded consequences. But its implementation uses a variant of the propagation operator T_P that is close to ours. GASP supports cardinality constraints while ASPeRiX does not, on the other hand GASP does not accept functional terms.

A deep description of ASPeRiX is out of the scope of this paper and some details about its main features and a partial evaluation of its performances are given in [12]. In the following we focus on some peculiarities of ASPeRiX that we consider as improvements for the ASP community. When we report some results about other ASP systems, we take into account the two steps of ASP computation (grounding and solving) and versions of systems used are: Gringo 2.0.3 + Clasp 1.2.1 (or their combination Clingo), Lparse 1.1.1 + Smodels 2.32, DLV Oct 11 2007 and GASP (june 2009).

First, we can cite that since each grounder has its own strategy, it imposes some particular syntactical restrictions about the occurrences of variables in rules. For Lparse the program has to be ω-restricted [22], for Gringo it has to be λ-restricted [8] and for DLV and GASP only safe rules are required. This point is not a theoretical one because if a problem is encoded following the syntax of a grounder, then it is also possible to encode it respecting the syntax of another grounder[4]. But, in practice, it leads to some difficulties to compare the efficiency of solvers because various encoding of a same problem can induce various sizes of the search space. Moreover, if we want to disseminate the ASP paradigm over various domains of applications, to encourage the development of portable ASP libraries and to promote the interoperability of systems, there is a need for the less restrictive language. For the encoding of a practical knowledge

---

[3] Iclingo [6] realizes an incremental grounding of programs that are parametrized by the size of the solution but it does not escape the grounding/solving separation.

[4] Recall that we restrict our attention to "classical" normal logic programs without aggregate functions or cardinality constraints.

representation or constraint satisfaction problem the constraint of safety is not seen as a drawback and is currently admitted. Since ASPeRiX requires only the safety of the program, we can say that our system is as general as possible.

But, for us, the main drawback of the preliminary grounding phase is that it leads to a lot of useless and counter-intuitive work in some situations.

Example 2. Let N be an integer. From, the following program

$$P_2 = \begin{cases} a \leftarrow not\ b., & p(1) \leftarrow ., \ldots, \ p(N) \leftarrow ., \\ b \leftarrow not\ a., ., & pa(X) \leftarrow a,\ p(X)., & aa(X,Y) \leftarrow pa(X),\ pa(Y)., \\ \leftarrow a., & pb(X) \leftarrow b,\ p(X)., & bb(X,Y) \leftarrow pb(X),\ pb(Y). \end{cases}$$

DLV, Gringo and Lparse[5] generate about $2 \times N^2$ ground rules.

Because of the constraint $\leftarrow a.$ that eliminates from the possible solutions every atom set containing a, it is easy to see that all N ground rules with a positive body containing a, like $pa(\_) \leftarrow a, p(\_).$, are useless since they can never contribute to generate an answer set of P_2. And then, the N^2 rules with head $aa(\_,\_)$ are useless too. In defense of the actual grounders, their inability to eliminate these particular rules is not surprising since the reason justifying this elimination is the consequence of a reasoning taking into account the stable model semantics. If we refer to Fig. 1 it is clear that this task is relevant to the solver box and not to the grounder box. Thus, if we want to limit as much as possible the number of rules and atoms to deal with, we have to not separate grounding and answer set computing as we propose in our methodology. The Table 1 gives few experimental results illustrating that ASPeRiX is the most efficient system for the computation of one stable model of P_2.

Table 1. Experimental results for P_2

		Lparse+Smodels	DLV	Clingo	GASP	ASPeRiX
$N = 200$	time in sec	5.4	2	1.8	3	0.7
	memory in MB	30	64	28	46	7
$N = 400$	time in sec	21.3	10.5	11.7	14.4	2.6
	memory in MB	100	240	103	54	19
$N = 600$	time in sec	48	29	42	40	6.6
	memory in MB	220	523	223	68	37

With respect to the efficiency of our new approach we mention that ASPeRiX appears to be the fastest ASP system to compute the unique stable model of a stratified program. It is known that for this class of programs, the computation of the stable model can be done polynomialy with a bottom-up greedy algorithm. This is exactly how does our system by using only the propagation step via γ_{pro} and generating no choice point since γ_{cho} is never used (see Fig. 2). This is achieved through a dependency graph on first order rules that allows

---

[5] For Lparse, $p(X)$ and $p(Y)$ must be inserted in the two last rules in order to respect the ω-restricted condition.

to treat all non monotonic rules as monotonic ones if the program is stratified. This property is also exploited very efficiently by ASPeRiX for locally stratified programs (details can be found in [12]).

Another important improvement of our methodology that integrates grounding of rules into their forward chaining is that it allows to manage functions in a very natural and easy way. The inherent difficulty with arithmetic in particular and functions in general in the framework of ASP is that it makes the Herbrand universe infinite in whole generality. Furthermore, various syntactic restrictions have already been exhibited and ASP solvers are able to deal with some restricted versions of functions. The reader will find in [2] many details about the treatment of functions in ASP and the introduction of a new class of programs with functions, for which answer set computation is always possible. This work has led to the system DLV-complex [2] that is an extension of DLV and certainly the most advanced ASP system available today to deal with functions in ASP.

But, once again, a lot of the difficulties arising with functions become from the fact that the pregrounding of the program is required for traditional ASP solvers. A very fast growing of the memory usage is much more possible than without function symbols. With our approach since we generate the Herbrand universe as we need we just have to fix an a priori limit on the biggest admissible integer and the maximum number of nested function symbols (DLV-complex does the same to deal with non finitely-ground programs [2]). By this way, our Theorem 2 is still valid since the computation always ends.

In Example 3 we give the simplest normal logic program that we can write with respect to the accepted language of ASPeRiX in order to compute the Fibonacci number F_k for a given k and not compute the sequence beyond this limit. Let us recall that $F_0 = 0$, $F_1 = 1$ and $\forall i > 0, F_{i+2} = F_i + F_{i+1}$.

Example 3. With the following program

$$P_3 = \left\{ \begin{array}{ll} fibo(0,0) \leftarrow ., & tocompute fibo(10) \leftarrow . \\ fibo(1,1) \leftarrow ., & \\ fibo(N+2, F1+F) \leftarrow fibo(N,F), fibo(N+1,F1), \\ \qquad tocompute fibo(K), K > N+1. \end{array} \right\}$$

ASPeRiX returns a stable model containing $fibo(10,55)$ representing $F_{10} = 55$.

Let us remark that this schema of recursive rule is not allowed by grounders Lparse and Gringo since the program is neither ω nor λ-restricted. For these systems, we need to modify the recursive rule as

$$fibo(N2, F2) \leftarrow p(N), p(F), p(F1), fibo(N,F), N1 = N+1, fibo(N1, F1),$$
$$tocompute fibo(K), K > N1, N2 = N1+1, F2 = F1+F.$$

and adding $p(0) \leftarrow ., \ldots p(2000) \leftarrow .$ for instance if we want to limit the possible used numbers to the interval $0, \ldots, 2000$. This syntactic constraint is very strong in context of arithmetic, since the predicate p enumerates a lot of useless numbers when the computation of the Fibonacci sequence needs only few ones. By a direct consequence, the pregrounding becomes very hard to do since a huge number of atoms are unnecessarily generated. For instance, to compute F_{10}, Clingo needs more than 2 minutes and Lparse does not find the solution after 30 minutes of CPU time and both systems consume more than 1 GB of memory. On its side

ASPeRiX computes immediately $F_{46} = 1836311903$ the biggest value that it is possible to compute under its limit of 2^{31} and DLV-complex does the same even if its limit is $F_{44} = 701408733$. GASP, with a slight syntactic adaptation, can compute $F_{21} = 10946$. This example illustrates that only the condition of safety for rules is admissible if we want to introduce non trivial arithmetic calculus in ASP. It is also an illustration that our approach is naturally well adapted to this kind of computation. The difference with DLV-complex is that the calculus are made only when they are necessary.

ASPeRiX is also able to manage symbolic functions occurring in rules like $p(X) \leftarrow p(f(X))$. or $p(f(X)) \leftarrow p(X)$. indifferently. The following commented program P_4 illustrates the use of symbolic functions by extending Example 3. It encodes a kind of generic system to compute every recursive arithmetic functions.

Example 4. From the following program

$$P_4 = \left\{ \begin{array}{l} \text{\% definition of fibonacci} \\ value(fibo(0),0) \leftarrow ., \qquad\qquad value(fibo(1),1) \leftarrow . \\ value(fibo(N+2), R+R1) \leftarrow value(fibo(N), R), value(fibo(N+1), R1), \\ \qquad\qquad\qquad\qquad\qquad not\ ok(fibo(N+1)). , \\ \text{\% the values to compute for Fibonacci} \\ tocompute(fibo(20)) \leftarrow ., \qquad\quad tocompute(fibo(10)) \leftarrow ., \\ \text{\% and the non maximum values to compute} \\ nmaxtocompute(fibo(N1)) \leftarrow tocompute(fibo(N1)), \\ \qquad\qquad\qquad\qquad tocompute(fibo(N2)), N2 > N1. , \\ \hline \text{\% definition of exponentiation} \\ value(expo(X,0),1) \leftarrow tocompute(expo(X,N)). , \\ value(expo(X,N+1), R*X) \leftarrow value(expo(X,N), R), not\ ok(expo(X,N)). , \\ \text{\% the values to compute for exponentiation} \\ tocompute(expo(2,10)) \leftarrow ., \qquad\quad tocompute(expo(2,15)) \leftarrow ., \\ tocompute(expo(4,10)) \leftarrow . \\ \text{\% and the non maximum values to compute} \\ nmaxtocompute(expo(X,N1)) \leftarrow tocompute(expo(X,N1)), \\ \qquad\qquad\qquad\qquad tocompute(expo(X,N2)), N2 > N1. , \\ \hline \text{\% the limit of computation for each recursive function Phi} \\ maxtocompute(Phi) \leftarrow tocompute(Phi), not\ nmaxtocompute(Phi). \\ \text{\% to detect when all necessary computations are done for one function Phi} \\ ok(Phi) \leftarrow maxtocompute(Phi), value(Phi, R). , \\ \text{\% to collect the results} \\ result(C,R) \leftarrow tocompute(C), value(C,R). , \end{array} \right.$$

ASPeRiX and also DLV-complex return (among some others) the atoms:
$result(fibo(10), 55) \qquad result(fibo(20), 6765)$
$result(expo(2, 10), 1024)\ result(expo(2, 15), 32768)\ result(expo(4, 10), 1048576)$
representing all the wanted calculus.

To end, we mention that the treatment of symbolic lists is also possible with our system by means of terms like $l(\_, l(\ldots, l(\_, nil)\ldots))$. But, because of the particular importance of lists in logic programming we think that a dedicated treatment of them is required as in `DLV-complex`. This point is not yet implemented in `ASPeRiX` but it poses no theoretical difficulties with respect to our general algorithm.

5   Conclusion

In this work, we have elaborated a new approach of answer set computation that escapes the preliminary phase of grounding. Our methodology deals with first order rules following a forward chaining with grounding process realized on the fly and has been implemented in a new ASP solver `ASPeRiX`.

One direct consequence of our new approach is that the use of symbolic functions in general and arithmetic calculus in particular inside ASP is greatly facilitated. Another point is that our methodology allows very good performances for definite, stratified and almost stratified programs. Again, we have shown that our approach escapes the bottleneck of pregrounding phase that is the only difficulty for some classes of programs.

With our present work, we wanted also to emphasize on the P (Programming) of ASP. Indeed, we think that we do not have to forget that ASP is in the field of Logic Programming, and not only SAT or CSP. Many application domains of ASP, let us cite only the web semantic, use rules that are chained ones with the others when this is not always the case for a program encoding a CSP. For this category of programs, we think that our approach may be of great interest.

Furthermore, computing the answer sets of a program is a fundamental goal but not an exclusive one. Debugging a program, controlling its behavior, introducing in it some features coming from other programming languages may be of great interest for ASP. We think that our methodology of answer set computing, guided by the rules of the program, is the good starting point towards these new goals.

References

1. Baral, C.: Knowledge Representation, Reasoning and Declarative Problem Solving. Cambridge University Press, Cambridge (2003)
2. Calimeri, F., Cozza, S., Ianni, G., Leone, N.: Computable functions in ASP: Theory and implementation. In: Garcia de la Banda, M., Pontelli, E. (eds.) ICLP 2008. LNCS, vol. 5366, pp. 407–424. Springer, Heidelberg (2008)
3. Calimeri, F., Perri, S., Ricca, F.: Experimenting with parallelism for the instantiation of ASP programs. Journal of Algorithms 63(1-3), 34–54 (2008)
4. Eiter, T., Lu, J.J., Subrahmanian, V.S.: Computing non-ground representations of stable models. In: Dix, J., Furbach, U., Nerode, A. (eds.) LPNMR 1997. LNCS, vol. 1265, pp. 198–217. Springer, Heidelberg (1997)
5. Ferraris, P., Lee, J., Lifschitz, V.: A new perspective on stable models. In: Proceedings of the 20th International Joint Conference on Artificial Intelligence (IJCAI 2007), pp. 372–379 (2007)
6. Gebser, M., Kaminski, R., Kaufmann, B., Ostrowski, M., Schaub, T., Thiele, S.: Engineering an incremental ASP solver. In: Garcia de la Banda, M., Pontelli, E. (eds.) ICLP 2008. LNCS, vol. 5366, pp. 190–205. Springer, Heidelberg (2008)

7. Gebser, M., Kaufmann, B., Neumann, A., Schaub, T.: Conflict-driven answer set solving. In: Proceedings of the 20th International Joint Conference on Artificial Intelligence (IJCAI 2007), pp. 386–392 (2007)
8. Gebser, M., Schaub, T., Thiele, S.: GrinGo: A new grounder for answer set programming. In: Baral, C., Brewka, G., Schlipf, J. (eds.) LPNMR 2007. LNCS (LNAI), vol. 4483, pp. 266–271. Springer, Heidelberg (2007)
9. Gelfond, M., Lifschitz, V.: The stable model semantics for logic programming. In: Kowalski, R.A., Bowen, K. (eds.) ICLP, pp. 1070–1080. MIT Press, Cambridge (1988)
10. Giunchiglia, E., Lierler, Y., Maratea, M.: Answer set programming based on propositional satisfiability. Journal of Automated Reasoning 36(4), 345–377 (2006)
11. Gottlob, G., Marcus, S., Nerode, A., Salzer, G., Subrahmanian, V.S.: A non-ground realization of the stable and well-founded semantics. Theoretical Computer Science 166(1-2), 221–262 (1996)
12. Lefèvre, C., Nicolas, P.: The first version of a new ASP solver: ASPeRiX. In: Erdem, E., Lin, F., Schaub, T. (eds.) LPNMR 2009. LNCS. Springer, Heidelberg (2009)
13. Leone, N., Pfeifer, G., Faber, W., Eiter, T., Gottlob, G., Perri, S., Scarcello, F.: The DLV system for knowledge representation and reasoning. ACM Transactions on Computational Logic 7(3), 499–562 (2006)
14. Lin, F., Zhao, Y.: ASSAT: computing answer sets of a logic program by SAT solvers. Artificial Intelligence 157(1-2), 115–137 (2004)
15. Lin, F., Zhou, Y.: From answer set logic programming to circumscription via logic of GK. In: Proceedings of the 20th International Joint Conference on Artificial Intelligence (IJCAI 2007), pp. 441–446 (2007)
16. Liu, L., Pontelli, E., Son, T.C., Truszczynski, M.: Logic programs with abstract constraint atoms: The role of computations. In: Dahl, V., Niemelä, I. (eds.) ICLP 2007. LNCS, vol. 4670, pp. 286–301. Springer, Heidelberg (2007)
17. Liu, L., Truszczynski, M.: Pbmodels - software to compute stable models by pseudoboolean solvers. In: Baral, C., Greco, G., Leone, N., Terracina, G. (eds.) LPNMR 2005. LNCS (LNAI), vol. 3662, pp. 410–415. Springer, Heidelberg (2005)
18. Niemelä, I.: Logic programs with stable model semantics as a constraint programming paradigm. Annals of Mathematics and Artificial Intelligence 25(3-4), 241–273 (1999)
19. Dal Palù, A., Dovier, A., Pontelli, E., Rossi, G.: Answer set programming with constraints using lazy grounding. In: Hill, P., Warren, D. (eds.) ICLP 2009. LNCS, vol. 5649. Springer, Heidelberg (2009)
20. Simons, P., Niemelä, I., Soininen, T.: Extending and implementing the stable model semantics. Artificial Intelligence 138(1-2), 181–234 (2002)
21. Syrjänen, T.: Implementation of local grounding for logic programs for stable model semantics. Technical report, Helsinki University of Technology (1998)
22. Syrjänen, T.: Omega-restricted logic programs. In: Eiter, T., Faber, W., Truszczyński, M. (eds.) LPNMR 2001. LNCS (LNAI), vol. 2173, pp. 267–279. Springer, Heidelberg (2001)

Knowledge Qualification through Argumentation

Loizos Michael and Antonis Kakas

Department of Computer Science
University of Cyprus
{loizosm,antonis}@cs.ucy.ac.cy

Abstract. We propose a framework that brings together two major forms of default reasoning in Artificial Intelligence: default property classification in static domains, and default property persistence in temporal domains. Emphasis in this work is placed on the *qualification problem*, central when dealing with default reasoning, and in any attempt to integrate different forms of such reasoning.

Our framework can be viewed as offering a semantics to two natural problems: *(i)* that of employing default static knowledge in a temporal setting, and *(ii)* the dual one of temporally projecting and dynamically updating default static knowledge.

The proposed integration is introduced through a series of example domains, and is then formalized through argumentation. The semantics follows a pragmatic approach. At each time-point, an agent predicts the next state of affairs. As long as this is consistent with the available observations, the agent continues to reason forward. In case some of the observations cannot be explained without appealing to some *exogenous* reason, the agent revisits and revises its past assumptions.

We conclude with some formal results, including an algorithm for computing complete admissible argument sets, and a proof of *elaboration tolerance*, in the sense that additional knowledge can be gracefully accommodated in any domain.

1 Introduction

An important aspect of intelligence is the ability to reason, and draw conclusions about properties of one's environment that are not directly visible. In the area of logic-based reasoning, it is assumed that such conclusions are drawn by applying some knowledge base comprised of logic rules. To account for the inherent barrier of representing all possible knowledge for any but the simplest domains, *default* logic rules can be used in a knowledge base, so that conclusions are still drawn in the absence of sufficient information, but can be retracted in the presence of evidence to the contrary.

Two major forms of default reasoning have been extensively studied on their own in Artificial Intelligence, but have rarely been addressed in the same formalism. These are *default property classification* as applied to inheritance systems [1,2], and *default persistence* central to temporal reasoning in theories of Reasoning about Actions and Change (RAC) [3,4,5]. Here we consider the question [6]: How can a formalism synthesize the reasoning encompassed within each of these two forms of default reasoning?

Central to these two (and indeed all) forms of default reasoning is the *qualification problem*: default conclusions are qualified by information that can block the application of the default inference. One aspect of the qualification problem is to express, within the theory, the knowledge required to properly qualify and block the default inference

E. Erdem, F. Lin, and T. Schaub (Eds.): LPNMR 2009, LNCS 5753, pp. 209–222, 2009.

under exceptional situations. This *endogenous* form of qualification is implicit in the theory, driven by auxiliary observations that enable the known qualifying information to be applied. For example, known exceptional classes in the case of default property inheritance, or known action laws (and their ramifications) in the case of default persistence, qualify, respectively, the static and temporal forms of default reasoning.

Completely representing, within a given theory, the qualification knowledge is impractical and indeed undesirable, as we want to jump to default conclusions based on a minimal set of information available. We, therefore, also need to allow for default conclusions to be qualified unexpectedly from observed information that is directly (or explicitly) contrary to them. In this *exogenous* form of qualification, the theory itself cannot account for the qualification of the default conclusion, but our observations tell us explicitly that this is so and we attribute the qualification to some unknown reason.

Recent work [7,8] has shown the importance for RAC theories to properly account for these two forms of qualification, so that exogenous qualification is employed only when observations cannot be accounted for by endogenous qualification of default persistence and the causal laws. When integrating default static and temporal theories, this means that we need to ensure that the two theories properly qualify each other endogenously, so that the genuine cases of exogenous qualification can be correctly recognized.

The mutual qualification of the default static and temporal theories can be understood in two dual ways: On the one hand, temporal reasoning is extended to include default static knowledge, which acts as a global, but defeasible, constraint that qualifies the temporal evolution of a domain. On the other hand, default static reasoning is extended to a temporal setting, where persistence, the effects of actions, and observations across the time-line can qualify the default laws that are used for reasoning.

In particular, we study how four different types of information present in such an integrated framework interact with, and can qualify, each other: *(i)* information generated by default persistence of fluents, *(ii)* action laws that can qualify default persistence, *(iii)* default static laws of fluent relationships that can qualify these action laws, and *(iv)* observations that can (exogenously) qualify any of these types of information. This hierarchy of information comes full circle, as the bottom layer of default persistence of observations (which carry the primary role of qualification) can, also, qualify the static theory. Hence, in our proposed integrated framework, temporal projection with the observations help to determine the admissible states of the default static theory. In turn, admissible states qualify the actions laws and the temporal projection they generate.

The semantics of the proposed integration is motivated and introduced through a series of example domains in Section 2, and formalized through an argumentation-based framework in Section 3. The framework follows a pragmatic approach. An agent holds certain beliefs about the state of the world at a certain time-point, executes some actions, and possibly makes some observations. It then predicts the next state of the world, by devising an admissible argument set, taking into account all the pieces of knowledge at its disposal, and resolving conflicts based on the strength of the corresponding arguments. If an admissible argument set that agrees with all available observations is found, the agent assumes that the computed state is the actual state of the world, and moves forward, repeating the process at the next time-point. This is repeated until the state predicted by the agent based solely on its knowledge, and without appealing to some

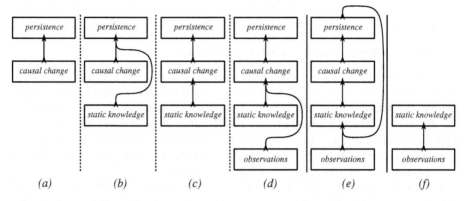

Fig. 1. Solutions to the knowledge qualification problem. Arrows point from the knowledge type that qualifies to the knowledge type being qualified. Root nodes in the graphs correspond to strict knowledge, and internal nodes correspond to default knowledge (qualified by its parent nodes).

exogenous reason, is in conflict with the observations. In such a case, the agent revisits its past assumptions, and revises them so as to resolve conflicts with the observations.

We establish some formal results in Section 4. From a computational point of view, we provide an algorithm that is guaranteed to compute a complete admissible argument set. From an epistemological point of view, we show that domains interpreted under our proposed semantics are *elaboration tolerant*, in the sense that they can be extended with, and gracefully accommodate, *arbitrary* pieces of knowledge. In particular, the semantics enjoys a *free-will* property, so called because an agent may choose to execute any action, without this causing an inconsistency and compromising its ability to reason. We conclude in Section 5, where we briefly discuss related and future work.

2   Knowledge Qualification

Through a series of examples, we present in this section the issues that arise when examining the qualification of knowledge, and place in context the various problems and solutions considered so far. We remark that we generally use the term *qualification* in a broader sense than that used in the context of Reasoning about Actions and Change.

For illustration purposes, we employ the syntax of the action description language $\mathcal{ME}$ [7] for temporal domain descriptions, and a pseudo-syntax based on that of propositional logic for representing static theories describing default or strict domain constraints. Strict static knowledge is represented in classical propositional logic. Default static knowledge is represented in terms of default rules of the form "$\phi \rightsquigarrow \psi$", where ϕ, ψ are propositional formulas. In this pseudo-syntax we specify the relative strength between two default rules by statements of the form *"rule (i) overrides rule (j)"*. Formulas which contain variables are a shorthand representation of all formulas obtained by substituting the variables over an appropriate finite domain of constants.

We do not reproduce here the formal syntax for these theories. In particular, the formal semantics of our approach, given in the next section, will not depend on the specific form of the static theories, and different frameworks such as Default Logic [2] or argumentation [9] can be used. For the example domains in this section, it is sufficient for the reader to use the informal reading of the static theories for their semantics.

2.1 Past Investigations of Knowledge Qualification

One of the first knowledge qualification problems formally studied in Artificial Intelligence relates to the *Frame Problem* (see, e.g., [5]) of how the causal change properly qualifies the default persistence; see Fig. 1(a). In the archetypical Yale Shooting Problem domain [3], a turkey named Fred is initially alive, and one asks whether it is still alive after loading a gun, waiting for some time, and then shooting Fred. The lapse of time cannot cause the gun to become unloaded. Default persistence is qualified only by known events and known causal laws linked to these events.

The consideration of richer domains gave rise to the *Ramification Problem* (see, e.g., [10]) of how indirect action effects are generated and qualify persistence; see Fig. 1(b). Static knowledge expressing relationships between different properties (i.e., domain constraints) was introduced to encode indirect effects. In early solutions to the Ramification Problem a direct action effect would cause this static knowledge to be violated, unless a minimal set of indirect effects were also assumed in order to maintain consistency [10,11]. Thus, given the static knowledge that "dead birds do not walk", the shooting action that causes Fred to be dead would also indirectly cause Fred to stop walking, thus qualifying the persistence of Fred walking.

Subsequent work examined default causal knowledge, bringing to focus the *Qualification Problem* (see, e.g., [8]) of how such default causal knowledge is qualified by domain constraints; see Fig. 1(c). In some solutions to the Qualification Problem, static knowledge within the domain description was identified as the knowledge that *endogenously* qualified causal knowledge, as opposed to as an aid to causal knowledge in qualifying persistence [7]. The Ramification Problem was now addressed by the explicit addition of causal laws, and the development of a richer semantics to account for their interaction. A typical domain is shown below.

Shoot(x) causes *FiredAt(x)*
FiredAt(x) causes ¬*Alive(x)*   *static theory:*
¬*Alive(x)* causes ¬*Walks(x)*
Alive(Fred) holds-at *1*   ¬(¬*Alive(x) and Walks(x))*
Walks(Fred) holds-at *1*   ¬(*GunBroken and FiredAt(x))*
Shoot(Fred) occurs-at *2*

Fix a model implying *"GunBroken* holds-at *1"*. Then we reason that the static theory (of domain constraints) qualifies the direct effect of the action *"Shoot(Fred)"* on *"FiredAt(Fred)"*, and hence it also prevents the indirect effect *"¬Walks(Fred)"* from being triggered. Thus, the default persistence of Fred walking is not qualified, and Fred keeps walking. If, on the other hand, a model implies *"¬GunBroken* holds-at *1"*, then no causal law is qualified by the static theory. Note that the effect *"¬Alive(Fred)"* is not qualified despite the observation *"Walks(Fred)* holds-at *1"*; the causal knowledge *"¬Alive(Fred)* causes ¬*Walks(Fred)"* provides an escape route to this qualification. Hence, the default persistence of *"Walks(Fred)"* is qualified, and Fred is not walking after time-point 2. Models derived according to either of the cases are valid.

Perhaps the next natural step was realizing that observations after causal change, also, qualify the latter when the two are in conflict, a problem known as the *Exogenous Qualification Problem* (see, e.g., [7]); see Fig. 1(d). Consider, for example, the last domain extended by the observation *"¬FiredAt(Fred)* holds-at *4"*. Even though the effect of the *"Shoot(Fred)"* is not, as we have seen, necessarily qualified by the static theory alone, the explicit observation that the action's direct effect is not produced leads us to conclude that it was necessarily qualified. The interaction with the endogenous qualification of the causal laws by the static theory comes from the fact that *"GunBroken"* together with the static theory qualifies the action law, and provides, thus, an explanation of the observed action failure. So, if we wish to minimize the unknown exogenous cases of qualification, we would conclude that *"GunBroken"* holds, as this is the only known way to endogenously account for the observed failure.

Independently of the study of qualification in a temporal setting, another qualification problem was examined in the context of *Default Static Theories* [2] that consider how observed facts qualify default static knowledge; see Fig. 1(f). In the typical domain, represented below, one asks whether a bird named Tweety has the ability to fly, when the *only* extra given knowledge is that Tweety is a bird.

static theory:

Bird(Tweety)

(1) Penguin(x) ⤳ ¬CanFly(x)
(2) Penguin(x) → Bird(x)
(3) Bird(x) ⤳ CanFly(x)
rule (1) overrides rule (3)

In the absence of any explicit information on whether Tweety has the ability to fly, the theory predicts *"CanFly(Tweety)"*. Once extended with the fact *"Penguin(Tweety)"*, however, *"CanFly(Tweety)"* is retracted. The same retraction happens if instead of the fact *"Penguin(Tweety)"*, the fact *"¬CanFly(Tweety)"* is added. In either case the static theory is qualified, and yields to explicit facts or stronger evidence to the contrary.

2.2 Putting Fred and Tweety in the Same Scene

In this paper we investigate how temporal domains can incorporate default static theories, or dually, how static theories should be revised when interpreted in a temporal setting, in the presence of default persistence, default causal change, and observations. The technical challenge lies in understanding how the four types of domain knowledge, three of which may now be default, interact and qualify each other; see Fig. 1(e).

We view observations as part of the non-defeasible part in default static theories, thus primarily taking the role of qualifying the static knowledge, which then in turn will qualify the causal knowledge as described above. Due to the temporal aspect of a domain, however, a point-wise interpretation of observations as facts in the default static theory is insufficient, *even* in domains with no causal laws and, thus, strict persistence. Consider a temporal domain with the observations *"Penguin(Tweety)* holds-at *1"* and *"Bird(Tweety)* holds-at *4"*, and a static theory as in the Tweety example above. By viewing each time-point in isolation, we can conclude that *"CanFly(Tweety)"* holds only at time-point 4, but not at time-point 1. This cannot be extended into a temporal model without violating the (strict) persistence. Rather, *"Penguin(Tweety)* holds-at *1"* should persist everywhere, as if *"Penguin(Tweety)"* were observed at every time-point. This persistence, then, qualifies the static theory at every time-point, and implies that *"¬CanFly(Tweety)"*. Analogously, if the observation *"CanFly(Tweety)* holds-at *7"* is included in the domain, the observation persists everywhere and qualifies the default conclusion of the static theory that the penguin Tweety cannot fly.

Assume, now, that observations and persistence have appropriately qualified the static theory *at each time-point T*, so that the theory's default extensions (models) determine the set of *admissible* states at T. Through these sets of admissible states, the qualified static knowledge then qualifies the change that the temporal part of the theory attempts to generate through its causal knowledge. Given a time point T, it is natural that causal knowledge will be qualified by admissible states as determined *immediately after T*. This type of qualification is illustrated in the next example domain.

ClapHands causes *Noise*	
Noise causes *Fly(x)*	*static theory:*
Noise causes ¬*Noise*	*(1) Penguin(x)* ↝ ¬*CanFly(x)*
Spell(x) causes *CanFly(x)*	*(2) Penguin(x)* → *Bird(x)*
Penguin(Tweety) holds-at *1*	*(3) Bird(x)* ↝ *CanFly(x)*
ClapHands occurs-at *3*	*rule (1) overrides rule (3)*
Spell(Tweety) occurs-at *5*	*(4) ¬CanFly(x)* → ¬*Fly(x)*
ClapHands occurs-at *7*	

The persistence of *"Penguin(Tweety)* holds-at *1"* implies that *"¬CanFly(Tweety)"* holds in each set of admissible states up to time-point 5. In particular, this conclusion holds immediately after *"ClapHands* occurs-at *3"*, and qualifies through the static theory the causal generation of *"Fly(Tweety)"* by the action *"ClapHands"*.

This domain illustrates also a new aspect of the qualification problem. Intuitively, we expect *"Spell(Tweety)* occurs-at *5"* to override the static theory's default conclusion *"¬CanFly(Tweety)"* from holding at time-points following time-point 5. Note, however, that up to now we have assumed that the default static theory is stronger than the causal knowledge, and that it qualifies any change implied by the latter. But this is not the case now, since we wish to specify that some causal information is stronger than the default static theory. How, then, can we ensure that the causal generation of *"CanFly(Tweety)"* by *"Spell(Tweety)"* will not be qualified in this particular case?

This requirement can be accommodated by interpreting the particular causal law of interest *"Spell(x)* causes *CanFly(x)"* as a default rule in the static theory, and giving it priority over other default rules of the static theory with the contrary conclusion.[1] This interpretation need not be explicated. It suffices to mark *strong* causal laws as such, and then let their effects qualify the static theory, much in the same way that observations and persistence do. Because of this qualification, then, *"CanFly(Tweety)"* will hold in the set of admissible states associated with the time-point immediately following the occurrence of the action *"Spell(Tweety)"*, allowing the action's effect to come about and override the static theory's usual default conclusion that *"¬CanFly(Tweety)"*.

---

[1] We remind the reader that our goal here is not to provide semantics for static theories, and that using an informal reading in all presented example domains suffices for their semantics.

Such strong actions[2] (like *"Spell(x)"*) take the world out of the normal default state of affairs (where penguins cannot fly) into an exceptional, from the point of view of the static theory, state (where Tweety, a penguin, can fly). The rest of the default conclusions of the static theory still apply in this exceptional state (following time-point 5), conditioned on the exception (that Tweety can fly) that the strong action has brought about. This exception holds by persistence until some later action occurrence (e.g., *"UndoSpell(Tweety)"*) brings the world back into its normal state. In our domain, the second occurrence of *"ClapHands"* is while in an exceptional state; thus, the causal change is not qualified, and Tweety (a penguin able to fly) flies after time-point 7.

Consider now replacing *"Spell(Tweety)* occurs-at *5"* in the domain above with the observation *"Fly(Tweety)* holds-at *5"*. By persistence, this observation qualifies the static theory so that *"Fly(Tweety)"* holds in each set of admissible states at time-points strictly after 3. This does not hold for time-points up to and including time-point 3, since the occurrence of the action *"ClapHands"* at time-point 3 can now account for the change from *"¬Fly(Tweety)"* by qualifying its persistence, as the static theory does not now qualify *"ClapHands* occurs-at *3"*. Note that the interpretation of the observation *"Fly(Tweety)* holds-at *5"* is that Tweety flies for some *exogenous* reason (e.g., it is on a plane), and thus it is not known how the static theory is qualified, but only that it is somehow exogenously qualified. If an action at time-point 6 were to cause Tweety to stop flying, then this would *release* the static theory's default conclusion that penguins do not fly, so that the subsequent action *"ClapHands* occurs-at *7"* would be qualified by the static theory, and would not cause Tweety to fly again.

A somewhat orthogonal question to the one of *when* causal knowledge is qualified by the static theory, is *how* this qualification happens. Consider the Fred meets Tweety domain [6] below, and assume we wish to know whether Fred is alive after firing at it. One concludes that Fred is dead from time-point 2 onwards, and also that Tweety is flying. What happens, however, if one were to observe *"¬Fly(Tweety)* holds-at *4"*? Could one still conclude that Fred is dead? Interestingly enough, the answer depends on why Tweety would not fly after Fred would be shot! The observation by itself does not explain why the causal laws that would normally cause Tweety to fly did not do so.

Shoot(x) causes *FiredAt(x)*
FiredAt(x) causes *¬Alive(x)*
Shoot(x) causes *Noise*
Noise causes *Fly(x)*
Noise causes *¬Noise*
Alive(Fred) holds-at *1*
Turkey(Fred) holds-at *1*
Bird(Tweety) holds-at *1*
Shoot(Fred) occurs-at *2*

static theory:

(1) Penguin(x) or Turkey(x) ⤳ *¬CanFly(x)*
(2) Penguin(x) or Turkey(x) → *Bird(x)*
(3) Bird(x) ⤳ *CanFly(x)*
rule (1) overrides rule (3)
(4) ¬CanFly(x) → *¬Fly(x)*

An endogenous explanation would be that Tweety is a penguin, and *"Fly(Tweety)"* is qualified from being caused. An exogenous explanation would be that Tweety could not fly due to exceptional circumstances (e.g., an injury). In either case, we would

---

[2] The set of strong actions is domain-dependent, and it is the domain designer's task to identify them and to mark them as such in the domain provided to an agent for reasoning.

presumably conclude that Fred is dead. However, Tweety might not have flown because the shooting action failed to cause a noise, or even because the action failed altogether. Different conclusions on Fred's status might be reached depending on the explanation.

3   Argumentation Semantics

Motivated by the discussion in Section 2, we propose in this section a formal semantics for the qualification problem in the context of integrating default static and temporal theories. Argumentation offers a natural framework for this purpose, as it allows the easy specification of different types of knowledge as arguments, and the specification of their relative strengths as preferences imposed over these arguments. This, in turn, provides a clean formalization for the non-monotonic nature of knowledge qualification. A list of numerous non-monotonic logics that have been (re-) formulated in terms of argumentation, and a discussion of how argumentation offers a uniform formalism for understanding non-monotonic reasoning, can be found in [12] and references therein.

We emphasize that our proposed semantics does not hinge on any particular syntax or semantics used in the previous and this section for illustration purposes. In particular, we take a black-box approach to the syntax and semantics of default static theories, and assume simply that we have access to their models, without concerning ourselves with how these models are derived. For the temporal part of our semantics, we follow a pragmatic approach. We first focus on defining how an agent can reason from what holds in the current state of affairs to what will hold in the subsequent one. We discuss later how this single-step approach can be extended across the entire time-line.

We assume a time structure defined over the non-negative integers. Fix a positive integer T, and a state of affairs $\mathcal{E}$ that is believed to hold at time-point $T - 1$. Given a domain $\mathcal{D}$ expressed in some syntax, and interpreted according to some semantics, one derives a set of arguments of what holds at time-point T. Again, we take a black-box approach here, and do not concern ourselves with how these arguments are derived.

Definition 1. *Denote by* $\mathcal{U}_{\mathcal{D},\mathcal{E},T}$ *the **argument universe for** domain $\mathcal{D}$ **at** time-point T **given** state $\mathcal{E}$. $\mathcal{U}_{\mathcal{D},\mathcal{E},T}$ comprises* stat, *and arguments of the form* argm(L), *namely* assm(L), pers(L), ngen(L), sgen(L), exog(L), *as determined by the causal, static, and narrative parts of the given domain $\mathcal{D}$, and assuming the state $\mathcal{E}$ holds at $T - 1$.*

Assumption arguments assm(L) are necessary only at time-point 0, where, in fact, they can be thought of as a special case of generation arguments. Beyond this, assumptions are useful only for ease of presentation, and perhaps from a computational point of view in abstracting the past by postulating that something holds without a proof. Persistence arguments pers(L) exist exactly if L holds in the current state of affairs $\mathcal{E}$.

Normal generation arguments ngen(L) exist when causal change is triggered by some action occurrence, associated with a set of causal laws.[3] These normal generation

---

[3] As already illustrated in Section 2, certain RAC frameworks follow the approach that direct action effects may trigger other indirect effects. For ease of presentation, we do not make this distinction here, and focus on the more fundamental problem of how action effects (direct and indirect alike) interact with other pieces of knowledge. We note, however, that conditional arguments could be introduced so as to properly accommodate for indirect effects.

arguments are assumed to be qualified by the static theory; intuitively, we think of static theory as a compiled form of normal causal knowledge. Strong generation arguments $\mathtt{sgen}(L)$ are similar to normal generation arguments, but they exist when the action effects are produced through strong causal laws, as these are defined in a domain. These strong generation arguments do not yield to the static theory, but, rather, override it.

Exogenous arguments $\mathtt{exog}(L)$ exist when L is observed to hold at time-point T. Observations in a domain do not capture causal or static knowledge explaining *why* the environment reaches a particular state. Instead, they postulate that something holds for reasons *exogenous* to the causal and static theory. As such, observations are linked to the exogenous arguments of our framework in a one-to-one correspondence.

The static argument $\mathtt{stat}$ serves to indicate that the static theory is to be taken into account. This may give rise to additional conclusions, but also opens up the possibility for some of these conclusions to be questioned in lieu of stronger counter-arguments.

The relative strengths of various types of knowledge are captured by imposing preferences between the corresponding arguments; see Definition 2. As expected, assumptions are qualified by every other type of knowledge, while observations (i.e., exogenous reasons) qualify all other types of knowledge. The static theory qualifies normal causal change, but it is qualified by strong causal change, persistence, and observations. Finally, persistence is qualified by causal change. We have made the working assumption that strong causal change is incomparable in strength to normal causal change; that is, their only difference is with respect to the static theory. This assumption is retractable, and does not affect any of the definitions or results that follow in any important way.

Definition 2. *Define a **preference** relation $\succ$ between pairs of arguments, so that for every literal L, the arguments on the left are preferred over those on the top in the table:*

	stat	$\mathtt{assm}(\overline{L})$	$\mathtt{ngen}(\overline{L})$	$\mathtt{pers}(\overline{L})$	$\mathtt{sgen}(\overline{L})$	$\mathtt{exog}(\overline{L})$
stat		$\succ$	$\succ$			
$\mathtt{assm}(L)$						
$\mathtt{ngen}(L)$		$\succ$		$\succ$		
$\mathtt{pers}(L)$	$\succ$	$\succ$				
$\mathtt{sgen}(L)$	$\succ$	$\succ$		$\succ$		
$\mathtt{exog}(L)$	$\succ$	$\succ$	$\succ$	$\succ$	$\succ$	

Since the static theory is qualified by other pieces of knowledge, we need to assume that the static theory is associated with a revision mechanism. Following our black-box approach, we make no assumptions on what this mechanism is, beyond its existence.

Definition 3 (Dynamic Revision of Static Theory). *It is assumed that there exists a fixed revision function $\mathtt{rev}(\cdot, \cdot)$ that given a static theory and a set of literals, revises the static theory so that it entails all literals in the set. Given an argument set $\mathcal{A} \subseteq \mathcal{U}_{\mathcal{D},\mathcal{E},T}$, define $\mathcal{Q}(\mathcal{A}) \triangleq \{L \mid \mathtt{exog}(L) \in \mathcal{A} \text{ or } \mathtt{sgen}(L) \in \mathcal{A} \text{ or } \mathtt{pers}(L) \in \mathcal{A}\}$. The **associated** static theory $\mathcal{S}_{\mathcal{D},\mathcal{A}}$ of an argument set $\mathcal{A} \subseteq \mathcal{U}_{\mathcal{D},\mathcal{E},T}$ **under** a domain $\mathcal{D}$ with a static theory $\mathcal{S}_{\mathcal{D}}$, is defined to be $\mathtt{rev}(\mathcal{S}_{\mathcal{D}}, \mathcal{Q}(\mathcal{A}))$.*

Those types of knowledge in an argument set that are preferred over the static theory are captured by $\mathcal{Q}(\mathcal{A})$, and are used to qualify the static theory. This revised static theory is then used to draw conclusions and possibly qualify other types of knowledge.

Definition 4 (Argument Set Entailment and Completeness). *An argument set* $\mathcal{A} \subseteq \mathcal{U}_{\mathcal{D},\mathcal{E},T}$ *entails a literal* L, *denoted* $\mathcal{A} \models L$, *if either* $\mathrm{argm}(L) \in \mathcal{A}$, *or* $\mathrm{stat} \in \mathcal{A}$ *and* L *holds in all those models of* $\mathcal{S}_{\mathcal{D},\mathcal{A}}$ *that are consistent with every literal* G *such that* $\mathrm{argm}(G) \in \mathcal{A}$. *An argument set* $\mathcal{A} \subseteq \mathcal{U}_{\mathcal{D},\mathcal{E},T}$ *minimally entails a literal* L *if* $\mathcal{A} \models L$ *and there exists no argument set* $\mathcal{A}' \subset \mathcal{A}$ *such that* $\mathcal{A}' \models L$.

An argument set $\mathcal{A} \subseteq \mathcal{U}_{\mathcal{D},\mathcal{E},T}$ *is* **complete for** *a fluent* F *if either* $\mathcal{A} \models F$ *or* $\mathcal{A} \models \overline{F}$. *An argument set* $\mathcal{A} \subseteq \mathcal{U}_{\mathcal{D},\mathcal{E},T}$ *is* **complete** *if* $\mathcal{A}$ *is complete for every fluent. A complete argument set* $\mathcal{A} \subseteq \mathcal{U}_{\mathcal{D},\mathcal{E},T}$ *entails a state* $\mathcal{E}'$ *if* $\mathcal{A}$ *entails every literal in* $\mathcal{E}'$.

Definitions for attacks and admissibility are given next, in a manner that closely follows corresponding definitions in the literature (see, e.g., [13]). We emphasize this point, since it allows one to use existing and well-studied argumentation frameworks, and exploit computational models that have been developed for those (see, e.g., [12]).

Definition 5 (Attacking Relation). *An argument set* $\mathcal{A}_1 \subseteq \mathcal{U}_{\mathcal{D},\mathcal{E},T}$ *attacks an argument set* $\mathcal{A}_2 \subseteq \mathcal{U}_{\mathcal{D},\mathcal{E},T}$ *(on the literal* L) *if* $\mathcal{A}_1 \models \overline{L}$ *and* $\mathcal{A}_2 \models L$, *and there exist argument sets* $\mathcal{A}_1^m \subseteq \mathcal{A}_1$ *and* $\mathcal{A}_2^m \subseteq \mathcal{A}_2$ *such that the following conditions hold:*

(i) $\mathcal{A}_1^m$ *minimally entails* $\overline{L}$ *and* $\mathcal{A}_2^m$ *minimally entails* L;
(ii) *if an argument in* $\mathcal{A}_2^m$ *is preferred over an argument in* $\mathcal{A}_1^m$, *then an argument in* $\mathcal{A}_1^m$ *is preferred over an argument in* $\mathcal{A}_2^m$.

Definition 6 (Admissibility). *An argument set* $\mathcal{A} \subseteq \mathcal{U}_{\mathcal{D},\mathcal{E},T}$ *is* **admissible** *if the following conditions hold:*

(i) $\mathcal{A}$ *does not attack itself;*
(ii) $\mathcal{A}$ *attacks every argument set* $\mathcal{A}' \subseteq \mathcal{U}_{\mathcal{D},\mathcal{E},T}$ *that attacks* $\mathcal{A}$.

We now have the necessary machinery to formalize the integration of default static and temporal theories for the single time-step case. For a domain $\mathcal{D}$, an agent starts with a state $\mathcal{E}$ at time-point $T - 1$, and a set of available arguments $\mathcal{U}_{\mathcal{D},\mathcal{E},T}$, and constructs a complete admissible argument set $\mathcal{A} \subseteq \mathcal{U}_{\mathcal{D},\mathcal{E},T}$. In turn, $\mathcal{A}$ entails a state that — according to the information available to the agent — is the state of affairs at time-point T. It is straightforward to extend this to multiple time-steps, where the prediction from a time-step serves as input to the next time-step. For convenience, and without loss of generality, we assume that there are no observations or causal effects at time-point 0.

Definition 7 (Pre-Models). *An* **interpretation** $\mathcal{M}$ *of a domain* $\mathcal{D}$ *is a total mapping from time-points to states, so that for every time-point* $T \geq 0$, $\mathcal{M}(T)$ *denotes the state associated with* T. *A* **pre-model** $\mathcal{M}$ *of a domain* $\mathcal{D}$ **supported by** *a mapping* α *is an interpretation of* $\mathcal{D}$ *such that* $\mathcal{M}(0)$ *is a model of* $\mathcal{S}_{\mathcal{D}}$, *and for every time-point* $T > 0$, *there is a complete admissible argument set* $\alpha(T) \subseteq \mathcal{U}_{\mathcal{D},\mathcal{M}(T-1),T}$ *that entails* $\mathcal{M}(T)$.

Recall that argument sets may contain exogenous arguments. Intuitively, these are used when the reason for which something holds is unknown, yet it is known that it does hold, since it was observed so. Appealing to such exogenous reasons should be minimized across the set of all argument sets used to construct a temporal model.

Definition 8 (Models). *A **model** $\mathcal{M}$ of a domain $\mathcal{D}$ is a pre-model of $\mathcal{D}$ supported by some α, such that there exists no pre-model $\mathcal{M}'$ of $\mathcal{D}$ supported by some α' that point-wise contains a subset of the exogenous arguments contained in α.*

In accordance to our pragmatic point of view, we propose that models of a domain be computed through a combination of forward and backward reasoning steps. Although it is beyond the scope of this work to devise a full computational procedure for the developed semantics, we briefly discuss how such a procedure would look like.

Initially the agent reasons forward, starting from some state at time-point 0, and computing the states at time-point 1, time-point 2, and so on. Whenever it executes an action, or makes an observation, it also reasons forward to compute the state of its environment at the next time-point. As long as the argument sets used in this reasoning process contain no exogenous arguments, the corresponding computed states are assumed to be part of some model (since point-wise they trivially minimize the use of exogenous arguments). If when computing the state at some time-point T, the need arises for an argument set to use exogenous arguments, then the agent enters the backward reasoning phase. It revisits the state at time-point $T - 1$, and examines what assumptions it has to change so that the state at T can be computed without appealing to exogenous arguments. In the process of doing so, the need may arise to employ exogenous arguments for the argument set that entails the state at time-point $T - 1$. If this is the case, the agent revisits the state at time-point $T - 2$, and so on, going backwards possibly until time-point 0. Once the exogenous arguments are eliminated, or some minimal use of exogenous arguments is found to be necessary, the forward reasoning resumes.

4   Formal Results

We now discuss some formal properties of our proposed formalism. An algorithm for constructing complete admissible arguments sets is first presented and shown correct.

Given $\mathcal{U}_{\mathcal{D},\mathcal{E},T}$, construct the argument set $\mathcal{A}$ according to the following steps:

(1) Set $\mathcal{A} := \emptyset$, and set $\mathcal{L}$ to be the set of all literals.

(2) While there is $L \in \mathcal{L}$ s.t. $\mathtt{pers}(L) \in \mathcal{U}_{\mathcal{D},\mathcal{E},T}$, and $\mathtt{exog}(\overline{L}), \mathtt{sgen}(\overline{L}), \mathtt{ngen}(\overline{L}) \notin \mathcal{U}_{\mathcal{D},\mathcal{E},T}$, set $\mathcal{A} := \mathcal{A} \cup \{\mathtt{pers}(L)\}$, and set $\mathcal{L} := \mathcal{L} \setminus \{\overline{L}\}$.

(3) While there is $L \in \mathcal{L}$ s.t. $\mathtt{sgen}(L) \in \mathcal{U}_{\mathcal{D},\mathcal{E},T}$ and $\mathtt{exog}(\overline{L}) \notin \mathcal{U}_{\mathcal{D},\mathcal{E},T}$, set $\mathcal{A} := \mathcal{A} \cup \{\mathtt{sgen}(L)\}$, and set $\mathcal{L} := \mathcal{L} \setminus \{\overline{L}\}$.

(4) While there is $L \in \mathcal{L}$ s.t. $\mathtt{exog}(L) \in \mathcal{U}_{\mathcal{D},\mathcal{E},T}$, set $\mathcal{A} := \mathcal{A} \cup \{\mathtt{exog}(L)\}$, and set $\mathcal{L} := \mathcal{L} \setminus \{\overline{L}\}$.

(5) Set $\mathcal{A} := \mathcal{A} \cup \{\mathtt{stat}\}$.

(6) While there is $L \in \mathcal{L}$ s.t. $\mathtt{pers}(L) \in \mathcal{U}_{\mathcal{D},\mathcal{E},T}$, and both $\mathcal{A} \models L$ and there exists $\mathcal{A}^m \subseteq \mathcal{A}$ that minimally entails L, set $\mathcal{A} := \mathcal{A} \cup \{\mathtt{pers}(L)\}$, and set $\mathcal{L} := \mathcal{L} \setminus \{\overline{L}\}$.

(7) While there is $L \in \mathcal{L}$ s.t. $\mathtt{ngen}(L) \in \mathcal{U}_{\mathcal{D},\mathcal{E},T}$, and either $\mathcal{A} \not\models \overline{L}$ or there exists no $\mathcal{A}^m \subseteq \mathcal{A}$ that minimally entails $\overline{L}$, set $\mathcal{A} := \mathcal{A} \cup \{\mathtt{ngen}(L)\}$, and set $\mathcal{L} := \mathcal{L} \setminus \{\overline{L}\}$.

(8) Return $\mathcal{A}$, and terminate.

Theorem 1 (Correctness of Construction). *For every domain* $\mathcal{D}$, *time-point* T, *and state* $\mathcal{E}$, *the algorithm above returns a complete admissible argument set* $\mathcal{A} \subseteq \mathcal{U}_{\mathcal{D},\mathcal{E},T}$.

Proof (sketch). Consider an argument set $\mathcal{A}' \subseteq \mathcal{U}_{\mathcal{D},\mathcal{E},T}$ such that for every literal L it holds that: *(i)* if $\mathtt{sgen}(L) \in \mathcal{U}_{\mathcal{D},\mathcal{E},T}$, then $\mathtt{pers}(L) \notin \mathcal{A}'$; and *(ii)* if $\mathtt{exog}(L) \in \mathcal{U}_{\mathcal{D},\mathcal{E},T}$, then $\mathtt{sgen}(L), \mathtt{pers}(L) \notin \mathcal{A}'$. By case analysis it can be shown that for every argument $\mathtt{argm}(G) \in \mathcal{A}'$, either $\mathtt{argm}(G) \in \mathcal{A}$, or $\mathcal{A}$ attacks $\mathcal{A}'$ on literal G. It follows that if $\mathcal{A}$ does not defend an attack from $\mathcal{A}'$, then $\mathcal{A}' \subseteq \mathcal{A}$, which leads to a contradiction.

The case of arbitrary argument sets can be reduced to the special case considered above. Overall, then, $\mathcal{A}$ is admissible. The completeness of $\mathcal{A}$ follows easily.   □

We now continue to establish an *elaboration tolerance* property: every domain has a model, as long as its static theory is not inconsistent to begin with.

Theorem 2 (Guaranteed Consistency of Domains). *For every domain* $\mathcal{D}$, *and every state* $\mathcal{E}$ *that is a model of* $\mathcal{S}_{\mathcal{D}}$, *there exists a model* $\mathcal{M}$ *of* $\mathcal{D}$ *such that* $\mathcal{M}(0) = \mathcal{E}$.

Proof (sketch). Theorem 1 immediately implies the existence of pre-models with the claimed property. This, then, implies the existence of a pre-model that minimizes the exogenous arguments, and is, thus, a model of the domain.   □

As a special type of elaboration tolerance, we show that our formalism enjoys a *free-will* property: an agent may attempt to execute any sequence of actions in the future, without requiring revision of any of its beliefs about the past. The need for such a property in the context of Reasoning about Actions and Change has been argued in [7].

Theorem 3 (Free-Will Property of Reasoning). *Consider any two domains* $\mathcal{D}_1, \mathcal{D}_2$ *for which the following conditions hold: (i) neither domain has observations at time-points after* T_0, *and (ii) the domains differ only on the occurrences of actions whose effects are brought about at time-points after* T_0. *For every model* $\mathcal{M}_1$ *of* $\mathcal{D}_1$ *there exists a model* $\mathcal{M}_2$ *of* $\mathcal{D}_2$ *such that for every time-point* $T \leq T_0$, $\mathcal{M}_1(T) = \mathcal{M}_2(T)$.

Proof (sketch). Let $\mathcal{M}_1$ be a pre-model of $\mathcal{D}_1$ supported by α_1, and let $\mathcal{M}_2$ be a pre-model of $\mathcal{D}_2$ supported by α_2. By Theorem 1, it can be shown that α_2 can be chosen so that for every time-point $T \leq T_0$, $\alpha_1(T) = \alpha_2(T)$. Since there are no observations at time-points after T_0, it follows that for every time-point $T > T_0$, neither $\alpha_1(T)$ nor $\alpha_2(T)$ contain any exogenous argument. Since $\mathcal{M}_1$ minimizes the exogenous arguments, so does $\mathcal{M}_2$. Thus, $\mathcal{M}_2$ is a model of $\mathcal{D}_2$.   □

Recall that one of the problems for which our proposed integration offers semantics is that of how to temporally project and dynamically update a static theory. We conclude this section by briefly reiterating the stance that our framework takes on this problem.

The original static theory $\mathcal{S}_{\mathcal{D}}$ is determined solely by a given domain $\mathcal{D}$. Since initially, at time-point 0, no temporal information (i.e., observations or causal effects) is present, this original static theory need not be revised. Indeed, according to our semantics (cf. Definition 7), the state of affairs at time-point 0 is consistent exactly with this original static theory $\mathcal{S}_{\mathcal{D}}$. As time progresses, however, observations and causal effects (from strong causal laws) become available. This information needs to be respected, even if it is not consistent with the original static theory $\mathcal{S}_\mathcal{D}$. Additionally, if something holds in a past state $\mathcal{E}$ and is not caused to stop, its persistence needs, also, to be respected. According to our semantics (cf. Definition 3), all these pieces of information are taken into account to construct the revised static theory $\mathcal{S}_{\mathcal{D},\mathcal{A}}$ for some $\mathcal{A} \subseteq \mathcal{U}_{\mathcal{D},\mathcal{E},T}$; the state of affairs at time-point T is consistent exactly with this revised static theory. Note further that since temporal knowledge might be non-deterministic (e.g., due to non-deterministic causal effects, or due to conflicting observations at the same time-point), so might be the revision of the static theory. Indeed, the choice of the argument set $\mathcal{A}$ corresponds to a choice of one of the possible temporal evolutions of the world, and this, then, determines the revised static theory $\mathcal{S}_{\mathcal{D},\mathcal{A}}$ that corresponds to this choice.

5   Concluding Remarks

We have proposed an integrated formalism for reasoning with both default static and default causal knowledge, two problems that have been extensively studied in isolation from each other. The semantics was developed through argumentation, and follows a pragmatic point of view that we feel is appropriate for use in real-world settings.

Our agenda for future research includes investigation of scenarios where it is appropriate for static knowledge to generate extra (rather than block) causal change, when the former qualifies the latter. We would also like to develop a full-fledged computational procedure, along the lines already discussed in the preceding sections.

Beyond the work that introduced the problem and discussed some early ideas [6], we are not aware of other previous work that explicitly addresses the problem of integrating default static and temporal reasoning. However, much work has been done on the use of default reasoning in inferring causal change. Of particular note in the context of the qualification problem are [14,8]. An interesting approach to distinguishing between default and non-default causal rules in the context of the Language $\mathcal{C}+$ is given in [15].

References

1. Horty, J., Thomason, R., Touretzky, D.: A Skeptical Theory of Inheritance in Nonmonotonic Semantic Networks. Artificial Intelligence 42(2-3), 311–348 (1990)
2. Reiter, R.: A Logic for Default Reasoning. Artificial Intelligence 13(1-2), 81–132 (1980)
3. Hanks, S., McDermott, D.: Nonmonotonic Logic and Temporal Projection. Artificial Intelligence 33(3), 379–412 (1987)
4. McCarthy, J., Hayes, P.: Some Philosophical Problems from the Standpoint of Artificial Intelligence. Machine Intelligence 4, 463–502 (1969)
5. Shanahan, M.: Solving the Frame Problem: A Mathematical Investigation of the Common Sense Law of Inertia. MIT Press, Cambridge (1997)
6. Kakas, A., Michael, L., Miller, R.: Fred meets Tweety. In: Proc. of the 18th European Conference on Artificial Intelligence (ECAI 2008), pp. 747–748 (2008)
7. Kakas, A., Michael, L., Miller, R.: Modular-E: An Elaboration Tolerant Approach to the Ramification and Qualification Problems. In: Baral, C., Greco, G., Leone, N., Terracina, G. (eds.) LPNMR 2005. LNCS (LNAI), vol. 3662, pp. 211–226. Springer, Heidelberg (2005)
8. Thielscher, M.: The Qualification Problem: A Solution to the Problem of Anomalous Models. Artificial Intelligence 131(1-2), 1–37 (2001)

9. Bondarenko, A., Dung, P., Kowalski, R., Toni, F.: An Abstract Argumentation-Theoretic Approach to Default Reasoning. Artificial Intelligence 93(1-2), 63–101 (1997)
10. Lin, F.: Embracing Causality in Specifying the Indirect Effects of Actions. In: Proc. of the 14th International Joint Conference on Artificial Intelligence (IJCAI 1995), pp. 1985–1991 (1995)
11. Lin, F., Reiter, R.: State Constraints Revisited. Journal of Logic and Computation 4(5), 655–678 (1994)
12. Kakas, A., Toni, F.: Computing Argumentation in Logic Programming. Journal of Logic and Computation 9, 515–562 (1999)
13. Kakas, A., Miller, R., Toni, F.: An Argumentation Framework for Reasoning about Actions and Change. In: Gelfond, M., Leone, N., Pfeifer, G. (eds.) LPNMR 1999. LNCS (LNAI), vol. 1730, pp. 78–91. Springer, Heidelberg (1999)
14. Doherty, P., Gustafsson, J., Karlsson, L., Kvarnström, J.: TAL: Temporal Action Logics Language Specification and Tutorial. Electronic Transactions on Artificial Intelligence 2(3-4), 273–306 (1998)
15. Chintabathina, S., Gelfond, M., Watson, R.: Defeasible Laws, Parallel Actions, and Reasoning about Resources. In: Proc. of the 8th International Symposium on Logical Formalizations of Commonsense Reasoning (Commonsense 2007), pp. 35–40 (2007)

Simple Random Logic Programs

Gayathri Namasivayam and Mirosław Truszczyński

Department of Computer Science, University of Kentucky, Lexington, KY
40506-0046, USA

Abstract. We consider random logic programs with two-literal rules
and study their properties. In particular, we obtain results on the probability that random "sparse" and "dense" programs with two-literal rules
have answer sets. We study experimentally how hard it is to compute
answer sets of such programs. For programs that are *constraint-free* and
purely negative we show that the easy-hard-easy pattern emerges. We
provide arguments to explain that behavior. We also show that the hardness of programs from the hard region grows quickly with the number of
atoms. Our results point to the importance of purely negative constraint-
free programs for the development of ASP solvers.

1 Introduction

The availability of a simple model of a random CNF theory was one of the
enabling factors behind the development of fast satisfiability testing programs
— *SAT solvers*. The model constrains the length of each clause to a fixed integer,
say k, and classifies k-CNF theories according to their *density*, that is, the ratio
of the number of clauses to the number of atoms. k-CNF theories with low
densities have few clauses relative to the number of atoms. Thus, most of them
have many solutions, and solutions are easy to find. k-CNF theories with high
densities have many clauses relative to the number of atoms. Thus, most of
them are unsatisfiable. Moreover, due to the abundance of clauses, proofs of
contradiction are easy to find. As theories in low- and high-density regions are
"easy," they played essentially no role in the development of SAT solvers.

There is, however, a narrow range of densities "in between," called the *phase
transition*, where random k-CNF theories change *rapidly* from most being satisfiable to most being unsatisfiable. Somewhere in that narrow range is a value d
such that random k-CNF theories with density d are satisfiable with the probability $1/2$. The problem of determining that value has received much attention.
For instance, for 3-CNF theories, the phase-transition density was found experimentally to be about 4.25 [1]. A paper by Achlioptas discusses recent progress
on the problem, including some lower and upper bounds on the phase transition
value [2]. A key property of 3-CNF theories from the phase transition region
is that they are hard.[1] Thus, we have the easy-hard-easy difficulty pattern as

---

[1] It should be noted that the low- and high-density regions also contain challenging
theories, but they are relatively rare [4]).

E. Erdem, F. Lin, and T. Schaub (Eds.): LPNMR 2009, LNCS 5753, pp. 223–235, 2009.
© Springer-Verlag Berlin Heidelberg 2009

the function of density. Moreover, deciding satisfiability of programs from the hard region is very hard indeed! Designing solvers that could solve random unsatisfiable 3-CNF theories with 700 atoms generated from the phase-transition region was one of grand challenges for SAT research posed by Selman, Kautz and McAllester [3]. It resulted in major advances in SAT solver technology.

As in the case of the SAT research, work on *random logic programs* is likely to lead to new insights into the properties of *answer sets* of programs, and lead to advances in *ASP solvers* — software for computing them. Yet, the question of models of random logic programs has received little attention so far, with the work of Zhao and Lin [5] being a notable exception. Our objective is to propose a model of simple random logic programs and investigate its properties.

As in SAT, we consider random programs with rules of the same length. For the present study, we further restrict our attention to programs with two-literal rules. These programs are simple, which facilitates theoretical studies. But despite their simplicity, they are of considerable interest. First, every problem in NP can be reduced in polynomial time to the problem of deciding the existence of an answer set of a program of that type [6]. Second, many problems of interest have a simple encoding in terms of such programs [7]. We study experimentally and analytically properties of programs with two-literal rules. We obtain results on the probability that random programs with two-literal rules, both "sparse" and "dense," have answer sets. We study experimentally how hard it is to compute answer sets of such programs. We show that for programs that are *constraint-free* and *purely negative* the easy-hard-easy pattern emerges. We give arguments to explain that phenomenon, and show that the hardness of programs from the hard region grows quickly with the number of atoms. Our results point to the importance of constraint-free purely negative programs for the development of ASP solvers, as they can serve as useful benchmarks when developing good search heuristics. However, unlike in the case of SAT, depending on the parameters of the model, we either do not observe the phase transition or, when we do, it is gradual not sudden.

Even relatively small programs from the hard region are very hard for the current generation of ASP solvers. Interestingly, that observation may also have implications for the design of SAT solvers. If P is a purely negative program, answer sets of P are models of its completion $comp(P)$, a certain propositional theory [8]. For programs with two-literal rules the completion is (essentially) a CNF theory. Our experiments showed that these theories are very hard for the present-day SAT solvers, despite the fact that most of their clauses are binary.

2   Preliminaries

Logic programs consist of *rules*, that is, of expressions of the form

$$a \leftarrow b_1, \ldots, b_m, not\ c_1, \ldots, not\ c_n \tag{1}$$

and

$$\leftarrow b_1, \ldots, b_m, not\ c_1, \ldots, not\ c_n, \tag{2}$$

where a, b_i and c_j are atoms. Rules (1) are called *definite*, and rules (2) — *constraints*. A rule is *proper* if no atom occurs in it more than once. A rule is *k-regular* if it consists of k literals (that is, it is a definite rule with $k-1$ literals in the body, or a constraint with k literals in the body).

If r is a rule of type (1) or (2), the expression $b_1, \ldots, b_m, not\ c_1, \ldots, not\ c_n$ (understood as the conjunction of its literals) is the *body* of r. We denote it by $bd(r)$. The set of atoms $\{b_1, \ldots, b_m\}$ is the *positive* body of r, denoted $bd^+(r)$, and the set of atoms $\{c_1, \ldots, c_n\}$ is the *negative* body of r, denoted $bd^-(r)$. In addition, the *head* of r, $hd(r)$, is defined as a, if r is of type (1), and as $\perp$, otherwise. A program P is *constraint-free* if it contains no constraints. A program P is *purely negative* if for every non-constraint rule $r \in P$, $bd^+(r) = \emptyset$.

A set of atoms M is an *answer set* of a program P if it is the least model of the *reduct* of P with respect to M, that is, the program P^M obtained by removing from P every rule r such that $M \cap bd^-(r) \neq \emptyset$, and by removing all literals of the form *not c* from all other rules of P.

Computing answer sets of propositional logic programs is the basic reasoning task of answer-set programming, and fast programs that can do that, known as *answer-set programming solvers* (*ASP solvers*, for short) have been developed in the recent years [9,10,11,12,13].

3   2-Regular Programs

We assume a fixed set of atoms $At = \{a_1, a_2, \ldots\}$. There are five types of 2-regular rules: $a \leftarrow not\ b$; $a \leftarrow b$; $\leftarrow not\ a, not\ b$; $\leftarrow a, not\ b$; $\leftarrow a, b$. Accordingly, we define five classes of programs, mR_n^-, mR_n^+, mC_n^-, $mC_n^\pm$, and mC_n^+, with atoms from $At_n = \{a_1, \ldots, a_n\}$ and consisting of m *proper* rules of each of these types, respectively. Without the reference to m, the notation refers to all programs with n atoms of the corresponding type (for instance, R_n^+ stands for the class of all programs over At_n consisting of proper rules of the form $a \leftarrow b$).

The maximum value of m for which mR_n^-, mR_n^+ and $mC_n^\pm$ are not empty is $n(n-1)$. The maximum value of m for which mC_n^- and mC_n^+ are not empty is $n(n-1)/2$. Let $0 \leq m_1, m_2, c_2 \leq n(n-1)$ and $0 \leq c_1, c_3 \leq n(n-1)/2$ be integers. By $[m_1 R^- + m_2 R^+ + c_1 C^- + c_2 C^\pm + c_3 C^+]_n$ we denote the class of programs P that are unions of programs from the corresponding classes. We refer to these programs as *components* of P. If any of the integers m_i and c_i is 0, we omit the corresponding term from the notation. When we do not specify the numbers of rules, we allow any programs from the corresponding classes. For instance, $[R^- + R^+ + C^- + C^\pm + C^+]_n$ stands for the class of all proper programs with atoms from At_n.

Given integers n and m, it is easy to generate uniformly at random programs from each class mR_n^-, mR_n^+, mC_n^-, $mC_n^\pm$, and mC_n^+. For instance, a random program from mR_n^- can be viewed as the result of a process in which we start with the empty program on the set of atoms At_n and then, in each step, we add a randomly generated proper rule of the form $a \leftarrow not\ b$, with repeating rules discarded, until m rules are generated. This approach generalizes easily to programs from other classes we consider, in particular, to programs from $[m_1R^- + m_2R^+ + c_1C^- + c_2C^\pm + c_3C^+]_n$. Our goal is to study properties of such random programs.

We start with a general observation. If $P \in [m_2R^+ + c_1C^- + c_2C^\pm + c_3C^+]_n$ ($m_1 = 0$), then either P has no answer sets (if $c_1 \neq 0$) or, otherwise, $\emptyset$ is a unique answer set of P. Thus, in order to obtain interesting classes of programs, we must have $m_1 > 0$. In other words, programs from R_n^- (proper purely negative and constraint-free) play a key role.

4   The Probability of a Program to Have an Answer Set

We study first the probability that a random program in the class $[m_1R^- + m_2R^+ + c_1C^- + c_2C^\pm + c_3C^+]_n$ has an answer set. In several places we use results from random graph theory [14,15]. To this end, we exploit graphs associated with programs. Namely, with a program $P \in [R^- + R^+ + C^\pm]_n$ we associate a *directed* graph $D(P)$ with the vertex set At_n, in which a is connected to b with a directed edge (a, b) if $b \leftarrow not\ a$, $b \leftarrow a$ or $\leftarrow b, not\ a$ is a rule of P. For $P \in [R^- + R^+]_n$, the graph $D(P)$ is known as the *dependency* graph of a program. Similarly, with a program $P \in [R^- + R^+ + C^- + C^\pm + C^+]_n$ we associate an undirected graph $G(P)$ with the vertex set At_n, in which a is connected to b with an *undirected* edge $\{a, b\}$ if a and b appear together in a rule of P. If $P \in [R^- + R^+ + C^\pm]_n$, then $D(P)$ may have fewer edges than P has rules (the rules $a \leftarrow not\ b$, $a \leftarrow b$ and $\leftarrow b, not\ a$ determine the same edge). A similar observation holds for $G(P)$.

These graphs contain much information about the underlying programs. For instance, it is well known that if $P \in [R^- + R^+]_n$ and $D(P)$ has no cycles then P has a unique answer set. Similarly, if $P \in [m_1R^- + m_2R^+ + c_1C^- + c_2C^\pm + c_3C^+]_n$ and M is an answer set of P then $\overline{M}$ is an independent set in the graph $G(P_1)$, where P_1 is the component of P from $m_1R_n^-$.

We denote by $\mathcal{AS}^+$ the class of all programs over At that have answer sets. We write $Prob(P \in \mathcal{AS}^+)$ for the probability that a random graph P from one of the classes defined above has an answer set. That probability depends on n (technically, it also depends on the numbers of rules of particular types but, whenever it is so, the relevant numbers are themselves expressed as functions of n). We are interested in understanding the behavior of $Prob(P \in \mathcal{AS}^+)$ for random programs P from the class $[R^- + R^+ + C^- + C^\pm + C^+]_n$ (or one of its subclasses). More specifically, we will investigate $Prob(P \in \mathcal{AS}^+)$ as n grows to infinity. If $Prob(P \in \mathcal{AS}^+) \to 1$ as $n \to \infty$, we say that P *asymptotically almost surely*, or *a.a.s* for short, has answer sets. If $Prob(P \in \mathcal{AS}^+) \to 0$ as $n \to \infty$, we say that P a.a.s. has no answer sets.

To ground our results in some intuitions, we first consider the probability that a program from mR_{150}^- has an answer set as a function of the density $d = m/150$ (or equivalently, the number of edges m). The graphs, shown in Figure 1, were obtained experimentally. For each value of d, we generated 1000 graphs from the set mR_{150}^-, where $m = 150d$. The graph on the left shows the behavior of the probability across the entire range of d. The graph on the right shows in more detail the behavior for small densities.

(a)   (b)

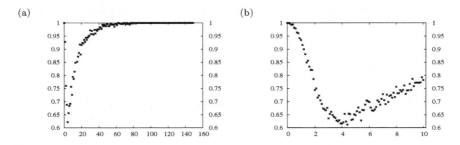

Fig. 1. The probability that a graph from mR_{150}^- ($m = 150d$) has an answer set, as a function of d

The graphs show that the probability is close to 1 for very small densities, then drops rapidly. After reaching a low point (around 0.6, in this case), it starts getting larger again and, eventually, reaches 1. We also note that the rate of drop is faster than the rate of ascent. We will now present theoretical results that quantify some of these observations. Our results concern the two extremes: programs of low density and graphs of high density.

We start with programs of low density and assume first that they do not have constraints. In this case, the results do not depend on whether or not we allow positive rules.

Theorem 1. *If $m_1 + m_2 = o(n)$ and $P \in [m_1 R^- + m_2 R^+]_n$, then P a.a.s has a unique answer set.*

Proof. (Sketch) Let P be a random program from $[m_1 R^- + m_2 R^+]_n$. The directed graph $D(P)$ can be viewed as a random directed graph with n vertices, and $m' = o(n)$ edges ($m' \leq m$, as different rules in P may map onto the same edge). Thus, $D(P)$ a.a.s. has no directed cycles (the claim can be derived from the property of random *undirected* graphs: a random undirected graph with n vertices and $o(n)$ edges a.a.s. has no cycles [15]). It follows that P a.a.s. has a unique answer set.   □

If there are constraints in the program, the situation changes. Even a single constraint of the form $\leftarrow not\ a, not\ b$ renders a sparse random program inconsistent.

Corollary 1. *If $c_1 \geq 1$, $m_1 + m_2 = o(n)$, and P is a random program from $[m_1 R^- + m_2 R^+ + c_1 C^-]_n$, then P a.a.s. has no answer sets.*

Proof. Let P be a random program from $[m_1 R^- + m_2 R^+ + c_1 C^-]_n$. Then, $P = P_1 \cup P_2$, where P_1 is a random program from $[m_1 R^- + m_2 R^+]_n$ and P_2 is a random program from $c_1 C_n^-$. By Theorem 1, P_1 a.s.s. has a unique answer set, say M. Since P_1 has $o(n)$ non-constraint rules, $|M| = o(n)$. The probability that a randomly selected constraint of the form $\leftarrow not\ a, not\ b$ is violated by M is given by the probability that $\{a, b\} \cap M = \emptyset$. That probability is $\binom{n-o(n)}{2} / \binom{n}{2}$ and it converges to 1 as $n \to \infty$. Thus, the assertion follows.   □

If we exclude such constraints then there again is a small initial interval of densities, for which random programs are consistent with high probability.

Corollary 2. *If $c_1 = 0$, $c_2 + c_3 \geq 1$, $(m_1 + m_2)c_2 = o(n)$, $(m_1 + m_2)^2 c_3 = o(n^2)$, and P is a random program from $[m_1 R^- + m_2 R^+ + c_2 C^\pm + c_3 C^+]_n$, then P a.a.s. has an answer set.*

Proof. (Sketch) Let P be a random program from $[m_1 R^- + m_2 R^+ + c_2 C^\pm + c_3 C^+]_n$. Thus, $P = P_1 \cup P_2 \cup P_3$, where P_1, P_2 and P_3 are random programs from $[m_1 R^- + m_2 R^+]_n$, $c_2 C_n^\pm$ and $c_3 C_n^+$, respectively. Since $c_2 > 0$ or $c_3 > 0$, $m_1 + m_2 = o(n)$. By Theorem 1, P_1 a.a.s. has a unique answer set, say M. Moreover, the size of M is at most $m_1 + m_2$. Under the assumptions of the corollary, one can show that a.a.s. each constraint $\leftarrow a, not\ b$ in P_2 has no atoms in M, and a.a.s. each constraint $\leftarrow a, b$ in P_3 has at most one atom in M. Thus, a.a.s. programs P_2 and P_3 are satisfied by M. Consequently, P a.a.s. has M as its unique answer set of P.   □

We move on to programs of high density. Our first result concerns programs from R_n^- (proper, purely negative and constraint-free programs with n atoms).

Theorem 2. *Let $0 < c < 1$ be a constant. For every fixed x, a random program from $m R_n^-$, where $m = \lfloor cN + x\sqrt{c(c-1)N} \rfloor$ and $N = n(n-1)$, a.a.s. has an answer set.*

Proof. (Sketch) To show the assertion, it is enough prove that a random directed graph with n vertices and m edges, where m is as in the statement of the theorem, a.a.s. has a kernel. It is known [16] that a.a.s. a random directed graph with n nodes drawn from the *binomial model* (edges are selected independently of each other and with the same probability c) has a kernel. Moreover, one can show that if $m' > m$, $m' = m + O(n)$, and G_m and $G_{m'}$ are random directed graphs with n nodes, and m and m' edges, respectively, then $Prob(G_m$ has a kernel$) \leq Prob(G_{m'}$ has a kernel$) + o(1)$. That property can be used instead of convexity in Theorem 2(ii) [14], which allows us to transform properties of graphs from the binomial model into properties of graphs from the *uniform* model that we are considering. Thus, the assertion follows.   □

Theorem 2 concerns only a narrow class of dense programs, its applicability being limited by the specific number of rules programs are to have ($m = \lfloor cN + x\sqrt{c(c-1)N} \rfloor$, where $N = n(n-1)$). It also does not apply to "very" dense graphs with $m = n^2 - o(n^2)$ rules. However, based on that theorem and on experimental results (Figure 1), we conjecture that for every $c > 0$, a program from $m R_n^-$, where $m \geq cn^2$, a.a.s. has an answer set.

We will now consider the effect of adding positive rules (rules of the form $a \leftarrow b$) and constraints. In fact, as soon as we have just slightly more than $n \log n$ positive rules in a random program that program a.a.s. has no answer sets.

Theorem 3. *For every $\epsilon > 0$, if $m_1 \geq 1$, $m_2 \geq (1 + \epsilon)n \log n$, and P is a random program from $[m_1 R^- + m_2 R^+ + c_1 C^- + c_2 C^\pm + c_3 C^+]_n$, then P a.a.s. has no answer sets.*

Proof. Let $P \in [m_1 R^- + m_2 R^+ + c_1 C^- + c_2 C^\pm + c_3 C^+]_n$, where $m_1 \geq 1$. Also, let P_2 be the component of P from $m_2 R^+$. If $D(P_2)$ contains a Hamiltonian cycle, then P has no answer sets. Indeed, $\emptyset$ is not an answer set due to the rule of the form $a \leftarrow not\ b$ that is present in P. Thus, if P has an answer set, say M, then $M \neq \emptyset$. Clearly, P^M contains P_2. By the assumption on $D(P_2)$, the least model of P^M contains all atoms in At_n. Thus, $M = At_n$. But then, P^M contains no atoms (all its rules are either from P_2 or are constraints of the form $\leftarrow a, b$) and so, the least model of P^M is $\emptyset$, a contradiction. Clearly, there is a precise correspondence between programs from $m_2 R^+$ and random directed graphs with n nodes and m edges (no loops). The assertion follows now from the result that states that a random directed graph with n nodes and at least $(1 + \epsilon)n \log n$ edges a.a.s. has a Hamiltonian cycle [14].   □

The presence of sufficiently many constraints of the form $\leftarrow a, b$ or $\leftarrow a, not\ b$ also eliminates answer sets. To see that, we recall that if M is an answer set of a program $P = P_1 \cup P_2$, where $P_1 \in R_n^-$ and $P_2 \in [R^+ + C^- + C^\pm + C^+]_n$, then M is the complement of an independent set in $G(P_1)$. The following property will be useful. For every real $c > 0$ there is a real $d > 0$ such that a.a.s. a graph with n vertices and $m \geq cn^2$ edges has no independent set with more than $d \log n$ elements [14]. Thus, we get the following result that provides a lower bound on the size of an answer set in a dense random logic program.

Theorem 4. *For every real $c > 0$, there is a real $d > 0$ such that a.a.s. the complement of every answer set of a random program $P = P_1 \cup P_2$, where $P_1 \in m R_n^-$, $P_2 \in [R^+ + C^- + C^\pm + C^+]_n$ and $m \geq cn^2$, has size at most $d \log n$.*

We now consider the effect of constraints of the form $\leftarrow a, b$ on the existence of answer sets in programs with many purely negative rules. Intuitively, even a small number of such constraints should suffice to "kill" all answer sets. Indeed, according to Theorem 4, these answer sets are large and contain "almost all" atoms. Formalizing this intuition, we get the following result.

Theorem 5. *For every $c > 0$ there is $d > 0$ such that if $m_1 \geq cn^2$, $c_3 \geq d \log n + 1$, and P is a random program from $[m_1 R^- + m_2 R^+ + c_1 C^- + c_2 C^\pm + c_3 C^+]_n$, then P a.a.s. has no answer sets.*

Constraints of the form $\leftarrow a, not\ b$ do not have such a dramatic effect. However, a still relatively small number of such constraints a.a.s. eliminates all answer sets.

Theorem 6. *For every $c > 0$, and for every $\epsilon > 0$, if $m_1 \geq cn^2$, $c_2 \geq n^{1+\epsilon}$, and P is a random program form $[m_1 R^- + m_2 R^+ + c_1 C^- + c_2 C^\pm + c_3 C^+]_n$, then a.a.s. P has no answer sets.*

Proof. We set $N = n(n - 1)$ and $N_i = i(n - i)$. Let $X \subseteq At_n$ consist of $n - i$ elements, where $0 < i < n$. We will first compute the probability that in a random directed graph with the set of vertices At_n and with m edges, there is no edge starting in $\overline{X}$ and ending in X. That probability is given by $\binom{N-N_i}{m} / \binom{N}{m}$. One can show that it can be bounded from above by $(1 - N_i/N)^m$. It follows that the probability that at least one $X \subseteq At_n$ such that $0 < |\overline{X}| \leq k$ has that property is bounded by $\sum_{i=1}^{k} \binom{n}{i}(1 - N_i/N)^m$. Let $d > 0$ be a constant. One can show that for every $\epsilon > 0$, if $k \leq d\log n$ and $m \geq n^{1+\epsilon}$, then $\sum_{i=1}^{k} \binom{n}{i}(1 - N_i/N)^m \to 0$ as $n \to \infty$. Let us interpret that result in terms of programs. Let d be a constant such that the complement of every answer set in a random program P from $[m_1 R^- + m_2 R^+ + c_1 C^- + c_2 C^\pm + c_3 C^+]_n$ has size at most $d\log n$ (such d exists by Theorem 4) and let ϵ be any fixed positive real. Let Q be the component from $c_2 C_n^\pm$ of P. Then $D(Q)$ has at least $n^{1+\epsilon}$ edges. Thus, a.a.s. for every set X such that $1 \leq |\overline{X}| \leq d\log n$, there is an edge (a, b) in $D(P)$ that originates in $\overline{X}$ and ends in X. Such edge corresponds to a constraint $\leftarrow b, not\ a$ in Q. Clearly, this constraint is violated by X. Thus, a.a.s. P has no non-empty answer sets. Since $X = At_n$ is not an answer set either (the reduct of P wrt At_n contains no atoms and so, it is inconsistent or its least model is empty), a.a.s. P has no answer sets.   □

The case of constraints $\leftarrow not\ a, not\ b$ is less interesting. Large answer sets (having at least $n - d\log n$ atoms) that arise for programs with dense component from R_n^- typically satisfy them and to "kill" all answer sets of such programs with high probability almost all constraints $\leftarrow not\ a, not\ b$ must be present.

5   Hardness of Programs

We will now study the hardness of programs from $[m_1 R^- + m_2 R^+ + c_1 C^- + c_2 C^\pm + c_3 C^+]_n$ for ASP solvers. The bulk of our experimental results concern programs in the class R_n^-. It turns out these programs (for appropriately chosen density) are especially challenging.

Unless stated otherwise, our experiments separate programs that have answer sets (are *consistent*) from those that do not (are *inconsistent*). For each experiment we generate a sample of instances of programs of each of these two types. In the previous section we provided evidence that programs in mR_n^-, where $m \geq cn^2$ (cf. Figure 1 and Theorem 2), a.a.s. have an answer set. Therefore, when experimenting with inconsistent programs, we restrict the number of rules in a program to values for which inconsistent programs appear with probability sufficiently larger than 0 (about 0.05) to allow for building samples of inconsistent programs of sizes large enough to justify drawing conclusions from experiments (typically 100 programs per sample).

In experiments, we used *smodels* (with lookahead) [9] and *clasp* [10]. We took the average number of choice points as reported by these systems as the measure of the *hardness* of a family of programs.

Our first observation is that as we increase m, programs from mR_n^- show the easy-hard-easy pattern. That is, low-density programs are easy for the two solvers. When m grows, programs get harder. Then, at some point, they start getting easier again. We illustrate that behavior in Figure 2 below. The two graphs show separately the results for consistent and inconsistent programs from the classes mR_{100}^-. Each figure shows together the results (average number of

(a)   (b)

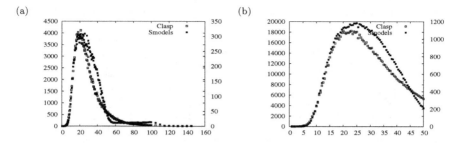

Fig. 2. Average number of choice points for consistent (graph (a)) and inconsistent (graph(b)) programs with 150 atoms; *smodels* (scale on the right) and *clasp* (scale on the left). The *x*-axis represents the density. Sample sizes are 500 for consistent programs, and 100 for inconsistent programs.

choice points) for *smodels* (the scale on the right) and *clasp* (the scale on the left). The *x*-axis shows the density, that is, the ratio of the number of rules to the number of atoms in a program. We stress that the scales differ. Thus, the figures are not meant to compare the performance of *smodels* and *clasp*. But they do show that for each solver a similar easy-hard-easy pattern emerges, and that the features of the pattern are remarkably similar for the two solvers.

We obtained the same type of a pattern in our experiments with programs with 125, 175 and 200 atoms. However, we observed some minor deviations from that pattern for *smodels* (but not for *clasp*) for programs with 100 atoms. Given our results for $n \geq 125$, it seems plausible that the irregular behavior arises only for some smaller numbers of atoms.

We used the term *hard region* above somewhat informally. To make that concept more precise, we define it now as the maximum interval $[u, v]$ such that for every density $d \in [u, v]$ the average number of choice points is at least 90% of the maximum (peak) average number of choice points. Table 1 shows the hard regions, the density for which the number of choice points reaches the maximum, and the number of choice points at the peak location for consistent and inconsistent instances with $n = 125, 150, 175$ and 200 atoms. The key observations are: (1) the location of the hard region does not seem to depend much on the solver; it is centered around the density of 19 for consistent programs, and 22 for inconsistent ones, (2) inconsistent programs are significantly harder than consistent ones, (3) the peak of hardness is not sharp or, in other words, the hard region extends over a sizable range of densities, and (4) the hardness of programs in the hard region grows very quickly.

We conclude with arguments to explain the presence of the easy-hard-easy pattern we observed for programs in the class R_n^-. First, we note that programs in mR^-, where $m = o(n)$, a.a.s. are stratified (Theorem 1). Computing answer sets for such programs is easy. As the density (the number of rules) grows, cycles in the graph $D(P)$ start appearing (that happens roughly when a program has as many rules as atoms). Initially, there are few cycles and the increase in hardness is slow. At some point, however, there are enough cycles in $D(P)$ to

Table 1. Hard region, peak location, and the number of choice points at the peak location for consistent and inconsistent programs. Results for *clasp* and *smodels*.

	Inconsistent programs					
	clasp			smodels		
n	hard region	peak	choice points at peak	hard region	peak	choice points at peak
125	[17.5 − 27]	22	5261	[17.5 − 24]	21	388
150	[18 − 27]	23	18639	[19 − 31]	24.5	1184
175	[18.5 − 27.5]	22	59704	[17.5 − 23.5]	20.5	3582
200	[18 − 28]	22	189576	[18 − 26]	22.5	14407
	Consistent programs					
125	[15.5 − 21.5]	17.5	1231	[16 − 25]	20	130
150	[16 − 23]	17.5	4033	[16 − 29.5]	20	308
175	[18.5 − 21.5]	20	14230	[17.5 − 21.5]	20	1110
200	[17.5 − 23]	19.5	43345	[18.5 − 24.5]	19.5	4232

make computing answer sets of P hard. To explain why the task gets easier again, we note the following property of binary trees.

Proposition 1. *Let T be a binary tree with m leaves, the height n, and with the number of left edges on any path from the root to a leaf bounded by k. Then $m \leq 2^k \binom{n}{k}$.*

Proof: Let $S(n,k)$ be the maximum number of leaves in such a tree. Then $S(n,k)$ is given by the recursive formula $S(n,k) = S(n-1,k) + S(n-1,k-1)$, for $n \geq k+1$ and $k \geq 1$, with the initial conditions $S(n,0) = 1$ and $S(n,n) = 2^n$, for $n \geq 0$. The assertion can now be proved by an easy induction.   □

We denote by $\mathcal{S}$ the class of complete solvers with the following three properties: (1) they compute answer sets (or determine that no answer set exists) by generating a sequence of partial assignments so that if an answer set exists then it occurs among the generated assignments; (2) they use boolean constraint propagation to force truth assignments on unassigned atoms and trigger backtracking if contradictions are found; and (3) the generated assignments can be represented by a binary tree, whose nodes are atoms, and where the left (right) edge leaving an atoms corresponds to assigning that atom *false* (*true*). This class of solvers includes in particular solvers that use chronological backtracking, as well as those that perform backjumping (we note that in that latter case, some nodes corresponding to decision atoms may have only one child).

Proposition 2. *Let $P \in R_n^-$ be such that the maximum size of an independent set in $G(P)$ equals β. Then, the number of assignments generated by any solver in the class $\mathcal{S}$ is $O((2n)^{\beta+1})$.*

Proof: The tree representing the space of assignments generated by a solver from $\mathcal{S}$ for P has height at most n and at most $\beta+1$ left edges on every path. Indeed, if there are ever $\beta+1$ left edges on a path in the tree, then $\beta+1$ atoms are set to false. Atoms in that set do not form an independent set in $G(P)$, and so for some two of them, say a and b, the rule $a \leftarrow not\ b$ is in P. Boolean propagation forces a or b to be true, while both of these atoms are false. Thus, a backtrack will occur (the current path will not be extended). The assertion follows now by Proposition 1, as $\binom{n}{k} \leq n^k$. □

We noted earlier that when $m \geq cn^2$, $\beta = O(\log n)$. Thus, when $m \geq cn^2$, the size of the search space is bounded by $n^{O(1)}2^{O(\log^2 n)}$, which is asymptotically much smaller than $O(2^n)$. Furthermore, with m getting closer to $n(n-1)$, β gets even smaller and so, the search space gets smaller, too.

Finally, we note (due to space limits, we do not discuss these results in detail) that adding even a small number of positive rules or constraints to programs from mR_n^- generally makes the resulting programs easier. These results suggest that from the perspective of benchmarking and insights into search heuristics, proper purely negative constraint-free programs are especially important.

6   Benchmarks for SAT Solvers

Deciding whether a logic program has an answer set is in the class NP. Thus, there are polynomial-time methods to reduce the task of computing stable models of a program to that of computing models of a CNF theory. Unfortunately, all known reductions lead to theories whose size is superlinear with respect to that of the original program [17].

However, linear-size reductions exist for programs that are *tight* [18]. Namely, answer sets of a tight program P are precisely models of the Clark's *completion* of P [8]. Purely negative programs are tight. In particular, programs in R_n^- are tight. Moreover, if $P \in R_n^-$, then the completion of P has especially simple form. It can be written as the collection of the following clauses: (1) $a \vee b$, where $a \leftarrow not\ b \in P$, and (2) $\neg a \vee \neg b_1 \vee \ldots \vee \neg b_k$, where $a \leftarrow not\ b_i$, $1 \leq i \leq k$, are *all* rules in P with a as the head.

Theories of that type obtained from programs form R_n^-, constitute an interesting class of benchmarks for SAT solvers. They are simple in that most of their clauses consist of two literals and all other clauses are disjunctions of atoms. Moreover, as the density grows, there is no phase transition, unlike in the case of the standard model. Instead, we observe the familiar easy-hard-easy

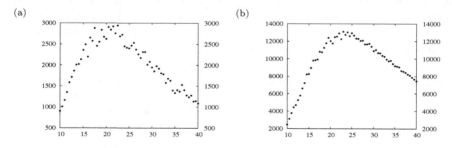

Fig. 3. Easy-hard-easy pattern shown by *minisat* [19] on the completions of programs from mR^-n, where $n = 150$, satisfiable theories in graph (a), inconsistent ones in graph (b)

property, with the hard region correlated well with the one we observed for *clasp* and *smodels* (Figure 3).

7 Discussion

We proposed and considered a model of random logic programs with fixed-length rules. We focused on the case of proper programs with two-literal rules. Our model is parameterized by the number of atoms and five integers that specify the numbers of rules of each possible type. Due to its simplicity, the model lends itself to theoretical investigations. To the best of our knowledge, our paper provides first non-trivial theoretical results on the properties of random programs. Our experimental results show that while simple, the model allows us to generate relatively small programs that are hard for the current ASP solvers. Computing answer sets of proper purely negative constraint-free programs with 600 atoms generated from the hard region seems to be infeasible at present. We also noted that completions of hard programs from our model are challenging benchmarks for SAT solvers. One of the main outcomes of our paper is the emergence of proper purely negative constraint-free programs as the core class for generating benchmarks and a key to theoretical studies of properties of random programs.

The model we proposed for the case of two-literal rules can be generalized to programs consisting of $k \geq 3$ rules. We believe that most properties we identified in this paper generalize, too. In particular, our preliminary experiments show the same easy-hard-easy pattern for proper purely negative constraint-free programs with three-literal rules. Moreover, programs from the hard region are harder than hard-region two literal ones.

There are several differences between our work and that of Zhao and Lin [5]. First, we consider the fixed rule-length model (and more narrowly, only the case of two-literal programs). Second, we can specify in our model the composition of programs in terms of the numbers of rules of particular types. That facilitates studies of the effect these rules have when added to the "core" consisting of proper purely negative constraint-free programs. Third, we focus on, what we believe, is the key class of random logic programs — the class of programs that are proper purely negative and constraint-free. Despite these differences, one of specializations of our model (that allows for constraints) is quite closely related to Zhao and Lin model and shows similar properties. We also note that the first and the last of the issues discussed above differentiate our approach from an unpublished work by Wong, Schlipf and Truszczyński [20].

Finally, we note that for the class R_n^-, as well as for several other classes of programs we can define in our framework, we do not observe the phase transition. That is, unlike in SAT, increasing the density (the number of rules) does not result in a sudden transition from consistent to inconsistent programs. In fact there is no density for which programs are a.a.s inconsistent. We believe it is due to the nonmonotonicity of the semantics of answer sets. For some classes of programs, namely those with sufficiently many constraints, a transition from consistent to inconsistent programs can be observed (Zhao and Lin's model shows such transition, too). However, the transition is relatively slow.

References

1. Mitchell, D., Selman, B., Levesque, H.: Hard and easy distributions of SAT problems. In: Proceedings of AAAI 1992, pp. 459–465. Morgan Kaufmann, San Francisco (1992)
2. Achlioptas, D.: Random satisfiability. In: Biere, A., Heule, M., van Maaren, H., Walsh, T. (eds.) Handbook of Satisfiability, pp. 245–270. IOS Press, Amsterdam (2009)
3. Selman, B., Kautz, H., McAllester, D.: Ten challenges in propositional reasoning and search. In: Proceedings of the 15th International Joint Conference on Artificial Inteligence, IJCAI 1997, pp. 50–54. Morgan Kaufmann, San Francisco (1997)
4. Gent, I., Walsh, T.: Easy problems are sometimes hard. Artificial Intelligence 70, 335–345 (1994)
5. Zhao, Y., Lin, F.: Answer set programming phase transition: a study on randomly generated programs. In: Palamidessi, C. (ed.) ICLP 2003. LNCS, vol. 2916, pp. 239–253. Springer, Heidelberg (2003)
6. Marek, W., Truszczyński, M.: Autoepistemic logic. Journal of the ACM 38, 588–619 (1991)
7. Huang, G.S., Jia, X., Liau, C.J., You, J.H.: Two-literal logic programs and satisfiability representation of stable models: a comparison. In: Cohen, R., Spencer, B. (eds.) Canadian AI 2002. LNCS (LNAI), vol. 2338, pp. 119–131. Springer, Heidelberg (2002)
8. Clark, K.: Negation as failure. In: Gallaire, H., Minker, J. (eds.) Logic and data bases, pp. 293–322. Plenum Press, New York (1978)
9. Simons, P., Niemelä, I., Soininen, T.: Extending and implementing the stable model semantics. Artificial Intelligence 138, 181–234 (2002)
10. Gebser, M., Kaufmann, B., Neumann, A., Schaub, T.: Clasp: A conflict-driven answer set solver. In: Baral, C., Brewka, G., Schlipf, J. (eds.) LPNMR 2007. LNCS (LNAI), vol. 4483, pp. 260–265. Springer, Heidelberg (2007)
11. Leone, N., Pfeifer, G., Faber, W., Eiter, T., Gottlob, G., Perri, S., Scarcello, F.: The dlv system for knowledge representation and reasoning. ACM Transactions on Computational Logic 7, 499–562 (2006)
12. Lin, F., Zhao, Y.: ASSAT: Computing answer sets of a logic program by SAT solvers. In: Proceedings of the 18th National Conference on Artificial Intelligence (AAAI 2002), pp. 112–117. AAAI Press, Menlo Park (2002)
13. Lierler, Y., Maratea, M.: Cmodels-2: Sat-based answer set solver enhanced to non-tight programs. In: Lifschitz, V., Niemelä, I. (eds.) LPNMR 2004. LNCS (LNAI), vol. 2923, pp. 346–350. Springer, Heidelberg (2004)
14. Bollobás, B.: Random Graphs. Academic Press, London (1985)
15. Janson, S., Luczak, T., Ruciński, A.: Random Graphs. Wiley-Interscience, Hoboken (2000)
16. de la Vega, W.F.: Kernels in random graphs. Discrete Math. 82, 213–217 (1990)
17. Janhunen, T.: Representing normal programs with clauses. In: de Mántaras, R.L., Saitta, L. (eds.) Proceedings of the 16th Eureopean Conference on Artificial Intelligence, ECAI 2004, pp. 358–362. IOS Press, Amsterdam (2004)
18. Fages, F.: Consistency of Clark's completion and existence of stable models. Journal of Methods of Logic in Computer Science 1, 51–60 (1994)
19. Eén, N., Sörensson, N.: An extensible SAT solver. In: Giunchiglia, E., Tacchella, A. (eds.) SAT 2003. LNCS, vol. 2919, pp. 502–518. Springer, Heidelberg (2003)
20. Wong, D., Schlipf, J., Truszczyński, M.: On the distribution of programs with stable models. Unpublished, presented at the Dagstuhl Seminar 05171, Nonmonotonic Reasoning, Answer Set Programming and Constraints (2005)

Max-ASP:
Maximum Satisfiability of Answer Set Programs*

Emilia Oikarinen and Matti Järvisalo

Helsinki University of Technology TKK
Department of Information and Computer Science
PO Box 5400, FI-02015 TKK, Finland
emilia.oikarinen@tkk.fi, matti.jarvisalo@tkk.fi

Abstract. This paper studies answer set programming (ASP) in the generalized
context of soft constraints and optimization criteria. In analogy to the well-known
Max-SAT problem of maximum satisfiability of propositional formulas, we introduce the problems of unweighted and weighted Max-ASP. Given a normal logic
program P, in Max-ASP the goal is to find so called optimal Max-ASP models,
which minimize the total cost of unsatisfied rules in P and are at the same time
answer sets for the set of satisfied rules in P. Inference rules for Max-ASP are
developed, resulting in a complete branch-and-bound algorithm for finding optimal models for weighted Max-ASP instances. Differences between the Max-ASP
problem and earlier proposed related concepts in the context of ASP are also discussed. Furthermore, translations between Max-ASP and Max-SAT are studied.

1 Introduction

Answer set programming (ASP) is a well-studied declarative programming paradigm
that has proven to be an effective approach to knowledge representation and reasoning
in various hard combinatorial problem domains. The task of answer set solvers is to
find answer sets of ASP programs, representing solutions to the underlying decision
problem instance at hand. However, it can often be the case that the problem instance
has no solutions since it may be over-constrained. While answer set solvers can in this
case prove the non-existence of answer sets, instead of a simple "no" answer, a "near-
solution" would be of interest, i.e., an interpretation that is optimal with respect to a
specific minimization or maximization criterion, such as the number of unsatisfied rules
in the program. For example, in debugging ASP programs (see e.g. [1] and references
therein), such an interpretation, or *optimal solution*, could give hints to the reasons for
the non-existence of answer sets through a minimal set of unsatisfied rules.

In the field of Boolean satisfiability (SAT), which has close connections to ASP especially from the viewpoint of solver technology, interest in methods for solving the
Max-SAT problem (the optimization variant (or generalization) of SAT) has risen especially during recent years [2,3,4]. Motivation for Max-SAT, where the interest is in

---

* This work is financially supported by Academy of Finland under the project *Methods for
Constructing and Solving Large Constraint Models* (grant #122399).

E. Erdem, F. Lin, and T. Schaub (Eds.): LPNMR 2009, LNCS 5753, pp. 236–249, 2009.
© Springer-Verlag Berlin Heidelberg 2009

optimal truth assignments with respect to the number of unsatisfied clauses, and especially its weighted variant, comes from the possibilities of expressing and solving various optimization and probabilistic reasoning tasks via (weighted) Max-SAT.

In this paper, we study the problem analogous to Max-SAT for normal logic programs under the stable model semantics, namely *Max-ASP*, or *maximum satisfiability of answer set programs*. In other words, given a normal logic program and integer weights for each rule in the program, in weighted Max-ASP the goal is to find *optimal Max-ASP models* that minimize the total cost (sum of weights) of unsatisfied rules in the program and are at the same time a stable model for remaining (satisfied) rules of the program.

Our contributions are many-fold. In addition to considering basic properties of (optimal) Max-ASP models, we develop various inference (or transformation) rules for reasoning about the optimal cost of Max-ASP instances. Based on the transformation rules, we present a complete branch-and-bound algorithm for determining the optimal cost and an associated optimal model for any Max-ASP instance, also in the weighted case. In fact, the algorithm can be viewed as a generalization of complete search methods proposed for ASP, as some of the presented transformation rules are in a sense generalizations of tableau rules [5] for ASP inference applied in ASP solvers. We also study the relation between Max-ASP and Max-SAT with translations which preserve the solutions between the problems. Furthermore, we discuss differences between Max-ASP and other generalizations [6,7,8,9] of answer sets/answer set programs which have a similar flavor. For example, in contrast to Max-ASP, often costs are assigned on literals instead of rules, penalizing the inclusion or exclusion of specific atoms in an answer set of the program at hand, and often answer sets for the whole program are still sought.

This paper is organized as follows. After necessary concepts related to ASP (Sect. 2), we define the Max-ASP problem and discuss properties of optimal Max-ASP models (Sect. 3). We then (Sect. 4) define various transformation rules which preserve the cost of all Max-ASP models and present a complete algorithm for determining the optimal cost of any weighted Max-ASP instance. Before conclusions, translations between Max-ASP and Max-SAT (Sect. 5), and the question of how earlier proposed related concepts can be expressed in Max-ASP (Sect. 6), are considered.

2   Preliminaries

We consider *normal logic programs* (NLPs) in the *propositional* case. A normal logic program Π consists of a finite set of rules of the form

$$r \ : \ h \leftarrow a_1, \ldots, a_n, \sim b_1, \ldots, \sim b_m, \tag{1}$$

where h, a_i's, and b_j's are propositional atoms. A rule r consists of a *head*, $\mathrm{head}(r) = h$, and a *body*, $\mathrm{body}(r) = \{a_1, \ldots, a_n, \sim b_1, \ldots, \sim b_m\}$. The symbol "$\sim$" denotes *default negation*. A *default literal* is an atom a, or its default negation $\sim a$. The set of atoms occurring in a program Π is $\mathrm{atom}(\Pi)$, and $\mathrm{dlits}(\Pi) = \{a, \sim a \mid a \in \mathrm{atom}(\Pi)\}$ is the set of default literals in Π. We use the shorthands $L^+ = \{a \mid a \in L\}$ and $L^- = \{a \mid \sim a \in L\}$ for a set L of default literals. Furthermore, we define $\mathrm{body}(\Pi) = \bigcup_{r \in \Pi}\{\mathrm{body}(r)\}$, and $\mathrm{def}(a, \Pi) = \{r \in \Pi \mid \mathrm{head}(r) = a\}$.

In ASP, we are interested in *stable models* [10] (or *answer sets*) of a program Π. An *interpretation* $M \subseteq \text{atom}(\Pi)$ defines which atoms of Π are true ($a \in M$) and which are false ($a \notin M$). An interpretation $M \subseteq \text{atom}(\Pi)$ satisfies a set L of literals, denoted $M \models L$, if and only if $L^+ \subseteq M$ and $L^- \cap M = \emptyset$; and M satisfies a rule $r \in \Pi$, denoted $M \models r$, if and only if $M \models \text{body}(r)$ implies $\text{head}(r) \in M$. An interpretation $M \subseteq \text{atom}(\Pi)$ is a *(classical) model* of Π, denoted by $M \models \Pi$, if and only if $M \models r$ for each rule $r \in \Pi$. A model M of a program Π is a stable model of Π if and only if there is no model $M' \subset M$ of Π_M, where $\Pi_M = \{\text{head}(r) \leftarrow \text{body}(r)^+ \mid r \in \Pi$ and $\text{body}(r)^- \cap M = \emptyset\}$ is called the *Gelfond-Lifschitz reduct* of Π with respect to M. The set of stable models of Π is denoted by $\text{SM}(\Pi)$.

Additional concepts relevant in this work are related to *loops*. The *positive dependency graph* of Π, denoted by $\text{Dep}^+(\Pi)$, is a directed graph with $\text{atom}(\Pi)$ and $\{\langle b, a \rangle \mid \exists r \in \Pi$ such that $b = \text{head}(r)$ and $a \in \text{body}(r)^+\}$ as the sets of vertices and edges, respectively. A non-empty set $L \subseteq \text{atom}(\Pi)$ is a loop in $\text{Dep}^+(\Pi)$ if for any $a, b \in L$ there is a path of non-zero length from a to b in $\text{Dep}^+(\Pi)$ such that all vertices in the path are in L; $\text{loop}(\Pi)$ denotes the set of all loops in $\text{Dep}^+(\Pi)$. The set of *external bodies* of a loop L in Π is

$$\text{eb}_\Pi(L) = \{\text{body}(r) \mid r \in \Pi, \text{ head}(r) \in L, \text{ body}(r)^+ \cap L = \emptyset\}.$$

3   Max-ASP

As a central starting point of this work, in this section we define unweighted and weighted Max-ASP and discuss some interesting properties of Max-ASP models.

Definition 1. *Given a NLP Π, an interpretation $M \subseteq \text{atom}(\Pi)$ is a* Max-ASP model *for Π, if M is a stable model for some subset-maximal $\Pi' \subseteq \Pi$ (there is no $\Pi'' \supset \Pi'$ such that $M \in \text{SM}(\Pi''))$. The* cost *of M is $|\Pi \setminus \Pi'|$. A Max-ASP model is* optimal *if it has minimum cost over all Max-ASP models for Π.*

In this work we are especially interested in finding optimal Max-ASP models. We denote the set of all optimal Max-ASP models of a NLP Π by $\text{MaxSM}(\Pi)$.

Example 1. Consider $\Pi = \{a \leftarrow \sim b.\ b \leftarrow \sim c.\ c \leftarrow \sim a\}$. Now $\text{SM}(\Pi) = \emptyset$, but $M = \{a\}$ is a Max-ASP model for Π, since $M \in \text{SM}(\Pi')$ for $\Pi' = \{a \leftarrow \sim b.\ c \leftarrow \sim a\}$. The cost of M is $|\{b \leftarrow \sim c\}| = 1$. Also $\emptyset$ (cost 3), $\{b\}$ (cost 1), and $\{c\}$ (cost 1) are Max-ASP models of Π. Thus $\text{MaxSM}(\Pi) = \{\{a\}, \{b\}, \{c\}\}$ since $\emptyset$ is not optimal.

Notice that Max-ASP models have the following basic properties:

1. Every NLP Π has a Max-ASP model; at least $\emptyset \subseteq \Pi$ trivially has a stable model.
2. If M is a Max-ASP model for Π such that $M \in \text{SM}(\Pi')$ for subset-maximal $\Pi' \subseteq \Pi$, then $M \not\models r$ for all $r \in \Pi \setminus \Pi'$.
3. The sets of optimal Max-ASP models and stable models for Π coincide if and only if Π has a stable model, i.e., $\text{MaxSM}(\Pi) = \text{SM}(\Pi)$ if and only if $\text{SM}(\Pi) \neq \emptyset$.
4. If $M \in \text{MaxSM}(\Pi)$ such that $M \in \text{SM}(\Pi')$ for subset-maximal $\Pi' \subseteq \Pi$, then $\text{SM}(\Pi'') = \emptyset$ for all $\Pi'' \supset \Pi'$.

3.1 Weighted Max-ASP

In analogy with weighted Max-SAT, Max-ASP allows for a natural extension to the weighted case where the rules can be weighted with integer costs.

Definition 2. *A weighted normal logic program is a pair* $\mathcal{P} = \langle \Pi, W \rangle$, *where* Π *is a NLP and* $W : \Pi \rightarrow \mathbb{N}$ *is a function that associates a nonnegative integer (a weight) with each rule in* Π.

We use the notation introduced in Section 2 in analogous way for weighted NLPs, e.g., atom($\mathcal{P}$) = atom(Π) for $\mathcal{P} = \langle \Pi, W \rangle$.

The concept of a *weighted* Max-ASP model is then naturally defined as follows.

Definition 3. *Given a weighted NLP* $\mathcal{P} = \langle \Pi, W \rangle$, *a Max-ASP model* M *for* Π *is a (weighted) Max-ASP model for* $\mathcal{P}$. *The cost of a Max-ASP model* M *which is a stable model for subset-maximal* $\Pi' \subseteq \Pi$ *is*

$$\sum_{r \in \Pi \setminus \Pi'} W(r).$$

A Max-ASP model is optimal *if it has minimum cost over all Max-ASP models for* $\mathcal{P}$.

We denote by MaxSM($\mathcal{P}$) the set of all optimal (weighted) Max-ASP models of $\mathcal{P}$. Notationally, we represent a weighted NLP $\mathcal{P} = \langle \Pi, W \rangle$ as a set of pairs

$$\{(r; w) \mid r \in \Pi \text{ and } w = W(r)\}.$$

Example 2. Consider $\mathcal{P} = \{(a \leftarrow \sim b; 5), (b \leftarrow \sim a; 5), (c \leftarrow a, b; 1), (d \leftarrow \sim c, \sim d; 1)\}$. Now, SM($\mathcal{P}$) = $\emptyset$, as the first two rules choose either a or b to be true, while the last rule requires that c is true. But to satisfy this, both a and b need to be true. The weights assigned for the rules imply that the mutual exclusion of a and b is more important than satisfaction of the constraint for c. Thus, MaxSM($\mathcal{P}$) = $\{\{a\}, \{b\}\}$ which both have cost 1, as the rule $d \leftarrow \sim c, \sim d$ is not satisfied. Notice that there is no Max-ASP model for $\mathcal{P}$ such that c is true, because no subset of the rules in $\mathcal{P}$ has such a stable model.

It is worth noticing that the properties of unweighted Max-ASP models discussed in Section 3 also hold in the weighted case.

The weighted variant of Max-ASP is in fact more expressive than the unweighted case. Namely, the decision problems of determining whether a (weighted) NLP has an optimal (weighted) Max-ASP model of cost less than a given value is in $\mathrm{P}^{\mathrm{NP}[\log n]}$ and P^{NP} for the unweighted and weighted case, respectively[1]. This is due to similar results for the well-known Max-SAT and weighted Max-SAT problems [11].

4   Branch-and-Bound for Max-ASP

In this section we present a branch-and-bound algorithm for finding the cost of an optimal Max-ASP model of a weighted NLP $\mathcal{P}$. The algorithm applies a set of equivalence-preserving transformation rules. For presenting these transformations and the branch-and-bound algorithm, we start with some additional notation.

---

[1] The unweighted case can be decided in deterministic polynomial time using logarithmic number of calls to an NP-oracle, while a linear number of calls are required for the weighted case.

We will assume that an explicit proper upper bound $\top$ is known for the cost of an optimal Max-ASP model for a weighted NLP $\mathcal{P} = \langle \Pi, W \rangle$. This is in analogy with [3], where a similar approach is applied in the context of Max-SAT. Notice that any value larger than the sum of the weights of all the rules of a program gives such an upper bound $\top$. Given a weighted NLP $\mathcal{P} = \langle \Pi, W \rangle$, an upper bound $\top$, and an interpretation $M \subseteq \mathrm{atom}(\mathcal{P})$, the cost of M in $\mathcal{P}$, denoted by $\mathrm{cost}(M, \mathcal{P}, \top)$, is $c = \sum_{r \in \Pi \setminus \Pi'} W(r)$ if $M \in \mathrm{SM}(\Pi')$ for subset maximal $\Pi' \subseteq \Pi$ such that $c < \top$, and $\top$ otherwise. Furthermore, M is a Max-ASP model for $\mathcal{P}$ if $\mathrm{cost}(M, \mathcal{P}, \top) < \top$, and M is optimal if it has minimum cost over all Max-ASP models for $\mathcal{P}$.

For a given upper bound $\top$, all rules which have weight $w > \top$ must necessarily be satisfied. Thus, such rules can be interpreted as *hard*, whereas rules with a weight less than $\top$ are *soft*. Without loss of generality, we can limit all the costs to the interval $[0 \ldots \top]$ and define $w_1 \oplus w_2 = \min(w_1 + w_2, \top)$. Finally, we use the symbol $\square$ to denote *falsity*, i.e., a rule that is always unsatisfied. Thus, if $(\square, w) \in \mathcal{P}$, then the cost of any Max-ASP model for $\mathcal{P}$ is at least w, and if $w = \top$, then $\mathcal{P}$ is *unsatisfiable*, i.e., has no Max-ASP models.

Remark 1. By setting $\top = 1$, the problem of finding a Max-ASP model for a weighted NLP $\mathcal{P} = \langle \Pi, W \rangle$ reduces to the problem of finding a stable model for Π.

Next, we will present the transformation rules which form a central part of our branch-and-bound algorithm for weighted Max-ASP.

4.1   Equivalence-Preserving Transformations

For presenting transformations preserving Max-ASP models, we begin by defining when two weighted NLPs are equivalent.

Definition 4. *Weighted NLPs $\mathcal{P}_1$ and $\mathcal{P}_2$ with a common upper bound $\top$ are equivalent, denoted by $\langle \mathcal{P}_1, \top \rangle \equiv \langle \mathcal{P}_2, \top \rangle$, if*

1. $\mathrm{atom}(\mathcal{P}_1) = \mathrm{atom}(\mathcal{P}_2)$, *and*
2. $\mathrm{cost}(M, \mathcal{P}_1, \top) = \mathrm{cost}(M, \mathcal{P}_2, \top)$ *for all $M \subseteq \mathrm{atom}(\mathcal{P}_1) = \mathrm{atom}(\mathcal{P}_2)$.*

Notice that in order $\mathcal{P}_1$ and $\mathcal{P}_2$ to be equivalent they need to have the same upper bound. Furthermore, notice that it is not sufficient that $\mathrm{MaxSM}(\mathcal{P}_1) = \mathrm{MaxSM}(\mathcal{P}_2)$ holds, but in addition, the cost of each interpretation has to be the same.

Remark 2. With $\top = 1$, i.e., when a stable model is sought, the relation $\equiv$ turns out to be the same as *ordinary* or *weak equivalence*, which requires that $\mathcal{P}_1$ and $\mathcal{P}_2$ have the same set of stable models. Notice that the additional condition $\mathrm{atom}(\mathcal{P}_1) = \mathrm{atom}(\mathcal{P}_2)$ can always be satisfied, e.g., by adding rules of the form $a \leftarrow a$.

Given weighted NLPs $\mathcal{P}, \mathcal{P}_1 \subseteq \mathcal{P}$, and $\mathcal{P}_2$, we use $\mathcal{P}_1 \equiv_{\mathcal{P}} \mathcal{P}_2$ as a shorthand for

$$\langle \mathcal{P}, \top \rangle \equiv \langle (\mathcal{P} \setminus \mathcal{P}_1) \cup \mathcal{P}_2, \top \rangle.$$

Finally, we use shorthands $(a; w)$ and $(\sim a; w)$ where $a \in \mathrm{atom}(\Pi)$ for weighted literals; and $(B; w)$ and $(\sim B; w)$ where $B \in \mathrm{body}(\Pi)$ for weighted bodies and their complements which can appear in a weighted NLP $\mathcal{P} = \langle \Pi, W \rangle$. These shorthands are easily presented with weighted rules, e.g., $(\{l_1, \ldots, l_n\}; w)$ is a shorthand for weighted rules $(f \leftarrow \sim f, \sim l; w)$ and $(l \leftarrow l_1, \ldots, l_n; \top)$; and $(\sim\{l_1, \ldots, l_n\}; w)$ for $(f \leftarrow \sim f, l; w)$ and $(l \leftarrow l_1, \ldots, l_n; \top)$, where f and l do not appear in atom(Π).

First we present transformation rules for *aggregation*, *hardening*, and *lower-bounding*.

Proposition 1. *Let $\mathcal{P} = \langle \Pi, W \rangle$ be a weighted NLP with an upper bound $\top$. If ψ is a rule $r \in \Pi$ (only in Items 1 and 2), a default literal $l \in$ dlits(Π), or a body $B \in$ body(Π) or its complement $\sim B$, then the following equivalences hold:*

1. *Aggregation:* $\{(\psi; w_1), (\psi; w_2)\} \equiv_{\mathcal{P}} \{(\psi; w_1 \oplus w_2)\}$
2. *Hardening:* $\{(\psi; w_1), (\Box; w_2)\} \equiv_{\mathcal{P}} \{(\psi; \top), (\Box; w_2)\}$, *if $w_1 \oplus w_2 = \top$*
3. *Lower-bounding:* $\{(\psi; \top), (\sim\psi; w)\} \equiv_{\mathcal{P}} \{(\psi; \top), (\Box; w)\}$

The hardening rule allows one to identify rules that are equivalent to hard counterparts: the violation of $(\psi; w_1)$ has cost $\top$. The lower-bounding rule makes the lower bound implied by the violation of a hard constraint explicit. The proof of Proposition 1 is omitted due to space constraints.

The next transformations are for inference between bodies and literals in the bodies.

Proposition 2. *Let $\mathcal{P} = \langle \Pi, W \rangle$ be a weighted NLP, $\top$ an upper bound, l and l' literals in* dlits(Π), *and $B \in$ body(Π). The following equivalences hold:*

4. *Forward true body (FTB):* $\{(l; \top) \mid l \in B\} \equiv_{\mathcal{P}} \{(l; \top) \mid l \in B\} \cup \{(B; \top)\}$
5. *Backward false body (BFB):*

$$\{(\sim B; \top)\} \cup \{(l; \top) \mid l \in B \setminus \{l'\}\}$$
$$\equiv_{\mathcal{P}} \{(\sim B; \top)\} \cup \{(l; \top) \mid l \in B \setminus \{l'\}\} \cup \{(\sim l'; \top)\}$$

6. *Forward false body (FFB):* $\{(\sim l; \top)\} \equiv_{\mathcal{P}} \{(\sim l; \top), (\sim B; \top)\}$, *if $l \in B$*
7. *Backward true body (BTB):* $\{(B; \top)\} \equiv_{\mathcal{P}} \{(B; \top), (l; \top))\}$, *if $l \in B$*

All these transformation rules require the constraints involved be hard, and this way a body is interpreted as a conjunction of its default literals. Without going into further details, we note that the correctness of these transformations follows from the similarity with sound ASP inference rules for normal logic programs presented in [5].

Finally, we have transformations relating head atoms with the rules defining them.

Proposition 3. *Let $\mathcal{P} = \langle \Pi, W \rangle$ be a weighted NLP and $\top$ an upper bound. The following equivalences hold:*

8. *Forward true atom (FTA):*
$$\{(B; \top), (h \leftarrow B; w)\} \equiv_{\mathcal{P}} \{(B; \top), (h \leftarrow B; 0), (h; w)\}$$
9. *Backward false atom (BFA):*
$$\{(\sim h; \top), (h \leftarrow B; w)\} \equiv_{\mathcal{P}} \{(\sim h; \top), (h \leftarrow B; 0), (\sim B; w)\}$$
10. *Forward false atom (FFA):*

$$\{(\sim B; \top) \mid B \in \text{body}(\text{def}(h, \Pi))\}$$
$$\equiv_{\mathcal{P}} \{(\sim B; \top) \mid B \in \text{body}(\text{def}(h, \Pi))\} \cup \{(\sim h; \top)\}$$

11. *Forward loop (FL): Let $h \in L$ and $L \in \mathsf{loop}(\Pi)$. Then*

$$\{(\sim B; \top) \mid B \in \mathsf{eb}_\Pi(L)\} \equiv_\mathcal{P} \{(\sim B; \top) \mid B \in \mathsf{eb}_\Pi(L)\} \cup \{(\sim h; \top)\}$$

12. *Backward true atom (BTA):*

$$\{(\sim B; \top) \mid B \in \mathsf{body}(\mathsf{def}(h, \Pi)) \setminus \{B'\}\} \cup \{(h; \top)\}$$
$$\equiv_\mathcal{P} \{(\sim B; \top) \mid B \in \mathsf{body}(\mathsf{def}(h, \Pi)) \setminus \{B'\}\} \cup \{(h; \top), (B'; \top)\}$$

For understanding these transformations, we note that since a Max-ASP model needs to be a stable model for some subset of rules, only hard constraints are inferred in *FFA*, *FL*, and *BTA*. On the other hand, the "soft" transformation rules *FTA* and *BFA* correspond to satisfaction of rules (the cost of a Max-ASP model M comes from the rules r such that $M \not\models r$). Moreover, the precise form of *FTA* and *BFA* is related to the global nature of *FFA*, *FL*, and *BTA*. Instead of removing the rule in question when applying *FTA* or *BFA*, we change its cost to zero. There is no cost involved if a rule with zero cost is unsatisfied, but nevertheless, the effect of such rules needs to be taken into account when applying *FFA*, *FL*, and *BTA*. We now consider the correctness of the transformation rules *FTA* and *FFA* in more detail. The other cases are similar.

FTA: Let

$$\mathcal{P} = \mathcal{P}' \cup \{(B; \top), (h \leftarrow B; w)\} \text{ and } \mathcal{P}'' = \mathcal{P}' \cup \{(B; \top)\}, (h \leftarrow B; 0), (h; w),$$

and consider arbitrary $M \subseteq \mathsf{atom}(\mathcal{P})$. If $M \not\models B$, then $\mathsf{cost}(M, \mathcal{P}, \top) = \top$ and $\mathsf{cost}(M, \mathcal{P}'', \top) = \top$. If $M \models B$, then $M \not\models h \leftarrow B$ if and only if $h \notin M$. Thus, either $\{(h \leftarrow B; w)\}$ and $\{(h; w), (h \leftarrow B; 0)\}$ are both satisfied in M, or neither of them is satisfied, and $\mathsf{cost}(M, \mathcal{P}, \top) = \mathsf{cost}(M, \mathcal{P}'', \top)$.

FFA: Let

$$\mathcal{P} = \langle \Pi, W \rangle = \mathcal{P}' \cup \{(\sim B; \top) \mid B \in \mathsf{body}(\mathsf{def}(h, \Pi))\} \text{ and}$$
$$\mathcal{P}'' = \mathcal{P}' \cup \{(\sim B; \top) \mid B \in \mathsf{body}(\mathsf{def}(h, \Pi))\} \cup \{(\sim h; \top)\},$$

and consider arbitrary $M \subseteq \mathsf{atom}(\mathcal{P})$. If $h \notin M$, then it holds that $\mathsf{cost}(M, \mathcal{P}'', \top) = \mathsf{cost}(M, \mathcal{P}, \top)$. If $h \in M$, then $\mathsf{cost}(M, \mathcal{P}'', \top) = \top$. Assume that cost $(M, \mathcal{P}, \top) < \top$. Then there is $\Pi' \subseteq \Pi$ such that $M = \mathsf{SM}(\Pi')$. However, $M \not\models B$ for all $B \in \mathsf{body}(\mathsf{def}(h, \Pi))$. This implies that there can be no rule r in Π' such that $\mathsf{head}(r) = h$ and $M \models \mathsf{body}(r)$, and furthermore, $h \notin M$. This is a contradiction, and thus, $\mathsf{cost}(M, \mathcal{P}, \top) = \top$.

4.2 Branch-and-Bound

We are now ready to introduce a depth-first branch-and-bound algorithm which, given a weighted NLP $\mathcal{P}$ and a cost upper bound $\top$, determines the cost of the optimal weighted Max-ASP models of $\mathcal{P}$ given that the optimal cost is less than $\top$. The method, presented as Algorithm 1, applies the equivalence-preserving transformations introduced in Propositions 1, 2, and 3 in PROPAGATE($\mathcal{P}, \top$) (Line 1). After applying the transformations, the algorithm makes *choices* by case analysis on $(l; \top)$, where $l \in \mathsf{dlits}(\mathcal{P})$, such that $(l; \top) \notin \mathcal{P}$ (represented by the choice heuristic SELECTLITERAL($\mathcal{P}$) on

---

Algorithm 1. MAXASP($\mathcal{P}, \top$)

---

1. $\mathcal{P} := \text{PROPAGATE}(\mathcal{P}, \top)$
2. **if** $(\square, \top) \in \mathcal{P}$ **then return** $\top$
3. **if** $\forall a \in \text{atom}(\mathcal{P})$ either $(a; \top) \in \mathcal{P}$ or $(\sim a; \top) \in \mathcal{P}$ **then**
   3a. **if** $(\square, w) \in \mathcal{P}$ **then return** w
   3b. **else return** 0
4. $l := \text{SELECTLITERAL}(\mathcal{P})$
5. $v := \text{MAXASP}(\mathcal{P} \cup \{(l; \top)\}, \top)$
6. $v := \text{MAXASP}(\mathcal{P} \cup \{(\sim l; v)\}, v)$
7. **return** v

Line 4). This leads to a complete search for determining the cost of Max-ASP models of weight less than $\top$. The algorithm can also easily be modified to also return an optimal model in addition to its cost (as demonstrated in Example 3). If there are no models with cost less than $\top$, the algorithm returns $\top$ (Line 2). Recall that the lower-bounding rule guarantees that $\{(a; \top), (\sim a; \top)\} \subseteq \mathcal{P}$ is impossible.

While a formal correctness proof is omitted here, the correctness is based on the fact that, once $(a; \top) \in \mathcal{P}$ or $(\sim a; \top) \in \mathcal{P}$ for all $a \in \text{atom}(\mathcal{P})$ (Line 2), the transformation rules are complete in the sense that they assure that the lower bound can not be tightened further. Since the transformation rules also guarantee that the current weighted NLP remains equivalent to the original one, the current lower bound is the optimal cost.

The following example illustrates a run of the algorithm.

Example 3. Consider $\mathcal{P} = \{(b \leftarrow a;\ 1), (a \leftarrow b;\ 2), (a \leftarrow \sim c;\ 3)\}$ with $\top = 7$. We start with $\text{PROPAGATE}(\mathcal{P}, \top)$. Since atom c has no defining rules, we can apply *FFA* and obtain $\mathcal{P} \cup \{(\sim c; \top)\}$. We continue by applying *FTB* and *FTA* and get $\mathcal{P}_1 = \{(b \leftarrow a;\ 1), (a \leftarrow b;\ 2), (a \leftarrow \sim c;\ 0), (\sim c; \top), (\{\sim c\}; \top), (a;\ 3)\}$. None of the transformation rules is applicable to $\mathcal{P}_1$, and the stopping criteria on Line 2 and Line 3 do not hold. Thus we make a choice. Let $\text{SELECTLITERAL}(\mathcal{P})$ return $\sim a$.

- We add $(\sim a; \top)$ to $\mathcal{P}_1$ and after propagation using *FFB*, *lower-bounding*, *FFA*, and *FFB*, we obtain

$$\mathcal{P}_2 = \{(b \leftarrow a;\ 1), (a \leftarrow b;\ 2), (a \leftarrow \sim c;\ 0), (\sim c; \top), (\{\sim c\}; \top),$$
$$(\sim a; \top), (\sim\{a\}; \top), (\square;\ 3), (\sim b; \top), (\sim\{b\}; \top)\}.$$

No further propagation is applicable, and the condition on Line 3 holds. Now, $\{a \in \text{atom}(\mathcal{P}) \mid (a; \top) \in \mathcal{P}_2\} = \emptyset$ and $\text{cost}(\emptyset, \mathcal{P}, \top) = \text{cost}(\emptyset, \mathcal{P}_2, \top) = 3$. Furthermore, $\emptyset$ is a (not necessarily optimal) Max-ASP model for $\mathcal{P}_2$, and thus also for $\mathcal{P}$. We set $\top = 3$ and continue the search by backtracking.

- After adding $(a; \top)$ to $\mathcal{P}_1$, we get $\{(b \leftarrow a; 0), (a \leftarrow b;\ 2), (a \leftarrow \sim c;\ 0), (\sim c; \top), (\{\sim c\}; \top), (a; \top), (\{a\}; \top), (b; 1)\}$ by application of *aggregation*, *FTB*, and *FTA*. Let $\text{SELECTLITERAL}(\mathcal{P})$ return b. After the use of *aggregation* and *FTB* we have

$$\mathcal{P}_3 = \{(b \leftarrow a;\ 0), (a \leftarrow b;\ 0), (a \leftarrow \sim c;\ 0), (\sim c; \top), (\{\sim c\}; \top), (a; \top),$$
$$(\{a\}; \top), (b;\ \top), (\{b\}; \top)\}.$$

Now, $M = \{a, b\} = \{a \in \mathsf{atom}(\mathcal{P}) \mid (a; \top) \in \mathcal{P}_3\}$ and $\mathsf{cost}(M, \mathcal{P}, \top) = \mathsf{cost}(M, \mathcal{P}_3, \top) = 0$. Thus we have found an optimal Max-ASP model, which in this case is a stable model for the NLP.

Remark 3. In the case $\top = 1$, the transformation rules in Propositions 2 and 3 resemble closely inference rules in tableau calculi for ASP proposed in [5]. In fact, Algorithm 1 can be viewed as a generalization of complete search methods proposed for ASP, as some of the presented transformation rules are generalizations of tableau rules [5]. However, for compactness we have intentionally left out additional transformation rules (related to *well-founded sets* and loops) which would generalize their counterparts in [5].

5   From Max-ASP to Max-SAT

In this section we analyze the relationship between Max-ASP and Max-SAT, giving solution-preserving translations between these problems. For this we first briefly go through necessary concepts related to Max-SAT.

Let X be a set of Boolean variables. Associated with every variable $x \in X$ there are two *literals*, the positive literal, denoted by x, and the negative literal, denoted by $\bar{x}$. A *clause* is a disjunction of distinct literals. We view a clause as a finite set of literals and a *CNF formula* as a finite set of clauses. A *truth assignment* τ associates a truth value $\tau(x) \in \{\mathsf{false}, \mathsf{true}\}$ with each variable $x \in X$. A truth assignment *satisfies* a CNF formula if and only if it satisfies every clause in it. A clause is satisfied if and only if it contains at least one satisfied literal, where a literal x ($\bar{x}$, respectively) is satisfied if $\tau(x) = \mathsf{true}$ ($\tau(x) = \mathsf{false}$, respectively). A CNF formula is *satisfiable* if there is a truth assignment that satisfies it, and *unsatisfiable* otherwise.

A *weighted CNF formula* is a pair $\mathcal{W} = \langle \mathcal{C}, W \rangle$, where $\mathcal{C}$ is a CNF formula and $W : \mathcal{C} \to \mathbb{N}$ is a function that associates a nonnegative integer with each clause in $\mathcal{C}$.

Definition 5. *Given a weighted CNF formula $\mathcal{W} = \langle \mathcal{C}, W \rangle$, an upper bound $\top$, and a truth assignment τ for $\mathcal{C}$, the cost of τ in $\mathcal{W}$, denoted by $\mathsf{cost}(\tau, \mathcal{W}, \top)$, is the sum of weights of the clauses not satisfied by τ. If $\mathsf{cost}(\tau, \mathcal{W}, \top) < \top$, τ is a Max-SAT model for $\mathcal{W}$, and τ is optimal if it has minimum cost over all Max-SAT models for $\mathcal{W}$.*

5.1   Max-ASP as Max-SAT

The translation from Max-ASP to Max-SAT is based on a typical translation from ASP to SAT, i.e., Clark's completion [12,13] with loop formulas [14]. In the following, we assume without loss of generality that $r_i \neq r_j$ for all rules $r_i, r_j \in \Pi$ in a weighted NLP $\mathcal{P} = \langle \Pi, W \rangle$. Furthermore, we use shorthands as follows: if l is a default literal, i.e., a or $\sim a$, then $x_l = x_a$ if $l = a$, and $x_l = \bar{x}_a$ if $l = \sim a$.

Definition 6. *Given a weighted NLP $\mathcal{P} = \langle \Pi, W \rangle$ with an upper bound $\top$, the translation $\mathsf{MaxSAT}(\mathcal{P}, \top)$ consists of the following weighted clauses:*

1. *For each $a \in \mathsf{atom}(\Pi)$:*
   $(\{x_a, \bar{x}_{B_1}\}; w_1), \ldots, (\{x_a, \bar{x}_{B_m}\}; w_m)$ and $(\{x_{B_1}, \ldots, x_{B_m}, \bar{x}_a\}; \top)$,
   where $B_i = \mathsf{body}(r_i)$ and $w_i = W(r_i)$ for each $r_i \in \mathsf{def}(a, \Pi)$.

2. *For each* $B = \{l_1, \ldots, l_n\} \in \mathsf{body}(\Pi)$:
   $(\{x_{l_1}, \bar{x}_B\}; \top), \ldots, (\{x_{l_n}, \bar{x}_B\}; \top)$, *and* $(\{x_B, \bar{x}_{l_1}, \ldots, \bar{x}_{l_n}\}; \top)$.
3. *For each* $L \in \mathsf{loop}(\Pi)$ *and each* $a \in L$:
   $(\{x_{B_1}, \ldots, x_{B_m}, \bar{x}_a\}; \top)$ *where* $\mathsf{eb}_\Pi(L) = \{B_1, \ldots, B_m\}$.

There is a bijective correspondence between the Max-ASP models of $\mathcal{P}$ and Max-SAT models of its translation $\mathsf{MaxSAT}(\mathcal{P}, \top)$.

Theorem 1. *Given a weighted NLP* $\mathcal{P} = \langle \Pi, W \rangle$, *an upper bound* $\top$, *and a Max-ASP model* M *for* $\mathcal{P}$ *with cost* $w < \top$, *the truth assignment*

$$\tau(x) = \begin{cases} \mathsf{true}, & \text{if } x = x_a \text{ for } a \in \mathsf{atom}(\Pi) \text{ and } a \in M, \\ \mathsf{true}, & \text{if } x = x_B \text{ for } B \in \mathsf{body}(\Pi) \text{ and } M \models B, \\ \mathsf{false}, & \textit{otherwise}. \end{cases}$$

is a Max-SAT model for $\mathsf{MaxSAT}(\mathcal{P}, \top)$ *with cost* w.

Proof. Let M be a Max-ASP model for $\mathcal{P} = \langle \Pi, W \rangle$ with cost w, τ defined as in Definition 6, and $\Pi' \subseteq \Pi$ subset-maximal such that $M \in \mathsf{SM}(\Pi')$.

Consider first the clauses in Item 2 of Definition 6. Assume that clause $(\{x_l, \bar{x}_B\}; \top)$ is not satisfied by τ, i.e., $\tau(x_l) = \mathsf{false}$ and $\tau(x_B) = \mathsf{true}$. But then $M \not\models l$ and $M \models B$ which leads to contradiction as $l \in B$. Similarly, τ satisfies all clauses of the form $(\{x_B, \bar{x}_{l_1}, \ldots, \bar{x}_{l_n}\}; \top)$; and furthermore, τ satisfies all clauses in Item 2.

Next, notice that the hard clauses in Item 3 are effectively the loop formulas of Π in clausal form [14]. Assume now that there is a clause $(\{x_{B_1}, \ldots, x_{B_m}, \bar{x}_a\}; \top)$ that τ does not satisfy, i.e, $\tau(x_{B_i}) = \mathsf{false}$ for all i and $\tau(x_a) = \mathsf{true}$. Thus $M \not\models B$ for all $B \in \mathsf{eb}_\Pi(L)$ and there is $a \in M \cap L$. Since $a \in M$ and $M = \mathsf{LM}((\Pi')_M)$, there is a rule $r \in (\Pi')_M$ such that $a = \mathsf{head}(r)$ and $M \models \mathsf{body}(r)$. Moreover, $\mathsf{body}(r) \cap L \neq \emptyset$, since $M \not\models B$ for all $B \in \mathsf{eb}_\Pi(L)$. Thus there is $b \in \mathsf{body}(r) \cap L \cap M$. Continuing this process, one notices that $M \models L$ and moreover, $L \in \mathsf{loop}(\Pi')$. Since $\Pi' \subseteq \Pi$, also $\mathsf{eb}_{\Pi'}(L) \subseteq \mathsf{eb}_\Pi(L)$. Therefore $M \not\models B$ for all $B \in \mathsf{eb}_{\Pi'}(L)$, and there is a loop formula of Π' not satisfied in M. This is a contradiction to $M \in \mathsf{SM}(\Pi')$, since a stable model of a program must satisfy all its loop formulas [14]. Thus, τ satisfies all the clauses in Item 3.

As regards the weighted clauses in Item 1, we notice that if $a \in M$, then there is a rule $a \leftarrow B \in \mathsf{def}(a, \Pi') \subseteq \mathsf{def}(a, \Pi)$ such that $M \models B$, i.e., $\tau(x_a) = \mathsf{true}$ and $\tau(x_B) = \mathsf{true}$ for some $B \in \mathsf{body}(\mathsf{def}(a, \Pi))$. Thus τ satisfies all weighted clauses in Item 1. On the other hand, if $a \notin M$, then we notice that $M \not\models \mathsf{body}(r)$ for all $r \in \mathsf{def}(a, \Pi) \cap \Pi'$ and $M \models \mathsf{body}(r)$ for all $r \in \mathsf{def}(a, \Pi) \setminus \Pi'$. The cost of clauses related to $\mathsf{def}(a, \Pi)$ which τ does not satisfy is the same as the cost of rules $r \in \mathsf{def}(a, \Pi)$ such that $M \not\models r$. Thus, the overall cost of violated clauses is exactly the cost of rules in $\Pi \setminus \Pi'$. $\square$

Remark 4. Note that if $\top = 1$, all clauses are hard, and $\mathsf{MaxSAT}(\mathcal{P}, \top)$ is a clausal form of Clark's completion of $\mathcal{P}$ with loop formulas of $\mathcal{P}$. Thus, $M \in \mathsf{SM}(\mathcal{P})$ if and only if τ satisfies $\mathsf{MaxSAT}(\mathcal{P}, 1)$ [14].

Theorem 2. *Given a weighted NLP* $\mathcal{P} = \langle \Pi, W \rangle$, *an upper bound* $\top$, *and a Max-SAT model* τ *with cost* $w < \top$ *for* $\mathsf{MaxSAT}(\mathcal{P}, \top)$, *the interpretation*

$$\{a \in \text{atom}(\Pi) \mid \tau(x_a) = \text{true}\}$$

is a Max-ASP model for $\mathcal{P}$ with cost w.

Proof. Assume that τ is a Max-SAT model for MaxSAT$(\mathcal{P}, \top)$ with cost $w < \top$, and $M = \{a \in \text{atom}(\Pi) \mid \tau(x_a) = \text{true}\}$. Define $\Pi' = \{r \in \Pi \mid M \models r\}$ and $\mathcal{P}' = \langle \Pi', W' \rangle$ such that $W'(r) = W(r)$ for all $r \in \Pi'$.

Let us consider MaxSAT$(\mathcal{P}', 1)$. Since $w < \top$, the clauses in Item 2 of Definition 6 of MaxSAT$(\mathcal{P}, \top)$ force that $\tau(x_B) = \text{true}$ if and only if $\tau(x_l) = \text{true}$ for all $l \in B$. Furthermore, since body$(\Pi') \subseteq$ body(Π), τ satisfies all the clauses in Item 2 of MaxSAT$(\mathcal{P}', 1)$ as well.

Consider arbitrary $a \in \text{atom}(\Pi)$. If $a \in M$, i.e., $\tau(x_a) = \text{true}$, then def$(a, \Pi') = $ def(a, Π). Then MaxSAT$(\mathcal{P}', 1)$ contains the same clauses related to def(a, Π') as MaxSAT$(\mathcal{P}, \top)$, and furthermore, τ satisfies all these clauses. If $a \notin M$, i.e., $\tau(x_a) = $ false, then for all $r \in $ def(a, Π') it holds that $M \not\models $ body(r). Thus τ satisfies all clauses related to def(a, Π').

Consider an arbitrary $L \in \text{loop}(\Pi') \subseteq \text{loop}(\Pi)$. If eb$_{\Pi'}(L) = $ eb$_{\Pi}(L)$, then τ satisfies all clauses in Item 3 of MaxSAT$(\mathcal{P}', 1)$. Assume eb$_{\Pi'}(L) \subset $ eb$_{\Pi}(L)$ and consider arbitrary $a \in L$. If $a \notin M$, i.e., $\tau(x_a) = $ false, then τ satisfies the clause in Item 3 related to a. If $a \in M$, i.e., $\tau(x_a) = $ true, then there is $B \in $ body$(\text{def}(a, \Pi'))$ such that $\tau(x_B) = $ true, i.e., $M \models B$, since τ satisfies the clause $(\{x_{B_1}, \ldots, x_{B_m}, \bar{x}_a\}; \top)$ for $\{B_1, \ldots, B_m\} = $ body$(\text{def}(a, \Pi'))$. If $B \in $ eb$_{\Pi'}(L)$ then τ satisfies all clauses in Item 3 related to L. If $B \notin $ eb$_{\Pi'}(L)$, then $B \cap L \neq \emptyset$, and thus $M \models B$ implies that there is $a' \in L \cap M$, i.e., $\tau(x_{a'}) = $ true. Again, there must be $B' \in $ body$(\text{def}(a', \Pi'))$ such that $\tau(x_{B'}) = $ true, i.e., $M \models B'$. If $B' \in $ eb$_{\Pi'}(L)$, we are done as τ satisfies all clauses in Item 3 related L. Assume now that $\tau(x_B) = $ false for all $B \in $ eb$_{\Pi'}(L)$. Continuing the process we notice that $a \in M$ for all $a \in L$. Since τ satisfies the clauses in Item 3 of MaxSAT$(\mathcal{P}, \top)$ there is $B \in $ eb$_{\Pi}(L) \setminus $ eb$_{\Pi'}(L)$ such that $\tau(x_B) = $ true. Thus, there is $r \in \Pi \setminus \Pi'$ such that $B = $ body(r) and head$(r) \in L$. Since $r \notin \Pi'$, it holds that head$(r) \notin M$. This is in contradiction with $L \subseteq M$, and therefore τ satisfies all clauses in Item 3 of MaxSAT$(\mathcal{P}', 1)$.

Now, τ is a Max-SAT model for MaxSAT$(\mathcal{P}', 1)$ with cost 0, i.e., M is a stable model of Π'. Finally, note that the sum of the weights of the rules in $\Pi \setminus \Pi'$ is w. □

5.2 Max-SAT as Max-ASP

There is a natural linear-size translation from CNF formulas to NLPs so that there is a bijective correspondence between the satisfying truth assignments of any CNF formula and stable models of its translation [15]. We give a generalization of the translation to establish a similar correspondence between Max-SAT and Max-ASP. If l is a literal, i.e., variable x or its negation $\bar{x}$, then $\sim a_l = a_x$ if $l = \bar{x}$ and $\sim a_l = \sim a_x$ if $l = x$. We assume without loss of generality that $C_i \neq C_j$ for all clauses $C_i, C_j \in \mathcal{C}$ in a weighted CNF formula $\mathcal{W} = \langle \mathcal{C}, W \rangle$.

Definition 7. *Given a weighted CNF formula $\mathcal{W} = \langle \mathcal{C}, W \rangle$ with an upper bound $\top$, the translation MaxASP$(\mathcal{W}, \top)$ consists of the following weighted rules:*

1. *For each clause $C_i = \{l_1, \ldots, l_n\} \in \mathcal{C}$:*
   $(f_i \leftarrow \sim f_i, \sim a_{l_1}, \ldots, \sim a_{l_n}; w_i)$, *where* $w_i = W(C_i)$.
2. *For each variable x in $\mathcal{C}$:* $(\hat{a}_x \leftarrow \sim a_x; \top)$, *and* $(a_x \leftarrow \sim \hat{a}_x; \top)$.

Theorem 3. *Let $\mathcal{W} = \langle \mathcal{C}, W \rangle$ be a weighted CNF formula, and $\top$ an upper bound.*

- *If τ is a Max-SAT model for $\mathcal{W}$ with cost $w < \top$, then*

$$\{a_x \mid \tau(x) = \mathsf{true}\} \cup \{\hat{a}_x \mid \tau(x) = \mathsf{false}\}$$

is a Max-ASP model for $\mathsf{MaxASP}(\mathcal{W}, \top)$ *with cost w.*

- *If M is a Max-ASP model for* $\mathsf{MaxASP}(\mathcal{W}, \top)$ *with cost $w < \top$, then*

$$\tau(x) = \begin{cases} \mathsf{true}, & \text{if } a_x \in M, \text{ and} \\ \mathsf{false}, & \text{if } \hat{a}_x \in M. \end{cases}$$

is a Max-SAT model for $\mathcal{W}$ with cost w.

6   Related Approaches in ASP

In this section we discuss the similarities and differences of Max-ASP and some extensions of stable models most closely related to Max-ASP models.

Simons et al. [7] consider an extended ASP language including cardinality and choice rules and, especially, *optimize statements*. An optimize statement is a *minimize statement* or its dual, a *maximize statement*. A minimize statement is of the form MINIMIZE $(a_1 = w_{a_1}, \ldots, a_n = w_{a_n}, \sim b_1 = w_{b_1}, \ldots, \sim b_m = w_{b_m})$ where a_i's and b_j's are atoms. First, recall that several optimize statements can be encoded into a single one with suitable weights [7]. Now, given a program Π and a minimize statement, the goal is to find a stable model M for Π such that $\sum_{a_i \in M} w_{a_i} + \sum_{b_i \notin M} w_{b_i}$ is minimized. For an NLP Π, we can view this optimization problem as a weighted NLP $\mathcal{P}$ where all the rules from Π are hard, and in addition, there are soft constraints $(f \leftarrow \sim f, a_i; w_{a_i})$ for each $1 \le i \le n$ and $(f \leftarrow \sim f, \sim b_i; w_{b_i})$ for each $1 \le i \le m$.

Buccafurri et al. [6] consider *strong and weak constraints* in *disjunctive Datalog programs*. A strong constraint is of the form $\leftarrow a_1, \ldots, a_n, \sim b_1, \ldots, \sim b_m$ which can be viewed as a rule $f \leftarrow \sim f, a_1, \ldots, a_n, \sim b_1, \ldots, \sim b_m$. Thus, a stable model must necessarily satisfy all the strong, i.e., hard constraints. A weak constraint is of the form $\Leftarrow a_1, \ldots, a_n, \sim b_1, \ldots, \sim b_m$, and it is possible to set different priorities for weak constraints. The goal is to minimize the number of unsatisfied weak constraints according to their priorities. In the case of NLPs, weak constraints can be represented in a natural way in Max-ASP by using weights for expressing priorities.

On the other hand, while (weighted) Max-ASP can be expressed with NLPs and either minimize statements or weak constraints (all three problems have the same complexity), we do not see an immediate and natural way of encoding Max-ASP in the other two formalisms.

Gebser et al. [9] introduce ι-stable models, which have similar properties as (unweighted) Max-ASP models. An interpretation $M \subseteq \mathsf{atom}(\Pi)$ is a ι-*stable model* for NLP Π, if $M = \mathsf{LM}(\Pi'_\emptyset)$ for some subset maximal $\Pi' \subseteq \Pi$ such that $\mathsf{body}^+(\Pi') \subseteq \mathsf{LM}(\Pi'_\emptyset)$ and $\mathsf{body}^-(\Pi') \cap \mathsf{LM}(\Pi'_\emptyset) = \emptyset$. It is worth noticing that a ι-stable model

M of Π is a stable model of Π if and only if $M \models \Pi$, and the same applies to Max-ASP models. It is straightforward to see that a ι-stable model M for Π is a Max-ASP model for Π. However, ι-stable models are not necessarily *optimal* Max-ASP models. For instance, $\Pi_1 = \{a \leftarrow \sim d.\ b \leftarrow \sim e.\ c \leftarrow a, b.\ e \leftarrow \sim a\}$ [9] has two ι-stable models, namely $\{e\}$, and $\{a, b, c\}$. Now, $\mathsf{SM}(\Pi_1) = \{\{a, b, c\}\}$ and $\{e\}$ is not optimal. On the other hand, a Max-ASP model of a NLP is not necessarily a ι-stable model. For instance, consider $\Pi_2 = \{a \leftarrow \sim c.\ a \leftarrow b, c.\ b \leftarrow a.\ b \leftarrow c.\ c \leftarrow a, b\}$ from [9, Example 6.3]. Now, $\mathsf{MaxSM}(\Pi_2) = \{\emptyset, \{a\}, \{a, b\}\}$ which each have cost one. However, $\{a, b\}$ is the unique ι-stable model of Π_2.

7   Conclusions

We study the unweighted and weighted Max-ASP problems from several different angles. Most importantly, we develop sound transformation rules for Max-ASP inference, based on which we present a complete algorithm for computing optimal weighted Max-ASP models. Translations between Max-ASP and Max-SAT are also developed, in addition to relating Max-ASP to related concepts in ASP.

Our branch-and-bound algorithm builds ground for implementations for Max-ASP search. We find that the study of inference and search methods, including study of additional transformation rules as well as solver development, for (weighted) Max-ASP could prove a fruitful direction for further study. Additionally, we aim to study translation-based approaches for solving Max-ASP as, e.g., Max-SAT, so that the already developed Max-SAT solvers can be exploited in analogy to SAT-based approaches to ASP such as [14].

References

1. Gebser, M., Pührer, J., Schaub, T., Tompits, H.: A meta-programming technique for debugging answer-set programs. In: Proc. AAAI 2008, pp. 448–453. AAAI Press, Menlo Park (2008)
2. Li, C., Manyá, F.: MaxSAT, hard and soft constraints. In: Handbook of Satisfiability. IOS Press, Amsterdam (2009)
3. Larrosa, J., Heras, F., de Givry, S.: A logical approach to efficient Max-SAT solving. Artificial Intelligence 172(2-3), 204–233 (2008)
4. Bonet, M., Levy, J., Manyà, F.: Resolution for Max-SAT. Artificial Intelligence 171(8-9), 606–618 (2007)
5. Gebser, M., Schaub, T.: Tableau calculi for answer set programming. In: Etalle, S., Truszczyński, M. (eds.) ICLP 2006. LNCS, vol. 4079, pp. 11–25. Springer, Heidelberg (2006)
6. Buccafurri, F., Leone, N., Rullo, P.: Enhancing disjunctive datalog by constraints. IEEE Transactions on Knowledge and Data Engineering 12(5), 845–860 (2000)
7. Simons, P., Niemelä, I., Soininen, T.: Extending and implementing the stable model semantics. Artificial Intelligence 138(1-2), 181–234 (2002)
8. Brewka, G., Niemelä, I., Truszczynski, M.: Answer set optimization. In: Proc. IJCAI 2003, pp. 867–872. Morgan Kaufmann, San Francisco (2003)
9. Gebser, M., Gharib, M., Mercer, R., Schaub, T.: Monotonic answer set programming. Journal of Logic and Computation (in press, 2009), doi:10.1093/logcom/exn040

10. Gelfond, M., Lifschitz, V.: The stable model semantics for logic programming. In: Proc. ICLP/SLP 1988, pp. 1070–1080. MIT Press, Cambridge (1988)
11. Papadimitriou, C.: Computational Complexity. Addison-Wesley, Reading (1995)
12. Clark, K.: Negation as failure. In: Readings in nonmonotonic reasoning, pp. 311–325. Morgan Kaufmann, San Francisco (1987)
13. Fages, F.: Consistency of Clark's completion and existence of stable models. Journal of Methods of Logic in Computer Science 1, 51–60 (1994)
14. Lin, F., Zhao, Y.: ASSAT: Computing answer sets of a logic program by SAT solvers. Artificial Intelligence 157(1-2), 115–137 (2004)
15. Niemelä, I.: Logic programs with stable model semantics as a constraint programming paradigm. Annals of Mathematics and Artificial Intelligence 25(3-4), 241–273 (1999)

Belief Revision with Bounded Treewidth[*]

Reinhard Pichler, Stefan Rümmele, and Stefan Woltran

Vienna University of Technology, Austria

Abstract. Problems arising from the revision of propositional knowledge bases have been intensively studied for two decades. Many different approaches to revision have thus been suggested, with the ones by Dalal or Satoh being two of the most fundamental ones. As is well known, most computational tasks in this area are intractable. Therefore, in practical applications, one requires sufficient conditions under which revision problems become efficiently solvable. In this paper, we identify such tractable fragments for the reasoning and the enumeration problem exploiting the notion of treewidth. More specifically, we present new algorithms based on dynamic programming for these problems in Dalal's setting and a tractability proof using Courcelle's Theorem for Satoh's approach.

1 Introduction

Since knowledge is continually evolving, there is a constant need to be able to revise a knowledge base as new information is received. In this paper, we restrict ourselves to *propositional knowledge bases*, i.e., they are given by propositional formulae. Problems arising from the revision of knowledge bases (referred to as *belief revision*) have been intensively studied for two decades. Formally, the problem of belief revision is usually specified as follows: Given a knowledge base (i.e., a formula) α and a formula β, find a revised knowledge base $\alpha \circ \beta$, such that β is true in all models of $\alpha \circ \beta$ and the change compared to the models of α is minimal. The following problems are of great interest:

- *Reasoning.* Given formulae α, β, and γ, decide if $\alpha \circ \beta \models \gamma$ holds.
- *Enumeration.* Given formulae α and β, compute the models of $\alpha \circ \beta$.

Several realizations for $\circ$ have been proposed in the literature and desired properties are formulated by the famous AGM-Postulates [1], or in terms of propositional logic and finite knowledge bases, by Katsuno and Mendelzon [2]. Two of the most fundamental approaches are due to Dalal [3] and Satoh [4]. Complexity results for the reasoning problem w.r.t. different $\circ$ operators are provided in [5] including $\Theta_2 P$- and $\Pi_2 P$-completeness for Dalal's, respectively Satoh's, operator.

An interesting approach to dealing with intractable problems is parameterized complexity theory. In fact, hard problems can become tractable if some problem parameter is bounded by a fixed constant. Such problems are called *fixed-parameter tractable*. One important parameter is treewidth, which measures the "tree-likeness" of a graph or, more generally, of a structure. By using a seminal result due to Courcelle [6], several fixed-parameter tractability (FPT) results in the area of AI and KR have been recently proven [7]. The goal of this work is to obtain tractability results also for belief revision.

---

[*] This work was supported by the Austrian Science Fund (FWF), project P20704-N18.

E. Erdem, F. Lin, and T. Schaub (Eds.): LPNMR 2009, LNCS 5753, pp. 250–263, 2009.
© Springer-Verlag Berlin Heidelberg 2009

Courcelle's Theorem states that any property of finite structures, that is definable in monadic second-order logic (MSO), becomes tractable over structures with bounded treewidth. In this work, we show FPT for the reasoning problem with Satoh's operator $\circ_S$ by giving an MSO definition of this problem, i.e., defining the property "$\alpha \circ_S \beta \models \gamma$" of a structure representing formulae α, β, and γ. In order to prove an analogous result for Dalal's operator, we have to make use of an extension [8] of Courcelle's Theorem.

The proof of Courcelle's Theorem and its extension in [8] is "constructive": It works by transforming the MSO evaluation problem into a tree language recognition problem, which is then solved via a finite tree automaton (FTA). However, the "algorithms" resulting from such an MSO-to-FTA transformation are usually not practical due to excessively large constants. Consequently, Niedermeier states that MSO "is a very elegant and powerful tool for quickly deciding about FPT, but it is far from any efficient implementation" [9]. We therefore present a novel algorithm for the reasoning problem with Dalal's operator $\circ_D$. This algorithm is based on dynamic programming and builds upon an algorithm of [10] for the #SAT problem (i.e., the problem of counting all models of a given propositional formula). Moreover, we extend our reasoning algorithm to an algorithm for the enumeration problem of "$\circ_D$". As far as the complexity is concerned, our algorithms work in linear time for the reasoning problem and with linear delay (i.e., the time needed for computing the first model and for any further model of $\alpha \circ_D \beta$). Due to lack of space, we only present such dedicated algorithms for Dalal's approach (Satoh's approach can be handled by a similar dynamic programming algorithm, which will be included in the full version of this paper).

Results. Our main contributions are as follows.

- A novel algorithm based on dynamic programming for deciding $\alpha \circ_D \beta \models \gamma$ in *linear time* provided that the treewidth of the formulae α, β, and γ is bounded by a constant (for a formal definition of treewidth, see Section 2).
- An extension of the algorithm for the reasoning problem, such that also the set of all models of $\alpha \circ_D \beta$ is computed. In paticular, our algorithm works with linear delay if the formulae α and β have bounded treewidth.
- We show that the property "$\alpha \circ_S \beta \models \gamma$" is definable in MSO and that "$\alpha \circ_D \beta \models \gamma$" can be defined in the extension of MSO from [8]. In case of $\circ_S$ we thus establish FPT w.r.t. the treewidth of the reasoning problem. In case of $\circ_D$ we thus get an alternative proof of FPT, which follows of course also from our dedicated algorithm via dynamic programming.

2 Background

Throughout the paper, we assume a universe U of propositional atoms. A literal is either an atom a or a negated atom $\bar{a}$. For a set A of atoms, $\overline{A} = \{\bar{a} : a \in A\}$. Clauses are sets of literals. An interpretation (or assignment) I is a set of atoms and we define, for a clause c and $O \subseteq U$, $Mod_O(c) = \{I \subseteq O : (I \cup \overline{(O \setminus I)}) \cap c \neq \emptyset\}$. For a set C of clauses, $Mod_O(C) = \bigcap_{c \in C} Mod_O(c)$. For $O = U$, we write $Mod(C)$ instead of $Mod_O(C)$ for the set of classical models of C. By $At(C)$ we denote the set of atoms occurring in C. In what follows, we use the term *formula* to refer to a set of clauses. As usual, $\alpha \models \beta$ iff each model of formula α is also a model of formula β.

Revision Operators. The approaches of revision we deal with here rely on so-called model-based change operators. Such operators usually utilize a model distance $M \Delta M'$

which yields the set of atoms differently assigned in interpretations M and M', i.e. in our notation, $M \Delta M' = (M \setminus M') \cup (M' \setminus M)$. Assuming that α is consistent (we tacitly make this assumption throughout the paper), the operators due to Satoh [4] ("$\circ_S$"), and respectively, Dalal [3] ("$\circ_D$"), can be defined as follows:

$$Mod(\alpha \circ_S \beta) = \{J \in Mod(\beta) \,:\, \exists I \in Mod(\alpha) \text{ s.t. } I \Delta J \in \Delta^{min}(\alpha, \beta)\};$$
$$Mod(\alpha \circ_D \beta) = \{J \in Mod(\beta) \,:\, \exists I \in Mod(\alpha) \text{ s.t. } |I \Delta J| = |\Delta|^{min}(\alpha, \beta)\};$$

where we use $\Delta^{min}(\alpha, \beta) = min_\subseteq(\{I \Delta J \,:\, I \in Mod(\alpha), J \in Mod(\beta)\})$ with $min_\subseteq$ selecting elements which are minimal w.r.t. set inclusion, and $|\Delta|^{min}(\alpha, \beta) = min(\{|I \Delta J| \,:\, I \in Mod(\alpha), J \in Mod(\beta)\}$.

Subsequently, we refer to a *revision scenario* as either a pair of formulae α, β (in case of the enumeration problem) or a triple α, β, γ (in case of the reasoning problem). For revision scenarios, we assume unless stated otherwise, that the universe U is given by the set of atoms occurring in the involved formulae. In particular, we thus have that $Mod(\alpha \circ \beta)$ with $\circ \in \{\circ_S, \circ_D\}$ refers to a set of models over $At(\alpha \cup \beta)$ (resp. over $At(\alpha \cup \beta \cup \gamma)$ in the context of a reasoning problem).

As shown in [5], given formulae α, β, γ, deciding $\alpha \circ_S \beta \models \gamma$ is $\Pi_2 P$-complete while deciding $\alpha \circ_D \beta \models \gamma$ is $\Theta_2 P$-complete. For both results, hardness holds even in case γ is a single atom.

Tree Decomposition and Treewidth. A *tree decomposition* of a graph $G = (V, E)$ is a pair (T, χ), where T is a tree and χ maps each node n of T (we use $n \in T$ as a shorthand below) to a *bag* $\chi(n) \subseteq V$, such that (1) for each $v \in V$, there is an $n \in T$, s.t. $v \in \chi(n)$; (2) for each $(v, w) \in E$, there is an $n \in T$, s.t. $v, w \in \chi(n)$; (3) for each $n_1, n_2, n_3 \in T$, s.t. n_2 lies on the path from n_1 to n_3, $\chi(n_1) \cap \chi(n_3) \subseteq \chi(n_2)$ holds.

A tree decomposition (T, χ) is *normalized* (or *nice*) [11], if (1) each $n \in T$ has ≤ 2 children; (2) for each $n \in T$ with two children n_1, n_2, $\chi(n) = \chi(n_1) = \chi(n_2)$; and (3) for each $n \in T$ with one child n', $\chi(n)$ and $\chi(n')$ differ in exactly one element.

The *width* of a tree decomposition is defined as the cardinality of its largest bag $\chi(n)$ minus one. It is known that every tree decomposition can be normalized in linear time without increasing the width [11]. The *treewidth* of a graph G, denoted as $tw(G)$, is the minimum width over all tree decompositions of G. For arbitrary but fixed $w \geq 1$, it is feasible in linear time to decide if a graph has treewidth $\leq w$ and, if so, to compute a tree decomposition of width w, see [12].

Tree Decompositions for Revision Scenarios. To build tree decompositions for revision scenarios (α, β, γ), we use incidence graphs[1] over $\Gamma = \alpha \cup \beta \cup \gamma$. Thus, for formulae α, β, γ, such a graph G is given by vertices $\Gamma \cup At(\Gamma)$ and has as edges all pairs (a, c) with an atom a appearing in a clause c of Γ. In case of normalized tree decompositions, we distinguish between six types of nodes: atom introduction (AI), clause introduction (CI), atom removal (AR), clause removal (CR), branch (B), and leaf (L) nodes. The first four types will be often augmented with the element e (either an atom or clauses) which is removed or added compared to the bag of the child node.

Example 1. Figure 1 shows the revision scenario $A \circ B$, which is used as a running example throughout the paper. Since the models of these formulae are $Mod(A) = \{\{x\}\}$ and $Mod(B) = \{\{\}, \{x, y, z\}\}$, it is easy to verify, that $|\Delta|^{min}(A, B) = 1$ and

---

[1] See [10] for justifications why incidence graphs are favorable over other types of graphs.

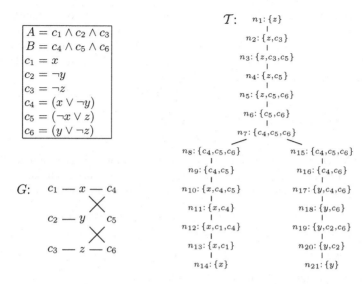

$$A = c_1 \wedge c_2 \wedge c_3$$
$$B = c_4 \wedge c_5 \wedge c_6$$
$$c_1 = x$$
$$c_2 = \neg y$$
$$c_3 = \neg z$$
$$c_4 = (x \vee \neg y)$$
$$c_5 = (\neg x \vee z)$$
$$c_6 = (y \vee \neg z)$$

Fig. 1. Revision scenario $A \circ B$; incidence graph G; and normalized tree decomposition $\mathcal{T}$ of G

$\Delta^{min}(A, B) = \{\{x\}, \{y, z\}\}$. Hence, $Mod(A \circ_D B) = \{\{\}\}$ and $Mod(A \circ_S B) = \{\{\}, \{x, y, z\}\}$. Figure 1 also shows the incidence graph G of this scenario, together with a normalized tree decomposition $\mathcal{T}$ of G having width 2. Actually, G cannot have a tree decomposition of width < 2, since only trees have treewidth $= 1$ and G contains cycles. Hence, the tree decomposition in Figure 1 is in fact optimal and we have $tw(G) = 2$. Examples for node types are n_{21} as (L) node, n_{20} as (c_2-CI) node, n_{16} as (y-AR) node, n_7 as (B) node, n_6 as (c_4-CR) node, and n_5 as (z-AI) node.   ⊣

3   Applying Courcelle's Theorem

An important tool for establishing FPT of a decision problem is Courcelle's Theorem [6], stating that any property of finite structures definable in monadic second-order logic (MSO) becomes tractable (in fact, even linear), if the treewidth of the structures is bounded by a constant. This result was extended to counting problems as well as extremum problems [8]. We recall that MSO extends first-order logic by the use of *set variables* (denoted by upper case letters), which range over sets of domain elements.

In order to show the FPT of the aforementioned belief revision problems using Courcelle's Theorem, we first have to define how the problem instances can be modeled as finite structures. Let formulae α, β, γ be given by a structure $\mathcal{A}$ with signature $\sigma = \{atom(\cdot), clause_\alpha(\cdot), clause_\beta(\cdot), clause_\gamma(\cdot), pos(\cdot, \cdot), neg(\cdot, \cdot)\}$. $\mathcal{A}$ has domain $A = \Gamma \cup At(\Gamma)$, where $\Gamma = \alpha \cup \beta \cup \gamma$. Moreover, for each relation symbol in σ, a relation over A is contained in $\mathcal{A}$ with the following intended meaning: $atom$ designates the set of atoms, $clause_\alpha$, $clause_\beta$ and $clause_\gamma$ denote the set of clauses of α, β and γ respectively. Furthermore $pos(a, c)$ denotes that atom a occurs positively in clause c. Negative literals are described by $neg(a, c)$. The treewidth of a structure $\mathcal{A}$ is defined as the treewidth of the graph that we get by taking the set of domain elements (in our case,

$A = \Gamma \cup At(\Gamma))$ as vertices and by considering two vertices (i.e., domain elements) as adjacent if these domain elements jointly occur in some tuple of the structure, i.e., the edges of this graph are of the form (a, c) where either $pos(a, c)$ or $neg(a, c)$ is contained in the structure. Hence, the treewidth of $\mathcal{A}$ is precisely the treewidth defined via the *incidence graph* of Γ as described in the previous section.

Models of a formula φ can then be stated by the MSO property (see also [7]):

$$mod_\varphi(I) \equiv \forall x[x \in I \rightarrow atom(x)] \wedge$$
$$\forall c[clause_\varphi(c) \rightarrow \exists a((pos(a, c) \wedge a \in I) \vee (neg(a, c) \wedge a \notin I))].$$

Towards an MSO-encoding for $\alpha \circ_S \beta$ we define three more helper formulae. The first one yields all models J of β together with the possible differences to models of α. The other two characterize valid triples $I \Delta J = K$, respectively proper subsets $X \subset Y$.

$$modD_{\alpha,\beta}(J, K) \equiv mod_\beta(J) \wedge \exists I[mod_\alpha(I) \wedge diff(I, J, K)],$$
$$diff(I, J, K) \equiv \forall a[a \in K \leftrightarrow ((a \in I \wedge a \notin J) \vee (a \notin I \wedge a \in J))],$$
$$sub(X, Y) \equiv \forall a(a \in X \rightarrow a \in Y) \wedge \exists b(b \in Y \wedge b \notin X).$$

We put things together to characterize the models of $\alpha \circ_S \beta$:

$$rev^S_{\alpha,\beta}(J) \equiv \exists K[modD_{\alpha,\beta}(J, K) \wedge \forall J' \forall K'(sub(K', K) \rightarrow \neg modD_{\alpha,\beta}(J', K'))].$$

It is now easy to see that the MSO formula $\forall J(rev^S_{\alpha,\beta}(J) \rightarrow Mod_\gamma(J))$ characterizes the reasoning problem $\alpha \circ_S \beta \models \gamma$ for Satoh's revision operator. We thus obtain via Courcelle's Theorem the following result.

Theorem 1. *The reasoning problem $\alpha \circ_S \beta \models \gamma$ is fixed-parameter linear w.r.t. the treewidth, i.e., it is solvable in time $\mathcal{O}(f(w) \cdot \|\alpha \cup \beta \cup \gamma\|)$, where f is a function depending only on the treewidth w of the revision scenario (α, β, γ).*

For Dalal's operator, we require additional machinery. Following the notation of [8], a model of $Mod(\alpha \circ_D \beta)$ can be described by a *linear extended monadic second-order extremum problem* $\min_{modD_{\alpha,\beta}(J,K)} |K|$, where $modD_{\alpha,\beta}(J, K)$ is the MSO-formula given above in the MSO-characterization of Satoh's revision operator. By using the extension of Courcelle's Theorem of [8], we thus get the following FPT result.

Theorem 2. *Assuming unit cost for arithmetic operations, the reasoning problem $\alpha \circ_D \beta \models \gamma$ is fixed-parameter linear w.r.t. the treewidth, i.e., it is solvable in time $\mathcal{O}(f(w) \cdot \|\alpha \cup \beta \cup \gamma\|)$, where f is a function depending only on the treewidth w of (α, β, γ).*

Proof. The key observation is that $\alpha \circ_D \beta \models \gamma$ holds if and only if $|\Delta|^{min}(\alpha, \beta \wedge \gamma) < |\Delta|^{min}(\alpha, \beta \wedge \neg\gamma)$. Moreover, both expressions $|\Delta|^{min}(\alpha, \beta \wedge \gamma)$ and $|\Delta|^{min}(\alpha, \beta \wedge \neg\gamma)$ can be characterized by linear extended MSO extremum problems. By Theorem 5.6 of [8], those can be evaluated in linear time if we assume unit cost for arithmetic operations and if the treewidth of $\alpha \cup \beta \cup \gamma$ is bounded by a fixed constant. $\square$

4   The Dynamic Programming Approach for $\circ_D$

In this section, we show how the theoretical results from Section 3 can be put to practice by dynamic programming. Due to the space restrictions we discuss here only the realization for the Dalal-revision operator $\circ_D$ in detail.

We start with an algorithm to decide $\alpha \circ_D \beta \models \gamma$. The very idea of such an algorithm is to associate certain objects (so-called bag assignments) to each node n of a tree decomposition for this problem, such that certain information about the subproblem represented by the subtree rooted at n remains available. Consequently, results for the entire problem can be read off the root of the tree decomposition.

We then make use of our algorithm also for the enumeration problem. Hereby, we traverse the tree decomposition a second time, but starting from the root, where we already have identified certain objects which will allow us to compute the models of the revised knowledge base. However, to guarantee that the enumeration does not provide duplicate models, some additional adjustments in the data structure will be necessary.

4.1   Reasoning Problem

For the problem $\alpha \circ_D \beta \models \gamma$, we restrict ourselves here to scenarios where γ is a single atom occurring in $At(\alpha \cup \beta)$ in order to keep the presentation simple. In what follows, we fix $\mathcal{T} = (T, \chi)$ to be a normalized tree decomposition of the incidence graph for $\alpha \cup \beta$. We refer to the root node of T as n_{root}, and we require that the bags of n_{root} and of all leaf nodes of T do not contain any clauses. Such a tree decomposition is easily obtained from a normalized one by suitably adding (CI)- and (CR)-nodes. Additionally, we require that γ appears in $\chi(n_{root})$. Finally, we assume $\alpha \cap \beta = \emptyset$ holds, thus for any clause $c \in \alpha \cup \beta$, its origin $o(c)$ is either α or β. Also recall that we fix $U = At(\alpha \cup \beta)$.

For a node $n \in T$, we denote by T_n the subtree of T rooted at n. For a set S of elements (either atoms or clauses), $n|_S$ is a shorthand for $\chi(n) \cap S$; moreover, $n\!\downarrow_S$ is defined as $\bigcup_{m \in T_n} m|_S$, and $n\!\Downarrow_S$ abbreviates $n\!\downarrow_S \setminus n|_S$.

Definition 1. *A tuple* $\vartheta = (n, M_\alpha, M_\beta, C)$, *where* $n \in T$, $M_\alpha, M_\beta \subseteq n|_U$, *and* $C \subseteq n|_{\alpha \cup \beta}$ *is called a* bag assignment *(for node n).*

Bag assignments for a node n implicitly talk about interpretations over $n\!\downarrow_U$. The following definition makes this more precise.

Definition 2. *For a bag assignment* $\vartheta = (n, M_\alpha, M_\beta, C)$ *and* $\varphi \in \{\alpha, \beta\}$, *define*

$$E_\varphi(\vartheta) = \big\{ K \subseteq n\!\downarrow_U : K \setminus (n\!\Downarrow_U) = M_\varphi;$$
$$(C \cap \varphi) \cup (n\!\Downarrow_\varphi) = \{ c \in n\!\downarrow_\varphi : K \in Mod_{n\downarrow_U}(c) \} \big\}.$$

In other words, we associate with a bag assignment $(n, M_\alpha, M_\beta, C)$ all interpretations K that extend M_α in such a way, that all clauses from α appearing in C and in bags below node n are satisfied. The same is done for β. Bag assignments for which such extended interpretations exist for both α and β are of particular interest for us.

Definition 3. *A bag assignments* ϑ *is called* bag model *iff* $E_\alpha(\vartheta) \neq \emptyset \neq E_\beta(\vartheta)$.

We next rephrase the main features of the definition of $\circ_D$ in terms of bag models and then show that bag models for the root node capture $\circ_D$ as expected.

Definition 4. *For any bag model* $\vartheta = (n, M_\alpha, M_\beta, C)$, *define*

$$\delta(\vartheta) = \min \big\{ |I_\alpha \Delta I_\beta| : I_\alpha \in E_\alpha(\vartheta), I_\beta \in E_\beta(\vartheta) \big\}; \quad and$$
$$\mathcal{E}(\vartheta) = \big\{ I_\beta \in E_\beta(\vartheta) : \exists I_\alpha \in E_\alpha(\vartheta), |I_\alpha \Delta I_\beta| = \delta(\vartheta) \big\}.$$

Theorem 3. *Let Θ be the set of all bag models ϑ for n_{root}, such that no bag model ϑ' for n_{root} with $\delta(\vartheta') < \delta(\vartheta)$ exists. Then, $Mod(\alpha \circ_D \beta) = \bigcup_{\vartheta \in \Theta} \mathcal{E}(\vartheta)$.*

Proof. ($\subseteq$): Let $J \in Mod(\alpha \circ_D \beta)$. Hence, $J \in Mod(\beta)$ and there exists an $I \in Mod(\alpha)$, such that $|I \Delta J| = |\Delta|^{min}(\alpha, \beta) = k$. Consider $\vartheta = (n_{root}, M_\alpha, M_\beta, \emptyset)$ where $M_\alpha = n_{root}|_I$ and $M_\beta = n_{root}|_J$. Since we assumed that no clauses are stored in $\chi(n_{root})$ and $n_{root} \downarrow_U = U$, $J \in Mod(\beta)$ yields that $J \in Mod_{n \downarrow_U}(c)$ holds for each $c \in n_{root} \downarrow_\beta = n_{root} \Downarrow_\beta$. The same argumentation applies to I and α. Hence, $I \in E_\alpha(\vartheta)$, $J \in E_\beta(\vartheta)$, and thus ϑ is a bag-model. To show $\vartheta \in \Theta$, it remains to show that no other $\vartheta' \in \Theta$ exists with $\delta(\vartheta') < \delta(\vartheta) \leq k$. Towards a contradiction, suppose such a $\vartheta' = (n_{root}, M'_\alpha, M'_\beta, C')$ exists. By definition, then there exists an $I'_\alpha \in E_\alpha(\vartheta')$ and an $I'_\beta \in E_\beta(\vartheta')$ with $|I'_\alpha \Delta I'_\beta| = \delta(\vartheta')$. Let $\varphi \in \{\alpha, \beta\}$. By definition of $E_\varphi(\cdot)$, we obtain $I'_\varphi \in Mod_{n_{root} \downarrow_U}((C' \cap \varphi) \cup (n_{root} \Downarrow_\varphi))$. Again $C' = \emptyset$ by our assumption for n_{root}, and thus $n_{root} \Downarrow_\varphi = \varphi$. We also know $U = n_{root} \downarrow_U$. $I'_\varphi \in Mod(\varphi)$ follows. Hence, we have found models I'_α, I'_β for α, and resp. β, such that $|I'_\alpha \Delta I'_\beta| < k$. A contradiction to our assumption that $|\Delta|^{min}(\alpha, \beta) = k$. The other direction holds by essentially the same arguments. ☐

We now put our concept of bag models to work also below the root node. Our goal is to characterize bag models ϑ without an explicit computation of $E_\varphi(\vartheta)$. To this end, first note that bag models for leaf nodes n are easily built from all pairs of interpretations over the atoms in the bag $\chi(n)$; also recall that we assumed that no clause is in $\chi(n)$. Thus, formally, the set of all bag models for a leaf node n is given by $\{(n, M, N, \emptyset) : M, N \subseteq n|_U\}$. For each such bag model $\vartheta = (n, M, N, \emptyset)$, $\delta(\vartheta) = |M \Delta N|$ is clear. Next, we define a relation $\prec_T$ between bag assignments, such that all bag models of a node are accordingly linked to bag models of the child(ren) node(s). We thus can propagate, starting from the leaves, bag models upwards the tree decomposition. Afterwards, we will show how $\delta(\vartheta)$ can be treated accordingly.

Definition 5. *For bag assignments $\vartheta = (n, M_\alpha, M_\beta, C)$ and $\vartheta' = (n', M'_\alpha, M'_\beta, C')$, we have $\vartheta' \prec_T \vartheta$ iff n has a single child n', and the following properties are satisfied, depending on the node type of n:*

1. *(c-CR): $M_\alpha = M'_\alpha$, $M_\beta = M'_\beta$, $C = C' \setminus \{c\}$, $c \in C'$;*
2. *(c-CI): $M_\alpha = M'_\alpha$, $M_\beta = M'_\beta$, and $C = C' \cup \{c\}$ if $M_{o(c)} \in Mod_{n|_U}(c)$; and $C = C'$ otherwise;*
3. *(a-AR): $M_\alpha = M'_\alpha \setminus \{a\}$, $M_\beta = M'_\beta \setminus \{a\}$, $C = C'$;*
4. *(a-AI): one of the following cases applies*
   - *$M_\alpha = M'_\alpha \cup \{a\}$, $N = M'_\beta \cup \{a\}$, $C = C' \cup \{c \in n|_{\alpha \cup \beta} : a \in c\}$;*
   - *$M_\alpha = M'_\alpha \cup \{a\}$, $M_\beta = M'_\beta$, $C = C' \cup \{c \in n|_\alpha : a \in c\} \cup \{d \in n|_\beta : \bar{a} \in d\}$;*
   - *$M_\alpha = M'_\alpha$, $M_\beta = M'_\beta \cup \{a\}$, $C = C' \cup \{c \in n|_\alpha : \bar{a} \in c\} \cup \{d \in n|_\beta : a \in d\}$;*
   - *$M_\alpha = M'_\alpha$, $M_\beta = M'_\beta$, $C = C' \cup \{c \in n|_{\alpha \cup \beta} : \bar{a} \in c\}$.*

For branch nodes, we extend (with slight abuse of notation) $\prec_T$ to a ternary relation.

Definition 6. *For bag assignments $\vartheta = (n, M_\alpha, M_\beta, C)$, $\vartheta' = (n', M'_\alpha, M'_\beta, C')$ and $\vartheta'' = (n'', M''_\alpha, M''_\beta, C'')$, we have $(\vartheta', \vartheta'') \prec_T \vartheta$ iff n has two children n', n'', $M_\alpha = M'_\alpha = M''_\alpha$, $M_\beta = M'_\beta = M''_\beta$, and $C = C' \cup C''$.*

Lemma 1. *Let $\vartheta, \vartheta', \vartheta''$ be bag assignments, such that $\vartheta' \prec_T \vartheta$ (resp. $(\vartheta', \vartheta'') \prec_T \vartheta$). Then, ϑ is a bag model iff ϑ' is a bag model (resp. both ϑ' and ϑ'' are bag models).*

Proof. For the proof, one has to distinguish between the node types. Here, we only show the case where ϑ is a bag assignment for a $(c\text{-CI})$ node n with child m. In this case, $\vartheta' \prec_T \vartheta$ holds exactly for assignments of the form $\vartheta = (n, M_\alpha, M_\beta, C)$ and $\vartheta' = (m, M_\alpha, M_\beta, C')$, where $C = C' \cup \{c\}$ if c appears in $\varphi \in \{\alpha, \beta\}$ and M_φ is a partial model of c (i.e., $M_\varphi \in Mod_{n|_U}(c)$); and $C = C'$ otherwise. Consider the case $c \in \alpha$ (the other case is symmetric). We show $E_\alpha(\vartheta) = E_\alpha(\vartheta')$ and $E_\beta(\vartheta) = E_\beta(\vartheta')$. The assertion then follows. There is only room to sketch a proof for $E_\alpha(\vartheta) = E_\alpha(\vartheta')$.

We have $V = n\!\downarrow_U = m\!\downarrow_U$, $W = n\!\Downarrow_U = m\!\Downarrow_U$. It is sufficient to show $(C \cap \alpha) \cup n\!\Downarrow_\alpha = \{d \in n|_\alpha : K \in Mod_V(d)\}$ iff $(C' \cap \alpha) \cup m\!\Downarrow_\alpha = \{d \in m|_\alpha : K \in Mod_V(d)\}$ for each $K \subseteq V$ s.t. $K \setminus W = M_\alpha$. Fix such a K and note that $n\!\Downarrow_\alpha = m\!\Downarrow_\alpha$. One can show $(C \cap \alpha) = \{d \in n|_\alpha : K \in Mod_V(d)\}$ iff $(C' \cap \alpha) = \{d \in m|_\alpha : K \in Mod_V(d)\}$ by observing that $K \in Mod_V(c)$ iff $M_\alpha \in Mod_{n|_U}(c)$. $\square$

Next, we define recursively a number assigned to bag models n and show that this number in fact matches the minimal distance $\delta(\vartheta)$ defined above.

Definition 7. *Let $\vartheta = (n, M_\alpha, M_\beta, C)$ be a bag model. We define*

$$
\rho(\vartheta) = \begin{cases}
|M_\alpha \Delta M_\beta| & \text{if } n \text{ is a leaf node} \\
\rho(\vartheta') & \text{if } n \text{ is type (CI) or (CR); and } \vartheta' \prec_T \vartheta \\
\min\{\rho(\vartheta') : \vartheta' \prec_T \vartheta\} & \text{if } n \text{ is type (AR)} \\
\min\{\rho(\vartheta') : \vartheta' \prec_T \vartheta\} & \text{if } n \text{ is type (a-AI) and } a \in M_\alpha \text{ iff } a \in M_\beta \\
\min\{\rho(\vartheta') : \vartheta' \prec_T \vartheta\} + 1 & \text{if } n \text{ is type (a-AI) and } a \notin M_\alpha \text{ iff } a \in M_\beta \\
\min\{\vartheta' \times \vartheta'' : (\vartheta', \vartheta'') \prec_T \vartheta\} & \text{if } n \text{ is type (B)}
\end{cases}
$$

where $\vartheta' \times \vartheta''$ stands for $\rho(\vartheta') + \rho(\vartheta'') - |M_\alpha \Delta M_\beta|$.

Lemma 2. *For any bag model ϑ, $\delta(\vartheta) = \rho(\vartheta)$.*

Example 2. In Fig. 2, we list all bag models ϑ of the tree decomposition from Example 1 together with values $\rho(\vartheta)$. For instance, leaf node n_{21} has bag models for all pairs of interpretations over $\{y\}$. If we go upwards the tree, we observe that bag models for n_{19} additionally contain clauses from $\{c_2, c_6\}$ satisfied by the respective assignments. In the next node n_{18} only those bag models survive where c_2 was contained, since n_{18} is a $(c_2\text{-CR})$ node. Due to space restrictions, we cannot discuss all steps in detail. Note that for the root, the bag model $\vartheta = (n_1, \emptyset, \emptyset, \emptyset)$ is the one with minimal $\rho(\vartheta)$. It can be checked that $\mathcal{E}(\vartheta) = \{\{\}\}$ as expected (recall that $Mod(\alpha \circ_D \beta) = \{\{\}\}$). $\dashv$

Theorem 4. *Assuming unit cost for arithmetic operations, $\alpha \circ_D \beta \models \gamma$ can be decided in time $O(f(w) \cdot \|\alpha \cup \beta\|)$, where f is a function depending only on the treewidth w of (α, β).*

Proof. Lemma 1 suggests the following algorithm: first, we establish the bag models ϑ for leaf nodes together with their value for $\delta(\vartheta) = \rho(\vartheta)$; then we compute all remaining bag models via $\prec_T$ in a bottom-up manner, and keep track of $\delta(\cdot)$ using the definition of ρ, which is indeed feasible thanks to Lemma 2. As soon as we have the

$(n_1, \emptyset, z, \emptyset): 2$	$(n_5, z, z, c_5c_6): 1$	$(n_8, \emptyset, \emptyset, c_4): 0$	$(n_{15}, \emptyset, \emptyset, c_6): 1$
$(n_1, \emptyset, \emptyset, \emptyset): 1$	$(n_5, z, z, c_5): 0$	$(n_8, \emptyset, \emptyset, c_5): 1$	$(n_{15}, \emptyset, \emptyset, c_4): 0$
$(n_2, z, z, \emptyset): 1$	$(n_5, z, \emptyset, c_6): 1$	$(n_9, \emptyset, \emptyset, c_4): 0$	$(n_{16}, \emptyset, \emptyset, c_6): 1$
$(n_2, z, \emptyset, \emptyset): 2$	$(n_5, z, \emptyset, c_5c_6): 2$	$(n_9, \emptyset, \emptyset, c_5): 1$	$(n_{16}, \emptyset, \emptyset, c_4): 0$
$(n_2, \emptyset, z, c_3): 2$	$(n_5, \emptyset, z, c_5c_6): 2$	$(n_{10}, x, x, c_4): 0$	$(n_{17}, \emptyset, y, c_6): 1$
$(n_2, \emptyset, \emptyset, c_3): 1$	$(n_5, \emptyset, z, c_5): 1$	$(n_{10}, x, \emptyset, c_5): 1$	$(n_{17}, \emptyset, \emptyset, c_4): 0$
$(n_3, z, z, c_5): 1$	$(n_5, \emptyset, \emptyset, c_6): 0$	$(n_{11}, x, x, c_4): 0$	$(n_{18}, \emptyset, y, c_6): 0$
$(n_3, z, \emptyset, \emptyset): 1$	$(n_5, \emptyset, \emptyset, c_5c_6): 1$	$(n_{11}, x, \emptyset, c_4): 1$	$(n_{18}, \emptyset, \emptyset, \emptyset): 0$
$(n_3, \emptyset, \emptyset, c_5): 2$	$(n_6, \emptyset, \emptyset, c_6): 1$	$(n_{12}, x, x, c_1c_4): 0$	$(n_{19}, y, y, c_6): 0$
$(n_3, \emptyset, z, c_5c_3): 2$	$(n_6, \emptyset, \emptyset, \emptyset): 0$	$(n_{12}, x, \emptyset, c_1): 1$	$(n_{19}, y, \emptyset, \emptyset): 0$
$(n_3, \emptyset, \emptyset, c_3): 0$	$(n_6, \emptyset, \emptyset, c_5): 1$	$(n_{12}, \emptyset, x, c_4): 1$	$(n_{19}, \emptyset, y, c_2c_6): 1$
$(n_3, \emptyset, \emptyset, c_5c_3): 1$	$(n_7, \emptyset, \emptyset, c_4c_6): 1$	$(n_{12}, \emptyset, \emptyset, \emptyset): 0$	$(n_{19}, \emptyset, \emptyset, c_2): 0$
$(n_4, z, z, c_5): 1$	$(n_7, \emptyset, \emptyset, c_5c_6): 2$	$(n_{13}, x, x, c_1): 0$	$(n_{20}, y, y, \emptyset): 0$
$(n_4, z, \emptyset, \emptyset): 1$	$(n_7, \emptyset, \emptyset, c_4): 0$	$(n_{13}, x, \emptyset, c_1): 1$	$(n_{20}, y, \emptyset, \emptyset): 0$
$(n_4, z, \emptyset, c_5): 2$	$(n_7, \emptyset, \emptyset, c_4c_5): 1$	$(n_{13}, \emptyset, x, \emptyset): 0$	$(n_{20}, \emptyset, y, c_2): 1$
$(n_4, \emptyset, z, c_5): 2$		$(n_{13}, \emptyset, \emptyset, \emptyset): 0$	$(n_{20}, \emptyset, \emptyset, c_2): 0$
$(n_4, \emptyset, \emptyset, \emptyset): 0$		$(n_{14}, x, x, \emptyset)\ 0$	$(n_{21}, y, y, \emptyset): 0$
$(n_4, \emptyset, \emptyset, c_5): 1$		$(n_{14}, x, \emptyset, \emptyset)\ 1$	$(n_{21}, y, \emptyset, \emptyset): 1$
		$(n_{14}, \emptyset, x, \emptyset)\ 1$	$(n_{21}, \emptyset, y, \emptyset): 1$
		$(n_{14}, \emptyset, \emptyset, \emptyset)\ 0$	$(n_{21}, \emptyset, \emptyset, \emptyset): 0$

Fig. 2. All bag models for the tree decomposition from Example 1

bag models for the root node together with their δ-values we know that bag models in Θ as defined in Theorem 3 characterize the models of $\alpha \circ_D \beta$. Due to our assumption that γ is just a single atom occurring in $\chi(n_{root})$, it remains to check whether for each $(n_{root}, M_\alpha, M_\beta, C) \in \Theta$, $\gamma \in M_\beta$ holds.

The effort needed for processing a leaf node as well as the transition from child to parent nodes only depends on the treewidth but not on $\|\alpha \cup \beta\|$. The size of $\mathcal{T}$ is linear bounded by the size of $\alpha \cup \beta$, thus the desired time bound for our algorithm follows.   □

4.2   Enumeration Problem

Our reasoning algorithm from Section 4.1 gathers the following information along the bottom-up traversal of $\mathcal{T}$: (1) all bag models $\vartheta = (n, M_\alpha, M_\beta, C)$ for all nodes n in $\mathcal{T}$, (2) the minimal distance $\delta(\vartheta)$ between the models $I_\alpha \in E_\alpha(\vartheta)$ and $I_\beta \in E_\beta(\vartheta)$, and (3) the relation $\prec_\mathcal{T}$ indicating which bag model(s) ϑ' at the child node n' (resp. at the two child nodes n' and n'') give rise to which bag model ϑ at a node n in $\mathcal{T}$. In principle, this is all the information needed to enumerate the models in $\alpha \circ_D \beta$ by starting with the bag models ϑ in Θ from Theorem 3 (i.e., the bag models ϑ at the root node n_{root}, s.t. no other bag model ϑ' at n_{root} with smaller value of $\delta(\cdot)$ exists) and determining $\mathcal{E}(\vartheta)$ for every $\vartheta \in \Theta$ by traversing $\mathcal{T}$ in top-down direction following the $\prec_\mathcal{T}$ relation in reversed direction. However, such an enumeration algorithm faces two problems:

(1) In Definition 7 (with $\rho(\vartheta) = \delta(\vartheta)$, by Lemma 2) we computed the minimum value attainable by $\rho(\vartheta)$ over all possible bag models ϑ' (resp. pairs $(\vartheta', \vartheta'')$) with $\vartheta' \prec_\mathcal{T} \vartheta$ (resp. $(\vartheta', \vartheta'') \prec_\mathcal{T} \vartheta$). Hence, when we now follow the $\prec_\mathcal{T}$ relation in the reversed direction, we have to make sure that, from any ϑ, we only continue with bag models ϑ' (resp. with pairs $(\vartheta', \vartheta'')$) that actually lead to the minimal value of $\rho(\vartheta)$.

(2) For distinct bag models ϑ, ϑ' for any node n, $\mathcal{E}(\vartheta) \cap \mathcal{E}(\vartheta') = \emptyset$ is *not* guaranteed. More precisely, suppose that two bag models $\vartheta = (n, M_\alpha, M_\beta, C)$ and $\vartheta' = (n, M'_\alpha, M'_\beta, C')$ fulfill the condition $M_\beta = M'_\beta$. Then it may well happen that some model I_β is contained both in $E_\beta(\vartheta)$ and $E_\beta(\vartheta')$, s.t. I_β has minimal distance $\delta(\vartheta)$ from some $I_\alpha \in E_\alpha(\vartheta)$ and also minimal distance $\delta(\vartheta')$ from some $I'_\alpha \in E_\alpha(\vartheta')$. However, for our enumeration algorithm we want to avoid the computation of duplicates since this would, in general, destroy the linear time upper bound on the delay.

The first problem is dealt with below by restricting the relation $\prec_\mathcal{T}$ to a subset $\ll_\mathcal{T}$ of $\prec_\mathcal{T}$. For the second problem, we shall extend the relation $\ll_\mathcal{T}$ on bag models to a relation on sets of bag models. We start with the definition $\ll_\mathcal{T}$ on bag models. Let us introduce some additional notation first: We identify the components of a bag assignment $\vartheta = (n, M_\alpha, M_\beta, C)$ as $\vartheta_{node} = n$; $\vartheta_\alpha = M_\alpha$; $\vartheta_\beta = M_\beta$; and $\vartheta_{clause} = C$. If Θ is a set of bag assignments, s.t. ϑ_β is identical for all $\vartheta \in \Theta$, then we write Θ_β to denote ϑ_β for any $\vartheta \in \Theta$. Finally, we write $\mathcal{E}(\Theta)$ as a short-hand for $\bigcup_{\vartheta \in \Theta} \mathcal{E}(\vartheta)$.

Definition 8. *Let ϑ, ϑ', and optionally, ϑ'' be bag models, s.t. ϑ'_{node} (and, optionally, also ϑ''_{node}) is a child of $n = \vartheta_{node}$. We define $\vartheta' \ll_\mathcal{T} \vartheta$ (resp. $(\vartheta', \vartheta'') \ll_\mathcal{T} \vartheta$) iff $\vartheta' \prec_\mathcal{T} \vartheta$ (resp. $(\vartheta', \vartheta'') \prec_\mathcal{T} \vartheta$) and one of the following conditions is fulfilled:*

(i) n is of type (CI) or (CR);
(ii) n is of type (AR) and $\delta(\vartheta) = \delta(\vartheta')$;
(iii) n is of type (a-AI), $a \in \vartheta_\alpha$ iff $a \in \vartheta_\beta$, and $\delta(\vartheta) = \delta(\vartheta')$;
(iv) n is of type (a-AI), $a \notin \vartheta_\alpha$ iff $a \in \vartheta_\beta$, and $\delta(\vartheta) = \delta(\vartheta') + 1$; or
(v) n is of type (B) and $\delta(\vartheta) = \delta(\vartheta') + \delta(\vartheta'') - |\vartheta_\alpha \Delta \vartheta_\beta|$.

We now extend $\ll_\mathcal{T}$ from a relation on bag models to a relation on sets Θ of bag models ϑ (with identical component ϑ_β). By slight abuse, we reuse the same symbol $\ll_\mathcal{T}$.

Definition 9. *Let $\Theta \neq \emptyset$ a set of bag models for $n \in \mathcal{T}$, s.t. for all $\vartheta, \vartheta' \in \Theta$, $\vartheta_\beta = \vartheta'_\beta$.*

(i) Suppose that n is either of type (CI), (CR) or of type (a-AI) with $a \notin \Theta_\beta$. Then we define $\Theta' \ll_\mathcal{T} \Theta$ for $\Theta' = \{\vartheta' : \vartheta'_\beta = \Theta_\beta \text{ and } \vartheta' \ll_\mathcal{T} \vartheta \text{ for some } \vartheta \in \Theta\}$.

(ii) Suppose that n is of type (a-AI) with $a \in \Theta_\beta$. Then we define $\Theta' \ll_\mathcal{T} \Theta$ for $\Theta' = \{\vartheta' : \vartheta'_\beta = \Theta_\beta \setminus \{a\} \text{ and } \vartheta' \ll_\mathcal{T} \vartheta \text{ for some } \vartheta \in \Theta\}$.

(iii) Suppose that n is of type (a-AR). Then we define $\Theta'_1 \ll_\mathcal{T} \Theta$ and $\Theta'_2 \ll_\mathcal{T} \Theta$ for $\Theta'_1 = \{\vartheta' : \vartheta'_\beta = \Theta_\beta \text{ and } \vartheta' \ll_\mathcal{T} \vartheta \text{ for some } \vartheta \in \Theta\}$ and $\Theta'_2 = \{\vartheta' : \vartheta'_\beta = \Theta_\beta \cup \{a\} \text{ and } \vartheta' \ll_\mathcal{T} \vartheta \text{ for some } \vartheta \in \Theta\}$.

(iv) Suppose that n is of type (B). Then we define $\Theta' \ll_\mathcal{T} \Theta$ for $\Theta' = \{\vartheta' : \exists \vartheta'' \text{ with } (\vartheta', \vartheta'') \ll_\mathcal{T} \vartheta \text{ for some } \vartheta \in \Theta\}$.

Moreover, for every $\hat{\Theta} \subseteq \Theta'$ with $\Theta' \ll_\mathcal{T} \Theta$, we define $(\hat{\Theta}, \Theta'') \ll_\mathcal{T} \Theta$, where $\Theta'' = \{\vartheta'' : \exists \vartheta \in \Theta \text{ and } \exists \vartheta' \in \hat{\Theta} \text{ with } (\vartheta', \vartheta'') \ll_\mathcal{T} \vartheta\}$.

The following lemma states that every model in $\mathcal{E}(\Theta)$ at node n can be computed via $\mathcal{E}(\Theta')$ (and optionally $\mathcal{E}(\Theta'')$) at the child node(s) of n and, conversely, that every element in $\mathcal{E}(\Theta')$ (and optionally $\mathcal{E}(\Theta'')$) can indeed be extended to a model of $\mathcal{E}(\Theta)$.

Lemma 3. *Let $n \in \mathcal{T}$ and Θ be a non-empty set of bag models for n, s.t. for all $\vartheta, \vartheta' \in \Theta$, $\vartheta_\beta = \vartheta'_\beta$. Then the following properties hold:*

(i) *Suppose n is of type (CI), (CR), (AR), or of type (a-AI), s.t. $a \notin \Theta_\beta$. Then $I \in \mathcal{E}(\Theta)$ iff $I \in \mathcal{E}(\Theta')$, s.t. $\Theta' \ll_T \Theta$.*

(ii) *Suppose n is of type (a-AI), s.t. $a \notin \Theta_\beta$. Then $I \in \mathcal{E}(\Theta)$ iff $(I \setminus \{a\}) \in \mathcal{E}(\Theta')$, s.t. $\Theta' \ll_T \Theta$.*

(iii) *Suppose n is of type (B). Then $I \in \mathcal{E}(\Theta)$ iff $I = I' \cup I''$ for some $I' \in \mathcal{E}(\hat{\Theta})$ and $I'' \in \mathcal{E}(\Theta'')$, where $\hat{\Theta} \subseteq \Theta'$, $\Theta' \ll_T \Theta$, and $(\hat{\Theta}, \Theta'') \ll_T \Theta$.*

For our enumeration algorithm, we start at the root node of $\mathcal{T}$ and first partition the relevant bag models ϑ according to ϑ_β. Formally, let Θ be the set of all bag models ϑ for n_{root}, such that no bag model ϑ' for n_{root} with $\delta(\vartheta') < \delta(\vartheta)$ exists. Then we partition Θ into $\Theta_1, \ldots, \Theta_n$, such that for each $\vartheta, \vartheta' \in \Theta_i$, $\vartheta_\beta = \vartheta'_\beta$, and for each $\vartheta \in \Theta_i$, $\vartheta' \in \Theta_j$ with $i \neq j$, $\vartheta_\beta \neq \vartheta'_\beta$. Clearly, the sets $\mathcal{E}(\Theta_i)$ are pairwise disjoint. Hence, no duplicates will be computed when we compute $\mathcal{E}(\Theta_1), \ldots, \mathcal{E}(\Theta_n)$ separately.

Given a set Θ of bag models, we compute $\mathcal{E}(\Theta)$ by implementing an appropriate *iterator* for every node n in $\mathcal{T}$. The iterator provides functions open, get_current, and get_next. In addition, other functions like close (to deallocate state information) are needed which we do not discuss here.

The open *function.* The open function serves to initialize the state information at each node of a given subtree of $\mathcal{T}$. For instance, it is convenient (in particular, for branch nodes) to store in a Boolean flag first whether get_next() has not yet been called since the initialization with the call of function open. Moreover, for an (AR) node, there can exist two sets Θ_1 and Θ_2 with $\Theta_i \ll_T \Theta$. We have to store in the state of the (AR) node, which one of these two sets is currently being processed at the child node.

The open function takes as input a set Θ of bag models ϑ with identical ϑ_β and recursively calls open(Θ') with $\Theta' \ll_T \Theta$. If the current node is of type (CI), (CR), or (AI), then Θ' is unique. Likewise, Θ' is unique for the first child of a branch node. In case of an (AR) node, Θ' corresponds to Θ_1 from Definition 9, case (iii), provided that it is non-empty. Otherwise, $\Theta' = \Theta_2$ is chosen. The children of a branch node are treated asymmetrically by the $\ll_T$-relation and, hence, also by the open function. In the first place, we only compute Θ' with $\Theta' \ll_T \Theta$ for the *first child* of every branch node. As we shall explain below, the function get_next() computes the set of assignments $\mathcal{E}(\Theta)$, returning one such assignment per call. For branch nodes, we thus compute for the first child the set $\mathcal{E}(\Theta')$ with $\Theta' \ll_T \Theta$. For each assignment I' thus returned, get_next() also yields $\Gamma' = \{\hat{\vartheta} : I' \in \mathcal{E}(\hat{\vartheta})\} \subseteq \Theta'$. Then the subtree rooted at the second child node is processed with Θ'', s.t. $(\Gamma', \Theta'') \ll_T \Theta$. Hence, for every assignment I', we have to compute Θ'' (which is uniquely determined by Γ' and Θ) and call open(Θ''), before we can retrieve the assignments I'' in $\mathcal{E}(\Theta'')$ with get_next().

The get_next *and* get_current *function.* Suppose that a node n in $\mathcal{T}$ has been initialized by a call of open(Θ) with $\vartheta_{node} = n$ for every $\vartheta \in \Theta$. Then we can call get_next() for this node in order to retrieve the first resp. the next assignment I in $\mathcal{E}(\Theta)$. In addition to the assignment I, the get_next function also provides a set $\Gamma \subseteq \Theta$ as output, s.t. $\Gamma = \{\hat{\vartheta} : I \in \mathcal{E}(\hat{\vartheta})\}$. As we have already seen, this set Γ of bag assignments is needed when we encounter a branch node on our way back to the root, in order to determine the set Θ'' for the second child. The get_current function is called (for the first child of a branch node) to retrieve once again the result from the previous call to get_next. If no next assignment exists, then get_next returns the value "Done".

Function get_next for a branch node n with child nodes n', n''
  Let Θ be the input parameter of the previous call of function open
  if first *then*
    first = False;
    $(I', \Gamma') = n'$.get_next();
    Let Θ'' s.t. $(\Gamma', \Theta'') \ll_{\mathcal{T}} \Theta$;
    n''.open(Θ'')
    $(I'', \Gamma'') = n''$.get_next()
  else
    $(I', \Gamma') = n'$.get_current()
    $(I'', \Gamma'') = n''$.get_next()
    if $(I'', \Gamma'') =$ undefined (i.e., the call of n''.get_next() returned "Done") *then*
      $(I', \Gamma') = n'$.get_next();
      if $(I', \Gamma') =$ undefined (i.e., the call of n'.get_next() returned "Done") *then*
        return "Done"
      endif
      Let Θ'' s.t. $(\Gamma', \Theta'') \ll_{\mathcal{T}} \Theta$;
      n''.open(Θ'')
      $(I'', \Gamma'') = n''$.get_next()
    endif
  endif
  return $(I' \cup I'', \{\vartheta \in \Theta \; : \; \exists \gamma' \in \Gamma'$ and $\exists \gamma'' \in \Gamma''$, s.t. $(\gamma', \gamma'') \ll_{\mathcal{T}} \vartheta\})$

Fig. 3. Function get_next() for a branch node

In order to compute the first resp. next assignment I, we traverse $\mathcal{T}$ downwards by recursive calls of get_next() until the leaves are reached. In the leaves, we start with the assignment $I = \Theta_\beta$ and also set $\Gamma = \Theta$. This assignment I and the set Γ are now updated on the way back to the root. The only modifications to I are in fact done when we are at an $(a\text{-}AI)$ node or at a (B) node. For (AI) nodes, we add a in case a is added to the respective Θ_β. For (B) nodes, we set $I = I' \cup I''$, where I' (resp. I'') is the assignment returned by the call of get_next() for the first (resp. second) child node. In Figure 3 we give the pseudo-code of the get_next function in case of a branch node. For the remaining node types, the implementation of get_next is even simpler.

Theorem 5. *Given formulae α and β, the models in $Mod(\alpha \circ_D \beta)$ can be computed with delay $O(f(w) \cdot |\alpha \cup \beta|)$, where f is a function depending only on the treewidth w of (α, β).*

Proof. By our definition of $\ll_{\mathcal{T}}$ on sets and by Lemma 3, we can be sure that (1) every assignment in $Mod(\alpha \circ_D \beta)$ is eventually computed by our iterator-based implementation via recursive calls of get_next and (2) no assignment is computed twice. Indeed, our set-based definition of the $\ll_{\mathcal{T}}$-relation groups together bag models ϑ with identical ϑ_β and, for any bag models ϑ' with $\vartheta_\beta \neq \vartheta'_\beta$, we trivially have $\mathcal{E}(\vartheta) \cap \mathcal{E}(\vartheta') = \emptyset$.

As far as the complexity is concerned, note that the recursive calls of the open function come down to a top-down traversal of $\mathcal{T}$. (In fact, by the asymmetric treatment of the children of a branch node, open is only called for the nodes along the left-most path in $\mathcal{T}$.) Similarly, each call of get_next and get_current leads to a single traversal of the subtree below the current node n. The work actually carried out inside each call is independent of the size of $\mathcal{T}$. Hence, in total, we end up with a time bound that is linear in the size of $\mathcal{T}$ and, hence, in the size of α and β.   □

5   Conclusion

The quest for (fixed-parameter) tractability has been pursued in many areas of KR and AI. However, in the context of belief revision, very few activities have been undertaken in this direction – apart from the work of Darwiche [13], which relies on compilation techniques. To the best of our knowledge, neither Courcelle's Theorem [6] (or one of its extensions as [8]) nor dynamic programming approaches (along the lines of tree decompositions) have been applied to belief revision problems, so far. For other KR formalisms though such approaches already proved to be successful (see, e.g., [7,14]).

In this work, we have identified new tractable classes of revision problems with respect to two of the most fundamental approaches [3,4]. Moreover, we provided novel dynamic programming algorithms for Dalal's revision operator [3] (i.e. for the problem of deciding $\alpha \circ_D \beta \models \gamma$, and enumerating the models of $\alpha \circ_D \beta$) which run in linear time (resp. with linear delay) if the treewidth of the revision scenario is bounded.

Future work includes to apply the methods used in this paper also to update operators. The approach due to Winslett [15], for instance, can be shown tractable by "plain" MSO, while the approach due to Forbus [16] is less accessible to such techniques since the concept of cardinality is "hidden" in the characterization (see, e.g. [17] for further discussions on this problem). Another direction of research is to apply our methods to approaches for iterated belief revision. This however calls for the additional requirement that the outcome of a single revision step has to be of bounded treewidth as well. It is an interesting research question of its own how to ensure such a property.

References

1. Alchourrón, C., Gärdenfors, P., Makinson, D.: On the logic of theory change: Partial meet functions for contraction and revision. Journal of Symbolic Logic 50, 510–530 (1985)
2. Katsuno, H., Mendelzon, A.: Propositional knowledge base revision and minimal change. Artificial Intelligence 52, 263–294 (1991)
3. Dalal, M.: Investigations into a theory of knowledge base revision. In: Proc. AAAI 1988, pp. 449–479. AAAI Press / The MIT Press (1988)
4. Satoh, K.: Nonmonotonic reasoning by minimal belief revision. In: Proceedings of the International Conference on Fifth Generation Computer Systems, pp. 455–462 (1988)
5. Eiter, T., Gottlob, G.: On the complexity of propositional knowledge base revision, updates, and counterfactuals. Artificial Intelligence 57, 227–270 (1992)
6. Courcelle, B.: Graph rewriting: An algebraic and logic approach. In: Handbook of Theoretical Computer Science, vol. B, pp. 193–242. Elsevier Science Publishers, Amsterdam (1990)
7. Gottlob, G., Pichler, R., Wei, F.: Bounded treewidth as a key to tractability of knowledge representation and reasoning. In: Proc. AAAI 2006, pp. 250–256. AAAI Press, Menlo Park (2006)
8. Arnborg, S., Lagergren, J., Seese, D.: Easy problems for tree-decomposable graphs. Journal of Algorithms 12, 308–340 (1991)
9. Niedermeier, R.: Invitation to Fixed-Parameter Algorithms. Oxford University Press, Oxford (2006)

10. Samer, M., Szeider, S.: Algorithms for propositional model counting. In: Dershowitz, N., Voronkov, A. (eds.) LPAR 2007. LNCS (LNAI), vol. 4790, pp. 484–498. Springer, Heidelberg (2007)
11. Kloks, T.: Treewidth. Computations and Approximations. Springer, Heidelberg (1994)
12. Bodlaender, H.L.: A linear-time algorithm for finding tree-decompositions of small treewidth. SIAM J. Comput. 25, 1305–1317 (1996)
13. Darwiche, A.: On the tractable counting of theory models and its application to truth maintenance and belief revision. Journal of Applied Non-Classical Logics 11, 11–34 (2001)
14. Jakl, M., Pichler, R., Woltran, S.: Answer-set programming with bounded treewidth. To appear in Proc. IJCAI (2009)
15. Winslett, M.: Reasoning about action using a possible models approach. In: Proc. (AAAI 1988), pp. 89–93. AAAI Press / The MIT Press (1988)
16. Forbus, K.: Introducing actions into qualitative simulation. In: Proc. IJCAI, pp. 1273–1278 (1989)
17. Szeider, S.: Monadic second order logic on graphs with local cardinality constraints. In: Ochmański, E., Tyszkiewicz, J. (eds.) MFCS 2008. LNCS, vol. 5162, pp. 601–612. Springer, Heidelberg (2008)

Casting Away Disjunction and Negation under a Generalisation of Strong Equivalence with Projection*

Jörg Pührer and Hans Tompits

Institut für Informationssysteme 184/3, Technische Universität Wien,
Favoritenstraße 9–11, A–1040 Vienna, Austria
{puehrer,tompits}@kr.tuwien.ac.at

Abstract. In answer-set programming (ASP), many notions of program equivalence have been introduced and formally analysed. A particular line of research in this direction aims at studying conditions under which certain syntactic constructs can be eliminated from programs preserving some given equivalence relation. In this paper, we continue this endeavour introducing novel conditions under which disjunction and negation can be eliminated from answer-set programs under *relativised strong equivalence with projection*. This notion is a generalisation of the usual strong-equivalence relation, as introduced by Lifschitz, Pearce, and Valverde, by allowing parametrisable context and output alphabets, which is an important feature in view of practical programming techniques like the use of local variables and modules. We provide model-theoretic conditions that hold for a disjunctive logic program P precisely when there is a program Q, equivalent to P under our considered notion, such that Q is either *positive*, *normal*, or *Horn*, respectively. Moreover, we outline how such a Q, called a *casting of P*, can be obtained, and consider complexity issues.

1 Introduction

An important area of research in answer-set programming (ASP) is devoted to the study of different notions of program equivalence. This particular field emerged with the seminal paper on *strong equivalence* by Lifschitz, Pearce, and Valverde [1]. In contrast to *ordinary equivalence*, which holds whenever two programs have the same answer sets, strong equivalence holds whenever two programs are ordinarily equivalent in *every context*. More formally, two programs P and Q are strongly equivalent iff, for every context program R, $P \cup R$ and $Q \cup R$ are ordinarily equivalent.

Strong equivalence circumvents a particular weakness of ordinary equivalence, viz. that the latter fails to yield a replacement property similar to the one of classical logic. That is to say, under ordinary equivalence, given a program P, replacing some subprogram $Q \subseteq P$ by an equivalent program R may yield an overall program $(P \setminus Q) \cup R$ which is not equivalent to P. Clearly, this is undesirable as far as modular programming and program optimisation is concerned. Strong equivalence does allow subprogram replacements, basically by definition, yet it is too restrictive for some purposes. In particular, strong equivalence does not take standard programming techniques like the use

---

* This work was partially supported by the Austrian Science Fund (FWF) under projects P18019 and P21698.

E. Erdem, F. Lin, and T. Schaub (Eds.): LPNMR 2009, LNCS 5753, pp. 264–276, 2009.

of local variables into account, which are ignored in the final output. Thus, it does not admit the *projection* of answer sets to a set of designated output letters.

A generalisation of strong equivalence taking this aspect into account is *relativised strong equivalence with projection*, defined as a special instance of a general framework for defining parameterised program-correspondence notions in ASP [2]. Relativised strong equivalence with projection extends the usual strong-equivalence relation via two parameters: one parameter specifies the alphabet of the context set and the other the alphabet of the output atoms. Thus, it is possible to specify an *input alphabet* and an *output alphabet*, allowing to view programs as black boxes computing some task with respect to a defined input/output behaviour. We note that if no projection is performed (i.e., if the output alphabet coincides with the overall program alphabet), then we arrive at the notion of *relativised strong equivalence*, first studied by Woltran [3].

In this paper, we are interested in the question whether a given disjunctive logic program P can be replaced by a program Q that is from a syntactically simpler class than P preserving relativised strong equivalence with projection (we refer to Q as a *casting* of P). In particular, we are interested in the questions whether a given program can be casted (i) to a program without disjunctions, (ii) to a program without negations, and (iii) to a program without both disjunctions and negations. In other words, we consider the question whether a program can be replaced (preserving relativised strong equivalence with projection) by a *normal*, *positive*, or *Horn program*, respectively. Our results follow a line of research dealing with analogous questions studied previously for ordinary, strong, and uniform equivalence [4,5] as well as for hyperequivalence [6], but they are the first in this direction to take the issue of projection into account.

The main results of our paper are the following. First of all, we introduce model-theoretic conditions which are necessary and sufficient for having positive answers of our casting questions. For each casting question, we actually provide two different conditions: one in terms of *minimal certificates* [2], and one in terms of *relativised SE-models* [3]. These concepts are the model-theoretic structures underlying relativised strong equivalence with and without projection, respectively, i.e., two programs are equivalent in one of these senses iff they have the same associated structures. Interestingly, our characterisations show that (i) in case the elimination of disjunction is possible, there is always a casting of the given program P just over the atoms $atm(P)$ of P, (ii) in case the elimination of both disjunction and negation is possible, a casting just over $atm(P)$ intersected with the input and output alphabet exists, while (iii) in case the elimination of negation is possible, a casting may introduce new atoms. Secondly, we provide upper complexity bounds for checking our casting questions. It turns out that the complexity of these tasks is not higher than the complexity of checking relativised strong equivalence with projection, which lies on the fourth level of the polynomial hierarchy. Thirdly, we outline how a casting can be obtained, in case a casting exists.

2   Preliminaries

Syntax and Semantics of Answer-Set Programs. We deal with finite propositional disjunctive logic programs containing rules (over a set At of atoms) of form

$$a_1 \vee \cdots \vee a_l \leftarrow b_1, \ldots, b_m, not\, b_{m+1}, \ldots, not\, b_n,$$

where $l \geq 0$, $n \geq m \geq 0$, all a_i, b_j are from At, and *not* denotes default negation. A rule r as above is *normal*, if $l \leq 1$; *positive*, if $m = n$; a *constraint*, if $l = 0$ and $m + n > 0$; and *Horn* if it is positive and normal. We define the *head* of r as $H(r) = \{a_1, \ldots, a_l\}$ and the *body* of r as $B(r) = \{b_1, \ldots, b_m, not\ b_{m+1}, \ldots, not\ b_n\}$. Furthermore, we also define $B^+(r) = \{b_1, \ldots, b_m\}$ and $B^-(r) = \{b_{m+1}, \ldots, b_n\}$. A *disjunctive logic program (over At)*, or simply a *program*, is a finite set of rules (over At). A program P is normal (resp., positive, Horn) if every rule in P is normal (resp., positive, Horn). We use N, P, and H to refer to the classes of normal, positive, and Horn programs, respectively. Finally, $atm(P)$ denotes the set of all atoms occurring in P.

Let I be an *interpretation*, i.e., a set of atoms. I *satisfies* a rule r, symbolically $I \models r$, iff $I \cap H(r) \neq \emptyset$ whenever $B^+(r) \subseteq I$ and $I \cap B^-(r) = \emptyset$ jointly hold. I is a *model* of a program P, symbolically $I \models P$, iff $I \models r$, for all $r \in P$. I is an *answer set* [7] of P iff I is a minimal model of P^I, where

$$P^I = \{H(r) \leftarrow B^+(r) \mid r \in P, B^-(r) \cap I = \emptyset\}$$

is the *reduct of P relative to I*. The set of all answer sets of P is denoted by $\mathcal{AS}(P)$.

Strong Equivalence and its Generalisations. Next, we introduce basic equivalence notions relevant for our purposes and provide model-theoretic characterisations for them. All of the equivalence notions discussed are special instances of a general framework for specifying parameterised program correspondence relations between answer-set programs [2].

We begin with *strong equivalence*, originally introduced by Lifschitz, Pearce, and Valverde [1]: Two programs, P and Q, are strongly equivalent, symbolically $P \equiv_s Q$, iff $\mathcal{AS}(P \cup R) = \mathcal{AS}(Q \cup R)$, for any program R (which is also referred to as a *context program*). Interestingly, strong equivalence corresponds to equivalence in the logic of here-and-there [8], HT, which, from a semantical point of view, is intuitionistic logic restricted to two worlds, "here" and "there". More specifically, two programs, viewed as logical theories, are strongly equivalent iff they have the same models in HT [1]. Emerging from this observation, Turner [9] characterised strong equivalence in terms of *SE-models* which directly correspond to models in HT: By an *SE-interpretation* we understand a pair (X, Y) of interpretations $X, Y \subseteq At$ such that $X \subseteq Y$. If $X = Y$, then (X, Y) is *total*, otherwise (X, Y) is *non-total*. In view of HT, the first component of an SE-interpretation is identified with the world "here", whilst the second component refers to the world "there". An SE-interpretation (X, Y) is an *SE-model* of a program P over At if $Y \models P$ and $X \models P^Y$. The set of all SE-models of P is denoted by $SE(P)$. It then holds that two programs P and Q are strongly equivalent iff $SE(P) = SE(Q)$. By limiting the context programs to be defined over particular alphabets only, we arrive at the notion of *relativised strong equivalence* [3]. Formally, given an alphabet $A \subseteq At$, programs P and Q over At are strongly equivalent *relative to A*, symbolically $P \equiv_s^A Q$, iff $\mathcal{AS}(P \cup R) = \mathcal{AS}(Q \cup R)$, for any program R over A. Model-theoretically, strong equivalence relative to A is captured in terms of *A-SE-models*: an SE-interpretation (X, Y) is an *A-SE-model* of a program P over At if (i) $Y \models P$, (ii) for all $Y' \subset Y$ with $Y' \cap A = Y \cap A$, it holds that $Y' \not\models P^Y$, and (iii) if $X \subset Y$, there is an $X' \subset Y$ with $X' \cap A = X$ such that $X' \models P^Y$. The set of all A-SE-models of P is denoted by $SE^A(P)$. For programs P and Q, it holds that $P \equiv_s^A Q$ iff $SE^A(P) = SE^A(Q)$.

A set $\mathcal{S}$ of SE-interpretations is *A-complete* if, for all $(X, Y) \in \mathcal{S}$ with $X \subset Y$, it holds that $(Y, Y) \in \mathcal{S}$ and $X \subset (Y \cap A)$, and, for all $(X, Y), (Z, Z) \in \mathcal{S}$ with $Y \subset Z$, it holds that $(X \cap A, Z) \in \mathcal{S}$. Note that $SE^A(P)$ is A-complete for every program P and, conversely, for every A-complete set $\mathcal{S}$ of SE-interpretations, there exists a program Q with $SE^A(Q) = \mathcal{S}$.

A further relaxation of strong equivalence is *relativised strong equivalence with projection* [2]: Given sets $A, O \subseteq At$, two programs P and Q over At are *strongly equivalent relative to A under projection to O*, or $\langle A, O \rangle$-*equivalent*, symbolically $P \equiv_{|_O}^A Q$, iff, for any program R over A, it holds that $\{I \cap O \mid I \in \mathcal{AS}(P \cup R)\} = \{I \cap O \mid I \in \mathcal{AS}(Q \cup R)\}$. For better readability, let us write $I|_O$ for $I \cap O$, for an interpretation I and an alphabet O, and define $\mathcal{S}|_O = \{I|_O \mid I \in \mathcal{S}\}$ for a set of interpretations $\mathcal{S}$. Then, we have that $P \equiv_{|_O}^A Q$ iff $\mathcal{AS}(P \cup R)|_O = \mathcal{AS}(Q \cup R)|_O$, for any program R over A.

Clearly, relativised strong equivalence with projection includes strong equivalence and relativised strong equivalence as special cases. Indeed, for all programs P, Q over At, $P \equiv_s Q$ iff $P \equiv_{|_{At}}^{At} Q$, and $P \equiv_s^A Q$ iff $P \equiv_{|_{At}}^A Q$. Certainly, we also have $P \equiv_s Q$ iff $P \equiv_s^{At} Q$.

In the spirit of the model-theoretic characterisations above, $\langle A, O \rangle$-equivalence can be characterised as follows [2]: Let $A, O \subseteq At$ be sets of atoms. A *certificate structure* is a pair (Ξ, Y), where Ξ is a set of interpretations and Y is an interpretation. For a program P over At, a certificate structure (Ξ, Y) is an $\langle A, O \rangle$-*certificate* of P if there is some $(Y', Y') \in SE^A(P)$ with $Y = Y'|_{A \cup O}$ and $\Xi = \{X \mid (X, Y') \in SE^A(P), X \subset Y'\}$. An $\langle A, O \rangle$-certificate (Ξ, Y) of P is *minimal* if, for any $\langle A, O \rangle$-certificate (Ξ', Y) of P, $\Xi' \subseteq \Xi$ implies $\Xi' = \Xi$. By $\mathcal{C}_{A,O}(P)$ we denote the set of all $\langle A, O \rangle$-certificates of P, and $\mathcal{C}_{A,O}^m(P)$ stands for the set of all minimal $\langle A, O \rangle$-certificates of P. Then, for two programs P and Q, $P \equiv_{|_O}^A Q$ holds iff $\mathcal{C}_{A,O}^m(P) = \mathcal{C}_{A,O}^m(Q)$.

For our later purposes, we need to characterise minimal $\langle A, O \rangle$-certificates directly in terms of A-SE-models. To this end, we introduce the following notion:

Definition 1. *Let $A, O \subseteq At$ be sets of atoms, $\mathcal{S}$ a set of SE-interpretations over At, and $(X, Y) \in \mathcal{S}$. Then, (X, Y) is $\langle A, O \rangle$-optimal in $\mathcal{S}$ if there is no $(Y', Y') \in \mathcal{S}$ with $Y|_{A \cup O} = Y'|_{A \cup O}$ and $\{U \mid (U, Y') \in \mathcal{S}, U \subset Y'\} \subset \{U \mid (U, Y) \in \mathcal{S}, U \subset Y\}$.*

We then have the following property:

Theorem 1. *Let $A, O \subseteq At$ be sets of atoms and P a program over At. Then, $(\Xi, Y) \in \mathcal{C}_{A,O}^m(P)$ iff there is some (Y', Y') that is $\langle A, O \rangle$-optimal in $SE^A(P)$ with $Y'|_{A \cup O} = Y$ and $\Xi = \{X \mid (X, Y') \in SE^A(P), X \subset Y'\}$.*

3   Main Results

In this section, we present necessary and sufficient conditions such that for a given disjunctive logic program P over At and given sets $A, O \subseteq At$, there is a program Q which is $\langle A, O \rangle$-equivalent to P, where Q is either normal, positive, or Horn. We call such a Q, if it exists, an $\langle A, O \rangle$-$\mathcal{C}$-*casting of* P, for $\mathcal{C} \in \{N, P, H\}$, referring to the normal, positive, or Horn case, respectively, or simply a *casting of* P if no ambiguity arises.

$r_1 : b \leftarrow p, not\ c$	$r_5 : p \vee j \leftarrow b, not\ d$	$r_9 : v \leftarrow j, not\ c, not\ d$
$r_2 : b \leftarrow p, not\ d$	$r_6 : \quad\quad p \leftarrow j, not\ c, not\ d$	$r_{10} : c \leftarrow j, not\ d, not\ v$
$r_3 : b \leftarrow j, not\ c$	$r_7 : \quad\quad p \leftarrow v$	$r_{11} : d \leftarrow j, not\ c, not\ v$
$r_4 : b \leftarrow j, not\ d$	$r_8 : \quad\quad \leftarrow p, d$	

Fig. 1. Program P_{ex}

In each case, we present two kinds of model-theoretic conditions—one based on certificate structures and one on SE-interpretations. As well, we approach our conditions by first addressing the case of relativised strong equivalence *without projection*, which can be directly obtained from previous results about casting under hyperequivalence [6].

We start with presenting a typical scenario for our casting questions, serving as a running example for our subsequent elaborations.

Example 1. Consider a party-attendance problem for determining who will attend a party based on given preferences and constraints of potential party guests. Assume the following circumstances: Our friend Betty only attends if Peter or Mary-Jane does and if there is no need to dance or no cheesy music playing. If Betty comes, in case there is no dancing, she will bring Peter or Mary-Jane along. As Peter needs to talk to Mary-Jane, he will definitely attend if Mary-Jane comes and there is neither dancing nor cheesy music playing. He also comes if there is only vegetarian food, but as he hates dancing he is not coming if dancing is required. Mary-Jane is a party-tiger and a die-hard vegetarian, well-known to force people to either dance, listen to cheesy music, or eat only vegetarian meals.

Program P_{ex} in Fig. 1 is an encoding of this information, where atoms b, p, and j represent the attendance of Betty, Peter, or Mary-Jane, respectively, and d, v, and c indicate dancing, vegetarian food, or cheesy music at the party. Note that P_{ex} is understood as representing partial information only, since we expect further preferences and constraints from Betty, Peter, or Mary-Jane concerning their attendance. Thus, P_{ex} will later be joined with further rules containing atoms from $A = \{b, p, j\}$.

As discussed below, no program Q exists not involving disjunction or negation that is strongly equivalent to P_{ex}, even when context programs are built from atoms in A only. However, assume we are only interested in who attends the party but we do not care about which party activities take place. Then, we do not mind if the answer sets of a casting disagree with those of P_{ex} on atoms in $\{d, v, c\}$. Hence, we are interested whether there is a casting Q of P_{ex} under relativised strong equivalence under projection. Our model-theoretic properties presented later on allow to answer this question.

3.1 Normal Logic Programs

For getting an intuition of the mechanisms that underlie our characterisations, it is helpful to deliberate how the syntactic class of a program influences its SE-models. Consider a normal program P. Then, for any interpretation Y, the reduct P^Y is a Horn program. Since the models of a Horn program are closed under intersection, it follows that for all SE-models $(X_1, Y), (X_2, Y)$ of P, $(X_1 \cap X_2, Y)$ must also be an SE-model of P, as both $X_1 \models P^Y$ and $X_2 \models P^Y$ holds. Following Eiter et al. [5], let us call a collection $\mathcal{S}$

of SE-interpretations *closed under here-intersection* if $(X_1, Y) \in \mathcal{S}$ and $(X_2, Y) \in \mathcal{S}$ implies $(X_1 \cap X_2, Y) \in \mathcal{S}$. So, our above argument shows that the set of all SE-models of a normal program is closed under here-intersection. Consequently, if, for a DLP P, there is some strongly equivalent normal program Q, then $SE(P)$ must be closed under here-intersection. However, as shown by Eiter et al. [5], closure under here-intersection is also a *sufficient* condition to guarantee the existence of a normal program Q being strongly equivalent to a given DLP P. Interestingly, this characterisation holds for strong equivalence relative to A as well, but using A-SE-models instead of SE-models.

Proposition 1 ([6]). *Given a set $A \subseteq At$ of atoms and a program P over At, a normal program Q over At exists with $P \equiv_s^A Q$ iff $SE^A(P)$ is closed under here-intersection.*

Example 2. Consider the program P_{ex} from Example 1 and $A = \{b, p, j\}$. Then,[1]

$$SE^A(P_{ex}) = \{(\emptyset, \emptyset), (\emptyset, bp), (bp, bp), (\emptyset, bjc), (bjc, bjc), (\emptyset, bjd), (b, bjd),$$
$$(bjd, bjd), (\emptyset, bpjv), (bp, bpjv), (bpjv, bpjv), (\emptyset, bpjc),$$
$$(bp, bpjc), (bj, bpjc), (bpjc, bpjc)\}.$$

As $(bp, bpjc), (bj, bpjc) \in SE^A(P_{ex})$ but $(b, bpjc) \notin SE^A(P_{ex})$, $SE^A(P_{ex})$ is not closed under here-intersection. Hence, there is no normal program Q with $P_{ex} \equiv_s^A Q$.

Turning to the case of relativised strong equivalence with projection, we now define our key properties for casting to normal programs—first for certificate structures and then for SE-interpretations.

Definition 2. *Let $A, O \subseteq At$ be sets of atoms. Then, a set $\mathscr{S}$ of certificate structures is $\langle A, O \rangle_{N,c}$-compliant if, for every $(\Xi, Y) \in \mathscr{S}$, Ξ is closed under intersection.*

Definition 3. *Let $A, O \subseteq At$ be sets of atoms. Then, a set S of SE-interpretations is $\langle A, O \rangle_{N,s}$-compliant if the set of $\langle A, O \rangle$-optimal SE-interpretations in S is closed under here-intersection.*

Consider a normal program P. Since, in view of Proposition 1, for every (X_1, Y), $(X_2, Y) \in SE^A(P)$, we have that $(X_1 \cap X_2, Y) \in SE^A(P)$, and, by Definition 1, either all or none of (X_1, Y), (X_2, Y), and $(X_1 \cap X_2, Y)$ are $\langle A, O \rangle$-optimal in $SE^A(P)$, it follows that $SE^A(P)$ is $\langle A, O \rangle_{N,s}$-compliant. Hence, if a DLP P is $\langle A, O \rangle$-equivalent to a normal program Q, $SE^A(P)$ must be $\langle A, O \rangle_{N,s}$-compliant. As the next result shows, the converse also holds, as well as similar relations for $\langle A, O \rangle_{N,c}$-compliance.

Theorem 2. *Let $A, O \subseteq At$ be sets of atoms and P a program over At. Then, the following statements are equivalent:*

1. *$\mathscr{C}_{A,O}^m(P)$ is $\langle A, O \rangle_{N,c}$-compliant;*
2. *$SE^A(P)$ is $\langle A, O \rangle_{N,s}$-compliant;*
3. *a normal program Q over $At \cup At'$ exists for some universe At' with $P \equiv_{|O}^A Q$;*
4. *a normal program Q over $atm(P)$ exists such that $P \equiv_{|O}^A Q$.*

---

[1] For brevity, in what follows, we omit braces and commas for interpretations, i.e., we write, e.g., bp instead of $\{b, p\}$.

Note that if a casting exists, then there is a casting Q that is built just of atoms from the input program P. Indeed, it is sufficient to remove all A-SE-models from $SE^A(P)$ that are not $\langle A, O \rangle$-optimal in order to construct the set S of A-SE-models of Q. We will see later on that for other casting questions a solution can be found only at the expense of introducing new atoms into the casting.

Example 3. The minimal $\langle A, O \rangle$-certificates of our running example P_{ex} for $A = O = \{b, p, j\}$ are given by $\mathscr{C}^m_{A,O}(P_{ex}) = \{(\emptyset, \emptyset), (\{\emptyset\}, bp), (\{\emptyset\}, bj), (\{\emptyset, bp\}, bpj)\}$. As $\emptyset, \{\emptyset\}$, and $\{\emptyset, bp\}$ are closed under intersection, $\mathscr{C}^m_{A,O}(P_{ex})$ is $\langle A, O \rangle_{N,c}$-compliant. Hence, there exists a normal program being $\langle A, O \rangle$-equivalent to P_{ex}. For instance, the following program is one in question:[2]

$$Q = \{b \leftarrow p; \quad b \leftarrow j; \quad j \leftarrow c; \quad p \leftarrow b, not\, c; \quad p \leftarrow b, not\, j; \quad p \leftarrow j, not\, c;$$
$$c \leftarrow b, not\, p; \quad c \leftarrow j, not\, p; \quad v \leftarrow j, not\, c\}.$$

3.2   Positive Logic Programs

A positive program P satisfies $P = P^I$, for any interpretation I. Hence, if $(X, Y) \in SE(P)$, then $(X, X) \in SE(P)$ as well, since $X \models P$ trivially implies $X \models P^I$. This motivates the following definition: A set S of SE-interpretations for which $(X, Y) \in S$ implies $(X, X) \in S$ is called *here-total*. Considering strong equivalence relative to A, here-totality of A-SE-models does not make sense, as the here-component of an A-SE-model is not a model of the program but only a *projection* of a model. Hence, we need to adapt the property as follows: A set S of SE-interpretations is A-*here-total* if, for all $(X, Y) \in S$ with $X \subset Y$, some $(X', X') \in S$ exists with $X'|_A = X$ and $X' \subset Y$.

Proposition 2 ([6]). *Let $A \subseteq At$ be a set of atoms and P a program over At. Then, there is a positive program Q over At such that $P \equiv^A_s Q$ iff $SE^A(P)$ is A-here-total.*

Example 4. Consider again our running example. Since $(b, bjd) \in SE^A(P_{ex})$, for $A = \{b, p, j\}$, but there is no $(X', X') \in SE^A(P_{ex})$ with $X'|_A = \{b\}$, $SE^A(P_{ex})$ is not A-here-total. Hence, there is no positive program Q with $P_{ex} \equiv^A_s Q$.

We continue with the characterising conditions for casting to positive programs.

Definition 4. *Let $A, O \subseteq At$ be sets of atoms. Then, a set $\mathscr{S}$ of certificate structures is $\langle A, O \rangle_{P,c}$-compliant if, for every $X \in \Xi$ where $(\Xi, Y) \in \mathscr{C}^m_{A,O}(P)$, there is some $(\Xi', X') \in \mathscr{C}^m_{A,O}(P)$ such that $X'|_O \subseteq Y|_O$, $X'|_A = X$, and $\Xi' \subseteq \Xi$.*

Definition 5. *Let $A, O \subseteq At$ be sets of atoms. Then, a set S of SE-interpretations is $\langle A, O \rangle_{P,s}$-compliant if, for every (X, Y) which is $\langle A, O \rangle$-optimal in S with $X \subset Y$, there is some $(X', X') \in S$ with $X'|_O \subseteq Y|_O$, $X'|_A = X$, and $(V, X') \in S$ with $V \subset X'$ implies $(V, Y) \in S$.*

Intuitively, the implication in the last part of the above condition—which does not have a pendant in the definition of A-here-totality—ensures A-completeness of some A-here-total variant S' of S that amounts to the set of A-SE-models of the desired casting.

Theorem 3. *Let $A, O \subseteq At$ be sets of atoms and P a program over At. Then, the following statements are equivalent:*

---

[2] Section 4 contains a description how to construct a casting.

1. $\mathscr{C}^m_{A,O}(P)$ is $\langle A, O \rangle_{\mathsf{P,c}}$-compliant;
2. $SE^A(P)$ is $\langle A, O \rangle_{\mathsf{P,s}}$-compliant;
3. a positive program Q over $At \cup At'$ exists with $P \equiv^A_{|_O} Q$, for some universe At'.

Example 5. It can be checked that $\mathscr{C}^m_{A,O}(P_{ex})$ is $\langle A, O \rangle_{\mathsf{P,c}}$-compliant for $A = O = \{b, p, j\}$ and hence there is a positive program being $\langle A, O \rangle$-equivalent to P_{ex}, e.g., $Q = \{p \lor c \leftarrow b; b \leftarrow c; b \leftarrow p; b \leftarrow j; j \leftarrow c\}$.

In contrast to Example 5, where Q contains only atoms from P_{ex}, unlike in the case of normal programs, building an $\langle A, O \rangle$-equivalent positive program might require atoms not occurring in the original program. The following program illustrates this point.

Example 6. Consider program P over $At = \{a, b, c, h\}$, given by the following rules:

$$
\begin{array}{llll}
a \leftarrow not\ b, not\ c; & \quad a \leftarrow not\ h; & \quad h \leftarrow not\ a, not\ c; \\
b \leftarrow not\ a, not\ c; & \quad b \leftarrow not\ h; & \quad h \leftarrow not\ b, not\ c.
\end{array}
$$

For $A = \{a, b, c\}$ and $O = \emptyset$, we obtain $\mathscr{C}^m_{A,O}(P) = \{(\emptyset, a), (\emptyset, b), (\emptyset, ab), (\{a, ab\}, abc)\}$. Although, $\mathscr{C}^m_{A,O}(P)$ is $\langle A, O \rangle_{\mathsf{P,c}}$-compliant, it can be shown that no $\langle A, O \rangle$-equivalent positive program exists containing only atoms from At.

3.3 Horn Programs

Horn programs are both normal and positive. However, it is not sufficient to combine the criteria of casting to normal and positive programs in order to obtain a characterisation for Horn programs. The reason is that the elimination of disjunction may introduce negation and vice versa. As stated earlier, the classical models of Horn theories are intersection closed. In terms of SE-models of such a program, this means that the there-components occurring in the SE-models, being the models of the program, are closed under intersection [4]. In analogy to closure under here-intersection, this property is called *closure under there-intersection*. More formally, a set S of SE-interpretations is closed under there-intersection iff, whenever $(X, X) \in S$ and $(Y, Y) \in S$, then $(X \cap Y, X \cap Y) \in S$. For a program P, a strongly equivalent Horn program Q exists iff $SE(P)$ is here-total and closed under there-intersection [4]. Note that closure under here-intersection follows automatically from these two conditions. For relativised strong equivalence, similar to here-totality for the positive case, closure under there-intersection has to be adapted with respect to the context alphabet: A set S of SE-interpretations is *A-closed under there-intersection* if, for all $(X, X), (Y, Y) \in S$, there is some $(Z, Z) \in S$ with $Z \subseteq (X \cap Y)$ and $Z|_A = (X \cap Y)|_A$.

Proposition 3 ([6]). *Let $A \subseteq At$ be a set of atoms and P a program over At. Then, there exists a Horn program Q over At such that $P \equiv^A_s Q$ iff $SE^A(P)$ is A-here-total and A-closed under there-intersection.*

The restriction of A-closure under there-intersection in interplay with A-completeness imposes an interesting side effect on the A-SE-models of Horn programs.

Theorem 4. *Let $A, O \subseteq At$ be sets of atoms and Q a Horn program over At. Then, $(Z_1, Z_1), (Z_2, Z_2) \in SE^A(Q)$ with $Z_1|_A = Z_2|_A$ only if $Z_1 = Z_2$.*

As a consequence of that, all A-SE-models of Q are $\langle A, O \rangle$-optimal ones, since for some $(X, Y) \in SE^A(Q)$ it holds that $(Y, Y) \in SE^A(Q)$, and hence, by Theorem 4, there cannot be any $(Y', Y') \in SE^A(Q)$ with $Y' \neq Y$ and $Y'|_{A \cup O} = Y|_{A \cup O}$, which is a requirement for violation of $\langle A, O \rangle$-optimality. Due to this restriction on the A-SE-models, in contrast to the case of positive programs, we do not need additional atoms for building a Horn casting Q, if one exists for some given program P. Indeed, Q can be built from atoms in $atm(P)|_{A \cup O}$ only.

We obtain the following conditions for casting to Horn programs:

Definition 6. *Let $A, O \subseteq At$ be sets of atoms. Then, a set $\mathscr{S}$ of certificate structures is $\langle A, O \rangle_{H,c}$-compliant if*

(i) *for all $(\Xi, Y) \in \mathscr{S}$ and all $X \in \Xi$ there is some $(\Xi', X') \in \mathscr{S}$ such that $X' \subset Y$ and $X'|_A = X$,*

(ii) *for all $(\Xi_1, Y_1), (\Xi_2, Y_2) \in \mathscr{S}$ there is some $(\Xi', Z) \in \mathscr{S}$ with $Z \subseteq (Y_1 \cap Y_2)$ and $Z|_A = (Y_1 \cap Y_2)|_A$, and*

(iii) *for all $(\Xi_Y, Y), (\Xi_Z, Z) \in \mathscr{S}$ with $Y \subseteq Z$ it holds that $\Xi_Y \subseteq \Xi_Z$. Moreover, if $Y \neq Z$, then $Y|_A \in \Xi_Z$.*

Definition 7. *Let $A, O \subseteq At$ be sets of atoms. Then, a set S of SE-interpretations is $\langle A, O \rangle_{H,s}$-compliant if*

(i) *for every (X, Y) which is $\langle A, O \rangle$-optimal in S with $X \subset Y$, some $(X', X') \in S$ exists such that $X'|_O \subseteq Y|_O$ and $X'|_A = X$,*

(ii) *for all $(Y_1, Y_1), (Y_2, Y_2)$ which are $\langle A, O \rangle$-optimal in S, some $(Z, Z) \in S$ exists with $Z|_O \subseteq (Y_1 \cap Y_2)|_O$ and $Z|_A = (Y_1 \cap Y_2)|_A$, and*

(iii) *for all $(X, Y), (Z, Z)$ which are are $\langle A, O \rangle$-optimal in S such that $X \subset Y$ and $Y|_{A \cup O} \subseteq Z_{A \cup O}$, $(X, Z) \in S$ holds. Moreover, if $Y \neq Z$, then $(Y|_A, Z) \in S$.*

The individual Subproperties (i), (ii), and (iii) directly correspond to the equally labelled ones from Definition 6. In both definitions, (i) and (ii) are the pendants of A-here-totality and A-closure under there-intersection for $\langle A, O \rangle$-equivalence, respectively. Note that (i) differs from $\langle A, O \rangle_{P,s}$-compliance in not having a subcondition for A-completeness. Instead, Subproperty (iii) corresponds to A-completeness of the $\langle A, O \rangle$-optimal SE-interpretations in S.

Theorem 5. *Let $A, O \subseteq At$ be sets of atoms and P a program over At. Then, the following statements are equivalent:*

1. $\mathscr{C}^m_{A,O}(P)$ *is $\langle A, O \rangle_{H,c}$-compliant;*
2. $SE^A(P)$ *is $\langle A, O \rangle_{H,s}$-compliant;*
3. *a Horn program Q over $At \cup At'$ exists, for some universe At', with $P \equiv^A_{|_O} Q$;*
4. *a Horn program Q over $atm(P)|_{A \cup O}$ exists such that $P \equiv^A_{|_O} Q$.*

Example 7. As we have seen, $\mathscr{C}^m_{A,O}(P_{ex})$ is $\langle A, O \rangle_{N,c}$-compliant and $\langle A, O \rangle_{P,c}$-compliant for $A = O = \{b, p, j\}$. However, it turns out that $\mathscr{C}^m_{A,O}(P_{ex})$ is not $\langle A, O \rangle_{H,c}$-compliant. Indeed, Condition (ii) of Definition 6 is violated as $(\{\emptyset\}, bp), (\{\emptyset\}, bj) \in \mathscr{C}^m_{A,O}(P_{ex})$ but there is no $(\Xi, Z) \in \mathscr{C}^m_{A,O}(P_{ex})$ with $Z \subseteq \{b\}$ and $Z|_A = \{b\}$. This means that, for preserving $\langle A, O \rangle$-equivalence, we can remove disjunctions from P_{ex} only by introducing negation and vice versa.

However, if we are not interested in who attends the party but whether we have to bear vegetarian food, dancing, or cheesy music, we can obtain a Horn program corresponding to P_{ex}: If we set $A' = O' = \{v, d, c\}$, we expect only input programs that mention party activities and we are only interested in output related to them. The minimal $\langle A', O' \rangle$-certificates of P_{ex} are given by $\mathscr{C}^m_{A',O'}(P_{ex}) = \{(\emptyset, \emptyset), (\{\emptyset\}, d), (\{\emptyset\}, v),$ $(\{\emptyset\}, c), (\{\emptyset, v, c\}, vc), (\{\emptyset, d, c\}, dc)\}$. All conditions of Definition 6 apply, therefore $\mathscr{C}^m_{A',O'}(P_{ex})$ is $\langle A', O' \rangle_{H,c}$-compliant. An example of a Horn program $\langle A', O' \rangle$-equivalent to P_{ex} is $Q = \{c \leftarrow v, d; \leftarrow d, v, c\}$.

3.4 Computational Complexity

We finally discuss the complexity of our casting questions. While checking strong equivalence is coNP-complete [10], deciding relativised strong equivalence and hyperequivalence is already Π^P_2-complete [11,12]. Checking whether, for a given program P, a hyperequivalent casting exists is Π^P_2-complete for normal castings and Π^P_3-complete for positive castings. Furthermore, for the case of Horn programs, the problem lies in Π^P_3 [6]. Although testing relativised strong equivalence with projection is presumably harder, viz. Π^P_4-complete [2], the next result shows that casting under this equivalence notion does not yield an additional source of complexity:

Theorem 6. *Given a program P over At and $A, O \subseteq At$, deciding whether $SE^A(P)$ is $\langle A, O \rangle_{\mathcal{C},s}$-compliant, for $\mathcal{C} \in \{N, P, H\}$, is in Π^P_4, as is deciding whether $\mathscr{C}^m_{A,O}(P)$ is $\langle A, O \rangle_{\mathcal{C},c}$-compliant.*

The major source of complexity is checking $\langle A, O \rangle$-optimality. Important for establishing the upper complexity bound is the observation that checking $\langle A, O \rangle$-optimality is not required in the consequent of any of the (sub)conditions of $\langle A, O \rangle_{\mathcal{C},s}$-compliance.

4   Proof Outline of Main Results

We briefly outline how Theorems 2, 3, and 5 can be shown, thereby providing methods for constructing castings. First of all, from the definitions of an A-SE-model, a minimal $\langle A, O \rangle$-certificate, and $\langle A, O \rangle$-optimality, the following lemma can be shown:

Lemma 1. *For any program P over At, any $A, O \subseteq At$, and any $\mathcal{C} \in \{N, P, H\}$, $\mathscr{C}^m_{A,O}(P)$ is $\langle A, O \rangle_{\mathcal{C},c}$-compliant iff $SE^A(P)$ is $\langle A, O \rangle_{\mathcal{C},s}$-compliant.*

The general proof schema, then, proceeds by showing, for a given program P over At and $\mathcal{C} \in \{N, P, H\}$ that

1. if P has a casting $Q \in \mathcal{C}$ over some $At' \supseteq At$, then $\mathscr{C}^m_{A,O}(P)$ is $\langle A, O \rangle_{\mathcal{C},c}$-compliant, and
2. if $\mathscr{C}^m_{A,O}(P)$ is $\langle A, O \rangle_{\mathcal{C},c}$-compliant, then there exists a casting $Q \in \mathcal{C}$ over some At'', where (i) $At'' = atm(P)$ for $\mathcal{C} = N$, (ii) $At'' \supseteq At$ for $\mathcal{C} = P$, and (iii) $At'' = atm(P)|_{A \cup O}$ for $\mathcal{C} = H$.

Showing Item 1 can be reduced to showing that for every program $Q' \in \mathcal{C}$ it holds that $\mathscr{C}^m_{A,O}(Q')$ is $\langle A, O \rangle_{\mathcal{C},c}$-compliant. Hence, given a casting $Q \in \mathcal{C}$ of P, $\mathscr{C}^m_{A,O}(Q)$ is $\langle A, O \rangle_{\mathcal{C},c}$-compliant and, in view of $\mathscr{C}^m_{A,O}(P) = \mathscr{C}^m_{A,O}(Q)$, so is $\mathscr{C}^m_{A,O}(P)$.

For proving Item 2, we devised algorithms that allow for obtaining a casting $Q \in \mathcal{C}$ over At'' whenever $\mathscr{C}^m_{A,O}(P)$ is $\langle A, O \rangle_{\mathcal{C},c}$-compliant. These involve three transformations, which have the following properties:

(i) $f^{\mathcal{C}}_{A,O}$ maps a program P over At to $f^{\mathcal{C}}_{A,O}(P) = SE^A(Q)$, where Q is a program over At'' such that $Q \in \mathcal{C}$, providing $SE^A(P)$ is $\langle A, O \rangle_{\mathcal{C},s}$-compliant. Furthermore, (X, Y) is $\langle A, O \rangle$-optimal in $SE^A(P)$ iff (X, Y) is $\langle A, O \rangle$-optimal in $f^{\mathcal{C}}_{A,O}(P)$.

(ii) $c^{\mathcal{C}}_A$ maps a set $\mathcal{S}$ of SE-interpretations over At'' to $c^{\mathcal{C}}_A(\mathcal{S}) = SE(Q)$, where Q is a program over At'' such that $Q \in \mathcal{C}$ and $SE^A(Q) = \mathcal{S}$. If such a Q does not exist, then $c^{\mathcal{C}}_A$ is undefined.

(iii) $p^{\mathcal{C}}$ maps a set $\mathcal{S}$ of SE-interpretations over At'' into $p^{\mathcal{C}}(\mathcal{S}) = Q$, where Q is a program over At'' such that $Q \in \mathcal{C}$ and $SE(Q) = \mathcal{S}$. If such a Q does not exist, then $c^{\mathcal{C}}_A$ is undefined.

Given a program P for which $SE^A(P)$ is $\langle A, O \rangle_{\mathcal{C},s}$-compliant, a casting Q of P is given by $Q = p^{\mathcal{C}}(c^{\mathcal{C}}_A(f^{\mathcal{C}}_{A,O}(P)))$. Indeed, by the properties of $f^{\mathcal{C}}_{A,O}$, the set of $\langle A, O \rangle$-optimal A-SE-models of P coincides with the set of $\langle A, O \rangle$-optimal elements in $f^{\mathcal{C}}_{A,O}(P)$. But, again by construction of $f^{\mathcal{C}}_{A,O}$, there exists some $Q' \in \mathcal{C}$ such that $SE^A(Q') = f^{\mathcal{C}}_{A,O}(P)$. Hence, $c^{\mathcal{C}}_A(f^{\mathcal{C}}_{A,O}(P))$ is defined and $c^{\mathcal{C}}_A(f^{\mathcal{C}}_{A,O}(P)) = SE(Q'')$, for some $Q'' \in \mathcal{C}$ such that $SE^A(Q'') = f^{\mathcal{C}}_{A,O}(P)$. By the latter condition, we obtain in turn that $p^{\mathcal{C}}$ is defined and $p^{\mathcal{C}}(c^{\mathcal{C}}_A(f^{\mathcal{C}}_{A,O}(P))) = Q$, where Q is a program such that $Q \in \mathcal{C}$ and $SE(Q) = c^{\mathcal{C}}_A(f^{\mathcal{C}}_{A,O}(P))$. Now, since $c^{\mathcal{C}}_A(f^{\mathcal{C}}_{A,O}(P)) = SE(Q'')$, we have that $SE(Q) = SE(Q'')$, and thus $SE^A(Q) = SE^A(Q'')$. But $SE^A(Q'') = f^{\mathcal{C}}_{A,O}(P)$, so $SE^A(Q) = f^{\mathcal{C}}_{A,O}(P)$. Consequently, the set of $\langle A, O \rangle$-optimal A-SE-models of P coincides with the set of $\langle A, O \rangle$-optimal elements of Q. Therefore, by Theorem 1, it follows that P and Q have the same minimal $\langle A, O \rangle$-certificates, and thus $P \equiv^A_{|o} Q$.

The specific definitions of the above functions are as follows: $c^{\mathcal{C}}_A$ and $p^{\mathcal{C}}$ are functions defined in previous work about casting under hyperequivalence [6]. Specifically, c^{N}_A, c^{P}_A, and c^{H}_A are the *completion transformations* for normal, positive, and Horn programs, respectively [6, Definitions 4,7, and 9], while p^{N}, p^{P}, and p^{H} are the functions given by the adaptions in that work of the technique of *canonical programs* [2]. The functions $f^{\mathcal{C}}_{A,O}$, for $\mathcal{C} = \{\mathsf{N}, \mathsf{P}, \mathsf{H}\}$ are novel, however, and are briefly discussed in what follows. To begin with, $f^{\mathsf{N}}_{A,O}$ and $f^{\mathsf{H}}_{A,O}$ are defined as follows:

$$f^{\mathsf{N}}_{A,O}(P) = \{(X, Y) \mid (X, Y) \text{ is } \langle A, O \rangle\text{-optimal in } SE^A(P)\};$$
$$f^{\mathsf{H}}_{A,O}(P) = \{(X|_{A \cup O}, Y|_{A \cup O}) \mid (X, Y) \text{ is } \langle A, O \rangle\text{-optimal in } SE^A(P)\}.$$

The transformations reflect the fact that a normal casting can be build from atoms in P only and a Horn casting from atoms that occur in $atm(P)|_{A \cup O}$ only.

Obtaining the transformation $f^{\mathsf{P}}_{A,O}(P)$ for positive castings requires a more sophisticated approach, however, for which we first need to build a rooted forest $F_{A,O}(P)$ from the A-SE-models of P as an auxiliary structure. The nodes are triples (h, Y, Ξ), where h is an atom, unique for every node, that is from a set At' of auxiliary atoms not occurring in At, Y is an interpretation, and Ξ is a set of interpretations. Each node corresponds to an $\langle A, O \rangle$-certificate of the casting Q to build, and each root node to a minimal $\langle A, O \rangle$-certificate of P. The intuition is to preserve the minimal $\langle A, O \rangle$-certificates of P, for retaining $\langle A, O \rangle$-equivalence to P, and add missing (non-minimal) $\langle A, O \rangle$-certificates of Q that ensure A-here-totality of the A-SE-models of Q in order for Q to be a positive program. Roughly, for every $(X, Y) \in f^P_{A,O}(P)$ with $X \subset Y$ that is built from a node N, a descendant of N guarantees that there is some $(X', X') \in f^P_{A,O}(P)$ with $X'|_A = X$ and $X' \subset Y$, as required by A-here-totality. $F_{\langle A, O \rangle}(P)$ is obtained by the following algorithm, which starts with $F_{\langle A, O \rangle}(P)$ being empty:

1. For every (Y, Y) that is $\langle A, O \rangle$-optimal in $SE^A(P)$, add (h, Y, Ξ) as a root node of $F_{\langle A, O \rangle}(P)$, where $h \in At'$ is not occurring elsewhere in $F_{\langle A, O \rangle}(P)$ and $\Xi = \{X \mid (X, Y) \in SE^A(P), X \subset Y\}$.
2. While there is a leaf node (h, Y, Ξ) such that $\Xi \neq \emptyset$ do:
   - for each $X \in \Xi$, find some (U, U) that is $\langle A, O \rangle$-optimal in $SE^A(P)$,[3] where $U|_A = X$, $U|_O \subseteq Y|_O$, and $(V, U) \in SE^A(P)$ with $V \subset U$ implies $V \in \Xi$, and add $(h', U|_{A \cup O}, \Xi')$ as child node of (h, Y, Ξ), where $h \in At'$ does not occur elsewhere in $F_{\langle A, O \rangle}(P)$ and $\Xi' = \{V \mid (V, U) \in SE^A(P), V \subset U\}$.

Having $F_{\langle A, O \rangle}(P)$ built, we then obtain $f^P_{\langle A, O \rangle}(P)$ as follows:

$$f^P_{\langle A, O \rangle}(P) = \{(Y', Y') \mid N = (h, Y, \Xi) \text{ is a node of } F_{\langle A, O \rangle}(P), Y' = Y \cup \{h\} \cup$$
$$\{h' \mid (h', Z, \Xi') \text{ is ancestor or descendant of } N\}\} \cup$$
$$\{(X, Y') \mid N = (h, Y, \Xi) \text{ is a node of } F_{\langle A, O \rangle}(P), X \in \Xi, Y' = Y \cup$$
$$\{h\} \cup \{h' \mid (h', Z, \Xi') \text{ is ancestor or descendant of } N\}\}.$$

When creating $f^P_{A,O}(P)$ from $F_{A,O}(P)$, atoms from At' are added in a way such that A-completeness of the A-SE-models of Q is ensured, by making Y_1 and Y_2 incomparable for all $(X, Y_1), (X, Y_2) \in SE^A(Q)$ which are not associated to nodes on the same path in the forest. Nodes along the same path cannot violate A-completeness per se.

5 Discussion

We provided necessary and sufficient semantical conditions that hold for a program P and sets A, O of atoms iff there exists another program Q that is normal, positive, or Horn, respectively, such that P and Q are strongly equivalence relative to A under projection to O. These conditions are parameterised in A and O and defined on the minimal $\langle A, O \rangle$-certificates of P and, alternatively, on the A-SE-models of P. Furthermore, we showed that deciding whether a casting exists is not computationally harder than checking equivalence under the considered notion, and provided methods for constructing a casting whenever one exists. Our results contribute to the understanding of problem settings in logic programming in the sense that they show in what scenarios the usage of certain constructs are superfluous or not.

---

[3] The existence of such a (U, U) is guaranteed when $SE^A(P)$ is $\langle A, O \rangle_{P,s}$-compliant.

An interesting equivalence notion related to the one studied here is *modular equivalence*, introduced by Oikarinen and Janhunen [13] for the purpose of modular programming, accommodating the specification of input, output, and hidden atoms. Compared to the use of projection, where answer sets that coincide on the projected part are treated as identical, modular equivalence distinguishes between answer sets with differing hidden atoms and thereby enforces corresponding modules to have the same number of answer sets. *Update equivalence* [14], on the other hand, has no provision of projection but generalises relativised strong equivalence by allowing also the *deletion* of certain program parts during comparison.

A natural next step is to consider casting questions for nonground programs. Towards this end, recent work provides nonground versions of the kind of generalised equivalences studied here [15], yet relativisation quickly leads to undecidable instances.

References

1. Lifschitz, V., Pearce, D., Valverde, A.: Strongly equivalent logic programs. ACM Transactions on Computational Logic 2(4), 526–541 (2001)
2. Eiter, T., Tompits, H., Woltran, S.: On solution correspondences in answer set programming. In: IJCAI 2005, pp. 97–102 (2005)
3. Woltran, S.: Characterizations for relativized notions of equivalence in answer set programming. In: Alferes, J.J., Leite, J. (eds.) JELIA 2004. LNCS (LNAI), vol. 3229, pp. 161–173. Springer, Heidelberg (2004)
4. Eiter, T., Fink, M., Tompits, H., Woltran, S.: Simplifying logic programs under uniform and strong equivalence. In: Lifschitz, V., Niemelä, I. (eds.) LPNMR 2004. LNCS (LNAI), vol. 2923, pp. 87–99. Springer, Heidelberg (2004)
5. Eiter, T., Fink, M., Tompits, H., Woltran, S.: On eliminating disjunctions in stable logic programming. In: KR 2004, pp. 447–458. AAAI Press, Menlo Park (2004)
6. Pührer, J., Tompits, H., Woltran, S.: Elimination of disjunction and negation in answer-set programs under hyperequivalence. In: Garcia de la Banda, M., Pontelli, E. (eds.) ICLP 2008. LNCS, vol. 5366, pp. 561–575. Springer, Heidelberg (2008)
7. Gelfond, M., Lifschitz, V.: Classical negation in logic programs and disjunctive databases. New Generation Computing 9, 365–385 (1991)
8. Heyting, A.: Die formalen Regeln der intuitionistischen Logik. Sitzungsberichte, physikalisch-mathematische Klasse, preußische Akademie der Wissenschaften (1930)
9. Turner, H.: Strong equivalence made easy: Nested expressions and weight constraints. Theory and Practice of Logic Programming 3(4-5), 602–622 (2003)
10. Lin, F.: Reducing strong equivalence of logic programs to entailment in classical propositional logic. In: KR 2002, pp. 170–176. Morgan Kaufmann, San Francisco (2002)
11. Eiter, T., Fink, M., Woltran, S.: Semantical characterizations and complexity of equivalences in answer set programming. ACM Transactions on Computational Logic 8(3), 1–53 (2007)
12. Woltran, S.: A common view on strong, uniform, and other notions of equivalence in answer-set programming. Theory and Practice of Logic Programming 8(2), 217–234 (2008)
13. Oikarinen, E., Janhunen, T.: Achieving compositionality of the stable model semantics for SMODELS programs. TPLP 8(5-6), 717–761 (2008)
14. Inoue, K., Sakama, C.: Equivalence of logic programs under updates. In: Alferes, J.J., Leite, J. (eds.) JELIA 2004. LNCS (LNAI), vol. 3229, pp. 174–186. Springer, Heidelberg (2004)
15. Oetsch, J., Tompits, H.: Program correspondence under the answer-set semantics: The nonground case. In: Garcia de la Banda, M., Pontelli, E. (eds.) ICLP 2008. LNCS, vol. 5366, pp. 591–605. Springer, Heidelberg (2008)

A Default Approach to Semantics of Logic Programs with Constraint Atoms

Yi-Dong Shen[1] and Jia-Huai You[2]

[1] State Key Laboratory of Computer Science, Institute of Software,
Chinese Academy of Sciences, Beijing 100190, China
ydshen@ios.ac.cn
[2] Department of Computing Science, University of Alberta, Edmonton, Alberta,
Canada T6G 2E8
you@cs.ualberta.ca

Abstract. We define the semantics of logic programs with (abstract) constraint atoms in a way closely tied to default logic. Like default logic, formulas in rules are evaluated using the classical entailment relation, so a constraint atom can be represented by an equivalent propositional formula. Therefore, answer sets are defined in a way closely related to default extensions. The semantics defined this way enjoys two properties generally considered desirable for answer set programming − *minimality* and *derivability*. The derivability property is very important because it guarantees free of self-supporting loops in answer sets. We show that when restricted to basic logic programs, this semantics agrees with the conditional-satisfaction based semantics. Furthermore, answer sets by the minimal-model based semantics can be recast in our approach. Consequently, the default approach gives a unifying account of the major existing semantics for logic programs with constraint atoms. This also makes it possible to characterize, in terms of the minimality and derivability properties, the precise relationship between them and contrast with others.

1   Introduction

In knowledge representation and reasoning, it is often desirable to embed into a general inference process/structure special methods for solving/querying a predefined relation. The goal is to make representation easier and reasoning more effective. The technique has been explored, under the name of *global constraints*, in constraint programming [1], and *aggregates* in logic programming [2].

More recently, a great deal of attention has been paid to incorporating constraint atoms (*c-atoms* for short, sometimes under the name of *aggregates*) into answer set programming (ASP) [3,4,5,6,7,8,9,10,11,12,13,14,15]. The intensive study on the subject is largely due to the unsettling question on the semantics for these programs. All of the major semantics proposed so far agree on logic programs with monotone c-atoms; when arbitrary c-atoms are allowed, disagreements exist. Among the recent proposals beyond monotone constraints, two approaches have attracted the most attention − the minimal-model based

E. Erdem, F. Lin, and T. Schaub (Eds.): LPNMR 2009, LNCS 5753, pp. 277–289, 2009.
© Springer-Verlag Berlin Heidelberg 2009

semantics [4] (with its extension [6]) and the conditional-satisfaction based semantics [15] (with the equivalent one [12]). Although example programs have been used to show their differences, since the ways in which they are defined are quite different, to the best of our knowledge, no precise relationship has been established.

In this paper, we propose a different approach to the semantics of logic programs with c-atoms, with a close tie to default logic [16]. We consider programs consisting of rules of the form $H \leftarrow G$, where H is a propositional atom and G an arbitrary formula built from atoms, c-atoms and standard connectives. Like default logic, formulas in rules are evaluated using the classical entailment relation, so a constraint atom can be represented by an equivalent propositional formula. Therefore, answer sets are defined in a way closely related to default extensions. The semantics defined this way enjoys two properties generally considered desirable for answer set programming — *minimality* and *derivability*. The derivability property is very important because it guarantees free of self-supporting loops in answer sets. Several major existing semantics, such as [4,6,9], lack this property, thus their answer sets may incur self-supporting loops.

We show that for basic logic programs, this semantics agrees with the conditional-satisfaction based semantics. This reveals that the latter can be viewed as a form of default reasoning without resorting to conditional satisfaction. Furthermore, answer sets by the minimal-model based semantics can be recast in our approach. This results in a generalization of the existing semantics to arbitrary propositional formulas. This generalization differs from the one to nested expressions [6] in their different underlying logics and the ways in which c-atoms are encoded by formulas. Our approach is like default logic where formulas are evaluated as in classical logic, while answer sets of nested expressions are minimal total models in the logic here-and-there [17].

Our approach provides a unifying framework for evaluating/comparing the major existing semantics. In particular, we can characterize their differences in terms of the minimality and derivability properties. We show that an answer set by the conditional-satisfaction based semantics is an answer set by the minimal-model based semantics, with the additional condition that it is *derivable* in the sense of default logic.

2   Preliminaries

We restrict attention to a propositional language, as we consider only Herbrand interpretations so that the first-order case reduces to a propositional one via grounding. We assume a propositional language determined by a fixed countable set Σ of propositional *atoms*.

Any subset I of Σ is an *interpretation*. We call $I^+ = I$ the *positive literals* of I, and $I^- = \{\neg a \mid a \in \Sigma \setminus I\}$ the *negative literals*.

A *c-atom* A is a pair (D, C), where D is a finite set of atoms in Σ and C is a collection of sets of atoms in D, i.e., $C \subseteq 2^D$ with 2^D being the powerset of D. For convenience, we use A_d and A_c to refer to the components D and C

of A, respectively. As a general abstract framework, a c-atom A can be used to represent any constraint with a finite set A_c of admissible solutions over a finite domain A_d, including various aggregates [10,11]. Therefore, in the sequel we use the aggregate notation and c-atoms exchangeably to express abstract constraints.

The *complement* of a c-atom A is the c-atom A' with $A'_d = A_d$ and $A'_c = 2^{A_d} \setminus A_c$.

Formulas are built from atoms and c-atoms using connectives $\neg$ (negation), $\wedge$ (conjunction), $\vee$ (disjunction), and $\supset$ (implication), as in propositional logic. A *literal* is an atom/c-atom or a negated atom/c-atom. A *theory* is a set of formulas. When a theory mentions no c-atoms, it is called an *ordinary theory*. An ordinary theory may be simply called a theory if it is clear from the context.

A (deductive) *rule* r is of the form $H \leftarrow B$, where H is an atom, B is a formula, and $\leftarrow$ is an *if-then* operator. The meaning of r is that "if the logic property B holds then H holds." We use $head(r)$ and $body(r)$ to refer to the head H and body B of r, respectively.

A *general logic program* (or *program*) P is a set of rules. P is called a *basic program* if the body of each rule is a conjunction of literals. A *normal program* is a basic program without c-atoms. A *positive program* is a normal program without negative literals.

Let $I \subseteq \Sigma$ be an interpretation. I *satisfies* an atom A if $A \in I$; $\neg A$ if $A \notin I$. I satisfies a c-atom A if $A_d \cap I \in A_c$; $\neg A$ if $A_d \cap I \notin A_c$. Therefore, it follows that I satisfies $\neg A$ if and only if I satisfies the complement of A.

The satisfaction of a formula F by an interpretation I is defined as in propositional logic. I satisfies a rule r if it satisfies $head(r)$ or it does not satisfy $body(r)$. I is a *model* of a theory W if it satisfies all formulas of W. I is a *model* of a program P if it satisfies all rules of P. A *minimal* model is a model none of whose proper subset is also a model. A model I of P is *supported* if for each $a \in I$, there is a rule in P of the form $a \leftarrow body(r)$ such that $body(r)$ is satisfied by I. For any expression E, we say E is *true* (resp. *false*) in I if and only if I satisfies (resp. does not satisfy) E.

For an (ordinary) theory W, we say that W is *consistent* (or *satisfiable*) if W has a model. We use $Cn(W)$ to denote the *deductive closure* of W as in propositional logic.

The *entailment relation* $\models$ is defined as a minor extension to the one in propositional logic. For any two formulas F and G (possibly containing c-atoms), F *entails* G, denoted $F \models G$, if G is true in all models of F. We write $F \equiv G$ if $F \models G$ and $G \models F$, in which case F and G are said to be *logically equivalent*. Note that when F and G involve no c-atoms, the entailment relation is exactly the classical one. The extension makes it convenient to talk about formulas involving c-atoms being logically equivalent.

For a set S of atoms, when S appears in a formula, it expresses a conjunction $\bigwedge_{a \in S} a$. Similarly, $\neg S$ expresses a conjunction $\bigwedge_{a \in S} \neg a$.

In this paper, the *standard ASP semantics* refers to the stable model semantics defined in [18].

3   Answer Sets for Programs without C-Atoms

We start by recalling the *rationality principle* – the spirit of the standard ASP semantics for normal programs, which says that *one shall not believe anything one is not forced to believe* [19,18]. The rationality principle instructs us that in the construction of an answer set I from a normal program P, we should make as many negative beliefs as possible, provided that no contradiction is derived. On the one hand, this means that the answer set I shall be a *minimal* model of P. On the other hand, this means that given the set $I^- = \{\neg a_1, ..., \neg a_m\}$ of negative beliefs, all positive beliefs I^+ in I shall be *derivable* by applying the deductive rules $H \leftarrow B$ in P in the way that if B holds, then derive H.

We apply the rationality principle to general logic programs. The derivability property is formally defined by means of deductive sets.

Definition 1. Let P be a program without c-atoms and W an (ordinary) theory. The *deductive set* $Th(P, W)$ of P and W is the smallest set of formulas satisfying the following two conditions: (1) $W \subseteq Th(P, W)$; (2) for each rule r in P, if $Th(P, W) \models body(r)$ then $head(r) \in Th(P, W)$.

We can alternatively use a fixpoint approach to define the deductive set. We first introduce the following one-step provability operator:

$$T_P(W) = \{H \mid P \text{ has a rule } H \leftarrow B \text{ such that } W \models B\}$$

Since the entailment relation $\models$ is defined as in propositional logic, $W \models B$ implies $V \models B$ if $W \subseteq V$. Then, T_P is monotone, i.e., for any theories W_1, W_2 such that $W_1 \subseteq W_2$, $T_P(W_1) \subseteq T_P(W_2)$. Therefore, for any (ordinary) theory W, the sequence $T_P^i(W)$, where $T_P^0(W) = W$ and $T_P^{i+1}(W) = W \cup T_P(T_P^i(W))$, is monotonically increasing and has a fixpoint $T_P^\delta(W)$ (i.e., there exists the first ordinal δ such that $T_P^\delta(W) = T_P^{\delta+1}(W)$). We then have this fixpoint theory $T_P^\delta(W)$ as the deductive set of P and W, i.e. $Th(P, W) = T_P^\delta(W)$.[1]

The above fixpoint definition shows that all positive literals in a deductive set are derived stratum by stratum via the fixpoint sequence. This means that the derivability property guarantees free of self-supporting loops.

Our definition of answer sets for general logic programs adheres to the rationality principle and builds directly on the derivability property.

Definition 2. Let P be a program without c-atoms and I an interpretation. I is an *answer set* of P if (1) $Th(P, I^-)$ is consistent; and (2) for each $a \in I$, $Th(P, I^-) \models a$.

Since the head of each rule is an atom, $Th(P, I^-)$ consists of I^- plus a set of atoms. In particular, we have

Theorem 1. *Let P be a program without c-atoms and I an interpretation. I is an answer set of P if and only if $Th(P, I^-) = I \cup I^-$.*

---

[1] When P is a normal program and W is a set of negative literals, $T_P^\delta(W)$ coincides with the fixpoint $T_{P'}^\delta(\emptyset)$ introduced in [20], where $P' = P \cup W$.

Example 1. Consider the following program:

$$P_1: \quad p(a) \leftarrow (p(a) \wedge \neg p(b)) \vee (p(b) \wedge \neg p(a)) \vee (p(a) \wedge p(b)).$$
$$p(b) \leftarrow \neg q. \qquad q \leftarrow \neg p(b).$$

Let $\Sigma = \{p(a), p(b), q\}$ and $I = \{p(a), p(b)\}$. Then $I^- = \{\neg q\}$. $T^0_{P_1}(I^-) = I^- = \{\neg q\}$, $T^1_{P_1}(I^-) = I^- \cup T_{P_1}(T^0_{P_1}(I^-)) = \{\neg q, p(b)\}$, $T^2_{P_1}(I^-) = I^- \cup T_{P_1}(T^1_{P_1}(I^-)) = \{\neg q, p(b), p(a)\}$, and $T^3_{P_1}(I^-) = T^2_{P_1}(I^-)$. Note that the body of the second rule is entailed by $\{\neg q\}$, and the body of the first rule is entailed by $\{\neg q, p(b)\}$. Therefore, $Th(P_1, I^-) = T^2_{P_1}(I^-) = \{\neg q, p(b), p(a)\}$. Since $Th(P_1, I^-) = I \cup I^-$, $I = \{p(a), p(b)\}$ is an answer set of P_1.

It is easy to check $I = \{q\}$ is also an answer set of P_1. No other interpretations are answer sets of P_1. For instance, for $I = \{p(b)\}$ we have the deductive set $Th(P_1, I^-) = \{\neg q, \neg p(a), p(b), p(a)\}$, which is inconsistent.   □

Example 2. Consider another program:

$$P_2: \quad c \leftarrow \neg((a \wedge \neg b) \vee (b \wedge \neg a)).$$
$$a \leftarrow c. \qquad b \leftarrow a.$$

Let $\Sigma = \{a, b, c\}$ and $I = \{a, b, c\}$. Then $I^- = \emptyset$. $T^0_{P_2}(I^-) = \emptyset$ and $T^1_{P_2}(I^-) = T^0_{P_2}(I^-)$, where no rule body is entailed by $\emptyset$. $Th(P_2, I^-) = \emptyset$ and thus I is not an answer set of P_2. No other interpretations are answer sets of P_2. Therefore, P_2 has no answer set.   □

The derivability property of the answer sets as defined by Definition 2 immediately follows from the definition. It is also easy to show that such answer sets satisfy the minimality property.

Theorem 2. *Answer sets by Definition 2 are supported, minimal models.*

A supported, minimal model is not necessarily an answer set. For example, let P consist of two rules: $d \leftarrow \neg a$ and $a \leftarrow a$. $I = \{a\}$ is a supported, minimal model of P, but it is not an answer set.

For normal programs, answer sets defined above coincide with answer sets under the standard ASP semantics.

Theorem 3. *For a normal program P, an interpretation I is an answer set by Definition 2 if and only if I is an answer set under the standard ASP semantics.*

Our definition of answer sets is closely tied to Reiter's default logic [16]. Recall that a *default theory* is a pair $\triangle = (D, W)$, where W is a theory, D is a set of *defaults* of the form $\frac{\alpha : \beta_1, ..., \beta_m}{\gamma}$, and $\alpha, \beta_1, ..., \beta_m, \gamma$ are formulas. For a theory E, let $\Gamma_\triangle(E)$ be the smallest deductively closed set of formulas satisfying the following two conditions: (1) $W \subseteq \Gamma_\triangle(E)$; (2) for each default r in D, if $\Gamma_\triangle(E) \models \alpha$ and $\neg\beta_1, ..., \neg\beta_m \notin E$, then $\gamma \in \Gamma_\triangle(E)$. E is called an *extension* of $\triangle$ if $E = \Gamma_\triangle(E)$.

Theorem 4. *Let P be a program without c-atoms and I an interpretation. Let $D_1 = \{\frac{body(r):}{a} \mid a \leftarrow body(r) \in P\}$, $D_2 = \{\frac{: \neg a}{\neg a} \mid a \in \Sigma\}$, and $\triangle = (D_1 \cup D_2, \emptyset)$. $Cn(Th(P, I^-)) = \Gamma_\triangle(I \cup I^-)$.*

By Theorem 1, I is an answer set of P if and only if $I \cup I^- = Th(P, I^-)$; then by Theorem 4, if and only if $Cn(I \cup I^-) = \Gamma_\triangle(I \cup I^-)$. Therefore, I is an answer set of P if and only if $Cn(I \cup I^-)$ is an extension of $\triangle$.

4   Answer Sets for Programs with C-Atoms

Since our semantics (Definition 2) is defined by deductive sets using the classical entailment relation, it has an advantage that replacing the body of a rule by a logically equivalent formula preserves answer sets. Therefore, to extend the semantics to programs with c-atoms, it suffices to represent each c-atom as an equivalent formula.

Recall that the satisfaction of a c-atom A is defined by means of propositional interpretations [11]: for any interpretation I, I satisfies A if $A_d \cap I \in A_c$, and I satisfies $\neg A$ if $A_d \cap I \notin A_c$. We want to have a suitable formula F with the property that for any interpretation I, I satisfies A if and only if I satisfies F. Such a formula has been introduced in [13], where it was used to justify an abstract representation of c-atoms.

Definition 3 ([13]). Let $A = (A_d, A_c)$ be a c-atom with $A_c = \{S_1, ..., S_m\}$. The *DNF* formula $C_1 \vee ... \vee C_m$ for A is defined as: each C_i is a conjunction $a_1 \wedge ... \wedge a_k \wedge \neg b_1 \wedge ... \wedge \neg b_l$ built from S_i such that $S_i = \{a_1, ..., a_k\}$ and $(A_d \setminus S_i) = \{b_1, ..., b_l\}$.

Proposition 1. *Let A be a c-atom and I an interpretation. Let $C_1 \vee ... \vee C_m$ be the DNF formula for A. I satisfies A if and only if I satisfies $C_1 \vee ... \vee C_m$. I satisfies $\neg A$ if and only if I satisfies $\neg(C_1 \vee ... \vee C_m)$.*

Let $body(r)$ be the body of a rule r, and $body(r')$ be $body(r)$ with all c-atoms replaced by their DNF formulas. By Proposition 1, an interpretation I satisfies $body(r)$ if and only if I satisfies $body(r')$. We then have the following ASP semantics for general logic programs.

Definition 4. Let P be a program and I an interpretation. I is an *answer set* of P if I is an *answer set* of P' as defined in Definition 2, where P' is P with all c-atoms replaced by their DNF formulas.

Since answer sets of P' satisfy both the minimality and the derivability property, by Proposition 1 answer sets of P satisfy the two properties as well.

Example 3. Consider the following program (borrowed from [21]):

$$P_3 : \quad p(a) \leftarrow COUNT(\{X \mid p(X)\}) > 0.$$
$$p(b) \leftarrow \neg q. \qquad q \leftarrow \neg p(b).$$

The aggregate notation $COUNT(\{X \mid p(X)\}) > 0$ corresponds to a c-atom A with $A_d = \{p(a), p(b)\}$ and $A_c = \{\{p(a)\}, \{p(b)\}, \{p(a), p(b)\}\}$. The DNF formula for A is $(p(a) \wedge \neg p(b)) \vee (p(b) \wedge \neg p(a)) \vee (p(a) \wedge p(b))$. By replacing A with this DNF formula, we obtain P_3' which is the same as P_1 in Example 1. Since $P_3' = P_1$ contains no c-atom, its answer sets can be computed applying Definition 2. P_1 has two answer sets: $I_1 = \{p(a), p(b)\}$ and $I_2 = \{q\}$. Therefore, by Definition 4 P_3 has the two answer sets.   □

Example 4. Consider the following program with a negated constraint:

$P_4:$ $c \leftarrow \neg 1\{a, b\}1.$
    $a \leftarrow c.$ $b \leftarrow a.$

The cardinality constraint $1\{a, b\}1$ corresponds to a c-atom A with $A_d = \{a, b\}$ and $A_c = \{\{a\}, \{b\}\}$. The DNF formula for A is $(a \wedge \neg b) \vee (b \wedge \neg a)$. By replacing A with this DNF formula, we obtain P_4' which is the same as P_2 in Example 2. P_2 has no answer set, thus P_4 has no answer set.   □

5   Relating to Other Approaches

Our approach provides a unifying framework for evaluating the major existing proposals for arbitrary c-atoms, including the conditional-satisfaction based semantics [15], the minimal-model based semantics [4,5], and the computation-based semantics [9]. In this section, we characterize the differences of these semantics in terms of the minimality and derivability properties for basic programs.

5.1   Conditional-Satisfaction Based Semantics

We show that when restricted to basic programs, Definition 4 defines the same answer sets as the ones by conditional-satisfaction [15].

To introduce the notion of conditional satisfaction, we need a simple program transformation. Any atom a can be expressed as an *elementary* c-atom $A = (\{a\}, \{\{a\}\})$; any negated atom $\neg a$ can be expressed as a c-atom $A = (\{a\}, \{\emptyset\})$; and any negated c-atom $\neg A$ can be replaced by the complement of A. Due to this, in this section we assume all basic programs have been rewritten so that their rules are of the form $H \leftarrow A_1 \wedge ... \wedge A_m$, where each A_i is a c-atom.

Definition 5. Let R and S be two sets of atoms and A a c-atom. We say R *conditionally satisfies* A w.r.t. S, denoted $R \models_S A$, if R satisfies A and for every S' such that $R \cap A_d \subseteq S'$ and $S' \subseteq S \cap A_d$, we have $S' \in A_c$. For a rule r in a basic program, $R \models_S body(r)$ if $R \models_S A_i$ for each A_i in $body(r)$.

An immediate consequence operator $\mathcal{T}_P(R, S)$ is defined in terms of the conditional satisfaction.

Definition 6. Let P be a basic program and R and S be two sets of atoms. Define

$$\mathcal{T}_P(R, S) = \{a \mid a \leftarrow body(r) \in P \text{ and } R \models_S body(r)\}$$

It is proved that when the second argument is a model of P, $\mathcal{T}_P$ is monotone w.r.t. its first argument [15]. Therefore, for any model M the monotone sequence $\mathcal{T}_P^i(\emptyset, M)$, where $\mathcal{T}_P^0(\emptyset, M) = \emptyset$ and $\mathcal{T}_P^{i+1}(\emptyset, M) = \mathcal{T}_P(\mathcal{T}_P^i(\emptyset, M), M)$, converges to a fixpoint $\mathcal{T}_P^\alpha(\emptyset, M)$.

The *conditional-satisfaction based ASP semantics* says that a model M is an answer set of P if M coincides with the fixpoint $\mathcal{T}_P^\alpha(\emptyset, M)$ [15]. Denecker et al. [22] and Shen and You [13] define the same semantics using different definitions.

Example 5. Assume P_4 of Example 4 has been rewritten to

$$P_4: \quad c \leftarrow (\{a,b\}, \{\emptyset, \{a,b\}\}).$$
$$a \leftarrow (\{c\}, \{\{c\}\}).$$
$$b \leftarrow (\{a\}, \{\{a\}\}).$$

Note that the c-atom in the body of the first rule is the complement of the original c-atom. This program has no answer set under the conditional-satisfaction based semantics. Take $M = \{a,b,c\}$ as an example, which is a model of P_4. Let $T_{P_4}^0(\emptyset, M) = \emptyset$. Although the c-atom in the body of the first rule is satisfied by $\emptyset$, it is not conditionally satisfied by $\emptyset$ w.r.t. M. As a result, we reach the fixpoint $T_{P_4}^1(\emptyset, M) = T_{P_4}(T_{P_4}^0(\emptyset, M), M) = \emptyset$. M does not agree with the fixpoint $\emptyset$, thus is not an answer set of P_4. □

The conditional-satisfaction based fixpoint and the deductive set are closely related.

Theorem 5. *Let P be a basic program, P' be P with all c-atoms replaced by their DNF formulas, and M be a model of P. We have $T_P^\alpha(\emptyset, M) = Th(P', M^-) - M^-$.*

The following corollary follows immediately, showing that the ASP semantics of Definition 4 and the conditional-satisfaction based semantics agree on basic programs.

Corollary 1. *Let P be a basic program and M a model of P. M is an answer set of P under the conditional-satisfaction based semantics if and only if M is an answer set defined by Definition 4.*

Since answer sets by Definition 4 respect the rationality principle, the conditional-satisfaction based semantics also satisfies the two properties for basic programs. Similar results have been established earlier [15], which shows that any answer set under the conditional-satisfaction based semantics is a minimal model for which a *level mapping* exists. The definition of a level mapping relies on identifying structural properties of a c-atom w.r.t. a given interpretation, in order to capture conditional satisfaction. Thus, the derivability property can be seen as an independent way to capture level mapping, without conditional satisfaction.

However, conditional satisfaction is not defined over disjunction of c-atoms. In fact, a direct application could cause unintuitive behaviors.

Example 6 (borrowed from [23]). Let A be the aggregate $SUM(\{X \mid p(X)\}) \neq -1$. In standard mathematics, this is equivalent to $SUM(\{X \mid p(X)\}) > -1$ or $SUM(\{X \mid p(X)\}) < -1$, because $A \equiv B \vee C$, where B and C denote the latter two aggregates, respectively.

Now suppose $S = \{p(1)\}$ and $M = \{p(1), p(2), p(-3)\}$. It can be verified that while $S \models_M A$, it is not the case that $S \models_M B$ or $S \models_M C$. □

In our approach, c-atoms are represented by propositional formulas, which are then evaluated as in propositional logic. As a result, logic equivalence guarantees the preservation of answer sets. In fact, it is not difficult to verify that, for any program P, replacing a formula in the body of a rule in P by a logically equivalent one results in a program strongly equivalent to P. This is an advantage of our approach, inherited from that of default logic.

5.2   Minimal-Model Based Semantics

Let P be a basic program and I an interpretation. To check if I is an answer set, this semantics first removes all rules whose body contains a literal that is not satisfied by I, then defines I to be an answer set if I is a minimal model of the simplified program P^I [4,5]. Ferraris [6] defines an ASP semantics in a different way, which agrees with the minimal-model based one on basic programs.

We now recast the minimal-model based semantics in our framework for general logic programs, where c-atoms are represented by their DNF formulas. We define the *reduct* P^I of P w.r.t. I as the result of removing all rules r from P such that $body(r)$ is not satisfied by I.

Definition 7. Let P be a general logic program and P' be P with all c-atoms replaced by their DNF formulas. An interpretation I is said to be a *weak answer set* of P if I is a minimal model of P'^I.

Theorem 6. *Weak answer sets of a program P are minimal models of P.*

For basic programs, by Proposition 1 weak answer sets coincide with answer sets under the minimal-model based semantics. Furthermore, we have

Theorem 7. *Let P be a general logic program and P' be P with all c-atoms replaced by their DNF formulas. An interpretation I is an answer set of P by Definition 4 if and only if I is a weak answer set of P such that for any $a \in I$, $Th(P', I^-) \models a$.*

It follows immediately that for basic programs, answer sets under the conditional-satisfaction based semantics are such answer sets under the minimal-model based semantics that are derivable via the deductive set.

The minimal-model based semantics does not satisfy the derivability property, even for basic programs.

Example 7. Consider the following basic program:

$$P_5: \quad p(1).$$
$$p(2) \leftarrow p(-1).$$
$$p(-1) \leftarrow SUM(\{X \mid p(X)\}) \geq 1.$$

The aggregate $SUM(\{X \mid p(X)\}) \geq 1$ corresponds to a c-atom A, where $A_d = \{p(-1), p(1), p(2)\}$ and $A_c = \{\{p(1)\}, \{p(2)\}, \{p(-1), p(2)\}, \{p(1), p(2)\}, \{p(-1), p(1), p(2)\}\}$. The DNF formula for A is

$$(p(1) \wedge \neg p(-1) \wedge \neg p(2)) \vee (\neg p(1) \wedge \neg p(-1) \wedge p(2)) \vee (\neg p(1) \wedge p(-1) \wedge p(2))$$
$$\vee (p(1) \wedge \neg p(-1) \wedge p(2)) \vee (p(1) \wedge p(-1) \wedge p(2))$$

which as in propositional logic, can be simplified to $(p(1) \wedge \neg p(-1)) \vee p(2)$.

Let $\Sigma = \{p(1), p(-1), p(2)\}$. P_5 has only one model, $\{p(1), p(-1), p(2)\}$. By Definition 4 or equivalently under the conditional-satisfaction based semantics, P_5 has no answer set, since the model is not derivable via the deductive set. However, $I = \{p(1), p(-1), p(2)\}$ is an answer set of P_5 under the minimal-model based semantics.   □

It should be pointed out that due to lack of derivability, the minimal-model based semantics ([4] and its extension [6]) may incur undesirable self-supporting loops in its answer sets. For instance, in the above example $I = \{p(1), p(-1), p(2)\}$ is the answer set of P_5, where $p(2)$ and $p(-1)$ can only be deduced via one of the following self-supporting loops:

$$p(2) \rightarrow SUM(\{X \mid p(X)\}) \geq 1 \rightarrow p(-1) \rightarrow p(2).$$
$$p(-1) \rightarrow p(2) \rightarrow SUM(\{X \mid p(X)\}) \geq 1 \rightarrow p(-1).$$

That is, in order to derive $p(2)$ and $p(-1)$ from P_5 we must assume either $p(2)$ or $p(-1)$ is true in advance.

5.3 Computation-Based Semantics

Liu et al. [9] define an answer set I for a basic program P to be the fixpoint of a *computation* $\langle I_i \rangle_{i=0}^{\infty}$ with $I_0 = \emptyset$ and $I_{i+1} = T_{Q_i}^{nd}(I_i)$, where Q_i is a subset of the rules in P whose bodies are satisfied by I_i and $T_{Q_i}^{nd}(I_i)$ consists of all heads of rules in Q_i, such that for each $i \geq 0$, $I_i \subseteq I_{i+1}$ and $Q_i \subseteq Q_{i+1}$.

Consider the following basic program:

$P_6:$ $p(1).$
　　　$p(-1) \leftarrow p(2).$
　　　$p(2) \leftarrow SUM(\{X \mid p(X)\}) \geq 1.$

As described in Example 7, the DNF formula for $SUM(\{X \mid p(X)\}) \geq 1$ is $(p(1) \wedge \neg p(-1)) \vee p(2)$. P_6 has two models, $\{p(1), p(-1)\}$ and $\{p(1), p(-1), p(2)\}$.

Let r_i refer to the i-th rule. We have a computation for P_6, where $I_0 = \emptyset$, $I_1 = \{p(1)\}$ with $Q_0 = \{r_1\}$, $I_2 = \{p(1), p(2)\}$ with $Q_1 = \{r_1, r_3\}$, and $I_3 = \{p(1), p(-1), p(2)\}$ with $Q_2 = \{r_1, r_2, r_3\}$. I_3 is the fixpoint of the computation, where $I_0 \subseteq \ldots \subseteq I_3$ and $Q_0 \subseteq Q_1 \subseteq Q_2$. Thus I_3 is an answer set of P_6 under the computation-based semantics.

Observe that I_3 is not a minimal model of P_6; neither is it derivable via the deductive set $Th(P_6, I_3^-)$. This shows that the computation-based semantics satisfies neither the minimality nor the derivability property.

Just like the minimal-model based semantics, due to lack of derivability the computation-based semantics may incur undesirable self-supporting loops in its answer sets. For instance, $I_3 = \{p(1), p(-1), p(2)\}$ is an answer set of P_6, where $p(2)$ can only be deduced via a self-supporting loop

$$p(2) \rightarrow SUM(\{X \mid p(X)\}) \geq 1 \rightarrow p(2).$$

$\{p(1), p(-1), p(2)\}$ is an answer set of P_6 under the computation-based semantics, but it is not under the minimal-model based one. On the contrary, $\{p(1), p(-1), p(2)\}$ is an answer set of P_5 under the minimal-model based semantics, but it is not under the computation-based one. This means that answer sets under the minimal-model based semantics are not necessarily answer sets under the computation-based semantics, and vice versa.

To get more restricted computations (answer sets), Liu et al. [9] propose to use a *sub-satisfiability* relation ▷ to replace the standard satisfiability with the property that for any interpretation I and c-atom A, $I \vartriangleright A$ implies I satisfies A. Then, different computations and answer sets can be obtained by embedding different sub-satisfiability relations into the definition of computation.

6   Summary and Discussion

Incorporating constraints into a general knowledge representation and reasoning (KR) system has proven to be a crucial step in gaining representation power and reasoning efficiency. When logic programs with constraint atoms are taken as the underlying KR language, the question on the semantics has raised great interest, with competing views and definitions of answer sets.

For normal logic programs, the stable model semantics has roots in different nonmonotonic formalisms, thus, as summarized by Lifschitz [24], leading to current twelve different definitions. As pointed out by Lifschitz, "there are reasons why each of them is valuable and interesting. A new characterization of stable models can suggest an alternative picture of the intuitive meaning of logic programs; ⋯ or it can be interesting simply because it demonstrates a relationship between seemingly unrelated ideas."

In this paper, we developed an alternative new approach to the semantics of logic programs with c-atoms, where formulas in rules are evaluated using the classical entailment relation and c-atoms are represented by equivalent propositional formulas. The resulting framework can be seen as one with c-atoms embedded into a fragment of default logic. It turns out that both the conditional-satisfaction based semantics and the minimal-model based semantics have a root in the default framework. The former can be viewed as a form of default reasoning without conditional satisfaction, while under exactly the same representation of c-atoms, the latter can be recast in the same framework. As a result, we are able to identify the precise relationship between the two major existing semantics and contrast with others such as the computation-based semantics.

The semantics defined by the default approach has two important properties – minimality and derivability. The derivability property is very useful; it guarantees free of self-supporting loops in answer sets. Our examples show that several major existing semantics, such as [4,6,9], lack this property, thus their answer sets may incur self-supporting loops.

Another advantage of the default approach is that replacing a constraint by a logically equivalent one preserves strong equivalence. This is an important feature, since constraint atoms are supposed to represent pre-defined/built-in/global constraints in constraint solving, and special constraint propagation rules for such a built-in constraint may need to be implemented or updated. In this case, one only needs to verify the preservation of satisfaction. This feature also provides a methodology of representing a constraint by some (logically equivalent) combination of constraints. For example, an aggregate $SUM(..) \neq k$ can be substituted by $SUM(..) > k \vee SUM(..) < k$, according to the standard

Y.-D. Shen and J.-H. You

mathematics while preserving strong equivalence. Since constraints are viewed as propositional formulas, this feature is applicable in a more general context: Given a program P, replacing a formula (possibly including constraint atoms) of the body of a rule in P by a logically equivalent one results in a program P' which is strongly equivalent to P.

As remarked earlier, our approach is different from that of [6], due to the different underlying logics and methods of representing c-atoms. In [6] rule bodies and c-atoms are nested expressions in the logic of here-and-there, while in our approach they are classical propositional formulas. We would like to argue that an intimate integration of classical formulas into logic programs, as done in our approach which is inherited from default logic, facilitates knowledge representation. The same point has been argued for a more general context by [25] and recent work on integrating ASP with ontologies and description logics for the Semantic Web (where classical formulas in rule bodies are interpreted as queries to a description logic knowledge base [26]).

Our language can be extended to accommodate c-atoms in rule heads like $A \leftarrow G$, where A is a c-atom and G an arbitrary formula. It can also be extended to logic programs whose rule heads may be a disjunction of atoms. As a concrete application, we are applying the default approach to characterizing the semantics of *description logic programs* for the Semantic Web [26].

Acknowledgments

We would like to thank the anonymous referees for their constructive comments. Yi-Dong Shen is supported in part by NSFC grants 60673103, 60721061 and 60833001, and by the National High-tech R&D Program (863 Program). The work by Jia-Huai You is supported in part by the Natural Sciences and Engineering Research Council of Canada.

References

1. Marriott, K., Stuckey, P.: Programming with Constraints. MIT Press, Cambridge (1998)
2. Kemp, D., Stuckey, P.: Semantics of logic programs with aggregates. In: Proc. Int'l Symposium on Logic Programming, pp. 387–401. MIT Press, Cambridge (1991)
3. Calimeri, F., Faber, W., Leone, N., Perri, S.: Declarative and computational properties of logic programs with aggregates. In: IJCAI 2005, pp. 406–411 (2005)
4. Faber, W., Leone, N., Pfeifer, G.: Recursive aggregates in disjunctive logic programs: Semantics and complexity. In: Alferes, J.J., Leite, J. (eds.) JELIA 2004. LNCS (LNAI), vol. 3229, pp. 200–212. Springer, Heidelberg (2004)
5. Faber, W., Pfeifer, G., Leone, N., Dell'Armi, T., Ielpa, G.: Design and implementation of aggregate functions in the dlv system. TPLP 8(5) (2008)
6. Ferraris, P.: Answer sets for propositional theories. In: Baral, C., Greco, G., Leone, N., Terracina, G. (eds.) LPNMR 2005. LNCS (LNAI), vol. 3662, pp. 119–131. Springer, Heidelberg (2005)
7. Lee, J., Lifschitz, V., Palla, R.: A reductive semantics for counting and choice in answer set programming. In: AAAI 2008, pp. 472–479 (2008)

8. Liu, L., Truszczynski, M.: Properties and applications of programs with monotone and convex constraints. JAIR 7, 299–334 (2006)
9. Liu, L., Pontelli, E., Son, T., Truszczynski, M.: Logic programs with abstract constraint atoms: the role of computations. In: Dahl, V., Niemelä, I. (eds.) ICLP 2007. LNCS, vol. 4670, pp. 286–301. Springer, Heidelberg (2007)
10. Marek, V.W., Remmel, J.B.: Set constraints in logic programming. In: Lifschitz, V., Niemelä, I. (eds.) LPNMR 2004. LNCS (LNAI), vol. 2923, pp. 167–179. Springer, Heidelberg (2004)
11. Marek, V.W., Truszczynski, M.: Logic programs with abstract constraint atoms. In: AAAI 2004, pp. 86–91 (2004)
12. Pelov, W., Denecker, M., Bruynooghe, M.: Well-founded and stable semantics of logic programs with aggregates. TPLP 7(3), 301–353 (2007)
13. Shen, Y.D., You, J.H.: A generalized Gelfond-Lifschitz transformation for logic programs with abstract constraints. In: AAAI 2007, pp. 483–488 (2007)
14. Simons, P., Niemela, I., Soininen, T.: Extending and implementing the stable model semantics. Artificial Intelligence 138(1-2), 181–234 (2002)
15. Son, T.C., Pontelli, E., Tu, P.H.: Answer sets for logic programs with arbitrary abstract constraint atoms. JAIR 29, 353–389 (2007)
16. Reiter, R.: A logic for default reasoning. Artificial Intelligence 13, 81–132 (1980)
17. Lifschitz, V., Pearce, D., Valverde, A.: Strongly equivalent logic programs. ACM Transactions on Computational Logic 2(4), 526–541 (2001)
18. Gelfond, M., Lifschitz, V.: The stable model semantics for logic programming. In: ICLP 1988, pp. 1070–1080 (1988)
19. Gelfond, M.: Answer sets. Handbook of Knowledge Representation. Elsevier, Amsterdam (2008)
20. van Gelder, A.: The alternating fixpoint of logic programs with negation. In: PODS, pp. 1–10 (1989)
21. Son, T.C., Pontelli, E.: A constructive semantic characterization of aggregates in answer set programming. TPLP 7(3), 355–375 (2007)
22. Denecker, M., Pelov, N., Bruynooghe, M.: Ultimate well-founded and stable semantics for logic programs with aggregates. In: Codognet, P. (ed.) ICLP 2001. LNCS, vol. 2237, pp. 212–226. Springer, Heidelberg (2001)
23. Liu, G.H., You, J.H.: Lparse programs revisited: Semantics and representation of aggregates. In: Garcia de la Banda, M., Pontelli, E. (eds.) ICLP 2008. LNCS, vol. 5366, pp. 347–361. Springer, Heidelberg (2008)
24. Lifschitz, V.: Twelve definitions of a stable model. In: Garcia de la Banda, M., Pontelli, E. (eds.) ICLP 2008. LNCS, vol. 5366, pp. 37–51. Springer, Heidelberg (2008)
25. Denecker, M., Vennekens, J.: Building a knowledge base sysem for an integration of logic programmng and classical logic. In: ICLP 2009, pp. 71–76 (2009)
26. Eiter, T., Ianni, G., Lukasiewicz, T., Schindlauer, R., Tompits, H.: Combining answer set programming with description logics for the semantic web. Artif. Intell. 172(12-13), 1495–1539 (2008)

The Complexity of Circumscriptive Inference in Post's Lattice*

Michael Thomas

Institut für Theoretische Informatik, Gottfried Wilhelm Leibniz Universität
Appelstr. 4, 30167 Hannover, Germany
thomas@thi.uni-hannover.de

Abstract. Circumscription is one of the most important formalisms for reasoning with incomplete information. It is equivalent to reasoning under the *extended closed world assumption*, which allows to conclude that the facts derivable from a given knowledge base are all facts that satisfy a given property. In this paper, we study the computational complexity of several formalizations of inference in propositional circumscription for the case that the knowledge base is described by a propositional theory using only a restricted set of Boolean functions. To systematically cover all possible sets of Boolean functions, we use Post's lattice. With its help, we determine the complexity of circumscriptive inference for all but two possible classes of Boolean functions. Each of these problems is shown to be either Π_2^p-complete, coNP-complete, or contained in L.

In particular, we show that in the general case, unless P = NP, only literal theories admit polynomial-time algorithms, while for some restricted variants the tractability border is the same as for classical propositional inference.

1 Introduction

Circumscription is a non-monotonic logic introduced by McCarthy for first-order theories to overcome the 'qualification problem' which is concerned with the impossibility of representing all conditions for the successful performance of an action [17]. Circumscription allows to conclude that the objects that can be shown to have a certain property P by reasoning from a given knowledge base Γ are all objects that satisfy P. Moreover, circumscription has been shown to coincide with reasoning under the *extended closed world assumption*, in which all formulae involving only propositions from P that cannot be derived from Γ are assumed to be false [12]. To date, circumscription has become one of the most well developed and extensively studied formalisms for non-monotonic reasoning.

Given a theory Γ containing a predicate P, circumscribing P amounts to selecting only the models of Γ in which P is assigned the value true on a minimal set of tuples. The key intuition behind this rationale is that minimal models have as few 'exceptions' as possible and, thus, embody common sense. In propositional

---

* Supported by DFG grant VO 630/6-1.

E. Erdem, F. Lin, and T. Schaub (Eds.): LPNMR 2009, LNCS 5753, pp. 290–302, 2009.
© Springer-Verlag Berlin Heidelberg 2009

logic, P is simply a set of propositions; whence propositional circumscription asks for the minimal models of Γ w.r.t. the coordinatewise partial order induced on P by $0 < 1$. The remaining propositions are partitioned into sets Q and Z where propositions in Q are fixed and propositions in Z are allowed to vary in minimizing the extent of P. We write $<_{(P,Q,Z)}$ for this order. In the literature, circumscription has also been studied in a restricted form, in which all propositions are subject to minimization (i.e., $Q = Z = \emptyset$). Following [19], we will call this restricted form *basic circumscription* and omit the (P, Q, Z)-subscript.

One of the most elementary tasks in logic is inference. The *circumscriptive inference problem* asks whether a formula ψ holds in all models of φ that are minimal w.r.t. $<_{(P,Q,Z)}$. As with most non-monotonic logics, inference is Π_2^p-complete [11,8] and, thus, strictly harder than the inference problem for classical propositional logic unless the polynomial hierarchy collapses. This negative result raises the question for fragments of lower complexity. One of the most natural ways to exhibit such fragments is to restrict the formulae allowed in the premise or the conclusion.

As one example, Schaefer studied the complexity of conjunctions of generalized clauses, i.e., conjunctions of arbitrary relations from a fixed set S [25]. He used this approach to classify the complexity of the satisfiability problem for all possible sets S. In this context, circumscriptive inference is well-studied from the computational complexity perspective: in 2005, Nordh proved a trichotomy classifying the complexity of circumscriptive inference to be either Π_2^p-complete, coNP-complete, or in P [19]. For basic circumscription, Kirousis and Kolaitis first established a dichotomy, classifying the inference problem to be either Π_2^p-complete or in coNP [15]. To also refine this result into a trichotomy for basic circumscription, Durand and Hermann showed that the inference problem for affine formulae is coNP-complete [9]. Finally, Durand *et al.* completed the classification of the complexity basic circumscriptive inference in Schaefer's framework by establishing a trichotomy [10].

However, Schaefer's framework is just one possibility of restricting the set of premises. Instead of restricting formulae to conjunctions of relations from a fixed set S, one may require that the premise formulae are written using a restricted set B of Boolean functions. One would likewise assume the complexity to drop into lower classes if the set B is very restrictive. This approach has first been taken by Lewis, who classified the complexity of the satisfiability problem w.r.t. to all finite sets of Boolean functions [16]. Many problems have been studied in this way since then (see [24,23,13,1,3,2,18], amongst others).

Though this approach bears similarity to Schaefer's framework, we stress that in general the results do not imply each other. Consider, e.g., the set of relations S definable using Horn formulae. The circumscriptive inference of a clause from a conjunction of generalized clauses from S is coNP-complete [8]. On the contrary, if the knowledge base is restricted to consist of formulae expressible using only Boolean implication $x \to y$, then circumscriptive inference is Π_2^p-complete (Theorems 3 and 12), albeit $\{(x, y) \mid x \to y\} \in S$.

The present paper takes Post's systematic approach to completely classify the complexity for several formalizations of the inference problem for circumscription. Let B be a finite set of Boolean functions and let a B-*formula* be a formula using functions from B as connectives only. We determine the complexity of inference of a B-formula from the circumscription of a set of B-formulae, written $\text{CircINF}(B)$, for all finite sets B, except for the case that only affine functions based on the ternary *exclusive-or* $\oplus$ are allowed. The complexity in both propositional circumscription and basic circumscription for all remaining sets B is trichotomic: inference in propositional circumscription is Π_2^p-complete whenever B can simulate non-monotone or non-affine Boolean functions, coNP-complete for all monotone B that implement the *or*-function $\vee$, and tractable otherwise (viz. if any B-formula is equivalent to a set of literals). The complexity of inference in basic circumscription becomes tractable in addition to the above cases if all functions from B can be defined using disjunctions only.

We also study the complexity of inferring a clause instead of a B-formula, and the restriction of the premise set to a singleton. In both cases, we completely classify the complexity for all finite sets B: (1) if a clause is to be inferred, then inference remains equivalent to the first formalization with the exception that for affine functions implementing the ternary $\oplus$, the problems becomes coNP-hard; (2) if the premise set is required to be a singleton, then inference becomes tractable for exactly the same clones as does the ordinary inference problem for B-formulae in propositional logic [2], i.e., for all sets of Boolean functions such that all its members can be expressed using either the *and*-function $\wedge$, $\vee$, or $\oplus$.

As a result, we obtain an almost complete classification for the complexity of propositional circumscription w.r.t. the Boolean functions allowed. Moreover, a tradeoff between the expressivity of sets of Boolean functions and the complexity of deciding the circumscriptive inference problem is exhibited, thus attaining a clear picture of the complexity inherent to each Boolean function w.r.t. minimal model semantics.

From the application point of view, our results imply that syntactically restricting the knowledge base does not lead to tractable, yet expressive fragments of circumscription. They furthermore allow the designer of nonmonotonic deduction systems to choose an appropriate representation, according to the required expressiveness: depending on whether syntactically restricting the knowledge base is possible or not, the complexity of the arising problem allows to select appropriate heuristics for solving the problem.

The paper is organized as follows. Section 2 contains preliminaries, while Sect. 3 introduces circumscription and the inference problems studied in this paper. Our main results are presented in Sect. 4 to 6, where we classify the complexity of circumscriptive inference problems w.r.t. the Boolean functions allowed to construct the formulae. In each of these sections, we first give the main theorems; the corresponding proofs are established from subsequent lemmas and propositions. Finally, Sect. 7 concludes with a discussion of the results and identifies interesting further work. In the interests of space, some proofs are omitted and will be included in the full version of this paper.

2   Preliminaries

In this paper we make use of standard notions of complexity theory. The arising complexity degrees encompass the classes $\oplus$L, P, NP, coNP, Σ_2^p and Π_2^p, where $\oplus$L forms the class of languages recognizable by logspace Turing machines with an odd number of accepting paths [5]. We also require the circuit complexity classes AC^0 and $AC^0[2]$. The class AC^0 denotes the class of languages recognizable using logtime-uniform Boolean circuits of constant depth and polynomial size over $\{\wedge, \vee, \neg\}$, where the fan-in of gates of the first two types is not bounded. The class $AC^0[2]$ is defined similarly, but in addition to $\{\wedge, \vee, \neg\}$ we also allow $\oplus$-gates of unbounded fan-in. For further information, we refer to [20,27].

For the hardness results we use *constant-depth* reductions, defined as follows: A language A is *constant-depth reducible* to a language B ($A \leq_{cd} B$) if there exists a logtime-uniform AC^0-circuit family $\{C_n\}_{n \geq 0}$ with unbounded fan-in $\{\wedge, \vee, \neg\}$-gates and oracle gates for B such that for all x, $C_{|x|}(x) = 1$ iff $x \in A$ (cf. [27]). Denote by MOD_2 the problem to decide whether a given string $w \in \{a, b\}^*$ contains an even number of occurrences of the letter a. It is easily verified that MOD_2 is complete for $AC^0[2]$ under $\leq_{cd}$-reductions.

We assume familiarity with propositional logic. The set of all propositional formulae is denoted by $\mathcal{L}$. For finite $\Gamma \subseteq \mathcal{L}$ and $\varphi \in \mathcal{L}$, we identify Γ with $\bigwedge \Gamma$ and write $\Gamma \models \varphi$ if all assignments satisfying all formulae in Γ also satisfy φ. For a formula φ, let $\varphi[\alpha/\beta]$ denote φ with all occurrences of α replaced by β, and let $\Gamma[\alpha/\beta] := \{\varphi[\alpha/\beta] : \varphi \in \Gamma\}$ for $\Gamma \subseteq \mathcal{L}$. A propositional formula using only functions from a finite set B of Boolean functions as connectives is called a B-formula. The set of all B-formulae is denoted by $\mathcal{L}(B)$. In order to cope with the infinitely many finite sets B of Boolean functions, we require some algebraic tools to classify the complexity of the infinitely many arising reasoning problems. A *clone* is a set B of Boolean functions that is closed under superposition, i. e., B contains all projections and is closed under arbitrary composition [21]. Let B be a clone. For ease of notation, we will also write $\mathcal{L}(B)$ and identify B with an arbitrary finite base for B. For an arbitrary finite set B of Boolean functions, we denote by $[B]$ the smallest clone containing B and call B a *base* for $[B]$. In [22], Post classified the lattice of all Boolean clones and found a finite base for each clone. In order to introduce the clones relevant to this paper, we define the following notions for n-ary Boolean functions f:

- f is *c-reproducing* if $f(c, \ldots, c) = c$, $c \in \{0, 1\}$.
- f is *monotone* if $a_1 \leq b_1, \ldots, a_n \leq b_n$ implies $f(a_1, \ldots, a_n) \leq f(b_1, \ldots, b_n)$.
- f is *self-dual* if $f \equiv \mathrm{dual}(f)$, where $\mathrm{dual}(f)(x_1, \ldots, x_n) = \neg f(\neg x_1, \ldots, \neg x_n)$.
- f is *affine* if $f \equiv x_1 \oplus \cdots \oplus x_n \oplus c$ for a constant $c \in \{0, 1\}$ and variables $x_1, \ldots, x_n$.

The clones relevant to this paper are listed in Tab. 1. The definition of all Boolean clones can be found, e. g., in [4].

Some of our results rely on the fact that the functions $\vee$, $\wedge$ and $\oplus$ can be implemented using 'short' formulae in order to avoid super-polynomial blowups. The following lemma states the relevant cases.

Table 1. A list of Boolean clones with definitions and bases

Name	Definition	Base
BF	All Boolean functions	$\{\wedge, \neg\}$
R_0	$\{f : f$ is 0-reproducing$\}$	$\{\wedge, \nrightarrow\}$
R_1	$\{f : f$ is 1-reproducing$\}$	$\{\vee, \rightarrow\}$
M	$\{f : f$ is monotone$\}$	$\{\vee, \wedge, 0, 1\}$
D_2	$\{f : f$ is self-dual and monotone$\}$	$\{(x \wedge y) \vee (y \wedge z) \vee (x \wedge z)\}$
L	$\{f : f$ is affine$\}$	$\{\oplus, 1\}$
L_1	$L \cap R_1$	$\{\equiv\}$
V	$\{f : f$ is a disjunction of variables or a constant$\}$	$\{\vee, 0, 1\}$
V_2	$V \cap R_0 \cap R_1$	$\{\vee\}$
E	$\{f : f$ is a conjunction of variables or a constant$\}$	$\{\wedge, 0, 1\}$
E_1	$E \cap R_1$	$\{\wedge, 1\}$
N	$\{f : f$ depends on at most one variable$\}$	$\{\neg, 0, 1\}$
I	$\{f : f$ is a projection or a constant$\}$	$\{\mathrm{id}, 0, 1\}$

Lemma 1. *Let B be a finite set of Boolean functions such that $[B] \in \{V, M, BF\}$ (resp. $[B] \in \{E, M, BF\}$). Then, there exists a B-function $f(x, y)$ such that $f(x, y) \equiv x \vee y$ (resp. $f(x, y) \equiv x \wedge y$) and x, y occur exactly once in f.*

3   Circumscription

Let $\varphi \in \mathcal{L}$ be a formula over a set X of propositions. We write $\varphi(X)$ to denote that the propositions occurring in φ are a subset of X. An *assignment* σ *of* X is a mapping $\sigma \colon X \to \{0, 1\}$. Slightly abusing notation, we will identify σ with the set $\{x \in X : \sigma(x) = 1\}$. We say σ is a *model of* φ if σ evaluates φ to 1, written $\sigma \models \varphi$. For a partition (P, Q, Z) of X, define the parametrized order $\leq_{(P,Q,Z)}$ on assignments as follows: for assignments $\sigma, \sigma' \colon X \to \{0, 1\}$, let $\sigma \leq_{(P,Q,Z)} \sigma'$ if $\sigma \cap P \subseteq \sigma' \cap P$ and $\sigma \cap Q = \sigma' \cap Q$. We write $\sigma <_{(P,Q,Z)} \sigma'$ if $\sigma \leq_{(P,Q,Z)} \sigma'$ and $\sigma \cap P \neq \sigma' \cap P$. We say that a model σ of φ is *minimal w. r. t.* (P, Q, Z) (or (P, Q, Z)-*minimal*) if there is no model σ' such that $\sigma <_{(P,Q,Z)} \sigma'$. Note that if $Q = Z = \emptyset$ then $<_{(P,Q,Z)}$ coincides with the coordinatewise partial order $<$ induced by $0 < 1$.

An assignment σ is a model of the circumscription of P in $\varphi(X)$, written $\sigma \models^{\mathrm{circ}}_{(P,Q,Z)} \varphi$, if σ is a (P, Q, Z)-minimal model. A formula $\psi(X)$ can be inferred from (the circumscription of P in) $\varphi(X)$, written $\varphi \models^{\mathrm{circ}}_{(P,Q,Z)} \psi$, if ψ holds in all (P, Q, Z)-minimal models of φ. Analogously define $\Gamma \models^{\mathrm{circ}}_{(P,Q,Z)} \psi$ for sets of formulae Γ. For any finite set B of Boolean functions, we define the *inference problem* for B-formulae in propositional circumscription as

> Problem: CircINF(B)
> Instance: $\Gamma(X) \subseteq \mathcal{L}(B)$, $\psi(X) \in \mathcal{L}(B)$, a partition (P, Q, Z) of X.
> Question: Does $\Gamma \models^{\mathrm{circ}}_{(P,Q,Z)} \psi$ hold?

Let B be a clone. Again, we will also write CircINF(B) to denote all problems CircINF(B), where B is a finite base for B. Whenever considering basic circumscription, i. e., whenever CircINF is restricted to $Q = Z = \emptyset$, we omit the (P, Q, Z) subscript and write BasicCircINF.

Regarding automated reasoning, it is useful to require formulae to be represented in (generalized) conjunctive normal form. It is hence natural to ask for inference of clauses instead of B-formulae. To compare our results to preliminary work, we will thus also consider the variant $\text{CircINF}_{cl}(B)$ of the inference problem, in which ψ is replaced by a clause over X. Moreover, we consider the restriction $\text{CircINF}^1(B)$, in which Γ is assumed to be a single formula.

For the standard case $B = \{\wedge, \vee, \neg\}$, Cadoli and Lenzerini showed that $\text{CircINF}(\{\wedge, \vee, \neg\})$ is contained in Π_2^P [7]. The matching lower bound was given by Eiter and Gottlob, who proved that $\text{CircINF}_{cl}(\{\wedge, \vee, \neg\})$ is Π_2^P-hard, even if ψ is restricted to be a single literal [11]. The upper bound can be easily generalized to arbitrary sets B and the reduction given by Eiter and Gottlob can be computed using constant-depth reductions indeed. We hence obtain the following proposition matching our formalization.

Proposition 2 ([7,11]). $\text{CircINF}(\mathsf{BF})$, $\text{CircINF}_{cl}(\mathsf{BF})$, and $\text{CircINF}^1(\mathsf{BF})$ are Π_2^P-complete, even if $Q = Z = \emptyset$.

4   Inference of B-Formulae

Theorem 3. Let B be a finite set of Boolean functions. Then $\text{CircINF}(B)$ is

1. Π_2^P-complete if $[B] \not\subseteq \mathsf{M}$ and $[B] \not\subseteq \mathsf{L}$,
2. coNP-complete if $[B] \subseteq \mathsf{M}$ and $[B] \not\subseteq \mathsf{E}$,
3. in coNP and $\oplus\mathsf{L}$-hard if $[B] \subseteq \mathsf{L}$ and $[B] \not\subseteq \mathsf{N}$,
4. $\text{AC}^0[2]$-complete if $[B] \subseteq \mathsf{N}$ and $[B] \not\subseteq \mathsf{I}$, and
5. in AC^0 in all other cases (i. e., if $[B] \subseteq \mathsf{E}$).

What is particularly interesting about Theorem 3 is that the complexity of circumscriptive inference differs for sets of Boolean functions being dual to each other: while in propositional logic the inference problem for V- and E-formulae is AC^0-complete under $\leq_{cd}$-reductions [2], it is not the case for circumscription. $\text{CircINF}(B)$ is coNP-hard if $\mathsf{V}_2 \subseteq [B]$ (i. e., if $\vee \in [B]$). Intuitively, this derives from the fact that $\Gamma \subseteq \mathcal{L}(\mathsf{V}_2)$ is identified with the conjunction of the contained formulae, i. e., $\Gamma \equiv \bigwedge_i \bigvee_j x_{ij}$, while the (P, Q, Z)-minimality of models for Γ allows for modelling atomic negations. Hence, for $\mathsf{V}_2 \subseteq [B]$, $\text{CircINF}(B)$ is as hard as the implication problem for arbitrary formulae; while for $B \subseteq \mathsf{E}$, where $\Gamma \equiv \bigwedge_i x_i$, the complexity of the problem remains in AC^0. This asymmetry neatly contrasts with the complexity of inference for basic circumscription.

Theorem 4. Let B be a finite set of Boolean functions. Then $\text{BasicCircINF}(B)$ is

1. Π_2^P-complete if $[B] \not\subseteq \mathsf{M}$ and $[B] \not\subseteq \mathsf{L}$,
2. coNP-complete if $[B] \subseteq \mathsf{M}$ and $[B] \not\subseteq \mathsf{E}$ and $[B] \not\subseteq \mathsf{V}$,
3. in coNP and $\oplus\mathsf{L}$-hard if $[B] \subseteq \mathsf{L}$ and $[B] \not\subseteq \mathsf{N}$,

4. $AC^0[2]$-*complete if* $[B] \subseteq N$ *and* $[B] \not\subseteq I$, *and*
5. *in* AC^0 *in all other cases (i. e., if* $[B] \subseteq V$ *or* $[B] \subseteq E$).

The proof of Theorems 3 and 4 will be established from the propositions in this section. To begin with, the following easy lemma reduces the number of clones to be considered.

Lemma 5. *Let B be a finite set of Boolean functions. Then* $\mathrm{CircINF}(B) \equiv_{cd}$ $\mathrm{CircINF}(B \cup \{1\})$, *and* $\mathrm{CircINF}(B) \equiv_{cd} \mathrm{CircINF}(B \cup \{0\})$ *if* $\neg \in [B]$ *or* $\vee \in [B]$. *The equivalences hold even if* $Z = Q = \emptyset$ *is assumed.*

From the proof of Lemma 5, we also obtain the following corollary that will be of use in Sect. 5.

Corollary 6. *Let B be a finite set of Boolean functions. Then* $\mathrm{CircINF}_{cl}(B) \equiv_{cd}$ $\mathrm{CircINF}_{cl}(B \cup \{0, 1\})$, *even if* $Z = Q = \emptyset$.

Recall that when writing CircINF(B) for a clone B, we identify B with an arbitrary, finite base for B. As a result of Lemma 5, to completely classify the complexity of $\mathrm{CircINF}(B)$ only the clones BF, M, V, E, E_1, L, L_1, N, I, and $I_1 = I \cap R_1$ have to be considered (see Fig. 1). For example, if $[B] \not\subseteq M$ and $[B] \not\subseteq L$ then $\mathrm{CircINF}(B) \equiv_{cd} \mathrm{CircINF}(BF)$. Similarly $\mathrm{CircINF}(B) \equiv_{cd} \mathrm{CircINF}(M)$ if $[B] \subseteq M$, $[B] \not\subseteq E$, and $[B] \not\subseteq V$. And if $[B] \subseteq N$ and $[B] \not\subseteq I$ (resp. $[B] \subseteq L$ and $[B] \not\subseteq N$) then $\mathrm{CircINF}(B)$ is equivalent to CircINF(N) (resp. CircINF(L) or CircINF(L_1), depending on B).

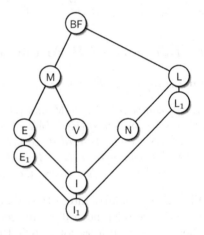

Fig. 1. Boolean clones relevant to the classification of $\mathrm{CircINF}(B)$

Proposition 7. $\mathrm{CircINF}(M)$ *is* coNP-*complete, even if* $Q = Z = \emptyset$.

Proof. For CircINF(M) $\in$ coNP, let B be a finite set of Boolean functions such that $[B] = M$, let (P, Q, Z) partition the set of propositions X, $\Gamma \subseteq \mathcal{L}(B)$ and $\psi \in \mathcal{L}(B)$. It holds that $\Gamma \models^{\mathrm{circ}}_{(P,Q,Z)} \psi$ iff for all (P, Q, Z)-minimal models σ of Γ, $\sigma \models \Gamma$ implies $\sigma \models \psi$.

For a monotone formula φ, we have that $\sigma \models \varphi$ implies $\sigma \cup \{x\} \models \varphi$ for all $x \in X$. Thus, a model $\sigma \colon X \to \{0, 1\}$ is (P, Q, Z)-minimal for Γ iff $(\sigma \cup Z) \setminus \{p\} \not\models \Gamma$ for all $p \in P$ with $\sigma(p) = 1$. One can hence check in polynomial time whether σ is a (P, Q, Z)-minimal model of Γ. Consequently, to prove that $\Gamma \not\models^{\mathrm{circ}}_{(P,Q,Z)} \psi$ it suffices to guess an assignment σ and then to check (in polynomial time according to the discussion above) that σ is a minimal model of Γ falsifying ψ. This shows that $\mathrm{CircINF}(B) \in$ coNP.

As for coNP-hardness, we give a reduction from TAUT, that is, the problem to decide whether a given formula in disjunctive normal form is a tautology.

Let $\varphi \in \mathcal{L}$ be in disjunctive normal form over propositions $X = \{x_1, \ldots, x_n\}$. Let $Y = \{y_1, \ldots, y_n\}$ be a set of propositions disjoint from X. Denote by φ' the formula derived from φ by replacing all negative literals $\neg x_i$ by y_i. By virtue of Lemma 1, we may w.l.o.g. assume that $\vee, \wedge \in B$. We can hence define the reduction function f as

$$f: \varphi \mapsto \left(\left\{ \bigwedge_{1 \leq i \leq n} (x_i \vee y_i) \right\}, \varphi', (X \cup Y, \emptyset, \emptyset) \right).$$

$\bigwedge_{1 \leq i \leq n}(x_i \vee y_i)$ and φ' are obviously monotone and can furthermore be constructed using AC^0-circuits. We show that $\varphi \in \mathrm{TAUT}$ if and only if $(\{\bigwedge_{1 \leq i \leq n}(x_i \vee y_i)\}, \varphi', (X \cup Y, \emptyset, \emptyset)) \in \mathrm{CircINF}(B)$.

First assume that $\varphi \in \mathrm{TAUT}$. Then $\sigma \models \varphi(X)$ for any assignment $\sigma \colon X \to \{0, 1\}$. We define σ' as the extension of σ to $X \cup Y$ defined by $\sigma'(y_i) = \neg\sigma(x_i)$ for all $1 \leq i \leq n$. As a result, for $\sigma_1, \sigma_2 \colon X \to \{0, 1\}$, the corresponding assignments σ'_1, σ'_2 are incomparable under $\leq_{(X \cup Y, \emptyset)}$.

Now assume that there exists a $(X \cup Y, \emptyset, \emptyset)$-minimal model σ' of the premise such that $\sigma'(x_i) = \sigma'(y_i) = 1$ for some $1 \leq i \leq n$. Then both assignments $\sigma' \setminus \{x_i\} <_{(P,Q,Z)} \sigma'$ and $\sigma' \setminus \{y_i\} <_{(P,Q,Z)} \sigma'$ still satisfy $\bigwedge_{1 \leq i \leq n}(x_i \vee y_i)$, a contradiction to σ' being minimal. Hence, any $(X \cup Y, \emptyset, \emptyset)$-minimal model σ' of the premise satisfies $\sigma'(x_i) \neq \sigma'(y_i)$ for all $1 \leq i \leq n$. As a result, any such model is an extension of some assignment $\sigma \colon X \to \{0, 1\}$ as defined above. As $\sigma \models \varphi$ by assumption, we obtain $\sigma' \models \varphi[\neg x_1/y_1, \ldots \neg x_n/y_n] = \varphi'$.

As for $\varphi \notin \mathrm{TAUT}$, there exists an assignment $\sigma \colon X \to \{0, 1\}$ falsifying $\sigma \models \varphi$. Let σ' again be defined as the extension of σ to $X \cup Y$ satisfying $\sigma'(y_i) = \neg\sigma(x_i)$. Then σ' is a $(X \cup Y, \emptyset, \emptyset)$-minimal model of $\bigwedge_{1 \leq i \leq n}(x_i \vee y_i)$ such that $\sigma' \not\models \varphi'$. Hence, σ' witnesses $\{\bigwedge_{1 \leq i \leq n}(x_i \vee y_i)\} \not\models^{\mathrm{circ}}_{(P,Q,Z)} \varphi'$. $\quad\square$

The following proposition re-proves coNP-hardness for a clone properly contained in M. However, it does so only for propositional circumscription; additionally, the proof of Proposition 7 will be required to establish Theorem 15.

Proposition 8. CircINF(V) *is coNP-complete, while BasicCircINF(V) is contained in* AC^0.

Proof (Sketch). Membership of CircINF(V) in coNP follows from Proposition 7. To prove the coNP-hardness of CircINF(V), let B be a finite set of Boolean functions such that $[B] = V$. We reduce the unsatisfiability problem for formulae in conjunctive normal form to CircINF(B). Given $\varphi = \bigwedge_{i=1}^{n} c_i$ over propositions in X, we map $\varphi \mapsto (\Gamma, z, (P, Q, Z))$, where $P := X \cup \{\overline{x} \mid x \in X\} \cup \{g_i \mid 1 \leq i \leq n\}$, $Q := \emptyset$, $Z := \{z\}$ with z being a fresh proposition, and Γ is defined as follows:

1. for each proposition $x \in X$, Γ contains the formula $x \vee \overline{x}$,
2. for each clause $c_i = x_{i1} \vee \cdots \vee x_{ik_i} \vee \neg x_{ik_i+1} \vee \cdots \vee \neg x_{in_i}$ of φ, Γ contains the $n_i + 1$ formulae $g_i \vee \overline{x_{i1}}, \ldots, g_i \vee \overline{x_{ik_i}}, g_i \vee x_{ik_i+1}, \ldots, g_i \vee x_{in_i}$ and $g_i \vee z$.

It holds that if φ is unsatisfiable then every (P, Q, Z)-minimal model σ of Γ sets $\sigma(g_i) = 0$ for at least one $1 \leq i \leq n$. As $g_i \vee z \in \Gamma$ for all $1 \leq i \leq n$, we conclude $\Gamma \models z$. On the other hand, let σ be a (P, Q, Z)-minimal model of Γ. If $\sigma(z) = 0$ then $\sigma \models \bigwedge_{i=1}^{n} g_i$, which in turn implies that $\sigma \cap X \models \varphi$. Hence, all (P, Q, Z)-minimal models of Γ satisfy z.

As the propositions g_i need not be numbered by i, but can rather be indexed according to their position in the input string, we conclude that the reduction can be computable using AC^0-circuits. Whence the coNP-hardness of $CircINF(B)$ under $\leq_{cd}$-reductions follows.

To prove that $BasicCircINF(B) \in AC^0$, let $X = \{x_1, \ldots, x_n\}$ denote the set of of all propositions and let $\Gamma \subseteq \mathcal{L}(B)$, $\psi \in \mathcal{L}(B)$ be given. Then $\Gamma \equiv c_1 \vee \bigwedge_{j=1}^{m} \bigvee_{i \in I_j} x_i$ and $\psi \equiv c_2 \vee \bigvee_{i \in J} x_i$, where $c_1, c_2 \in \{0, 1\}$ and $I_1, \ldots, I_m, J \subseteq \{1, \ldots, n\}$. It is readily observed that these representations can be computed using AC^0-circuits. Assume w. l. o. g. that $\psi \not\equiv 1$ and let $\sigma' \colon \{x_i \mid i \in J\} \to \{0, 1\}$ be the partial assignment defined by $\sigma'(x_i) = 0$ for all $i \in J$. Then $\Gamma \models^{circ} \psi$ iff σ' can not be extended to a minimal model of φ. This is the case iff $\Gamma[X_J/0] := c_1 \vee \bigwedge_{j=1}^{m} \bigvee_{i \in I_j \setminus J} x_i$ is unsatisfiable. This is equivalent to not being satisfiable by the all-1 assignment, because $\Gamma[X_J/0]$ is monotone. As $\vee$-formulae can further be evaluated in AC^0 [26], $BasicCircINF^1(B) \in AC^0$ follows. $\square$

An argument similar to the above can now be used to show that $CircINF(E) \in AC^0$ for both propositional and basic circumscription.

Proposition 9. $CircINF(E)$ *is contained in* AC^0.

Proposition 10. $CircINF(L)$ *and* $CircINF(L_1)$ *are contained* coNP *and* $\oplus L$-*hard, even if* $Q = Z = \emptyset$.

Proof (Sketch). It is well known that minimality of models for sets of affine formulae can be decided in polynomial time [14]. To prove that $\Gamma \not\models^{circ}_{(P,Q,Z)} \psi$, we hence guess an assignment and verify that it is a (P, Q, Z)-minimal model of Γ falsifying ψ.

For the $\oplus L$-hardness, observe that the classical inference problem for affine formulae is hard for $\oplus L$ under $\leq_{cd}$-reductions [2]. Hence, mapping an instance (Γ, ψ) over propositions $X = \{x_1, \ldots, x_n\}$ to $(\Gamma \cup \Delta, \psi, (X \cup Y, \emptyset, \emptyset))$ with $Y = \{y_1, \ldots, y_n\}$ and $\Delta = \{x_i \oplus y_i \mid 1 \leq i \leq n\}$ yields the desired reduction. $\square$

Proposition 11. $CircINF(N)$ *is* $AC^0[2]$-*complete, even if* $Q = Z = \emptyset$.

Proof (Sketch). Let B be a finite set of Boolean functions such that $[B] = N$. Let $\Gamma(X) \subseteq \mathcal{L}(B)$, $\psi \in \mathcal{L}(B)$, and let (P, Q, Z) be a partition of the set of propositions X. Then ψ is equivalent to a literal and, similarly, $\Gamma \equiv \bigwedge_{\varphi \in \Gamma} \ell_\varphi$. Consequently, all (P, Q, Z)-minimal models σ of Γ satisfy $\sigma(x) = 1 \iff \varphi \equiv x$ for some $\varphi \in \Gamma$, for all $x \in P \cup Z$. We can thus compute the above representation of Γ and accept iff $\psi \equiv \varphi$ for some $\varphi \in \Gamma$. Membership in $AC^0[2]$ follows from the fact that N-formulae can be evaluated in $AC^0[2]$ (cf. [26]).

To establish the $AC^0[2]$-hardness, we reduce MOD_2 to of $CircINF(B)$ by mapping the letters a and b to the B-representation of the negation and the identity, resp. The reduction can be computed using AC^0-circuits, as one may w.l.o.g. assume that the variable is the last symbol in both functions.   □

5   Inference of Clauses

We will now consider the problem $CircINF_{cl}(B)$, where inference of a clause c rather than a B-formula is to be checked. This variant of the inference problem is close to the formalization of circumscriptive inference in Schaefer's framework. Indeed some upper and lower bounds can be derived from, e.g., [10]; however, these do not suffice to obtain a complete classification in Post's lattice.

Theorem 12. *Let B be a finite set of Boolean functions. If $[B] \notin \{L, L_1\}$ then $CircINF_{cl}(B)$ and $BasicCircINF_{cl}(B)$ are $\leq_{cd}$-equivalent to $CircINF(B)$. Otherwise, if $[B] = L$ or $[B] = L_1$ then $CircINF_{cl}(B)$ and $BasicCircINF_{cl}(B)$ are coNP-complete.*

Proof (Sketch). By Corollary 6, it suffices to consider the clones BF, M, V, L, N, E and I. The Π_2^p-completeness of $CircINF_{cl}(BF)$ follows from Proposition 2. For the coNP-complete fragments, we distinguish the following two cases.

Clones M and V: Membership in coNP follows from a straightforward adaptation of the algorithm given in the proof of Proposition 7. Note that the proof of Proposition 8 shows that $CircINF(B)$ is coNP-hard, even if ψ is restricted to a single proposition. We adapt the reduction to show hardness for basic circumscription, too. Therefore, we drop z from the set of propositions, remove from Γ the formulae $g_i \vee z$, $1 \leq i \leq n$, and set the clause to be inferred to $\neg g_1 \vee \cdots \vee \neg g_n$. The modified reduction now satisfies that the input formula φ is unsatisfiable iff $\Gamma \models^{circ} \bigvee_{i=1}^n \neg g_i$.

Clone L: Durand and Hermann showed that $CircINF_{cl}(\{\oplus\})$ is coNP-complete under polynomial time many-one reductions, even if $Q = Z = \emptyset$ [9]. Their reduction can indeed be computed using AC^0-circuits. By virtue of Corollary 6, it follows that $CircINF_{cl}(B)$ remains coNP-complete under $\leq_{cd}$-reductions for all $L_2 \subseteq [B] \subseteq L$, where $L_2 = [\{x \oplus y \oplus z\}]$.

To show the $AC^0[2]$-completeness of $CircINF_{cl}(N)$, we follow the proof of Proposition 11. Let B be a finite set of Boolean functions such that $[B] = N$ and let (P, Q, Z) partition the set of all propositions X. Further let $\Gamma \subseteq \mathcal{L}(B)$ and let c be a clause. Then the set of (P, Q, Z)-minimal models σ for Γ is characterized by $\sigma(x) = 1$ iff $x \equiv \varphi$ for $x \in X$ and some $\varphi \in \Gamma$. Now, $\Gamma \models^{circ}_{(P,Q,Z)} c$ iff $\sigma \models c$, which can be tested using AC^0-circuits. Hence, $CircINF_{cl}(B)$ is solvable in $AC^0[2]$. For the $AC^0[2]$-hardness, note that the reduction given holds for $CircINF_{cl}$, too.

It now remains to show $CircINF_{cl}(B) \in AC^0$ for all B such that $[B] \subseteq E$. Analogously to the proof of Proposition 9, we have that $\Gamma \equiv \bigwedge_{x \in X'}$ for some set $X' \subseteq X$ and that all (P, Q, Z)-minimal models σ of Γ are uniquely determined on $X' \cup P$ by the partial assignment σ' defined by $\sigma(x) = 1$ iff $x \in X'$. It thus remains to test whether all such σ satisfy the given clause c. This is again the case iff c does not include propositions from $X \setminus X'$ □

Note that while $\text{CircINF}(B) \equiv_{cd} \text{CircINF}_{cl}(B)$ for all finite sets B such that $[B] \notin \{\mathsf{L}, \mathsf{L_1}\}$, the restriction of CircINF_{cl} to $Q = Z = \emptyset$ does not decrease the complexity of the problem for V-formulae. Informally, this derives from the fact that the all-0 assignment falsifying a V-formula is well-behaved w.r.t. minimiza- tion, whereas for CircINF_{cl} an arbitrary assignment σ may falsify the clause at hand. It hence remains necessary to determine whether σ is among the minimal models of the premise set.

6 Inference from Singletons

As the last formalization of circum- scriptive inference, we will consider the restriction CircINF^1, where Γ is required to be a singleton set (i.e., a formula φ). Due to this restriction, the constant 1 can no longer be gen- erated as in the proof of Lemma 5; the following restricted version holds nevertheless.

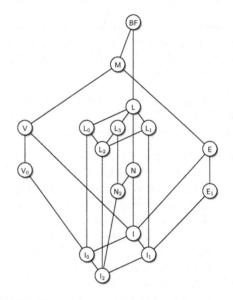

Lemma 13. *Let B be a finite set of Boolean functions. If $\vee \in [B]$ or $\neg \in [B]$, then $\text{CircINF}^1(B) \equiv_{cd} \text{CircINF}^1(B \cup \{0\})$. If $\wedge \in [B]$, then $\text{CircINF}^1(B) \equiv_{cd} \text{CircINF}^1(B \cup \{1\})$. The equivalences hold even if $Z = Q = \emptyset$ is assumed.*

Fig. 2. Boolean clones relevant to the clas- sification of $\text{CircINF}^1(B)$

Lemma 14. *Let B be a finite set of Boolean functions such that $\mathsf{D_2} \subseteq [B]$. Then $\text{CircINF}^1(B) \equiv_{cd} \text{CircINF}^1 (B \cup \{0\})$.*

As a result of Lemmas 13 and 14, it suffices to consider the complexity of $\text{CircINF}^1(B)$ for the clones depicted in Fig. 2.

The following theorem summarizes the complexity of $\text{CircINF}^1(B)$. As a re- sult of the restriction to singleton sets, the inference problem for both V- and L-formulae becomes tractable. Additionally restricting the problem to basic cir- cumscription does not lower the complexity of the problem anymore.

Theorem 15. *Let B be a finite set of Boolean functions. Then $\text{CircINF}^1(B)$ and $\text{BasicCircINF}_1(B)$ are*

1. *Π_2^p-complete if $[B] \nsubseteq \mathsf{M}$ and $[B] \nsubseteq \mathsf{L}$,*
2. *coNP-complete if $[B] \subseteq \mathsf{M}$ and $[B] \nsubseteq \mathsf{E}$ and $[B] \nsubseteq \mathsf{V}$,*
3. *$\text{AC}^0[2]$-complete if $[B] \subseteq \mathsf{L}$ and $[B] \nsubseteq \mathsf{I}$,*
4. *in AC^0 in all other cases (i.e., if $[B] \subseteq \mathsf{V}$ or $[B] \subseteq \mathsf{E}$).*

7 Concluding Remarks

In this paper we almost completely classified the complexity of the inference problem for circumscription for different formalizations arising in the context of Post's lattice. The obtained complexity results form a trichotomy for all problems studied: depending on the set of Boolean connectives allowed, the inference problem is either Π_2^p-complete, coNP-complete, or contained in P (more precisely, complete for $AC^0[2]$ or contained in AC^0, both of which form strict subclasses of L). However, we have to leave open the complexity of $CircINF(L)$ and $CircINF(L_1)$, for which we were only able to obtain $\oplus$L-hardness and membership in coNP.

Our complexity results exhibit the trade-off between expressivity and tractability in the worst-case: restricting the set of Boolean connectives allowed does not lower the complexity of the circumscriptive inference problem unless all functions in B are either monotone or affine—a significant restriction to its expressivity; yet, the complexity of those fragments remains coNP-complete in general and decreases only if further restrictions are imposed (cf. Tab. 2). This supports previous results stating that non-monotonic reasoning is more complex due to a 'super-compact' knowledge representation as compared to classical propositional logic [6].

Non-monotonic reasoning had been introduced in hope of increasing the performance of automated reasoning. Unfortunately, the increased computational complexity of those problems seems to tell another story. But these results take into consideration only the worst-case, they give no indication on the correlation of nonmonotonicity and computational complexity. We thus believe that a finer analysis beyond the usual worst-case measures is an interesting topic for future work.

Table 2. The complexity of deciding circumscriptive inference for selected clones

	$[B] \subseteq E$	$[B] = V$	$[B] = M$	$[B] = M$
$CircINF_{cl}(B)$	$\in P$	coNP-complete	coNP-complete	coNP-complete
$CircINF(B)$	$\in P$	coNP-complete	?	coNP-complete
$CircINF^1(B)$	$\in P$	$\in P$	$\in P$	coNP-complete

References

1. Bauland, M., Schneider, T., Schnoor, H., Schnoor, I., Vollmer, H.: The complexity of generalized satisfiability for linear temporal logic. In: Seidl, H. (ed.) FOSSACS 2007. LNCS, vol. 4423, pp. 48–62. Springer, Heidelberg (2007)
2. Beyersdorff, O., Meier, A., Thomas, M., Vollmer, H.: The complexity of propositional implication. ACM CoRR, arXiv:0811.0959v1 [cs.CC] (2008)
3. Beyersdorff, O., Meier, A., Thomas, M., Vollmer, H.: The complexity of reasoning for fragments of default logic. In: Kullmann, O. (ed.) SAT 2009. LNCS, vol. 5584, pp. 51–64. Springer, Heidelberg (2009)
4. Böhler, E., Creignou, N., Reith, S., Vollmer, H.: Playing with Boolean blocks, part I: Post's lattice with applications to complexity theory. SIGACT News 34(4), 38–52 (2003)

5. Buntrock, G., Damm, C., Hertrampf, U., Meinel, C.: Structure and importance of logspace MOD-classes. Mathematical Systems Theory 25, 223–237 (1992)
6. Cadoli, M., Donini, F., Schaerf, M.: On compact representations of propositional circumscription. Theor. Comput. Sci. 182, 205–216 (1995)
7. Cadoli, M., Lenzerini, M.: The complexity of closed world reasoning and circumscription. In: Proc. 8th AAAI, pp. 550–555 (1990)
8. Cadoli, M., Lenzerini, M.: The complexity of propositional closed world reasoning and circumscription. J. Comput. Syst. Sci. 48(2), 255–310 (1994)
9. Durand, A., Hermann, M.: The inference problem for propositional circumscription of affine formulas is coNP-complete. In: Alt, H., Habib, M. (eds.) STACS 2003. LNCS, vol. 2607, pp. 451–462. Springer, Heidelberg (2003)
10. Durand, A., Hermann, M., Nordh, G.: Trichotomies in the complexity of minimal inference (2008), http://www.ida.liu.se/~gusno/
11. Eiter, T., Gottlob, G.: Propositional circumscription and extended closed world reasoning are Π_2^p-complete. Theor. Comput. Sci. 114(2), 231–245 (1993); Addendum in TCS 118, p. 15 (1993)
12. Gelfond, M., Przymusinska, H., Przymusinski, T.C.: On the relationship between circumscription and negation as failure. Artificial Intelligence 38(1), 75–94 (1989)
13. Kirousis, L.M., Kolaitis, P.: The complexity of minimal satisfiability in Post's lattice. Unpublished notes (2001)
14. Kirousis, L.M., Kolaitis, P.G.: The complexity of minimal satisfiability problems. Inf. Comput. 187(1), 20–39 (2003)
15. Kirousis, L.M., Kolaitis, P.G.: A dichotomy in the complexity of propositional circumscription. Theory Comput. Syst. 37(6), 695–715 (2004)
16. Lewis, H.: Satisfiability problems for propositional calculi. Mathematical Systems Theory 13, 45–53 (1979)
17. McCarthy, J.: Circumscription – A form of non-monotonic reasoning. Artificial Intelligence 13, 27–39 (1980)
18. Meier, A., Mundhenk, M., Thomas, M., Vollmer, H.: The complexity of satisfiability for fragments of CTL and CTL*. In: Proc. 2nd Workshop on Reachability Problems, pp. 201–213 (2008)
19. Nordh, G.: A trichotomy in the complexity of propositional circumscription. In: Baader, F., Voronkov, A. (eds.) LPAR 2004. LNCS (LNAI), vol. 3452, pp. 257–269. Springer, Heidelberg (2005)
20. Papadimitriou, C.H.: Computational Complexity. Addison-Wesley, Reading (1994)
21. Pippenger, N.: Theories of Computability. Cambridge University Press, Cambridge (1997)
22. Post, E.: The two-valued iterative systems of mathematical logic. Annals of Mathematical Studies 5, 1–122 (1941)
23. Reith, S.: On the complexity of some equivalence problems for propositional calculi. In: Rovan, B., Vojtáš, P. (eds.) MFCS 2003. LNCS, vol. 2747, pp. 632–641. Springer, Heidelberg (2003)
24. Reith, S., Wagner, K.: The complexity of problems defined by Boolean circuits. In: Proc. MFI 1999, pp. 141–156. World Science Publishing (2005)
25. Schaefer, T.J.: The complexity of satisfiability problems. In: Proc. 10th STOC, pp. 216–226. ACM Press, New York (1978)
26. Schnoor, H.: The complexity of the Boolean formula value problem. Technical report, Theoretical Computer Science, University of Hannover (2005)
27. Vollmer, H.: Introduction to Circuit Complexity. Springer, Heidelberg (1999)

Trichotomy Results on the Complexity of Reasoning with Disjunctive Logic Programs

Mirosław Truszczyński

Department of Computer Science, University of Kentucky, Lexington, KY 40506, USA

Abstract. We present trichotomy results characterizing the complexity of reasoning with disjunctive logic programs. To this end, we introduce a certain definition schema for classes of programs based on a set of allowed arities of rules. We show that each such class of programs has a finite representation, and for each of the classes definable in the schema we characterize the complexity of the existence of an answer set problem. Next, we derive similar characterizations of the complexity of skeptical and credulous reasoning with disjunctive logic programs. Such results are of potential interest. On the one hand, they reveal some reasons responsible for the hardness of computing answer sets. On the other hand, they identify classes of problem instances, for which the problem is "easy" (in P) or "easier than in general" (in NP).

1 Introduction

It is well known that the problem to decide whether a disjunctive logic program has an answer set (the *EAS problem*, for short) is Σ_2^P-complete [1]. It is also well known that putting restrictions on the input instances may affect the complexity. For example, the EAS problem for normal logic programs is NP-complete [2].

In this paper we study the complexity of the EAS problem for classes of disjunctive logic programs that can be defined by sets of program rule *arities*. We show that in each case the problem is either in P, is NP-complete or is Σ_2^P-complete, and we characterize the classes of programs that fall into each category. We extend this result to establish similar characterizations for the problems of skeptical and credulous reasoning with disjunctive logic programs. Such results are of potential interest. On the one hand, they reveal some reasons responsible for the hardness of computing answer sets; cf. Lemmas 4 and 5. On the other hand, they identify classes of problem instances, for which the problem is "easy" (in P) or "easier than in general" (in NP); cf. Lemmas 1 and 2.

Our results can be regarded as *trichotomy* results for the complexity of reasoning tasks in disjunctive logic programming. Similar results are known for the complexity of reasoning in other formalisms: propositional satisfiability [3,4,5], reasoning with minimal models [6], default logic [7], and abductive reasoning [8]. There is however, an important distinction between those earlier papers and our approach. The results contained there are concerned with the setting in which formulas are conjunctions of Boolean relations, and the set of models of a formula is the intersection of the sets of models of its constituent relations (that, in particular, implies the monotonicity of inference from such formulas). The basic results concern the complexity of the satisfiability problem

E. Erdem, F. Lin, and T. Schaub (Eds.): LPNMR 2009, LNCS 5753, pp. 303–315, 2009.

for classes of formulas determined by sets of Boolean relation allowed as formula conjuncts. It turns out that there is a simple characterization of all those classes, for which the problem is in P; moreover for all other classes the problem is NP-complete [3,4,5]. This result can be exploited to characterize the complexity of reasoning with systems, in which basic reasoning tasks reduce to series of satisfiability tests [6,7,8]. In the setting of disjunctive logic programs, these earlier results seem of little help. It is well known that the answer-set semantics is *nonmonotone* and so, logic programs under the answer-set semantics are *not* conjunctions of their rules. Thus, it is unclear that defining classes of programs in terms of semantic properties of individual rules could yield any useful insights.

2   Preliminaries

We fix an infinite countable set At of propositional variables. A *disjunctive program* (or simply, program) over the set of atoms At is a collection of *disjunctive logic program rules*, that is, expressions of the form

$$r = a_1|\ldots|a_k \leftarrow b_1,\ldots,b_m, not\ c_1,\ldots, not\ c_n, \tag{1}$$

where a_i, b_i and c_i are atoms from At. The disjunction $a_1|\ldots|a_k$ is the *head* of r and the conjunction $b_1,\ldots,b_m, not\ c_1,\ldots, not\ c_n$ is the *body* of r. Further, we call the triple $[k, m, n]$ the *arity* of r. We denote the arity of r by $a(r)$ and write $a_1(r)$, $a_2(r)$ and $a_3(r)$ for the three components of $a(r)$. Thus, under this notation, $a(r) = [a_1(r), a_2(r), a_3(r)]$. If $a_1(r) \geq 1$, we call r *proper*. Otherwise, $a_1(r) = 0$ and r is a *constraint*. Programs consisting of proper rules are *proper* programs. Similarly, programs consisting of constraints only are *constraint* programs.

We recall that given a program P and a set of atoms M (an interpretation), the *reduct* of P with respect to M, P^M, is the program obtained by removing all rules with a literal *not c*, where $c \in M$, in the body and, then, removing all negative literals (negated atoms) from the bodies of all remaining rules. A set of atoms M is an *answer set* of a disjunctive program P if M is a minimal model of P^M [9].

For a program P, we denote by $\overline{P}$ and $\overline{\overline{P}}$ the programs consisting of all proper rules and of all constraints in P, respectively. The following result is well known.

Theorem 1. *A set $M \subseteq At$ is an answer set of a program P if and only if M is an answer set of $\overline{P}$ and a model of $\overline{\overline{P}}$.*

One can define classes of logic programs by specifying arities of rules. For instance, the set $\{[1, 1, 0], [0, 1, 0]\}$ defines the set of all Horn programs such that each rule has at most one atom in the body. Some classes of programs do not have such a finitary representation. For instance, the class of all Horn programs with constraints can be defined by the set $\{[k, m, 0]\ |\ k \leq 1, 0 \leq m\}$, but there is no *finite* set of arities that could be used instead. To handle such cases, we introduce now a general representation schema for classes of programs defined by sets of arities.

Let $U = \{0, 1,\ldots\} \cup \{\infty\}$. We consider U to be ordered by the relation $\leq$ (the standard $\leq$ ordering relation on non-negative integers, extended by $i \leq \infty$, for every

$i = 0, 1 \ldots)$. Next, we define $\mathcal{T} = \{[k, m, n] \mid k, m, n \in U\}$. Thus, $\mathcal{T}$ contains all arities, as well as *additional* triples — those containing at least one occurrence of ∞. We refer to triples of that latter sort as *superarities*. We emphasize that superarities are *not* arities as we do not consider infinitary rules. If $\alpha \in \mathcal{T}$, we write α_1, α_2 and α_3 for the components of α.

Let $\alpha, \beta \in \mathcal{T}$. We define $\alpha \preceq \beta$ if (1) $\alpha_i \leq \beta_i$, for $i = 1, 2, 3$ and, (2) if $\alpha_1 = 0$ then $\beta_1 = 0$. We write $\alpha \prec \beta$ when $\alpha \preceq \beta$ and $\alpha \neq \beta$.

If $\Delta \subseteq \mathcal{T}$, then we define $\mathcal{F}(\Delta)$ to be the set of all *finite* programs P that satisfy the following condition: for every rule $r \in P$ there is $\alpha \in \Delta$ such that $a(r) \preceq \alpha$. The condition (2) in the definition of $\preceq$ allows us to distinguish between classes of proper programs and classes of programs with constraints. Indeed, without the condition (2), every class of programs of the form $\mathcal{F}(\Delta)$ would contain constraints. With the condition (2), we can specify constraint-free classes of programs by means of sets Δ such that for every $\alpha \in \Delta$, $\alpha_1 \geq 1$. Including in Δ elements α with $\alpha_1 = 0$ yields classes of programs with constraints. As there are classes of proper programs that are of interest (Horn programs and normal logic programs are typically defined as consisting of proper rules only), the distinction is needed and motivates the condition (2) in the definition of $\preceq$.

Using this schema we can define several important classes of programs. For instance, the class of proper Horn programs can be described as $\mathcal{F}(\{[1, \infty, 0]\})$ and the class of normal logic programs with constraints as $\mathcal{F}(\{[1, \infty, \infty], [0, \infty, \infty]\})$.

Our main goal in this paper is to determine the complexity of the EAS problem when input programs come from classes $\mathcal{F}(\Delta)$, for $\Delta \subseteq \mathcal{T}$.

3   The Case of Finite Δ

In this section we tackle the case when Δ is finite. We note that given a finite set $\Delta \subseteq \mathcal{T}$, the problem to decide the membership of a program in the class $\mathcal{F}(\Delta)$ is in P.

We start by establishing the upper bounds on the complexity of the EAS problem for classes $\mathcal{F}(\Delta)$ given by some particular finite sets Δ of arities. Our first result is concerned with the following classes: $\mathcal{F}(\{[\infty, \infty, 0]\})$ — the class of proper positive disjunctive programs; $\mathcal{F}(\{[1, \infty, 0], [0, \infty, \infty]\})$ — the class of programs whose every rule is either a proper Horn rule or a constraint; $\mathcal{F}(\{[\infty, 1, 0], [0, 1, 0]\})$ — the class of *dual Horn programs*, that is, programs whose every rule, when viewed as a proposi-tional clause, is a dual Horn clause; and $\mathcal{F}(\{[i, j, 0] \in \mathcal{T} \mid i + j \leq 2\})$ — the class of positive programs whose every rule consists of at most two literals. For each of these classes of programs, the EAS problem is easy (that is, in P).

Lemma 1. *If Δ is one of:*

1. $\{[\infty, \infty, 0], [0, 0, 0]\}$
2. $\{[1, \infty, 0], [0, \infty, \infty]\}$
3. $\{[\infty, 1, 0], [0, 1, 0]\}$
4. $\{[i, j, 0] \in \mathcal{T} \mid i + j \leq 2\}$

then the EAS problem for $\mathcal{F}(\Delta)$ is in P.

Proof: We omit the proofs of the first two statements, which are quite straightforward.

Thus, let us assume that $P \in \mathcal{F}(\{[\infty, 1, 0], [0, 1, 0]\})$ or $P \in \mathcal{F}(\{[i, j, 0] \in \mathcal{T} \mid i + j \leq 2\})$. Then P is a dual Horn program, or P is positive and every clause in P consists of two literals. In each case, one can decide in polynomial time whether P has a model. If the answer is "no," then P has no answer sets. Otherwise, P has a model, say M. Since M is a model of $\overline{P}$, there is a subset M' of M such that M' is a minimal model of $\overline{P}$. We have $\overline{P}^{M'} = \overline{P}$. Thus, M' is an answer set of $\overline{P}$. Since M satisfies $\overline{\overline{P}}$ and each rule in $\overline{\overline{P}}$ is of the form $\leftarrow a$ or $\leftarrow a, b$, M' satisfies $\overline{\overline{P}}$, too. Thus, M' is an answer set of P. Again, the assertion follows. $\qquad \square$

The second result establishes sufficient conditions for the EAS problem to be in the class NP. It turns out to be the case for the following three classes of programs: $\mathcal{F}(\{[1, \infty, \infty], [0, \infty, \infty]\})$ — the class of normal logic programs with constraints; $\mathcal{F}(\{[\infty, 1, \infty], [0, \infty, \infty]\})$ — the class of programs whose reducts consist of proper dual Horn rules and constraints; and $\mathcal{F}(\{[\infty, \infty, 0], [0, \infty, 0]\})$ — the class of positive programs.

Lemma 2. *If Δ is one of:*

1. $\{[1, \infty, \infty], [0, \infty, \infty]\}$
2. $\{[\infty, 1, \infty], [0, \infty, \infty]\}$
3. $\{[\infty, \infty, 0], [0, \infty, 0]\}$

then the EAS problem for $\mathcal{F}(\Delta)$ is in NP.

Proof: If $\Delta = \{[1, \infty, \infty], [0, \infty, \infty]\}$, $\mathcal{F}(\Delta)$ consists of normal logic programs with constraints. In this case, the result is well known [2].

Next, let $\Delta = \{[\infty, 1, \infty], [0, \infty, \infty]\}$ and $P \in \mathcal{F}(\Delta)$. To prove the assertion it is enough to show that there is a polynomial time algorithm for deciding whether a set of atoms $M \subseteq At(P)$ is an answer set of P. To this end, we note that M is a minimal model of $\overline{P}^M$ if and only if for every $a \in M$, the program $\overline{P}^M \cup \{\leftarrow a\} \cup \{\leftarrow b \mid b \in At(P) \setminus M\}$ does not have a model. Since $\overline{P}^M \cup \{\leftarrow a\} \cup \{\leftarrow b \mid b \in At(P) \setminus M\}$ is dual Horn, verifying whether M is a minimal model of $\overline{P}^M$ can be accomplished in polynomial time. In addition, checking that M is a model of $\overline{\overline{P}}$ can also be done in polynomial time, too. Thus, in this case, the assertion follows.

Finally, if $\Delta = \{[\infty, \infty, 0], [0, \infty, 0]\}$ and $P \in \mathcal{F}(\Delta)$, then deciding whether P has an answer set is equivalent to deciding whether P has a model. Since the problem to decide whether P has a model is in NP, the assertion follows. $\qquad \square$

Next, we will prove several lower-bound results. We will first exhibit classes of programs of the form $\mathcal{F}(\Delta)$ for which the EAS problem is NP-hard. To this end, we need a lemma establishing the NP-hardness of the SAT problem for some simple classes of CNF theories. For the most part, the result is folklore. We sketch an argument for the sake of completeness.

Lemma 3. *The SAT problem restricted to each of the following classes of CNF theories is NP-hard:*

1. *the class of all CNF formulas ψ such that each clause of ψ is a disjunction of two negated atoms, or of at most three atoms*

2. *the class of all CNF formulas ψ such that each clause of ψ consists of at most two literals, or is a disjunction of two atoms and one negated atom*
3. *the class of all CNF formulas ψ such that each clause of ψ consists of at most two atoms, or of one negated atom, or is a disjunction of an atom and two negated atoms*
4. *the class of all CNF formulas ψ such that each clause of ψ is a disjunction of two atoms, or of at most three negated atoms.*

Proof: We will only prove the case (3). The argument in all other cases is similar.

Let φ be a CNF formula whose every clause has three literals, and let X be a set of atoms occurring in φ. For each atom $z \in X$ we introduce a fresh atom z'. Next, in each clause c we replace some of its positive literals a with $\neg a'$, and some of its negative literals $\neg b$ with b' so that the resulting clause, we will denote it by $\hat{c}$, is the disjunction of exactly one atom and two negated atoms. Such replacements can always be found.

Finally, we introduce one more fresh atom, say f, and define $F(\varphi)$ as follows:

$$F(\varphi) = \{z \vee z' \mid z \in X\} \cup \{f \vee \neg z \vee \neg z' \mid z \in X\} \cup \{\neg f\} \cup$$
$$\{\hat{c} \mid c \text{ is a clause in } \varphi\}$$

It is evident that $F(\varphi)$ is in the class of theories under consideration.

We will show that φ has a model if and only if $F(\varphi)$ has a model. To this end, we note that models of $F(\varphi)$ (if exist) are of the form $M \cup \{z' \mid z \in X \setminus M\}$, where $M \subseteq X$. It is now easy to see that M is a model of φ if and only if $M \cup \{z' \mid z \in X \setminus M\}$ is a model of $F(\varphi)$. Thus, the claim and, consequently, the assertion, follows.

As we noted, the argument for the remaining classes is similar. We only need to change the definition of $\hat{c}$ and use clauses $\neg z \vee \neg z'$ instead of $f \vee \neg z \vee \neg z'$ (there is no need to introduce f, as clauses being the disjunctions of two negated atoms are allowed in formulas in each of the classes considered in (1), (2) and (4)).   □

We will now use Lemma 3 to establish the NP-hardness of the EAS problem for $\mathcal{F}(\Delta)$ for several simple sets $\Delta \subseteq \mathcal{T}$.

Lemma 4. *If Δ is any of:*

1. $\{[1, 0, 1]\}$
2. $\{[2, 0, 0], [0, 0, 1]\}$
3. $\{[3, 0, 0], [0, 2, 0]\}$
4. $\{[2, 1, 0], [0, 2, 0]\}$
5. $\{[2, 0, 0], [1, 2, 0], [0, 1, 0]\}$
6. $\{[2, 0, 0], [0, 3, 0]\}$

then the EAS problem for $\mathcal{F}(\Delta)$ is NP-hard.

Proof (sketch): (1) The proof of the NP-completeness of the EAS problem for normal logic programs given by Marek and Truszczyński [2] establishes the assertion (1).

(2) We will construct a reduction from the SAT problem concerning the class considered in Lemma 3(4). Let φ be a CNF of the appropriate form. We denote by $pos(\varphi)$ the set of all clauses in φ that are of the form $a \vee b$, where $a, b \in At$. We denote by $neg(\varphi)$ the

set of all remaining clauses in φ (all of them are disjunctions of at most three negative literals).

For every clause $c = \neg y_1 \vee \ldots \vee \neg y_k$ in $neg(\varphi)$, we introduce a fresh atom x_c. Next, we define

$$P(\varphi) = \{a | b \leftarrow \; | \; a \vee b \in pos(\varphi)\} \cup$$
$$\{x_c | y_i \leftarrow \; | \; c \in neg(\varphi),$$
$$c = \neg y_1 \vee \ldots \vee \neg y_k, 1 \leq i \leq k\},$$
$$Q(\varphi) = \{\leftarrow \; not \; x_c \; | \; c \in neg(\varphi)\}, \text{ and}$$
$$R(\varphi) = P(\varphi) \cup Q(\varphi).$$

One can now show that φ is satisfiable if and only if $R(\varphi)$ has an answer set (we omit the details). Since $R(\varphi) \in \mathcal{F}(\Delta)$, the assertion follows by Lemma 3(4).

(3)-(6) In all the remaining cases, we exploit the fact that $P \in \mathcal{F}(\Delta)$ has an answer set if and only if P has a model (the same argument that we used in the proof of Lemma 2 applies). The latter problem for each of the cases (3)-(6) is equivalent to the satisfiability problem for the classes considered in Lemma 3(1)-(4), respectively. In each of these cases the problem is NP-hard (Lemma 3), and so the assertion follows. □

The next lemma establishes conditions guaranteeing Σ_2^P-hardness of the EAS problem. Eiter and Gottlob [1] proved that given $P \in \mathcal{F}(\{[2,0,0],[1,3,0],[1,0,1]\})$, it is Σ_2^P-hard to decide whether P has an answer set. The proof can be modified to the case when the class of input programs is restricted to $\mathcal{F}(\{[2,0,0],[1,2,0],[1,0,1]\})$, as clauses of the arity $[1,3,0]$ can be simulated by clauses of arity $[1,2,0]$. Moreover, in the construction provided by Eiter and Gottlob, the only rule of the arity $[1,0,1]$ used is a constraint and it can be simulated by a rule of the arity $[0,0,1]$. Thus, the Σ_2^P-hardness holds also for the class $\mathcal{F}(\{[2,0,0],[1,2,0],[0,0,1]\})$ of programs. We omit the details and state the result only.

Lemma 5. *If Δ is any of:*

1. $\{[2,0,0],[1,2,0],[0,0,1]\}$
2. $\{[2,0,0],[1,2,0],[1,0,1]\}$

then the EAS problem for $\mathcal{F}(\Delta)$ is Σ_2^P-hard.

We will now derive the main result of this section. It provides a complete characterization of the complexity of the EAS problem for the class $\mathcal{F}(\Delta)$. To state the result we introduce one more piece of notation. Given $\Delta, \Theta \subseteq \mathcal{T}$, we write $\Delta \preceq \Theta$ if for every $\alpha \in \Delta$ there is $\beta \in \Theta$ such that $\alpha \preceq \beta$.

Theorem 2. *Let $\Delta \subseteq \mathcal{T}$ be finite.*

(A) If

1. $\Delta \preceq \{[\infty,\infty,0],[0,0,0]\}$, *or*
2. $\Delta \preceq \{[1,\infty,0],[0,\infty,\infty]\}$, *or*
3. $\Delta \preceq \{[\infty,1,0],[0,1,0]\}$, *or*
4. $\Delta \preceq \{[i,j,0] \in \mathcal{T} \; | \; i+j \leq 2\}$,

then the EAS problem for $\mathcal{F}(\Delta)$ is in P.

(B) Otherwise, if

1. $\Delta \preceq \{[1, \infty, \infty], [0, \infty, \infty]\}$, *or*
2. $\Delta \preceq \{[\infty, 1, \infty], [0, \infty, \infty]\}$, *or*
3. $\Delta \preceq \{[\infty, \infty, 0], [0, \infty, 0]\}$,

then the EAS problem for $\mathcal{F}(\Delta)$ is NP-*complete.*

(C) Otherwise, the EAS problem for $\mathcal{F}(\Delta)$ is Σ_2^P-complete.

Proof: The claim (A) follows directly from Lemma 1. Thus, let us assume that Δ does not fall under the scope of (A) and satisfies the assumptions of (B). By Lemma 2, the latter implies that the EAS problem for $\mathcal{F}(\Delta)$ is in NP.

If $\{[1, 0, 1]\} \preceq \Delta$ or $\{[2, 0, 0], [0, 0, 1]\} \preceq \Delta$, the NP-hardness of the EAS problem for $\mathcal{F}(\Delta)$ follows from Lemma 4, parts (1) and (2), respectively. Thus, let us assume that $\{[1, 0, 1]\} \npreceq \Delta$ and $\{[2, 0, 0], [0, 0, 1]\} \npreceq \Delta$.

Since $\{[1, 0, 1]\} \npreceq \Delta$, we have $\Delta \preceq \{[\infty, \infty, 0], [0, \infty, \infty]\}$. Since Δ does not satisfy the condition (A2), $\{[2, 0, 0]\} \preceq \Delta$. Since $\{[2, 0, 0], [0, 0, 1]\} \npreceq \Delta$, $\{[0, 0, 1]\} \npreceq \Delta$. Thus, $\Delta \preceq \{[\infty, \infty, 0], [0, \infty, 0]\}$.

Since Δ does not satisfy the condition (A1), $\{[0, 1, 0]\} \preceq \Delta$. Similarly, since Δ does not satisfy the condition (A3), $\{[1, 2, 0]\} \preceq \Delta$ or $\{[0, 2, 0]\} \preceq \Delta$. We also have that Δ does not satisfy the condition (A4). Thus, there is $\alpha \in \Delta$ such that $\alpha_1 + \alpha_2 \geq 3$. Since we already proved that $\{[2, 0, 0]\} \preceq \Delta$, it follows that at least one of the following conditions holds: $\{[3, 0, 0], [0, 2, 0]\} \preceq \Delta$, $\{[2, 1, 0], [0, 2, 0]\} \preceq \Delta$, $\{[2, 0, 0], [1, 2, 0], [0, 1, 0]\} \preceq \Delta$, or $\{[2, 0, 0], [0, 3, 0]\} \preceq \Delta$. Thus, the NP-hardness of the EAS problem for $\mathcal{F}(\Delta)$ follows again from Lemma 4 and completes the proof of (B).

To prove (C), we observe that if Δ does not fall under the scope of (B), then $\{[2, 0, 0], [1, 2, 0], [0, 0, 1]\} \preceq \Delta$. Indeed, since Δ does not satisfy (B1), $\{[2, 0, 0]\} \preceq \Delta$. Similarly, since Δ does not satisfy (B2), $\{[1, 2, 0]\} \preceq \Delta$. Finally, since Δ does not satisfy (B3), $\{[0, 0, 1]\} \preceq \Delta$ or $\{[1, 0, 1]\} \preceq \Delta$. Thus, the Σ_2^P-hardness follows by Lemma 5. Since the EAS problem is in Σ_2^P even without any restrictions on the class of programs, both (C) and the assertion of the lemma follows.   □

4   The Case of Infinite Δ

The question we study now is whether there are interesting classes of programs of the form $\mathcal{F}(\Delta)$, when Δ is infinite. The main result of this section is that by allowing Δ to be infinite, we do not obtain *any* new classes of programs. In other words, for every class of programs of the form $\mathcal{F}(\Delta)$ there is a finite set $\Delta' \subseteq \mathcal{T}$ such that $\mathcal{F}(\Delta) = \mathcal{F}(\Delta')$.

A sequence $\{\alpha^k\}_{k=1}^{\infty}$ is *monotone* (*strictly* monotone) if for every k, $\alpha^k \preceq \alpha^{k+1}$ ($\alpha^k \prec \alpha^{k+1}$, respectively). Let $\{\alpha^k\}_{k=1}^{\infty}$ be a *monotone* sequence of elements of $\mathcal{T}$. We define the *limit* of this sequence as $\alpha^{\infty} = [(\alpha^{\infty})_1, (\alpha^{\infty})_2, (\alpha^{\infty})_3]$, where $(\alpha^{\infty})_i = \sup\{(\alpha^k)_i \mid k = 1, 2 \ldots, \}$, for $i = 1, 2, 3$. Let $\Delta \subseteq \mathcal{T}$. A monotone sequence $\{\alpha^k\}_{k=1}^{\infty}$ of elements of Δ is *maximal* if there is *no* $\alpha \in \Delta$ such that $\alpha^{\infty} \prec \alpha$. We define $A(\Delta)$ to be the set of the limits of maximal sequences in Δ.

We have the following two lemmas (we omit the proof of the first one as it is evident).

Lemma 6. *Let $\{\beta^k\}_{k=1}^{\infty}$ be a strictly monotone sequence of elements from T. For every $\alpha \in T$, if $\alpha \prec \beta^{\infty}$, then there is k such that $\alpha \prec \beta^k$.*

Lemma 7. *Let $\Delta \subseteq T$ and $\alpha \in \Delta$. Then, there is $\alpha' \in A(\Delta)$ such that $\alpha \preceq \alpha'$.*

Proof: Let $X = \{\beta \in \Delta \mid \alpha \preceq \beta\}$. If X has a maximal element, say γ, a sequence with each term equal to γ is maximal. Its limit, also equal to γ, clearly satisfies $\gamma \in A(\Delta)$ and $\alpha \preceq \gamma$. Thus, the assertion follows.

Otherwise, X has no maximal elements. Let α^1 be any element in X (we note that $X \neq \emptyset$, as $\alpha \in X$). Let $k \geq 1$ and let $\langle \alpha^1, \ldots, \alpha^k \rangle$ be a strictly monotone sequence of k elements in X, for some $k \geq 1$. Since X has no maximal elements, X contains elements that are strictly greater than α^k. Let us select as α^{k+1} an element $\beta \in X$ such that $\alpha^k \prec \beta$ and $\alpha_i^k < \beta_i$ on as many positions $i = 1, 2, 3$ as possible. An infinite sequence we define in that way, we will denote it by α, is strictly monotone. Let us assume that there is $\beta \in \Delta$ such that $\alpha^{\infty} \prec \beta$. It follows that there is j, $1 \leq j \leq 3$, such that $(\alpha^{\infty})_j < \beta_j$. Thus, $(\alpha^{\infty})_j = m$, for some integer m, and there is n such that $(\alpha^n)_j = m$. Since $\alpha^{n+1} \preceq \beta$ and $(\alpha^n) = (\alpha^{n+1})_j = m$, the number of positions i such that $(\alpha^n)_i < (\alpha^{n+1})_i$ is strictly smaller than the number of positions i such that $(\alpha^n)_i < \beta_i$. Since $\beta \in X$, that contradicts the way we constructed the sequence α.

It follows that the sequence $\{\alpha^k\}_{k=1}^{\infty}$ is maximal for Δ and so, the assertion follows in this case, too. $\square$

We now have the following properties. We provide a proof for the first of them and omit proofs, rather direct and technical, of the remaining two.

Proposition 1. *For every $\Delta \subseteq T$, $\mathcal{F}(\Delta) = \mathcal{F}(A(\Delta))$.*

Proof: To prove the assertion, it is enough to show that for every *arity* α (no occurrence of ∞), $\{\alpha\} \preceq \Delta$ if and only if $\{\alpha\} \preceq A(\Delta)$. Let us first assume that $\{\alpha\} \preceq \Delta$. It follows that there is an element $\alpha' \in \Delta$ such that $\alpha \preceq \alpha'$. By Lemma 7, there is $\alpha'' \in A(\Delta)$ such that $\alpha' \preceq \alpha''$. Thus, $\alpha \preceq \alpha''$ and so, $\{\alpha\} \preceq A(\Delta)$.

Conversely, let $\{\alpha\} \preceq A(\Delta)$. It follows that there is $\beta \in A(\Delta)$ such that $\alpha \preceq \beta$. Since $\beta \in A(\Delta)$, there is a monotone sequence $\{\beta^k\}_{k=1}^{\infty}$ of elements of Δ such that its limit is β. Wlog we can assume that either starting with some k_0, the sequence $\{\beta^k\}_{k=1}^{\infty}$ is constant, or the sequence β is strictly monotone. In the first case, $\alpha \preceq \beta = \beta^{k_0}$. Since $\beta^{k_0} \in \Delta$, $\{\alpha\} \preceq \Delta$. In the second case, Lemma 6 implies that there is k such that $\alpha \prec \beta^k$, and again $\{\alpha\} \preceq \Delta$ follows. $\square$

Proposition 2. *For every Δ, $A(\Delta)$ is an antichain.*

Proposition 3. *Every antichain in the partially ordered set $\langle T, \preceq \rangle$ is finite.*

These properties imply the main result of this section. It asserts that every class of programs $\mathcal{F}(\Delta)$ can be defined by means of a finite set Δ' that is an antichain in $\langle T, \preceq \rangle$.

Theorem 3. *For every set $\Delta \subseteq T$ there is a finite subset $\Delta' \subseteq T$ such that Δ' is an antichain and $\mathcal{F}(\Delta) = \mathcal{F}(\Delta')$.*

Proof: Let us define $\Delta' = A(\Delta)$. By Propositions 2 and 3, Δ' is a finite antichain in $\langle T, \preceq \rangle$, and by Proposition 1, $\mathcal{F}(\Delta) = \mathcal{F}(\Delta')$. Thus, the theorem follows. $\square$

5   The Complexity of Skeptical and Credulous Reasoning

The EAS problem is just one example of a reasoning task that arises in the context of disjunctive logic programs with the answer-set semantics. There are several other tasks that are of interest, too. They concern deciding whether a program nonmonotonically entails a literal, that is an atom, say a, or its negation $\neg a$. For a disjunctive logic program P and a literal l we say that

1. P *skeptically entails* l, written $P \models_s l$, if $M \models l$, for every answer set M of P (we recall that if M is a model (a set of atoms) and a is an atom, $M \models a$ if $a \in M$, and $M \models \neg a$ if $a \notin M$)
2. P *credulously entails* l, written $P \models_c l$, if there is an answer set M of P such that $M \models l$.

We note that $P \models_s l$ if and only if $P \not\models_c \bar{l}$, where $\bar{l}$ is l's *dual* literal. Thus, to establish fully the complexity of deciding nonmonotonic entailment it is enough to focus on deciding whether $P \models_c \neg a$ and $P \models_s \neg a$, where a is an atom. These two decision tasks were studied by Eiter and Gottlob [1], who proved that, in general, the first one is Σ_2^P-complete and the second one is Π_2^P-complete.

Reasoning with answer sets is related to circumscription and closed-world reasoning with propositional theories. A detailed study of the complexity of those forms of reasoning was conducted by Cadoli and Lenzerini [10]. Using Theorem 2 and one of the results from that paper (which we state in the proof below), one can characterize in terms of our definition schema the complexity of deciding, given a program P and an atom a, whether $P \models_c \neg a$ and $P \models_s \neg a$. The two problems are addressed in the following two theorems.

Theorem 4. *Let $\Delta \subseteq \mathcal{T}$ be finite.*

(A) If

1. $\Delta \preceq \{[1, \infty, 0], [0, \infty, \infty]\}$*, or*
2. $\Delta \preceq \{[\infty, 1, 0], [0, 1, 0]\}$*, or*
3. $\Delta \preceq \{[i, j, 0] \in \mathcal{T} \mid i + j \leq 2\}$,

then the problem to decide whether $P \models_c \neg a$, where $P \in \mathcal{F}(\Delta)$ and a is an atom, is in P.

(B) Otherwise, if

1. $\Delta \preceq \{[1, \infty, \infty], [0, \infty, \infty]\}$*, or*
2. $\Delta \preceq \{[\infty, 1, \infty], [0, \infty, \infty]\}$*, or*
3. $\Delta \preceq \{[\infty, \infty, 0], [0, \infty, 0]\}$,

then the problem to decide whether $P \models_c \neg a$, where $P \in \mathcal{F}(\Delta)$ and a is an atom, is NP-*complete.*

(C) Otherwise, the problem to decide whether $P \models_c \neg a$, where $P \in \mathcal{F}(\Delta)$ and a is an atom, is Σ_P^2-complete.

Proof: It is well known that P has an answer set M such that $M \models \neg a$ (that is, $a \notin M$) if and only if $P \cup \{ \leftarrow a\}$ has an answer set. Let $\Delta \subseteq \mathcal{T}$ be finite and let us define $\Delta' = \Delta \cup \{[0, 1, 0]\}$. Clearly, if $P \in \mathcal{F}(\Delta)$, then $P \cup \{ \leftarrow a\} \in \mathcal{F}(\Delta')$. Moreover, if Δ falls under the scope of (A) ((A) or (B), respectively) of this theorem then Δ' falls under the scope of (A) ((A) or (B), respectively) of Theorem 2. Consequently, the upper bound follows by Theorem 2.

The proof of hardness exploits earlier results on the hardness of the EAS problem. The reductions are provided by the following constructions. For a program P we define $P' = P \cup \{a|b\}$, $P'' = P \cup \{a \leftarrow not\ b; b \leftarrow not\ a\}$, and $P''' = \overline{P} \cup \{a \leftarrow bd(r) \mid r \in \overline{\overline{P}}\}$, where a and b are fresh atoms. Clearly, P has an answer set if and only if $P' \models_c \neg a$ ($P'' \models_c \neg a$, $P''' \models_c \neg a$, respectively). We omit the details due to space limits. $\square$.

Theorem 5. *Let $\Delta \subseteq \mathcal{T}$ be finite.*

(A) If $\Delta \preceq \{[1, \infty, 0], [0, \infty, \infty]\}$, then the problem to decide whether $P \models_s \neg a$, where $P \in \mathcal{F}(\Delta)$ and a is an atom, is in P.

(B) Otherwise, if

1. *$\Delta \preceq \{[1, \infty, \infty], [0, \infty, \infty]\}$, or*
2. *$\Delta \preceq \{[\infty, 1, \infty], [0, \infty, \infty]\}$*

then the problem to decide whether $P \models_s \neg a$, where $P \in \mathcal{F}(\Delta)$ and a is an atom, is coNP-*complete.*

(C) Otherwise, the problem to decide whether $P \models_s \neg a$, where $P \in \mathcal{F}(\Delta)$ and a is an atom, is Π_2^P-complete.

Proof: It is well known that P has an answer set such that $M \not\models \neg a$ (that is, $a \in M$) if and only if $P \cup \{ \leftarrow not\ a\}$ has an answer set. That observation implies all upper bound results (by a similar argument as that used in the proof of the previous theorem).

We will now prove the lower bounds for the cases (B) and (C). Let us assume that Δ does not satisfy (A) but falls under the scope of (B). If Δ satisfies (B1), then $\{[1, 0, 1]\} \preceq \Delta$. Let $P \in \mathcal{F}(\Delta)$ and let a and a' be fresh atoms. We note that P has an answer set if and only if $P \cup \{a \leftarrow not\ a'\}$ has an answer set M such that $a \in M$ or, equivalently, if and only if $P \cup \{a \leftarrow not\ a'\} \not\models_s \neg a$. Thus, the hardness follows (cf. Lemma 4(1)).

Let us assume then that Δ does not satisfy (B1). Then, we have that $\{[2, 0, 0]\} \preceq \Delta$. It follows from the results of Cadoli and Lenzerini [10] that it is NP-complete to decide whether a given 2CNF theory, whose every clause is a disjunction of two atoms, has a minimal model that contains a given atom a. As minimal models of such theories are precisely answer sets of the corresponding disjunctive program, it follows that given a program $P \in \mathcal{F}(\Delta)$ and an atom a, it is coNP-complete to decide whether $P \models_s \neg a$.

Finally, let us assume that neither (A) nor (B) apply to Δ. Then $[1, 2, 0] \in \Delta$ and $[2, 0, 0] \in \Delta$. Eiter and Gottlob [1] proved that if $P \in \mathcal{F}(\{[2, 0, 0], [1, 3, 0]\})$ and a is an atom then it is Π_2^P-hard to decide whether $P \models_s \neg a$. That result can be strengthened to the case when $P \in \mathcal{F}(\{[2, 0, 0], [1, 2, 0]\})$, as clauses of the arity $[1, 3, 0]$ can be simulated by clauses of arity $[1, 2, 0]$ (cf. the comments preceding Lemma 5). $\square$

We note that credulous reasoning is simple (in P or in NP) for several classes of programs. In contrast, there are fewer classes of programs, for which skeptical reasoning is simple (in P or coNP). The main reason behind this asymmetry is that in the cases (A2), (A3) and (B3) of Theorem 4 (positive programs) answer sets and minimal models coincide. Thus, in these cases, credulous reasoning asks for the existence of a *minimal* model that does not contain an atom a, which is equivalent to the existence of a model (not necessarily minimal) that does not contain a. In other words, the requirement of minimality becomes immaterial (one source of complexity disappears). This is not so with skeptical reasoning, where not having a in any minimal model is not the same as not having a in any model. A similar comparison of skeptical and credulous reasoning for positive programs was offered by Eiter and Gottlob for the coarser setting of classes of programs they considered [1].

6 Another Representation Schema

Finally, we consider briefly an alternative way, in which classes of programs could be described by means of arities of rules. When defining the class $\mathcal{F}(\Delta)$, we view each element $\alpha \in \Delta$ as a shorthand for the set of all arities β such that $\beta \preceq \alpha$. In other words, Δ is an *implicit* representation of the set of all allowed arities: not only those arities that are explicitly listed in Δ are legal but also those that are "dominated" by them.

There is another, more direct (more explicit), way to use arities to define classes of programs. Let $\Delta \subseteq \mathcal{T}$ be a set of *arities*, that is, we now do not allow superarities in Δ. We define $\mathcal{G}(\Delta)$ to consist of all finite programs P such that for every rule $r \in P$, $a(r) \in \Delta$. Thus, when defining the class $\mathcal{G}(\Delta)$, Δ serves as an *explicit* specification of the set of allowed arities.

One can show that the results of Section 3 can be adapted to the setting of classes of the form $\mathcal{G}(\Delta)$, where Δ is a *finite* set of arities. In particular, we have the following result.

Theorem 6. *Let* $\Delta \subseteq \mathcal{T}$ *be a finite set of arities. If there are no* $k, m \geq 1$ *such that* $\{[k, 0, 0]\} \in \Delta$ *or* $\{[k, 0, m]\} \in \Delta$, *then the EAS problem for* $\mathcal{G}(\Delta)$ *is in* P. *Otherwise:*

(A) If

1. $\Delta \preceq \{[\infty, \infty, 0], [0, 0, 0]\}$, *or*
2. $\Delta \preceq \{[1, \infty, 0], [0, \infty, \infty]\}$, *or*
3. $\Delta \preceq \{[\infty, 1, 0], [0, 1, 0]\}$, *or*
4. $\Delta \preceq \{[i, j, 0] \in \mathcal{T} \mid i + j \leq 2\}$,

then the EAS problem for $\mathcal{G}(\Delta)$ *is in* P.

(B) Otherwise, if

1. $\Delta \preceq \{[1, \infty, \infty], [0, \infty, \infty]\}$, *or*
2. $\Delta \preceq \{[\infty, 1, \infty], [0, \infty, \infty]\}$, *or*
3. $\Delta \preceq \{[\infty, \infty, 0], [0, \infty, 0]\}$,

then the EAS problem for $\mathcal{G}(\Delta)$ *is NP-complete.*

(C) Otherwise, the EAS problem for $\mathcal{G}(\Delta)$ *is* Σ_P^2-*complete.*

Proof (sketch): Let us first assume that there are no $k, m \geq 1$ such that $\{[k, 0, 0]\} \in \Delta$ or $\{[k, 0, m]\} \in \Delta$, and let $P \in \mathcal{G}(\Delta)$. Then every rule in $\overline{P}$ has at least one positive atom in the body and so, $M = \emptyset$ is the unique answer set of $\overline{P}$. It can be verified in polynomial time whether $M = \emptyset$ is a model of $\overline{\overline{P}}$. Thus, the EAS problem for programs in $\mathcal{G}(\Delta)$ can be decided in polynomial time.

To prove the remaining part of the assertion, we note that the upper bound is implied directly by Theorem 2 (as $\mathcal{G}(\Delta) \subseteq \mathcal{F}(\Delta)$). To prove the lower bounds, we observe that if there are $k, m \geq 1$ such that $\{[k, 0, 0]\} \in \Delta$ or $\{[k, 0, m]\} \in \Delta$, then the EAS problem for $\mathcal{F}(\Delta)$ can be reduced to the EAS problem for $\mathcal{G}(\Delta)$. Indeed, let $P \in \mathcal{F}(\Delta)$ and let $r \in P$. Then there is $\alpha \in \Delta$ such that $a(r) \preceq \alpha$. Having $\{[k, 0, 0]\} \in \Delta$ or $\{[k, 0, m]\} \in \Delta$, where $k, m \geq 1$, allows us to "simulate" the effect of r with a rule r' of arity α obtained by repeating atoms in the head of r, and by inserting an atom a and a negated atom $not\ b$, where a and b are fresh, as many times as necessary in the body of r to "reach" the arity α. We also add the rule $a|\ldots|a \leftarrow$ or $a|\ldots|a \leftarrow not\ a', \ldots, not\ a'$, where a' is another fresh atom and a and $not\ a'$ are repeated k, or k and m times, respectively.   □

Thus, as long as Δ is finite, our approach and results obtained earlier apply to the classes $\mathcal{G}(\Delta)$, as well. In particular, given a finite Δ, there is a polynomial-time (in the size of Δ) algorithm to decide which of the cases of Theorem 6 applies. In addition, there is a polynomial-time algorithm to decide whether $P \in \mathcal{G}(\Delta)$.

If Δ is infinite, the situation is different. For some infinite sets Δ there is no problem. For instance, when Δ is specified by means of a finite set Δ' of arities and superarities and consists of all arities α such that $\{\alpha\} \preceq \Delta$, then we have $\mathcal{G}(\Delta) = \mathcal{F}(\Delta')$. Thus, Theorem 2 can be used to determine the complexity of the EAS problem for programs from the class $\mathcal{G}(\Delta)$. However, in general, each finitary representation schema for the classes $\mathcal{G}(\Delta)$ would need to be studied separately and, possibly, it might not lend itself easily to reductions to Theorem 2.

7   Discussion

In the paper, we studied classes of programs defined in terms of "legal" arities of rules. Specifically, we focused on classes of programs of the form $\mathcal{F}(\Delta)$, where $\Delta \subseteq \mathcal{T}$. We proved that each such class has a finite representation and, for each finite set Δ, we determined the complexity of reasoning tasks for programs from $\mathcal{F}(\Delta)$. We also considered a related family of classes of programs, namely those of the form $\mathcal{G}(\Delta)$, where $\Delta \subseteq \mathcal{T}$ consists of arities only. For classes of programs of that form we established the complexity of the EAS problem, as well as that of the credulous and skeptical reasoning. In each case, the complexity is given by one of three complexity classes (P, NP-complete, and Σ_2^P-complete; or P, coNP-complete, and Π_2^P-complete, depending on the type of the reasoning task).

As we noted, our trichotomy results have some similarity to the dichotomy result by Schaefer, and its corollaries for other logic formalisms: the abductive reasoning [8], reasoning with minimal models [6] and, reasoning in default logic [7]. The classes of theories and formulas considered in those papers are defined in terms of boolean relations that are allowed in the language [3,4,5]. That definition schema satisfies the dichotomy property: for every class of formulas definable in that schema, the satisfiability problem is in P, or is NP-complete. The monotonicity of the propositional logic (the set of models of the conjunction of two formulas is the intersection of the sets of models of the conjuncts) is a fundamental property required by that result. Since logic programs with the answer-set semantics do not satisfy the monotonicity property, it is unclear how to extend that formalism the approach originated by Schaefer. Thus, we based our approach on a different definition schema developed specifically for programs, and related to the "complexity" of rules as measured by the numbers of atoms in the head, and positive and negative literals in the body.

It turns out though, that some classes of programs/theories appear prominently in both settings (for instance: Horn programs and Horn theories; positive programs with no more than two literals per rule and 2CNF theories). It is then an interesting problem whether a result based on the classification in terms of types of boolean relations can be obtained for disjunctive logic programs. One possibility might be to consider a more general setting of answer-set programs in the language of propositional logic under the semantics of equilibrium models [11].

Acknowledgments

This work was partially supported by the NSF grant IIS-0325063 and the KSEF grant KSEF-1036-RDE-008.

References

1. Eiter, T., Gottlob, G.: On the computational cost of disjunctive logic programming: propositional case. Annals of Mathematics and Artificial Intelligence 15, 289–323 (1995)
2. Marek, W., Truszczyński, M.: Autoepistemic logic. Journal of the ACM 38, 588–619 (1991)
3. Schaefer, T.: The complexity of satisfiability problems. In: Proceedings of the 10th Annual ACM Symposium on Theory of Computing, STOC 1978, pp. 216–226 (1978)
4. Bulatov, A.A., Jeavons, P., Krokhin, A.A.: Classifying the complexity of constraints using finite algebras. SIAM J. Comput. 34, 720–742 (2005)
5. Creignou, N., Khanna, S., Sudan, M.: Complexity Classifications of Boolean Constraint Satisfaction Problems. SIAM, Philadelphia (2001)
6. Cadoli, M.: The complexity of model checking for circumscriptive formulae. Information Processing Letters 44, 113–118 (1992)
7. Chapdelaine, P., Hermann, M., Schnoor, I.: Complexity of default logic on generalized conjunctive queries. In: Baral, C., Brewka, G., Schlipf, J. (eds.) LPNMR 2007. LNCS (LNAI), vol. 4483, pp. 58–70. Springer, Heidelberg (2007)
8. Nordh, G., Zanuttini, B.: What makes propositional abduction tractable. Artificial Intelligence 172, 1245–1284 (2008)
9. Gelfond, M., Lifschitz, V.: Classical negation in logic programs and disjunctive databases. New Generation Computing 9, 365–385 (1991)
10. Cadoli, M., Lenzerini, M.: The complexity of propositional closed world reasoning and circumscription. Journal of Computer and System Sciences 48, 255–310 (1994)
11. Ferraris, P., Lifschitz, V.: Mathematical foundations of answer set programming. In: We Will Show Them! Essays in Honour of Dov Gabbay, pp. 615–664. College Publications (2005)

Belief Logic Programming: Uncertainty Reasoning with Correlation of Evidence[*]

Hui Wan and Michael Kifer

State University of New York at Stony Brook
Stony Brook, NY 11794, USA

Abstract. Belief Logic Programming (BLP) is a novel form of quantitative logic programming in the presence of uncertain and inconsistent information, which was designed to be able to combine and correlate evidence obtained from non-independent information sources. BLP has non-monotonic semantics based on the concepts of belief combination functions and is inspired by Dempster-Shafer theory of evidence. Most importantly, unlike the previous efforts to integrate uncertainty and logic programming, BLP can correlate structural information contained in rules and provides more accurate certainty estimates. The results are illustrated via simple, yet realistic examples of rule-based Web service integration.

1 Introduction

Quantitative reasoning has been widely used for dealing with uncertainty and inconsistency in knowledge representation, and, more recently, on the Semantic Web. A less explored issue in quantitative reasoning is combining correlated pieces of information. Most works disregard correlation or assume that all the information sources are independent. Others make an effort to take some forms of correlation into account, but only in an ad hoc manner.

Among the models of uncertainty, probabilistic logic programming is particularly popular: [4,13,14,17,19,20]—just to name a few. However, when these approaches are used to combine evidence from different sources, the usage of probabilistic models becomes questionable, as discussed in Section 6. Another well-established way of dealing with uncertainty is Fuzzy Logic [29]. It has been successful in many application domains, but remains controversial due to some of its properties [6,8]. For example, if S' is a complement of the fuzzy set S, then $S \cap S' \neq \varnothing$ and even $S \subset S'$ are possible. This property is problematic for some applications.

Dempster-Shafer theory of evidence [5,22] has also been central to many approaches to quantitative reasoning [1,2,3,15,21,23,27]. This theory is based on belief functions [22]—a generalization of probability distributions, which represents degrees of belief in various statements. If these beliefs come from different

---

[*] This work is part of the SILK (Semantic Inference on Large Knowledge) project sponsored by Vulcan, Inc.

E. Erdem, F. Lin, and T. Schaub (Eds.): LPNMR 2009, LNCS 5753, pp. 316–328, 2009.

sources, the belief functions must be combined in order to obtain more accurate information. The difficult problem here is that these sources might not be independent.

Yet another line of work is based on deductive database methodology without dedicating to any particular theory of modeling uncertainty [10,11,12,15,24].

To the best of our knowledge, all existing works avoid correlating belief derivation paths, which often leads to counter-intuitive behavior when such correlation is essential for correctness. Most approaches simply restrict logic dependencies to avoid combining sources that are not independent [10,23]. Those that do not, might yield incorrect or inaccurate results when combining sources that are correlated due to overlapping belief derivation paths. For example, consider two rules A :- $B \wedge C$ and A :- $B \wedge D$, each asserting its conclusion with certainty 0.5. The approach in [15] would directly combine the certainty factors for A derived from the two rules as if they are independent, assigning A a combined certainty, which is likely going to be too high. Clearly, the independence assumption does not hold here, as both rules rely on the same fact B. A few notable exceptions include Baldwin's [1,2], Lakshmanan's [12] and Kersting's [9] approaches. Kersting et al. provide a very general framework, which could, in principle, be used to handle correlation. However, combination of two inconsistent conclusions is hard to explain in probability theory. Both Baldwin's and Lakshmanan's methods assume that every pair of rules with the same head have the same correlation. Consequently their methods are inadequate for scenarios such as the one described in Sections 2 and 7.

This paper introduces a novel form of quantitative reasoning, called *Belief Logic Programming* (BLP). BLP was designed specifically to account for correlation of evidence obtained from non-independent and, possibly, contradictory information sources. BLP has non-monotonic semantics based on belief combination functions and inspired by Dempster-Shafer theory of evidence. The BLP theory is orthogonal to the choice of a particular method of combining evidence and, in fact, several different methods can be used simultaneously for different pieces of uncertain information.[1] Most importantly, unlike the previous efforts to integrate uncertainty and logic programming, BLP can correlate structural information contained in rules and provides more accurate certainty estimates. The framework and the results are illustrated using simple, yet realistic examples of rule-based integration of Web services that deal with uncertain information.

This paper is organized as follows. Section 2 presents a motivating example, which is revisited in Section 7 from a technical standpoint. Section 3 provides background on Dempster-Shafer theory of evidence. Then, we define the syntax of BLP in Section 4 and the semantics in Section 5. Section 6 discusses several aspects of BLP and relates it to some other theories of non-monotonic reasoning. Section 8 concludes the paper.

---

[1] The reader should not confuse Dempster-Shafer theory of evidence with Dempster's combination rule. BLP does not depend on that combination rule, but can use it for modeling beliefs.

2   Motivating Example

A group of Stony Brook students is planning a trip to see a Broadway musical. Normally, it takes 1.5 hours by car to get to Manhattan, but the students know that Long Island Expressway is not called by the locals "the longest parking lot in America" for nothing. The students consult a traffic service, which integrates information from several independent information sources to provide traffic advisory along various travel routes. Let us assume that these sources are:

- weather forecast (rain, snow, fog)
- social activity (parades, motorcades, marathons)
- police activity (accidents, emergencies)
- roadwork

The service uses the following rules (which are simplified for this example) to generate advisories:

1. If the weather is bad, and there is roadwork along the route, the likelihood of a delay is 0.9.
2. If there is roadwork and social activities along the route, the likelihood of a delay is 0.8.
3. If there is roadwork and police activity along the route, the likelihood of a delay is 0.99.

These rules are expressed in BLP as shown below, where $?r$ is the variable that represents the travel route.

$$[0.9, 1] \quad delay(?r) \; :- \; roadwork(?r) \land bad\_weather(?r)$$
$$[0.8, 1] \quad delay(?r) \; :- \; roadwork(?r) \land social\_act(?r)$$
$$[0.99, 1] \quad delay(?r) \; :- \; roadwork(?r) \land police\_act(?r)$$

The service generates advisories expressed as the likelihood of delays along the routes of interest. The students do not want to miss the show due to traffic, but they also have conference deadlines and so do not want to leave too early. They decide that if the advisory says that the likelihood of delays is between 0.2 and 0.4 then they add one extra hour to the trip time. If the likelihood is between 0.4 and 0.6, then they add two hours, and if the likelihood is over 0.6 then they take a train.

The key observation here is that the three rules used in generating the advisory are not independent—they all rely on the roadwork information from Department of Transportation. Our intuition suggests that predictions based on the independence assumption might cost our student a Broadway show, a few hours of sleep, or a conference paper.

As mentioned in the introduction, the novelty of our approach is that it does not assume that the information sources are independent and instead properly correlates inferences obtained using the rules that represent these sources. In Section 7, we will return to this example and show that our approach improves the quality of the advisory and could help the students avoid unnecessary grief.

3   Preliminaries

In BLP, uncertainty is represented using belief functions of Dempster-Shafer Theory [5,22].

In Probability Theory, a probability distribution function assigns probabilities to mutually exclusive events. In Dempster-Shafer Theory, a mass function assigns evidence (also known as degree of belief, certainty, or support) to *sets of* mutually exclusive states. For example, the state $\{A, B, C\}$ may have an associated degree of belief 0.4, which means that either A or B or C is true with certainty 0.4. This statement does *not* imply anything about the individual truth of A, B, or C, or about any of the sets $\{A, B\}$, $\{A, C\}$, or $\{B, C\}$.

Let $\mathcal{U}$ be the **universal set**—a set of all possible mutually exclusive states under consideration. The power set $\mathbb{P}(\mathcal{U})$ is the set of all possible sub-sets of $\mathcal{U}$, including the empty set, $\varnothing$.

A **mass function** is a mapping $\mathtt{mass} : \mathbb{P}(\mathcal{U}) \longrightarrow [0, 1]$ such that $\mathtt{mass}(\varnothing) = 0$ and $\sum_{S \in \mathbb{P}(\mathcal{U})} \mathtt{mass}(S) = 1$. $\mathtt{mass}(S)$ expresses the proportion of all relevant and available evidence that supports the claim that the actual true state belongs to the set $S \subseteq \mathcal{U}$ and to no known proper subset of S. If it is also known that the actual state belongs to a subset S' of S then $\mathtt{mass}(S')$ will also be non-zero.

The **belief** associated with the set S is defined as the sum of all the masses of S's subsets: $\mathtt{belief}(S) = \sum_{S' \subseteq S} \mathtt{mass}(S')$.

Dempster's combination rule [5,22] addresses the issue of how to combine two independent sets of mass assignments. It emphasizes the agreement between multiple sources and ignores correlation and conflict through a normalization factor. As it turns out, ignoring these aspects leads to unexpected derivations [30]. To avoid this problem, BLP supports a family of combination methods, e.g., the rules in [21,28], and does not commit to any particular one. As a special case, Dempster's combination rule and its extensions can be used when appropriate.

4   Syntax of BLP

A **belief logic program** (or a *blp*, for short) is a set of annotated rules. Each **annotated rule** has the following format:

$$[v, w] \ X \ \text{:-} \ Body$$

where X is a positive atom and *Body* is a Boolean combination of atoms, i.e., a formula composed out of atoms by conjunction, disjunction, and negation. We will use capital letters to denote positive atoms, e.g., A, and a bar over such a letter will denote negation, e.g., $\overline{A}$. The annotation $[v, w]$ is called a **belief factor**, where v and w are real numbers such that $0 \leq v \leq w \leq 1$.

The informal meaning of the above rule is that if *Body* is true, then this rule supports X to the degree v and $\overline{X}$ to the degree $1 - w$. The difference, $w - v$, is the *information gap* (or the degree of ignorance) with regard to X.

Note that, in keeping with the theory of evidence, BLP uses what is known as explicit negation (or, strong negation) [18] rather than negation as failure. That is, if nothing is known about A, it only means that there is no evidence that A holds; it does not mean that the negation of A holds.

An annotated rule of the form $[v, w]\ X$:- *true* is called an **annotated fact**; it is often written simply as $[v, w]\ X$. In the remainder of this paper we will deal only with annotated rules and facts and refer to them simply as rules and facts.

Definition 1. *Given a blp P, an atom X is said to **depend** on an atom Y*

- *directly, if X is the head of a rule R and Y occurs in the body of R;*
- *indirectly, if X is dependent on Z, and Z depends on Y.*   □

We require that in a ground blp no atom depends on itself. So, there can be no cyclic dependency among ground atoms. Most other works in this area, e.g., [3,19,20], make the same assumption. The extension of BLP that allows cyclic dependency is future work and is beyond the scope of this paper.

5   Semantics of BLP

We begin with the concept of *combination functions*.

5.1   Combination Functions

Definition 2. *Let D be the set of all sub-intervals of $[0, 1]$, and $\Phi : D \times D \to D$ be a function. Let us represent $\Phi([v_1, w_1], [v_2, w_2])$ as $[V(v_1, w_1, v_2, w_2), W(v_1, w_1, v_2, w_2)]$. We say that Φ is a **belief combination function** if Φ is associative and commutative.*   □

A useful common-sense restriction on combination functions is that the functions V and W above are monotonically increasing in each of their four arguments, but this is not required for our results. Due to the associativity of Φ, we can extend it from two to three and more arguments as follows:

$$\Phi([v_1, w_1], ..., [v_k, w_k]) = \Phi\big(\Phi([v_1, w_1], ..., [v_{k-1}, w_{k-1}]), [v_k, w_k]\big)$$

For convenience, we also extend Φ to the nullary case and the case of a single argument as follows: $\Phi() = [0, 1]$ and $\Phi([v, w]) = [v, w]$. Note that the order of arguments in a belief combination function is immaterial, since such functions are commutative, so we often write such functions as functions on multisets of intervals, e.g., $\Phi(\{[v_1, w_1], ..., [v_k, w_k]\})$.

As mentioned earlier, there are many ways to combine evidence and so there are many useful belief combination functions. Different functions can be used for different application domains and even for different types of data within the same domain. Our examples will be using the following three popular functions:

- *Dempster's combination rule:*
  - $\Phi^{DS}([0, 0], [1, 1]) = [0, 1]$.
  - $\Phi^{DS}([v_1, w_1], [v_2, w_2]) = [v, w]$ if $\{[v_1, w_1], [v_2, w_2]\} \neq \{[0, 0], [1, 1]\}$, where $v = \frac{v_1 \cdot w_2 + v_2 \cdot w_1 - v_1 \cdot v_2}{K}$, $w = \frac{w_1 \cdot w_2}{K}$, and $K = 1 + v_1 \cdot w_2 + v_2 \cdot w_1 - v_1 - v_2$. In this case, $K \neq 0$ and thus v and w are well-defined.
- *Maximum:* $\Phi^{max}([v_1, w_1], [v_2, w_2]) = [max(v_1, v_2), max(w_1, w_2)]$.
- *Minimum:* $\Phi^{min}([v_1, w_1], [v_2, w_2]) = [min(v_1, v_2), min(w_1, w_2)]$.

5.2   Semantics

Given a blp P, the definitions of *Herbrand Universe* U_P and *Herbrand Base* B_P of P are the same as in the classical case. As usual in logic programming, the easiest way to define a semantics is by considering *ground* (i.e., variable-free) rules. We assume that each atom $X \in B_P$ has an associated belief combination function, denoted Φ_X. Intuitively, Φ_X is used to help determine the *combined* belief in X accorded by the rules in P that support X.

Definition 3. *A **truth valuation** over a set of atoms α is a mapping from α to $\{\mathbf{t}, \mathbf{f}, \mathbf{u}\}$. The set of all possible valuations over α is denoted as $TVal(\alpha)$.*

*A **truth valuation** I for a blp P is a truth valuation over B_P. Let $TVal(P)$ denote the set of all the truth valuations for P, so $TVal(P) = TVal(B_P)$.* □

It is easy to see that $TVal(P)$ has $3^{|B_P|}$ truth valuations. If α is a set of atoms, we will use $Bool(\alpha)$ to denote the set of all Boolean formulas constructed out of these atoms (i.e., using $\wedge$, $\vee$, and negation).

Definition 4. *Given a truth valuation I over a set of atoms α and a formula $F \in Bool(\alpha)$, $I(F)$ is defined as in Lukasiewicz's three-valued logic: $I(A \vee B) = max\big(I(A), I(B)\big)$, $I(A \wedge B) = min\big(I(A), I(B)\big)$, and $I(\overline{A}) = \neg I(A)$, where $\mathbf{f} < \mathbf{u} < \mathbf{t}$ and $\neg \mathbf{t} = \mathbf{f}$, $\neg \mathbf{f} = \mathbf{t}$, $\neg \mathbf{u} = \mathbf{u}$. We say that $I \models F$ if $I(F) = \mathbf{t}$.* □

Definition 5. *A **support function** for a set of atoms α is a mapping m_α from $TVal(\alpha)$ to $[0,1]$ such that $\sum_{I \in TVal(\alpha)} m_\alpha(I) = 1$.*

*The atom-set α is called the **base** of m_α. A **support function** for a blp P is a mapping m from $TVal(P)$ to $[0,1]$ such that $\sum_{I \in TVal(P)} m(I) = 1$.* □

Support functions, defined above, are always associated with mass functions of Dempster-Shafer theory, as discussed in Section 6. In Dempster-Shafer theory, every mass function has a corresponding belief function and, similarly, BLP support functions are associated with belief functions, defined next.

Definition 6. *Recall that $Bool(B_P)$ denotes the set of all Boolean formulas composed out of the atoms in B_P. A mapping $\mathtt{bel} : Bool(B_P) \longrightarrow [0,1]$ is said to be a **belief function** for P if there exists a support function m for P, so that for all $F \in Bool(B_P)$*

$$\mathtt{bel}(F) = \sum_{I \in TVal(P) \ such \ that \ I \models F} m(I)$$ □

Belief functions can be thought of as *interpretations* of belief logic programs. However, as usual in deductive database and logic programming, we are interested not just in interpretations, but in models. We define this next.

Definition 7. *Given a blp P and a truth valuation I, we define P's **reduct** **under** I to be $P_I = \{R \mid R \in P, I \models Body(R)\}$, where $Body(R)$ denotes the body of the rule R.*

Let $P(X)$ denote the set of rules in P with the atom X in the head. P's **reduct under I with X as head** is defined as $P_I(X) = P_I \cap P(X)$. Thus, $P_I(X)$ is simply that part of the reduct P_I, which consists of the rules that have X as their head.   □

We now define a measure for the degree by which I is supported by $P(X)$.

Definition 8. *Given a blp P and a truth valuation I for P, for any $X \in B_P$, we define $s_P(I, X)$, called the P-support for X in I, as follows.*

1. *If $P_I(X) = \phi$, then*
   - *If $I(X) = \mathbf{t}$ or $I(X) = \mathbf{f}$, then $s_P(I, X) = 0$;*
   - *If $I(X) = \mathbf{u}$, then $s_P(I, X) = 1$.*
2. *If $P_I(X) = \{R_1, \ldots, R_n\}$, $n > 0$, let $[v, w]$ be the result of applying Φ_X to the belief factors of the rules $R_1, \ldots, R_n$. Then*
   - *If $I(X) = \mathbf{t}$, then $s_P(I, X) = v$;*
   - *If $I(X) = \mathbf{f}$, then $s_P(I, X) = 1 - w$;*
   - *If $I(X) = \mathbf{u}$, then $s_P(I, X) = w - v$.*   □

Informally, $I(X)$ represents what the possible world I believes about X. The above interval $[v, w]$ produced by the Φ_X represents the combined support accorded by the rule set $P_I(X)$ to that belief. $s_P(I, X)$ measures the degree by which a truth valuation I is supported by $P(X)$. If X is true in I, it is the combined belief in X supported by P given the truth valuation I. If X is false in I, $s_P(I, X)$ is the combined disbelief in X. Otherwise, it represents the combined information gap about X.

It is easy to see that the case of $P_I(X) = \varnothing$ in the above definition is just a special case of $P_I(X) = \{R_1, \ldots, R_n\}$, since $\Phi_X(\varnothing)$ is $[0, 1]$, by Definition 2.

We now introduce the notion of P-support for I as a whole. It is defined as a cumulative P-support for all atoms in the Herbrand base.

Definition 9. *If I is a truth valuation for a blp P, then*

$$\hat{m}_P(I) = \prod_{X \in B_P} s_P(I, X)$$ □

Theorem 1. *For any blp P, $\sum_{I \in TVal(P)} \hat{m}_P(I) = 1$.*   □

In other words, $\hat{m}_P$ is a support function. This theorem is crucial, as it makes the following definition well-founded.

Definition 10. *The **model** of a blp P is the following belief function:*

$$\mathtt{model}(F) = \sum_{I \in TVal(P) \text{ such that } I \models F} \hat{m}_P(I), \quad \text{where } F \in \mathcal{B}ool(B_P).$$ □

The belief function $\mathtt{model}(F)$ measures the degree by which F is supported by P. It is easy to see that every blp has a *unique* model. The rationale for the above definition is expressed by the following theorem:

Theorem 2. *Let P be a blp and A an atom. For any rule R, let $Body(R)$ denote its body. Let S be a subset of $P(A)$ that satisfies (i) $\mathtt{model}(\bigwedge_{R\in S} Body(R)) > 0$; and (ii) S is maximal: if $S' \supseteq S$ is another subset of $P(A)$ that satisfies (i) then $S' = S$. Let $[v_R, w_R]$ denote the belief factor associated with the rule R and suppose $\Phi_A(\{[v_R, w_R]\}_{R\in S}) = [v, w]$ ($\Phi_A(\{[v_R, w_R]\}_{R\in S})$ is the result of applying Φ_A to the belief factors in S). Then*

$$\frac{\mathtt{model}\big(A \wedge \bigwedge_{R\in S} Body(R)\big)}{\mathtt{model}\big(\bigwedge_{R\in S} Body(R)\big)} = v \qquad \frac{\mathtt{model}\big(\overline{A} \wedge \bigwedge_{R\in S} Body(R)\big)}{\mathtt{model}\big(\bigwedge_{R\in S} Body(R)\big)} = 1 - w. \quad \square$$

In other words, $\mathtt{model}$ is a "correct" (and unique) belief function that embodies the evidence that P provides for each atom. In other words, $\mathtt{model}$ supports each atom in the Herbrand base with precisely the expected amount of support. In contrast, all other works that we are aware of either do not account for combined support provided by multiple rules deriving the same atom or do not have a clear model-theoretic account for that phenomenon.

It is not hard to see that the BLP semantics is non-monotonic.[2] To see that, suppose rule r_1 has the form $[0.4, 0.4]$ X and rule r_2 is $[0.8, 0.8]$ X. Let P_1 be $\{r_1\}$, P_2 be $\{r_2\}$, P_3 be $\{r_1, r_2\}$, and $\mathtt{bel}_i$ be the model of P_i, $i = 1, 2, 3$. For any combination function Φ, let $[v, w] = \Phi([0.4, 0.4], [0.8, 0.8])$, since $v \leq w$, either $v < 0.8$ is true, or $w > 0.4$ is true, or both. If $v < 0.8$, then $\mathtt{bel}_3(X) < 0.8 = \mathtt{bel}_2(X)$. Thus, adding r_1 to P_2 reduces the support for X. If $w > 0.4$, then $\mathtt{bel}_3(\overline{X}) < 0.6 = \mathtt{bel}_1(\overline{X})$, meaning that adding r_2 to P_1 reduces the support for $\overline{X}$. Non-monotonicity of Dempster-Shafer theory was also discussed in [16].

Also, under the BLP semantics, the support for A provided by a rule of the form $[v, w]$ A :- $B_1 \vee B_2$ might differ from the support for A provided by the pair of rules $[v, w]$ A :- B_1 and $[v, w]$ A :- B_2, if $\Phi_A([v, w], [v, w]) \neq [v, w]$.

A direct implementation of the semantics would have high complexity. A much more efficient query answering algorithm is presented in [26].

6   Discussion

First, one might be wondering whether the combination functions in BLP are really necessary and whether the same result could not be achieved without the use of combination functions. The answer is that combination functions *can* be dispensed with. However, this requires an extension of BLP with default negation and, more importantly, causes an exponential blowup of the program (making it an unlikely tool for knowledge engineering). This theme will be elaborated upon in a full version of this paper.

Next we discuss the relationship of BLP to Dempster-Shafer belief functions and defeasible reasoning.

---

[2] However, monotonicity holds under certain conditions, for instance, if every belief factor is of the form $[v, 1]$ *and* every combination function $\Phi([v_1, 1], [v_2, 1])$ is monotonically increasing in v_1 and in v_2. Under these conditions, the belief in any negated literal is 0.

Probability vs. belief in combination of evidence. Probability theory has been widely used for reasoning with uncertainty. However, several aspects of the application of this theory to modeling uncertainty has been criticized [22,31], especially when it comes to combining evidence obtained from different sources.

To illustrate, consider two mutually exclusive states A and $\overline{A}$. Suppose this distribution is provided by two different sources. Source 1 may assert that $prob(A) = 0.8$, $prob(\overline{A}) = 0.2$, meaning that the probability of A is 0.8. Source 2 may assert that $prob(A) = 0.6$, $prob(\overline{A}) = 0.4$, meaning that the probability of A is 0.6. There is no obvious way to combine information from these two sources because probability is objective. Some approaches take the maximum or the minimum of the two probability values; others take the average. However, none of these has any probabilistic justification.

In some frameworks, e.g. [4,15], probability intervals are used to model uncertainty. Suppose source 1 asserts $0.8 \leq prob(A) \leq 1$ and source 2 asserts $0.6 \leq prob(A) \leq 0.7$. Some approaches [4] compute the intersection of the two intervals, yielding $\varnothing$ (thus concluding nothing). Some other approaches [15] simply combine the uncertainty ranges, for instance, $[min(0.8, 0.6), max(1, 0.7)]$. Again, no probabilistic justification exists for either of these rules of combination, so probability theory is used here in name but not in substance.

In contrast, Dempster-Shafer theory [5,22] gives up certain postulates of the probability theory in order to provide an account for the phenomenon of combined evidence.

Relationship to Dempster-Shafer theory of evidence. We now relate our semantics to Dempster-Shafer theory.

Definition 11. *A **complete valuation** over a set of atoms α is a mapping from α to $\{\mathbf{t}, \mathbf{f}\}$. The set of all complete valuations over α is denoted as $\mathcal{U}(\alpha)$. A **complete valuation** I for a blp $\mathbf{P}$ is a complete valuation over $B_\mathbf{P}$. Let $\mathcal{U}(\mathbf{P})$ denote the set of all the complete valuations for $\mathbf{P}$, so $\mathcal{U}(\mathbf{P}) = \mathcal{U}(B_\mathbf{P})$.* □

Complete valuations correspond to interpretations in classical 2-values logic programs. A complete valuation J can be viewed as a state, and it is clear that two different complete valuations represent two mutually exclusive states. $\mathcal{U}(\alpha)$ is a universal set of mutually exclusive states.

A truth valuation I over α (see Definition 3) can be uniquely mapped to a set of complete valuations: $\Psi_I = \{J \in \mathcal{U}(\alpha) \mid \forall A \in \alpha, \ J(A) = \mathbf{t} \text{ if } I(A) = \mathbf{t}, J(A) = \mathbf{f} \text{ if } I(A) = \mathbf{f}\}$. In other words, each truth valuation I represents a subset Ψ_I of $\mathcal{U}(\alpha)$, and hence an element Ψ_I of $\mathbb{P}(\mathcal{U}(\alpha))$, the power set of $\mathcal{U}(\alpha)$. Obviously, $\{\Psi_I | I \in \mathcal{TVal}(\alpha)\} \subset \mathbb{P}(\mathcal{U}(\alpha))$.

Dempster-Shafer's mass function, $\mathtt{mass}$, is a mapping from $\mathbb{P}(\mathcal{U}(\alpha))$ to $[0, 1]$ such that $\sum_{S \in \ \mathbb{P}(\mathcal{U}(\alpha))} \mathtt{mass}(S) = 1$. According to Definition 5, our support function m is a mapping from $\mathcal{TVal}(\alpha)$ to $[0, 1]$ such that $\sum_{I \in \ \mathcal{TVal}(\alpha)} m(I) = 1$. Given any support function m, it can be related to a unique mass function, $\mathtt{mass}$, as follows.

$$\begin{aligned} \mathtt{mass}(\Psi_I) &= m(I), && \forall I \in \mathcal{TVal}(\alpha) \\ \mathtt{mass}(S) &= 0, && \text{if } S \neq \Psi_I, \forall I \in \mathcal{TVal}(\alpha) \end{aligned} \qquad (1)$$

Based on the above correspondence between Dempster-Shafer's mass function and BLP support function introduced in Definition 5, we establish a correspondence between Dempster-Shafer's belief function and BLP's belief function of Definition 6.

Any BLP formula $F \in \mathcal{B}ool(B_P)$ (defined in Section 5.2) uniquely corresponds to a set of complete valuations: $\Theta_F = \{J \in \mathcal{U}(B_P) \mid J \models F\}$. Clearly, $\Theta_F \in \mathbb{P}(\mathcal{U}(B_P))$, and it can be shown that $\{\Theta_F \mid F \in \mathcal{B}ool(B_P)\} = \mathbb{P}(\mathcal{U}(B_P))$.

Theorem 3. *Let m be a support function and* mass *its corresponding mass function as in (1). Also let* bel *be the belief function of Definition 6 constructed from m, and let* belief *be the Dempster-Shafer's belief function based on* mass: belief$(S) = \sum_{\forall S' \subseteq S}$ mass(S'). *Then, for any $F \in \mathcal{B}ool(B_P)$, the following holds:* bel$(F) =$ belief(Θ_F). □

In other words, any BLP belief function corresponds to a Dempster-Shafer's belief function.

Relationship to defeasible logic programs and explicit negation. There is an interesting correspondence between the treatment of contradictory information in BLP and a form of defeasible reasoning called Courteous Logic Programming [7] and, more generally, Logic Programming with Courteous Argumentation Theories (LPDA) [25]. Here we consider only LPDA without default negation. A defeasible LPDA rule has the form

$$@r \quad H \; \text{:-} \; B_1 \wedge \cdots \wedge B_n \tag{2}$$

where r is a label, H and B_i, $1 \le i \le n$, are atoms or explicitly negated atoms. As before, we use $\overline{A}$ to represent explicit negation of A.

For any atom A, let $\lambda(A) = [1,1]\,A$ and $\lambda(\overline{A}) = [0,0]\,A$. We extend λ to rules of the form (2) so that $\lambda(@r\; H \; \text{:-} \; B_1 \wedge \cdots \wedge B_n)$ is $\lambda(H) \; \text{:-} \; B_1 \wedge \cdots \wedge B_n$. Finally, λ is extended to programs so that $\lambda(\Pi) = \{\lambda(R) \mid R \in \Pi\}$. Note that $\lambda(\Pi)$ is a blp.

Let $\Pi \models_{LPDA} F$ denote that Π entails F under the semantics of LPDA [25] with one of the courteous argumentation theories and let the combination function Φ_1 be such that $\Phi_1([0,0],[1,1]) = [0,1]$, $\Phi_1([0,0],[0,0]) = [0,0]$, $\Phi_1([1,1],[1,1]) = [1,1]$.

Theorem 4. *Let Π be an acyclic LPDA program that consists of the rules of the form (2) and let* bel$_{\lambda(\Pi)}$ *be the model of the blp $\lambda(\Pi)$ with the combination function Φ_1. Assume, in addition, that none of the rules in Π defines or uses the special predicates* overrides *and* opposes, *which in LPDA provide information about conflicting rules and their defeasibility properties. Then, for any formula $F \in \mathcal{B}ool(B_\Pi)$, $\Pi \models_{LPDA} F$ if and only if* bel$_{\lambda(\Pi)}(F) = 1$. □

In both theories, the presence of A and $\overline{A}$ (in LPDA) or of $[1,1]\,A$ and $[0,0]\,A$ (in BLP) implies that A's truth value is undefined. That is, inconsistent information is self-defeating. In contrast, Pearce and Wagner's logic programs with strong negation, as defined in [18], handle inconsistent information by explicitly declaring a contradiction. Thus, if Π in Theorem 4 is a program with strong negation then $\Pi \models_{LPSN} F$ is equivalent to bel$_{\lambda(\Pi)}(F) = 1$ only if Π is *consistent*. Here $\models_{LPSN}$ denotes the entailment in logic programing with strong negation [18].

In the opposite direction, the connection is more complicated and we do not have the space to describe it here.

7   Motivating Example (Cont'd)

Returning to the example in Section 2, suppose that our information sources predict 50% chance of bad weather, parades with 50% certainty, roadwork along the Long Island Expressway (henceforth LIE) with certainty 80%, and police activity due to accidents with the likelihood of 40%. This information is expressed in BLP as follows.

$$
\begin{array}{ll}
[0.8, 0.8] \;\; roadwork(LIE) & [0.5, 0.5] \;\; bad\_weather(LIE) \\
[0.5, 0.5] \;\; social\_act(LIE) & [0.4, 0.4] \;\; police\_act(LIE)
\end{array}
\tag{3}
$$

The traffic service fetches the above information from four different information sources and integrates them using these rules:[3]

$$
\begin{array}{ll}
[0.9, 1] & delay(?r) \; :- \; roadwork(?r) \wedge bad\_weather(?r) \\
[0.8, 1] & delay(?r) \; :- \; roadwork(?r) \wedge social\_act(?r) \\
[0.99, 1] & delay(?r) \; :- \; roadwork(?r) \wedge police\_act(?r)
\end{array}
$$

Suppose the atom $delay(?r)$ is associated with the combination function Φ^{max} defined in Section 5.1. When correlation is not taken into account, as in [1,12], the belief factor of $delay(LIE)$ is $[0.36, 1]$, which means that the available information predicts traffic delay with certainty 0.36 and smooth traffic with certainty 0. Based on this advisory, the students would decide to drive and leave one hour earlier than normal (see Section 2 for the explanation of how the students make their travel plans). It is not hard to see that this advisory might cost our students a show. The information from the weather forecast and Department of Transportation alone (the first rule) is enough to predict traffic delays with certainty 0.36. Taking into account the possibilities of parades and accidents, it is reasonable to up the expectation of delays. In contrast, BLP computes the belief factor for traffic delays to be $[0.63, 1]$, which means that our students will shell out a little extra for the train but will make it to the show.

One may argue that the problem was the Φ^{max} combination function and a different function, such as the Dempster's combination rule Φ^{DS} (Section 5.1), may do just fine even without BLP. While this might be true for the traffic conditions in (3), a wrong advice will be given in the following scenario:

$$
\begin{array}{ll}
[0.2, 0.2] \;\; roadwork(LIE) & [0.8, 0.8] \;\; bad\_weather(LIE) \\
[0.9, 0.9] \;\; social\_act(LIE) & [0.3, 0.3] \;\; police\_act(LIE)
\end{array}
$$

where we assume that $delay(?r)$ is associated with the combination function Φ^{DS}. Without taking correlation into account, the belief factor of $delay(LIE)$ becomes $[0.31, 1]$—again suggesting to add one extra hour to the trip. However,

---

[3] Note that although the semantics is defined for ground rules, query answering algorithms work with non-ground rules [26].

this advisory errs on the cautious side. Here all three rules make their predictions building the same roadwork factor into their decision, so this factor is counted multiple times. In contrast, BLP recognizes that the three predictions are not independent and its calculated certainty factor is $[0.18, 1]$. Thus, the students will allocate no extra time for the eventualities and will get that badly needed extra hour of sleep before their conference deadlines.

We thus see that, by correlating rules, BLP is able to better predict certainty factors of the combined information.

8 Conclusions

We introduced a novel logic theory, Belief Logic Programming, for reasoning with uncertainty. BLP is based on the concept of belief function and is inspired by Dempster-Shafer theory, but it is not simply integration of Dempster-Shafer theory and logic programming. First, unlike the previous efforts in applying Dempster-Shafer theory in logic programming [1,2,15], BLP can correlate structural information contained in derivation paths for beliefs, as illustrated in the motivating example in Section 2 and Section 7. Secondly, BLP is not restricted to any particular combination rule, instead any number of reasonable combination rules can be used. Apart from traditional uses in expert systems, such a language can be used to integrate semantic Web services and information sources, such as sensor networks and forecasts, which deal with uncertain data.

For future work, we plan to extend the algorithms to deal with non-ground rules and queries, and to make them optimized based on belief factors given in the query. Another important extension is to allow cyclic dependency among ground atoms.

References

1. Baldwin, J.F.: Support logic programming. Intl. Journal of Intelligent Systems 1, 73–104 (1986)
2. Baldwin, J.F.: Evidential support logic programming. Fuzzy Sets and Systems 24(1), 1–26 (1987)
3. Bergsten, U., Schubert, J.: Dempster's rule for evidence ordered in a complete directed acyclic graph. Int. J. Approx. Reasoning 9, 37–73 (1993)
4. Dekhtyar, A., Subrahmanian, V.S.: Hybrid probabilistic programs. J. of Logic Programming 43, 391–405 (1997)
5. Dempster, A.P.: Upper and lower probabilities induced by a multi-valued mapping. Ann. Mathematical Statistics 38 (1967)
6. Elkan, C.: The paradoxical success of fuzzy logic. IEEE Expert, 698–703 (1993)
7. Grosof, B.N.: A courteous compiler from generalized courteous logic programs to ordinary logic programs. Technical Report Supplementary Update Follow-On to RC 21472, IBM (July 1999)
8. Halpern, J.Y.: Reasoning About Uncertainty. MIT Press, Cambridge (2003)
9. Kersting, K., De Raedt, L.: Bayesian logic programs. Technical report, Albert-Ludwigs University at Freiburg (2001)
10. Kifer, M., Li, A.: On the semantics of rule-based expert systems with uncertainty. In: Gyssens, M., Van Gucht, D., Paredaens, J. (eds.) ICDT 1988. LNCS, vol. 326, pp. 102–117. Springer, Heidelberg (1988)

11. Kifer, M., Subrahmanian, V.S.: Theory of generalized annotated logic programming and its applications. J. of Logic Programming 12(3,4), 335–367 (1992)
12. Lakshmanan, L.V.S., Shiri, N.: A parametric approach to deductive databases with uncertainty. IEEE Trans. on Knowledge and Data Engineering 13(4), 554–570 (2001)
13. Lukasiewicz, T.: Probabilistic logic programming with conditional constraints. ACM Trans. on Computational Logic 2(3), 289–339 (2001)
14. Muggleton, S.: Learning stochastic logic programs. Electron. Trans. Artif. Intell. 4(B), 141–153 (2000)
15. Ng, R.T.: Reasoning with uncertainty in deductive databases and logic programs. Intl. Journal of Uncertainty, Fuzziness and Knowledge-Based Systems 5(3), 261–316 (1997)
16. Ng, R.T., Subrahmanian, V.S.: Relating dempster-shafer theory to stable semantics. In: International Symposium on Logic Programming, pp. 551–565 (1991)
17. Ng, R.T., Subrahmanian, V.S.: A semantical framework for supporting subjective probabilities in deductive databases. J. of Automated Reasoning 10(2), 191–235 (1993)
18. Pearce, D., Wagner, G.: Logic programming with strong negation. In: Eriksson, L.-H., Hallnäs, L., Schroeder-Heister, P. (eds.) ELP 1991. LNCS, vol. 596, pp. 311–326. Springer, Heidelberg (1991)
19. Poole, D.: The independent choice logic and beyond. In: De Raedt, L., Frasconi, P., Kersting, K., Muggleton, S.H. (eds.) Probabilistic Inductive Logic Programming. LNCS (LNAI), vol. 4911, pp. 222–243. Springer, Heidelberg (2008)
20. De Raedt, L., Kersting, K.: Probabilistic inductive logic programming. In: De Raedt, L., Frasconi, P., Kersting, K., Muggleton, S.H. (eds.) Probabilistic Inductive Logic Programming. LNCS (LNAI), vol. 4911, pp. 1–27. Springer, Heidelberg (2008)
21. Ruthven, I., Lalmas, M.: Using dempster-shafers theory of evidence to combine aspects of information use. J. of Intelligent Systems 19, 267–301 (2002)
22. Shafer, G.: A Mathematical Theory of Evidence. Princeton University Press, Princeton (1976)
23. Shenoy, P.P., Shafer, G.: Axioms for probability and belief-function propagation. In: Uncertainty in Artificial Intelligence, pp. 169–198. North-Holland, Amsterdam (1990)
24. Subrahmanian, V.S.: On the semantics of quantitative logic programs. In: International Symposium on Logic Programming, pp. 173–182 (1987)
25. Wan, H., Grosof, B.N., Kifer, M., Fodor, P., Liang, S.: Logic programming with defaults and argumentation theories. In: Intl. Conf. on Logic Programming (2009)
26. Wan, H., Kifer, M.: Belief logic programming and its extensions. Technical report, Stony Brook University (2009), http://www.cs.sunysb.edu/~hwan/BLP_TR.html
27. Yager, R.R.: Decision making under dempster-shafer uncertainties. In: Classic Works on the Dempster-Shafer Theory of Belief Functions, pp. 619–632. Springer, Heidelberg (2008)
28. Yamada, K.: A new combination of evidence based on compromise. Fuzzy Sets System 159(13), 1689–1708 (2008)
29. Zadeh, L.A.: Fuzzy sets. Information Control 8, 338–353 (1965)
30. Zadeh, L.A.: A review of (a mathematical theory of evidence. g. shafer, princeton university press, princeton, nj, 1976). The AI Magazine (1984)
31. Zadeh, L.A.: Fuzzy sets as a basis for a theory of possibility. Fuzzy Sets Syst. 100(supp.), 9–34 (1999)

Weight Constraint Programs with Functions

Yisong Wang[1,2], Jia-Huai You[2], Li-Yan Yuan[2], and Mingyi Zhang[3]

[1] Department of Computer Science, Guizhou University, Guiyang, China
[2] Department of Computing Science, University of Alberta, Canada
[3] Guizhou Academy of Sciences, Guiyang, China

Abstract. In this paper we consider a new class of logic programs, called weight constraint programs with functions, which are lparse programs incorporating functions over non-Herbrand domains. We define answer sets for these programs and develop a computational mechanism based on loop completion. We present our results in two stages. First, we formulate loop formulas for lparse programs (without functions). Our result improves the previous formulations in that our loop formulas do not introduce new propositional variables, nor there is a need of translating lparse programs to nested expressions. Building upon this result we extend the work to weight constraint programs with functions. We show that the loop completion of such a program can be transformed to a Constraint Satisfaction Problem (CSP) whose solutions correspond to the answer sets of the program, hence off-the-shelf CSP solvers can be used for answer set computation. We show some preliminary experimental results.

1 Introduction

Logic programming based on stable model/answer set semantics, called *answer set programming* (ASP), has been considered a promising paradigm for declarative problem solving. The general idea is to encode a problem by a logic program such that the answer sets of the program correspond to the solutions of the problem. The class of logic programs with weight constraints, typically called *lparse programs* [15,19], are among the most commonly used forms of programs in ASP. A number of ASP systems are implemented according to the semantics of lparse programs defined in [15,19], which include SMODELS, CMODELS, and CLASP.

The idea of loop formulas [11] makes it possible to compute answer sets of logic programs using SAT solvers or SAT solving techniques. ASSAT and CMODELS are such implementations. The mechanism of loop completion (completion plus loop formulas) has been extensively studied for some general classes of logic programs, including disjunctive logic programs [9], logic programs with nested expressions [6], logic programs with arbitrary propositional formulas [4], arbitrary first-order sentences [8] and logic programs with abstract constraint atoms [12,20].

To deal with weight rules, CMODELS implemented a translation defined in [6], where an lparse program is translated into a logic program with nested expressions [7]. In general, the translation defined in [6] may generate an exponentially larger program. Liu and Truszcynski studied logic programs with monotone and convex constraints and showed that an lparse program can be transformed to a set of pseudo-boolean

E. Erdem, F. Lin, and T. Schaub (Eds.): LPNMR 2009, LNCS 5753, pp. 329–341, 2009.

constraints, based on its loop completion [12]. They implemented the system called PBMODELS for lparse programs in which negative literals are not allowed in weight constraints. The restriction can be removed by a translation using new propositional variables.

In another development, ASP has recently been extended to incorporate functions. The approach presented in [10] aims at economically and naturally encoding problems in ASP, by allowing functions over non-Herbrand domains. With loop completion, a normal logic program with functions can be translated to an instance of the Constraint Satisfaction Problem (CSP). Thus, off-the-shelf CSP solvers can be employed as black boxes for answer set computation.

It is important to notice that the goal of [10] is different from most of the other approaches to adding functions into ASP (see, e.g., [1,3,18]). Functions in these other approaches, just like Horn clause logic programming, are interpreted by fixed mappings and are used to define recursive data structures over infinite Herbrand domains. While these approaches aim at increasing the expressive power of ASP, the approach of [10] aims at economic and natural representation of knowledge and efficient computation.

In this paper, we consider adding functions into lparse programs along the line of [10]. We present our work in two stages. First, we show that loop completion can be formulated directly for lparse programs with arbitrary weight constraints, without introducing new symbols. This improves previous results in that we need not transform an lparse program into a program with nested expressions, as defined in [6] and implemented in CMODELS [7], nor do we need to require weight constraints to contain only positive literals, as in [12].

More importantly, we consider a new class of logic programs, called *weight (constraint) programs with functions*. We define answer sets for these programs, generalize the definition of completion and loop formulas to this class of programs, and show that answer sets can be characterized by loop completion. This provides a basis for extending FASP[1], an implementation for normal logic programs with functions, to compute answer sets of weight programs with functions. This approach possesses two main advantages. First, CSP facilities such as *global constraints* can be readily brought into the ASP language, since the programs in this language are translated into CSP instances. In this sense, this approach can be seen as another attempt to integrate CSP with ASP (cf. [14]). Another advantage of this approach is that the sizes of grounded programs sometimes can be substantially smaller than those of the standard ASP encodings.

In a preliminary experiment, we tested our extended implementation on the magic N-square problem encoded by a weight program with functions. We use two top ranked solvers from the 3rd CSP Solver Competition[2], as black boxes to FASP. Our experiment shows that FASP outperforms all the above mentioned state-of-the-art ASP solvers.

In the next section, we propose loop formulas for lparse programs, while in Section 3 we add functions to lparse programs, define the semantics, and formulate loop formulas for these programs. We report preliminary experimental results in Section 4, and conclude the paper with final remarks in Section 5.

---

[1] http://www.cse.ust.hk/fasp/
[2] http://cpai.ucc.ie/08/

2 Lparse Programs

2.1 Preliminary Definitions

We assume an underlying propositional language $\mathcal{L}^P$. As given in [15], a *rule element* is an atom (positive rule element) or an atom prefixed with *not* (negative rule element). A *weight constraint* is an expression of the form

$$l \leq \{c_1 : w_1, \ldots, c_n : w_n\} \leq u \tag{1}$$

where

- each of l and u is a real number or one of the symbols $+\infty$, $-\infty$, if l (resp. u) is $-\infty$ (resp. $+\infty$) then "$l \leq$" (resp. "$\leq u$") can be omitted.
- $c_1, \ldots, c_n$ are rule elements different from each other, and
- $w_1, \ldots, w_n$ are nonnegative real numbers[3].

We denote the weight constraint (1) by $l \leq S \leq u$, where l or u may be omitted as defined above and $S = \{c_1 : w_1, \ldots, c_n : w_n\}$.

Let $C = l \leq S \leq u$ be a weight constraint. We denote $\mathit{Atoms}^+(C) = \{a \mid (a : w_a) \in S\}$ and $\mathit{Atoms}(C) = \mathit{Atoms}^+(C) \cup \{b \mid (not\, b : w_b) \in S\}$. Let $\mathcal{K}$ be a set of weight constraints. By $\mathit{Atoms}^+(\mathcal{K})$, we mean the set $\bigcup_{C \in \mathcal{K}} \mathit{Atoms}^+(C)$ and by $\mathit{Atoms}(\mathcal{K})$ we mean the set $\bigcup_{C \in \mathcal{K}} \mathit{Atoms}(C)$.

Let Z be a set of atoms, c a rule element and C a weight constraint of the form (1). Z *satisfies* c, denoted by $Z \models c$, if $c \in Z$ whenever c is an atom, and $c \notin Z$ otherwise. The set Z *satisfies* the weight constraint C, written $Z \models C$, if the sum of the weights w_j for all j such that Z satisfies c_j is not less than l and not greater than u.

An *lparse program* (or just a *logic program*) is a finite set of *weight rules* of the form

$$C_0 \leftarrow C_1, \ldots, C_n \tag{2}$$

where C_i $(0 \leq i \leq n)$ are weight constraints. Let r be a rule of the form (2). C_0 and $\{C_1, \ldots, C_n\}$ are called its *Head* and *Body*, respectively. Alternatively, we also write r in the form of *Head $\leftarrow$ Body*, for convenience.

We denote by $\mathit{Atoms}(P)$ the union of $\mathit{Atoms}(C)$ where C is a weight constraint occurring in P. A weight rule is a *Horn rule* if its head is a positive rule element and every member of its body has the form $l \leq S$, where S does not mention any negative rule elements.

In the following, let Z be a set of atoms.

Z *satisfies* a logic program P if, for every rule of the form (2) in P, Z satisfies C_0 whenever Z satisfies $C_1, \ldots, C_n$.

The *reduct* of a weight constraint $C = l \leq S \leq u$ with respect to Z, denoted C^Z, is the weight constraint $l^Z \leq S'$, where

- S' is obtained from S be dropping all pairs $(not\, b : w)$, and
- l^Z is l minus the sum of the weights w for all pairs $(not\, b : w)$ in S such that $b \notin Z$.

---

[3] We consider only nonnegative weights in the paper, as negative weights can be replaced by negative rule elements [13,19].

Since the reduct of a weight constraint $l \leq S \leq u$ no longer mentions u, we will simply write $(l \leq S)^Z$ instead.

The *reduct* of a weight rule

$$C_0 \leftarrow l_1 \leq S_1 \leq u_1, \ldots, l_n \leq S_n \leq u_n \tag{3}$$

with respect to Z is the set of rules

$$\{a \leftarrow (l_1 \leq S_1)^Z, \ldots, (l_n \leq S_n)^Z \mid a \in Atoms^+(C_0) \cap Z, Z \models S_i \leq u_i \text{ for all } i\}.$$

The *reduct* P^Z of a logic program P with respect to Z is the union of the reducts of the rules of P with respect to Z. Clearly, P^Z consists of Horn rules only.

If an lparse program P consists of Horn rules then it has a unique minimal set S of atoms such that $S \models P$. The set is called the *deductive closure* of P and denoted by $cl(P)$. S is an *answer set* of P if $S \models P$ and $S = cl(P)$. Given an lparse program P and a set of atoms S, S is an *answer set* of P if and only if $S \models P$ and $cl(P^S) = S$.

2.2   Completion and Loop Formulas

Completion. Following [12], to define the completion we first introduce an extension $\mathcal{L}^{Pwc}$ of the language $\mathcal{L}^P$. A *formula* in $\mathcal{L}^{Pwc}$ is an expression built from weight constraints by means of boolean connectives $\wedge$, $\vee$, $\supset$, and $\neg$. The notion of a model of a formula extends in a standard way to the class of formulas in $\mathcal{L}^{Pwc}$.

Let P be an lparse program. The *completion* of P, denoted $COMP(P)$, is defined as follows:

1. For every rule $Head \leftarrow Body$ in P we include in $COMP(P)$ an $\mathcal{L}^{Pwc}$ formula

$$\left(\bigwedge Body \right) \supset Head \tag{4}$$

2. For every atom $a \in Atoms(P)$, we include in $COMP(P)$ an $\mathcal{L}^{Pwc}$ formula

$$a \supset \bigvee_{1 \leq i \leq n} \left(\bigwedge Body_i \right) \tag{5}$$

where $(Head_1 \leftarrow Body_1), \ldots, (Head_n \leftarrow Body_n)$ are the rules of P such that $a \in Atoms^+(Head_i)$, for all i $(1 \leq i \leq n)$. Please note that, $\bigvee \emptyset = \perp$ ("false") and $\bigwedge \emptyset = \top$ ("true").

Loops and Loop Formulas. Let P be an lparse program. The *positive dependency graph* of P, written G_P, is the directed graph (V, E), where

- $V = Atoms(P)$,
- $(a, b) \in E$ if there a rule of the form (2) in P such that $a \in Atoms^+(C_0)$ and $b \in Atoms^+(C_i)$ for an $i(1 \leq i \leq n)$.

A nonempty subset L of $Atoms(P)$ is a *loop* of P if there is a non-zero length cycle in G_P that goes through only and all the nodes in L.

Let $C = l \leq S \leq u$ be a weight constraint and L a set of atoms. The *restriction* of C to L, written $C_{|L}$, is the following $\mathcal{L}^{Pwc}$ formula

$$l \leq S' \wedge S \leq u \tag{6}$$

where S' is obtained from S by removing every pair $(c_i : w_i)$ from S if $c_i \in L$.

The *loop formula* of a loop L, written $LF(L, P)$, is the following $\mathcal{L}^{Pwc}$ formula

$$\bigvee L \supset \bigvee_{1 \leq i \leq n} \bigwedge_{C \in Body_i} C_{|L}$$

where $(Head_1 \leftarrow Body_1), \ldots, (Head_n \leftarrow Body_n)$ are the rules of P such that, for each i $(1 \leq i \leq n)$, $Atoms^+(Head_i) \cap L \neq \emptyset$.

Theorem 1. *Let P be an lparse program. M is an answer set of P if and only if M is a model of $COMP(P) \cup LF(P)$, where $LF(P)$ is the set of loop formulas of P.*

Since a literal l can be regarded as the weight constraint $1 \leq \{l = 1\} \leq 1$, normal logic programs can be seen as special cases of lparse programs. Note also that, suppose L is a set of atoms and l a literal, if l is an atom in L then $l_{|L}$ is $1 \leq \{\} \leq 1$ which is equivalent to $\perp$, and l itself otherwise. In this way, the above definitions of loops and loop formulas can be regarded as a generalization of those for normal logic programs.

3   Weight Programs with Functions

3.1   Syntax and Semantics

We consider the class of weight programs with functions. Following [10], we assume a many-sorted first-order language $\mathcal{L}$ that may have pre-interpreted symbols such as the standard arithmetic functions " $+$ ", " $-$ " and so on. By an *atom* we mean an atomic formula that does not mention equality, an *equality atom* is a formula of the form $t = t'$ where t and t' are terms of $\mathcal{L}$. Unless stated otherwise, by functions we mean *proper functions* (whose arities are greater than 0). Rule elements are extended to be atoms and equality atoms (positive rule elements), and their negations (negative rule elements). For convenience, we write $t \neq t'$ for *not* $t = t'$.

A *weight constraint with functions* is an expression of the form

$$l \leq \{c_1 : w_1, \ldots, c_n : w_n\} \leq u \tag{7}$$

where each c_i $(1 \leq i \leq n)$ is a rule element, l and u are real numbers, w_i $(1 \leq i \leq n)$ are nonnegative real numbers. A *weight rule* now is an expression

$$C_0 \leftarrow C_1, \ldots, C_n \tag{8}$$

where each C_i $(1 \leq i \leq n)$ is a weight constraint with functions. Again, C_0 and $\{C_1, \ldots, C_n\}$ are its *Head* and *Body*, respectively. A *weight (constraint) program with functions* P is a set of weight rules together with a set of *type definitions*, one for each type τ used in the rules of P. A type definition takes the form

$$\tau : D \qquad (9)$$

where D is a finite nonempty set of elements, called the *domain* of τ. In particular, if a constant c of type τ occurs in a rule of P, then the domain D of τ as specified in the type definitions of P must contain c. In addition, the notations about sets of atoms labeled by the symbols $\mathcal{A}toms^+$ and $\mathcal{A}toms$ are extended to the case of weight programs with functions (note that they denote sets of non-equality atoms).

Let P be a weight program with functions. An atom $p(c_1, \ldots, c_k)$ is said to *reside in* P if p is a predicate of type $\tau_1 \times \cdots \times \tau_k$ in P, and $c_i \in D_i$, where D_i is the domain of the type τ_i, for each i. We denote by $\mathcal{A}t(P)$ the set of atoms residing in P.

Example 1. The magic N-square problem is to construct an $N \times N$ array using each integer in $\{1, \ldots, N^2\}$ as an entry in the array exactly once in such a way that entries in each row, each column, and either of two main diagonals sum up to $N(N^2+1)/2$. Borrowing the familiar notation of *conditional literals* [15] (below, we use "|" instead of ":" for conditional literals for syntactic distinction), we can formulate these requirements by the following weight rules:

$$1 \leq \{square(X,Y) = W : 1 \mid num(X;Y)\} \leq 1 \leftarrow value(W) \qquad (10)$$
$$w \leq \{square(X,Y) = W : W \mid num(X), value(W)\} \leq w \leftarrow num(Y) \qquad (11)$$
$$w \leq \{square(X,Y) = W : W \mid num(Y), value(W)\} \leq w \leftarrow num(X) \qquad (12)$$
$$w \leq \{square(X,X) = W : W \mid num(X), value(W)\} \leq w \qquad (13)$$
$$w \leq \{square(X,N-X) = W : W \mid num(X), value(W)\} \leq w \qquad (14)$$

where $square$ is a function of the type $num \times num \rightarrow value$, the domain of num is the set of numbers $\{1, 2, \ldots, N\}$, the domain of $value$ is $\{1, 2, \ldots, N^2\}$, and $w = N \times (N^2+1)/2$. Rule (10) guarantees that all cells $square(X,Y)$ have distinct values from $\{1, \ldots, N^2\}$. An instance of the problem is specified by a given number N.

Let P be a weight program with functions. The grounding of P consists of type definitions in P and the rules that are obtained by replacing variables in the rules of P with elements in their respective domains.

As noted in [10], a ground rule may have symbols not in the original language $\mathcal{L}$. We let $\mathcal{L}_P$ be the language that extends $\mathcal{L}$ by introducing a new constant for each element that is in the domain of a type, but not a constant in $\mathcal{L}$. These new constants will have the same type as their corresponding elements. In this case, the fully instantiated rules will be in the language $\mathcal{L}_P$. In the following, unless otherwise stated, we shall equate a weight program with functions with its grounding in the extended language $\mathcal{L}_P$. Note that the language $\mathcal{L}_P$ associates with the given program P.

Given a weight program with functions P, an *interpretation* I of P is a first-order structure that defines a mapping such that:

- The domains of I are those specified in the type definitions of P.
- A constant is mapped to itself.

- If R is a relation of arity $\tau_1 \times \cdots \times \tau_n$ and the type definitions $\tau_i : D_i, 1 \leq i \leq n$, are in P, then $R^I \subseteq D_1 \times \cdots \times D_n$.
- If f is a function of type $\tau_1 \times \cdots \times \tau_n \to \tau_{n+1}, n \geq 1$, and the type definitions $\tau_i : D_i, 1 \leq i \leq n+1$, are in P, then f^I is a function from $D_1 \times \cdots \times D_n$ to D_{n+1}. A pre-interpreted function should follow its standard interpretation.

By I^a, we denote the set of atoms that are true under I, i.e., $\{p(c) \mid p^I(c) \ holds\}$, where c is a tuple of constants matching the arity of relation symbol p. Note that an interpretation always associates with a weight program with functions. The valuation of a term t under an interpretation I, denoted by t^I, is a constant defined as:

- $t^I = c$ if $t = c$;
- $t^I = c'$ if $t = f(s)$ and $f^I(s^I) = c'$, where s is a tuple of terms matching the type of f, say $s = (t_1, \ldots, t_n)$, and s^I stands for $(t_1^I, \ldots, t_n^I)$.

Let I be an interpretation. I *satisfies* an atom $p(t)$ if $t^I \in p^I$, i.e., $p(t^I) \in I^a$. The interpretation I *satisfies* an equality atom $t_1 = t_2$ if $t_1^I = t_2^I$. The *satisfaction* (and *model*), written by $\models$, for literals, weight constraints with functions and weight programs with functions can be straightforwardly defined accordingly.

Let P be a weight program with functions, I an interpretation for P, and $l \leq S \leq u$ a weight constraint occurring in P. The *reduct* of $l \leq S \leq u$ with respect to I, written $[l \leq S]^I$ (again, since the reduct does not mention u, we will omit u), is the weight constraint $l^{(I,S)} \leq S^I$, where

- S^I is obtained from S by
  - replacing each functional term $f(t)$ in a rule by d if $f^I(t^I) = d$;
  - removing all pairs $(c : w)$ whenever c is of the forms "*not a*" or "$t = t'$" where a is an atom or an equality atom, t and t' are terms.
- $l^{(I,S)}$ is the following expression:

$$l - \sum_{\substack{c:w_c \in S \\ I \models c}} w_c$$

where c is a negative rule element or an equality atom.

The *reduct* of a rule of the form

$$l_0 \leq S_0 \leq u_0 \leftarrow l_1 \leq S_1 \leq u_1, \ldots, l_n \leq S_n \leq u_n$$

in P, with respect to an interpretation I, is the set of rules

$$\{p(c) \leftarrow [l_1 \leq S_1]^I, \ldots, [l_n \leq S_n]^I \mid p(c) : w \in S_0^I, I \models p(c), I \models S_i \leq u_i \text{ for all } i\}.$$

The *reduct* of P under the interpretation I, written P^I, is the union of the reducts of the rules in P with respect to I.

Clearly, P^I is an lparse program consisting of Horn rules only. The answer set semantics for such logic programs is already defined; i.e., a set of atoms Z is an answer set of P^I if $Z \models P^I$ and $Z = cl(P^I)$. We call the interpretation I for P an *answer set* of P if I^a is an answer set of P^I.

Example 2. Consider the following weight program with functions P:

$$f : \tau \to \tau, \quad p : \tau, \quad \tau : \{0, 1\},$$
$$1 \leq \{(f(0) \neq f(1)) : 1, p(0) : 2\} \leq 2 \leftarrow 1 \leq \{not\, p(f(1)) : 2, p(f(0)) : 1\} \leq 2.$$

Let's consider the interpretation I for P such that $f^I(0) = f^I(1) = 0$, $p^I(0)$ holds and $p^I(1)$ does not. The reduct P^I consists of a single rule: $p(0) \leftarrow 1 \leq \{p(0) : 1\}$. Since $\emptyset$ is the answer set of P^I which different from $I^a = \{p(0)\}$, it follows that I is not an answer set of P.

Note that functions in our weight programs are not semantically necessary, since they can be replaced by predicates while preserving the semantics (cf. [10]).

3.2   Completion and Loop Formulas

Similar to Section 2.2, to accommodate logic formulas that may contain weight constraints with functions, we extend the language $\mathcal{L}$ to $\mathcal{L}^{wc}$, in which a weight constraint with functions is regarded as an atomic formula of $\mathcal{L}^{wc}$.

Let P be a weight program with functions. The *completion* of P, written $COMP(P)$, consists of the following formulas of the language $\mathcal{L}^{wc}$:

1. For every rule *Head* $\leftarrow$ *Body* in P, we include the following $\mathcal{L}^{wc}$ formula in $COMP(P)$

$$(\bigwedge Body) \supset Head.$$

2. For each atom $p(c) \in At(P)$, we include the following formula in $COMP(P)$

$$p(c) \supset \bigvee_{1 \leq i \leq n} \left[\bigwedge Body_i \wedge \left(\bigvee_{p(t) \in Atoms^+(Head_i)} t = c \right) \right]$$

where $Head_i \leftarrow Body_i (1 \leq i \leq n)$ are the rules of P, $t = c$ stands for the formula $t_1 = c_1 \wedge \cdots \wedge t_n = c_n$ whenever $t = (t_1, \ldots, t_n)$ and $c = (c_1, \ldots, c_n)$.

Let P be a weight program with functions. The *positive dependency graph* of P, written G_P, is the directed graph (V, E), where

- $V = At(P)$ and
- $(p(c), q(d)) \in E$ if P has a rule of the form (8) such that, $p(t) \in Atoms^+(C_0)$, $q(s) \in Atoms^+(C_i)$, for some i $(1 \leq i \leq n)$, and there exists an interpretation I for P with $t^I = c$ and $s^I = d$.

A non-empty subset L of $At(P)$ is a *loop* of P if there is a non-zero length cycle in G_P that goes through only and all the nodes in L.

In the following, to define loop formulas, we further extend the language $\mathcal{L}^{wc}$ to $\mathcal{L}_o^{wc}$ such that, for any predicate p and a non-empty subset L of $At(P)$, $\mathcal{L}_o^{wc}$ contains a predicate p_L that has the same arity as that of p. The interpretations for P can be similarly extended. Let $p(t)$ be an atom and $L \subseteq At(P)$. The *L-irrelevant formula* of $p(t)$, written $IL(p(t), L)$, is defined as

$$p_L(t) \equiv \left(p(t) \wedge \bigwedge_{p(c) \in L} \neg(c = t) \right) \tag{15}$$

Intuitively, given an interpretation I for P, $p_L^I(t^I)$ holds if and only if $p^I(t^I)$ holds and $p(t^I) \notin L$. Note that, suppose t mentions only constants, we then have $p_L(t) \equiv p(t)$ if $p(t) \notin L$, and $p_L(t) \equiv \bot$ otherwise.

Let P be a weight program with functions, $L \subseteq At(P)$ and C a weight constraint $l \le S \le u$. The *restriction* of C to L, written $C_{\|L}$, is the following $\mathcal{L}_o^{wc}$ formula:

$$(l \le S') \wedge (S \le u) \wedge \left(\bigwedge_{p(t) \in Atoms^+(C)} IL(p(t), L) \right) \tag{16}$$

where S' is obtained from S by replacing each atom $p(t) \in Atoms^+(C)$ by its L-irrelevant formula $p_L(t)$. In particular, if C mentions only literals that contain no function symbols, then $C_{\|L}$ coincides with $C_{|L}$ in the language $\mathcal{L}^{wc}$.

Let L be a loop of a weight program with functions P. The *loop formula* of L with respect to P, written $LF(L, P)$, is the following $\mathcal{L}_o^{wc}$ formula:

$$\bigvee L \supset \bigvee_{1 \le i \le n} \left[\left(\bigvee_{\substack{p(c) \in L \\ p(t) \in Atoms^+(Head_i)}} t = c \right) \wedge \left(\bigwedge_{C \in Body_i} C_{\|L} \right) \right] \tag{17}$$

where $Head_i \leftarrow Body_i (1 \le i \le n)$ are the rules of P.

Theorem 2. *Let P be a weight program with functions. An interpretation I for P is an answer set of P if and only if $o(I)$ is a model of $COMP(P) \cup LF(P)$ where $LF(P)$ is the set of loop formulas of P and $o(I)$ is the extension of I according to (15).*

3.3   Translation to CSP

In order to translate a weight program with functions into a CSP instance, similar to [10], we need to assume a certain "normal form" for functional terms.

Let P be a weight program with functions. We say P is *free of functions in arguments* if all terms that can be evaluated independently of interpretations have been replaced by constants in $\mathcal{L}_P$, and none of the predicates or functions that are not pre-interpreted have a functional term in their arguments. Given a weight program with functions P, we can translate it into one that is free of functions in arguments using the following procedure:

(1) evaluate all terms that mention only constants and pre-interpreted functions to constants;

(2) for each rule in P, repeatedly replace every occurrence of a term $f(u_1, \ldots, u_n)$ in any argument of a predicate or a function that is not pre-interpreted in the rule by a new fresh variable v of the same type as the range of f, and add $f(u_1, \ldots, u_n) = v$ to the body of the rule, where u_i is a simple term; and

(3) ground the rules obtained in the above step.

It is evident that the original program and the transformed one are equivalent in the sense that they have the same answer sets. In the following, without loss of generality, we assume that weight programs with functions are free of functions in arguments. In this case, we can replace $C_{\|L}$ with $C_{|L}$ in the equation (17).

Accordingly, using completion and loop formulas, we can translate a weight program with functions to a CSP [17], denoted $\mathcal{R}(P) = \langle \mathcal{X}, \mathcal{D}, \mathcal{C} \rangle$, as follows. The set $\mathcal{X}$ of variables and their domains are

- for each atom p in $\mathcal{At}(P)$, there is a variable for it whose domain is $\{0, 1\}$, and
- for each functional term $f(u_1, \ldots, u_n)$ occurring in P such that f is not pre-interpreted, there is a variable for it whose domain is the range of f.

The set $\mathcal{C}$ of the constraints is: for each formula ϕ in $\mathcal{COMP}(P) \cup LF(P)$, there is a constraint $c(\phi) = \langle S, R \rangle$ in $\mathcal{C}$, where R is the constraint obtained from ϕ by replacing atoms and functional terms in it by their corresponding variables, and S is the set of variables occurring in R.

Theorem 3. *Let P be a weight program with functions that is free of functions in arguments, and I an interpretation of P. Then I is an answer set of P if and only if $v(I)$ is a solution of $\mathcal{R}(P)$, where $v(I)$ is defined as follows:*

- *if $x \in X$ corresponds to an atom p, then $v(I)$ assigns x the value 1 iff p is true under I; and*
- *if $x \in X$ corresponds to a term $f(u_1, \ldots, u_n)$, then $v(I)$ assigns x the value u iff $f^I(u_1, \ldots, u_n) = u$.*

4   Implementation and Experiments

Implementation. By Theorem 3, we can compute the answer sets of weight programs with functions using the same algorithm proposed in [10]. As we know, the current CSP encoding formalism under the name of XCSP 2.1[4], which is designed for the third CSP solvers competition, allows global constraints, such as *weightSum*, *allDifferent*, *among*, *atleast*, *atmost*, *cumulative*, etc. Our approach of using CSP solvers to compute answer sets allows us to make use of these facilities. Therefore, we can incorporate global constraints into the ASP language for weight programs with functions, by introducing appropriate denotations. For example, we use the following notation

$$f : \tau_1 \times \ldots \times \tau_n \to \tau[\textit{allDifferent}]$$

---

[4] http://www.cril.univ-artois.fr/~lecoutre/research/benchmarks/benchmarks.html#format

Table 1. The magic N-square problem

N=	smodels	cmodels	clasp	pbmodels		FASP		Size	
?				pbs	satzoo	cpHydra	Mistral	lparse	FASP
4	0.340	11.390	0.030	1.4041	3.0162	0.0720	0.0480	55K	781B
5	–	71.120	8.060	11.4487	32.510	0.0560	0.0320	130K	798B
6	–	–	6.260	113.8152	118.0994	10.7452	0.0240	263K	1.1K
7	–	–	92.880	–	420.9823	10.3772	1.2401	479K	2.1K
8	–	–	–	–	–	10.2013	2.3602	808K	2.6K
9	–	–	–	–	–	0.3360	0.4000	1.3M	3.2K

Legends: – – No result in 0.5 hours. lparse - the LPARSE encoding. FASP - the encoding using functions.

to express not only the input/output types of the function symbol f but also the requirement that the function $f(t)$ should produce different values for different t's. Accordingly, the rule (10) in Example 1 can be omitted by specifying

$$square : num \times num \to value[allDifferent]$$

for the definition of the function $square$. In the current version of FASP, invocation to the CSP global constraint *allDifferent* has been implemented.

Notice further that, in the rules (11–14) of Example 1, the expression "$square(X, Y) = W : W$" means if the value of $square(X, Y)$ is W then the weight of "$square(X, Y) = W$" is W, i.e., the value of $square(X, Y)$. Thus, we can simply write $square(X, Y)$ instead of "$square(X, Y) = W : W$" in those rules. In this case, a weight constraint with functions can be encoded by a CSP constraint in terms of linear inequations. That is, the values W of $square(X, Y)$ are not explicitly enumerated in the generation of the CSP instance. As a result, the size of the ground programs is in the order of $O(n^2)$.

Experiments. In our experiments, we used two CSP solvers, cpHydra-k-10 and Mistral 1.331, which were two top CSP solvers in the 2008 CSP competition.

We tested our implementation with the benchmark magic N-square problem and compared the time cost in seconds with the ASP solvers:[5] SMODELS, CMODELS, CLASP, and PBMODELS. The experiments were done on a PC with Intel Pentium(R) Dual-Core CPU (2.50GHz), 1GB RAM running Fedora Linux Core 6.0 and the results are summarized in Table 1.

The encoding for the magic N-square problem in LPARSE is adopted from the website of PBMODELS as follows:

```
#weight wt(I) = I.          num(1..n).           1{sqr(I,J,D):num(I;J)}1 :- data(D).
data(1..n*n).               1{sqr(I,J,D):data(D)}1 :- num(I;J).
n(n*2+1)/2 [ sqr(I, J, A)=A : num(J): data(A) = weight(wt(A)) ] n(n*2+1)/2 :- num(I).
n(n*2+1)/2 [ sqr(I, J, A)=A : num(I): data(A) = weight(wt(A)) ] n(n*2+1)/2 :- num(J).
n(n*2+1)/2 [ sqr(I, I, A)=A : num(I): data(A) = weight(wt(A)) ] n(n*2+1)/2.
n(n*2+1)/2 [ sqr(I, n+1-I, A)=A : num(I;n+1-I): data(A) = weight(wt(A)) ] n(n*2+1)/2.
```

---

[5] Solving time only, not including the grounding time, nor the time for generating CSP instances.

Grounding this program may produce a ground program of the size $O(n^4)$.

We tested PBMODELS using two engines, PBS and SATZOO, which are considered the best for this benchmark [12]. It is clear that FASP outperforms all the ASP solvers for this benchmark. For comparison, we also output the sizes of these ground programs in LPARSE and in our FASP. The ground programs in LPARSE are generated using *lparse* with the "-t" option. Table 1 shows that, using functions may significantly reduce the size of the ground program.

5   Final Remarks

In this paper we formulated completion and loop formulas for lparse programs directly, and generalized them to weight (constraint) programs with functions that are interpreted over non-Herbrand domains. We show that the completion and loop formulas capture the answer sets of these programs. This enables us to extend FASP for computing answer sets of a weight program with functions. We apply our extended implementation to the magic N-square problem, and the experimental results show a clear advantage of our approach over the traditional ASP solvers.

For adding functions into ASP, a majority of recent work support functions over the Herbrand interpretations [1,3,18]. Quantified Equilibrium Logic (QEL) [16] and the General Theory of Stable Models [5] allow non-Herbrand interpretations. In QEL, the equilibrium models are Kripke structures while the answer set semantics for the General Theory of Stable Models is defined by translating a sentence into a second-order one. On the other hand, the work of [2] is closely related to ours, where a functional logic programming language is defined, in which the valuation of functions can be partial while it must be total in our case.

Our past experience shows that, typically for scheduling benchmarks, CSP-based solvers tend to be more efficient, sometimes they can be orders of magnitude faster than the traditional ASP solvers. We hope that our fine tuning of FASP and further experiments on it can close the gap.

Acknowledgment. We thank the reviewers for their detailed comments, which helped improve the presentation of this paper. Yisong Wang and Mingyi Zhang were partially supported by the Natural Science Foundation of China under grants 90718009 and 60703095, the Fund of Guizhou Science and Technology: 2008[2119], the Fund of Education Department of Guizhou Province: 2008[011] and Scientific Research Fund for talents recruiting of Guizhou University 2007[042].

References

1. Baselice, S., Bonatti, P.A., Criscuolo, G.: On finitely recursive programs. In: Dahl, V., Niemelä, I. (eds.) ICLP 2007. LNCS, vol. 4670, pp. 89–103. Springer, Heidelberg (2007)
2. Cabalar, P.: A functional action language front-end. In: (presented at) The Third International Workshop on Answer Set Programming: Advances in Theory and Implementation, Bath, UK (July 2005), http://www.dc.fi.udc.es/~cabalar/asp05_C.pdf

3. Calimeri, F., Cozza, S., Ianni, G., Leone, N.: Computable functions in ASP: Theory and implementation. In: Garcia de la Banda, M., Pontelli, E. (eds.) ICLP 2008. LNCS, vol. 5366, pp. 407–424. Springer, Heidelberg (2008)
4. Ferraris, P.: Answer sets for propositional theories. In: Baral, C., Greco, G., Leone, N., Terracina, G. (eds.) LPNMR 2005. LNCS (LNAI), vol. 3662, pp. 119–131. Springer, Heidelberg (2005)
5. Ferraris, P., Lee, J., Lifschitz, V.: A new perspective on stable models. In: Proceedings of the 20th International Joint Conference on Artificial Intelligence, pp. 372–379 (2007)
6. Ferraris, P., Lifschitz, V.: Weight constraints as nested expressions. Theory and Practice of Logic Programming 5(1-2), 45–74 (2005)
7. Giunchiglia, E., Lierler, Y., Maratea, M.: Answer set programming based on propositional satisfiability. Journal of Automated Reasoning 36(4), 345–377 (2006)
8. Lee, J., Meng, Y.: On loop formulas with variables. In: Proceedings of Eleventh International Conference on Principles of Knowledge Representation and Reasoning, Sydney, Australia, pp. 444–453 (2008)
9. Lee, J., Lifschitz, V.: Loop formulas for disjunctive logic programs. In: Palamidessi, C. (ed.) ICLP 2003. LNCS, vol. 2916, pp. 451–465. Springer, Heidelberg (2003)
10. Lin, F., Wang, Y.: Answer set programming with functions. In: Proceedings of Eleventh International Conference on Principles of Knowledge Representation and Reasoning, Sydney, Australia, pp. 454–464 (2008)
11. Lin, F., Zhao, Y.: ASSAT: computing answer sets of a logic program by sat solvers. Artificial Intelligence 157(1-2), 115–137 (2004)
12. Liu, L., Truszczynski, M.: Properties and applications of programs with monotone and convex constraints. Journal of Artificial Intelligence Research 27, 299–334 (2006)
13. Marek, V., Niemelä, I., Truszczyński, M.: Logic programs with monotone abstract constraint atoms. Theory and Practice of Logic Programming 8(2), 167–199 (2008)
14. Mellarkod, V.S., Gelfond, M., Zhang, Y.: Integrating answer set programming and constraint logic programming. Annals of Mathematics and Artificial Intelligence 53(1-4), 251–287 (2008)
15. Niemelä, I., Simons, P.: Extending the smodels system with cardinality and weight constraints. In: Logic-Based Artificial Intelligence, pp. 491–521. Kluwer Academic Publishers, Dordrecht (2000)
16. Pearce, D., Valverde, A.: Towards a first order equilibrium logic for nonmonotonic reasoning. In: Alferes, J.J., Leite, J. (eds.) JELIA 2004. LNCS (LNAI), vol. 3229, pp. 147–160. Springer, Heidelberg (2004)
17. Rina, D.: Constraint Processing. Morgan Kaufmann, San Francisco (2003)
18. Simkus, M., Eiter, T.: FDNC: Decidable non-monotonic disjunctive logic programs with function symbols. In: Dershowitz, N., Voronkov, A. (eds.) LPAR 2007. LNCS, vol. 4790, pp. 514–530. Springer, Heidelberg (2007)
19. Simons, P., Niemelä, I., Soininen, T.: Extending and implementing the stable model semantics. Artificial Intelligence 138(1-2), 181–234 (2002)
20. You, J.-H., Liu, G.: Loop formulas for logic programs with arbitrary constraint atoms. In: Proceedings of the Twenty-Third AAAI Conference on Artificial Intelligence, Chicago, Illinois, USA, pp. 584–589. AAAI Press, Menlo Park (2008)

Bridging the Gap between High-Level Reasoning and Low-Level Control

Ozan Caldiran, Kadir Haspalamutgil, Abdullah Ok, Can Palaz,
Esra Erdem, and Volkan Patoglu

Faculty of Engineering and Natural Sciences, Sabancı University, Istanbul, Turkey

Abstract. We present a formal framework where a nonmonotonic formalism (the action description language $\mathcal{C}+$) is used to provide robots with high-level reasoning, such as planning, in the style of cognitive robotics. In particular, we introduce a novel method that bridges the high-level discrete action planning and the low-level continuous behavior by trajectory planning. We show the applicability of this framework on two LEGO MINDSTORMS NXT robots, in an action domain that involves concurrent execution of actions that cannot be serialized.

1 Introduction

As robotics technology is broadening its applications from factory to more general-purpose applications in public use, demands from robots shift from speed and precision towards safety and cognition. New levels of robustness, physical dexterity, and cognitive capability are necessitated from robots that can perform in dynamic environments involving humans. While traditional robotics design and construct extremely rigid robots with high position control gains, cognitive robotics [1] is concerned with providing robots with higher level cognitive functions that involve reasoning about goals, perception, actions, the mental states of other agents, collaborative task execution, etc., so that they can give high-level decisions to act intelligently in a dynamic world. This paper is an attempt to close the gap between traditional robotics and cognitive robotics, to meet the demands of various applications from robots.

There have been various studies to close the gap between traditional robotics and cognitive robotics, by implementing high-level robot control systems based on different families of formalisms for reasoning about actions and change. For instance, [2] describes a system, LEGOLOG[1], that controls a LEGO MINDSTORMS RIS robot using the high-level control language GOLOG [3] based on the situation calculus [4,5]. [6] presents an execution monitoring system for GOLOG and the RHINO control software which operates on RWI B21 and B14 mobile robots. [7] studies coordination of soccer playing robots, using an extension of GOLOG. In the WITAS Unmanned Aerial Vehicle Project[2] temporal action logic [8], features and fluents [9], and cognitive robotics logic [10] are used for representing the actions and the events, as a part of a helicopter control system [11]. [12] describes how event calculus [13,14] can be used to provide

---

[1] http://www.cs.toronto.edu/cogrobo/Legolog
[2] http://www.ida.liu.se/ext/witas

E. Erdem, F. Lin, and T. Schaub (Eds.): LPNMR 2009, LNCS 5753, pp. 342–354, 2009.

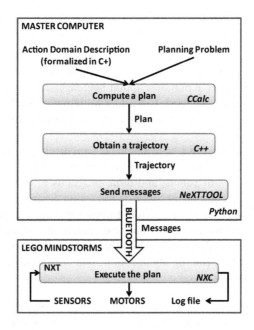

Fig. 1. The overall system architecture

high-level control for a Khepera robot. The agent programming language FLUX [15], based on the fluent calculus [16], has also been used to control the execution of some robots.[3] For instance, [17] presents how FLUX can be used for monitoring the execution of a plan, on a Pioneer 2 mobile robot.

We continue this line of research to provide traditional robotics with high-level reasoning in the style of cognitive robotics, with a different formalism (i.e., the action description language $\mathcal{C}+$ [18]), a different reasoner (i.e., CCALC[4]), a different robot (i.e., LEGO MINDSTORMS NXT), and most importantly with a different method, which bridges the high-level discrete action planning and the low-level continuous behavior with "trajectory planning".

Also we consider high-level reasoning in a different sort of action domain that involves concurrent execution of actions that cannot be serialized. In particular, we study planning problems that require two robots to pick up and carry a payload from an initial location to a goal location, on a maze, while avoiding obstacles. The idea is for the robots to automatically generate a plan, and then execute it collaboratively (Fig. 1). In this domain, the robots can follow complex paths (not necessarily a straight path marked a priori, as in LEGOLOG) avoiding obstacles; this is why our system has trajectory planning between the high-level discrete action planning and the low-level continuous behavior. We can describe this action domain (in particular, the frame problem, the ramification problem, the qualification problem, nonserializable concurrency) in $\mathcal{C}+$ in a straightforward way.

---

[3] http://www.fluxagent.org/projects.htm
[4] http://www.cs.utexas.edu/users/tag/cc

Our main contributions can be summarized in two parts:

- Action description languages [19] are well-studied for various sorts of high-level reasoning about actions and change. On the other hand, unlike the other formalisms mentioned above, it has not been shown on a real robot how high-level reasoning performed within an action description language can be useful for traditional robotics. In that sense, our work is the first to demonstrate the use of action description languages for high-level reasoning and control of robots, in the style of cognitive robotics.
- LEGO MINDSTORMS NXT is available at a relatively low price and is widely available all over the world compared to more sophisticated robots. It allows one to build various kinds of robots, and write programs to control them. Also, the high-level reasoning component based on the action description language $C+$ can be replaced by one based on a different formalism. These features of the overall system enable the reproduction and improvement of our work for educational and research purposes by other researchers who study action description languages, other formalisms for reasoning about actions and change, cognitive robotics, and also traditional robotics.

In the rest of the paper, first we describe the overall system shown in Fig. 1. After we describe the particular action domain and the kind of planning problems we are interested in, we formalize them in the language of CCALC. After that, we explain how a plan computed by CCALC is executed by a LEGO MINDSTORMS NXT robot. We conclude with a discussion on the results and the challenges, as well as the future work.

2   The Overall System Architecture

The overall architecture of our high-level reasoning and control platform is illustrated in Fig. 1.

We start with a description of an action domain in the action description language $C+$ [18]. The idea is, based on this description, to plan the actions of two LEGO MINDSTORMS NXT robots to achieve a common goal. For that, we use the reasoner CCALC. Given an initial state and goal conditions, CCALC computes a plan to reach a goal state, and displays the complete history (including the state information). From such a history, we extract the trajectories of the robots (including the positions and the orientations of the joints of the robots) using inverse kinematics; these trajectories are obtained from a history automatically with a C++ program. After that, we pass these trajectories to the robots, by means of messages via Bluetooth communication, using the program NeXT-Tool. All these tasks are automatically performed on a PC using a Python program.

The brain of a LEGO MINDSTORMS NXT robot is NXT—an embedded controller (with an ARM7 microprocessor) capable of processing messages via the Bluetooth communication, and sending signals to three motors. In our work, two motors are used for movements of the robot on a plane; a third motor is used for the rotation of the robot arm. Since gripping would require an additional degree of freedom, a permanent magnet is used as the end-effector; by this way, a payload with metal endpoints can be grabbed by the robots. Several methods and languages exist for programming NXTs.

Due to its documentation and relative ease of use, we use the programming language NXC to control the movements of the robots according to the received messages.

3   Example: Two Robots and a Payload

Consider two robots, and a payload (a long metal stick) on a platform. Suppose that each robot has a magnet at its end-effector so that it can hold the payload only at one end. None of the robots can carry the payload alone; they have to hold the payload at both ends to be able to carry it. The goal is to place the payload at a specified goal position on the platform.

3.1   Action Domain Description

We view the platform as a maze. We represent the robots by the constants `r1` and `r2`. We describe the payload by its end points, and denote them by the constants `pl1` and `pl2`.

We characterize each robot by its end-effector, and describe its position by a grid point on the maze. The location `(X,Y)` of a robot `R` is specified by two functional fluents, `xpos(R)=X` and `ypos(R)=Y`. Similarly, the location `(X,Y)` of an end point `P1` of the payload is specified by two fluents, `xpay(P1)=X` and `ypay(P1)=Y`. Movements of a robot `R` in some direction `D` are described by actions of the form `move(R,D)`. Each such action has an attribute that specifies the number of steps to be taken by the robot.

In the following, suppose that `R` denotes a robot, `P1` and `P2` denote the end points of the payload, `N` and `N1` range over nonnegative integers 1, ..., `maxN`, and `D` and `D1` range over all directions, `up`, `down`, `right`, `left`. Also suppose that `X1`, `X2`, `Y1`, `Y2` range over nonnegative integers 1, ..., `maxXY`.

We present the causal laws in the language of CCALC.

Direct effects of actions. We describe the effect of a robot's moving right, by the causal laws

```
move(R,right) causes xpos(R)=X2 if steps(R,right)=N & xpos(R)=X1
    where X2=X1+N & X2 =< maxN.
```

Similarly, we describe the effects of moving in other directions.

Ramifications. If a robot `R` is at the same location as an end point `P1` of the payload, the end-effector of that robot attracts that end point:

```
caused on(R,P1) if xpos(R)=xpay(P1) & ypos(R)=ypay(P1).
```

Then the location of the payload is determined by the locations of the robots:

```
caused xpay(P1)=X1 if on(R,P1) & xpos(R)=X1.
caused ypay(P1)=Y1 if on(R,P1) & ypos(R)=Y1.
```

Preconditions of actions. We describe that a robot cannot move in opposite directions by the causal laws

```
nonexecutable move(R,up) & move(R,down).
nonexecutable move(R,left) & move(R,right).
```

We describe each robot's range of motion, taking into account the Pythagorean Theorem, by the causal laws

```
nonexecutable move(R,D) & move(R,D1)
   if D @< D1 & steps(R,D)=N & steps(R,D1)=N1
   where N*N+N1*N1 > maxN*maxN.
```

The robots can carry the payload only if both of them hold the payload at its end points.

```
nonexecutable move(R,D) if -canCarry & on(R,P1).
```

The conditions under which two robots can carry the payload are described by canCarry:

```
caused canCarry if on(r1,P1) & on(r2,P2) & P1\=P2
   after on(r1,P1) & on(r2,P2) & P1\=P2.
```

Note that it is required by the causal laws above that the robots wait for one step immediately after they hold the payload at both ends.

Constraints. We make sure that a payload cannot move places unless it is carried by the causal laws

```
caused false if xpay(P1)=X1 & X1\=X2
   after -canCarry & xpay(P1)=X2.
caused false if ypay(P1)=Y1 & Y1\=Y2
   after -canCarry & ypay(P1)=Y2
```

Since CCALC can only deal with integers, we cannot keep track of the exact locations of the payload. (Consider, for instance, moving one end of the horizontally-situated payload up by 2 steps.) Therefore, we allow the payload's length change with a small tolerance for a more flexible motion. Suppose that linklengthsq denotes the square of the length of the payload; and tolerance denotes the maximum change allowed in the payload's length. The following laws ensure that the payload's length cannot increase/decrease more than tolerance:

```
caused false
   if xpay(pl1)=X1 & xpay(pl2)=X2 & ypay(pl1)=Y1 & ypay(pl2)=Y2
   where
      (X2-X1)*(X2-X1)+(Y2-Y1)*(Y2-Y1) < (linklengthsq-tolerance) ++
      (X2-X1)*(X2-X1)+(Y2-Y1)*(Y2-Y1) > (linklengthsq+tolerance).
```

To take care of obstacles on the platform (to prevent collisions), we add the following causal laws:

```
caused false if xpos(R)=X1 & ypos(R)=Y1
   after xpos(R)=X2 & ypos(R)=Y2
   where collision(X1,Y1,X2,Y2) &
      between(X2-maxN,X2+maxN,X1) & between(Y2-maxN,Y2+maxN,Y1).
caused false
   if xpay(pl1)=X1 & ypay(pl1)=Y1 & xpay(pl2)=X2 & ypay(pl2)=Y2
   where collision(X1,Y1,X2,Y2).
```

Here `collision` is an external function defined in C++, and `between` is an external SWI Prolog function; both are evaluated in SWI Prolog while grounding the causal laws. The first law above prevents the robot end-effectors from moving to a position occupied by an obstacle. The second law ensures that at every state of the world the payload cannot collide with an obstacle.

3.2 Collision Detection

The constraints above ensure at each step that the length of a payload does not change more than a specified tolerance, and the robot end-effectors and the payload do not collide with an obstacle. However, during the plan execution, between any two steps of the plan, the length of a payload can change out of the specified range, and there may be collisions. To ensure collision-free trajectories for the robot end-effectors and the payload, a collision detection algorithm is required.

Such a collision detection algorithm can be implemented in C++ as a function, which takes as inputs the end-effector coordinates of a robot (or the end point coordinates of the payload) at the current state and the next state, and returns "true" if the path is free of collisions. Let's call this function `trajectoryCollision`. After that, we can prevent collision by adding to the description in Section 3.1 the causal laws

```
caused false
  if xpay(pl1)=X1 & ypay(pl1)=Y1 & xpay(pl2)=X2 & ypay(pl2)=Y2
  after
    xpay(pl1)=X3 & ypay(pl1)=Y3 & xpay(pl2)=X4 & ypay(pl2)=Y4
  where trajectoryCollision(X1,Y1,X3,Y3,X2,Y2,X4,Y4) &
    between(X3-maxN,X3+maxN,X1) & between(Y3-maxN,Y3+maxN,Y1) &
    between(X4-maxN,X4+maxN,X2) & between(Y4-maxN,Y4+maxN,Y2).
```

However, there are too many of such causal laws (due to the 8 schematic variables). (Grounding the schematic law above takes more than an hour if we reduce the grid size and restrict the collision detection check to a small set of possible positions of the robot end-effectors.) We can modify `trajectoryCollision` so that it takes 6 variables instead (the positions of one of the end points and the orientation of the payload at the current state and the next state) to uniquely determine the current and the next positions of the robot end-effectors and the payload; add new definitions for the orientations; and modify the schematic causal law above. However, these modifications do not reduce the grounding time sufficiently.

Therefore, instead of adding to the action domain description $\mathcal{D}$ in Section 3.1 causal laws with too many variables, we apply Algorithm 1. The idea is to compute a plan with $\mathcal{D}$ using CCALC, and then check whether such a plan could lead to a trajectory collision. If such a collision is detected between Steps i and $i + 1$, then we extract the location L of the payload and the action A executed at Step i and ask CCALC for a different plan that does not execute A at a state where the payload is located at L. It is important to note that CCALC grounds the action domain only once at the very first iteration of the algorithm; after that, no grounding is done to compute a collision-free plan.

Algorithm 1. PLAN

Input: An action domain description $\mathcal{D}$, a planning problem $\mathcal{P}$
Output: A collision-free plan P of length at most n, if exists
  plan := *false*;   // no collision-free plan is computed so far
  while ¬*plan* **do**
    plan, P, H ← Compute a plan P of length at most n, within a history H, using CCALC
    with $\mathcal{D}$ and $\mathcal{P}$, if there exists such a plan;
    if *plan* **then**
      collision := *false*;   // no trajectory collision is detected so far
      $i := 0$;
      while ¬*collision* AND $i \le |P|$ **do**
        Δ ← Extract the relevant parameters from the history H to uniquely identify the
        positions of the robot end-effectors at Steps i and $i + 1$;
        // Extract the location L of the payload and the action A executed at Step i, if a collision
        is detected
        collision, L, A ← *trajectoryCollision*(Δ);
        $i + +$;
      end while
      if ¬*collision* **then**
        return P
      else
        $\mathcal{P}$ ← Modify the planning problem $\mathcal{P}$ to compute a plan that does not execute A at a
        state where the payload is located at L;
        plan := *false*;
      end if
    end if
  end while

---

4 Finding a Collision-Free Plan

Suppose that initially the robots r1 and r2 are at (1,1) and (2,1) respectively, and
the end points of the payload are at (4,1) and (9,1). The goal is to move the payload
to a location so that its end points are at (4,9) and (9,9). This planning problem can
be described in the language of CCALC by means of a "query" as follows:

```
:- query
maxstep :: 0..infinity;
0: -canCarry, xpos(r1)=1, ypos(r1)=1, xpos(r2)=1, ypos(r2)=1,
    xpay(pl1)=4, ypay(pl1)=1, xpay(pl2)=9, ypay(pl2)=1;
maxstep: xpay(pl1)=9, ypay(pl1)=9, xpay(pl2)=4, ypay(pl2)=9.
```

CCALC then computes the following plan (Plan 1) for this problem:

```
0: move(r1,up,steps=2) move(r1,right,steps=2) move(r2,up,steps=3)
1: move(r1,up,steps=3) move(r1,right,steps=2) move(r2,up,steps=2)
   move(r2,right,steps=3)
2: move(r1,down,steps=3) move(r1,right,steps=2) move(r2,down,steps=2)
   move(r2,right,steps=3)
3: move(r1,down,steps=2) move(r1,left,steps=3) move(r2,down,steps=3)
```

```
    move(r2,right,steps=2)
4:
5: move(r2,up,steps=3) move(r2,left,steps=1)
6: move(r1,up,steps=2) move(r1,right,steps=2)
   move(r2,up,steps=3) move(r2,right,steps=1)
7: move(r1,up,steps=2) move(r1,right,steps=3)
   move(r2,up,steps=2) move(r2,left,steps=3)
8: move(r1,up,steps=4) move(r2,left,steps=2)
```

However, while executing this plan, between time units 7 and 8, the payload collides with the obstacle as illustrated in Fig. 2. Therefore, from the history CCALC computed, we extract the position L of the payload at Step 7:

```
xpay(pl1)=6 xpay(pl2)=9 ypay(pl1)=3 ypay(pl2)=7
```

and the actions A executed at Step 7:

```
move(r1,up,steps=2) move(r1,right,steps=3)
move(r2,up,steps=2) move(r2,left,steps=3)
```

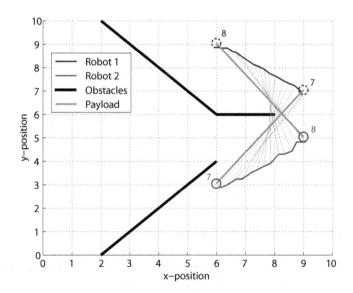

Fig. 2. This figure illustrates the execution of Plan 1 on the robot system, at time step 7. Colors blue, red, green and black are associated with the Robots 1 and 2, the payload, and the obstacles, respectively. Circles and their labels indicate the positions of the robot end-effectors, while the thick green lines denote the position of the payload at each step according to the history calculated by CCALC. For instance, according to the computed history, at Step 7, the end-effectors of Robots 1 and 2 are located at (6,3) and (9,7) respectively, holding the end points of the payload. The thinner green lines denote the payload configuration constructed from the motor encoder data. Observe that, although at time steps 7 and 8 the payload does not collide with the obstacles, between time steps 7 and 8 it does collide with the obstacles. Also the length of the payload changes more than the allowable tolerance.

After that, we ask CCALC to find a different plan that does not execute the actions A at a state where the payload is located at L, by modifying the query above as follows:

```
:- query
label::5;
maxstep :: 9..infinity;
0: -canCarry, xpos(r1)=1, ypos(r1)=1, xpos(r2)=1, ypos(r2)=1,
    xpay(pl1)=4, ypay(pl1)=1, xpay(pl2)=9, ypay(pl2)=1;
maxstep: xpay(pl1)=9, ypay(pl1)=9, xpay(pl2)=4, ypay(pl2)=9;
T<maxstep ->> (
    ((T: xpay(pl1)=6) && (T: ypay(pl1)=3) &&
     (T: xpay(pl2)=9) && (T: ypay(pl2)=7)) ->>
    -((T: move(r1,up)) && (T: steps(r1,up)=2) &&
      (T: move(r1,right)) && (T: steps(r1,right)=3) &&
      (T: move(r2,up)) && (T: steps(r2,up)=2) &&
      (T: move(r2,left)) && (T:steps(r2,left)=3) )).
```

Then CCALC computes the following plan (Plan 2)

```
0: move(r1,up,steps=2)  move(r1,right,steps=3)
   move(r2,up,steps=2)  move(r2,right,steps=3)
1: move(r1,up,steps=2)  move(r1,right,steps=3)
   move(r2,up,steps=2)  move(r2,right,steps=3)
2: move(r1,down,steps=1)  move(r1,right,steps=2)
   move(r2,down,steps=3)  move(r2,left,steps=1)
3: move(r1,down,steps=3)  move(r2,down,steps=1)  move(r2,left,steps=2)
4:
5: move(r1,up,steps=4)  move(r2,up,steps=1)  move(r2,right,steps=1)
6: move(r1,up,steps=4)  move(r2,up,steps=2)  move(r2,right,steps=3)
7: move(r1,left,steps=3)  move(r2,up,steps=1)  move(r2,right,steps=1)
8: move(r1,left,steps=2)  move(r2,up,steps=4)
```

According to this plan, for instance, at Step 7, Robot 1 moves left by 3 units, and Robot 2 moves up by 1 unit and right by 1 unit.

5   Executing a Plan on LEGO Robots

Once CCALC computes a plan for a given problem, it logs the complete history (including the state information). From such a plan, the positions of the robot end-effectors at each time step can be extracted. The simplest approach for executing the plan would be to convert these state values into motor angles and use these values as set-point references for the motors. However, set-point tracking does not guarantee a linear motion of the end-effector, and may cause collisions with the obstacles. To obtain more straight trajectories, a simplified trajectory tracking controller is implemented by introducing intermediate steps to the plan using linear interpolation. Then, these intermediate points are mapped to robot joint variables.

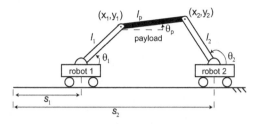

Fig. 3. Schematic representation of two robots carrying a payload

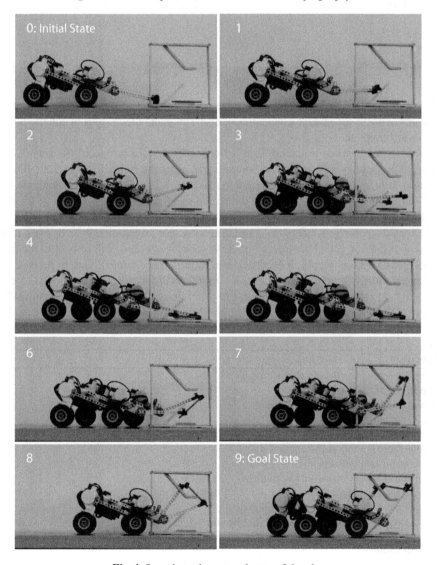

Fig. 4. Snapshot taken at each step of the plan

Fig. 3 depicts a schematic representation of two planar robots carrying a payload. For each robot i, its end-effector is located at a grid point (x_i, y_i) and its corresponding joint variables are denoted as (s_i, θ_i). The forward kinematics of each robot maps its joint variables to its end-effector coordinates and reads as

$$x_i = s_i + l_i \cos(\theta_i) \tag{1}$$
$$y_i = l_i \sin(\theta_i) \tag{2}$$

while the inverse kinematics maps the end-effector coordinates to the joint variables and is given as

$$s_i = x_i \pm \sqrt{l_i^2 - y_i^2} \tag{3}$$
$$\theta_i = atan2\left(\pm\sqrt{l_i^2 - y_i^2}, y_i\right) \tag{4}$$

where l_i represents the length of each robot arm. One can observe that two feasible solutions exist for the inverse kinematics of each robot and the $\pm$ signs in equations (3) and (4) are coupled.

After the joint space trajectories are calculated, they are passed to the robots, by means of messages via Bluetooth communication, using the NeXTTool program. Based on these joint space trajectories, the computed plan is executed by the robots via an NXC program. Algorithm 2 presents the structure of the NXC program used for the low level control of the robots.

To locate a robot at a reference configuration within an acceptable error margin, it is essential that the actual configuration of the robot is checked with respect to the reference configuration. Hence, a feedback controller is necessitated. Due to its ease of implementation, a proportional feedback controller (P-controller) is employed to ensure a robust tracking of the robots in the joint space. The P-controller continually compares the reference and actual joint variables and compensates for the error term by

---

Algorithm 2. NXC Program

Input: Trajectories (a list of reference angles)
Output: Log file
   Check for the Bluetooth communication
   Go to the initial configuration
   Wait for the start signal
   while There is a trajectory to follow **do**
      Read the reference angles
      while Not at the reference angles **do**
         Read the motor angles
         Calculate the error in motor position
         Rotate the motor to compensate for the error
         Record the motor angles
      end while
   end while

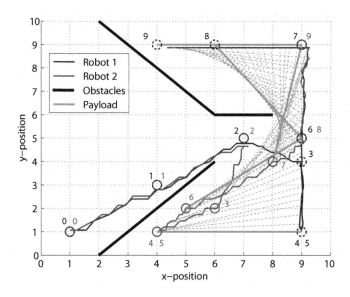

Fig. 5. This figure illustrates the execution of Plan 2 on the robot system. Colors blue, red, green and black are associated with the Robots 1 and 2, the payload, and the obstacles, respectively. Circles and their labels indicate the positions of the robot end-effectors, while the thick green lines denote the position of the payload at each step according to the history calculated by CCALC. For instance, according to the computed history, initially, the end-effectors of robots are located at the grid point (1,1), and the payload lies between the points (4,1) and (9,1); at Step 6, the end-effectors of Robots 1 and 2 are located at (9,5) and (5,2) respectively, holding the end points of the payload. Blue and red lines represent the end-effector trajectories of each robot, while thinner green lines denote the payload configuration as constructed from the motor encoder data. The black lines represent the obstacles.

commanding a counteracting motion that is proportional to the magnitude of the error signal. The P-controller gain is tuned empirically to achieve acceptably low overshoot and steady state error of the motor response.

For instance, consider the planning problem described in the previous section. After CCALC computes a collision-free plan (Plan 2) for the problem as described in the previous sections, the intermediate points are interpolated and mapped to the robot joint space as explained above. Then, the LEGO robots trace these trajectories as in Fig.s 4 and 5. Fig. 4 presents snapshots taken at each step of the plan, while Fig. 5 depicts the trajectories of the robot end-effectors and the payload.

6 Discussion

We have demonstrated with some planning problems in a sample action domain, how the logic-based formalism $\mathcal{C}+$ can be used to endow two LEGO MINDSTORMS NXT robots with high-level reasoning in the style of cognitive robotics.

In these experiments, we encountered many challenges. For instance, that CCALC can handle integers only, caused some difficulties in calculating the exact positions of the robots. To deal with this problem, we assumed that the length of the payload might increase/decrease within a specified tolerance. We also faced control challenges: Lack of floating point operations in NXC; low encoder resolution, high friction and backlash of the LEGO motors; and the flexible robot structure due to plastic parts. To address these challenges we have to upgrade the hardware/software of LEGO MINDSTORMS NXT robots. The modification of the overall architecture to include monitoring of the plan execution is a part of the ongoing work.

References

1. Levesque, H., Lakemeyer, G.: Cognitive robotics. In: Handbook of Knowledge Representation. Elsevier, Amsterdam (2007)
2. Levesque, H.J., Pagnucco, M.: Legolog: Inexpensive experiments in cognitive robotics. In: Proc. of CogRob, pp. 104–109 (2000)
3. Levesque, H.J., Reiter, R., Lin, F., Scherl, R.B.: GOLOG: A logic programming language for dynamic domains. JLP 31 (1997)
4. McCarthy, J.: Situations, actions, and causal laws. Technical report, Stanford University (1963)
5. Levesque, H.J., Pirri, F., Reiter, R.: Foundations for the situation calculus. ETAI 2, 159–178 (1998)
6. Hähnel, D., Burgard, W., Lakemeyer, G.: GOLEX - bridging the gap between logic (GOLOG) and a real robot. In: Herzog, O. (ed.) KI 1998. LNCS, vol. 1504, pp. 165–176. Springer, Heidelberg (1998)
7. Ferrein, A., Fritz, C., Lakemeyer, G.: Using GOLOG for deliberation and team coordination in robotic soccer. Künstliche Intelligenz 1 (2005)
8. Doherty, P., Gustafsson, J., Karlsson, L., Kvarnström, J.: Tal: Temporal action logics language specification and tutorial. ETAI 2, 273–306 (1998)
9. Sandewall, E.: Features and Fluents: A Systematic Approach to the Representation of Knowledge about Dynamical Systems. Oxford University Press, Oxford (1994)
10. Sandewall, E.: Cognitive robotics logic and its metatheory: Features and fluents revisited. ETAI 2, 307–329 (1998)
11. Doherty, P., Granlund, G., Kuchcinski, K., Sandewall, E., Nordberg, K., Skarman, E., Wiklund, J.: The WITAS unmanned aerial vehicle project. In: Proc. of ECAI, pp. 747–755 (2000)
12. Shanahan, M., Witkowski, M.: High-level robot control through logic. In: Castelfranchi, C., Lespérance, Y. (eds.) ATAL 2000. LNCS (LNAI), vol. 1986, pp. 104–121. Springer, Heidelberg (2001)
13. Kowalski, R., Sergot, M.: A logic-based calculus of events. New Gen. Comput. 4(1), 67–95 (1986)
14. Miller, R., Shanahan, M.: The event calculus in classical logic - alternative axiomatisations. ETAI 3(A), 77–105 (1999)
15. Thielscher, M.: FLUX: A logic programming method for reasoning agents. TPLP 5(4-5), 533–565 (2005)
16. Thielscher, M.: Introduction to the fluent calculus. ETAI 2, 179–192 (1998)
17. Fichtner, M., Großmann, A., Thielscher, M.: Intelligent execution monitoring in dynamic environments. In: Proc. of Workshop on Issues in Designing Physical Agents for Dynamic Real-Time Environments: World modeling, planning, learning, and communicating, Acapulco, Mexico (2003)
18. Giunchiglia, E., Lifschitz, J.L.V.: Nonmonotonic causal theories. AIJ 153 (2004)
19. Gelfond, M., Lifschitz, V.: Action languages. ETAI 2, 193–210 (1998)

A General Approach to the Verification of Cryptographic Protocols Using Answer Set Programming

James P. Delgrande, Torsten Grote, and Aaron Hunter

School of Computing Science,
Simon Fraser University,
Burnaby, B.C.,
Canada V5A 1S6
{jim,tga14,hunter}@cs.sfu.ca

Abstract. We introduce a general approach to cryptographic protocol verification based on answer set programming. In our approach, cryptographic protocols are represented as extended logic programs where the answer sets correspond to traces of protocol runs. Using queries, we can find attacks on a protocol by finding the answer sets for the corresponding logic program. Our encoding is modular, with different modules representing the message passing environment, the protocol structure and the intruder model. We can easily tailor each module to suit a specific application, while keeping the rest of the encoding constant. As such, our approach is more flexible and elaboration tolerant than related formalizations. The present system is intended as a first step towards the development of a compiler from protocol specifications to executable programs; such a compiler would make verification a completely automated process. This work is also part of a larger project in which we are exploring the advantages of explicit, declarative representations of protocol verification problems.

1 Introduction

A cryptographic protocol is a sequence of encrypted messages that is used to exchange information and achieve communicative goals over an insecure network. Proving that a cryptographic protocol is secure is difficult, because many attacks are subtle and difficult to find. The problem is further complicated by the fact that protocol goals are often specified imprecisely, which makes it difficult to know exactly what might constitute an attack. In this paper, we use Answer Set Programming (ASP) to encode relevant information involved in a cryptographic protocol, including agent capabilities and the message passing environment, and to automatically detect attacks.

A wide range of methods have been previously employed for the verification of cryptographic protocols, including encodings in many different formal logics. Our work is distinguished from most of these methods in that we use a declarative formalism for reasoning about action effects. We specify the steps of a cryptographic protocol in a logic program where the answer sets correspond to sequences of messages exchanged between agents. After translating a given protocol into our encoding, it is straightforward to automatically generate attacks by using an answer set solver.

E. Erdem, F. Lin, and T. Schaub (Eds.): LPNMR 2009, LNCS 5753, pp. 355–367, 2009.

This paper makes several contributions to work on cryptographic protocol verification. First, given our declarative approach, all aspects of protocol specification, including agent capabilities, the message-passing environment, and the goals of the protocol, are explicitly specified. As a result also, our model of message passing is more general and more flexible than other models used in protocol verification. For example, in our approach it is straightforward to modify the capabilities of the intruder for a specific application. Second, our work provides a methodology for compiling an arbitrary cryptographic protocol into a formal encoding suitable for automated analysis. This is a significant improvement over purely logical representations of cryptographic protocols, where the translation from formalism to implementation may be non-trivial.

We proceed as follows. In §2, we provide an illustrative example, related work, and motivation for our approach. In §3, we present the details of our formal model. In §4, we discuss the advantages of our approach as compared with related formalisms. We then offer some concluding remarks and directions for future work.

2   Background and Motivation

2.1   Illustrative Example: The Needham Schroeder Protocol

We introduce the well-known Needham Schroeder protocol [15]. This is an *authentication protocol* involving two agents A and B, which is to say that the goal is for each agent to establish that they are communicating with the other. In the protocol specification, the notation $\{M\}_K$ is used to denote the message M encrypted with the key K. In the given protocol, N_A and N_B denote random numbers, called *nonces*, generated by A and B respectively. The key K_A is the public key for A and the key K_B is the public key for B. In public key cryptography, each agent also has a corresponding private key that is used for decryption. We treat encryption as a 'black box' operation by assuming that an encrypted message can never be decrypted without the corresponding private key. Hence, a message encrypted with the public key of some agent can only be decrypted by that particular agent. Using these conventions, the protocol is specified as follows.

The Needham Schroeder Protocol
  1. $A \rightarrow B : \{N_A, A\}_{K_B}$
  2. $B \rightarrow A : \{N_A, N_B\}_{K_A}$
  3. $A \rightarrow B : \{N_B\}_{K_B}$

Each line in the specification indicates that a message is sent from one agent to the other, as indicated by the arrow. Hence, the first step in the protocol is for A to send B the nonce N_A as well as an identifier for A, both encrypted with B's public key. In the second step, B responds with the nonce N_A as well as a new nonce, both encrypted with A's public key. Intuitively, the idea is that A can conclude that B received the first message, because only B has the private key for decrypting this message. The protocol concludes with A responding to B's nonce, appropriately encrypted.

An attack on the Needham Schroeder protocol was discovered by Lowe [15]. Following convention, we describe the attack using the same "arrow" notation used for specifying protocols. The symbol I denotes a dishonest agent called the *intruder*. In order to clarify the attack, we use the notation I_A in situations where I is pretending to be A, either by intercepting a message intended for A or by sending a message that A would be expected to send. In the attack, A initiates a run of the protocol with I, but then I uses the nonce generated by A to start a new run with B.

Needham Schroeder Protocol Attack

$$
\begin{aligned}
&1. && A \to I && : \{N_A, A\}_{K_I} \\
&1'. && I_A \to B && : \{N_A, A\}_{K_B} \\
&2'. && B \to I_A && : \{N_A, N_B\}_{K_A} \\
&2. && I \to A && : \{N_A, N_B\}_{K_A} \\
&3. && A \to I && : \{N_B\}_{K_I} \\
&3'. && I_A \to B && : \{N_B\}_{K_B}
\end{aligned}
$$

By interweaving protocol runs, in this attack I is able to get A to decrypt the challenge nonce sent by B. Hence, at the conclusion of this trace, B believes that it has been communicating with A, when it has not.

2.2 Logic-Based Verification

Logical methods have been used to prove the correctness of cryptographic protocols. The basic idea is to encode the steps of a protocol as formulas in a logic, and also to formalize the goals of the protocol in the same logic. A protocol is then shown to be secure by proving that the goal is logically entailed by the formulas that correspond to the steps of the protocol. The first logic defined explicitly for protocol verification was the BAN logic of Burrows, Abadi and Needham [4]. This work has been highly influential for two main reasons. First, it illustrates that protocol verification can be reduced to formal reasoning in a logic. Second, it demonstrates the significance of knowledge and belief when proving protocol correctness. Although BAN logic itself has several known flaws, it has lead directly to the development of several more sophisticated protocol logics. In many cases, such logics are based on the well-known "runs and systems" model of message passing [7].

Virtually all logic-based approaches to protocol verification use the *Dolev-Yao intruder model*, which specifies that an intruder can read, intercept and forward the messages sent between honest agents [6]. In this model, messages are received with no proof of authorship, which means that a message recipient is never aware of the sender, except possibly via the contents of the sent message. Encryption is assumed to be unbreakable, so an intruder can never decrypt an encrypted message without the appropriate key.

In contrast to the bulk of the work in protocol verification, our approach emphasises a declarative representation. One declarative formalism that has been employed for protocol verification in the past is the situation calculus [14]. The first work in this area consisted of a direct encoding of message passing in the presence of a Dolev-Yao intruder [13]. In this model, message interception and forwarding was modeled by means of non-deterministic effects for message sending. More recently, in a companion to the present paper, we have introduced a new situation calculus model of cryptographic protocols [5]. In this approach, we introduce an explicit intruder model, in which message interception and forwarding are actions executed by an intruder. This approach is discussed further in the final section.

As well, there has been previous work in using ASP for protocol verification. Two distinct, yet similar, ASP models have been proposed in [1,16]. The basic idea in these approaches is to provide a template for defining a logic program that represents a specific protocol. Answer sets correspond to sequences of exchanged messages, which can then be analysed with respect to a set of explicitly defined attacks. Since attacks are specified in advance, this approach cannot readily be used to find "new" or "unexpected" attacks on a protocol. In contrast, in our approach, we define attacks through user-specified queries over the set of protocol runs. Another distinguishing feature of our approach is that we present modular definitions of message passing, intruder capabilities, and protocol structure. Hence, we define a formalism that is more flexible and elaboration tolerant.

2.3 Motivation

The standard specification of protocols, as illustrated by the Needham-Schroeder Protocol above, is both incomplete and imprecise. It is incomplete, in part, because the goal of the protocol is not explicitly given. Without a precise goal for a protocol, it is difficult to determine if a given trace constitutes an attack. The specification of protocols is ambiguous because it is not clear exactly what $A \rightarrow B$ actually means; clearly it does not mean that A successfully sends the message to B. Instead, it seems to conflate two things:

1. A *intends* to send the message to B.
2. A actually sends the message *to someone*.

This interpretation suggests that the intentions of a sender are relevant for protocol verification. Nevertheless, intentions are often left implicit in many approaches to protocol verification. One of the main motivations of this paper is to provide a more precise and explicit description of what is intended in a cryptographic protocol specification. An advantage of our approach is that we specify protocols and protocol goals in a declarative manner, which allows us to critically examine exactly what constitutes an attack.

One problem with existing formal models for protocol verification is that they implicitly assume that honest agents do not do anything irresponsible or dangerous. For example, honest agents do not send plain text messages that compromise secret keys. While this example is obvious, it may not be clear which other actions are irresponsible. By specifying attacks in terms of flexible queries on the set of traces satisfying a protocol, we are able to discover which actions should be avoided by an honest agent.

Lastly, we are interested in comparing alternative declarative paradigms for cryptographic protocol analysis, in particular, and multi-agent message passing, in general. Consequently we have been developing in parallel an ASP approach along with a situation calculus encoding. It is not immediately obvious, however, how to choose between ASP and a formalism such as the situation calculus. In the discussion, we briefly consider the relative merits of each approach.

3 Approach

Our aim is to define a logic program in which the answer sets correspond to sequences of exchanged messages between agents. We will refer to the honest agents as *principals*, in contrast to the *intruder* whose goal is to disrupt communication between principals. The formalization consists of three essentially independent modules. One module includes generic information about message passing and protocols; one module includes a specification of the intruders capabilities; and the third includes the structure of a specific protocol. This approach makes the formalization very elaboration tolerant, as we are free to manipulate the message passing environment, the protocol structure, or the intruder to suit different applications. The ability to change the intruder is particularly novel in the protocol verification community, where the Dolev-Yao intruder is so entrenched. Nevertheless, there are practical examples where the power of the intruder may be altered by external issues, such as network topology.

We assume the reader is familiar with ASP, as described in [11], for example. We also make extensive use of constrained *choice rules* involving expressions of the form

```
i {p, q, r} j
```

Such expressions are understood to indicate that at least i of the enclosed atoms are true, but not more than j are true. Another kind of expression we make use of is conditions of the form $p(X)$: $q(X)$. These expressions are used for instantiating the variable X with the domain predicate $q(X)$ to collections of terms within a single rule. For example, if there are the facts $q(1), q(2)$ and $q(3)$ in the program, then i $\{p(X)$: $q(X)\}$ j will be grounded to i $\{$ $p(1)$, $p(2)$, $p(3)$ $\}$ j. Also, our program is λ-restricted as described in [10] in order to avoid problems with function symbols.

3.1 Protocol Module

The protocol module sets up a general message passing framework, the principals' holdings and capabilities, as well as some high-level auxiliary predicates. It also specifies different types for variables.[1]

Keys and Nonces: Principals need keys for encryption. Our approach supports both asymmetric and symmetric key encryption. The protocol module is responsible for setting up the key infrastructure. It specifies keys for every agent and distributes the public or the shared keys to the appropriate principals. In order to guarantee the freshness of messages, certain protocols require nonces or timestamps in messages. Due to the propositional nature of our formalism, it is not possible to have every agent create an unlimited number of nonces on demand. Instead, a sufficient number n of nonces is assigned to each principal. For performance reasons, n defaults to 1, but can also be set as a command line parameter. In the initial state all nonces are fresh at the outset; and no messages have been sent or received.

---

[1] Of course these declarations are eventually grounded, but it is useful to think of the variables as ranging over different classes of entities.

Actions: Protocol analysis is focused on analyzing sequences of actions. Clearly, *send* and *receive* actions are central, although these need not be the only actions. Actions occur at some point in time, and the notion of time is abstracted into slices or time steps. In order to keep the search space manageable, internal actions related to message composition and encryption are implemented through state constraints. This means that a single time step is used for one agent to receive a message, decrypt its contents, compose a reply, encrypt and send it. The reply may then be received by a recipient in the following time step. The maximum number of time steps is also set as a parameter. Sending and receiving of messages is modelled as events or actions that are triggered once all preconditions are satisfied. A message is sent as soon as a principal has the message and wants to send it.[2]

```
send(A, B, M, T) :- wants(A, send(A, B, M), T),
                    has(A, M, T).
```

If the message does not get intercepted by the intruder, it will be received in the next time step.

```
receive(A, B, M, T+1) :- send(B, A, M, T),
                         not intercept(M, T+1).
```

The sending action effectively acts as a precondition for receiving, while an effect of the receive action is that the recipient possesses the message afterwards.

```
has(A, M, T) :- receive(A, B, M, T).
```

If a principal receives a message as part of a particular protocol, then they want to send the appropriate response. The notion of "appropriateness" is modelled by the `fit` predicate that matches certain message components.

```
{ wants(A, send(A, B, M), T) } 1 :- receive(A, B, M2, T),
    fit(msg(J, B, A, M2), msg(J+1, A, B, M)).
```

Auxiliary predicates. In order to be able to specify flexible goal and attack conditions in the instance module, we provide several auxiliary predicates. Some of these predicates are straightforward, such as the predicate `talked` to indicate that two agents have successfully communicated:

```
talked(A, B, T+1) :- send(A, B, M, T),
                     receive(B, A, M, T+1).
```

In contrast, the predicate `authenticated`, which is used as part of a goal specification, is much more complex. We say that *A* is *authenticated* with *B* only if *A* has sent a fresh "challenge nonce" encrypted with an appropriate key to *B*; *B* has to have replied to *A*'s challenge with the same nonce, again encrypted with a key so that only *A* can decrypt it. As well, *A* received *B*'s reply and all events happened in the right order. The predicate is given as follows:

---

[2] Note that this and following code examples may not be complete. On occasion, details, such as domain restrictions, are omitted for readability and brevity.

```
authenticated(A, B, T) :-
    send(A, B, enc(M1, K1), T1),
    fresh(A, nonce(A, Na), T1),
    part_m(nonce(A, Na), M1),
    key_pair(K1, Kinv1), has(A, K1, T1),
    has(B, Kinv1, T1),
    not has(C, Kinv1, T1) : agent(C) : C != B,
    send(B, A, enc(M2, K2), T2),
    receive(A, B, enc(M2, K2), T),
    part_m(nonce(A, Na), M2),
    key_pair(K2, Kinv2), has(B, K2, T1),
    has(A, Kinv2, T1),
    not has(C, Kinv2, T1) : agent(C) : C != A,
    T1 < T2, T2 < T.
```

Authentication is generally a difficult concept to define, and different notions of authentication may be required for different protocols. One advantage of our approach is that it is easy to change the definition of authentication to suit a particular application or a particular audience. Also, the protocol module can easily be extended by other auxiliary predicates that might be useful for a wide range of protocols.

3.2 Intruder Module

The intruder module specifies all aspects of the intruder. As noted, an advantage of our approach is that the intruder model is flexible and easy to modify. If our system is run without the intruder module, then it simply computes all valid protocol runs between honest principals within the given time bound. If an intruder module is given, then the system can be used to find attacks on the protocol that may be carried out by the intruder.

Holdings: By default, the intruder holds the public key of every agent. The intruder also holds a public-private key pair. It is important that the intruder has this pair, since it allows the intruder to "pretend" to be an honest agent that initiates protocol runs with principals. Since the intruder module can be modified, it is straightforward to model, for example, the situation in which an intruder has obtained a key belonging to someone else. This is an important improvement over existing approaches to verification, as it allows us to give customized proofs of security. For example, we may be able to prove that a certain protocol is secure, even if one of the participants has a compromised private or shared key.

Capabilities: The intruder module specifies the capabilities and limitations of the intruder. In general, we specify a Dolev-Yao type intruder, because this makes it easier to compare our approach with existing approaches. Hence, we specify that the intruder can intercept messages and that it receives the messages that it intercepts:

```
0 { intercept(M, T+1) } 1 :- send(A, B, M, T).
receive(I, A, M, T+1) :- send(A, B, M, T),
                         intercept(M, T+1).
```

The intruder can also send messages whenever it wants to. However, it can not send messages that it does not have:

```
:- send(I, A, M, T), not has(I, M, T).
```

Of course, the intruder can also fake the sender name of messages it sends:

```
1 { receive(A, B, M, T+1) : principal(B) } 1 :-
    send(I, A, M, T).
```

This will make the receive action of principal A look like it obtained the message from a principal B.

Again, the capabilities we have listed here are examples that are useful for many types of protocols. Using interface predicates from the protocol module and standard ASP syntax, it is possible to specify a wide range of different capabilities. For example, one might want to specify that an intruder can only eavesdrop on messages that arrive at or originate from one specific principal.

3.3   Instance Modules

In order to verify a particular protocol, we need to provide an encoding of that protocol in ASP. This encoding is done in the instance module. Currently, we have encoded two protocols: the Needham Schroeder Protocol and the Challenge Response Protocol. The structure of the ASP encoding is straightforward; indeed the next step in the project is to automate the process, so that, given a simple "arrows" specification like that for the Needham-Schroeder protocol, a corresponding instance module in ASP is produced. At present, however, the encodings are done by hand.

In the beginning of the instance module, certain options such as the maximum number of concurrent protocol runs have to be set. The principals participating have to be added, too. This is done by simply including several interface predicates of the form `principal(a)`. The instance module for a specific protocol also has to state which kind of encryption should be used. For example, public-key encryption is activated if the fact `set(pub_key_enc)` is added. All the details about the encryption are handled by the protocol module and is of no concern to the instance module.

The instance module has four main components: a specification of valid messages, completion conditions, the goal of the protocol, and attack conditions. We illustrate how each component is implemented, using the Needham Schroeder protocol as an example.

Messages: In order to model message passing in ASP, it is useful to bound the number of messages that can be exchanged. In practice, the set of messages is of course not bounded – given a composition operator and an encryption operation, it is possible to define an infinite set of messages. However, if we fix a specific protocol, then it is easy to specify the set of messages that can be used in a valid run. In the Needham Schroeder protocol, for example, there are three basic message forms that can be sent, one for each line of the protocol. However, each message can involve any nonce that is available and it can involve the public key of any agent on the network. In our framework there are only a finite number of nonces and agents, so it is easy to delimit the class of messages.

The protocol itself can then be described by a valid subset of possible messages:

```
msg(1,A,B,enc(m(nonce(C,Na),principal(A)),pub_key(B))).
msg(2,B,A,enc(m(nonce(C,Na),nonce(D,Nb)),pub_key(A))).
msg(3,A,B,enc(m(nonce(D,Nb)),pub_key(B))).
```

These rules correspond to the first, second and third message of the protocol.

Completion Conditions: Normally, a protocol run is completed if all of the messages have been sent and received. We define the completion conditions at the agent level, which means we need to specify when each participating agent *believes* the protocol is completed. To achieve this, we introduce a general `believes` predicate that is also used to represent other beliefs of the principals. If a principal has sent and received the appropriate messages in the right order, it will believe that it successfully completed a valid run of the protocol with another principal. Unfortunately, it was not possible to generalize this condition for every protocol in the protocol module. However a compiler could easily create this rule from the protocol specification.

Goals: As we saw, the protocol module includes a set of high-level meta-predicates that can be used to define specific goals. In the Needham Schroeder protocol instance, the goal is that both principals should believe that the other one is authenticated and that they actually are.

```
goal(A, B, T) :- authenticated(A, B, T),
                 believes(A, authenticated(A, B), T),
                 authenticated(B, A, T),
                 believes(B, authenticated(B, A), T).
```

This is essentially a definition of "mutual authentication", following the BAN tradition in which we only consider nested beliefs to depth two. As noted earlier, the concept of authentication is difficult to define and there might be disagreements about the definition used here. However, it is easy to change the definition of mutual authentication while leaving the rest of the encoding and the other modules unchanged.

Attacks: An attack on a protocol is a run of the protocol that causes an agent to believe the goal is true, when in fact the goal is not true. In the case of the Needham Schroeder protocol, an attack is specified in terms of a principal believing that it is authenticated, but is in fact not authenticated.

```
attack :- believes(A, authenticated(A, B), T),
          not authenticated(A, B, T).
```

In this case, the principal's belief is false and must have been created somehow. Our system can give an explanation of how this happened by showing the protocol trace that led to the false belief. This is done by including the integrity constraint `:- not attack` in the protocol instance. Then all answer sets of our system will represent an attack, because traces that are not an attack are excluded.

3.4 Protocol Verification

Given the three modules of our ASP encoding, an ASP solver can be used to find attacks on a given protocol. For our implementation, we ground our encoding using *GrinGo*

[10] and compute the answer sets using *clasp* [9]. For simplicity, we use the hybrid program *Clingo*[3] to perform both tasks in one step. Also, in order to format the display of the output, we employ a Python script output.py that makes the results more readable. Therefore, at the command line, we can verify the Needham Schroeder protocol by executing the following command.

```
clingo protocol.lp intruder.lp needham_schroeder.lp \
    -c n=1 -c t=6 0 | output.py
```

Note that you have to specify the number of nonces n that each agents has initially and the maximum trace length t as parameters. We also provide run scripts, so users do not need to remember the entire command line. These scripts just take n and t as optional parameters that are initialised with reasonable default values.

The above command will output attacks on the Needham Schroeder protocol in the following format:

```
send(a,i,enc(m(nonce(a,1),principal(a)),pub_key(i)),0)

receive(i,a,enc(m(nonce(a,1),principal(a)),pub_key(i)),1)
send(i,b,enc(m(nonce(a,1),principal(a)),pub_key(b)),1)

receive(b,a,enc(m(nonce(a,1),principal(a)),pub_key(b)),2)
send(b,a,enc(m(nonce(a,1),nonce(b,1)),pub_key(a)),2)

receive(i,b,enc(m(nonce(a,1),nonce(b,1)),pub_key(a)),3)
send(i,a,enc(m(nonce(a,1),nonce(b,1)),pub_key(a)),3)

receive(a,i,enc(m(nonce(a,1),nonce(b,1)),pub_key(a)),4)
send(a,i,enc(m(nonce(b,1)),pub_key(i)),4)

receive(i,a,enc(m(nonce(b,1)),pub_key(i)),5)
send(i,b,enc(m(nonce(b,1)),pub_key(b)),5)

believes(b,authenticated(b,a),6)
believes(b,completed(b,a),6)
receive(b,a,enc(m(nonce(b,1)),pub_key(b)),6)
```

The attack can be read directly from this output. To improve readability, it would be straightforward to incorporate an additional script to transform this output into the usual "arrow" syntax enriched with the believes predicates.

Our system employs a minimize statement by default to always return the shortest attack. But it can easily be reconfigured to return all attacks that are possible in the given trace length. Note that our system is not actually able to prove a protocol is correct, as it will simply continue looking for successively longer attacks. However, it is possible to prove that there is no attack of involving less than k actions, for a pre-specified number k.

---
[3] http://potassco.sf.net

It addition to Lowe's attack on the Needham Schroeder protocol, our system discovered several other attacks that become possible if the agents' capabilities are not circumscribed. The simplest such attack is what we have called the "Stupidity Attack" in which one principal sends an unencrypted nonce to the intruder. A more sophisticated attack we found was dubbed the "Bad Memory Attack", because it only works if the principals are unable to remember the current state of a protocol run.

Bad Memory Attack

1. $A \rightarrow B : \{N_A, A\}_{K_B}$
2. $B \rightarrow I_A : \{N_A, N_B\}_{K_A}$
2'. $I \rightarrow A : \{N_A, N_B\}_{K_A}$
3'. $A \rightarrow I : \{N_B\}_{K_I}$
3. $I_A \rightarrow B : \{N_B\}_{K_B}$

In this attack, A starts the protocol regularly with B. The second message is intercepted by the intruder who uses this message to send it as challenge to A in line with the second step of the protocol. A has a bad memory and does not realize that he did not initiate the protocol with I. So A replies to the challenge, effectively decrypting the nonce N_B for the intruder, who is then able to use it to fool B.

Our system produces attacks for the Challenge Response and Needham Schroeder protocol in less than two seconds. If each agent is given more than one nonce, the runtime increases significantly.[4] This seems to be partly caused by the grounding process. Nonces are involved in almost every ASP rule of our encoding and are responsible for the creation of exponentially many ground rules. Another reason for the increased overall runtime of our system when using more nonces or a larger plan length is the bigger search space. The option of minimizing the number of sent messages forces Clasp to almost traverse the entire search space to make sure that no smaller trace exists. Incremental grounding and solving as described in [8] might be useful for coping with the blowup caused by the plan length and the search for a minimal trace length of an attack. This is a topic for future work.

4 Discussion

4.1 Comparison with Related Formalisms

There has been previous work on declarative approaches to protocol verification in the situation calculus [13], in abductive logic programming [2], and in ASP [1,16]. Our approach is more flexible than these approaches in the sense that different aspects of the model can be modified, while maintaining a consistent overall framework. Our work is also distinguished by the fact that we can detect unexpected "attacks" on a protocol.

As well we mentioned a companion paper that formulates cryptographic protocol verification in terms of a situation calculus theory [5]. Both of our approaches are, or are in the process of, being implemented. The situation calculus approach allows a more procedural specification, in that agent actions are defined and "executed". As well, it allows a more refined specification of control and message passing. On the other hand,

---

[4] We obtained the following times with the Needham Schroeder protocol: 3 nonces: 2 sec, 6 nonces: 13 sec, 9 nonces: 51 sec, 12 nonces: 190 sec.

the situation calculus approach will require more work with regards to an implementation; we anticipate making use of more procedural constructs as found in ConGolog. For the ASP approach described here, implementation was more straightforward and a prototype was readily obtained[5]. The biggest issue in the ASP implementation was controlling the state explosion that resulted from grounding. Performance issues arise when the number of agents, nonces or concurrent protocol runs is increased.

Last, it is interesting to note that our original intention was to formulate our theory not directly in ASP, but rather in the action language C [12]. Unfortunately, C turned out to be problematic for several reasons, notably the fact that action effects must be Markovian. In reasoning about protocol exchanges, it is inevitably necessary to express effects due to a certain sequence of actions. Neither the description nor the query language of C support this. As a result, we found that it was easier and more straightforward to encode protocols directly in ASP.

4.2 Conclusion

We have described a formulation and implementation of cryptographic protocol analysis and verification in ASP. A primary advantage of using ASP is that it provides a declarative formalism for which efficient implementations exist. As a result we were able to specify a framework that is flexible, declarative, and elaboration tolerant. The approach provides for flexible models of the intruder and principals, and the specification of goals at an intensional level in terms of an agent's beliefs. The implementation discovered not just the standard known attacks, but also other possible attacks, among them the Bad Memory Attack and the Stupidity Attack. Both of these latter attacks are easily addressed; however the detection of such attacks for a simple protocol suggests that, in general, there may be other, more subtle attacks that a declarative formalism, with a full specification of agent actions, may be able to detect.

There have been challenges in the implementation, primarily in getting the implementation to scale up. Increasing the number of agents, the number of available nonces, and the suite of actions available to an agent all provide challenges. For future work, we will first continue to develop our implementations. Concurrently we are also working on a compiler, both for ASP and the situation calculus approaches, that will take an arbitrary protocol and produce an executable theory. Hence, in a broader potential contribution, this project will provide a testbed for comparing approaches implemented in a constraint formalism such as ASP, with a more planning-oriented approach such as the situation calculus.

References

1. Aiello, L.C., Massacci, F.: Verifying security protocols as planning in logic programming. ACM Trans. Comput. Logic 2(4), 542–580 (2001)
2. Alberti, M., Chesani, F., Gavanelli, M., Lamma, E., Mello, P., Torroni, P.: Security protocols verification in abductive logic programming: A case study. In: Dikenelli, O., Gleizes, M.-P., Ricci, A. (eds.) ESAW 2005. LNCS (LNAI), vol. 3963, pp. 106–124. Springer, Heidelberg (2006)

---

[5] The prototype is available for download from
http://www.cs.sfu.ca/~cl/software/ASP.SP/

3. Baral, C., Brewka, G., Schlipf, J. (eds.): LPNMR 2007. LNCS (LNAI), vol. 4483. Springer, Heidelberg (2007)
4. Burrows, M., Abadi, M., Needham, R.: A logic of authentication. ACM Transactions on Computer Systems 8(1), 18–36 (1990)
5. Delgrande, J.P., Hunter, A., Grote, T.: Modelling cryptographic protocols in a theory of action. In: Proceedings of the Ninth International Symposium on Logical Formalizations of Commonsense Reasoning (2009)
6. Dolev, D., Yao, A.: On the security of public key protocols. IEEE Transactions on Information Theory 29(2), 198–208 (1983)
7. Fagin, R., Halpern, J.Y., Moses, Y., Vardi, M.Y.: Reasoning about Knowledge. MIT Press, Cambridge (1995)
8. Gebser, M., Kaminski, R., Kaufmann, B., Ostrowski, M., Schaub, T., Thiele, S.: Engineering an incremental ASP solver. In: Garcia de la Banda, M., Pontelli, E. (eds.) ICLP 2008. LNCS, vol. 5366, pp. 190–205. Springer, Heidelberg (2008)
9. Gebser, M., Kaufmann, B., Neumann, A., Schaub, T.: clasp: A conflict-driven answer set solver. In: Baral et al [3], pp. 260–265
10. Gebser, M., Schaub, T., Thiele, S.: Gringo: A new grounder for answer set programming. In: Baral et al [3], pp. 266–271
11. Gelfond, M.: Answer sets. In: Handbook of Knowledge Representation, pp. 285–316. Elsevier, Amsterdam (2007)
12. Gelfond, M., Lifschitz, V.: Action languages. Electronic Transactions on AI 3 (1998)
13. Hernández-Orallo, J., Pinto, J.: Formal modelling of cryptographic protocols in situation calculus (Published in Spanish as: Especificación formal de protocolos criptográficos en Cálculo de Situaciones, *Novatica*, 143, pp. 57–63, 2000) (1997)
14. Levesque, H.J., Pirri, F., Reiter, R.: Foundations for the situation calculus. Linköping Electronic Articles in Computer and Information Science 3(18) (1998)
15. Lowe, G.: Breaking and fixing the needham-schroeder public-key protocol using FDR. In: Margaria, T., Steffen, B. (eds.) TACAS 1996. LNCS, vol. 1055, pp. 147–166. Springer, Heidelberg (1996)
16. Wang, S., Zhang, Y.: A logic programming based framework for security protocol verification. In: An, A., Matwin, S., Ras, Z.W., Slezak, D. (eds.) Foundations of Intelligent Systems. LNCS (LNAI), vol. 4994, pp. 638–643. Springer, Heidelberg (2008)

An ASP-Based System for e-Tourism

Salvatore Maria Ielpa[1], Salvatore Iiritano[2], Nicola Leone[1], and Francesco Ricca[1]

[1] Dipartimento di Matematica, Università della Calabria, 87030 Rende, Italy
{s.ielpa,leone,ricca}@mat.unical.it
[2] Exeura Srl, Via Pedro Alvares Cabrai - C.da Lecco 87036 Rende (CS), Italy
salvatore.iiritano@exeura.com

Abstract. In this paper we present the IDUM system, a successful application of logic programming to e-tourism. IDUM exploits two technologies that are based on the state-of-the-art ASP system DLV: (i) a system for ontology representation and reasoning, called OntoDLV; and, (ii) H𝑖L𝜀X a semantic information extraction tool. The core of IDUM is an ontology which models the domain of the touristic offers. The ontology is automatically populated by extracting the information contained in the touristic leaflets produced by tour operators. A set of specifically devised ASP programs is used to reason on the information contained in the ontology for selecting the holiday packages that best fit the customer needs. An intuitive web-based user interface eases the task of interacting with the system for both the customers and the operators of a travel agency.

1 Introduction

In the last few years, the tourism industry has strongly modified marketing strategies with the diffusion of e-tourism portals in the Internet. Big tour operators are exploiting new technologies, such as web portals and e-mails, in order to both simplify their advertising strategies and reduce the selling costs. The efficacy of e-tourism solutions is also witnessed by the continuously growing community of e-buyers that prefer to surf the Internet for buying holiday packages. On the other hand, traditional travel agencies are undergoing a progressive loss of marketing competitiveness. This is partially due to the presence of web portals, which basically exploit a new market. Indeed, Internet surfers often like to be engaged in self-composing their holiday by manually searching for flights, accommodation etc. Instead, the traditional selling process, whose strength lies in both direct contact with customer and knowledge about customer habits, is experiencing a reduced efficiency. This can be explained by the increased complexity of matching demand and offer. Indeed, travel agencies receive thousand of e-mails per day from tour operators containing new pre-packaged offers. Consequently, the employees of the agency cannot know all the available holiday packages (since they cannot analyze all of them). Moreover, customers are more demanding than in the past (e.g. the classic statement "I like the sea" might be enriched by "I like snorkeling", or "please find an hotel in Cortina" might be followed by "featuring beauty and fitness center") and they often do not declare immediately all their preferences and/or constraints (like, e.g., budget limits, preferred transportation mean or accommodation etc.). This knowledge of customers preferences plays a central role in the traditional selling process; However,

E. Erdem, F. Lin, and T. Schaub (Eds.): LPNMR 2009, LNCS 5753, pp. 368–381, 2009.

the task of matching this information with the large unstructured e-mail database is both difficult to carry out in a precise way and is time consuming. Consequently, the seller is often unable to find the best possible solution to the customer needs in a short time.

The goal of the IDUM project is to devise a system that addresses the above-mentioned causes of inefficiency by offering:

- (i) an automatic extraction and classification of the incoming touristic offers (so that they are immediately available for the seller), and
- (ii) an "intelligent" search that combines knowledge about users preferences with geographical information, and matches user needs with the available offers.

We could achieve the goal by exploiting Answer Set Programming (ASP) [1]. ASP is a powerful logic programming language, which is very expressive in a precise mathematical sense; in its general form, allowing for disjunction in rule heads and nonmonotonic negation in rule bodies, in a fully declarative way ASP can represent *every* problem in the complexity class Σ_2^P and Π_2^P (under brave and cautious reasoning, respectively) [2]. In particular, the core functionalities of IDUM were based on two technologies[1] relying on the state-of-the-art ASP system DLV [3]:

- OntoDLV [4,5] a powerful ASP-based ontology representation and reasoning system; and,
- H$\imath$LεX [6,7,8], an advanced tool for semantic information-extraction from unstructured or semi-structured documents.

More in detail, in the IDUM system, behind the web-based user interface (that can be used by both employees of the agency and customers), there is an "intelligent" core that exploits an OntoDLV ontology for both modeling the domain of discourse (i.e., geographic information, user preferences, and touristic offers, etc.) and storing the available data. The ontology is automatically populated by extracting the information contained in the touristic leaflets produced by tour operators. It is worth noting that, offers are mostly received by the travel agency in a dedicated e-mail account. Moreover, the received e-mails are human-readable, and the details are often contained in email-attachments of different format (plain text, pdf, gif, or jpeg files) and structure that might contain a mix of text and images. The H$\imath$LεX system allows for automatically processing the received contents, and to populate the ontology with the data extracted from touristic leaflets. Once the information is loaded on the ontology, the user can perform an "intelligent" search for selecting the holiday packages that best fit the customer needs. IDUM tries to mimic the behavior of the typical employee of a travel agency by exploiting a set of specifically devised logic programs that "reason" on the information contained in the ontology.

In the remainder of the paper, we first introduce the employed ASP-based technologies; then, in Section 3, we describe how the crucial tasks have been implemented. We show the architecture of the IDUM system in Section 4, and we draw the conclusion in Section 6.

---

[1] Both systems are developed by Exeura srl, a technology company working on analytics, data mining, and knowledge management, that is working on their industrialization finalized to commercial distribution.

2   Underlying ASP-Based Technologies

The core functionalities of the e-tourism systems IDUM were based on two technologies relying on the DLV system [3]: OntoDLV [4,5] a powerful ASP-based ontology representation and reasoning system; and, HιLεX [6,7,8], an advanced tool for semantic information-extraction from unstructured or semi-structured documents.

In the following we briefly describe both OntoDLV and HιLεX, the reader interested in a more detailed description is referred to [4,5] and [6,7,8], respectively.

2.1   The OntoDLV System

Traditional ASP in not well-suited for ontology specifications, since it does not directly support features like classes, taxonomies, individuals, etc. Moreover, ASP systems are a long way from comfortably enabling the development of industry-level applications, mainly because they lack important tools for supporting programmers. All the above-mentioned issues were addressed in OntoDLV [4,5] a system for ontologies specification and reasoning. Indeed, by using OntoDLV, domain experts can create, modify, store, navigate, and query ontologies; and, at the same time, application developers can easily develop their own knowledge-based applications on top of OntoDLV, by exploiting a complete Application Programming Interface [9]. OntoDLV implements a powerful logic-based ontology representation language, called OntoDLP, which is an extension of (disjunctive) ASP with all the main ontology constructs including classes, inheritance, relations, and axioms. In OntoDLP, a *class* can be thought of as a collection of individuals who belong together because they share some features. An individual, or *object*, is any identifiable entity in the universe of discourse. Objects, also called class instances, are unambiguously identified by their object-identifier (oid) and belong to a class. A class is defined by a name (which is unique) and an ordered list of attributes, identifying the properties of its instances. Each attribute has a name and a type, which is, in truth, a class. This allows for the specification of *complex objects* (objects made of other objects). Classes can be organized in a specialization hierarchy (or data-type taxonomy) using the built-in *is-a* relation (*multiple inheritance*). Relationships among objects are represented by means of *relations*, which, like classes, are defined by a (unique) name and an ordered list of attributes (with name and type). OntoDLP relations are strongly typed while in ASP relations are represented by means of simple flat predicates. Importantly, OntoDLP supports two kind of classes and relations: (base) classes and (base) relations, that correspond to basic facts (that can be stored in a database); and *collection* classes and *intensional* relations, that correspond to facts that can be inferred by logic programs; in particular, *collection classes* are mainly intended for object reclassification (i.e., for repeatedly classifying individuals of an ontology). For instance, the following statement declares a class modeling customers, which has six attributes, namely: *firstName, lastName*, and *status* of type string; *birthDate* of type Date; a positive integer *childNumber*, and *job* which contains an instance of another class called Job.

```
class Customer (firstName: string, lastName: string,
    birthDate: Date, status: string,
    childNumber: positive integer, job: Job).
```

As in ASP, logic programs are sets of logic rules and constraints. However, OntoDLP extends the definition of logic atom by introducing class and relation predicates, and complex terms (allowing for a direct access to object properties). This way, the OntoDLP rules merge, in a simple and natural way, the declarative style of logic programming with the navigational style of the object-oriented systems. In addition, logic programs are organized in *reasoning modules*, to take advantage of the benefits of modular programming. For example, with the following program we single out the pairs of customers having the same age

```
module (CustomersWithTheSameAge) {
  sameAge(C1,C2,D)  :- C1: Customer(birthDate:D),
                       C2: Customer(birthDate:D).
}
```

The core of OntoDLV is a rewriting procedure [5] that translates ontologies, and reasoning modules to an equivalent standard ASP program which, in the general case, runs on state-of-the art ASP system DLV [3].

OntoDLV features an advanced persistency manager allows one to store ontologies transparently both in text files and internal relational databases; while powerful type-checking routines are able to analyze ontology specifications and single out consistency problems. Importantly, if the rewritten program is stratified and non-disjunctive [1,10,11] (and the input ontology resides in relational databases) then the evaluation is carried out directly in mass memory by exploiting a specialized version of the same system, called DLV ^{DB} [12]. Note that, since class and relation specifications are rewritten into stratified and non-disjunctive programs, queries on ontologies can always be evaluated by exploiting a DBMS. This makes the evaluation process very efficient, and allows the knowledge engineer to formulate queries in a language more expressive than SQL. Clearly, more complex reasoning tasks (whose complexity is NP/co-NP, and up to Σ_2^P/Π_2^P) are dealt with by exploiting the standard DLV system instead.

2.2 The H$\imath$LεX System

H$\imath$LεX [6,7,8] is an advanced system for ontology-based information extraction from semi-structured and unstructured documents, that has been already exploited in many relevant real-world applications. The H$\imath$LεX system implements a semantic approach to the information extraction problem based on a new-generation semantic conceptual model by exploiting:

- ontologies as knowledge representation formalism;
- a general document representation model able to unify different document formats (html, pdf, doc, ...);
- the definition of a formal attribute grammar able to describe, by means of declarative rules, objects/classes w.r.t. a given ontology.

Most of the existing information extraction approaches do not work in a semantical way and they are not independent of the specific type of document they process. Contrariwise, the approach implemented in H$\imath$LεX confirms that it is possible recognize, extract and structure relevant information form heterogeneous sources.

H*i*L*ε*X is based on OntoDLP for describing ontologies, since this language perfectly fits the definition of semantic extraction rules.

Regarding the unified document representation, the idea is that a document (unstructured or semi-structured) can be seen as a suitable arrangement of objects in a two-dimensional space. Each object has its own semantics, is characterized by some attributes and is located in a two-dimensional area of the document called *portion*. A portion is defined as a rectangular area univocally identified by four cartesian coordinates of two opposite vertices. Each portion "contains" one or more objects and an object can be recognized in different portions.

The language of H*i*L*ε*X is founded on the concept of *ontology descriptor*. A "descriptor" looks like a production rule in a formal attribute grammar, where syntactic items are replaced by ontology elements, and where extensions for managing two-dimensional objects are added. Each descriptor allows us to describe: (i) an ontology object in order to recognize it in a document; or (ii) how to "generate" a new object that, in turn, may be added in the original ontology.

Note that an object may also have more than one descriptor, thus allowing one to recognize the same kind of information when it is presented in different ways.

3 The IDUM System

In this section we describe the core of the IDUM system and its innovative features based on ASP. IDUM is an e-tourism system, conceived for classifying and driving the search of touristic offers for both travel agencies operators and their customers.

IDUM, like other existing portals, has been equipped with a proper (web-based) user interface; but, behind the user interface there is an "intelligent" core that exploits knowledge representation and reasoning technologies based on ASP. In IDUM (see Figure 1, the information regarding the touristic offers provided by tour operators is received by the system as a set of e-mails. Each e-mail might contain plain text and/or a set of leaflets, usually distributed as pdf or image files which store the details of the offer (e.g., place, accommodation, price etc.). Leaflets are devised to be human-readable,

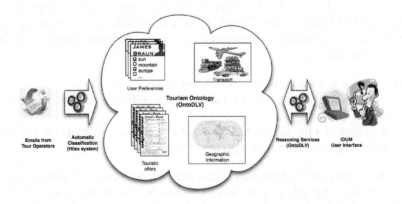

Fig. 1. The IDUM System

might mix text and images, and usually do not have the same layout. E-mails (and their content) are automatically processed by using the HɪLεX system,[2] and the extracted data about touristic offers is used to populate an OntoDLP ontology that models the domain of discourse: the *"tourism ontology"*. The resulting ontology is then analyzed by exploiting a set of reasoning modules (ASP programs) combining the extracted data with the knowledge regarding places (geographical information) and users (user preferences) that we inserted in the tourism ontology. The system mimics the typical deductions made by a travel agency employee for selecting the most appropriate answers to the user needs.

It is worth pointing out that, the core of the system is based on ASP technologies, since both the two most innovative and important features: automatic extraction of touristic offers and intelligent search, were implemented by exploiting OntoDLV and HɪLεX.

In the following sections, we briefly describe the tourism ontology and the implementation of the above-mentioned ASP-based features.

3.1 The Tourism Ontology

The "tourism ontology" has been specified by analyzing the nature of the input (we studied the characteristics of several touristic leaflets) with the cooperation of the staff of a real touristic agency, who were repeatedly interviewed. In this way, we could model the key entities that describe the process of organizing and selling a complete holiday package: in particular, the "tourism ontology" models all the required information, such as user profile, geographic information, kind of holiday, transportation means, etc. In Figure 2, we report some of the most relevant classes and relations that constitute the tourism ontology. In detail, the class *Customer* allows us to model the personal information of each customer, while a number of relations is used to model user preferences, like *CustomerPrefersTrip* and *CustomerPrefersMeans*, which associate each customer to his preferred kind of trip, and his preferred transportation means, respectively. The kind of trip is represented by using a class *TripKind*. Examples of *TripKind* instances are: *safari, sea_holiday*, etc. In the same way, e.g. *airplane, train*, etc. are instances of the class *TransportationMean*. Geographical information is modeled by means of the class *place*, which has been populated with the information regarding more than a thousand touristic places. Moreover, with each place is associated a kind of trip by means of the relation *PlaceOffer* (e.g. Kenya offers safari, Sicily offers both sea and sightseeing). Importantly, the natural a *part-of* hierarchy of places is easily modeled by using the intensional relation *Contains*. This allowed us to assert manually only some basic facts and to obtain all the basic inclusions in a simple yet efficient way. Indeed, the full hierarchy is computed by evaluating a rule (which, basically, encodes the transitive closure).

The mere geographic information is, then, enriched by other information that is usually exploited by travel agency employees for selecting a travel destination. For instance, one might suggest avoiding sea holidays in winter, or going to India during the

---

[2] Note that, e-mails are the main source of touristic offers, but IDUM can deal with sources different from e-mails; indeed, e-mails are "unwrapped": attachments are analyzed, enclosed external links are followed and corresponding web pages analized.

```
class Customer (firstName: string, lastName: string,
    birthDate: Date, status: string,
    childNumber: positive integer, job: Job).
relation CustomerPrefersTrip ( cust:Customer, kind: TripKind ).
relation CustomerPrefersMeans ( cust:Customer,
    means: TransportationMean ).  ...

class Place ( description:string ).
intentional relation Contains ( pl1:place, pl2:place )
{ Contains(P1,P2) :- Contains(P1,P3), Contains(P3,P2).
 Contains('Europe', 'Italy'). Contains('Italy', 'Sicily').
 Contains('Sicily', 'Palermo'). ... }

relation PlaceOffer( place: place, kind: tripKind ).
relation SuggestedPeriod ( place:place, period: positive integer ).
relation BadPeriod ( place:place, period: positive integer ).

class TouristicOffer( start: Place, destination: Place,
    kind: TripKind, means: TransportationMean,
    cost: positive integer, fromDay: Date, toDay: Date,
    maxDuration: positive integer, deadline: Date,
    uri: string, tourOperator: TourOperator ).

class TransportationMean ( description: string ).
class TripKind ( description: string ).
...
```

Fig. 2. Main entities of the touristic ontology

wet monsoon period; whereas, one should be recommended a visit to Sicily in summer. This was encoded by means of the two relations *SuggestedPeriod* and *BadPeriod*.

Finally, the *TouristicOffer* class contains an instance for each available holiday package. The instances of this class are added either automatically, by exploiting the H*i*L*ε*X system (see next section), or manually by the personnel of the agency.

3.2   Automatic Extraction of Touristic Offers

Touristic offers are mainly available in digital format and they are received via e-mail. It is usual that more than a hundred emails per day crowd the mail folder of a travel agency, and often the personnel cannot even analyze the entire in-box. This causes a loss of efficiency in the selling process, because some interesting offers might be ignored. Note that, most of the information is contained in pdf, gif or jpeg files attached to the e-mail messages, and this strongly limits the efficacy of standard search tools like, e.g., the ones provided by e-mail clients.

To deal with this problem, the IDUM system has been equipped with an automatic classification system based on H*i*L*ε*X. Basically, after some pre-processing steps, in which e-mails are automatically read from the inbox, and their attachments are properly handled (e.g. image files are analyzed by using OCR software), the input is sent to H*i*L*ε*X. In turn, H*i*L*ε*X is able to both extract the information contained in the e-mails

Fig. 3. Extracting offer information

and populate the *TouristicOffer* class. This has been obtained by encoding several ontology descriptors (actually, we provided several descriptors for each kind of file received
by the agency). For instance, the following descriptor:

```
<TouristicOffer (Destination, Period)> ->
    <X:place(XX)>{Destination:=X;}
    <X:date(XX)>{Period:=X;}
    SEPBY <X:separator()>.
```

extracts from the leaflet in Figure 3(a) that the proposed holiday package regards trips
to both the Caribbean islands and Brazil. Moreover, it also extracts the period in which
this trip is offered. The extracted portions are outlined in Figure 3(b). The result of the
application of this descriptor are two new instances of the TouristicOffer class.

3.3 Personalized Trip Search

The second crucial task carried out in the IDUM system is the personalized trip search.
This feature has been conceived to make simpler the task of selecting the holiday packages that best fit the customer needs. We tried to "simulate" the deductions made by an
employee of the travel agency in the selling process by using a set of specifically devised logic programs. In a typical scenario, when a customer enters the travel agency, an
employee tries to understand his current desires and preferences at first; then, the seller
has to match the obtained information with a number of pre-packaged offers. Actually, a
number of candidate offers are proposed to the customer and, then, the employee has to
understand his needs by interpreting the preferences of the customer. Customer preferences depend on his personal information (age, gender, marital status, lifestyle, budget,
etc.), but also on his holiday habits (e.g. he prefers going to the mountains, or he has
already been to Italy). Actually, this information has to be elicited by the seller by inter-
viewing the customer, but most of it might be already known by the employee if he is
serving an old customer. In this process, what has to be clearly understood (for properly
selecting a holiday package fitting the customer needs) is summerized in the following
four words: *where, when, how,* and *budget.* Indeed, the seller has to understand *where*

the customer desires to go; *when* he can/wishes to leave; how much time he will dedicate to his holiday; which is the preferred transportation means (*how*); and, finally, the available budget. However, the customer does not directly specify all this information, for example, he can ask for a sea holiday in January but he does not specify a precise place, or he can ask for a kind of trip that is unfeasible in a given period (e.g. winter holiday in Italy in August). In this case the seller has to exploit his knowledge of the world for selecting the right destination and a good offer in a huge range of proposals. This is exactly what the IDUM system does when a user starts a new search. Current needs are specified by filling an appropriate search form (see Figure 4), where some of the key information has to be provided (i.e. where and/or when and/or available money and/or how). Note that, the tourism ontology models both the knowledge of the seller (geographic information and preferred places), and the profile of customers (clearly, in the case of new customers the seller has to fill the ontology with profile information, whereas the ontology already contains the information regarding old customers); moreover, the extraction process continuously populates the ontology with new touristic offers. Thus, the system, by running a specifically devised reasoning module, combines the specified information with the one available in the ontology, and shows the holiday packages that best fit the customer needs. For example, suppose that a customer specifies the kind of holiday and the period, then the following (simplified) module creates a selection of holiday packages:

```
module(kindAndPeriod) {
  %detect possible and suggested places
  possiblePlace(P)  :- askFor(tripKind:K),
                       PlaceOffer(place:P, kind:K).

  suggestPlace(P)  :- possiblePlace(P), askFor(period:D),
                      SuggestedPeriod(place:P1, period:D),
                      not BadPeriod(place:P1, period:D).

  %select possible, alternative and suggestible packages
  possibleOffer(O)  :- O:TouristicOffer(destination:P),
                       possiblePlace(P).

  alternativeOffer(O)  :- O:TouristicOffer(destination:P),
                          suggestPlace(P).

  suggestOffer(O)  :- O:TouristicOffer(destination:P, mean:M),
                      suggestPlace(P), askFor(cust:C),
                      CustomerPrefersMean(cust:C, mean:M).
}
```

The first two rules select: possible places (i.e., the ones that offer the kind of holiday in input); and places to be suggested (because they offer the required kind of holiday in the specified period). Finally, the remaining three rules search in the available holiday packages the ones that: offer an holiday that matches the original input (possible offer); are good alternatives in suggested places (alternativeOffer); or match the customer's preferred means and can be suggested.

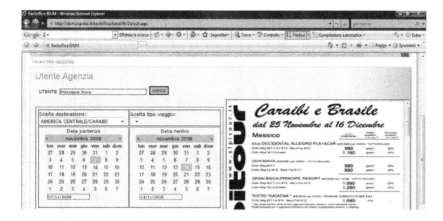

Fig. 4. The IDUM User Interface

The example above reports one of the several reasoning modules that have been devised for implementing the intelligent search. In the development process, we exploited many of the advanced features of the language like negation as failure and aggregates. The declarative nature of ASP allowed us to design effective solutions and to tune rapidly our modules by following the suggestions of the domain experts.

4   System Architecture

The architecture of the IDUM system, which is depicted in Figure 5, is made up of four layers: *Data Layer, Information Layer, Knowledge Layer*, and *Service Layer*. In the Data Layer the input sources are dealt with. In particular, the system is able to store and handle the most common kind of sources: e-mails, plain text, pdf, gif, jpeg, and HTML files. The Information Layer provides ETL (Extraction, Transformation and Loading) functionalities, in particular: in the loading step the documents to be processed are stored in an auxiliary database (that also manages the information about the state of the extraction activities); whereas, in the Transformation step, semi-structured or non-structured documents are manipulated. First the document format is normalized; then, the "bi-dimensional logical representation" is generated (basically, the $H\imath L\varepsilon X$ portions are identified); finally, $H\imath L\varepsilon X$ descriptors are applied in the Semantic Extraction step and ontology instances are recognized within processed documents. The outcome of this process is a set of concept instances, that are recognized by matching semantic patterns, and stored in the core knowledge base of the system where the tourism ontology resides (Knowledge Layer). Domain ontology and extracted information are handled by exploiting the Persistency manager of the OntoDLV system (see Section 2.1). The Services Layer features the profiling service and the intelligent search (see Section 3.3) which implements the reasoning on the core ontology by evaluating in the OntoDLV system a set of logic programs. The Graphical User Interface (GUI) can access the system features by interacting with a set of web-services.

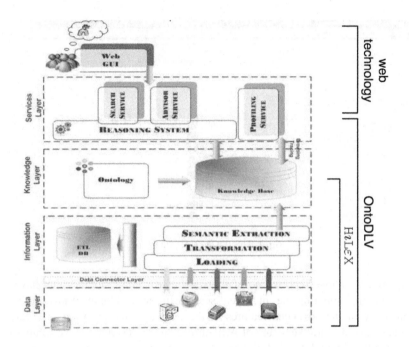

Fig. 5. System Architecture

5   Related Work

The usage of ontologies for developing e-tourism applications was already studied in the literature [13,14,15,16,17], and the potential of the application of semantic technology recognized [18,19].

The architecture of an e-tourism system able to create a touristic package in a dynamic way is presented in [14]. This system permits the customer to specify a set of preferences for a vacation and dynamically access and query a set of information sources to find component such as accommodation, car rental, and leisure activities in real time. It is based on an ontology written in OWL-DL [20]. The ontology exploited in [15,14] encodes the same key concepts of ours, but does not include information about user preferences. Another advantage of the our approach is the possibility of developing ASP programs that reason on the data contained in the ontology for developing complex searches, while there is no accepted solution for combining logic rules and OWL-DL ontologies.

The SPETA system [21], which is based on the ontology of [14], acts as an advisor for tourists. Fundamentally, SPETA follows people who need advising when visiting a new place, and who consequently do not know what is interesting to visit. Here the ontology is enriched with user profile information for determining the common characteristics of the previously visited places and the user behavior. In this way the system recommends attractions which are likely to fit the user expectations. It exploits GPS technology to know user position and it gets user data from previous users history and also from social networks. Both SPETA and IDUM exploit an ontology for building personalized solutions for the users, but the goal of SPETA is different from that of IDUM. Indeed, the former was conceived for offering assistance and information to the user when they already are on a place, while the goal of IDUM is to assist the users in the selection of an holiday.

The E-Tourism Working Group at DERI [22] is developing e-tourism solutions based on the Semantic Web technology. Their goal is to develop an advanced eTourism Semantic Web portal which will connect the customers and virtual travel agents from anywhere at anytime with any needs and requests. In [23], they present OnTour an information retrieval system that exploits an RDF engine for storing the data regarding accommodation facilities of different types. DERI also developed a content management system: OnTourism [24]. The solution is very similar to IDUM, but it is based on the use of Lixto Software [25] which makes information extraction from web pages. Information about current events is crawled from several web sources and it is rendered in a machine-accessible semantic.

6   Conclusion and Market Perspective

In this paper we have described a successful example of commercial and practical use of logic programming: the e-tourism system IDUM.

The core of IDUM is an ontology modeling the domain of the touristic offers, which is automatically populated by extracting the information contained in the e-mails sent by tour operators; and an intelligent search tool based on answer set programming is able to search the holiday packages that best fits the customer needs.

The system has been developed under project "IDUM: Internet Diventa Umana" (project n. 70 POR Calabria 2000/2006 Mis. 3.16 Azione D Ricerca e Sviluppo nella Imprese Regionali - Modulo B Voucher Tecnologici) funded by the Calabrian Region. The project team involved five organizations: the Department of Mathematics of the University of Calabria (that has ASP as one of the principal research area), the consortium Spin, Exeura srl (a company working on knowledge management), Top Class srl (a travel agency), and ASPIdea (a software farm specialized in the development of web applications). The members exploited their specific knowledge for developing the innovative features of the system. The strong synergy among partners made it possible to push the domain knowledge of the travel agency TopClass in both the ontology and in the reasoning modules. The result is a system that mimics the behavior of a seller of the agency and it is able to search in a huge database of automatically classified offers. IDUM combines the speed of computers with the knowledge of a travel agent for improving the efficiency of the selling process.

The IDUM system was initially conceived for solving the specific problems of a travel agency, and it is currently employed by one of the project partners: Top Class srl. We are working on an enterprise version of the system conceived for offering its advanced services to several travel agencies. The enhancements of IDUM will be developed under another technology-transfer PIA (Pacchetti Integrati di Agevolazione industria, artigianato e servizi) project funded by the Calabrian region. The value of the system was confirmed by the good position obtained by the proposal in the project evaluation ranking (IDUM occupies the 2nd position over more than 400 competing proposals). Moreover, we received very positive feedbacks from the market, indeed many travel agents are willing to use the system, and the potential of IDUM has been recognized also by the chair of the Italian touring club, which is the most important Italian association of tour operators.

References

1. Gelfond, M., Lifschitz, V.: Classical Negation in Logic Programs and Disjunctive Databases. NGC 9, 365–385 (1991)
2. Eiter, T., Gottlob, G., Mannila, H.: Disjunctive Datalog. ACM TODS 22(3), 364–418 (1997)
3. Leone, N., Pfeifer, G., Faber, W., Eiter, T., Gottlob, G., Perri, S., Scarcello, F.: The DLV System for Knowledge Representation and Reasoning. ACM TOCL 7(3), 499–562 (2006)
4. Ricca, F., Gallucci, L., Schindlauer, R., Dell'Armi, T., Grasso, G., Leone, N.: OntoDLV: an ASP-based system for enterprise ontologies. Journal of Logic and Computation (2009)
5. Ricca, F., Leone, N.: Disjunctive Logic Programming with types and objects: The DLV$^+$ System. Journal of Applied Logics 5(3), 545–573 (2007)
6. Manna, M.: Semantic Information Extraction: Theory and Practice. PhD thesis, Dipartimento di Matematica, Universitá della Calabria, Rende, Cosenza Italia (2008)
7. Ruffolo, M., Manna, M.: HiLeX: A System for Semantic Information Extraction from Web Documents. In: ICEIS (Selected Papers). LNBIP, vol. 3, pp. 194–209. Springer, Heidelberg (2008)
8. Ruffolo, M., Leone, N., Manna, M., Saccà, D., Zavatto, A.: Exploiting ASP for Semantic Information Extraction. In: Proceedings ASP 2005 - Answer Set Programming: Advances in Theory and Implementation, Bath, UK, pp. 248–262 (2005)
9. Gallucci, L., Ricca, F.: Visual Querying and Application Programming Interface for an ASP-based Ontology Language. In: Proc. of (SEA 2007), pp. 56–70 (2007)
10. Gelfond, M., Leone, N.: Logic Programming and Knowledge Representation – the A-Prolog perspective. AI 138(1-2), 3–38 (2002)
11. Minker, J.: Overview of Disjunctive Logic Programming. AMAI 12, 1–24 (1994)
12. Terracina, G., Leone, N., Lio, V., Panetta, C.: Experimenting with recursive queries in database and logic programming systems. TPLP 8, 129–165 (2008)
13. Maedche, A., Staab, S.: Applying semantic web technologies for tourism information systems. In: Proc. of ENTER 2002 (2002)
14. Jorge, C.: Combining the semantic web with dynamic packaging systems. In: Proc. of AIKED 2006, pp. 133–138. World Scientific and Engineering Academy and Society, Singapore (2006)
15. Cardoso, J.: Developing an owl ontology for e-tourism. In: Semantic Web Services, Processes and Applications, pp. 247–282 (2006)
16. Martin, H., Katharina, S., Daniel, B.: Towards the semantic web in e-tourism: can annotation do the trick? In: Proc. of the 14th ECIS 2006 (2006)
17. Martin, H., Katharina, S., Daniel, B.: Towards the semantic web in e-tourism: Lack of semantics or lack of content? In: Proc. of European Semantic Web Conference (ESWC 2006)
18. Dogac, A., Kabak, Y., Laleci, G., Sinir, S., Yildiz, A., Kirbas, S., Gurcan, Y.: Semantically enriched web services for the travel industry. SIGMOD Rec. 33(3), 21–27 (2004)
19. Joe, B., Carole, G.: Tourist: the application of a description logic based semantic hypermedia system for tourism. In: Proc. of HYPERTEXT 1998, pp. 132–141. ACM, New York (1998)
20. Smith, M.K., Welty, C., McGuinness, D.L.: OWL web ontology language guide. W3C Candidate Recommendation (2003), http://www.w3.org/TR/owl-guide/

21. Angel, G.C., Javier, C., Ismael, R., Myriam, M., Ricardo, C.P., Miguel, G.B.J.: SPETA: Social pervasive e-tourism advisor. Telemat. Inf. 26(3), 306–315 (2009)
22. DERI: Digital Enterprise Research Institute. Technikerstrae 21a. Innsbruck, A., `http://e-tourism.deri.at//`
23. Siorpaes, K., Bachlechner, D.: OnTour: Tourism information retrieval based on YARS. In: Proceedings of ESWC 2006 (2006)
24. Ding, Y., Prantner, K., Luger, M., Herzog, C.: OnTourism: Semantic etourism portal. In: Proc. of Scientific Conference of the e-Business Forum Business in Travel, Tourism and Hospitality, Athens, Greece (2008)
25. Lixto, `http://www.lixto.at//`

ccT on Stage: Generalised Uniform Equivalence Testing for Verifying Student Assignment Solutions*

Johannes Oetsch[1], Martina Seidl[2], Hans Tompits[1], and Stefan Woltran[1]

[1] Institut für Informationssysteme, Technische Universität Wien,
Favoritenstraße 9-11, A-1040 Vienna, Austria
{oetsch,tompits}@kr.tuwien.ac.at
woltran@dbai.tuwien.ac.at
[2] Institut für Softwaretechnik, Technische Universität Wien,
Favoritenstraße 9-11, A-1040 Vienna, Austria
seidl@big.tuwien.ac.at

Abstract. The tool ccT is an implementation for testing various parameterised notions of program correspondence between logic programs under the answer-set semantics, based on reductions to quantified propositional logic. One such notion is *relativised uniform equivalence with projection*, which extends standard uniform equivalence via two additional parameters: one for specifying the input alphabet and one for specifying the output alphabet. In particular, the latter parameter is used for *projecting* answer sets to the set of designated output atoms, i.e., ignoring auxiliary atoms during answer-set comparison. In this paper, we discuss an application of ccT for verifying the correctness of students' solutions drawn from a laboratory course on logic programming, employing relativised uniform equivalence with projection as the underlying program correspondence notion. We complement our investigation by discussing a performance evaluation of ccT, showing that discriminating among different back-end solvers for quantified propositional logic is a crucial issue towards optimal performance.

1 Introduction

This paper deals with a system for testing various refined notions of program correspondence for nonmonotonic logic programs under the answer-set semantics, called ccT (standing for "correspondence-checking tool") [1]. It belongs to a current line of research in answer-set programming (ASP) about questions of program equivalence relevant for different software engineering tasks like optimisation, modular programming, and verification. This research was for the most part initiated by the seminal work of Lifschitz, Pearce, and Valverde [2] about *strong equivalence*, which is defined to hold between two programs P and Q iff $P \cup R$ and $Q \cup R$ are ordinarily equivalent, i.e., have the same answer sets, for every program R (here, R is called *context*). Albeit strong equivalence circumvents the failure of ordinary equivalence to yield a replacement property similar to the one of classical logic, it is however too restrictive for certain applications. This led to the investigation of more liberal notions, chiefly among

---

* This work was partially supported by the Austrian Science Fund (FWF) under projects P18019 and P21698.

E. Erdem, F. Lin, and T. Schaub (Eds.): LPNMR 2009, LNCS 5753, pp. 382–395, 2009.

them *uniform equivalence* [3], which is defined similar to strong equivalence except that context programs are restricted to be sets of *facts*. In any case, both strong and uniform equivalence do not take standard programming techniques like the use of local (auxiliary) variables into account, which may occur in some subprograms but which are ignored in the final solutions. In other words, these notions do not admit the *projection* of answer sets to a set of dedicated output atoms. To accommodate issues like the above, Eiter et al. [4] introduced a general framework for specifying parameterised notions of program correspondence, allowing both answer-set projection as well as the specification which kind of context class should be used for program comparison. Thus, these notions generalise not only strong and uniform equivalence but also *relativised* versions thereof [5] (where relativisation refers to the possibility of specifying the alphabet of the context class).

The system ccⲦ was developed as a checker for specific correspondence problems belonging to the framework of Eiter et al. [4], based on reductions to the satisfiability problem of quantified propositional logic.[1] Such a reduction approach is motivated by two aspects: (i) the complexity of the considered problems—lying on the third and fourth level of the polynomial hierarchy, respectively—is captured by certain classes of quantified propositional formulas, and (ii) the availability of advanced solvers for quantified propositional logic.

Here, we are interested in specific correspondence problems computable by ccⲦ, viz. *propositional query equivalence problems* (PQEPs) [6], which generalise uniform equivalence amounting to *relativised uniform equivalence with projection*.[2] In particular, we discuss how PQEPs can be used to verify the correctness of solutions provided by students as part of their assignments for a laboratory course on knowledge-based systems at our university, relative to a reference solution. The assignments are taken from the domain of model-based diagnosis and use the diagnosis front-end of the well-known ASP solver DLV [7] as underlying reasoning engine. The main difficulty for verifying the students' solutions is that PQEPs deal with propositional programs only whilst the solution programs are non-ground. A naive grounding would not be feasible, so we resorted to a special technique restricting the domain to admissible inputs as well as employing the intelligent grounder of DLV. It turned out that verifying the solutions in this way yielded less false positives than with a test script currently in use, which is based on a collection of sample test cases.

As ccⲦ admits the use of different QBF solvers as back-end engines, we also report about an experimental evaluation of the tool using a set of benchmark problems showing the runtime behaviour of the system depending on a chosen solver. The experiments were based on a set of parameterisable benchmarks stemming from the hardness proof of the complexity analysis of the corresponding equivalence problems [6]. These benchmarks have the particular advantage that they can be used to easily verify the correctness not only of ccⲦ but also of the employed QBF solvers. This proved to be very

---

[1] Recall that quantified propositional logic is an extension of ordinary propositional logic allowing quantifications over atomic formulas. Following custom, we refer to formulas of quantified propositional logic as *quantified Boolean formulas* (QBFs).

[2] The name PQEP stems from taking a database point of view in which programs are considered as queries over databases.

helpful during the development of the system. The experiments show that discriminating among different back-end QBF solvers is crucial towards optimal performance.

The paper is organised as follows. In Section 2, we recapitulate the relevant aspects from ASP and correspondence checking, as well as from quantified propositional logic. Afterwards, in Section 3, we review the theoretical basis of $cc\top$, including some optimisations employed in the system. This is followed by Section 4 containing a discussion of the experimental results. Section 5 discusses the application of $cc\top$ for verifying students' solutions. The paper concludes with a brief summary and outlook in Section 6.

2 Background

Answer-set semantics. We are concerned with *disjunctive logic programs* (DLPs) which are finite sets of safe rules of form

$$a_1 \vee \cdots \vee a_l \leftarrow a_{l+1}, \ldots, a_m, \text{not } a_{m+1}, \ldots, \text{not } a_n, \tag{1}$$

where $n \geq m \geq l \geq 0$, all a_i are atoms from some fixed vocabulary $\mathcal{U}$, and "not" denotes *default negation*. Recall that safety means that all variables occurring in the head or negative body also occur in the positive body. A rule or program containing no variables is *ground*. The *grounding* of a program P relative to a set C of constants is defined as usual and denoted by $grd(P, C)$. Programs containing only atoms of arity 0 are called *propositional*. Following custom, we will identify ground programs with propositional ones. The set of all atoms occurring in a program P is denoted by $At(P)$. By a *fact*, we understand a rule of form $a \leftarrow$, usually just written a. Note that facts must be ground. Given a finite set A of ground atoms, the power set, 2^A, of A thus represents the set of all programs containing facts from A only.

Following Gelfond and Lifschitz [8], an interpretation I (i.e., a set of ground atoms) is an *answer set* of a ground program P iff it is a minimal model of the *reduct* P^I, resulting from P by (i) deleting all rules containing a default-negated atom not a such that $a \in I$, and (ii) deleting all default-negated atoms in the remaining rules. The answer sets of a non-ground program are given by the answer sets of the grounding over its Herbrand universe. The set of all answer sets of a program P is denoted by $\mathcal{AS}(P)$.

We continue with recapitulating the relevant program correspondence notions. For this, we consider propositional programs only in what follows as $cc\top$ deals with just these kinds of programs. To begin with, for collections $\mathcal{S}, \mathcal{S}'$ of sets of ground atoms, a set B of ground atoms, and $\odot \in \{\subseteq, =\}$, we define $\mathcal{S} \odot_B \mathcal{S}'$ as $\{Y \cap B \mid Y \in \mathcal{S}\} \odot \{Y \cap B \mid Y \in \mathcal{S}'\}$. Following Oetsch et al. [6], a *propositional query inclusion problem*, or *PQIP*, is a tuple of form $(P, Q, 2^A, \subseteq_B)$ and a *propositional query equivalence problem*, or *PQEP*, is a tuple of form $(P, Q, 2^A, =_B)$, where P and Q are two programs and A and B are sets of atoms, intuitively referring to sets of input and output atoms, respectively. We say that $(P, Q, 2^A, \odot_B)$ *holds*, for $\odot \in \{\subseteq, =\}$, iff, for each set of facts $F \in 2^A$, $AS(P \cup F) \odot_B AS(Q \cup F)$. For a PQEP $\Pi = (P, Q, 2^A, =_B)$, the PQIPs $\Pi^{\rightarrow} = (P, Q, 2^A, \subseteq_B)$ and $\Pi^{\leftarrow} = (Q, P, 2^A, \subseteq_B)$ are *associated* with Π. Clearly, Π holds iff both $\Pi^{\rightarrow}$ and $\Pi^{\leftarrow}$ hold.

PQEPs express *ordinary equivalence*, *uniform equivalence* [3], and *strong equivalence* [2] as follows: If P and Q are formed over alphabet $\mathcal{U}$, then P and Q are (i) ordinarily equivalent iff the PQEP $(P, Q, \{\emptyset\}, =_{\mathcal{U}})$ holds, (ii) uniformly equivalent iff

$(P, Q, 2^{\mathcal{U}}, =_{\mathcal{U}})$ holds, and (iii) strongly equivalent iff $(P, Q, \mathcal{P}_{\mathcal{U}}, =_{\mathcal{U}})$ holds, where $\mathcal{P}_{\mathcal{U}}$ is the set of all programs over $\mathcal{U}$.

Concerning the complexity of PQIPs and PQEPs, as shown previously [6], given programs $P, Q \in \mathcal{P}_{\mathcal{U}}$, sets $A, B \subseteq \mathcal{U}$ of atoms, and $\odot \in \{\subseteq, =\}$, deciding whether $(P, Q, 2^A, \odot_B)$ holds is Π_3^P-complete. Moreover, the problem is Π_2^P-complete in case $B = \mathcal{U}$. Both hardness results hold even for arbitrary but fixed A.

Quantified propositional logic. The complexity results above show that PQIPs and PQEPs can be efficiently reduced to *quantified propositional logic*, an extension of classical propositional logic in which formulas are permitted to contain quantifications over propositional variables. Such formulas are also called *quantified Boolean formulas* (QBFs); we denote them by upper-case Greek letters. Similar to predicate logic, $\exists$ and $\forall$ are used as symbols for existential and universal quantification, respectively.

For an interpretation I and a QBF Φ, the relation $I \models \Phi$ is defined analogously as in classical propositional logic, with the additional conditions that $I \models \exists p \Psi$ iff $I \models \Psi[p/\top]$ or $I \models \Psi[p/\bot]$, and $I \models \forall p \Psi$ iff $I \models \Psi[p/\top]$ and $I \models \Psi[p/\bot]$, for $\Phi = Qp\Psi$ with $Q \in \{\exists, \forall\}$, where $\Psi[p/\phi]$ denotes the QBF resulting from Ψ by replacing each free occurrence of p in Ψ by ϕ.[3] Satisfiability and validity of a QBF are defined analogously as for formulas in classical propositional logic. Note that for closed QBFs it holds that the notions of satisfiability and validity coincide.

Given a finite set P of atoms, $QP\Psi$ stands for any QBF $Qp_1 Qp_2 \ldots Qp_n \Psi$ such that $P = \{p_1, \ldots, p_n\}$. A QBF Φ is said to be in *prenex normal form* (PNF) iff it is closed and of the form $Q_n P_n \ldots Q_1 P_1 \phi$, where $n \geq 0$, ϕ is a propositional formula, and $Q_i \in \{\exists, \forall\}$ such that $Q_i \neq Q_{i+1}$ for $1 \leq i \leq n - 1$. Moreover, if ϕ is in conjunctive normal form, then Φ is in *prenex conjunctive normal form* (PCNF), and if ϕ is in disjunctive normal form, then Φ is in *prenex disjunctive normal form* (PDNF). A QBF $\Phi = Q_n P_n \ldots Q_1 P_1 \phi$ is also referred to as an (n, Q_n)-*QBF*. Any closed QBF Φ is easily transformed into an equivalent QBF in prenex normal form such that each quantifier occurrence from Φ corresponds to a quantifier occurrence in the prenex normal form.

Well-known complexity results for the evaluation problem of QBFs imply that PQIPs and PQEPs can be efficiently reduced to $(3, \forall)$-QBFs. These reductions are the central theoretical basis for ccT and are discussed next.

3 Underlying Translations of ccT

In this section, we recapitulate the basic encodings for mapping PQIPs and PQEPs into QBFs [6] and introduce a slightly simplified encoding.

We start with some notation and ancillary definitions. Given a set V of atoms, we assume (pairwise) disjoint copies $V^i = \{v^i \mid v \in V\}$, for every $i \geq 1$. Furthermore, we define $(V^i \leq V^j)$ as $\bigwedge_{v \in V}(v^i \rightarrow v^j)$, $(V^i < V^j)$ as $(V^i \leq V^j) \wedge \neg(V^j \leq V^i)$, and $(V^i = V^j)$ as $(V^i \leq V^j) \wedge (V^j \leq V^i)$. Loosely speaking, these operators allow to compare different subsets of atoms from a common set V under subset inclusion, proper-subset inclusion, and equality, respectively.

---

[3] The notion of a free variable occurrence is defined similarly as in predicate logic.

We use superscripts as a general renaming scheme for formulas and rules. That is, for each $i \geq 1$, α^i expresses the result of replacing each occurrence of an atom v in α by v^i, where α is any formula or rule. For a rule r of form (1), we define $H(r) = a_1 \vee \cdots \vee a_l$, $B^+(r) = a_{l+1} \wedge \cdots \wedge a_m$, and $B^-(r) = \neg a_{m+1} \wedge \cdots \wedge \neg a_n$. We identify empty disjunctions with $\bot$ and empty conjunctions with $\top$.

In order to express properties of logic programs and respective program reducts in the language of quantified propositional logic, we introduce the following concept: Given a propositional program P, define $P^{\langle i,j \rangle} = \bigwedge_{r \in P} \left((B^+(r^i) \wedge B^-(r^j)) \to H(r^i) \right)$. Then, for any propositional program P with $At(P) = V$, any interpretation I, and any $X, Y \subseteq V$ such that, for some $i, j \geq 0$, $I \cap V^i = X^i$ and $I \cap V^j = Y^j$, it holds that $X \models P^Y$ iff $I \models P^{\langle i,j \rangle}$ [9].

We are now in a position to state the encoding due to Oetsch et al. [6].

Proposition 1. *Let $\Pi = (P, Q, 2^A, \subseteq_B)$ be a PQIP, $At(P \cup Q) = V$, $A, B \subseteq V$, and*

$$\mathsf{S}[\Pi] = \forall V^1 \forall A^2 \neg \left(\Phi_\Pi \wedge \forall V^4 ((B^4 = B^1) \to \Psi_\Pi) \right), \text{ where}$$
$$\Phi_\Pi = P^{\langle 1,1 \rangle} \wedge (A^2 \leq A^1) \wedge \forall V^3 \left(((A^2 \leq A^3) \wedge (V^3 < V^1)) \to \neg P^{\langle 3,1 \rangle} \right) \text{ and}$$
$$\Psi_\Pi = \left((Q^{\langle 4,4 \rangle} \wedge (A^2 \leq A^4)) \to \exists V^5 (((A^2 \leq A^5) \wedge (V^5 < V^4)) \wedge Q^{\langle 5,4 \rangle}) \right).$$

Then, Π holds iff $\mathsf{S}[\Pi]$ is valid. Moreover, a PQEP $\Omega = (P, Q, 2^A, =_B)$ holds iff $\mathsf{S}[\Omega^\rightarrow] \wedge \mathsf{S}[\Omega^\leftarrow]$ is valid.

Besides the above encoding $\mathsf{S}[\cdot]$, ccT implements a slightly adapted version which we introduce next. The key observation for the subsequent adaption is that we use a *fixed assignment* for atoms in view of the subformula $B^4 = B^1$ of $\mathsf{S}[\cdot]$. Hence, for the quantifier block $\forall V^4$, it is sufficient to take only atoms from $V^4 \setminus B^4$ into account and replace all occurrences of atoms $v^4 \in B^4$ by v^1 within the remaining part of the formula. We thus obtain:

Theorem 1. *Let $\Pi = (P, Q, 2^A, \subseteq_B)$ be a PQIP, $At(P \cup Q) = V$, and $A, B \subseteq V$. Then, Π holds iff $\mathsf{T}[\Pi] = \forall V^1 \forall A^2 \neg (\Phi_\Pi \wedge \forall (V^4 \setminus B^4) \Psi_\Pi [B^4/B^1])$ is valid, where Φ_Π and Ψ_Π are the QBFs from Proposition 1 and $\Psi_\Pi [B^4/B^1]$ is the result of replacing all occurrences of atoms $v^4 \in B^4$ in Ψ_Π by v^1.*

Obviously, all encodings introduced so far, are (i) always linear in the size of P, Q, A, and B, and (ii) possess at most two quantifier alternations in any branch of the formula tree. The latter shows that any such encoding is easily translated into a $(3, \forall)$-QBF. Thus, the complexity of evaluating these QBFs is not harder than the complexity of the encoded decision problems, which shows the adequacy of the encodings in the sense of Besnard et al. [10]. The benefit of the refined encodings is, however, that the number of universally quantified variables is reduced—in fact, in some specific cases, one quantifier block even vanishes. This guarantees adequacy also for some special cases of query problems with lower complexity. Note that by a proper parameterisation of a PQIP (resp., PQEP) also some important special cases of correspondence checking can be realised, e.g., uniform equivalence and ordinary equivalence. It can easily be verified that all special cases without projection have in common that the resulting encodings based on $\mathsf{T}[\cdot]$ yield QBFs with at most one quantifier alternation in each branch of the formula tree, witnessing their Π_2^P-membership.

4 Dress Rehearsal: Some Preliminary Performance Evaluation

In this section, we present a preliminary experimental evaluation of our implementation, in order to assess the behaviour of cc⊤ under different QBF solvers, different encodings, and different problem settings in terms of runtime performance.

In the spirit of previous experiments with cc⊤ [1], we use a reduction from QBFs to PQIPs as given by the Π_3^P-hardness proof for deciding PQIPs [6]. This provides us with a class of random benchmark problems for cc⊤ which is easily parameterisable and which reflects, in some sense, the inherent hardness of the problem. More precisely, the method is as follows: (i) generate a random $(3, \forall)$-QBF Φ in PDNF; (ii) reduce Φ to a PQIP $\Pi_\Phi = (P, Q, 2^A, \subseteq_B)$ such that Φ is valid iff Π_Φ holds [6]; and (iii) apply cc⊤ to derive the corresponding encoding $S[\Pi_\Phi]$ or $T[\Pi_\Phi]$. A particular advantage of this method is that it allows in a straightforward way to verify the correctness of the overall system: just check whether Φ and one of $S[\Pi_\Phi]$ or $T[\Pi_\Phi]$ have the same truth value. Indeed, with the help of this feature, we were able to find errors in some QBF solvers.

Our benchmark set consists of 1000 instances. The randomly generated QBFs of Step (i) contain 24 different atoms each. From those 24 atoms, each quantifier block binds 8 of them. Each term in the PDNF contains 4 atoms which are selected randomly among the 24 atoms and are negated with probability 0.5. The whole formula consists of 38 terms. From the 1000 instances, 506 evaluate to true and 494 evaluate to false. Thus, the ratio between true and false instances is close to 1 which indicates that the instances are neither under-constrained nor over-constrained. From each Φ, we construct a PQIP $\Pi_\Phi = (P, Q, 2^A, \subseteq_B)$ such that Φ is true iff Π_Φ holds. Note that P, Q, and B are determined by the reduction but the context A can be chosen arbitrarily.

For our experiments, we use three different settings, viz. the empty context $A = \emptyset$, the full context $A = \mathcal{U}$, and an in-between setting $\emptyset \subseteq A \subseteq \mathcal{U}$. For the last setting, each atom occurring in one of the two programs P and Q is in A with probability 0.5. We consider both encodings from PQIPs to QBFs, $S[\cdot]$ and $T[\cdot]$, together with the three settings for the context. We compare the QBF solvers semprop [11] (release 24/02/02), qube-bj [12] (v1.2), quantor [13] (release 25/01/04), and qpro [14], all of them showed to be competitive in previous QBF evaluations. The solvers qpro, qube-bj, and semprop are based on the standard DPLL decision procedure extended by special learning techniques whereas quantor implements a combination of resolution and variable expansion. All solvers except qpro require the input to be in prenex conjunctive normal form. Thus, for those solvers, an intermediate prenexing step is necessary. All experiments were carried out on a 3.0 GHz Dual Intel Xeon workstation, with 4 GB of RAM and Linux version 2.6.8.

Figure 1 summarises the results of the comparison. The different QBF solvers, encodings ($S[\cdot]$, $T[\cdot]$), and settings for the context (empty, half-full, full, respectively) are given on the abscissa, and the median runtimes in seconds are depicted on the ordinate.

Observe that the alternative encoding $T[\cdot]$ does not achieve faster runtimes for all solvers, although it uses less variables. For qpro and qube-bj, QBFs from $T[\cdot]$ are solved—as one would expect—faster. This is not the case for semprop and quantor, where semprop solves QBFs from $S[\cdot]$ slightly faster and quantor solves them much faster. The median runtime for quantor with full context and encoding $T[\cdot]$ is even greater than 100 seconds. Concerning the influence of the context parameterisation on

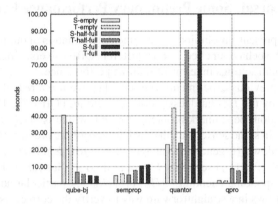

Fig. 1. Median runtimes for different solvers, encodings, and problem settings

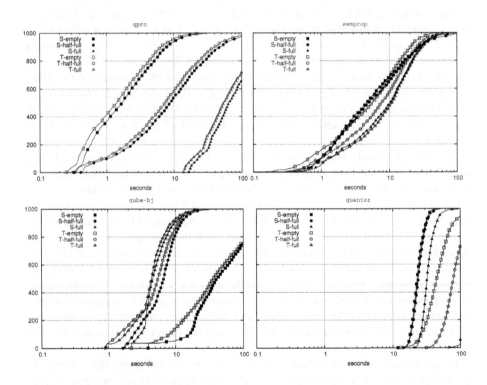

Fig. 2. Runtime distribution for qpro, semprop, qube-bj, and quantor

the runtimes, the non-normal-form solver qpro achieves best results for the empty context but rather poor results for the full context. For qube-bj the contrary is true, however, i.e., it achieves best results for the full context but poor results for the empty context—a quite surprising observation. Finally, the most robust solver in this aspect is semprop. Recall that each of the derived PQIPs $(P, Q, 2^A, \subseteq_B)$ either holds for any A, or does not hold for any A. The assignments of atoms from X^1 in our encodings which "guess" context-program candidates are thus irrelevant for the truth value of the QBFs. As qpro does not implement any heuristics concerning the selection of atoms, it is not surprising that runtimes scale exponentially with respect to the context. Interestingly, the runtimes for qube-bj even worsens without those "decoy" variables.

The results in Fig. 2 provide some deeper insights concerning the runtime behaviour of the non-normal-form solver qpro and the normal-form solvers semprop, qube-bj, and quantor, respectively. For those graphs, the abscissa gives the runtime in seconds (scaled logarithmically) and the ordinate gives the number of solved problem instances. This means that for each runtime in the data we depict how many instances were solved with runtime less than or equal to that time. The different curves correspond to the different combinations of the chosen encoding and context parameterisation. For better legibility, different symbols are attached to the curves.

5   cc⊤ on Stage: A Verification Application

We next discuss an application of cc⊤ for verifying the correctness of certain programs. In particular, these programs represent the solutions of students as part of their assignments for a laboratory course on knowledge-based systems at our university. We compare these solutions relative to a reference program based on verifying certain PQEPs. As the involved programs are non-ground, we need special techniques to take this into account. Hence, our results demonstrate also how our reduction approach to QBFs can be applied to non-ground programs as well.

One of the objectives of the course is to model a simple air-conditioning system by means of logic programs and, based on this model, to solve Reiter-style diagnosis tasks [15] with the dedicated diagnosis front-end of DLV [7]. The programs we consider here should represent the correct behaviour of the components of the air-condition system and are taken from three installments of the course between 2006 and 2008. The problem description slightly changed from year to year, yet Fig. 3 prototypically illustrates the specification of such an air-conditioning system. This system consists of four components, viz. a heater (h), a cooler (c), a switch (s), and a valve (v). They

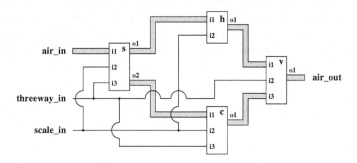

Fig. 3. Overall architecture of one particular air-conditioning system considered for a course

are connected by air lines (grey bars) and data lines (ordinary lines). The students' task is to model each component of the system as well as the connections between the components and some additional constraints required for diagnosing by respective programs. The problem description provides detailed specifications of the system and its components and defines the predicates to be used for the input and output of the single components and the whole system. The system's input airstream (air_in) is modeled by a temperature value, ranging from 0 to 60, and a value specifying whether or not air is streaming (on or off). The same holds for the output airstream (air_out). The input values of threeway_in is one of cool, heat, and off. Performance is regulated by the input value of scale_in which can be 0, 1, 2, or 3. The following specification determines the normal behaviour of the heater component:

> The heater warms-up the incoming airstream by three times the value set on the data line but only to a maximum value of 45. If the incoming airstream is already warmer than 45, then it is propagated unaltered to the heater's output. Air is streaming at the component's output iff it is streaming at the component's input. If it is not streaming at the output, the temperature is set to the value defined by a dedicated predicate ambient_t($\cdot$).

The following program represents this specification in DLV syntax. Note that ab($\cdot$) denotes the special predicate employed by DLV referring to a defective component.

```
% relates output airstream (on, off) with input airstream
s(H,o1,X) :- heater(H), s(H,i1,X), not ab(H).
% warming-up of the airstream according to the specification
t(H,o1,X) :-  heater(H), t(H,i1,Y), d(H,i2,Z), s(H,o1,on),
   A = Z*3, X = Y + A, X <= 45, not ab(H).
t(H,o1,45) :- heater(H), t(H,i1,Y), d(H,i2,Z), s(H,o1,on),
   A = Z*3, X = Y + A, X > 45, t(H,i1,T), T <= 45, not ab(H).
t(H,o1,T) :-  heater(H), t(H,i1,Y), d(H,i2,Z), s(H,o1,on),
   A = Z*3, X = Y + A, X > 45, t(H,i1,T), 45 < T, not ab(H).
% temperature of output airstream when air is not streaming
t(H,o1,X) :- heater(H), s(H,o1,off), ambient_t(X), not ab(H).
```

We will not go into further details but turn instead our attention to the considered verification tasks determining the correctness of the students' solutions.

Verification with ccT. Let us denote by σ a specification that has to be represented by a logic program. Furthermore, let $Stud(\sigma)$ be a student's attempt to represent σ while $Ref(\sigma)$ is the reference solution. The possible input for the program specified by σ is assumed to be defined over a fixed set $I(\sigma)$ of ground predicates, and, similarly, the output is fixed by a set $O(\sigma)$ of ground predicates. Usually, specifications make further (implicit) assumptions concerning the input, e.g., some predicates need to be defined always or are restricted to be functional in some argument. We call a set $A(\sigma) \subseteq I(\sigma)$ satisfying such assumptions *admissible with respect to* σ.

Definition 1. *A program* $Stud(\sigma)$ *is* correct *with respect to a specification* σ *iff, for any admissible set* $A(\sigma) \subseteq I(\sigma)$, $AS(Stud(\sigma) \cup A(\sigma)) =_{O(\sigma)} AS(Ref(\sigma) \cup A(\sigma))$.

Note that both $Stud(\sigma)$ and $Ref(\sigma)$ are non-ground programs. For instance, for the heater specification from above, σ_H, the set $I(\sigma_H)$ contains the atoms heater(C), ambient_t(T), d(C, i2, Z), t(C, i1, T), s(C, i1, on), and s(C, i1, off), while the set $O(\sigma_H)$ can be fixed to atoms t(C, o1, T), s(C, o1, on), and s(C, o1, off), where T

ranges from 0 to 60, $Z \in \{0, 1, 2, 3\}$, and $C \in \{\mathtt{s}, \mathtt{h}, \mathtt{c}, \mathtt{v}\}$. Any subset $A(\sigma_H) \subseteq I(\sigma_H)$ is admissible if $A(\sigma_H)$ contains exactly one predicate $\mathtt{ambient\_t}(\cdot)$, and predicates $\mathtt{t}(\cdot, \cdot, \cdot)$ and $\mathtt{s}(\cdot, \cdot, \cdot)$ are functional in their third argument.

In order to apply ccT for our verification purpose, the overall strategy is to ground programs $Stud(\sigma)$ and $Ref(\sigma)$ and then to reduce the problem of program correctness to a PQEP. The reason why a reduction to standard uniform equivalence or ordinary equivalence (with additional guessing rules) is not feasible, is the necessity of answer-set projection which has two sources: first, programmers usually employ auxiliary atoms which are not considered as output predicates, and second, new atoms are sometimes added by the grounding procedure (we return to this point in a moment). In terms of complexity theory, projection is the reason why deciding PQEPs is exponentially harder than to decide problems of ordinary equivalence or uniform equivalence.

First, we outline how to handle the restriction to consider only admissible inputs. We express admissibility conditions, as exemplified above, by constraints. For instance, the admissibility conditions for $\mathtt{ambient\_t}(\cdot)$ for the heater can be encoded by the following program:

```
def_ambient_t :- ambient_t(T).              :- not def_ambient_t.
:- ambient_t(T1), ambient_t(T2), T1 <> T2.
```

In general, we denote by $C(\sigma)$ the program representing admissibility constraints on the input according to specification σ. The following result establishes the connection between program correctness and PQEPs:

Theorem 2. *$Stud(\sigma)$ is correct with respect to σ iff the PQEP $(P, Q, 2^A, =_B)$ holds, where $P = grd(Stud(\sigma) \cup C(\sigma), D)$, $Q = grd(Ref(\sigma) \cup C(\sigma), D)$, $A = I(\sigma)$, $B = O(\sigma)$, and D is a finite set containing all constants in $Stud(\sigma) \cup Ref(\sigma) \cup C(\sigma) \cup I(\sigma)$.*

Verifying students' solutions by following the above theorem and then applying ccT is in principle possible since our domain is finite, but the resulting programs would be prohibitively large. So, instead of applying a naive grounding by strictly following the definition, we make use of the *intelligent grounding component* of DLV. This means that several optimisations are performed, e.g., input rewriting, deletion of rules whose body is always false, and semi-naive evaluation. The choice of enabled options has significant impact on the runtimes of the subsequently employed QBF solver, however. We also remark that some optimisations, e.g., the input rewriting, introduce new atoms. Thus, not only auxiliary atoms used by a programmer but also such new atoms stemming from the grounding request the use of projection in equivalence tests. Note that by using DLV's intelligent grounder, we can use strong negation as well as integer arithmetics and comparison predicates in the programs. The grounder translates these constructs such that they do not occur in the ground programs.

However, the optimisations of the intelligent grounder may be too excessive. For example, the grounding of the program for the heater above would result in the empty program since there are no facts, and therefore the bodies of the rules are always false. Our concrete method to ground programs is as follows: Let P be a program and σ its underlying specification. First, augment P by rules $a \leftarrow a'$ and $a' \vee a''$ for any $a \in I(\sigma)$, where a' and a'' are globally new atoms. Then, ground the augmented version of P. Finally, delete all rules containing primed or double-primed atoms from the resulting program. This method guarantees that the semantics of the ground program, possibly joined with atoms from $I(\sigma)$, is correctly preserved under the conservative assumption that the grounder only preserves ordinary equivalence.

However, the resulting programs are still too large. We thus sacrifice completeness of the verification problems by restricting the sets $I(\sigma)$ to contain only certain relevant predicates. For the heater specification σ_H from above for example, we restrict $I(\sigma_H)$ in such a way that not all temperature values from 0 to 60 are considered but only an interval around 45 since it is very likely that if a student program is not correct, then it will diverge from the specification on input from this interval.

Results of the verification. As already mentioned, we considered student data from three semesters. All experiments were carried out on a 3.0 GHz Quad Core Intel Xeon workstation, with 33 GB of RAM and SuSE Linux version 10.3. We used the QBF solver qpro with encoding $T[\cdot]$, as it turned out that all other solvers mentioned in the previous section showed a runtime behaviour several orders of magnitude worse than qpro's. Concerning the setting for the grounder, we achieved best performance when the option for input rewriting was disabled. The reason is that this optimisation introduces new atoms which seems to be disadvantageous for qpro.

We also compared the outcomes of the equivalence tests with results from a test approach currently used in the course. In the latter, test cases (admissible subsets of input predicates) are individually specified, and then it is tested whether a student's program and our reference program yield the same answer sets when joined with the test cases. Such sets of test cases usually comprise 10 to 20 instances. As it will turn out, many errors were undetected by our current approach, thus it is rather prone to false-positives with respect to the verification task.

Table 1. Outcomes of the program verification

semester	component	number of instances	classified as correct		runtimes	
			current approach	ccT approach	average	median
ws2006	c	50	44	38	0.9	1.0
	h	50	39	32	1.0	1.1
	s	50	29	22	0.4	0.1
	v	50	40	34	5.1	5.6
	all	50	42	32	70.2	103.0
ws2007	c	78	67	56	0.8	0.8
	h	78	69	59	0.6	0.6
	s	78	52	0	1.4	1.5
	v	78	48	8	4.5	2.9
	all	78	60	39	491.4	894.0
ws2008	c	100	54	40	1.3	2.3
	h	100	70	13	0.2	0.2
	s	100	59	28	1.4	3.0
	v	100	53	25	0.6	1.1
	all	100	52	19	132.3	72.5

Table 1 summarises the results of our experiments. We provide the year of the semester a course took place, the name of the component we considered, the number of instances of that component, the number of instances classified as correct by the current approach, the number of instances classified as correct by our reduction approach, the average runtime in seconds, as well as the median runtime for solving the QBFs. Components c, h, s, v denote the cooler, heater, switch, and valve as before, while 'all' refers to the overall program consisting of all components, the encoding of the connections between them, and additional constraints required for diagnosing. The ground programs for the component tests contain up to 985 rules. The number of variables in the resulting QBFs ranges from 229 to 623. For the overall tests, programs contain up to 4818 rules, and the QBFs contain 949 to 3143 variables. Note that whenever a program is classified as not correct by the current approach, then it is classified as not correct by the ccT approach as well. Hence, the difference between the numbers of programs classified as correct by the two approaches is the number of false positives for the current approach. Table 1 shows that runtimes for the solved QBFs keep in reasonable bounds. Coming as no surprise, the ccT approach reveals significantly more incorrect solutions than the current approach. The reason that the number of correct overall programs is not smaller than the minimum number of its correct components is mainly due to different restrictions on what admissible input means. The two significantly small numbers of correct solutions for the switch and the valve component in the 'ws2007' test set is because of subtle differences between the reference and the student solutions in case some input values are missing. If this is considered to be too strict, one can simply exclude such cases by changing the admissibility constraints accordingly.

6 Conclusion

In this paper, we discussed how correspondence problems which allow to restrict the alphabet of the context class and which facilitate the removal of auxiliary atoms in the comparison—two important concepts for program equivalence in practice—can be used in a concrete scenario. Moreover, though ccT processes propositional programs only, it can still be employed for program comparisons of non-ground programs. We recapitulated some details of the tool, based on an efficient reduction to QBFs, discussed one particular optimisation, and analysed experiments with different QBF solvers on a random benchmark set which reveals interesting differences of the solvers depending on the particular problem parameterisation and the choice of the encoding. More relevantly, we considered an application concerning the verification of programs.

There remain many issues for future work. For model-based diagnosis, native concepts of equivalence, directly defined in terms of a diagnosis problem, would be useful. In case programs are not equivalent, a counterexample that gives information why the programs are not equivalent would be of great value. ccT can be used to generate QBFs such that assignments for the open variables correspond to such counterexamples. However, few solvers can compute such assignments—to extend qpro in this way is future work. Also, often non-ground programs are formulated over a language with an infinite domain. An important topic is to single out at least sufficient conditions when we can restrict this domain to a finite subdomain such that program equivalence over this subdomain implies equivalence over the unrestricted domain.

Concerning related work, we mention the system DLPEQ [16] for deciding ordinary equivalence, which is based on a reduction to logic programs, and the system SELP [17] for checking strong equivalence, which is based on a reduction to classical logic. Strong equivalence between non-ground programs can be decided by a dedicated system that is based on a reduction to a decidable fragment of first-order logic [18].

References

1. Oetsch, J., Seidl, M., Tompits, H., Woltran, S.: ccT: A Tool for Checking Advanced Correspondence Problems in Answer-Set Programming. In: Proc. CIC 2006, pp. 3–10. IEEE, Los Alamitos (2006)
2. Lifschitz, V., Pearce, D., Valverde, A.: Strongly Equivalent Logic Programs. ACM Transactions on Computational Logic 2(4), 526–541 (2001)
3. Eiter, T., Fink, M.: Uniform Equivalence of Logic Programs under the Stable Model Semantics. In: Palamidessi, C. (ed.) ICLP 2003. LNCS, vol. 2916, pp. 224–238. Springer, Heidelberg (2003)
4. Eiter, T., Tompits, H., Woltran, S.: On Solution Correspondences in Answer Set Programming. In: Proc. IJCAI 2005, pp. 97–102 (2005)
5. Woltran, S.: Characterizations for Relativized Notions of Equivalence in Answer Set Programming. In: Alferes, J.J., Leite, J. (eds.) JELIA 2004. LNCS (LNAI), vol. 3229, pp. 161–173. Springer, Heidelberg (2004)
6. Oetsch, J., Tompits, H., Woltran, S.: Facts do not Cease to Exist Because They Are Ignored: Relativised Uniform Equivalence with Answer-Set Projection. In: Proc. AAAI 2007, pp. 458–464. AAAI Press, Menlo Park (2007)
7. Eiter, T., Faber, W., Leone, N., Pfeifer, G.: The Diagnosis Frontend of the DLV System. AI Communications 12(1-2), 99–111 (1999)
8. Gelfond, M., Lifschitz, V.: Classical Negation in Logic Programs and Disjunctive Databases. New Generation Computing 9, 365–385 (1991)
9. Tompits, H., Woltran, S.: Towards Implementations for Advanced Equivalence Checking in Answer-Set Programming. In: Gabbrielli, M., Gupta, G. (eds.) ICLP 2005. LNCS, vol. 3668, pp. 189–203. Springer, Heidelberg (2005)
10. Besnard, P., Schaub, T., Tompits, H., Woltran, S.: Representing Paraconsistent Reasoning via Quantified Propositional Logic. In: Bertossi, L., Hunter, A., Schaub, T. (eds.) Inconsistency Tolerance. LNCS, vol. 3300, pp. 84–118. Springer, Heidelberg (2005)
11. Letz, R.: Lemma and Model Caching in Decision Procedures for Quantified Boolean Formulas. In: Egly, U., Fermüller, C. (eds.) TABLEAUX 2002. LNCS (LNAI), vol. 2381, pp. 160–175. Springer, Heidelberg (2002)
12. Giunchiglia, E., Narizzano, M., Tacchella, A.: Backjumping for Quantified Boolean Logic Satisfiability. Artificial Intelligence 145, 99–120 (2003)
13. Biere, A.: Resolve and Expand. In: Hoos, H.H., Mitchell, D.G. (eds.) SAT 2004. LNCS, vol. 3542, pp. 59–70. Springer, Heidelberg (2005)
14. Egly, U., Seidl, M., Woltran, S.: A Solver for QBFs in Negation Normal Form. Constraints 14(1), 38–79 (2009)
15. Reiter, R.: A Theory of Diagnosis from First Principles. Artificial Intelligence 32(1), 57–95 (1987)
16. Oikarinen, E., Janhunen, T.: Verifying the Equivalence of Logic Programs in the Disjunctive Case. In: Lifschitz, V., Niemelä, I. (eds.) LPNMR 2004. LNCS (LNAI), vol. 2923, pp. 180–193. Springer, Heidelberg (2004)

17. Chen, Y., Lin, F., Li, L.: SELP - A System for Studying Strong Equivalence Between Logic Programs. In: Baral, C., Greco, G., Leone, N., Terracina, G. (eds.) LPNMR 2005. LNCS (LNAI), vol. 3662, pp. 442–446. Springer, Heidelberg (2005)
18. Eiter, T., Faber, W., Traxler, P.: Testing Strong Equivalence of Datalog Programs - Implementation and Examples. In: Baral, C., Greco, G., Leone, N., Terracina, G. (eds.) LPNMR 2005. LNCS (LNAI), vol. 3662, pp. 437–441. Springer, Heidelberg (2005)

Translating Preferred Answer Set Programs to Propositional Logic

Vernon Asuncion and Yan Zhang

Intelligent Systems Laboratory
School of Computing and Mathematics
University of Western Sydney, Australia
{vernon,yan}@scm.uws.edu.au

Abstract. We consider the problem of whether a given preferred answer set program can be reduced to a propositional formula. Research on this topic is of both theoretical and practical interests: on one hand, it will shed new insights to understand the expressive power of preferred answer set programs; on the other hand, it may also lead to efficient implementations for computing preferred answer sets of logic programs. In this paper, we focus on Brewka and Eiter's preferred answer set programs. We propose a translation from preferred answer set programs to propositional logic and show that there is one-to-one correspondence between the preferred answer sets of the program to the models of the resulting propositional theory. We then link this result to Brewka and Eiter's weakly preferred answer set semantics.

Keywords: answer set semantics, prioritized logic programs, answer set computations.

1 Introduction

In recent years, Answer Set Programming (ASP) has become one of the most effective approaches for declarative problem solving in knowledge representation and reasoning. One important research in this area is to translate various answer set programs, such as normal logic programs and disjunctive logic programs, into propositional logic, so that the answer sets of these logic programs are precisely captured by the models of corresponding propositional theories, e.g. [5,7]. Research on this topic is of both theoretical and practical interests because it has not only provided new insights for a better understanding of the expressive power of answer set programming, but also been led to some efficient computations for answer set programming [7].

On the other hand, preferred answer set programming is a promising method for dealing with conflict resolution in nonmonotonic reasoning. Over the years, a number of various preferred answer set program (also called prioritized logic program) frameworks have been developed, which extended traditional answer set semantics by integrating proper priorities into the underlying logic programs, e.g. [3,4,6,8].

However, it remains as an unaddressed question whether a similar translation can be achieved between preferred answer set programs and propositional logic. This paper provides a positive answer to this question. We focus on Brewka and Eiter's preferred

E. Erdem, F. Lin, and T. Schaub (Eds.): LPNMR 2009, LNCS 5753, pp. 396–401, 2009.

answer set programs and propose a translation between preferred answer set programs and propositional logic. In particular, we prove that given a preferred logic program, the models of a propositional theory, which consists of the completion, loop formulas, and preference formula of the program, precisely are the same as the preferred answer sets of the underlying program.

The rest of the paper is organized as follows. Section 2 overviews the basic notions and concepts of Brewka and Eiter's preferred and weakly preferred answer set programs. Section 3 presents our translation from the preferred answer set program to propositional logic and proves a one-to-one correspondence between the preferred answer sets of the program and the models of the translated propositional theory. Section 4 then links this result to weakly preferred answer set programs. Finally, section 5 concludes the paper with some discussions.

2 Brewka and Eiter's Preferred Answer Set Semantics: An Overview

Consider a propositional language $\mathcal{L}$ which consists of a set of propositional atoms. A *fully prioritized logic program* on the language $\mathcal{L}$ is a pair $(\Pi, <)$, where Π is a (finite) normal logic program and $<$ is a strict *total order* on Π. Since $<$ is a total order on the set Π, the rules in Π corresponds to a unique ordinal number, and thus to an enumeration $r_1, \cdots, r_\alpha, \cdots, r_{|\Pi|}$ of the elements of Π. Therefore we use the notion $\{r_\alpha\}_<$ to represent $(\Pi, <)$.

A ground rule r is *defeated* by a set of atoms S if there exist some atom $a \in S$ such that a appears in the negative body of r, i.e., "*not a*" is a part of r's body. We use $Head(r)$, $Pos(r)$ and $Neg(r)$ to denote the head atom, the set of atoms occurring in the positive body, and the set of atoms occurring in the negative body of rule r, respectively. Given a set of atoms S and a rule r, if $Head(r) \notin S$ and $Pos(r) \subseteq S$ then we refer to r as a *zombie* rule with respect to S or simply "zombie rule" when it is clear from the context. Intuitively, a zombie rule r is a rule which is assured to be *non-generating* with respect to S as $Head(r) \notin S$.

Definition 1. *[2] Let $\Pi_< = (\Pi, <)$ be a fully prioritized grounded (normal) logic program and X a set of ground atoms. Let $^X\Pi_{<'} = (^X\Pi, <')$ be the fully grounded prioritized logic program such that $^X\Pi$ is the set of rules obtained from Π by*

1. *deleting every rules having an atom p in its positive body where $p \notin X$, and*
2. *removing from each remaining rules their positive body,*

and $<'$ is inherited from $<$ by the map $f : ^X\Pi \longrightarrow \Pi$, i.e., $r_1' <' r_2'$ iff $f(r_1') < f(r_2')$, where $f(r') = r$ is the first rule in Π with respect to $<$ such that r' results from r by step 2.

Note that $<'$ is also a strict total order on $^X\Pi$. For a fully prioritized program $\Pi_<$, a set of atoms S, and $^S\Pi$ defined as in above, the preferred answer set semantics of $\Pi_<$ is defined through an operator $C_{S\,\Pi_{<'}} : 2^{Atoms(^S\Pi_{<'})} \longrightarrow 2^{Atoms(^S\Pi_{<'})}[1]$ such that

---

[1] Here $Atoms(^S\Pi_{<'})$ denotes the set of all atoms occurring in $^S\Pi_{<'}$.

an answer set A of Π satisfies the priorities if and only if $C_{A\Pi_{<'}}(A) = A$. The formal definition is given below.

Definition 2. *[2] For a fully prioritized grounded (normal) logic program $\Pi_< = (\Pi, <)$ and a set S of atoms, let $^S\Pi_{<'} = (^S\Pi, <') = \{r_\alpha\}_{<'}$ where $^S\Pi$ and $<'$ is defined as in Definition 1. The sequence S_α is defined as follows:*

$$S_0 = \emptyset$$

for $\alpha = 0$, and

$$S_{\alpha+1} = \begin{cases} S_\alpha & \text{if } r_{\alpha+1} \text{ is defeated by } S_\alpha \text{ or} \\ & Head(r_{\alpha+1}) \in S \text{ and } r_{\alpha+1} \text{ defeated by } S, \\ S_\alpha \cup \{Head(r_{\alpha+1})\}, & \text{otherwise.} \end{cases}$$

for $0 < \alpha < |\Pi|$. Then $C_{S\Pi_{<'}}(S) = S_{|\Pi|}$.

For a fully prioritized grounded logic program $\Pi_< = (\Pi, <)$ and an answer set A of Π, A is a *preferred answer set* of $\Pi_<$ if and only if $C_{A\Pi_{<'}}(A) = A$.

Obviously, there are some fully prioritized grounded logic programs that may have no preferred answer set. In [2], this problem was addressed by a proposed relaxation that gives preferred answer sets whenever they exist and an approximation called *weakly preferred answer sets*, in the other cases. For a formal definition, the notion of *inversion* is first introduced.

Definition 3. *[2] Let $<_1$ and $<_2$ be two well-orderings on set S. We define $Inv_S(<_1, <_2)$ (inversions of $<_2$ in $<_1$) as*

$$Inv_S(<_1, <_2) = \{(b, a) \mid a, b \in S, a <_2 b, b <_1 a\}.$$

The idea behind weakly preferred answer sets is linked to counting those inversions from a full prioritization $<_1$ of a grounded logic program Π to another full prioritization $<_2$ of the program. To formally define this, the notion of distance is introduced.

Definition 4. *[2] Let $<_1$ and $<_2$ be well-orderings of a finite set S. The distance from $<_1$ to $<_2$, denoted $d_S(<_1, <_2)$, is defined as*

$$d_S(<_1, <_2) = |Inv_S(<_2, <_1)|.$$

From the above definition, the notion of *preference violation degree*, denoted *pvd* is defined as follows.

Definition 5. *[2] Let $\Pi_< = (\Pi, <)$ be a finite fully prioritized grounded (normal) logic program. For an answer set A of Π, define $pvd_{\Pi_<}(A)$ (preferrence violation degree of A in $\Pi_<$) as*

$$pvd_{\Pi_<}(A) = min\{d_\Pi(<, <') \mid <' \text{ is any full prioritization of } \Pi \text{ such that} \\ A \text{ is a preferred answer set of } \Pi_{<'} = (\Pi, <')\}.$$

Intuitively, $pvd_{\Pi_<}(A)$ is the minimum distance possible from the full prioritization of $\Pi_<$ to any fully prioritized rule base $\Pi_{<'} = (\Pi, <')$ such that A is a preferred answer set of $\Pi_{<'}$. From the above definitions, the semantics of weakly preferred programs can be formally defined as follows.

Definition 6. *[2] Let $\Pi_< = (\Pi, <)$ be a finite fully prioritized grounded (normal) logic program. We define*

$$pvd(\Pi_<) = min\{pvd_{(\Pi_<)}(A) \mid A \text{ is an answer set of } \Pi\}.$$

Then A is a weakly preferred answer set *of $\Pi_<$ iff $pvd_{\Pi_<}(A) = pvd(\Pi_<)$.*

Informally A is a weakly preferred answer set of a fully prioritized grounded (normal) logic program $\Pi_<$ if there exist a full prioritization $<_1$ of Π such that A is a preferred answer set of $\Pi_{<_1}$, and for any other full prioritization $<_2$ of Π where there exist a preferred answer set A' of $\Pi_{<_2}$, we have $d_\Pi(<, <_1) \leq d_\Pi(<, <_2)$.

3 Preference Formulas and the Translation

In this section, we propose a translation from the preferred answer set semantics to propositional logic, such that a one-to-one correspondence exist for this translation.

Definition 7. *For a finite fully prioritized grounded (normal) logic program $\Pi_< = (\Pi, <)$, we define the* preference formula $PF(\Pi_<)$ *of $\Pi_<$ as follows:*

$$PF(\Pi_<) = \bigwedge_{r \in \Pi_<, Head(r) = a} (\neg a \wedge \bigwedge_{b \in Pos(r)} b \supset$$
$$\bigvee_{r' \in def_\Pi(r), r' < r} (\bigwedge_{c \in Pos(r')} c \wedge \bigwedge_{d \in Neg(r')} \neg d))$$

where $def_\Pi(r) = \{r' \mid r' \in \Pi, Head(r') \in Neg(r), Head(r') \neq Head(r) \text{ or } Neg(r') \neq Neg(r)\}$.

Informally, $def_\Pi(r)$ is the set of rules in Π that defeat r in a sense that each rule r' in $def_\Pi(r)$ is different from r with regards to either $Head(r)$ or $Neg(r)$, and $Head(r') \in Neg(r)$.

Theorem 1. *For a finite fully prioritized grounded (normal) logic program $\Pi_< = (\Pi, <)$ and an answer set A of Π, $A \models PF(\Pi_<)$ iff A is a preferred answer set of $\Pi_<$.*

In [7], Lin and Zhao proposed a translation of finite normal logic programs to propositional formulas without the need of extra variables. The translation is of the form $Comp(\Pi) \wedge LF(\Pi)$ where $Comp(\Pi)$ is the completion of the logic program Π and $LF(\Pi)$ is the conjunction of all loop formulas associated with Π. The loop formulas are a way of strengthening the completion of Π such that a set of atoms A is an answer set of Π iff A is a model of $Comp(\Pi) \wedge LF(\Pi)$. Using Lin and Zhao's result, we are able to translate prioritized normal logic programs to propositional formulas via the following theorem. Moreover, the models of the resulting propositional formula are in a one-to-one correspondence with the preferred answer sets of the logic program.

Theorem 2. *For a finite fully prioritized grounded (normal) logic program $\Pi_< = (\Pi, <)$, $A \models Comp(\Pi) \wedge LF(\Pi) \wedge PF(\Pi_<)$ iff A is a preferred answer set of $\Pi_<$.*

4 Linking Weakly Preferred Answer Set Programs

We now try to link the semantics of *weakly preferred answer sets* by extending our previously defined preference formula. Intuitively, we model the weakly preferred criterion by encoding the inversions between two full prioritizations of the rules of a given program. To achieve this, we introduce two classes of new atoms of the form (r_1, r_2) and $X_{(r_1, r_2)}$ where r_1 and r_2 are rule names of the given program.

Definition 8. *For a finite fully prioritized grounded (normal) logic program $\Pi_< = (\Pi, <)$, we define the* weak preference formula $WPF(\Pi_<)$ *as follows:*

$$WPF(\Pi_<) =$$

$$\bigwedge_{r \in \Pi, Head(r)=a} (\neg a \wedge \bigwedge_{b \in Pos(r)} b \supset \bigvee_{r' \in def_\Pi(r)} (\bigwedge_{c \in Pos(r')} c \wedge \bigwedge_{d \in Neg(r')} \neg d \wedge (r', r))) \tag{1}$$

$$\wedge \bigwedge_{r_1, r_2 \in \Pi, r_1 \neq r_2} (((r_1, r_2) \vee (r_2, r_1)) \wedge ((r_1, r_2) \supset \neg(r_2, r_1))) \tag{2}$$

$$\wedge \bigwedge_{r_1, r_2, r_3 \in \Pi, r_1 \neq r_2 \neq r_3} ((r_1, r_2) \wedge (r_2, r_3) \supset (r_1, r_3)) \tag{3}$$

$$\wedge \bigwedge_{r_1, r_2 \in \Pi_<, r_2 < r_1} (((r_1, r_2) \supset X_{(r_1, r_2)}) \wedge (X_{(r_1, r_2)} \supset (r_1, r_2))) \tag{4}$$

The formula $WPF(\Pi_<)$ is a conjunction of the four subformulas (1), (2), (3), and (4). As can be seen, formula (1) is similar to $PF(\Pi_<)$ except that a rule that defeats a zombie rule does not necessarily have to be more preferred with respect to $<$ and that if a rule r' defeats a zombie rule r then the atom (r', r) (i.e. encodes r' is more preferred than r but not necessarily with respect to $<$) should be satisfied (i.e. in the model satisfying $WPF(\Pi_<)$). Basically, the conjunction of the two formulas (2) and (3) encodes a full prioritization of the rules in Π (not necessarily $<$) where the prioritization relations are represented by the atoms of the form (r_1, r_2). The last formula (4) encodes the inversions of $<$ from the *other* full prioritization relations (that are represented by the atoms (r_1, r_2)), which are then represented by the atoms of the form $X_{(r_1, r_2)}$ (i.e. it indicates that (r_1, r_2) is an inversion of $r_2 < r_1$).

Before we present our main theorem of this section, we need to first introduce a useful notion. Let $\mathcal{L}_1$ and $\mathcal{L}_2$ be two propositional languages and $\mathcal{L}_1 \subseteq \mathcal{L}_2$, and A an interpretation of $\mathcal{L}_1$ (i.e. a subset of atoms of $\mathcal{L}_1$), an interpretation B of $\mathcal{L}_2$ is called an *extension* of A on $\mathcal{L}_2$, denoted as $B = ext(A)_{\mathcal{L}_2}$, if $A \subseteq B$, and A and B agree on the truth values of all propositional atoms of $\mathcal{L}_1$.[2]

---

[2] Intuitively, $\mathcal{L}_2$ is a *superset* of $\mathcal{L}_1$.

Theorem 3. *For a finite fully prioritized grounded (normal) logic program $\Pi_< = (\Pi, <)$ and an interpretation A of language $\mathcal{L} = Atom(\Pi)$, A is a weakly preferred answer set of $\Pi_<$ iff there exist an extension $ext(A)_P$ of A, where $P = Atoms(\Pi) \cup \{(r_i, r_j) \mid r_i, r_j \in \Pi\} \cup \{X_{(r_i, r_j)} \mid r_i, r_j \in \Pi_<, r_j < r_i\}^3$, such that $ext(A)_P \models Comp(\Pi) \wedge LF(\Pi) \wedge WPF(\Pi_<)$ and for all models M of $Comp(\Pi) \wedge LF(\Pi) \wedge WPF(\Pi_<)$, $|ext(A)_P \restriction_X| \leq |M \restriction_X|$.*

5 Conclusions

In this paper, we have proposed a translations between Brewka and Eiter's preferred answer set programs and propositional logic. We have also proved a one-to-one correspondence theorem for the translation. Moreover, we also provided a link between the weakly preferred answer sets and propositional logic. We believe that our work will be of practical values to serve as an alternative approach for current preferred answer set programming implementations [6]. Currently we are considering to implement a SAT based preferred answer set solver based on the work developed in this paper. From an implementation viewpoint, since our defined preference and weak preference formulas remain in a polynomial size of the underlying program, techniques of ASSAT [7] may be used to optimize the computation of preferred answer sets. For future work, we consider the possibility of applying similar methods to capture the preferred answer set framework in [3] which allows the specification of *dynamic* orderings such that *static* orderings are a trivial restriction of the more general dynamic case.

References

1. Ben-Eliyahu, R., Dechter, R.: Propositional semantics for disjunctive logic programs. Annals of Mathematics and Artificial Intelligence 12, 53–87 (1994)
2. Brewka, G., Eiter, T.: Preferred answer sets for extended logic programs. Artificial Intelligence 109, 297–356 (1999)
3. Delgrande, J., Schaub, T., Tompits, H.: A framework for compiling preferences in logic programs. Theory and Practice of Logic Programming 3, 129–187 (2003)
4. Delgrande, J., Schaub, T., Tompits, H., Wang, K.: A classification and survey of preference handling approaches in nonmonotonic reasoning. Computational Intelligence 20, 308–334 (2004)
5. Ferraris, P., Lee, J., Lifschitz, V.: A generalization of the Lin-Zhao theorem. Annals of Mathematics and Artificial Intelligence 47, 79–101 (2006)
6. Grell, S., Konczak, K., Schaub, T.: nomore++: A system for computing preferred answer sets. In: Baral, C., Greco, G., Leone, N., Terracina, G. (eds.) LPNMR 2005. LNCS (LNAI), vol. 3662, pp. 394–398. Springer, Heidelberg (2005)
7. Lin, F., Zhao, Y.: ASSAT: computing answer sets of a logic program by SAT solvers. Artificial Intelligence 157, 115–137 (2004)
8. Zhang, Y.: Logic program based updates. ACM Transactions on Computational Logic 7, 421–472 (2006)

---

^{3} $Atoms(\Pi)$ plus the relation and inversion 'indicator' atoms respectively.

CR-Prolog as a Specification Language for Constraint Satisfaction Problems

Marcello Balduccini

Intelligent Systems, OCTO
Eastman Kodak Company
Rochester, NY 14650-2102 USA
marcello.balduccini@gmail.com

Abstract. In this paper we describe an approach for integrating CR-Prolog and constraint programming, in which CR-Prolog is viewed as a specification language for constraint satisfaction problems. Differently from other methods of integrating ASP and constraint programming, our approach has the advantage of allowing the use of off-the-shelf, unmodified ASP solvers and constraint solvers, and of global constraints, which substantially increases practical applicability.

1 Introduction

Particular interest has been recently devoted to the integration of Answer Set Programming (ASP) [1] with Constraint Logic Programming (CLP) (see [2,3]), aimed at combining the ease of knowledge representation of ASP with the powerful support for numerical computations of CLP. Such approaches are mostly based on extending the ASP language and on using answer set and constraint solvers modified so that they can work together. Although the combination of ASP and CLP showed substantial performance improvements over ASP alone, the restriction of using ad-hoc ASP and CLP solvers limits the practical applicability of the approach. In fact, programmers can no longer select the solvers that best fit their needs (most notably, SMODELS, DLV, SWI-Prolog and SICStus Prolog), as is instead commonly done in ASP. Another limitation is the general lack of specific support for global constraints. Without global constraints, applications' performance is often heavily impacted by the combinatorial explosion of the underlying search space.

In [4] we have presented a method for integrating ASP and constraint programming. In this paper we extend the approach and integrate constraint programming with an extension of ASP, called CR-Prolog [5]. CR-Prolog introduces in ASP the notion of consistency restoring rule, which is particularly useful to represent unlikely events, less-desired choices, etc. Our technique consists in viewing CR-Prolog as a specification language for constraint satisfaction problems. CR-Prolog programs are written in such a way that their answer sets encode the desired constraint satisfaction problems; the solutions to those problems are found using constraint satisfaction techniques. Both the answer sets and the solutions to the constraint problems can be computed with arbitrary off-the-shelf solvers, as long as a (relatively simple) translation procedure is defined from the ASP encoding of the constraint problems to the input language of the

E. Erdem, F. Lin, and T. Schaub (Eds.): LPNMR 2009, LNCS 5753, pp. 402–408, 2009.

constraint solver selected. Moreover, our approach allows the use of the global constraints available in the selected constraint solver. Compared to the other approaches to the integration of ASP and CLP, our technique allows programmers to exploit the full power of the underlying state-of-the-art solvers when tackling industrial-size problems. Finally, although space restrictions prevent us from discussing it here, our experiments have also shown that our technique produces programs that are arguably more compact and easy to understand than those written in CLP alone, but with comparable performance.

2 Background

The syntax of CR-Prolog is defined as follows. A *regular rule* is a statement of the form[1] $h \leftarrow l_1, \ldots, l_m, \text{not } l_{m+1}, \ldots, \text{not } l_n$, where h and l_i's are literals (defined as usual). The intuitive meaning of the statment is that a reasoner who believes $\{l_1, \ldots, l_m\}$ and has no reason to believe $\{l_{m+1}, \ldots, l_n\}$, has to believe h. A *consistency-restoring rule* (or *cr-rule*) is a statement of the form: $r : h \stackrel{+}{\leftarrow} l_1, \ldots, l_m, \text{not } l_{m+1}, \ldots, \text{not } l_n$, where r, called the cr-rule's *name*, is a (possibly compound) term uniquely identifying the cr-rule. The intuitive reading of the statement is that a reasoner who believes $\{l_1, \ldots, l_m\}$ and has no reason to believe $\{l_{m+1}, \ldots, l_n\}$, *may possibly* believe h. The implicit assumption is that this possibility is used as little as possible, only when the reasoner cannot otherwise form a non-contradictory set of beliefs. A preference order on the use of cr-rules can also be given by means of atoms of the form $prefer(r_1, r_2)$ [5]. By rule we mean a regular rule or a cr-rule. Given rule ρ, we call $\{l_1, \ldots, \text{not } l_n\}$ its *body* $(body(\rho))$. Given cr-rule name r, $body(r)$ denotes the body of the corresponding cr-rule. A *program* is a set of rules. As usual, a non-ground program is viewed as a shorthand for the program consisting of the ground instances of its rules. Given a program Π, the *regular part* of Π is the set of its regular rules, and is denoted by $reg(\Pi)$. The set of its cr-rules is denoted by $cr(\Pi)$. The semantics of CR-Prolog can be found in [5].

Let us now turn our attention to Constraint Programming. The definition of constraint satisfaction problem that follows is adapted from [6]. A *Constraint Satisfaction Problem (CSP)* is a triple $\langle X, D, C \rangle$, where $X = \{x_1, \ldots, x_n\}$ is a set of variables, $D = \{D_1, \ldots, D_n\}$ is a set of domains, such that D_i is the domain of variable x_i (i.e. the set of possible values that the variable can be assigned), and C is a set of constraints. Each constraint $c \in C$ is a pair $c = \langle \sigma, \rho \rangle$ where σ is a list of variables and ρ is a subset of the Cartesian product of the domains of such variables. An *assignment* is a pair $\langle x_i, a \rangle$, where $a \in D_i$, whose intuitive meaning is that variable x_i is assigned value a. A *compound assignment* is a set of assignments to distinct variables from X. A *complete assignment* is a compound assignment to all the variables in X. A constraint $\langle \sigma, \rho \rangle$ specifies the acceptable assignments for the variables from σ. We say that such assignments *satisfy* the constraint. A *solution* to a CSP $\langle X, D, C \rangle$ is a complete assignment satisfying every constraint from C. Constraints can be represented either *extensionally*, by specifying the pair $\langle \sigma, \rho \rangle$, or *intensionally*, by specifying an expression involving variables, such as $x < y$. In this paper we focus on constraints represented intensionally.

---

[1] For simplicity we focus on non-disjunctive programs. Our results extend to disjunctive programs in a natural way.

A *global constraint* is a constraint that captures a relation between a non-fixed number of variables [7], such as $sum(x, y, z) < w$ and $all\_different(x_1, \ldots, x_k)$. One should notice that the mapping of an intensional constraint specification into a pair $\langle \sigma, \rho \rangle$ depends on the *constraint domain*. For example, the expression $1 \le x < 2$ corresponds to the constraint $\langle \langle x \rangle, \{\langle 1 \rangle\} \rangle$ if the finite domain is considered, while it corresponds to $\langle \langle x \rangle, \{\langle v \rangle \mid v \in [1, 2)\} \rangle$ in a continuous domain. For this reason, in this paper we assume that a CSP includes the specification of the intended constraint domain.

3 Encoding Constraint Problems in CR-Prolog

Our approach consists in writing CR-Prolog programs whose answer sets encode the desired constraint satisfaction problems (CSPs). The solutions to the CSPs are then computed using constraint satisfaction techniques.

CSPs are encoded in CR-Prolog using the following three types of statements: (1) a constraint domain declaration is a statement of the form $cspdomain(\mathcal{D})$, where $\mathcal{D}$ is a constraint domain such as fd, q, or r; informally, the statement says that the CSP is over the specified constraint domain, thereby fixing an interpretation for the intensionally specified constraints; (2) a constraint variable declaration is a statement of the form $cspvar(x, l, u)$, where x is a ground term denoting a variable of the CSP (CSP variable or constraint variable for short), and l and u are numbers from the constraint domain; the statement says that the domain of x is $[l, u]$;[2] (3) a constraint statement is a statement of the form $required(\gamma)$, where γ is an expression that intensionally represents a constraint on (some of) the variables specified by the *cspvar* statements; intuitively the statement says that the constraint intensionally represented by γ is required to be satisfied by any solution to the CSP. For the purpose of specifying global constraints, we allow γ to contain expressions of the form $[\delta/k]$. If δ is a function symbol, the expression intuitively denotes the sequence of all variables formed from function symbol δ and with arity k, ordered lexicographically. For example, given CSP variables $v(1), v(2), v(3)$, $[v/1]$ denotes the sequence $\langle v(1), v(2), v(3) \rangle$. If δ is a relation symbol and $k \ge 1$, the expression intuitively denotes the sequence $\langle e_1, e_2, \ldots, e_n \rangle$ where e_i is the last element of the ith k-tuple satisfying relation δ, according to the lexicographic ordering of such tuples. For example, given a relation r' defined by $r'(a, 3), r'(b, 1), r'(c, 2)$, the expression $[r'/2]$ denotes the sequence $\langle 3, 1, 2 \rangle$.

Example 1. A simple CSP is encoded by $A_1 = \{cspdomain(fd),\ cspvar(v(1), 1, 3),$ $cspvar(v(2), 2, 5),$ $cspvar(v(3), 1, 4),$ $required(v(1) + v(2) \le 4),$ $required(v(2) - v(3) > 1),\ required(sum([v/1]) \ge 4)\}$.

In the rest of this paper, we consider signatures that contain: relations *cspdomain*, *cspvar*, *required*; constant symbols for the constraint domains $\mathcal{FD}$, $\mathcal{Q}$, and $\mathcal{R}$; suitable symbols for the variables, functions and relations used in the CSP; the numerical constants needed to encode the CSP.

Let A be a set of atoms formed from relations *cspdomain*, *cspvar*, and *required*. We say that A is a *well-formed CSP definition* if: A contains exactly one constraint

---

[2] As an alternative, the domain of the variables could also be specified using constraints. We use a separate statement for similarity with CLP languages.

domain declaration; the same CSP variable does not occur in two or more constraint variable declarations of A; every CSP variable that occurs in a constraint statement from A also occurs in a constraint variable declaration from A. Let A be a well-formed CSP definition. The CSP *defined* by A is the triple $\langle X, D, C \rangle$ such that: $X = \{x_1, x_2, \ldots, x_k\}$ is the set of all CSP variables from the constraint variable declarations in A; $D = \{D_1, D_2, \ldots, D_k\}$ is the set of domains of the variables from X, where the domain D_i of variable x_i is given by arguments l and u of the constraint variable declaration of x_i in A, and consists of the segment between l and u in the constraint domain specified by the constraint domain declaration from A; C is a set containing a constraint γ' for each constraint statement $required(\gamma)$ of A, where γ' is obtained by: (1) replacing the expressions of the form $[f/k]$, where f is a function symbol, by the list of variables from X formed by f and of arity k, ordered lexicographically; (2) replacing the expressions of the form $[r/k]$, where r is a relation symbol and $k \geq 1$, by the sequence $\langle e_1, \ldots, e_n \rangle$, where, for each i, $r(t_1, t_2, \ldots, t_{k-1}, e_i)$ is the ith element of the sequence, ordered lexicographically, of atoms from A formed by relation r; (3) interpreting the resulting intensionally specified constraint w.r.t. the constraint domain specified by the constraint domain declaration from A.

Example 2. Set A_1 from Example 1 defines the CSP:

$$\langle \{v(1), v(2), v(3)\}, \left\{ \begin{array}{c} \{1,2,3\}, \{2,3,4,5\}, \\ \{1,2,3,4\} \end{array} \right\}, \left\{ \begin{array}{c} v(1) + v(2) \leq 4,\ v(2) - v(3) > 1, \\ sum(v(1), v(2), v(3)) \geq 4 \end{array} \right\} \rangle.$$

Let A be a set of literals. We say that A *contains* a well-formed CSP definition if the set of atoms from A formed by relations $cspdomain$, $cspvar$, and $required$ is a well-formed CSP definition. We also say that a CSP is defined by a set of literals A if it is defined by the well-formed CSP definition contained in A. Notice that, if a set A of literals does not contain a well-formed CSP definition, A does not define any CSP. For simplicity, in the rest of the discussion we omit the term "well-formed" and simply talk about CSP definitions.

Definition 1.

- *A pair $\langle A, \alpha \rangle$ is an* extended answer set *of cr-rule free program Π iff A is an answer set of Π and α is a solution to the CSP defined by A.*
- *Let Π be a program, and R be a set of names of cr-rules from Π, and $prefer^*$ be the transitive closure of relation $prefer$. $\mathcal{V} = \langle A, R \rangle$ is an* extended view *of Π if: (1) A is an extended answer set of $reg(\Pi) \cup \theta(R)$; (2) for every r_1, r_2, if $S \models prefer^*(r_1, r_2)$, then $\{r_1, r_2\} \not\subseteq R$; (3) for every r in R, $body(r)$ is satisfied by S.*
- *An extended view $\mathcal{V}$ is an* extended candidate answer set *of Π if, for every view $\mathcal{V}'$ of Π, $\mathcal{V}'$ does not dominate $\mathcal{V}$.*[3]
- *A is an* extended answer set *of Π if: (1) there exists a set R of names of cr-rules from Π such that $\langle A, R \rangle$ is a candidate answer set of Π, and (2) for every extended candidate answer set $\langle A', R' \rangle$ of Π, $R' \not\subseteq R$.*

---

[3] The notion of dominance extends to extended views in a natural way.

Example 3. Consider set A_1 from Example 1. An extended answer set of A_1 is $\langle A_1,$ $\{(v(1), 1), (v(2), 3), (v(3), 1)\}\rangle$. Consider program P_1 below and a corresponding extended answer set. Notice that the cr-rule is used to say that the sum of the CSP variables should be less than 20 *if at all possible*.

$$P_1 = \begin{cases} i(1).\ \dots i(4).\ \ cspdomain(fd). \\ cspvar(v(I), 1, 10) \leftarrow i(I). \\ required(v(I1) - v(I2) \geq 3) \leftarrow \\ \quad i(I1),\ i(I2),\ I2 = I1 + 1. \\ required(sum([v/1]) \geq 20) \leftarrow \\ \quad not\ can\_violate. \\ r :\ can\_violate \overset{+}{\leftarrow} . \end{cases}$$

Extended answer set:
$\langle \{i(1), \dots, i(4), cspdomain(fd),$
$\quad cspvar(v(1), 1, 10), \dots,$
$\quad cspvar(v(4), 1, 10),$
$\quad required(v(1) - v(2) \geq 3), \dots,$
$\quad required(v(3) - v(4) \geq 3)\},$
$\{(v(1), 10), (v(2), 7), (v(3), 4), (v(4), 1)\}\rangle$

To compute the extended answer sets of a cr-rule free program, we combine the use of answer set solvers and constraint solvers (see Algorithm 1). As discussed in [4], step (1) of Algorithm 1 relies on the correctness of the translation from the CSP definition to the encoding for the constraint solver. Soundness and completeness results for the algorithm can be found in [4]. The extended answer sets of arbitrary CR-Prolog programs can be computed by extending the CRMODELS algorithm from [8]. The complete algorithm is shown below (see Algorithm 2). We refer the reader to [8] for the definition of operators γ_i, τ, λ, ν and hr. To compute extended answer sets, the original algorithm is modified to use a new function $\epsilon_1(\Pi)$, which returns an arbitrary element of $\epsilon(\Pi)$ (and replaces function α_1 as used in the original algorithm). Soundness and completeness of the algorithm follow from soundness and completeness of Algorithm 1 and from the results in [8].

Algorithm 1. ϵ

Input: Program Π
Output: The set of extended answer sets of Π
1   $\mathcal{E} := \emptyset$
2   Let $\mathcal{A}$ be the set of answer sets of Π containing a CSP definition.
3   **for each** $A \in \mathcal{A}$ **do**
4   Select solver $solve_\mathcal{D}$ for constraint domain $\mathcal{D}$ as specified by $cspdomain(\mathcal{D}) \in A$.
5   Translate the CSP definition from A into an encoding $\chi_A^\mathcal{D}$ suitable for $solve_\mathcal{D}$.
6   Let $\mathcal{S} = \{\alpha_1, \dots, \alpha_k\}$ be the set of solutions returned by $solve_\mathcal{D}(\chi_A^\mathcal{D})$.
7   **for each** $\alpha \in \mathcal{S}$ **do** $\mathcal{E} := \mathcal{E} \cup \langle A, \alpha \rangle$.
8   **end**
9   **return** $\mathcal{E}$

---

4   Related Work

The *clingcon* system [3] integrates the answer set solver Clingo and the constraint solver Gecode. The system thus differs significantly from ours in that programmers cannot arbitrarily select the most suitable ASP and constraint solvers for the task at hand.

Algorithm 2. CRMODELS-CSP

Input: A CR-Prolog program Π

Output: The extended answer sets of Π

1 $C := \emptyset$; $\mathcal{A} := \emptyset$; $i := 0$
2 **while** $i \leq |cr(\Pi)|$ **do**
3     $C' := \emptyset$
4   **repeat**
5   **if** $\gamma_i(\Pi) \cup C$ *is inconsistent* **then** $M := \bot$
6   **else**
7           $\langle M, \alpha \rangle := \epsilon_1(\gamma_i(\Pi) \cup C)$
8   **if** $\tau(M, \Pi)$ *is inconsistent* **then**
9              $\mathcal{A} := \mathcal{A} \cup \{\langle M \cap \Sigma(\Pi), \alpha \rangle\}$
10             $C' := C' \cup \{ \leftarrow \lambda(M \cap atoms(appl, hr(\Pi))). \}$
11   **end**
12          $C := C \cup \{ \leftarrow \lambda(M), \nu(M). \}$
13   **end**
14   **until** $M = \bot$
15    $C := C \cup C'$; $i := i + 1$ *[now consider views obtained with one more cr-rule]*
16 **end**
17 **return** $\mathcal{A}$

The approach proposed in [2] is based on an extension, called $\mathcal{AC}(\mathcal{C})$, of CR-Prolog, allowing the use of CSP-style constraints in the body of the rules. The assignment of values to the constraint variables is denoted by means of special atoms occurring in the body of the rules. Such atoms are treated as abducibles, and their truth determined by solving a suitable CSP. The following result connects our approach and $\mathcal{AC}(\mathcal{C})$.

Theorem 1. *An $\mathcal{AC}_0$ program Π can be translated into a CR-Prolog program whose extended answer sets are in one-to-one correspondence with the answer sets of Π.*

References

1. Gelfond, M., Lifschitz, V.: Classical negation in logic programs and disjunctive databases. New Generation Computing, 365–385 (1991)
2. Mellarkod, V.S., Gelfond, M., Zhang, Y.: Integrating Answer Set Programming and Constraint Logic Programming. Annals of Mathematics and Artificial Intelligence (2008)
3. Gebser, M., Ostrowski, M., Schaub, T.: Constraint Answer Set Solving. In: 25th International Conference on Logic Programming (ICLP 2009). LNCS, vol. 5649. Springer, Heidelberg (2009)
4. Balduccini, M.: Representing Constraint Satisfaction Problems in Answer Set Programming. In: ICLP 2009 Workshop on Answer Set Programming and Other Computing Paradigms, ASPOCP 2009 (2009)
5. Balduccini, M., Gelfond, M.: Logic Programs with Consistency-Restoring Rules. In: Doherty, P., McCarthy, J., Williams, M.A. (eds.) International Symposium on Logical Formalization of Commonsense Reasoning, March 2003. AAAI 2003 Spring Symposium Series, pp. 9–18 (2003)

408   M. Balduccini

6. Smith, B.M.: 11. Modelling. Foundations of Artificial Intelligence. In: Handbook of Constraint Programming, pp. 377–406. Elsevier, Amsterdam (2006)
7. Katriel, I., van Hoeve, W.J.: 6. Global Constraints. Foundations of Artificial Intelligence. In: Handbook of Constraint Programming, pp. 169–208. Elsevier, Amsterdam (2006)
8. Balduccini, M.: CR-MODELS: An Inference Engine for CR-Prolog. In: Baral, C., Brewka, G., Schlipf, J. (eds.) LPNMR 2007. LNCS (LNAI), vol. 4483, pp. 18–30. Springer, Heidelberg (2007)

Modeling Multi-agent Domains in an Action Languages: An Empirical Study Using $\mathcal{C}$

Chitta Baral[1], Tran Cao Son[2], and Enrico Pontelli[2]

[1] Dept. Computer Science & Engineering, Arizona State University
chitta@asu.edu
[2] Dept. Computer Science, New Mexico State Univ.
{tson,epontell}@cs.nmsu.edu

Abstract. In the last two decades there has been a lot of research on action languages and reasoning about actions. Most of this research assume a domain with a single agent and possibly the environment. In this short paper we explore the relevance of this research vis-a-vis modeling multi-agent domains. We use the action language $\mathcal{C}$ and show that with minimal extensions it can capture several multi-agent domains from the literature.

1 Introduction and Motivation

There is now a large body of research on multi-agent systems. A lot of it is independent of research in reasoning about actions and action languages where the modeling is usually about a single agent in an environment. Often various papers in multi-agent systems introduce novel languages (with its syntax and semantics) that capture specific examples introduced in that paper. In this short paper we explore how far the action languages that were developed for a single agent (in an environment scenario) can be used and adapted for multi-agent domains. Our starting point is a well-studied and well-understood single agent action language—the language $\mathcal{C}$ [2]. We choose this language because it provides a number of features that appear necessary to handle multi-agent domains, such as concurrent interacting actions. The language is used to formalize a number of examples drawn from the multi-agent literature, describing different types of problems that can arise when dealing with multiple agents. Whenever necessary, we identify weaknesses of $\mathcal{C}$ and introduce simple extensions that are adequate to model these domains. The resulting language can be used as a foundation for different forms of reasoning in multi-agent domains (e.g., projection, validation of plans), which are formalized in the form of a query language.

Before we continue, let us discuss the desired features and the assumptions that we place on the target multi-agent systems. In this paper, we consider MAS domains as environments in which multiple agents can execute actions to modify the overall state of the world. We assume that: (1) Agents can execute actions concurrently; (2) Each agent knows its own capabilities; (3) Actions executed by different agents can interact; (4) Agents can communicate to exchange knowledge; and (5) Knowledge can be private to an agent or shared among groups of agents. The questions that we are interested in answering in a MAS domain involve both *hypothetical reasoning*, e.g., what happens if

E. Erdem, F. Lin, and T. Schaub (Eds.): LPNMR 2009, LNCS 5753, pp. 409–415, 2009.

agent A executes the action a; what happens if A executes a_1 and B executes b_1 at the same time; etc, and *planning/capability*, e.g., can a specified group of agents achieves a certain goal from a given state of the world. Variations of the above types of questions will also be considered. For example, what happens if the agents do not have complete information, if the agents do not cooperate, if the agents have preferences, etc.

2   Action Language $\mathcal{C}$ and Using It for Multi-agent Domains

The starting point of our investigation is the action language $\mathcal{C}$ [2]—an action description language originally developed to describe single agent domains, where the agent is capable of performing non-deterministic and concurrent actions. A domain description in $\mathcal{C}$ builds on a language signature $\langle \mathcal{F}, \mathcal{A} \rangle$, where $\mathcal{F}$ is a finite collection of fluent names and $\mathcal{A}$ is a finite collection of action names. Both the elements of $\mathcal{F}$ and $\mathcal{A}$ are viewed as propositional variables, and they can be used in formulae constructed using the traditional propositional operators. A propositional formula over $\mathcal{F} \cup \mathcal{A}$ is referred to simply as a *formula*, while a propositional formula over $\mathcal{F}$ is referred to as a *state formula*. A fluent literal is of the form f or $\neg f$ for any $f \in \mathcal{F}$.

A domain description D in $\mathcal{C}$ is a finite collection of axioms of the following forms:

> **caused** ℓ **if** F (*static causal axiom*)
> **caused** ℓ **if** F **after** G (*dynamic causal axiom*)

where ℓ is a fluent literal, F is a state formula, while G is a formula. The language also allows the ability to declare properties of fluents; in particular **non_inertial** ℓ declares that the fluent literal ℓ is to be treated as a non-inertial literal. A problem specification is obtained by adding an initial state description $\mathcal{I}$ to a domain D, composed of axioms of the form **initially** ℓ, where ℓ is a fluent literal.

$\mathcal{C}$ supports a query language (called $\mathcal{P}$ in [2]). This language allows queries of the form **necessarily** F **after** $A_1, \ldots, A_k$, where F is a state formula and $A_1, \ldots, A_k$ are sets of actions (called a *plan*). Intuitively, the query asks whether each state s reached after executing $A_1, \ldots, A_k$ from the initial state $\mathcal{I}$ has the property $s \models F$. This is denoted by $(D, \mathcal{I}) \models$ **necessarily** F **after** $A_1, \ldots, A_k$.

We now discuss some small modifications of $\mathcal{C}$ necessary to enable modeling MAS domains, through sample MAS problems drawn from the MAS literature. We will describe each domain from the perspective of someone (the modeler) who has knowledge of everything, including the capabilities and knowledge of each agent. Note that this is *only a modeling perspective*—it does not mean that we expect agents to have knowledge of everything, we only expect the *modeler* to have such knowledge.

We associate to each agent an element of a set of *agent identifiers*, $\mathcal{AG}$. We will describe a MAS domain over a set of signatures $\langle \mathcal{F}_i, \mathcal{A}_i \rangle$ for each $i \in \mathcal{AG}$, with the assumption that $\mathcal{A}_i \cap \mathcal{A}_j = \emptyset$ for $i \neq j$. Observe that $\bigcap_{i \in S} \mathcal{F}_i$ may be not empty for some $S \subseteq \mathcal{AG}$. This represents common knowledge between the agents in the group S of agents. The result is a $\mathcal{C}$ domain over the signature $\langle \bigcup_{i=1}^{n} \mathcal{F}_i, \bigcup_{i=1}^{n} \mathcal{A}_i \rangle$. We will require the following condition to be met: if **caused** ℓ **if** F **after** G is an axiom and $a \in \mathcal{A}_i$ appears in G, then the literal ℓ belongs to $\mathcal{F}_i$.

The Prison Domain. [5] In this example, we have two prison guards 1 and 2 who control two gates, the inner gate and the outer gate, by operating the four buttons a_1, b_1, a_2, and b_2. Agent 1 controls a_1 and b_1, while agent 2 controls a_2 and b_2. If either a_1 or a_2 is pressed, then the state of the inner gate is toggled. The outer gate, on the other hand, toggles if both b_1 and b_2 are pressed. In $\mathcal{C}$, this domain can be represented as follows. The set of agents is $\mathcal{AG} = \{1, 2\}$. For agent 1, we have:

$$\mathcal{F}_1 = \{in\_open, out\_open, pressed(a_1), pressed(b_1)\}.$$

Here, in_open and out_open represent the fact that the inner gate and outer gate are open respectively. $pressed(X)$ says that the button X where $X \in \{a_1, b_1\}$ is pressed. We have $\mathcal{A}_1 = \{push(a_1), push(b_1)\}$. This indicates that guard 1 can push buttons a_1 and b_1. Similarly, for agent 2, we have $\mathcal{F}_2 = \{in\_open, out\_open, pressed(a_2), pressed(b_2)\}$ and $\mathcal{A}_2 = \{push(a_2), push(b_2)\}$. We assume that the buttons do not stay pressed—thus, $pressed(X)$, for $X \in \{a_1, b_1, a_2, b_2\}$, is a non-inertial fluent with the default value *false*.

The axioms in the domain specification (D_{prison}) are:

> **non_inertial** $\neg pressed(X)$
> **caused** $pressed(X)$ **after** $push(X)$
> **caused** in_open **if** $pressed(a_1) \wedge \neg in\_open$
> **caused** $\neg in\_open$ **if** $pressed(a_1) \wedge in\_open$
> **caused** out_open **if** $pressed(b_1) \wedge pressed(b_2) \wedge \neg out\_open$
> **caused** $\neg out\_open$ **if** $pressed(b_1) \wedge pressed(b_2) \wedge out\_open$

where $X \in \{a_1, b_1, a_2, b_2\}$. The first axiom declares that $pressed(X)$ is non-inertial and has *false* as its default value. The second axiom describes the effect of the action $push(X)$. The rest is a collection of static law axioms representing the relationships between the fluents in $\mathcal{F}_1 \cup \mathcal{F}_2$.

Let us now consider queries from [5] and see how they can be answered using D_{prison}. In the first situation, both gates are closed, 1 presses a_1 and b_1, and 2 presses b_2. The question is whether the gates are open or not after the execution of these actions The initial situation is specified by the initial state description $\mathcal{I}_1$ containing

$$\mathcal{I}_1 = \{ \ \textbf{initially} \ \neg in\_open, \quad \textbf{initially} \ \neg out\_open \ \}$$

In this situation, there is only one initial state $s_0 = \{\neg\ell \mid \ell \in \mathcal{F}_1 \cup \mathcal{F}_2\}$. We can show that $(D_{prison}, \mathcal{I}_1) \models$ **necessarily** $out\_open \wedge in\_open$ **after** $\{push(a_1), push(b_1), push(b_2)\}$. If the outer gate is initially closed, i.e., $\mathcal{I}_2 = \{ \textbf{initially} \ \neg out\_open\}$, then the set of actions $A = \{push(b_1), push(b_2)\}$ is both necessary and sufficient to open it, i.e., the following queries hold: **necessarily** out_open **after** X and **necessarily** $\neg out\_open$ **after** Y where $A \subseteq X$ and $A \backslash Y \neq \emptyset$.

Adding Priority between Actions. Let us now present a small extension of $\mathcal{C}$ that allows for the encoding of competitive behavior between agents in MAS. For each domain specification D, we assume the presence of a function $Pr_D : 2^{\mathcal{A}} \rightarrow 2^{\mathcal{A}}$ with $Pr_D(A)$ denotes the set of actions whose effects will be accounted for when A is executed.

In the rocket domain [8], we have a rocket, a cargo, and the agents 1, 2, and 3. The rocket or the cargo are either in *london* or *paris*. The rocket can be moved by 1 and 2 between the two locations. The cargo can be loaded (unloaded) into the rocket by 1 and 3 (2 and 3). Agent 3 can refill the rocket if the tank is not full.

We will use the fluents $r(london)$ and $r(paris)$ to denote the location of the rocket. Likewise, $c(london)$ and $c(paris)$ denote the location of the cargo. in_rocket says that the cargo is inside the rocket and $tank\_full$ states that the tank is full. The signatures for the agents can be defined as follows.

$$\mathcal{F}_1 = \{\, in\_rocket, r(london), r(paris), c(london), c(paris) \,\}$$
$$\mathcal{F}_2 = \{\, in\_rocket, r(london), r(paris), c(london), c(paris) \,\}$$
$$\mathcal{F}_3 = \{\, in\_rocket, r(london), r(paris), c(london), c(paris), tank\_full \,\}$$

and $\mathcal{A}_2 = \{unload(2), move(2)\}$, $\mathcal{A}_3 = \{load(3), refill\}$, and $\mathcal{A}_1 = \{load(1), unload(1), move(1)\}$. This domain has a special feature—there are priorities among the actions. It states that $load$ or $unload$ will have no effect if $move$ is executed. The effects of two $load$ actions is the same as that of a single $load$ action. Likewise, two $unload$ actions have the same result as one $unload$ action. We define Pr_D as follows:

- $Pr_D(X) = \{move(a)\}$ if $\exists a.\, move(a) \in X$.
- $Pr_D(X) = \{load(a)\}$ if $move(x) \notin X$ for every $x \in \{1, 2, 3\}$ and $load(a) \in X$.
- $Pr_D(X) = \{unload(a)\}$ if $move(x) \notin X$ and $load(x) \notin X$ for every $x \in \{1, 2, 3\}$ and $unload(a) \in X$.
- $Pr_D(X) = X$ otherwise.

The domain specification D_{rocket} includes, among the others, and the following axioms:

caused in_rocket **after** $load(i)$	$(i \in \{1, 3\})$
caused $tank\_full$ **if** $\neg tank\_full$ **after** $refill$	
caused $\neg tank\_full$ **if** $tank\_full$ **after** $move(i)$	$(i \in \{1, 2\})$
caused $r(london)$ **if** $r(paris) \wedge tank\_full$ **after** $move(i)$	$(i \in \{1, 2\})$
caused $c(paris)$ **if** $r(paris) \wedge in\_rocket$	

Let $\mathcal{I}_4$ consist of the following axioms: **initially** $tank\_full$, **initially** $r(paris)$, **initially** $c(london)$, and **initially** $\neg in\_rocket$. Then the following query holds:

$$(D_{rocket}, \mathcal{I}_4) \models \textbf{necessarily } c(paris) \textbf{ after } \{move(1)\}, \{load(3)\}, \{refill\}, \{move(3)\}.$$

Adding Reward Strategies. The next example illustrates the need to handle numbers and optimization to represent reward mechanisms. The extension of $\mathcal{C}$ is simply the introduction of *numerical fluents*—i.e., fluents that, instead of being simply true or false, have a numerical value. For this purpose, we introduce a new variant of the necessity query

$$\textbf{necessarily max } F \textbf{ for } \varphi \textbf{ after } A_1, \ldots, A_n$$

where F is a numerical expressions involving only numerical fluents, φ is a state formula, and $A_1, \ldots, A_n$ is a plan. Given a domain specification D and an initial state description $\mathcal{I}$, we can define for each fluent numerical expression F and plan α:

$$value(F, \alpha) = \max\{s(F) \mid s \in \Phi^*(\alpha, s_0), s_0 \text{ is an initial state w.r.t. } \mathcal{I}, D\}$$

where $s(F)$ is the value of the expression F in state s and $\Phi^*(\alpha, s_0)$ are the states reached from the execution of the sequence of actions α. This allows us to define the following notion of entailment: **necessarily max** F **for** φ **after** $A_1, \ldots, A_n$ holds if:

- **necessarily** φ **after** $A_1, \ldots, A_n$ holds, and
- for every other plan $B_1, \ldots, B_m$ such that $(D, \mathcal{I}) \models$ **necessarily** φ **after** $B_1, \ldots,$ B_m we have that $value(F, [A_1, \ldots, A_n]) \geq value(F, [B_1, \ldots, B_m])$.

The following example has been derived from [1]. There are three agents. Agent 0 is a normative system that can play one of two strategies—either st_0 or $\neg st_0$. Agent 1 plays a strategy st_1, while agent 2 plays the strategy st_2. The reward system is described in the following tables (the first is for st_0 and the second one is for $\neg st_0$).

st_0	st_1	$\neg st_1$
st_2	1, 1	0, 0
$\neg st_2$	0, 0	−1, −1

$\neg st_0$	st_1	$\neg st_1$
st_2	1, 1	0, 0
$\neg st_2$	0, 0	1, 1

The signatures used by the agents are: $\mathcal{F}_0 = \{st_0, reward\}$, $\mathcal{F}_1 = \{st_1, reward_1\}$, $\mathcal{F}_2 = \{st_2, reward_2\}$, $\mathcal{A}_0 = \{play\_0, play\_not\_0\}$, $\mathcal{A}_1 = \{play\_1, play\_not\_1\}$, and $\mathcal{A}_2 = \{play\_2, play\_not\_2\}$. The domain specification D_{gam} consists of:

caused st_0 **after** $play\_0$ **caused** $\neg st_0$ **after** $play\_not\_0$
caused st_1 **after** $play\_1$ **caused** $\neg st_1$ **after** $play\_not\_1$
caused st_2 **after** $play\_2$ **caused** $\neg st_2$ **after** $play\_not\_2$
caused $reward\_1 = 1$ **if** $\neg st_0 \wedge st_1 \wedge st_2$ **caused** $reward\_2 = 1$ **if** $\neg st_0 \wedge st_1 \wedge st_2$
caused $reward\_1 = 0$ **if** $\neg st_0 \wedge st_1 \wedge \neg st_2$ **caused** $reward\_2 = 0$ **if** $\neg st_0 \wedge st_1 \wedge \neg st_2$
$\ldots$

caused $reward = a + b$ **if** $reward_1 = a \wedge reward_2 = b$

Assuming that $\mathcal{I} = \{$ **initially** $st_0 \}$ we can show that the following query holds:
$$(D_{gam}, \mathcal{I}) \models \text{ \textbf{necessarily max} } reward \text{ \textbf{after} } \{play_1, play_2\}.$$

3 Reasoning and Properties

We discuss various types of reasoning that are directly enabled by the semantics of $\mathcal{C}$.

Let us start exploring queries aimed at capturing the capabilities of agents. We will use the generic form **can** X **do** φ, where φ is a state formula and $X \subseteq \mathcal{AG}$ where $\mathcal{AG}$ is the set of agent identifiers of the domain. The intuition is to validate whether the group of agents X can guarantee that φ is satisfied.

If $X = \mathcal{AG}$ then the semantics of the capability query is simply expressed as $(D, \mathcal{I}) \models$ **can** X **do** φ iff $\exists k. \exists A_1, \ldots, A_k$ such that

$$(D, \mathcal{I}) \models \text{ \textbf{necessarily} } \varphi \text{ \textbf{after} } A_1, \ldots, A_k.$$

If $X \neq \{1, \ldots, n\}$, then we can envision different variants of this query.

The **Capability queries with non-interference and complete knowledge** are used to verify whether the agents in X can achieve φ when operating in an environment that includes *all* the agents, but the agents $\mathcal{AG} \setminus X$ are simply providing their knowledge and not performing actions or interfering. We will denote this type of queries as

$\mathbf{can}_g^n \ X \ \mathbf{do} \ \varphi$ (n: not interference, g: availability of all knowledge). The semantics of this type of queries can be formalized as follows: $(D, \mathcal{I}) \models \mathbf{can}_g^n \ X \ \mathbf{do} \ \varphi$ if there is a sequence of sets of actions $A_1, \ldots, A_k$ with the following properties: for each $1 \leq i \leq k$ we have that $A_i \subseteq \bigcup_{j \in X} \mathcal{A}_j$ (we perform only actions of agents in X), and $(D, \mathcal{I}) \models \mathbf{necessarily} \ \varphi \ \mathbf{after} \ A_1, \ldots, A_k$.

The **Capability queries with non-interference and projected knowledge** assume that not only the other agents ($\mathcal{AG} \setminus X$) are passive, but they also are not willing to provide knowledge to the active agents. We will denote this type of queries as $\mathbf{can}_l^n \ X \ \mathbf{do} \ \varphi$.

Let us refer to the *projection* of $\mathcal{I}$ w.r.t. X (denoted by $proj(\mathcal{I}, X)$) as the set of all the **initially** declarations that build on fluents of $\bigcup_{j \in X} \mathcal{F}_j$. The semantics of $\mathbf{can}_l^n$ type of queries can be formalized as follows: $(D, \mathcal{I}) \models \mathbf{can}_l^n \ X \ \mathbf{do} \ \varphi$ if there is a sequence of sets of actions $A_1, \ldots, A_k$ such that: for each $1 \leq i \leq k$ we have that $A_i \subseteq \bigcup_{j \in X} \mathcal{A}_j$, and $(D, proj(\mathcal{I}, X)) \models \mathbf{necessarily} \ \varphi \ \mathbf{after} \ A_1, \ldots, A_k$ (i.e., the objective will be reached irrespective of the initial configuration of the other agents).

The **Capability queries with interference** take into account the possible interference from other agents in the system. Intuitively, the query with interference, denoted by $\mathbf{can}^i \ X \ \mathbf{do} \ \varphi$, implies that the agents X will be able to accomplish X in spite of other actions performed by the other agents.

The semantics is as follows: $(D, \mathcal{I}) \models \mathbf{can}^i \ X \ \mathbf{do} \ \varphi$ if there is a sequence of sets of actions $A_1, \ldots, A_k$ such that: for each $1 \leq i \leq k$ we have that $A_i \subseteq \bigcup_{j \in X} \mathcal{A}_j$, and for each sequence of sets of actions $B_1, \ldots, B_k$, where $\bigcup_{j=1}^k B_j \subseteq \bigcup_{j \notin X} \mathcal{A}_j$, we have that $(D, \mathcal{I}) \models \mathbf{necessarily} \ \varphi \ \mathbf{after} \ (A_1 \cup B_1), \ldots, (A_k \cup B_k)$.

It is possible to design additional classes of queries to explore properties of a theory such as

- **Agent redundancy:** agent redundancy is a property of $(D, \mathcal{I})$ which indicates the ability to remove an agent to accomplish a goal.
- **Agent Necessity:** agent necessity is symmetrical to redundancy—it denotes the inability to accomplish a property φ if an agent is excluded.
- **Compositionality:** The formalization of multi-agent systems in $\mathcal{C}$ enables exploring the effects of composing domains; this is an important property, that allows us to model dynamic MAS systems (e.g., where new agents can join an existing coalition).

4 Conclusion

In this paper, we presented an investigation of the use of the $\mathcal{C}$ action language to model MAS domains. We discussed several interesting features that are necessary for modeling MAS, and showed how such features can be encoded in $\mathcal{C}$—either directly or with simple extensions of the action language. We mentioned several forms of reasoning that are naturally supported by the proposed language.

However, existing research in action languages reach their limits with respect the multi-agency when one needs to represent and reason about agents knowledge about other agents knowledge. An example of this is the *muddy children problem*. We explore this in detail in a sequel. Additional research directions include, adapting the more advanced forms of reasoning and implementation proposed for $\mathcal{C}$ to the case of MAS domains and investigating the use of the proposed extension of $\mathcal{C}$ in formalizing distributed systems.

References

1. Boella, G., van der Torre, L.: Enforceable social laws. In: AAMAS 2005, pp. 682–689. ACM, New York (2005)
2. Gelfond, M., Lifschitz, V.: Action languages. ETAI 3(6) (1998)
3. Gerbrandy, J.: Logics of propositional control. In: AAMAS 2006, pp. 193–200. ACM, New York (2006)
4. Herzig, A., Troquard, N.: Knowing how to play: uniform choices in logics of agency. In: AAMAS 2006, pp. 209–216 (2006)
5. Sauro, L., Gerbrandy, J., van der Hoek, W., Wooldridge, M.: Reasoning about action and cooperation. In: AAMAS 2006, pp. 185–192. ACM, New York (2006)
6. Son, T.C., Baral, C., Tran, N., McIlraith, S.: Domain-dependent knowledge in answer set planning. ACM Trans. Comput. Logic 7(4), 613–657 (2006)
7. Spaan, M., Gordon, G.J., Vlassis, N.A.: Decentralized planning under uncertainty for teams of communicating agents. In: AAMAS 2006, pp. 249–256 (2006)
8. van der Hoek, W., Jamroga, W., Wooldridge, M.: A logic for strategic reasoning, pp. 157–164. ACM, New York (2005)

Computing Weighted Solutions
in Answer Set Programming

Duygu Çakmak, Esra Erdem, and Halit Erdoğan

Faculty of Engineering and Natural Sciences, Sabancı University, Istanbul, Turkey

Abstract. For some problems with many solutions, like planning and phylogeny reconstruction, one way to compute more desirable solutions is to assign weights to solutions, and then pick the ones whose weights are over (resp. below) a threshold. This paper studies computing weighted solutions to such problems in Answer Set Programming. We investigate two sorts of methods for computing weighted solutions: one suggests modifying the representation of the problem and the other suggests modifying the search procedure of the answer set solver. We show the applicability and the effectiveness of these methods in phylogeny reconstruction.

1 Introduction

In Answer Set Programming (ASP) [1,8], a computational problem is described as an ASP program whose answer sets correspond to solutions, and answer sets for this program are computed using answer set solvers. Some problems, like planning and phylogeny reconstruction, have many solutions. Moreover, the correspondence between the answer sets and the solutions may not be one-to-one; there may be many answer sets that denote the same solution. For such problems, one way to compute more desirable solutions is to assign weights to solutions, and then pick the distinct solutions whose weights are over (resp. below) a threshold. For example, in a planning problem, we can define the weight of a plan in terms of the costs of actions (or action sequences), and then compute the distinct plans whose weights are less than a given value. In puzzle generation, we can define the weight of a puzzle instance by means of some difficulty measure, and then generate difficult puzzles whose weights are over a given value. Motivated by such applications, we study the problem of computing weighted solutions in ASP and show the applicability of our approach in phylogeny reconstruction (i.e., computing leaf-labeled trees, called phylogenies, to model the evolutionary history of a set of species).

We study two sorts of methods for computing weighted solutions: the representation-based methods and the search-based methods. In the former, the idea is to modify the ASP representation of the problem, to compute weighted solutions. In particular, we are interested in elaboration tolerant representations, where the weight of a solution is defined as an ASP program and added to the ASP representation of the problem. The latter, on the other hand, do not modify the ASP representation of the problem, but define the weight function externally (e.g., as a C++ program) and modify the search algorithm of the answer set solver to compute solutions over (resp. below) a given threshold. In this paper, we introduce such a search-based method for computing weighted solutions, and implement it by modifying the search algorithm of the answer set solver CLASP [7].

E. Erdem, F. Lin, and T. Schaub (Eds.): LPNMR 2009, LNCS 5753, pp. 416–422, 2009.

We apply these methods to phylogeny reconstruction. Reconstructing phylogenies for a given set of taxonomic units is important for various research such as historical linguistics, zoology, anthropology, archeology, etc.. For example, a phylogeny of parasites may help zoologists to understand the evolution of human diseases [4]; a phylogeny of languages may help scientists to better understand human migrations [11]. In this study, we define the weight of a phylogeny for the family of Indo-European languages studied in [2], in such a way as to reflect its plausibility and importance. Using this weight function, we show the applicability and effectiveness of the methods above (for computing weighted solutions) in reconstructing plausible phylogenies for Indo-European languages. No existing phylogenetic system has such utilities for experts to compute more plausible phylogenies, so our methods provide a useful tool for phylogenetics. Likewise, our methods provide a useful tool for various other applications of ASP.

2 Computing Weighted Solutions

We are interested in the following sorts of computational problems for computing weighted solutions:

> AT LEAST (resp. AT MOST) w-WEIGHTED SOLUTION: Given an ASP program $\mathcal{P}$ that formulates a computational problem P, a weight measure ω that maps a solution for P to a nonnegative integer, and a nonnegative integer w, decide whether a solution S exists for P such that $w(S) \geq w$ (resp. $w(S) \leq w$).

For instance, suppose that $\mathcal{P}$ describes the phylogeny reconstruction problem for Indo-European languages, and that ω describes the total weight of the characters compatible with the to-be-reconstructed phylogeny and takes into account some domain-specific information. Then finding phylogenies whose weights are at least 45 is an instance of the problem above.

We study two sorts of methods, representation-based and search-based, to compute at least/most w-weighted solutions in ASP.

The idea behind the representation-based methods is to modify the representation of the problem, to compute weighted solutions. For an elaboration tolerant representation $\mathcal{P}$, such modifications are done by means of adding some rules $\mathcal{W}$ describing the weight of a solution and some constraints $\mathcal{C}$ on the weights of the solutions (Fig. 1). Then we can compute at least (resp. most) w-weighted solutions by computing answer sets for the ASP program $\mathcal{P} \cup \mathcal{W} \cup \mathcal{C}$. In some cases, we do not have to define the weight of a solution explicitly; we can use aggregates (e.g., sum, count, times) to compute the weight, in the sense of [9,6,10]. Some other problems require an explicit definition of the weight of a solution. Phylogeny reconstruction problems we consider are in the latter group: the weight of a phylogeny does not only depend on the weights of some parts of the phylogeny but also some domain-specific information.

Search-based methods (as outlined in Fig. 2) on the other hand do not modify the ASP representation of the problem, but define the weight function externally (e.g., as a C++ program) and modify the search algorithm of the answer set solver to compute solutions over (resp. below) a given threshold. There is no answer set solver that

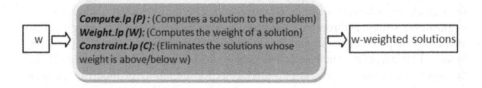

Fig. 1. Computing at most/least w-weighted solutions with representation-based method

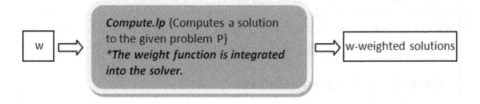

Fig. 2. Computing at most/least w-weighted solutions with search-based method

can compute weighted solutions in such a way. Therefore, in our studies, we consider the solver CLASP [7], and modify its search algorithm to implement this search-based method. We call the new algorithm as CLASP-W. CLASP-W is shown in Algorithm 1: the parts in red denote the modifications we have made over CLASP's algorithm (i.e., if we remove the red parts, then we get CLASP's algorithm). Note that, compared to CLASP, CLASP-W has a new function called WEIGHT-ANALYZE. At each step of the search, CLASP has a partial solution and tries to complete it to find a solution to the given problem. The function WEIGHT-ANALYZE computes an upper bound (resp. lower bound) for the weight of a completion of the partial solution, so that CLASP-W does not perform redundant search towards a complete solution. WEIGHT-ANALYZE function is domain-specific; therefore, in order to use CLASP-W, we need to implement this function (in a separate file) according to the given weight measure for the particular problem. We do not need to modify CLASP-W for different problems.

3   Computing Weighted Phylogenies for Indo-European Languages

The evolutionary relations between species (or "taxonomic unit") based on their shared traits can be modeled as a phylogeny (or a phylogenetic tree). The problem of phylogeny reconstruction asks for "plausible" phylogenies for a given set of taxonomic units. There have been various studies to compute plausible phylogenies (see [2] for a discussion). In the following, we will consider a character-based cladistics with respect to the compatibility criterion, as in [2]. Our goal is to find a phylogeny with a small number of incompatible characters. The problem of reconstructing a phylogeny with at most k incompatible characters (let us call this problem as k-CP) is NP-hard [5].

Algorithm 1. CLASP-W

Input: An ASP program Π and a nonnegative integer w
Output: An answer set for Π, that describes an at least (resp. at most) w-weighted solution
  $A \leftarrow \emptyset$ // current assignment of literals
  $\triangledown \leftarrow \emptyset$ // set of conflicts
  while A does not represent an answer set **do**
    // propagate according to the current assignment and conflicts;update the current assignment
    NOGOOD-PROPAGATION($\Pi, A, \triangledown$)
    // compute an upper (resp. lower) bound for the weight of a solution that contains A
    $weight \leftarrow$ WEIGHT-ANALYZE(A)
    // if the upper bound $weight$ is less than the desired weight value w
    // then no need to continue search to find an at least w-weighted solution
    if There is a conflict in unit-propagation OR $weight < w$ **then**
      RESOLVE-CONFLICT($\Pi, A, \triangledown$)   // learn and update the conflict set and do backtracking
    end if
    if Current assignment does not yield an answer set **then**
      SELECT($\Pi, A, \triangledown$)   // select a literal to continue search
    else
      return A
    end if
  end while
  return false

While reconstructing phylogenies, some characters may give more information than the others. For instance, to model the evolutionary history of a family of languages, morphological/phonological characters are more informative than lexical characters. In order to emphasize the role of such characters in reconstructing a phylogeny, we define the concept of a weighted phylogeny.

A *weighted phylogeny* is a phylogeny along with a weight function Φ that maps every character $i \in I$ to a nonnegative integer. The *weight of a phylogeny* can be defined in various ways with respect to Φ; in the following, we consider the weight of a phylogeny as the sum of the weights of all characters that are compatible with that phylogeny.

With such a weight measure and an ASP program describing phylogeny reconstruction (like the one in [2,3]), we can compute weighted phylogenies for the family of Indo-European languages described in [2], using the representation-based method or the search-based method described in Section 2.

To get more plausible phylogenies, we also incorporate further domain-specific information in the weight measure. It is told us by historical linguist Don Ringe that it is least likely that Greco-Armenian languages be siblings with Balto-Slavic languages. Similarly, but not as least likely as Greco-Armenian and Balto-Slavic, is the grouping of Greco-Armenian with Germanic languages. If the to-be-reconstructed phylogenies have such odd groupings of languages, we reduce some amount from the total weight of the phylogeny making sure that the weight of a phylogeny is not negative.

4   Experimental Results

We applied the computational methods described above (i.e., the representation-based method, and the search-based method) to reconstruct weighted phylogenies for Indo-European languages, as described in the previous section, with the dataset and the ASP program used in [2].

Let us consider computing an at least w-weighted phylogeny with at most c incompatible characters. In Table 1, for each problem, for each method, we present the computation time (CPU seconds), the size of the ground program (the number of atoms, and the number of rules), and the size of the memory (MB) used in computation.[1] For instance, let us consider computing a phylogeny with at most 17 incompatible characters, and whose weight is at least 45. With the representation-method, CLASP takes 15.32 CPU sec.s to compute such a phylogeny; the ground program has 79229 atoms and 1585419 rules; the computation of the phylogeny consumes 369 MB of memory. On the other hand, with the search-based method, CLASP-W takes 1.30 CPU sec.s to compute such a phylogeny; the ground program has 3744 atoms and 55219 rules; the computation of the phylogeny consumes 22 MB of memory.

Observe in Table 1 that in terms of both computation time and the memory used, the search-based method performs better than the representation-method. These results conforms with our expectations. The representation-based method explicitly defines the weight function, and thus the program/memory size is larger. The search-based method deals with the time consuming computation of weights of phylogenies, not at the representation level but at the search level, so it does not require an ASP representation of the weight function but requires a modification of the solver to guide it find a plausible phylogeny, and hence the smaller the program/memory size and the computation time.

Table 1. The representation-based method vs. the search-based method: computing a phylogeny with at most c incompatible characters, and whose weight is at least w.

# of incompatible characters (c)	weight (w)	method	time (CPU sec.s)	program size	memory size (MB)
16	45	representation-based	15.52	# of atoms: 79229 # of rules: 1585419	369
		search-based	1.34	# of atoms: 3744 # of rules: 55219	22
17	45	representation-based	15.32	# of atoms: 79229 # of rules: 1585419	369
		search-based	1.30	# of atoms: 3744 # of rules: 55219	22
18	45	representation-based	15.47	# of atoms: 79229 # of rules: 1585419	369
		search-based	1.10	# of atoms: 3744 # of rules: 55219	22

---

[1] All CPU times are in seconds, for a workstation with a 1.5GHz Xeon processor and 4x512MB RAM, running Red Hat Enterprise Linux (Version 4.3).

In [2], after computing all 45 phylogenies, the authors examine them manually, and identify 14 of them as plausible and with at most 18 incompatible characters. With the weight measure defined in Section 3, and the representation/search-based methods described above for computing weighted phylogenies, we could automatically compute all plausible phylogenies with at most 18 incompatible characters in 22.35 CPU seconds.

5 Discussion

We studied the problem of computing weighted solutions in ASP, where the weight of a solution should be defined explicitly. We introduced a search-based method that implements the weight of a solution as a C++ program, and modifies the search algorithm of the ASP solver CLASP to compute weighted solutions with respect to that weight program. We call the modified version of CLASP as CLASP-W.

We showed the applicability and effectiveness of the search-based method in reconstructing phylogenies for Indo-European languages, where the weight of a phylogeny takes into account domain-specific information to characterize the plausibility of phylogenies. In particular, by computing at least w-weighted phylogenies, we could compute more plausible and less number of phylogenies for Indo-European languages. We observed that the search-based method (with CLASP-W) is better than the representation-based method (with CLASP) in terms of computation time and space.

Since no existing phylogenetic system can compute weighted phylogenies, our search-based methods (including the weight measure and the solver CLASP-W) provide a useful tool for experts to automatically analyze phylogenies online. There are many appealing ASP applications (e.g., product configuration, planning) for which finding weighted solutions could be useful; in this sense, our methods are useful also for ASP.

Acknowledgments

Thanks to Martin Gebser and Benjamin Kaufmann for their help with CLASP. This work has been supported by TUBITAK Grant 107E229.

References

1. Baral, C.: Knowledge Representation, Reasoning and Declarative Problem Solving. Cambridge University Press, Cambridge (2003)
2. Brooks, D.R., Erdem, E., Erdogan, S.T., Minett, J.W., Ringe, D.: Inferring phylogenetic trees using answer set programming. JAR 39(4), 471–511 (2007)
3. Brooks, D.R., Erdem, E., Minett, J.W., Ringe, D.: Character-based cladistics and answer set programming. In: Hermenegildo, M.V., Cabeza, D. (eds.) PADL 2004. LNCS, vol. 3350, pp. 37–51. Springer, Heidelberg (2005)
4. Brooks, D.R., McLennan, D.A.: Phylogeny, Ecology, and Behavior: A Research Program in Comparative Biology. University of Chicago Press, Chicago (1991)
5. Day, W.H.E., Sankoff, D.: Computational complexity of inferring phylogenies by compatibility. Systematic Zoology 35(2), 224–229 (1986)

6. Faber, W., Pfeifer, G., Leone, N., Dell'Armi, T., Ielpa, G.: Design and implementation of aggregate functions in the dlv system. TPLP 8(5-6), 545–580 (2008)
7. Gebser, M., Kaufmann, B., Neumann, A., Torsten Schaub, T.: Conflict-driven answer set solving. In: Proc. of IJCAI, pp. 386–392. MIT Press, Cambridge (2007)
8. Lifschitz, V.: What is answer set programming? In: Proc. of AAAI (2008)
9. Simons, P., Soininen, T.: Stable model semantics of weight constraint rules. In: Gelfond, M., Leone, N., Pfeifer, G. (eds.) LPNMR 1999. LNCS (LNAI), vol. 1730, pp. 317–331. Springer, Heidelberg (1999)
10. Son, T.C., Pontelli, E.: A constructive semantic characterization of aggregates in answer set programming. TPLP 7(3), 355–375 (2007)
11. White, J.P., O'Connell, J.F.: A Prehistory of Australia, New Guinea, and Sahul. Academic, San Diego (1982)

Representing Multi-agent Planning in CLP

Agostino Dovier[1], Andrea Formisano[2], and Enrico Pontelli[3]

[1] Università di Udine
dovier@dimi.uniud.it
[2] Università di Perugia
formis@dipmat.unipg.it
[3] New Mexico State University
epontell@cs.nmsu.edu

Abstract. We explore the use of Constraint Logic Programming (CLP) as a platform for experimenting with planning domains in presence of multiple interacting agents. We develop a novel *constraint-based* action language, $\mathcal{B}^{\text{MAP}}$, that enables the declarative description of large classes of multi-agent and multi-valued domains. $\mathcal{B}^{\text{MAP}}$ supports several complex features, including combined effects, concurrency constraint, interacting actions, and delayed effects.

1 Introduction

Representing and programming intelligent and cooperating agents that are able to *acquire, represent,* and *reason* with knowledge is a challenging problem for which an extensive literature exists, presenting different languages for the description of planning domains (see, e.g., [11, 10, 8]) in the context of single-agent domains. Logic programming has been extensively used in this context, and *Answer Set Programming* [1] has been one of the paradigms of choice.

Constraint Logic Programming over Finite Domains has been recently shown to be another viable paradigm for reasoning about actions and change (e.g., [17, 18, 22, 6]). [6] makes a strong case for the use of constraint programming, demonstrating the flexibility of constraints in modeling several extensions of action languages, necessary to address real-world planning domains. Our goal is to develop a high-level description language to represent domains with multiple interacting agents, along with other properties relevant for representing complex domains (e.g., multi-valued fluents and constraints). Each agent can have different capabilities (it can perform different actions). The actions of the agents can also be *cooperative*—i.e., their cumulative effects are required to apply a change to the world—or *conflicting*—i.e., some actions may exclude other actions from being executed. Each agent maintains its own view of the world, but groups of agents may share knowledge of certain features of the world (through shared fluents).

The starting point of our project is represented by the design of a novel action language, named $\mathcal{B}^{\text{MAP}}$, for encoding multi-agent action domains. $\mathcal{B}^{\text{MAP}}$ adopts the perspective, shared by many other researchers (e.g., [3, 16, 20]), of viewing a multi-agent system from a *centralized* perspective, where a centralized

E. Erdem, F. Lin, and T. Schaub (Eds.): LPNMR 2009, LNCS 5753, pp. 423–429, 2009.

description defines the modifications to the world derived from the agents' action executions (even though the individual agents may not be aware of that). This perspective is important to investigate properties of a domain (e.g., existence of solutions, plan validation) and it represents the underlying semantics for the development of efficient centralized or distributed planning algorithms.

We have implemented a Prolog tool that maps a $\mathcal{B}^{MAP}$ action description into a constraint satisfaction problem and supports the process of validating the existence of plans to solve planning problems. Due to lack of space, we omit here the precise semantics of $\mathcal{B}^{MAP}$ and the g implementation details (see [7]).

Related work. Various authors have explored the use of logic programming, such as normal logic programs and abductive logic programs, to address cooperation between agents (e.g., [15, 19, 9]). To the best of our knowledge, the use of CLP technology for *modeling* multi-agent domains is novel. On the other hand, the use of constraints in modeling single-agent domains has been successfully validated by several authors. CLP has been used to support the development of the *fluent calculus* [22]. In [2] the authors analyze different encodings of sequential planning as CSP. Another interesting work on using constraints to guide planning in domain with actions with durations has been proposed in [23].

CLP has also been used to implement the centralized store of distributed programming platforms (e.g., OCP [13]). Some aspects of concurrency have been formalized and addressed also in the context of existing action languages (e.g., $\mathcal{C}, \mathcal{C}+, \mathcal{CARD}$ [10, 12, 5]) and in the area of multi-agent planning (e.g., [4, 3]). A recent effort along similar lines as ours, modeling an action language for multi-agent systems, has been proposed by Son and Sakama [21], relying on the use of answer set programming (with consistency restore rules).

2   Syntax of the Language $\mathcal{B}^{MAP}$

The signature of the $\mathcal{B}^{MAP}$ language consists of the following sets:

- $\mathcal{G}$: *agent* names, used to identify the agents participating in the domain.
- $\mathcal{F}$: *fluent* names; we assume that $\mathcal{F} = \bigcup_{a \in \mathcal{G}} \mathcal{F}_a$, where $\mathcal{F}_a$ are the fluents used to describe the knowledge of agent a. We assume that $a \neq a' \rightarrow \mathcal{F}_a \cap \mathcal{F}_{a'} = \emptyset$.
- $\mathcal{A}$: *action* names.
- $\mathcal{V}$: values for the fluents in $\mathcal{F}$. In the following, we assume $\mathcal{V} = \mathbb{Z}.$[1]

Agents and Fluents. An assertion (*agent declaration*) of the type `agent(a)`, where $a \in \mathcal{G}$, states the existence of the agent named a. The fluents that can be used by agent a to describe its own knowledge, are described by axioms: `fluent`$(a, f, \{v_1, \ldots, v_k\})$ with $a \in \mathcal{G}$ and $f \in \mathcal{F}_a$. This statement also determines the set of admissible values for f, i.e., $\{v_1, \ldots, v_k\} \subseteq \mathcal{V}.$[2]

---

[1] We will use a, b for agent names, f, g for fluent names, and x, y for action names.
[2] We admit the notation `fluent`(a, f, v_1, v_2) for `fluent`$(a, f, \{v_1, v_1 + 1, \ldots, v_2\})$.

Fluents can be used in *Fluent Expressions* (FE), which are defined inductively as follows (where $n \in \mathcal{V}$, $t \in \mathbb{Z}$, $\oplus \in \{+, -, *, /, \mathrm{mod}\}$, $f \in \mathcal{F}$, and $r \in \mathbb{N}$):

$$\mathrm{FE} ::= n \mid f^t \mid f \, @ \, r \mid \mathrm{FE}_1 \oplus \mathrm{FE}_2 \mid -(\mathrm{FE}) \mid \mathrm{abs}(\mathrm{FE}) \mid \mathrm{rei}(C)$$

Fluent expressions are evaluated with respect to an "history" of evolution of the world. Given an integer number t, an expression of the form f^t is an *annotated fluent expression*. Intuitively, for $t < 0$ ($t > 0$), the expression refers to the value f had $-t$ steps in the past (will have t steps in the future). f is a shorthand for f^0, the current value. Annotated expressions refer to points in time, *relatively* to the current state. The ability to create formulae that refer to different time points along the evolution of the world enables the encoding of non-Markovian processes. An expression of the form $f \, @ \, r$ denotes the value f has at the rth step in the evolution of the world (i.e., it refers to an *absolutely* specified point in time). The last alternative (reified expression) requires the notion of fluent constraint C (defined next). The intuitive semantics is that an expression $\mathrm{rei}(C)$ assumes a Boolean value depending on the truth of C.

A *primitive fluent constraints* (PC) is a formula $\mathrm{FE}_1 \ \mathbf{op} \ \mathrm{FE}_2$, where FE_1 and FE_2 are fluent expressions, and $\mathbf{op} \in \{=, \neq, \geq, \leq, >, <\}$ is a relational operator. *Fluent constraints* are propositional combinations of primitive constraints:

$$\mathrm{PC} ::= \mathrm{FE}_1 \ \mathbf{op} \ \mathrm{FE}_2 \qquad\qquad C ::= \mathrm{PC} \mid \neg C \mid C_1 \wedge C_2 \mid C_1 \vee C_2$$

To represent shared knowledge between agents, we assume the existence of an equivalence relation $\equiv_{\mathcal{F}} \subseteq \mathcal{F} \times \mathcal{F}$. Given two agents a and b, if $\mathcal{F}_a \ni f \equiv_{\mathcal{F}} f' \in \mathcal{F}_b$, then the two fluents f and f' represent the same property of the world.

Actions Description. An axiom of the form $\mathrm{action}(Ag, x)$, where $Ag \subseteq \mathcal{G}$ and $x \in \mathcal{A}$, declares that x is meant to be executed collectively by the set of agents Ag. The action is said to be *individual* if $|Ag| = 1$, *collective* if $|Ag| > 1$ and *exogenous* if $|Ag| = 0$.

For each axiom $\mathrm{action}(Ag, x)$, we introduce the expression $\mathrm{actocc}(Ag, x)$, called *action flag*, to denote the execution of that action. Action flags are intended to be Boolean-valued expressions, which will be used in combination with fluent constraints. *Action-fluent expressions* (AFE) extend the structure of fluent expressions by allowing propositions related to action occurrences:

$$\mathrm{AFE} ::= n \mid f^t \mid f \, @ \, r \mid \mathrm{actocc}(Ag, x)^t \mid \mathrm{actocc}(Ag, x) \, @ \, r \mid$$
$$\mathrm{AFE}_1 \oplus \mathrm{AFE}_2 \mid -(\mathrm{AFE}) \mid \mathrm{abs}(\mathrm{AFE}) \mid \mathrm{rei}(C)$$

where $n \in \mathcal{V}$, $t \in \mathbb{Z}$, $r \in \mathbb{N}$, $f \in \mathcal{F}$, $x \in \mathcal{A}$, $Ag \subseteq \mathcal{G}$, and $\oplus \in \{+, -, *, /, \mathrm{mod}\}$. Time-annotated action-fluent expressions allow us to refer to the occurrences of actions at any point during the evolution of the world. Action-fluent-expressions can be used to form *action-fluent constraints*, as done for fluent constraints.

An axiom of the form: $\mathrm{executable}(Ag, x, C)$, where $Ag \subseteq \mathcal{G}$, $x \in \mathcal{A}$, and C is an action-fluent constraint, states that C has to be entailed by the current state for x to be executable by the set of agents Ag. We assume that an executability axiom is present for each pair Ag, x such that $\mathrm{action}(Ag, x)$ is defined.

An axiom of the form `causes(Eff, Prec)` encode the effects of dynamic causal laws. `Prec` is an action-fluent constraint, called the *precondition constraint*, that implies the truth (w.r.t. the current state) of at least one action flag. `Eff` is a fluent constraint, called the *effect constraint*. The axiom asserts that if `Prec` is `true`, then `Eff` must hold in the next state. If `Prec` implies the conjunction of two (or more) action flags, then the effect refers to a *compound action*. Thus, a compound action is a concomitant execution of independently modeled actions.

Static causal laws and other State Constraints. Static causal laws can be expressed by axioms of the form `caused(`C_1, C_2`)`, stating that the action-fluent constraint $C_1 \rightarrow C_2$ must be entailed in any state encountered. Specific classes of static causal laws are commonly encountered. Due to lack of space, we introduce syntactic sugar for only on one of them. The specification of dynamic causal laws allows us to deal with effects derived from the concurrent execution of actions. Similarly, we may encounter situations where certain actions cannot be executed concurrently by different agents. This constraint can be enforced using the notions of action-fluent expressions and constraints. The axiom `concurrency_control(`C`)` states that the action-fluent constraint C must hold. It must contain at least two action flags. It is a syntactic sugar for the static causal law `caused(true, `C`)`. For instance, the fact that two agents can walk through a revolving door only one at the time can be modeled by

$$\texttt{concurrency\_control(actocc(}\{a\}, \texttt{walk\_through)} +$$
$$\texttt{actocc(}\{b\}, \texttt{walk\_through)} \leq 1).$$

Costs. In $\mathcal{B}^{\text{MAP}}$ it is possible to specify information about the *cost* of each action and about the global cost of a plan. In particular:

- `action_cost(`Ag, x, Val`)`, where $Ag \subseteq \mathcal{G}$, $x \in \mathcal{A}$, and Val specifies the cost of executing the action described by the axiom `action(`Ag, x`)` (otherwise, a default cost of 1 is assigned).
- `state_cost(`FE`)` specifies the cost of a generic state as the result of the evaluation of the fluent expression FE, built using the fluents present in the state (otherwise, a default cost of 1 is assumed).

Action Domains. An *action domain description* $\mathcal{D}$ is a collection of axioms of the forms described earlier. A specific instance of a planning problem is a tuple $\langle \mathcal{D}, \mathcal{I}, \mathcal{O} \rangle$, where $\mathcal{D}$ is an action domain description, $\mathcal{I}$ is a collection of axioms of the form `initially(`C`)` (describing the initial state of the world), and $\mathcal{O}$ is a collection of axioms of the form `goal(`C`)`, where C is a fluent constraint.

3   Experiments

An interpreter of the language $\mathcal{B}^{\text{MAP}}$, realized in SICStus Prolog, is available at `www.dimi.uniud.it/dovier/CLPASP/MAP` along with some planning domains. Once the action description is translated into a constraint satisfaction problem, solution's search starts following some heuristics. There is an overall "leftmost"

Table 1. Timing is in seconds. The symbol '–' denotes no answer within 1 hour.

Single-agent	Plan length	leftmost	ffc	ffcd	Multi-agent	Plan length	leftmost	ffc	ffcd
Three Barrels	11	0.12	0.11	0.07	Bob&Mary	5	0.01	0.01	0.01
Goat-Wolf etc.	23	0.14	0.04	0.28	Social Game	2	0.04	0.04	0.06
Gas diffusion	6	34.9	34.9	9.65	Dining phil.	9	339	439	–
15-Puzzle	15	62.0	64.7	4.55	Fuel&Cars	10	736	743	0.48
Peg Solitaire	31	–	–	44.7	Robots&Table	7	316	461	118

strategy that implements progression, i.e., we first considers the transition $\langle s_0, t_1, s_1 \rangle$ from the initial state s_0, then the transition $\langle s_1, t_2, s_2 \rangle$, etc. The programmer can choose the labeling strategy to be locally applied in these steps. The main options are `leftmost`, `ff` (first-fail), `ffc` (first-fail with a choice on the most constrained variable), or `ffcd` (which combines `ffc` with a downward selection of values for constrained variables). In most experiments, the latter strategy gives the best performance.

We run the system on some classical single-agent domains, such as the three barrels (12-7-5), a *Sam Lloyd's* puzzle, the goat-cabbage-wolf problem, *the gas diffusion problem* [6], and the peg-solitaire—csplib 037, also in the 2008 planning competition. Note that the peg-solitaire problem is solved in [14] in 388s, using a complex encoding that uses operations research techniques. We solve it in less than 45 seconds with a simple $\mathcal{B}^{\text{MAP}}$ encoding.

We have also tested the $\mathcal{B}^{\text{MAP}}$ implementation on the suite of peg-solitaire instances used in IPC08 for the "sequential satisficing track". The competition imposed these restrictions: the plan has to be produced within 30 minutes, by using at most 2GB of memory. The suite is composed of 30 problems. The $\mathcal{B}^{\text{MAP}}$ planner found the optimal plan for 24 problems.

Finally, we have tested the interpreter on some inherently concurrent domains, such as the dining philosophers—with the traditional rules, and when one eat he will be alive for 10 seconds; we seek a plan that ensures that all philosophers to be alive at a certain time—a problem of cars and fuels (EATCS bulletin N. 89, page 183—by Laurent Rosaz—tested with four cars), a *social-game* invented by us (that required 6 actions if solved by one agent), and two problems described in previous works on concurrency and knowledge representation. The first problem (Bob and Mary) is adapted from the working example of [4] and is related to the need of cooperation for opening a door. We have modeled it either using collective actions or using compound actions (for opening the door). The second problem is instead adapted from the working example of [3] and it is related to two agents that can use a table to carry sets of blocks from one room to another.

4 Conclusions

We presented a constraint-based action description language, $\mathcal{B}^{\text{MAP}}$, that extends the previously proposed language $\mathcal{B}^{MV}$ [6]. The new language retains all the features of $\mathcal{B}^{MV}$, e.g., multi-valued fluents and the possibility of referring to fluents in any different state of the trajectory. The major novelty of $\mathcal{B}^{MAP}$ consists of allowing declarative formalization of agent domains in presence of multiple interacting agents. Each agent can have a different (partial) view of the world and a different collection of executable actions. Moreover, preconditions, as well as effects, of the actions it performs, might interact with those performed by other agents. Concurrency and cooperation are easily modeled by means of static and dynamic causal laws, that might involve constraints referring to action occurrences, even performed by different agents in different points in time.

References

[1] Baral, C.: Knowledge representation, reasoning and declarative problem solving. Cambridge University Press, Cambridge (2003)
[2] Barták, R., Toropila, D.: Reformulating Constraint Models for Classical Planning. In: Proc. of FLAIRS 2008, pp. 525–530. AAAI Press, Menlo Park (2008)
[3] Boutilier, C., Brafman, R.: Partial order planning with concurrent interacting actions. J. of Artif. Intel. Research 14, 105–136 (2001)
[4] Brenner, M.: Planning for Multiagent Environments: From individual perceptions to coordinated execution. In: Proc. of ICAPS 2005 Workshop on Multiagent Planning and Scheduling (2005)
[5] Chintabathina, S., Gelfond, M., Watson, R.: Defeasible laws, parallel actions, and reasoning about resources. In: Proc. of CommonSense 2007. AAAI Press, Menlo Park (2007)
[6] Dovier, A., Formisano, A., Pontelli, E.: Multivalued Action Languages with Constraints in CLP(FD). In: Dahl, V., Niemelä, I. (eds.) ICLP 2007. LNCS, vol. 4670, pp. 255–270. Springer, Heidelberg (2007)
[7] Dovier, A., Formisano, A., Pontelli, E.: Multi-Agent Planning in CLP. In: Proc. of CILC 2009 (2009)
[8] Eiter, T., et al.: Answer Set Planning Under Action Costs. J. of Artif. Intel. Research 19, 25–71 (2003)
[9] Gelfond, G., Watson, R.: Modeling Cooperative Multi-Agent Systems. In: Proc. of ASP 2007 (2007)
[10] Gelfond, M., Lifschitz, V.: Action languages. Electr. Trans. on AI 2, 193–210 (1998)
[11] Gerevini, A., Long, D.: Plan Constraints and Preferences in PDDL3. Technical Report 2005-08-47, Dip. di Elettronica per l'Automazione, Univ. di Brescia (2005)
[12] Giunchiglia, E., et al.: Nonmonotonic Causal Theories. Artif. Intell. 153 (2004)
[13] Jaffar, J., Yap, R.H.C., Zhu, K.Q.: Coordination of Many Agents. In: Gabbrielli, M., Gupta, G. (eds.) ICLP 2005. LNCS, vol. 3668, pp. 98–112. Springer, Heidelberg (2005)
[14] Jefferson, C., et al.: Modelling and solving English Peg solitaire. Comput. Oper. Res. 33(10), 2935–2959 (2006)
[15] Kakas, A., Torroni, P., Demetriou, N.: Agent Planning, Negotiation and Control of Operation. In: Proc. of ECAI 2004. IOS Press, Amsterdam (2004)
[16] Knoblock, C.: Generating Parallel Execution Plans with a Partial-Order Planner. In: Proc. of AIPS 1994, pp. 98–103. AAAI Press, Menlo Park (1994)
[17] Lopez, A., Bacchus, F.: Generalizing graphplan by formulating planning as a CSP. In: Proc. of IJCAI 2003. Morgan Kaufmann, San Francisco (2003)

[18] Reiter, R.: Knowledge in Action: Logical Foundations for Describing and Implementing Dynamical Systems. MIT Press, Cambridge (2001)

[19] Sadri, F., Toni, F.: Abductive Logic Programming for Communication and Negotiation Amongst Agents. ALP Newsletter (2003)

[20] Sauro, L., Gerbrandy, J., van der Hoek, W., Wooldridge, M.: Reasoning about Action and Cooperation. In: Proc. of AAMAS 2006. ACM, New York (2006)

[21] Son, T.C., Sakama, C.: Reasoning and Planning with Cooperative Actions for Multiagents Using Answer Set Programming. In: Proc. of DALT 2009. LNCS (LNAI). Springer, Heidelberg (2009)

[22] Thielscher, M.: Programming of Reasoning and Planning Agents with FLUX. In: Proc. of KR 2002. Morgan Kaufmann, San Francisco (2002)

[23] Vidal, V., Geffner, H.: Branching and Pruning: An Optimal Temporal POCL Planner Based on Constraint Programming. Artif. Intell. 170 (2006)

Prototypical Reasoning with Low Complexity Description Logics: Preliminary Results

Laura Giordano[1], Valentina Gliozzi[2], Nicola Olivetti[3], and Gian Luca Pozzato[2]

[1] Dip. di Informatica - Univ. Piemonte O. "A. Avogadro"
laura@mfn.unipmn.it
[2] Dip. di Informatica - Università di Torino
{gliozzi,pozzato}@di.unito.it
[3] LSIS-UMR CNRS 6168 Univ. "P. Cézanne"
nicola.olivetti@univ-cezanne.fr

Abstract. We present an extension $\mathcal{EL}^{+^\perp}\mathbf{T}$ of the description logic $\mathcal{EL}^{+^\perp}$ for reasoning about prototypical properties and inheritance with exceptions. $\mathcal{EL}^{+^\perp}\mathbf{T}$ is obtained by adding to $\mathcal{EL}^{+^\perp}$ a typicality operator $\mathbf{T}$, which is intended to select the "typical" instances of a concept. In $\mathcal{EL}^{+^\perp}\mathbf{T}$ knowledge bases may contain inclusions of the form "$\mathbf{T}(C)$ is subsumed by P", expressing that typical C-members have the property P. We show that the problem of entailment in $\mathcal{EL}^{+^\perp}\mathbf{T}$ is in CO-NP.

1 Introduction

In Description Logics (DLs) the need of representing prototypical properties and of reasoning about defeasible inheritance of such properties naturally arises. The traditional approach is to handle defeasible inheritance by integrating some kind of nonmonotonic reasoning mechanism. This has led to study nonmonotonic extensions of DLs [2,3,4,5,6,12]. However, finding a suitable nonmonotonic extension for inheritance with exceptions is far from obvious.

In this work we introduce a defeasible extension of the description logic $\mathcal{EL}^{+^\perp}$ called $\mathcal{EL}^{+^\perp}\mathbf{T}$, continuing the investigation started in [7], where we extended the logic $\mathcal{ALC}$ with a typicality operator $\mathbf{T}$. The intended meaning of the operator $\mathbf{T}$ is that, for any concept C, $\mathbf{T}(C)$ singles out the instances of C that are considered as "typical" or "normal". Thus assertions as "typical football players love football" are represented by $\mathbf{T}(FootballPlayer) \sqsubseteq FootballLover$. The semantics of the typicality operator $\mathbf{T}$ turns out to be strongly related to the semantics of nonmonotonic entailment in KLM logic $\mathbf{P}$ [11].

In our setting, we assume that the TBox element of a KB comprises, in addition to the standard concept inclusions, a set of inclusions of the type $\mathbf{T}(C) \sqsubseteq D$ where D is a concept not mentioning $\mathbf{T}$. For instance, a KB may contain: $\mathbf{T}(Dog) \sqsubseteq Affectionate$; $\mathbf{T}(Dog) \sqsubseteq CarriedByTrain$; $\mathbf{T}(Dog \sqcap PitBull) \sqsubseteq NotCarriedByTrain$; $CarriedByTrain \sqcap NotCarriedByTrain \sqsubseteq \perp$, corresponding to the assertions: typically dogs are affectionate, normally dogs can be transported by train, whereas typically a dog belonging to the race of pitbull cannot

E. Erdem, F. Lin, and T. Schaub (Eds.): LPNMR 2009, LNCS 5753, pp. 430–436, 2009.

(since pitbulls are considered as reactive dogs); the fourth inclusion represents the disjointness of the two concepts *CarriedByTrain* and *NotCarriedByTrain*. Notice that, in standard DLs, replacing the second and the third inclusion with *Dog* $\sqsubseteq$ *CarriedByTrain* and *Dog* $\sqcap$ *PitBull* $\sqsubseteq$ *NotCarriedByTrain*, respectively, we would simply get that there are not pitbull dogs, thus the KB would collapse. This collapse is avoided as we do not assume that $\mathbf{T}$ is monotonic, that is to say $C \sqsubseteq D$ does not imply $\mathbf{T}(C) \sqsubseteq \mathbf{T}(D)$.

By the properties of $\mathbf{T}$, some inclusions are entailed by the above KB, as for instance $\mathbf{T}(Dog \sqcap CarriedByTrain) \sqsubseteq Affectionate$. In our setting we can also use the $\mathbf{T}$ operator to state that some domain elements are typical instances of a given concept. For instance, an ABox may contain either $\mathbf{T}(Dog)(fido)$ or $\mathbf{T}(Dog \sqcap PitBull)(fido)$. In the two cases, the expected conclusions are entailed: *CarriedByTrain(fido)* and *NotCarriedByTrain(fido)*, respectively.

In this work, we present some preliminary results on *low complexity* Description Logics extended with the typicality operator $\mathbf{T}$. In particular we focus on the logic $\mathcal{EL}^{+^{\perp}}$ of the well known $\mathcal{EL}$ family. The logics of the $\mathcal{EL}$ family allow for conjunction ($\sqcap$) and existential restriction ($\exists R.C$). Despite their relatively low expressivity, a renewed interest has recently emerged for these logics. Indeed, theoretical results have shown that $\mathcal{EL}$ has better algorithmic properties than its counterpart $\mathcal{FL}_0$, which allows for conjunction and value restriction ($\forall R.C$). Also, it has turned out that the logics of the $\mathcal{EL}$ family are relevant for several applications, in particular in the bio-medical domain; for instance, medical terminologies, such as GALEN, SNOMED, and the Gene Ontology used in bioinformatics, can be formalized in small extensions of $\mathcal{EL}$.

We present some results about the complexity of $\mathcal{EL}^{+^{\perp}} \mathbf{T}$. We show that, given an $\mathcal{EL}^{+^{\perp}} \mathbf{T}$ KB, if it is satisfible, then there is a *small* model whose size is polynomial in the size of KB. The construction of the model exploits the facts that (1) it is possible to reuse the same domain element (instance of a concept C) to fulfill existential formulas $\exists r.C$ w.r.t. domain elements; (2) we can restrict our attention to a class of models in which the preference relation $<$ is multi-linear and polynomial, that is it determines a set of disjoint chains of elements of polynomial length. The construction of the model allows us to conclude that the problem of deciding entailment in $\mathcal{EL}^{+^{\perp}} \mathbf{T}$ is in CO-NP.

Technical details and proofs can be found in the accompanying report [10].

2   The Logic $\mathcal{EL}^{+^{\perp}} \mathbf{T}$

We consider an alphabet of concept names $\mathcal{C}$, of role names $\mathcal{R}$, and of individuals $\mathcal{O}$. The language $\mathcal{L}$ of the logic $\mathcal{EL}^{+^{\perp}} \mathbf{T}$ is defined by distinguishing *concepts* and *extended concepts* as follows: (Concepts) $A \in \mathcal{C}$, $\top$, and $\perp$ are *concepts* of $\mathcal{L}$; if $C, D \in \mathcal{L}$ and $r \in \mathcal{R}$, then $C \sqcap D$ and $\exists r.C$ are *concepts* of $\mathcal{L}$. (Extended concepts) if C is a concept, then C and $\mathbf{T}(C)$ are extended concepts of $\mathcal{L}$. A knowledge base is a pair (TBox,ABox). TBox contains (i) a finite set of GCIs $C \sqsubseteq D$, where C is an extended concept (either C' or $\mathbf{T}(C')$), and D is a concept, and (ii) a finite set of role inclusions (RIs) $r_1 \circ r_2 \circ \cdots \circ r_n \sqsubseteq r$. ABox contains expressions of the form $C(a)$ and $r(a, b)$ where C is an extended concept, $r \in \mathcal{R}$, and $a, b \in \mathcal{O}$. In order to provide a semantics to the operator $\mathbf{T}$, we extend the definition of a model used in "standard" terminological logic $\mathcal{EL}^{+^\perp}$:

Definition 1 (Semantics of T). *A model $\mathcal{M}$ is any structure $\langle \Delta, <, I \rangle$, where Δ is the domain; $<$ is an irreflexive and transitive relation over Δ, and satisfies the following* Smoothness Condition*: for all $S \subseteq \Delta$, for all $a \in S$, either $a \in Min_<(S)$ or $\exists b \in Min_<(S)$ such that $b < a$, where $Min_<(S) = \{a : a \in S$ and $\nexists b \in S$ s.t. $b < a\}$. I is the extension function that maps each extended concept C to $C^I \subseteq \Delta$, and each role r to a $r^I \subseteq \Delta^I \times \Delta^I$. For concepts of $\mathcal{EL}^{+^\perp}$, C^I is defined in the usual way. For the $\mathbf{T}$ operator: $(\mathbf{T}(C))^I = Min_<(C^I)$. A model satisfying a KB (TBox,ABox) is defined as usual. Moreover, we assume the unique name assumption.*

Notice that the meaning of $\mathbf{T}$ can be split into two parts: for any a of the domain Δ, $a \in (\mathbf{T}(C))^I$ just in case (i) $a \in C^I$, and (ii) there is no $b \in C^I$ such that $b < a$. In order to isolate the second part of the meaning of $\mathbf{T}$, we introduce a new modality $\square$. The basic idea is simply to interpret the preference relation $<$ as an accessibility relation. By the Smoothness Condition, it turns out that $\square$ has the properties as in Gödel-Löb modal logic of provability G. The interpretation of $\square$ in $\mathcal{M}$ is as follows: $(\square C)^I = \{a \in \Delta \mid$ for every $b \in \Delta$, if $b < a$ then $b \in C^I\}$. We have that a is a typical instance of C ($a \in (\mathbf{T}(C))^I$) iff $a \in C^I$ and, for all $b < a$, $b \notin C^I$, namely we have that $a \in (\mathbf{T}(C))^I$ iff $a \in (C \sqcap \square \neg C)^I$. From now on, we consider $\mathbf{T}(C)$ as an abbreviation for $C \sqcap \square \neg C$. The Smoothness Condition ensures that typical elements of C^I exist whenever $C^I \neq \emptyset$, by preventing infinitely descending chains of elements.

3   Complexity of $\mathcal{EL}^{+^\perp}\mathbf{T}$

In order to give a complexity upper bound for the logic $\mathcal{EL}^{+^\perp}\mathbf{T}$, we show that, given a model $\mathcal{M} = \langle \Delta, <, I \rangle$ of a KB, we can build a *small* model of KB whose size is polynomial in the size of the KB.

Theorem 1 (Small model theorem). *Let KB=(TBox,ABox) be an $\mathcal{EL}^{+^\perp}\mathbf{T}$ knowledge base. For all models $\mathcal{M} = \langle \Delta, <, I \rangle$ of KB and all $x \in \Delta$, there exists a model $\mathcal{N} = \langle \Delta^\circ, <^\circ, I^\circ \rangle$ of KB such that (i) $x \in \Delta^\circ$, (ii) for all $\mathcal{EL}^{+^\perp}\mathbf{T}$ concepts C, $x \in C^I$ iff $x \in C^{I^\circ}$, and (iii) $\mid \Delta^\circ \mid$ is polynomial in the size of KB.*

Due to space limitations, here we only give a sketch of the proof, whose details can be found in [10]. The construction comprises three steps.

(step A) First of all, in order to reduce the size of the model, we cut a portion of it that includes x. We build a model $\mathcal{M}'$ by means of the following construction. For each atomic concept $C \in \mathcal{C}$ and for each role $r \in \mathcal{R}$ we let $S(C)$ and $R(r)$ be the mappings computed by the algorithm defined in [1] to compute subsumption by means of completion rules. As usual, for a given individual a in the ABox, we write a^I to denote the element of Δ corresponding to the extension of a in $\mathcal{M}$. We make use of three sets of elements: Δ_0 will be part of the domain of the model being constructed, and it contains a portion of the domain Δ of the initial model. All elements introduced in the domain must be processed in order to satisfy the existential formulas. *Unres* is used to keep track of not yet processed elements. Finally, Δ_1 is a set of elements that will belong to the domain of the constructed model. Each element w_C of Δ_1 is created for a corresponding atomic concept C and is used to satisfy any existential formula $\exists r.C$ throughout the model. In the following by w_C we mean the domain element of Δ_1 which is added for the atomic concept C. We provide an algorithmic description of the construction of model $\mathcal{M}'$ from the given model $\mathcal{M}$. Observe that $\mathcal{M}$ can be an infinite model.

1. $\Delta_0 := \{x\} \cup \{a^I \in \Delta \mid a$ occurs in the ABox $\}$
2. *Unres*$:=\{x\} \cup \{a^I \in \Delta \mid a$ occurs in the ABox $\}$
3. $\Delta_1:=\emptyset$
4. **while** *Unres* $\neq \emptyset$ **do**
5.   extract one y from *Unres*
6.   **for each** $\exists r.C$ occurring in KB s.t. $y \in (\exists r.C)^I$ **do**
7.   **if** $\nexists w_C \in \Delta_1$ **then**
8.   choose $w \in \Delta$ s.t. $(y,w) \in r^I$ and $w \in C^I$
9. $\Delta_0 := \Delta_0 \cup \{w\}$
10.   *Unres*$:=$ *Unres* $\cup \{w\}$
11.   create a new element w_C associated with C
12. $\Delta_1 := \Delta_1 \cup \{w_C\}$
13.   add $w <' w_C$
14.   add (y, w_C) to $r^{I'}$
15.   **else**
16.   add (y, w_C) to $r^{I'}$
17.   **for each** $y_i \in \Delta$ such that $y_i < y$ **do**
18. $\Delta_0 := \Delta_0 \cup \{y_i\}$
19.   *Unres*$:=$ *Unres* $\cup\{y_i\}$
20. **for each** $w_C, w_D \in \Delta_1$ with $C \neq D$ **do**
21.   **if** $(C, D) \in R(r)$ **then** add (w_C, w_D) to $r^{I'}$

The model $\mathcal{M}' = \langle \Delta', <', I' \rangle$ is defined as follows:

- $\Delta' = \Delta_0 \cup \Delta_1$

- we extend $<'$ computed by the algorithm by adding $u <' v$ if $u < v$, for each $u, v \in \Delta'$;

- the extension function I' is defined as follows: • for all atomic concepts $C \in \mathcal{C}$, for all domain elements in Δ', we define: for each $u \in \Delta_0$, we let $u \in C^{I'}$ if $u \in C^I$; for each $w_D \in \Delta_1$, we let $w_D \in C^{I'}$ if $C \in S(D)$. • for all roles r, we extend $r^{I'}$ constructed by the algorithm by means of the following role closure rules: for all inclusions $r \sqsubseteq s \in$ TBox, if $(u,v) \in r^{I'}$ then add (u,v) to $s^{I'}$; for all inclusions $r_1 \circ r_2 \sqsubseteq s \in$ TBox, if $(u,v) \in r_1^{I'}$ and $(v,w) \in r_2^{I'}$ then add (u,w) to $s^{I'}$. • I' is extended so that it assigns a^I to each individual a in the ABox.

$\mathcal{M}'$ is not guaranteed to have polynomial size in the KB because in line 18 we add an element y_i for each $y_i < y$, then the size of Δ_0 may be arbitrarily large.

(step B) We refine our construction in order to obtain from $\mathcal{M}'$ a multi-linear model with a polynomial number of chains. Intuitively, a model is *multi-linear* if the relation $<$ forms a set of chains of domain elements, that is, for every u, v, z of the domain, we have that: (i) if $u < z$ and $v < z$ and $u \neq v$, then $u < v$ or $v < u$; (ii) if $z < u$ and $z < v$ and $u \neq v$, then $u < v$ or $v < u$. From $\mathcal{M}'$ we can obtain a multilinear model $\mathcal{M}''$ that preserves the interpretation of atomic concepts with respect to common elements of the domain and has a polynomial number of chains.

(step C) We finally construct a model $\mathcal{N}$ from $\mathcal{M}''$ whose domain has polynomial size in the size of KB. The idea is as follows. Let us consider a chain $w_0, w_1, w_2, \ldots$ in the multi-linear model. We can observe that, given w_i and w_j in the chain such that $w_i < w_j$, the set of negated box formulas $\neg\Box\neg C$ of which w_i is an instance is a subset of the set of negated box formulas of which w_j is an instance. We can thus shrink each chain by retaining only the elements w_i, w_j such that $w_i < w_j$ implies there exists a formula $\neg\Box\neg C$ such that w_j is an instance of $\neg\Box\neg C$ and w_i is not an instance of $\neg\Box\neg C$. As there is only a polynomial number of such box formulas $\neg\Box\neg C$, each chain will contain only a polynomial number of elements. Since the number of chains is polynomial in itself (by step B), the resulting model $\mathcal{N}$ has a polynomial size.

Given Theorem 1 above, when evaluating the entailment, we can restrict our consideration to small models, namely, to polynomial multi-linear models of the KB. We write KB $\models \alpha$ to say that a query α holds in all the models of the KB. A query α is either a formula of the form $C(a)$ or a subsumption relation $C \sqsubseteq D$. We write KB $\models_s \alpha$ to say that α holds in all polynomial multi-linear models of the KB. It holds that KB $\models \alpha$ if and only if KB $\models_s \alpha$. As a consequence, we can give an upper bound on the complexity of $\mathcal{EL}^{+^\perp}\mathbf{T}$:

Theorem 2. *In $\mathcal{EL}^{+^\perp}\mathbf{T}$, the problem of deciding whether KB $\models \alpha$ is in* CO-NP. *The problems of satisfiability of a KB and of concept satisfiability are in* NP. *The problems of subsumption and of instance checking are in* CO-NP.

4   Conclusions and Future Issues

We have presented the description logic $\mathcal{EL}^{+^\perp}\mathbf{T}$, that is $\mathcal{EL}^{+^\perp}$ extended by a tipicality operator $\mathbf{T}$ intended to select the "most normal" instances of a concept. Whereas for $\mathcal{ALC} + \mathbf{T}$ deciding satisfiability (subsumption) is EXPTIME complete (see [9]), we have shown here that for $\mathcal{EL}^{+^\perp}\mathbf{T}$ the complexity is significantly smaller, namely it reduces to NP for satisfiability (and CO-NP for subsumption). This result is obtained by a "small" model property (of a particular kind: multi-linear) that fails for the whole $\mathcal{ALC} + \mathbf{T}$ as well as for $\mathcal{ALC}$. We believe that this bound is also a lower bound, but we have not proved it so far. Although validity/satisfiability for KLM logic $\mathbf{P}$ is known to be (co)NP hard, in

$\mathcal{EL}^{+^\perp}\mathbf{T}$, we can only directly encode nonmonotonic assertions $A \mathrel{\vert\!\sim} B$ where A is a conjunction of atoms and B is either an atom or $\perp$. As far as we know, the complexity of this fragment of $\mathbf{P}$ is unknown. Thus a lower bound for $\mathcal{EL}^{+^\perp}\mathbf{T}$ cannot be obtained from known results about KLM logic $\mathbf{P}$.

The logic $\mathcal{EL}^{+^\perp}\mathbf{T}$ in itself is not sufficient for prototypical reasoning and inheritance with exceptions, in particular we need a stronger (nonmonotonic) mechanism to cope with the problem known as *irrelevance*. Concerning the example of the Introduction, we would like to conclude that typical red dogs are affectionate, since the color of a dog is irrelevant with respect to the property of being affectionate. However, as the property of being red is not a property neither of all dogs, nor of typical dogs, in $\mathcal{EL}^{+^\perp}\mathbf{T}$ we are not able to conclude $\mathbf{T}(Dog \sqcap Red) \sqsubseteq Affectionate$. One possibility is to consider a stronger (nonmonotonic) entailment relation $\mathcal{EL}^{+^\perp}\mathbf{T}_{min}$ determined by restricting the entailment of $\mathcal{EL}^{+^\perp}\mathbf{T}$ to "minimal models", as defined in [8] for $\mathcal{ALC} + \mathbf{T}$. Intuitively, minimal models are those that maximise "typical instances" of a concept. As shown in [8], for $\mathcal{ALC} + \mathbf{T}_{min}$, minimal entailment can be decided in $\mathrm{CO\text{-}NExP}^{\mathrm{NP}}$. We believe that for $\mathcal{EL}^{+^\perp}\mathbf{T}_{min}$ we can obtain a smaller complexity upper bound on the base of the results presented here.

Acknowledgements. The work was partially supported by Regione Piemonte, Project "ICT4Law - *ICT Converging on Law: Next Generation Services for Citizens, Enterprises, Public Administration and Policymakers*".

References

1. Baader, F., Brandt, S., Lutz, C.: Pushing the $\mathcal{EL}$ envelope. In: Proc. of IJCAI 2005, Professional Book Center, pp. 364–369 (2005)
2. Baader, F., Hollunder, B.: Embedding defaults into terminological knowledge representation formalisms. J. Autom. Reasoning 14(1), 149–180 (1995)
3. Baader, F., Hollunder, B.: Priorities on defaults with prerequisites, and their application in treating specificity in terminological default logic. J. of Automated Reasoning (JAR) 15(1), 41–68 (1995)
4. Bonatti, P.A., Lutz, C., Wolter, F.: Description logics with circumscription. In: Proc. of KR, pp. 400–410 (2006)
5. Donini, F.M., Nardi, D., Rosati, R.: Description logics of minimal knowledge and negation as failure. ACM Trans. Comput. Log. 3(2), 177–225 (2002)
6. Eiter, T., Lukasiewicz, T., Schindlauer, R., Tompits, H.: Combining answer set programming with description logics for the semantic web. In: KR 2004, pp. 141–151 (2004)
7. Giordano, L., Gliozzi, V., Olivetti, N., Pozzato, G.L.: Preferential Description Logics. In: Dershowitz, N., Voronkov, A. (eds.) LPAR 2007. LNCS (LNAI), vol. 4790, pp. 257–272. Springer, Heidelberg (2007)
8. Giordano, L., Gliozzi, V., Olivetti, N., Pozzato, G.L.: Reasoning About Typicality in Preferential Description Logics. In: Hölldobler, S., Lutz, C., Wansing, H. (eds.) JELIA 2008. LNCS (LNAI), vol. 5293, pp. 192–205. Springer, Heidelberg (2008)

9. Giordano, L., Gliozzi, V., Olivetti, N., Pozzato, G.L.: On Extending Description Logics for Reasoning About Typicality: a First Step. Technical Report 116/09, Dip. di Informatica, Univ. di Torino (2009)

10. Giordano, L., Gliozzi, V., Olivetti, N., Pozzato, G.L.: Reasoning About Typicality in Low Complexity Description Logics: Preliminary Results. Technical Report 121/09, Dip. di Informatica, Univ. di Torino (2009)

11. Kraus, S., Lehmann, D., Magidor, M.: Nonmonotonic reasoning, preferential models and cumulative logics. Artificial Intelligence 44(1-2), 167–207 (1990)

12. Straccia, U.: Default inheritance reasoning in hybrid kl-one-style logics. In: Proc. of IJCAI, pp. 676–681 (1993)

AQL : A Query Language for Action Domains Modelled Using Answer Set Programming

Luke Hopton, Owen Cliffe, Marina De Vos, and Julian Padget

Department of Computer Science
University of Bath, BATH BA2 7AY, UK
lch21@bath.ac.uk, {occ,mdv,jap}@cs.bath.ac.uk

Abstract. We present a new general purpose query and abduction language for reasoning about action domains that allows the processing of simultaneous actions, definition of conditions and reasoning about fluents and actions. AQL provides a simple declarative syntax for the specification of constraints on the histories (the combination of action traces and state transitions) within the modelled domain. Its semantics, provided by the translation of AQL queries into Ans-$Prolog$, acquires the benefits of the reasoning power provided by Answer Set Programming (ASP). The answer sets obtained from combining the query and the domain description correspond to those histories of the domain changing over time that satisfy the query. The result is a simple, high-level query and constraint language that builds on ASP. Through the synthesis of features it offers a more flexible, versatile and intuitive approach compared to existing languages. Due to the use of ASP, AQL can also be used to reason about partial histories.

1 Introduction

Action domains are a useful mechanism for modelling a variety of domains such as planning, protocol definition and normative frameworks. Given an action description we can use established computational techniques, such as Answer Set Programming (ASP) or SAT solvers to verify or examine model properties. It is desirable that such a system should allow designers to specify model properties with a high degree of flexibility while offering qualitative properties of succinctness and human readability.

Action languages[2, 5] are a way of formally describing the effects of actions on a domain using a subset of natural language. Transition systems are central to action languages: with every action (or the combined effects of simultaneous actions) the environment changes. Traditionally action languages have two distinct parts: (i) a description language to capture the effects of actions, thus defining the transition system, and (ii) a query language to write queries or reason about the transition system.

In this paper, we present a new action query and abduction language AQL whose semantics is provided by ASP [1], a logic programming language used for knowledge representation and reasoning. AQL can be used in two ways: for transition system path selection and for model-checking. AQL extends existing query language to allow for simultaneous actions and the definition of conditions that can then form part of more complex queries. Furthermore, AQL does not rely on the use of any particular action description language but can be used on top of a $AnsProlog$ description or in conjunction with any action language description that maps to $AnsProlog$.

E. Erdem, F. Lin, and T. Schaub (Eds.): LPNMR 2009, LNCS 5753, pp. 437–443, 2009.

2   Action Domains

The purpose of action domain [4, 5] modelling is to be able to observe the effect of actions on the environment. A light switch offers a simple example: in one state it is on, the action of flicking a switch changes the state to off, etc. State transitions systems provide an effective representation mechanism, in which each state is identified by a set of fluents that are true in that state. Anything not in the set is not true (for the model). Actions are modelled by transition to a new state, wherein some fluents are added and others removed with respect to the previous state. Here we only consider deterministic domains in which there is exactly one next state resulting from the cumulative effect of a sequence of (sets of) actions—known as a *trace*—on the current state to build a history: $s_0 \xrightarrow{a_0} s_1 \xrightarrow{\{a_1,a_2\}} \ldots \xrightarrow{a_m} s_n$, where states are denoted $s_i, i \in 0 \ldots n$ and actions $a_j, j \in 0 \ldots m$. Planning [7] is probably the best known example of the use of action domains and reasoning over such domains, where the final state is the goal and the trace is the sequence of actions to achieve that goal.

In this paper, we use answer set programming to represent and reason about action domains, their traces and histories. As a result, histories are output as answer sets. However, rather than modelling domains directly in ASP, we put forward the abstraction of an *action language*—with subsequent translation to $AnsProlog$ —because this enables the user to focus on domain specifics, rather than generic details—such as inertia—that are common to every domain. Several such action language have been developed, with general languages like A [5], C [6] and DLV-K [3] and numerous domain-specific ones.

3   AQL

AQL is a query language that can be used directly with an $AnsProlog$ program representing the action domain or with any action language.

Given an action domain $\mathcal{M}$, we use $\mathcal{A_M}$[1] to denote the set of all actions in the action domain $\mathcal{M}$ while $\mathcal{F_M}$ is the set of all available fluents. When modelling histories and traces, we need the monitor the domain over a period of time (or a sequence of states). We assume that they are modelled using instant(I). The ordering of instances is established by next(I1, I2), with the final instance defined as final(I). Following convention, we assume that the truth of a fluent $F \in \mathcal{F}$ at a given state instance I is represented as holdsat(F, I), while an action $A \in \mathcal{A}$ is modelled as occurred(A, I).

AQL has two basic concepts: (i) *constraint:* an assertion of a property that must be satisfied by a valid trace (e.g. a restriction on which traces are considered), and (ii) *condition:* a specification of properties that can may hold for a given trace. Conditions can be declared in relation to other conditions and constraints can involve declared conditions. Table 1 summarises the syntax of the language, while the remainder of this section discusses in detail the elements of the language and their semantics.

Basic Constructs. AQL provides (Table 1: AQL-1–AQL-5) various forms of names: variable, variable_list, name, param_list and identifier. An *identifier* is

---

[1] The action domain will be omitted when it is clear from the context.

Table 1. A*QL* Syntax

Expression		Definition	
<variable>	::=	[A-Z][a-zA-Z0-9_]*	(AQL-1)
<variable_list>	::=	<variable> , <variable_list> | <variable>	(AQL-2)
<name>	::=	[a-z][a-zA-Z0-9_]*	(AQL-3)
<param_list>	::=	(<variable_list>)	(AQL-4)
<identifier>	::=	<name> <param_list> | <name>	(AQL-5)
<predicate>	::=	happens(<identifier>) | holds(<identifier>)	(AQL-6)
<literal>	::=	not <predicate> | <predicate>	(AQL-7)
<while_expr>	::=	<literal> while <while_expr> | <literal>	(AQL-8)
<after>	::=	after(<integer>) | after*l*	(AQL-9)
<after_expr>	::=	<while_expr> <after> <after_expr> |;	(AQL-10)
		<while_expr>	(AQL-11)
<condition_literal>	::=	not <identifier> | <identifier>	(AQL-12)
<term>	::=	<after_expr> | <condition_literal>	(AQL-13)
<conjunction>	::=	<term> and <conjunction> | <term>	(AQL-14)
<disjunction>	::=	<term> or <disjunction> | <term>	(AQL-15)
<condition_decl>	::=	condition <identifier> : <disjunction>; |	
		condition <identifier> : <conjunction>;	(AQL-16)
<constraint>	::=	constraint <disjunction> ; |	
		condition <identifier> : <conjunction>;	(AQL-17)

an arbitrary name which may have variable parameters, that enables the parameterisation of actions and fluents. A*QL* defines (Table 1: AQL-6) two *predicates* that form the basis of all queries. The first is `happens(Action)`, meaning that the specified action should occur at some point during history.The second is `holds(Fluent)`, which means that the specified fluent is true at some point. Negation (as failure) is provided by the unary operator `not` (Table 1: AQL-7). To construct complex queries, it is often easier to break them up into sub-queries, or in A*QL* terminology, sub-conditions. We can then join these with other criteria. Sub-conditions may be referenced within rules as *condition literals* (Table 1: AQL-12). The building block of query conditions is the *term* (Table 1: AQL-13). Query conditions are either after expressions or literals. *Terms* may be grouped and connected by the connectives `and` and `or` which provide logical conjunction and disjunction (Table 1: AQL-14 and AQL-15). On their own, they do not allow us create arbitrary combinations of *predicates* so additionally we provide the means to declare conditions. Table 1: AQL-16 defines a `condition` with the specified name to have a value equal to the specified `disjunction` or `conjunction`. This allows the `condition` name to be used as a `condition_literal`. A*QL* also allows for the specification of constraints (Table 1: AQL-17). They specify properties of the trace which must be true.

Example queries. To illustrate how this language is used to form queries, consider a simple light bulb action domain. The fluent `on` is true when the bulb is on. The action `switch` turns the light on or off. We can require that at some point the light is on:

```
constraint holds(on);
```

We can require that the light is never on:

```
condition light_on: holds(on);
constraint not light_on;
```

There is some subtlety here in that `light_on` is true if at any instant `on` is true. Therefore, if `light_on` is not true, there cannot be an instant at which `on` was true. And what if the bulb is brokenThis can be expressed as:

```
| constraint not light_on and happens(switch);
```
Using condition names, we can create arbitrary logical expressions. The statement that action $a1$ and either action $a2$ or $e3$ should occur can be expressed as follow:
```
| condition disj: happens(a2) or happens(a3);
| condition conj: happens(a1) and disj;
```

Query Semantics. In this paper we define the semantics of AQL in terms of its mapping to $AnsProlog$. The semantics of an AQL query is defined by the translation function T which translates AQL into $AnsProlog$. This function takes a fragment of AQL and generates a set of (partial) $AnsProlog$ rules. The semantics of predicates are defined as: $T(\texttt{happens(e)}) = \texttt{occurred(e, I)}$ and $T(\texttt{holds(f)}) = \texttt{holdsat(f, I)}$.

For a literal of the form $\texttt{not P}$ (where $\texttt{P}$ is a predicate) the semantics is: $T(\texttt{not } P) = \texttt{not } T(\texttt{P})$. For a condition literal they are: $T(\texttt{conditionName}) = \texttt{conditionName}$ and $T(\texttt{not conditionName}) = \texttt{not conditionName}$ and a conjunction of terms is: $T(t_1 \texttt{ and } t_2 \texttt{ and} \cdots \texttt{ and } t_n) = T(t_1), T(t_2), \ldots, T(t_n)$. Each disjunction is translated into multiple rules. The translation itself depends if depending on whether the disjunction is within a condition declaration or a constraint:

$$T(\texttt{condition conditionName}: \ t_1 \texttt{ or } t_2 \texttt{ or } \cdots \texttt{or } t_n;) =$$
$$\{\texttt{conditionName} \leftarrow T(t_i). \mid 1 \leq i \leq n\}$$
$$T(\texttt{constraint} \qquad\qquad t_1 \texttt{ or } t_2 \texttt{ or } \cdots \texttt{or } t_n;) =$$
$$\{\texttt{newName} \leftarrow T(t_i). \mid 1 \leq i \leq n\}\cup$$
$$\{\bot \leftarrow \texttt{not newName.}\}$$

Note that the term $\texttt{newName}$ denotes any identifier that is unique within the $AnsProlog$ program that is the combination of the query and the action program. In addition, each time instant $\texttt{I}$ generated in the translation of a predicate represents a name for a time instant that is unique within the AQL query.

Concurrent Actions and Fluents. We may wish to talk about actions occurring at the same time as one or more fluents are true, simultaneous occurrence of actions or combinations of fluents being simultaneously true (and/or false). For this situation, AQL has the keyword $\texttt{while}$ (Table 1: AQL-8) to indicate that literals are true *simultaneously*. Such $\texttt{while}$ expressions are only defined over literals constructed from predicates.

Returning to the light bulb example, we can now specify that we want only traces where the light was turned off at some point:
```
| constraint happens(switch) while holds(on);
```
Or that at some point the light was turned left on:
```
| constraint holds(on) while not happens(switch);
```
The semantics for $\texttt{while}$ is:
$$T(L_1 \texttt{ while } L_2 \texttt{ while } \cdots \texttt{ while } L_n) = T(L_1), T(L_2), \ldots, T(L_n), \texttt{instant(I)}.$$

Action and Fluent Ordering. The language allows for the expression of orderings over actions. This is done with the $\texttt{after}$ keyword (Table 1: AQL-9). In some cases we need to say not only that a given literal holds after some other literal, but that this is the case after a precise number of time instants (n). So, for a fluent that does (or does not) hold at time instant t_i or an action that occurs between t_i and t_{i+1}, we can talk about literals that hold at t_{i+n} or occur between t_{i+n} and t_{i+n+1}. An $\texttt{after}$ expression may contain only the $\texttt{after}$ operator or the $\texttt{after(n)}$ operator, depending on how precisely the gap between the two operands is to be specified.

Once again returning to the light bulb example, we can now specify a query which requires the light to be switched twice (or more):

```
constraint happens(switch) after happens(switch);
```

Or that once that light has is on, it cannot be switched off again:

```
condition switch_off: happens(switch) after holds(on);
constraint not switch_off;
```

We give the semantics for the binary operator after(n): $T(W_i \text{ after(n) } W_j) = T(W_i)$, $T(W_j)$, after(t_i, t_j, n) This can easily be generalised for after expressions built of sequences of after(n) operators mixed with after operators.

4 Reasoning with A*QL*

We now illustrate how A*QL* can be used to perform three common tasksin computational reasoning: prediction, postdiction and planning.

Prediction is the problem of ascertaining the resulting state for a given (partial) sequence of actions and initial state. That is, suppose some transition system is in state S and a (partial) sequence A of actions occurs. Then the prediction problem (S, A) is to decide the set of states $\{S'\}$ which may result. Postdiction is the converse problem: if a system is in state S' and we know that sequence A has occurred, then the problem (A, S') is to decide the set $\{S\}$ of states that could have held before A. The planning problem (S, S') is to decide which sequence(s) of actions, $\{A\}$, will bring about state S' from state S.

Identifying States. A state $S = \{f_1, \ldots, f_n\}$ is described by the set of fluents f_i that are true. In A*QL* states can be represented using using the while:

```
holds(f_1) while ... while holds(f_n) while
not holds(g_1) while ... while not holds(g_k)
```

where $f_{1\ldots k}$ are fluents that hold in S and $g_{1\ldots k}$ those that do not.

Describing Action Ordering. A sequence of actions $A = a_1, \ldots, a_n$ may be encoded as an after expression. In this case of complete information, we can express A as follows:

```
happens(a_n) after(1) ... after(1) happens(a_1)
```

If we do not have complete information, but we know that a_{i+1} happens later than a_i than we can express this as:

```
happens(a_i+1) after happens(a_i)
```

We can combine these cases throughout the formulation of A to represent the amount of information available.

The Prediction Problem. Given an initial state S and a sequence of actions A, the prediction problem (S, A) can be expressed in A*QL* as:

```
constraint A after(1) S;
```

This query limits traces to those in which at some point S holds after which the actions of A occur in sequence. The answer sets that satisfy this query will then contain $\{S'\}$.

The Postdiction Problem. Given a sequence of actions A and a resulting state S', the postdiction problem (A, S') can be expressed as:

```
| constraint S after(1) A;
```

This requires S to hold in the next instant following the final action of A.

The Planning Problem. Given a pair of states S and S' the planning problem (S, S') can be expressed in A$\mathcal{QL}$ as:

```
| constraint S' after S;
```

This allows any non-empty sequence of actions to bring about the transition from S to S'. If we want to consider plans of length k (i.e. $E = a_1, \ldots, a_k$) then we can express this using after(k) instead.

5   Discussion

In [5], the authors present four query languages: $\mathcal{P}, \mathcal{Q}, \mathcal{Q}_n, \mathcal{R}$. The action query language $\mathcal{P}$ has just two constructs: now L and necessarily F after A1, ..., An, where L refers to a fluent or its negation, F is a fluent and where Ai are actions. These queries can equally be encoded in A$\mathcal{QL}$ using the techniques discussed in Section 4. The same is true for the query languages $\mathcal{Q}, \mathcal{Q}_n$ and $\mathcal{R}$. Given the action ordering technique used, we can assign specific times to each of the fluents. A$\mathcal{QL}$ can express all the same kinds of queries as the query languages above, but in addition A$\mathcal{QL}$ is capable of modelling simultaneous actions and fluents, permits the expression of complex queries using disjunctions and conjunctions of conditions and, above all, allows reasoning with incomplete information, thus fully exploiting the reasoning power of answer set programming.

The Causal Calculator (CCALC)[4] is a versatile tool mainly used for modelling action domains. While CCALC can also support querying, AQL has been designed specifically as a query language, providing constructs to make specifying queries more intuitive and versatile. Relative ordering of actions or states is much more difficult in CCALC then it is AQL. Furthermore, CCALC also does not allow for the formulation of composite queries (condition literals).

As it stands A$\mathcal{QL}$ offers an intuitive and versatile query and abduction language for action domains, offering improvements in both query semantics and expressibility over existing approaches. The language is succinct and compact, comprising a small core of orthogonal operators (that is none can be expressed in terms of the others). However, from a software engineering point of view, there is a case for improving usability by providing commonly used constructs as part of the language.

References

[1] Baral, C.: Knowledge Representation, Reasoning and Declarative Problem Solving. Cambridge University Press, Cambridge (2003)
[2] Calvanese, D., Vardi, M.Y.: Reasoning about actions and planning in LTL action theories. In: Proc. KR 2002 (2002)
[3] Eiter, T., Faber, W., Leone, N., Pfeifer, G., Polleres, A.: The DLV^K Planning System. In: Flesca, S., Greco, S., Leone, N., Ianni, G. (eds.) JELIA 2002. LNCS (LNAI), vol. 2424, pp. 541–544. Springer, Heidelberg (2002)

[4] Giunchiglia, E., Lee, J., Lifschitz, V., McCain, N., Turner, H.: Nonmonotonic causal theories. Artificial Intelligence 153, 49–104 (2004)

[5] Gelfond, M., Lifschitz, V.: Action languages. Electron. Trans. Artif. Intell. 2, 193–210 (1998)

[6] Giunchiglia, E., Lifschitz, V.: An action language based on causal explanation: preliminary report. In: AAAI 1998/IAAI 1998, pp. 623–630. American Association for Artificial Intelligence (1998)

[7] Vladimir, L.: Action languages, answer sets and planning. In: The Logic Programming Paradigm: a 25-Year perspective, pp. 357–373. Springer, Heidelberg (1999)

Level Mapping Induced Loop Formulas for Weight Constraint and Aggregate Programs

Guohua Liu

Department of Computing Science
University of Alberta
Edmonton, Alberta, Canada
guohua@cs.ualberta.ca

Abstract. We improve the formulations of loop formulas for weight constraint and aggregate programs by investigating the level mapping characterization of the semantics for these programs. First, we formulate a level mapping characterization of the stable model semantics for weight constraint programs, based on which we define loop formulas for these programs. This approach makes it possible to build loop formulas for programs with arbitrary weight constraints without introducing new atoms. Secondly, we further use level mapping to characterize the semantics and propose loop formulas for aggregate programs. The main result is that for aggregate programs not involving the inequality comparison operator, the dependency graphs can be built in polynomial time. This compares to the previously known exponential time method.

1 Introduction

Logic programming under *stable model* semantics has been extended to incorporate a variety of constraints to facilitate knowledge representation and reasoning. These constraints include weight constraints [12], aggregates [1,11,14] and abstract constraints [10,13]. We refer to logic programs with these constraints as weight constraint, aggregate and abstract constraint programs, respectively.

Lin and Zhao [6] propose to compute stable models of normal logic program as the model of the loop completion of the program. The loop completion consists of the loop formulas and the formulas of the completion of the program. Liu and Truszczyński [8] extend the approach to weight constraint programs where the weight constraints contain only positive literals and weights. However, in order to transform an arbitrary weight constraint to a weight constraint with only positive literals and weights, new propositional atoms are needed [8,9]. In theory, new atoms enlarge the search space for stable model computation. An interesting question is whether the loop formulas for arbitrary weight constraint programs can be formulated without extra atoms.

The method of level mapping has been studied to characterize stable models [2,4] of normal programs. We observe that such a characterization is closely related to the formulation of loop formulas. We present level mapping characterization of the stable models of weight constraint programs. The characterization leads to a formulation of loop formulas for arbitrary weight constraint programs without introducing extra atoms.

E. Erdem, F. Lin, and T. Schaub (Eds.): LPNMR 2009, LNCS 5753, pp. 444–449, 2009.

Aggregate programs are closely related to weight constraint programs, since many aggregates can be encoded by weight constraints [7]. There are different semantics proposed for aggregate programs [3,5,14]. Among them, the semantics based on conditional satisfaction is considered the most *conservative* [13], in the sense that any answer set under this semantics is an answer set under others, but the reverse may not hold. We are interested in the formulation of loop formulas for this semantics. To distinguish from the stable model semantics of weight constraint programs, we call the semantics *answer set* semantics.

Loop formulas for answer set semantics are presented in [15]. In the approach, given a program, the construction of the dependency graph requires computing what is called "local power set" for the constraints in the program, to capture conditional satisfaction. The process takes exponential time in the size of the program. We investigate the level mapping characterization of answer sets and find that, for aggregates, the conditional satisfaction checking can be reduced to polynomial time standard satisfaction checking. Based on this finding, we define the levels of aggregates. The definition induces a formulation of loop formulas, where local power sets are not needed and the exponential process to construct the dependency graph is avoided.

2   Level Mapping Induced Loop Formulas for Weight Constraint Programs

A *weight constraint* is of the form

$$l\,[a_1{=}w_{a_1}, ..., a_n{=}w_{a_n}, \texttt{not}\ b_1{=}w_{b_1}, ..., \texttt{not}\ b_m{=}w_{b_m}]\,u \tag{1}$$

where each a_i, b_j is an atom. Atoms $a_i's$ and not-atoms $\texttt{not}\ b_i's$ are also called *literals* (*positive* and *negative* literals, respectively). We denote by $lit(W)$ the set of literals in a weight constraint. Each literal in a constraint is associated with a *weight*[1]. The numbers l and u give the lower and upper bounds of the constraint, respectively. The weights and bounds are real numbers. Either of the bounds may be omitted in which case the missing lower bound is taken to be $-\infty$ and the missing upper bound ∞.

A set of atoms M satisfies a weight constraint W of the form (1), denoted $M \models W$, if (and only if) $l \leq w(W, M) \leq u$, where $w(W, M) = \sum_{a_i \in M} w_{a_i} + \sum_{b_i \notin M} w_{b_i}$. M satisfies a set of weight constraints Π if $M \models W$ for every $W \in \Pi$.

A *weight constraint program* is a finite set of rules of the form

$$W_0 \leftarrow W_1, ..., W_n \tag{2}$$

where each W_i is a weight constraint. We use $hd(r)$ and $bd(r)$ to denote W_0 and $\{W_1, ..., W_n\}$, respectively. $Atom(P)$ denotes the set of the atoms appearing in program P. For the semantics of weight constraint programs, we refer the reader to [12].

Notations. In the rest of this paper, we will use the following notations: Given a weight constraint W of the form (1) and a set of atoms M, we define $M_a(W) = \{a_i \in$

---

[1] The weights of literals could be negative. It is pointed out that negative weights can be eliminated by a transformation [12]. We assume that weights are non-negative if not indicated otherwise.

$M \mid a_i \in lit(W)\}$ and $M_b(W) = \{b_i \in M \mid \text{not } b_i \in lit(W)\}$. Since W is always clear by context, we will simply write M_a and M_b.

In general, an atom may appear both positively and negatively in a weight constraint. We call such an atom a *dual* atom, e.g. atom a is a dual atom in $1[a = 1, \text{not } a = 2]1$.

Following the notation in [13], for a set of atoms X and a mapping λ from X to positive integers, we define $H(X) = max(\{\lambda(a) \mid a \in X\})$. For the empty set $\emptyset$, we define $max(\emptyset) = 0$ and $min(\emptyset) = \infty$.

2.1 Level Mapping Characterization of Stable Models

Given a set of atoms X, a *level mapping* of X is a function λ from the atoms in X to positive integers. Let W be a weight constraint of the form (1), M a set of atoms and λ a level mapping of M. The *level* of W w.r.t. M, denoted $L(W, M)$, is defined as:

$$L(W, M) = min(\{H(X_a) \mid X \subseteq M, \text{ and } w(W, X_a) \geq l + \sum_{b_i \in M \setminus X_a} w_{b_i}\}). \quad (3)$$

Proposition 1. *Let W be a weight constraint of the form (1), M and X be two sets of atoms. $w(W^M, X_a) \geq l^M$ iff $w(W, X_a) \geq l + \sum_{b_i \in M \setminus X_a} w_{b_i}$, where W^M is the reduct of W w.r.t. M as defined in [12] and l^M is the lower bound of W^M.*

Intuitively, the level of W w.r.t. M depends on the levels of atoms in M that are necessary to satisfy W^M and positive in W.

Definition 1. *Let P be a weight constraint program and M a set of atoms. M is said to be* level mapping justified *by P if there is a level mapping λ of M satisfying that for each $b \in M$, there is a rule $r \in P$ such that $b \in lit(hd(r))$, $M \models bd(r)$, and for each $W \in bd(r)$, $\lambda(b) > L(W, M)$.*

By Proposition 1, the following theorem can be proved.

Theorem 1. *Let P be a weight constraint program and M a set of atoms. M is a stable model of P iff M is a model of P and level mapping justified by P.*

2.2 Loop Formulas for Weight Constraint Programs

To characterize stable models by loop formulas, we need the concept of *completion* of a program, whose models are the supported models of the program. The definition of completion can be found in [8]. For a program P, we denote its completion $Comp(P)$.

The formulation of loop formulas consists of two steps: constructing a dependency graph and then establishing a formula for each loop in the graph.

Let P be a weight constraint program. The *dependency graph* of P, denoted $G_P = (V, E)$, is a directed graph, where (i). $V = Atoms(P)$ and (ii). (u, v) is a directed edge from u to v in E, if there is a rule of the form (2) in P, such that $u \in lit(W_0)$ and $v \in lit(W_i)$ for some i $(1 \leq i \leq n)$. Let $G = (V, E)$ be a directed graph. A set $L \subseteq V$ is a *loop* in G if the subgraph of G induced by L is strongly connected.

Let W be a weight constraint and L be a set of atoms. The *restriction* of W w.r.t. L, denoted $W_{|L}$, is a conjunction of weight constraints $W_{l|L} \wedge W_{u|L}$, where

– $W_{l|L}$ is obtained by removing the upper bound and all positive literals in L and their weights from W;
– $W_{u|L}$ is obtained by removing the lower bound from W.

Let P be a weight constraint program and L be a loop in G_P. The *loop formula* for L, denoted $LF(P, L)$, is defined as

$$LF(P, L) = \bigvee L \rightarrow \bigvee \{ \bigwedge_{W \in bd(r)} W_{|L} \mid r \in P, L \cap lit(hd(r)) \neq \emptyset \} \qquad (4)$$

Let P be a weight constraint program. The *loop completion* of P denoted $LComp(P)$ is defined as $LComp(P) = Comp(P) \cup \{LF(P, L) \mid L \text{ is a loop in } G_P\}$.

Theorem 2. *Let P be a weight constraint program and M a set of atoms. M is a stable model of P iff M is a model of $LComp(P)$.*

3   Level Mapping Induced Loop Formulas for Aggregate Programs

An aggregate is a constraint on sets taking the form $aggr(\{X \mid p(X)\})$ op *Result*, where $aggr$ is an *aggregate function*. The standard aggregate functions are those in {SUM, COUNT, AVG, MAX, MIN}. The relational operator *op* is from $\{=, \neq, <, >, \leq, \geq\}$ and *Result* is either a variable or a numeric constant.

An aggregate program is a set of rules of the form

$$h \leftarrow A_1, ..., A_n \qquad (5)$$

where h is an atom and $A_1, ..., A_n$ are aggregates. The semantics of aggregate programs bases on the notion of *conditional satisfaction*. We refer the reader to [14] for the details.

All of the standard aggregates (without the operator "$\neq$"[2]) can be encoded by weight constraints as shown in [7]. In this section, we focus on programs with aggregate SUM only[3]. For an aggregate A, we denote its weight constraint encoding $W(A)$. A property of aggregates is that their weight constraint encoding contain no dual atoms. This is useful to prove the proposition later.

3.1   Level Mapping Characterization of Answer Sets

Let A be an aggregate, M a set of atoms and λ a level mapping of M. The *answer set level* of A w.r.t. M, denoted $L^*(A, M)$, is defined as:

$$L^*(A, M) = min(\{H(X) \mid X \subseteq M, w(W(A), X_a) \geq l + \sum_{b_i \in M} w_{b_i}, \qquad (6)$$

$$and \ w(W(A), X_b) \leq u - \sum_{a_i \in M} w_{a_i}\}),$$

---

[2] We only study the aggregates without the operator "$\neq$", since for programs with $SUM(.) \neq k$, the *answer set existence problem* is at a level higher than NP-Completeness in the complexity hierarchy, according to Son and Pontelli [14].
[3] Note: Aggregates $COUNT$ and AVG are special cases of SUM. Aggregates MAX and MIN can be encoded by SUM [7].

where l and u are the lower and upper bounds of $W(A)$ respectively.

Intuitively, the level of A, w.r.t. M depends on the level of atoms in M that are necessary to conditionally satisfies A, w.r.t. M.

Proposition 2. *Let A be an aggregate and X and M two sets of atoms such that $X \subseteq M$. X conditionally satisfies A w.r.t. M iff $w(W(A), X_a) \geq l + \sum_{b_i \in M} w_{b_i}$ and $w(W(A), X_b) \leq u - \sum_{a_i \in M} w_{a_i}$, where l and u are the lower and upper bounds of $W(A)$, respectively.*

Definition 2. *Let P be an aggregate program and M a set of atoms. M is said to be strongly level mapping justified by P if there is a level mapping λ of M satisfying that for each $b \in M$, there is a rule $r \in P$ such that $b = hd(r)$, $M \models bd(r)$, and for each $A \in bd(r)$, $\lambda(b) > L^*(A, M)$.*

Using Proposition 2, we can prove the following theorem.

Theorem 3. *Let P be an aggregate program and M a set of atoms. M is an answer set of P iff M is a model of P and strongly level mapping justified by P.*

3.2 Loop Formulas for Aggregate Programs

The completion of aggregate programs consists of the same set of formulas as that of weight constraint programs, except that the weight constraints in the formulas are weight constraint encoding of aggregates.

Let P be an aggregate program. The *dependency graph* of P, denoted $G_P^* = (V, E)$, is a directed graph, where (i). $V = Atom(P)$ and (ii). (u, v) is a directed edge from u to v in E, if there is a rule of the form (5) in P such that $u = hd(r)$ and either v or not $v \in lit(W(A_i))$ for some i ($1 \leq i \leq n$).

Now we give the *strong restriction* of an aggregate w.r.t. a loop by defining the strong restriction of a weight constraint. Let W be a weight constraint and L a set of atoms. The strong restriction of W, w.r.t. L, denoted $W_{|L}^*$, is a conjunction of weight constraints $W_{l|L}^* \wedge W_{u|L}^*$, where

- $W_{l|L}^*$ is obtained by removing from W the upper bound, all positive literals that are in L and their weights;
- $W_{u|L}^*$ is obtained by removing from W the lower bound, not $b_i = w_{b_i}$ for each $b_i \in L$, and changing the upper bound to be $u - \sum_{b_i \in L} w_{b_i}$.

The strong restriction of an aggregate A w.r.t. a loop L is defined as the strong restriction of its weight constraint encoding w.r.t. the loop $W_{|L}^*(A)$.

Let P be an aggregate program and L a loop in G_P^*. The loop formula for L, denoted $LF^*(P, L)$, is defined as

$$LF^*(P, L) = \bigvee L \to \bigvee \{ \bigwedge_{A \in bd(r)} W(A)_{|L}^* \mid r \in P, hd(r) \in L \} \qquad (7)$$

Let P be an aggregate program. The loop completion of P, denoted $LComp^*(P)$, is defined as $LComp^*(P) = Comp(P) \cup \{LF^*(P, L) \mid L \text{ is a loop in } G_P^*\}$.

Theorem 4. *Let P be an aggregate program and M a set of atoms. M is an answer set of P iff it is a model of $LComp^*(P)$.*

4   Conclusion and Future Work

We present level mapping characterizations of semantics for weight constraint programs and aggregate programs, respectively. Based on the level mapping characterizations, we improve the formulation of loop formulas for these programs. For arbitrary weight constraint programs, we propose a formulation of loop formulas that does not require introducing any new atoms; For aggregate programs, we show that the dependency graph can be constructed in time polynomial in the size of programs.

The level mapping and loop formulas are defined on aggregate SUM. For the aggregates MAX and MIN, direct definition may be desired for intuitiveness. We leave this for the future work.

References

1. Dell'Armi, T., Faber, W., Lelpa, G., Leone, N.: Aggregate functions in disjunctive logic programming: semantics, complexity, and implementation in DLV. In: Proc. IJCAI 2003, pp. 847–852 (2003)
2. Erdem, E., Lifschitz, V.: Tight logic programs. Theory and Practice of Logic Programming 3(4), 499–518 (2003)
3. Faber, W., Leone, N., Pfeifer, G.: Recursive aggregates in disjunctive logic programs. In: Alferes, J.J., Leite, J. (eds.) JELIA 2004. LNCS (LNAI), vol. 3229, pp. 200–212. Springer, Heidelberg (2004)
4. Fages, F.: Consistency of clark's completion and existence of stable models. Journal of Methods of Logic in Computer Science 1, 51–60 (1994)
5. Ferraris, P.: Answer sets for propositional theories. In: Baral, C., Greco, G., Leone, N., Terracina, G. (eds.) LPNMR 2005. LNCS (LNAI), vol. 3662, pp. 119–131. Springer, Heidelberg (2005)
6. Lin, F., Zhao, Y.: ASSAT: Computing answer sets of a logic program by SAT solvers. Artificial Intelligence 157(1-2), 115–137 (2004)
7. Liu, G., You, J.: Lparse programs revisited: semantics and representation of aggregates. In: Garcia de la Banda, M., Pontelli, E. (eds.) ICLP 2008. LNCS, vol. 5366, pp. 347–361. Springer, Heidelberg (2008)
8. Liu, L., Truszczyński, M.: Properties and applications of programs with monotone and convex constraints. Journal of Artificial Intelligence Research 7, 299–334 (2006)
9. Marek, V., Niemelä, I., Truszczyński, M.: Logic programs with monotone abstract constraint atoms. Theory and Practice of Logic Programming 8(2), 167–199 (2008)
10. Marek, V.W., Remmel, J.B.: Set constraints in logic programming. In: Lifschitz, V., Niemelä, I. (eds.) LPNMR 2004. LNCS (LNAI), vol. 2923, pp. 167–179. Springer, Heidelberg (2004)
11. Pelov, N., Denecker, M., Bruynooghe, M.: Well-founded and stable semantics of logic programs with aggregates. Theory and Practice of Logic Programming 7, 301–353 (2007)
12. Simons, P., Niemelä, I., Soininen, T.: Extending and implementing the stable model semantics. Artificial Intelligence 138(1-2), 181–234 (2002)
13. Son, T.C., Pontelli, E., Tu, P.H.: Answer sets for logic programs with arbitrary abstract constraint atoms. Journal of Artificial Intelligence Research 29, 353–389 (2007)
14. Son, T.C., Pontelli, E.: A constructive semantic characterization of aggregates in answer set programming. Theory and Practice of Logic Programming 7, 355–375 (2006)
15. You, J., Liu, G.: Loop formulas for logic programs with arbitrary constraint atoms. In: Proc. AAAI 2008, pp. 584–589 (2008)

Layer Supported Models of Logic Programs

Luís Moniz Pereira and Alexandre Miguel Pinto

Centro de Inteligência Artificial (CENTRIA)
Universidade Nova de Lisboa, 2829-516 Caparica, Portugal
{lmp,amp}@di.fct.unl.pt

Abstract. Building upon the 2-valued Layered Models semantics for normal programs, we introduce a refinement — the Layer Supported Models semantics — which, besides keeping all of LMs' properties, furthermore respects the Well-Founded Model.

Keywords: Stable Models, Relevance, Semantics, Layering.

1   Introduction

In [5] we presented the Layered Models semantics, a 2-valued semantics for Normal Logic Programs (NLPs) which guarantees model existence, enjoys relevance and cumulativity, and is a conservative extension of the Stable Models (SMs) semantics [3]. Although the LMs proposed in [5] already enforces minimality of its models, it does not ensure compatibility with the Well-Founded Model (WFM). The refinement to the LMs we now propose — Layer Supported Models (LSMs) semantics — obeys the proviso that each model respects the WFM. To achieve this we simply refine the notion of layering original of [5]. Intuitively, a program is conceptually partitioned into "layers" which are subsets of its rules. Rules forming loops are placed in the same layer, while rules not forming loops are placed in different layers — higher layer rules depending on lower layer rules, but not vice-versa. An atom is then considered *true* in a model if there is some rule for it, at some layer, where all literals in its body supported by rules of lower layers are also *true*. Otherwise the atom is *false*.

The core reason SM semantics fails to guarantee model existence for every NLP is that the stability condition it imposes on models is impossible to be complied with by Odd Loops Over Negation (OLONs) — like $a \leftarrow not\ a, X$[1]. In fact, the SM semantics community uses such inability as a means to write Integrity Constraints (ICs).

The LSM semantics provides a semantics to all NLPs. In the $a \leftarrow not\ a, X$ example above, whenever X is *true*, the only LSM is $\{a, X\}$. For LSM semantics OLONs are not ICs. ICs are implemented with rules for reserved atom $falsum$, of the form $falsum \leftarrow X$, where X is the body of the IC we wish to prevent being true. This does not prevent $falsum$ from being in some models. From a theoretical standpoint this means that the LSM semantics does not include an IC compliance enforcement mechanism. ICs must be dealt with in two possible ways: either by 1) a syntactic post-processing step, as a "test phase" after the model generation "generate phase"; or by 2) embedding the IC

---

[1] OLON is a loop with an odd number of default negations in its circular call dependency path.

E. Erdem, F. Lin, and T. Schaub (Eds.): LPNMR 2009, LNCS 5753, pp. 450–456, 2009.

compliance in a query-driven (partial) model computation, where such method can be a top-down query-solving one *a la* Prolog, since the LSM semantics enjoys relevance. In this second case, the user must conjoin query goals with $not\ falsum$. If inconsistency examination is desired, like in the 1) case above, models including $falsum$ can be discarded *a posteriori*. This is how LSM semantics separates OLON semantics from IC compliance.

After notation and background definitions, we present the formal definition of LSM semantics and overview its properties. The applications afforded by LSMs are all those of SMs plus those requiring OLONs for model existence, and those where OLONs are actually employed for problem representation. Work under way concerns the efficient implementation of the LSM semantics.

2 Background Notation and Definitions

Definition 1. *Logic Rule.* *A Logic Rule r has the general form*
$H \leftarrow B_1, \ldots, B_n, not\ C_1, \ldots, not\ C_m$ *where H, the B_i and the C_j are atoms.*

We call H the head of the rule — also denoted by $head(r)$. And $body(r)$ denotes the set $\{B_1, \ldots, B_n, not\ C_1, \ldots, not\ C_m\}$ of all the literals in the body of r. Throughout this paper we will use 'not' to denote default negation. When the body of the rule is empty, we say the head of rule is a fact and we write the rule just as H. A Logic Program (LP for short) P is a (possibly infinite) set of ground Logic Rules of the form in Definition 1. In this paper we focus mainly on NLPs, those whose heads of rules are positive literals, i.e., simple atoms; and there is default negation just in the bodies of the rules. Hence, when we write simply "program" or "logic program" we mean an NLP.

3 Layering of Logic Programs

The well-known notion of stratification of LPs has been studied and used for decades now. But the common notion of stratification does not cover all LPs, i.e., there are some LPs which are non-stratified. The usual syntactic notions of dependency are mainly focused on atoms. They are based on a dependency graph induced by the rules of the program. Useful these notions might be, for our purposes they are insufficient as they leave out important structural information about the call-graph of P. To cover that information we also define below the notion of a rule's dependency. Indeed, layering puts rules in layers, not atoms. An atom B directly depends on atom A in P iff there is at least one rule with head B and with A or $not\ A$ in the body. An atom's dependency is just the transitive closure of the atom's direct dependency. A rule directly depends on atom B iff any of B or $not\ B$ is in its body. A rule's dependency is just the transitive closure of the rule's direct dependency. The Relevant part of P for some atom A, represented by $Rel_P(A)$, is the subset of rules of P with head A plus the set of rules of P whose heads the atom A depends on, cf. [2]. Likewise for the relevant part for an atom A notion [2], we define and present the notion of relevant part for a rule r. The Relevant part of P for rule r, represented by $Rel_P(r)$, is the set containing the rule r itself plus the set of rules relevant for each atom r depends on.

Definition 2. *Parts of the body of a rule.* *Let* $r = H \leftarrow B_1, \ldots, B_n, not\ C_1, \ldots, not$ C_m *be a rule of P. Then,* $r^l = \{B_i, not\ C_j : B_i\ depends\ on\ H \wedge C_j\ depends\ on\ H\}$. *Also,* $r^B = \{B_i : B_i \in (body(r) \setminus r^l)\}$, *and* $r^C = \{C_j : not\ C_j \in (body(r) \setminus r^l)\}$.

Definition 3. *HighLayer function.* *The HighLayer function is defined over a set of literals: its result is the highest layer number of all the rules for all the literals in the set, or zero if the set is empty.*

Definition 4. *Layering of a Logic Program* P. *Given a logic program P a layering function L/1 is just any function defined over the rules of P', where P' is obtained from P by adding a rule of the form H ← falsum for every atom H with no rules in P, assigning each rule r ∈ P' a positive integer, such that:*

- $L(r) = 0$ *if* $falsum \in body(r)$, *otherwise*
- $L(r) \geq max(HighLayer(r^l), HighLayer(r^B), (HighLayer(r^C) + 1))$

A layering of program P is a partition $P^1, \ldots, P^n$ *of P such that* P^i *contains all rules r having* $L(r) = i$, *i.e., those which depend only on the rules in the same layer or layers below it.*

This notion of layering does not correspond to any level-mapping [4], since the later is defined over atoms, and the former is defined over rules. Also, due to the definition of dependency, layering does not coincide with stratification [1], nor does it coincide with the layer definition of [6]. However, when the program at hand is stratified (according to [1]) it can easily be seen that its respective layering coincides with its stratification. In this sense, layering, which is always defined, is a generalization of the stratification.

Amongst the several possible layerings of a program P we can always find the least one, i.e., the layering with least number of layers and where the integers of the layers are smallest. In the remainder of the paper when referring to the program's layering we mean such least layering (easily seen to be unique).

4   Layer Supported Models Semantics

The Layer Supported Models semantics we now present is the result of the two new notions we introduced: the layering, formally introduced in section 3, which is a generalization of stratification; and the layered support, as a generalization of classical support. These two notions are the means to provide the desired 2-valued semantics which respects the WFM, as we will see below.

An interpretation M of P is classically supported iff every atom a of M is classically supported in M, i.e., *all* the literals in the body of some rule for a are *true* under M in order for a to be supported under M.

Definition 5. *Layer Supported interpretation.* *An interpretation M of P is layer supported iff every atom a of M is layer supported in M, and this holds iff a has a rule r where all literals in* $(body(r) \setminus r^l)$ *are true in M. Otherwise, it follows that a is false.*

Theorem 1. *Classical Support implies Layered Support.* *Given an NLP P, an interpretation M, and an atom a such that $a \in M$, if a is classically supported in M then a is also layer supported in M.*

Proof. Trivial from the definitions of classical support and layered support.   □

In programs without odd loops layered supported models are classically supported too.

Example 1. **Layered Unsupported Loop.** Consider program P:

$$c \leftarrow not\ a \qquad a \leftarrow c, not\ b \qquad b$$

The only rule for b is in the first layer of P. Since it is a fact it is always *true* in the WFM. Knowing this, the body of the rule for a is *false* because unsupported (both classically and layered). Since it is the only rule for a, its truth value is *false* in the WFM, and, consequently, c is *true* in the WFM. This is the intuitively desirable semantics for P, which corresponds to its LSM semantics. LM and the LSM semantics differences reside both in their layering notion and the layered support requisite of Def. 5. In this example, if we used LM semantics, which does not exact layered support, there would be two models: $LM_1 = \{b, a\}$ and $LM_2 = \{b, c\}$. $\{b\}$ is the only minimal model for the first layer and there are two minimal model extensions for the second layer, as a is not necessarily false in the LM semantics because Def. 5 is not adopted. Lack of layered support lets LM semantics fail to comply with the WFM. Note that adding a rule like $b \leftarrow a$ would not affect the semantics of the program, according to LSM. This is so because, such rule would be placed in the same layer with the rules for a and c, but leaving the fact rule b in the strictly lower layer.

Intuitively, the minimal layer supported models up to and including a given layer, respect the minimal layer supported models up to the layers preceding it. It follows trivially that layer supported models are minimal models, by definition. This ensures the truth assignment to atoms in loops in higher layers is consistent with the truth assignments in loops in lower layers and that these take precedence in their truth labeling. As a consequence of the layered support requirement, layer supported models of each layer comply with the WFM of the layers equal to or below it. Combination of the (merely syntactic) notion of layering and the (semantic) notion of layered support makes the LSM semantics.

Definition 6. *Layer Supported Model of P.* *Let $P^1, \ldots, P^n$ be the least layering of P. A layer supported interpretation M is a Layer Supported Model of P iff*

$$\forall_{1 \leq i \leq n} M|_{\leq i} \text{ is a minimal layer supported model of } \cup_{1 \leq j \leq i} P^j$$

where $M|_{\leq i}$ denotes the restriction of M to heads of rules in layers less or equal to i:

$$M|_{\leq i} \subseteq M \cap \{head(r) : L(r) \leq i\}$$

The Layer Supported semantics of a program is just the intersection of all of its Layer Supported Models.

Layered support is a more general notion than that of perfect models [7], with similar structure. Perfect model semantics talks about "least models" rather than "minimal models" because in strata there can be no loops and so there is always a unique least model which is also the minimal one. Layers, as opposed to strata, may contain loops and thus there is not always a least model, so layers resort to minimal models, and these are guaranteed to exist (it is well known every NLP has minimal models).

It is worth noting that atoms with no rules and appearing in the bodies of some rule are necessarily "placed" in the lowest layer: an atom a having no rules is equivalent to having the single rule $a \leftarrow falsum$. Any minimal model of this layer will consider the heads of such rules to be false. This ensures compliance with the Closed World Assumption (CWA).

In [5] the authors present an example (7) where three alternative joint vacation solutions are found by the LM semantics, for a vacation problem modeled by an OLON. The solutions found actually coincide with those found by the LSM semantics. We recall now a syntactically similar example (from [8]) but with different intended semantics, and show how it can be attained in LSM by means of ICs.

Example 2. **Working example [8].**

$$tired \leftarrow not\ sleep \qquad sleep \leftarrow not\ work \qquad work \leftarrow not\ tired$$

As in the example 7 of [5], the LSM semantics would provide three solutions for the above program: $\{work, tired\}, \{work, sleep\}, \{sleep, tired\}$. Although some (or even all!) of these solutions might be actually plausible in a real world case, they are not, in general, the intended semantics for this example. With the LSM semantics, the way to prune away some (or all) of these solutions is by means of ICs. For example, to eliminate the $\{work, sleep\}$ solution we would just need to add the IC $falsum \leftarrow work, sleep$.

The principle used by LSMs to provide semantics to any NLP — whether with OLONs or not — is to accept all, and only, the minimal models that are layer supported, i.e., that respect the layers of the program. The principle used by SMs to provide semantics to only some NLPs is a "stability" (fixed-point) condition imposed on the SMs by the Gelfond-Lifschitz operator.

5   Properties of the Layer Supported Models Semantics

Although the definition of the LSMs semantics is different from the LMs of [5], this new refinement enjoys the desirable properties of the LMs semantics, namely: guarantee of model existence, relevance, cumulativity, and being a conservative extension of the SMs semantics. Moreover, and this is the main contribution of the LSMs semantics, it respects the Well-Founded Model.

Due to lack of space, the complexity analysis of this semantics is left out of this paper. Nonetheless, a brief note is due. Model existence is guaranteed for every NLP, hence the complexity of finding if one LSM exists is trivial, when compared to SMs semantics. Brave reasoning — finding if there is any model of the program where some atom a is true — is an intrinsically NP-hard task from the computational complexity point of view. But since LSM semantics enjoys relevance, the computational scope of this task can be reduced to consider only $Rel_P(a)$, instead of the whole P. From a practical standpoint, this can have a significant impact in the performance of concrete applications. Cautious reasoning (finding out if some atom a is in all models) in the LSM semantics should have the same computational complexity as in the SMs, i.e., coNP-complete.

5.1   Respect for the Well-Founded Model

Definition 7. *Interpretation* M *of* P *respects the WFM of* P. *An interpretation* M *respects the WFM of* P *iff* M *contains the set of all the* true *atoms of the WFM of* P, *and it contains no* false *atoms of the WFM of* P.

Theorem 2. *Layer Supported Models respect the WFM.* Let P be an NLP, and $P^{\leq i}$ denote $\bigcup_{1 \leq j \leq i} P^j$, where P^j is P's j layer. Each LSM $M|_{\leq i}$ of $P^{\leq i}$, where $M \supseteq M|_{\leq i}$, respects the WFM of $P^i \cup M|_{<i}$.

Proof. By hypothesis, each $M|_{\leq i}$ is a full LSM of $P^{\leq i}$. Consider P^1. Any $M|_{\leq 1}$ contains the facts of P, and their direct positive consequences, since the rules for all of these are necessarily placed in the first layer in the least layering of P. Hence, $M|_{\leq 1}$ contains all the *true* atoms of the WFM of P^1. Layer 1 also contains whichever loops that do not depend on any other atoms besides those which are the heads of the rules forming the loop. Any such loops having no negative literals in the bodies are deterministic and, therefore, the heads of the rules forming the loop will be all *true* or all *false* in the WFM of P^1, depending on whether the bodies are fully supported by facts in the same layer, or not, and, in the latter case, if the rules are not involved in other types of loop making their heads undefined. In any case, an LSM of this layer will by necessity contain all the *true* atoms of the WFM of P^1. On the other hand, assume there is some atom b which is *false* in the WFM of P^1. b being *false* in the WFM means that either b has no rules or that every rule for b has an unsatisfiable body in P^1. In the first case, by definition 6 we know that b cannot be in any LSM. In the second case, every unsatisfiable body is necessarily unsupported, both classically and layered. Hence, b cannot be in any LSM of P^1. This means that any LSM contains no atoms *false* in the WFM of P^1, and, therefore, that they must respect the WFM of P^1.

By hypothesis $M|_{\leq i+1}$ is an LSM of $P^{\leq i+1}$ iff $M|_{\leq i+1} \supseteq M|_{\leq i}$, for some LSM $M|_{\leq i}$ of $P^{\leq i}$, which means the LSMs $M|_{\leq i+1}$ of $P^{\leq i+1}$ are exactly the LSMs $M|_{\leq i+1}$ of $P^{i+1} \cup M|_{\leq i}$. Adding the $M|_{\leq i}$ atoms as facts imposes them as *true* in the WFM of $P^{i+1} \cup M|_{\leq i}$. The then deterministically *true* consequences of layer $i+1$ — the *true* atoms of the WFM of $P^{i+1} \cup M|_{\leq i}$ — become necessarily present in every minimal model of $P^{i+1} \cup M|_{\leq i}$, and therefore in its every LSM $M|_{\leq i+1}$. On the other hand, every atom b *false* in the WFM of $P^{i+1} \cup M|_{\leq i}$ has now unsatisfiable bodies in all its rules (up to this layer $i+1$). Hence, b cannot be in any LSM of $P^{i+1} \cup M|_{\leq i}$. Therefore, every $M|_{\leq i+1}$ respects the WFM of $P^{i+1} \cup M|_{\leq i}$. Hence, more generally, every $M|_{\leq i}$ respects the WFM of $P^i \cup M|_{<i}$

$\square$

References

1. Apt, K.R., Blair, H.A.: Arithmetic classification of perfect models of stratified programs. Fundam. Inform. 14(3), 339–343 (1991)
2. Dix, J.: A Classification-Theory of Semantics of Normal Logic Programs: I, II. Fundamenta Informaticae XXII(3), 227–255, 257–288 (1995)
3. Gelfond, M., Lifschitz, V.: The stable model semantics for logic programming. In: ICLP/SLP, pp. 1070–1080. MIT Press, Cambridge (1988)

4. Hitzler, P., Wendt, M.: A uniform approach to logic programming semantics. TPLP 5(1-2), 93–121 (2005)
5. Pereira, L.M., Pinto, A.M.: Layered models top-down querying of normal logic programs. In: Gill, A., Swift, T. (eds.) PADL 2009. LNCS, vol. 5418, pp. 254–268. Springer, Heidelberg (2009)
6. Przymusinski, T.C.: Every logic program has a natural stratification and an iterated least fixed point model. In: PODS, pp. 11–21. ACM Press, New York (1989)
7. Przymusinski, T.C.: Perfect model semantics. In: ICLP/SLP, pp. 1081–1096 (1988)
8. Przymusinski, T.: Well-founded and stationary models of logic programs. Annals of Mathematics and Artificial Intelligence 12, 141–187 (1994)

Applying ASP to UML Model Validation

Mario Ornaghi, Camillo Fiorentini, Alberto Momigliano, and Francesco Pagano

Dipartimento di Scienze dell'Informazione, Università degli Studi di Milano, Italy
{ornaghi,fiorenti,momiglia,pagano}@dsi.unimi.it

Abstract. We apply ASP to model validation in a CASE setting, where models are UML class diagrams and object diagrams are called "snapshots". We present the design and implementation of MSG, a snapshot generator for UML models that employs DLV-Complex as a generator engine, the answer sets representing the legal snapshots.

1 Introduction

The object of this research is the application of ASP to model validation in a CASE setting, in particular evaluating the "correctness" of formal specifications (or *models*) with respect to their requirements. Here, models are UML class diagrams [8] with constraints, typically written in OCL: a diagram should represent an abstraction of the problem domain; the objects populating a system state should represent a "snapshot" of a corresponding counterpart in the modeled world. In the UML, snapshots are represented by object diagrams. The legal snapshots are those satisfying the constraints that can be attached to the model to better specify the desired properties. In this context, tools for snapshot generation (SG) are an important part of the "weaponry" of light-weight formal methods. In fact, the relevance of SG for validation and testing in OO software development is widely acknowledged and a relevant part of the recently branded field of "Model-Based Testing" [4]. The latter ranges from model animation to ways of establishing partial certification such as model consistency and constraints independence checking.

This paper presents the design and implementation of a snapshot generator for UML models called "MSG" (read as "Message" and standing for "Milano Snapshot Generator", cooml.dsi.unimi.it/msg), which employs DLV-Complex [3] as a generator engine, the answer sets representing the legal snapshots. This is integrated in a system that takes as input any UML class diagram in XMI format and eventually displays back to the user the answers, i.e. the snapshots in the same format. The main theoretical contribution consists in a specialized representation of UML class diagrams into DLV-Complex, tailored to the *fully automatic* generation of *non isomorphic* snapshots The representation makes essential use of DLV-Complex's external functions, but still requires the introduction of an intermediate language (DLVExi) adding polymorphic types and existential quantification [5].

Background on UML. The Unified Modeling Language (UML) comprises a variety of model types for describing system properties, both static (e.g. class models, object models) and dynamic (e.g. state-machines, activity graphs). One of the more prominent

E. Erdem, F. Lin, and T. Schaub (Eds.): LPNMR 2009, LNCS 5753, pp. 457–463, 2009.
© Springer-Verlag Berlin Heidelberg 2009

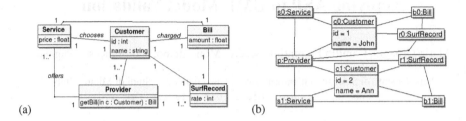

Fig. 1. A class diagram of an Internet Service Provider (a) and a snapshot for it (b)

model types is the class model (visualized as a class diagram) used to represent the underlying data model of a system in an object-oriented manner. A class diagram consists of a set of classes (rectangles) and relations among them, in particular associations (lines connecting rectangles), as shown in our running example (Fig. 1 (a)), a simplified version of the ISP example from http://www.brucker.ch/projects/hol-ocl/. In this scenario a Provider offers some Service(s) at a certain price. A Customer chooses one of these services and she is charged a Bill according to her SurfRecord and the download rate. According to the UML "type-instance" dichotomy [8], classes, associations and class diagrams represent "types". Classes instantiate over "objects", associations over "links" and class diagrams over object diagrams, called "snapshots". Fig. 1 (b) shows a snapshot, where a provider p (precisely, an object p of class Provider) offers two services s0, s1 and has two clients c0, c1, where c0 chooses s0, and so on. Association ends are decorated with various multiplicities. For example, the multiplicity 1..* at the Service-end of the association offers denotes that a provider may offer one or more services, while 1 at the Provider-end indicates that a service is offered by one provider.

We remark that we can omit *oids* (object identifiers) in a snapshot. One way to look at this is considering oids as abstractions of memory addresses, which are transparent in OO programming. Thus, two snapshots of the same class diagram should be considered equal if they are *isomorphic*, where, roughly speaking, a snapshot isomorphism is a bijective map among oids that preserves the navigations. For example, if in the ISP snapshot we map c0 into c1 and c1 into c0, we get an isomorphic copy, i.e., morally, the "same" snapshot.

2   Model Validation and Snapshot Generation

Since requirements are informal, model validation can be only empirical, i.e., it is performed by comparing the formal model with the user's expectations. In this context, a snapshot generation tool (SGT) plays an important role. A SGT has two inputs: a model M and a set G of *generation requests* (GR), needed to make the number of snapshots finite. SGT output the set of legal snapshots that satisfy G. It allows us to perform various "experiments". To name one, the mere existence of a snapshot ensures that the model is *consistent* with its constraints.

Fig. 2 shows the architecture and the data-flow of MSG. We start with the open source UML tool BOUML (http://bouml.free.fr/) to design diagrams and to generate the corresponding XMI representation. At the end of SG, BOUML displays the snapshots produced by the *as2xmi* module from the XMI format. The translation from XMI to DLV has been divided in two phases, using an intermediate language DLVExi. The latter allows us to decouple the representation of XMI models in logic from the definition and implementation of the generation request language. The Java component XMI2DLVEXI translates an XMI model M into a DLVExi program E_M, which is a faithful representation of M in the following sense: every legal snapshot of M is represented by an "answer set" of E_M and every "answer set"

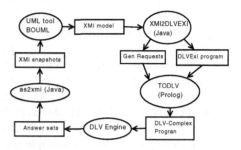

Fig. 2. The data flow view of the system

of E_M represents a legal snapshot of M. The component TODLV translates the program E_M and the generation requirements G into a DLV-Complex program $P_{M,G}$. The answer sets of $P_{M,G}$ are the answer sets of E_M that satisfy G. DLV-Complex [2] is used as the generator engine.

3   Representing UML into ASP

In this section we discuss some encoding techniques for UML+OCL class diagrams in DLV-Complex. We do this via an intermediate language, DLVExi, which can be seen as an extension of DLV-Complex with a ML-like polymorphic type system and allowing conjunction and existential quantification in the head of clauses, as shown by the following fragment of trel, one of the DLVExi encodings of the UML with which we have experimented.

```
type obj(C) --> o(C).   type typeId(X) --> tid.
type mult --> m(int,int); star(int); union(mult,mult).
type association(C1,C2) --> ass(assoc_name). ...
pred object(obj(C)), is_association(association(C1,C2)), is_class(class(C)),
    link(association(C1,C2),obj(C1),obj(C2)), att_rec(obj(C),T),
    mLeft(association(C1,C2),obj(C2),int),
    mRight(association(C1,C2),obj(C1),int),
    leftMult(association(C1,C2),mult),
    rightMult(association(C1,C2),mult),   violates(int,mult), ...
trel(C1:type, C2:type, X:C1, Y:C2, A:association(C1,C2)) isunit {
    object([C1],o(X)) v neg(object([C1],o(X))) if is_class(tid([C1])),    %g1
    link(A,o(X),o(Y)) v neg(link(A,o(X),o(Y))) if                         %g2
            is_association(A) & object(o(X)) & object(o(Y)),
    exi([v], att_rec(o(X),v)) if  object(o(X)),                          %ec
    false if leftMult(A,M)&object(o(Y))&mLeft(A,o(Y),N)&violates(N,M),    %t1
    false if rightMult(A,M)&object(o(X))&mRight(A,o(X),N)&violates(N,M)   %t2}.
```

Before explaining trel, some brief comments on the DLVExi language are in order. Types are expressed analogously to datatype declarations in functional programming languages, where --> productions introduce polymorphic types by listing the

type constructors (also called generators). For example, mult (multiplicity) is generated by m(int,int), star(int) and, recursively, union(mult,mult). The ground
mult-terms represent multiplicities, for example union(m(1,2),star(5)) represents
1..2, 5..*. One use of (polymorphic) types is as wrappers, abstracting away from
the types of the specific UML model. In particular, obj(C) is the type of the oids for
a class type C and association(C1,C2) the type of the associations between class C1
(left hand side) and C2 (right hand side). For the sake of type safe grounding [5], every
ground term must have a unique type. To this aim, [5] introduces annotated functions
$f_J(\ldots)$ and predicates $p_J(\ldots)$. In our concrete syntax, the annotations J are enclosed between square brackets. For example, tid([C1]) is the concrete syntax of tid_{C1}, while
object([C1],o(X)) is $object_{C1}(o(X))$. Annotations may be left understood and are
reconstructed by the system. If multiple annotations are possible, the system produces
an error message. Polymorphic types allow us to decouple the general representation
choices from the signature of the specific UML model. In this way, we can represent a
UML model M by a DLVExi theory $T_M = R \cup E_M$, where R is a general "representation theory", and an "encoding theory" E_M representing M in R. The theory R does not
depend on M, but only on the representation choices and the generation strategy. For
example, the above trel corresponds to a *relational representation* of the associations,
by means of the predicates object(O) ("O is a live object") and link(A,O1,O2) ("O1
and O2 are linked by the association A"). According to the "guess and test" methodology of DLV [9], live objects and links are *guessed* by the rules %g1 and %g2, while %t1
and %t2 "test" the multiplicity constraints. The XMI2DLVEXI component translates M
into the encoding theory E_M. For example, part of the ISP-encoding (Fig. 1 (a)) is:

```
type customer. type bill.  ...  % the types for the ISP classes
type customer_atb --> rec(id:int, name:string). % the types for the
type bill_atb --> rec(amount:float).             % attribute-records
... ISP isunit {
  is_class(tid([bill])) if true, % "bill" identifies a class-type
  ...
  is_association(ass([customer,bill],charged)) if true,
  % "charged" is the name of an association between customer and bill
  ...
  leftMult(ass([customer,bill],charged),m(1,1)) if true,
  % the multiplicity on the customer association end of "charged" is 1
  % ....}
```

Inputs to the TODLV component are the theory $T_M = R \cup E_M$ and a set GR of generation
requests. The output is a DLV-Complex program. The generation requests suggest a
finite set of possible object identifiers and a finite set of attribute values, in order to get
finitely many models. Examples of GR for customer and bill are:

```
type customer --> c1; c2.  type bill --> b1; b2.           %i
att_rec([customer], c1,V) if V=rec(0,ted),                 %ii
att_rec([customer], c2,V) if V=rec(1,mary), att_rec([bill],B,V)
if member(V,[rec(12.3), rec(10.5)]), ...
```

By %i, we fix a finite set of possible oids, while %ii gives a finite set of "witness-choices"
for the existential variable v of the clause %ec. The TODLV component replaces the

existential formula with a disjunction over the witness-choices, as shown in the clauses
%c4 of the following DLV-Complex program:

```
of(obj(C),o(X)) :- is_class([C],tid([C])), of(C,X). ...                    %c1
object([C],o(X)) v -object([C],o(X)) :- of(C,X),is_class([C],tid([C])).
link([C1,C2], A, o(X), o(Y)) v -link([C1,C2],A, o(X), o(Y)))) :-
        is_association([C1,C2],A), object([C1],o(X)), object([C2],o(Y)).
...
is_class([bill],tid([bill])). ...                                          %c2
is_association([customer,bill], ass([customer,bill],charged)).
....
of(customer,c1). of(customer,c2). of(bill,b1). of(bill,b2).                %c3
....
att_rec([customer],c1,rec(0,john)). att_rec([customer],c2,rec(1,mary)).%c4
att_rec([bill], B, rec(12.3)) v att_rec([bill], B, rec(10.5)) :-
        object([bill],o(B)). ...
```

Clauses c1 are the DLV-Complex translation of trel (we use of(C,X) as "X is of type
C"), excluding %ec. Clauses c2 comes from the ISP-encoding, c3 from the generation
requests for the oids and c4 from the existential clause %ec and the related generation requests. We remark that TODLV does not perform grounding, which is left to
DLV-Complex. Type and annotation reconstruction play a central role, since they enforce the correct grounding of polymorphic clauses. For example, the annotations of
is_association clauses of c2 are used to instantiate the type-variables C1, C2 in the
link-clause.

Finally we launch DLV-Complex with the above program, and we get back a set
of stable models that represent the possible snapshots. Such models can be visualized as object diagrams using BOUML or a graphical tool. Here we show a snapshot
that is consistent with the given specifications, yet does not fit with our expectations.
This would suggest some problem in the
modeling phase. Here, customer c_0 is associated with b_0, and r_1, while c_1 is associated with b_1 and r_0. This is surprising
since, for instance, r_1 is the surfrecord of
c_0, but r_1 and c_0 refer to different bills.
This brings about very well the usefulness of lightweight formal methods and
model validation in particular. One could
do all sort of heavy functional verification via interactive theorem proving only

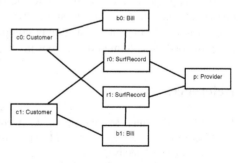

to discover that the initial model was under-specified and required further constraints.
Without some clever notion of proof reuse, this would have meant a lot of wasted effort.

We conclude with a final comment concerning the intermediate language DLVExi.
It has been introduced to enhance the expressive power of DLV-Complex, in order to
define different representations of UML, of which trel is an example, minimizing
the impact on the XMI2DLVEXI and TODLV components and on the GR language
and its semantics. In particular, we have developed a functional representation tfun of

the associations, not explained here for lack of space. The tfun representation drastically reduces the number of the generated isomorphic snapshots. For example, without generating the attributes, with 1 provider and at most 2 customers we have 100 snapshots with trel, while only 6 with tfun. The non isomorphic snapshots are 4.

4   Related and Future Work

Animation tools for UML diagrams such as state-chart, activity etc. are a commercial enterprise. Among academics, the USE tool [6] claims to be the only one supporting automatic SG; differently from us, SG requires the user to write Pascal-like procedures in a dedicated language. The issue of isomorphic models does not seem to be addressed and the performances of USE are very sensitive to the *order* of objects and attribute assignments [1]. Other animation and validation tools support different languages. Alloy [7] is based on first-order relational logic. The Alloy Analyzer compiles a formula in the Alloy language into quantifier-free booleans and feed to a SAT solver. According to [1], the Alloy Analyzer is the leading system for generation of instances of invariants, animation of the execution of operations and checking of user-specified properties. However, Alloy is not formally object-oriented, nor does it support UML and OCL.

We have described the design and implementation of MSG, a tool using ASP for MBT in the context of model validation of UML+OCL class diagrams. While the system is not yet ready to be released our preliminary experiments have shown that it compares favourably with the functionalities and the statistics reported in [1] w.r.t. our main "competitor", USE. Our main theoretical contribution has been the introduction of an intermediate language and of a representation of UML class diagrams tailored to the fully automatic generation of non isomorphic snapshots.

Future work include engineering the implementation, but also improve the representation, especially w.r.t. cyclic structures: the functional encoding yields rational terms. Possible approaches are *coinductive* techniques or identifying isomorphic graphs via classes of equivalence w.r.t. oid and link names, using a *nameless* representation. We plan to integrate one of the available compilers for OCL and address validation of pre/post conditions of methods supporting both *forward* and *backward* animation.

References

1. Aydal, E.G., Utting, M., Woodcock, J.: A comparison of state-based modelling tools for model validation. In: Paige, R.F., Meyer, B. (eds.) TOOLS (46). LNBIP, vol. 11, pp. 278–296. Springer, Heidelberg (2008)
2. Calimeri, F., Cozza, S., Ianni, G.: External sources of knowledge and value invention in logic programming. Ann. Math. Artif. Intell. 50(3-4), 333–361 (2007)
3. Calimeri, F., Cozza, S., Ianni, G., Leone, N.: DLV-Complex,
   http://www.mat.unical.it/dlv-complex
4. Dalal, S.R., et al.: Model-based testing in practice. In: ICSE 1999, pp. 285–294 (1999)
5. Fiorentini, C., Momigliano, A., Ornaghi, M.: Towards a type discipline for answer set programming. In: Berardi, S., Damiani, F., de Liguoro, U. (eds.) TYPES 2008 Post-Proceedings. LNCS, vol. 5497, pp. 117–135. Springer, Heidelberg (2009)

6. Gogolla, M., Bohling, J., Richters, M.: Validating UML and OCL models in USE by automatic snapshot generation. Software and System Modeling 4(4), 386–398 (2005)
7. Jackson, D.: Software Abstractions: Logic, Language, and Analysis. MIT Press, Cambridge (2006)
8. Larman, C.: Applying UML and Patterns: An Introduction to Object-Oriented Analysis and Design and Iterative Development. Prentice Hall, Upper Saddle River (2004)
9. Leone, N., et al.: The DLV system for knowledge representation and reasoning. ACM TOCL 7(3), 499–562 (2006)

The Logical Consequence Role in LPNMR: A Parameterized Computation Approach

Mauricio Osorio Galindo[1] and Simone Pascucci[2]

[1] Universidad de las Américas - Puebla
osoriomauri@gmail.com
[2] Universitá degli studi "La Sapienza" - Roma
cxjepa@yahoo.it

Abstract. We present results about the logical consequence test under classical logic w.r.t. the Theory of Parameterized Complexity and Computation [1]. We show how a normal logic program P can partitioned in subset of clauses such that we can define an algorithm proving sets of atoms which complexity is bounded by a relation exponential in terms of a fixed parameter k and polynomial on the original size of the problem, namely the size of P. As example of application we study the model checking problem w.r.t. the P-Stable semantics.

1 Introduction

In this paper, we study the P-Stable model checking problem to provide efficient methods for its implementation. We present some new results based on the application of modular evaluation techniques to our approach, which allow for further efficiency improvements by splitting the process of P-Stable model checking. Our approach is motivated by the theory of parameterized computation (TPC), for an extensive survey of the argument refer for example to the work of Chen [1]. A previous result which relates Logic Programming, Stable semantics and TPC can be found in the work of Gottlob et al. [11]. We do not provide an introduction about logic programming semantics and their relations with different types of logics, the interested reader can find an extensive survey of the subject in the following papers [5,3,10,2,6,7].

2 Background

A vocabulary $\mathcal{L}$ is a set of elements that we call atoms. A literal is an atom, a, or the negation of an atom $\neg a$. Given a set of atoms $\{a_1, ..., a_n\}$, we write $\neg\{a_1, ..., a_n\}$ to denote the set of atoms $\{\neg a_1, ..., \neg a_n\}$. A normal clause, C, is denoted: $a \leftarrow l_1, \ldots, l_n$, where $n \geq 0$, a is an atom, and each l_i is a literal. When $n = 0$ the clause is an abbreviation of the atom a. Sometimes, we denote a clause C by $a \leftarrow \mathcal{B}^+, \neg\mathcal{B}^-$, where $\mathcal{B}^+$ contains all the positive body atoms and $\mathcal{B}^-$ contains all the negative body atoms. We also use $body(C)$ to denote $\mathcal{B}^+ \cup \neg\mathcal{B}^-$.

E. Erdem, F. Lin, and T. Schaub (Eds.): LPNMR 2009, LNCS 5753, pp. 464–469, 2009.

A normal program P is a finite set of normal clauses, formally a normal program is a conjunction of its normal clauses.

The following is a program example that we call EXA:

$$e \leftarrow c. \qquad\qquad\qquad\qquad c \leftarrow e.$$

$$a \leftarrow \neg b,\, c. \qquad\qquad\qquad b \leftarrow \neg a, \neg e.$$

$$d \leftarrow b.$$

We denote by $\mathcal{L}_P$ the the set of atoms that occur in P. We denote by $HEAD(P)$ the set $\{a \mid a \leftarrow \mathcal{B}^+,\, \neg\mathcal{B}^- \in P\}$. Let P be a program and M a set of atoms. We write $P \models M$ if the conjunction of the atoms in M is a logical consequence of P. When M is the empty set the conjunction is interpreted as the formula $\top$ (the true constant). Note that $P \models \emptyset$ is true for every program P.

We will make a general assumption. Let P be program, then for every atom x that occurs in $\mathcal{L}_P$ then $x \in HEAD(P)$. We can always add some tautologies of the form $x \leftarrow x$, for certain atoms x, in order to ensure this property.

Given $\mathcal{L}$, an interpretation $I_{\mathcal{L}}$ is a boolean function with domain $\mathcal{L}$. As usual, we use interpretations to evaluate logical formulas. Sometimes, when the domain is understood, we drop the subscript of the interpretation. A model M of a program P is an interpretation, such that each formula in P evaluates to true. Given two interpretations $I_{\mathcal{L}_1}$, $J_{\mathcal{L}_2}$ such that $\mathcal{L}_1 \subseteq \mathcal{L}_2$, we say that $J_{\mathcal{L}_2}$ extends $I_{\mathcal{L}_1}$ if for every $x \in \mathcal{L}_1$, $I(x) = J(x)$. In this case, we also say that $I_{\mathcal{L}_1}$ is the restriction of $J_{\mathcal{L}_2}$ w.r.t $I_{\mathcal{L}_1}$. Since interpretations are functions, and functions are sets (a set of pairs) we can apply the union of two interpretations. Clearly if both interpretations are defined over disjoint domains, then the union is an interpretation with domain equal to the union of the domains of their respective interpretations. Finally, given a set of atoms M and a vocabulary $\mathcal{L}$ where $M \subseteq \mathcal{L}$. We define the interpretation $I_{\mathcal{L}}^M$ as:

$$I_{\mathcal{L}}^M(x) = \begin{cases} 1 \text{ if } x \in M \\ 0 \text{ if } x \in \mathcal{L} \setminus \mathcal{M} \end{cases}$$

The P-Stable semantics is defined in terms of a fixed point operator and classical logic [3].

Definition 1. *Let P be a normal program and M be a set of atoms. We define:*

$$RED(P, M) = \{h \leftarrow B^+, \neg(B^- \cap M) \mid h \leftarrow B^+, \neg B^- \in P\}$$

Definition 2. *Let P be a normal program and M be a set of atoms. We say that M is a p-stable model of P if $RED(P, M) \models M$ and $I_{\mathcal{L}}^M$ is a model of P.*

Let us consider again our example EXA, and let M be $\{b, d\}$. Then one can verify that M is a p-stable model of P. First, observe that M is a model of P. Note that $RED(P, M)$ is:

$$e \leftarrow c.$$
$$a \leftarrow \neg b, c.$$
$$d \leftarrow b.$$

$$c \leftarrow e.$$
$$b \leftarrow .$$

It is easy to check that $RED(P, M) \models b \wedge d$.

3   Reductions

In this section we state a set of results $w.r.t.$ programs proving atoms. These results constitute the basis for the formulation of a general algorithm for the test of logical consequence which will be given in the next section.

Definition 3. *Let P be a program and M, N be set of atoms. We define:*
$$REDp(P, M) = \{a \leftarrow \mathcal{B}^+ \setminus M, \ \neg \mathcal{B}^- : a \leftarrow \mathcal{B}^+, \ \neg \mathcal{B}^- \in P, \mathcal{B}^- \cap M = \emptyset\}.$$
$$REDn(P, M) = \{a \leftarrow \mathcal{B}^+, \ \neg \mathcal{B}^- : a \leftarrow \mathcal{B}^+, \ \neg \mathcal{B}^- \in P, \mathcal{B}^+ \cap M = \emptyset\}.$$
$$REDU(P, M, N) = REDn(REDp(P, M), N).$$

Consider again example EXA. Let $M = \{a, c\}$ and $N = \{b, e\}$. Then, we have

$$REDp(EXA, M) : e \leftarrow .$$
$$c \leftarrow e.$$
$$a \leftarrow \neg b.$$
$$d \leftarrow b.$$
$$REDU(P, M, N) : e \leftarrow .$$
$$a \leftarrow \neg b.$$

$$REDn(EXA, N) : e \leftarrow c.$$
$$a \leftarrow \neg b, c.$$
$$b \leftarrow \neg a, \neg e.$$

For reason of space we present in this section only the major result of our study.

Theorem 1. *Let P_1, P_2 two normal programs. Let $M = \{w : P_1 \models w\}$ and $N = \{w \in \mathcal{L}(P_1) : P_1 \not\models w\})$. Suppose the following conditions holds:*

1. *For every atom x, $x \in \mathcal{L}(P_1)$ then $x \notin Head(P_2)$.*
2. *For every atom z that occurs negated in P_1 then $z \in M$.*
3. *For every atom z that occurs negated in P_2 and $z \in \mathcal{L}(P_1)$ then $z \in M$.*
4. *Let y be any atom.*

Then $P_1 \cup P_2 \models y$ iff $P_1 \cup REDU(P_2, M, N) \models y$.

In the following, we assume the following notation by default. If P_1, P_2 are programs, M, N sets of atoms. Then, $M_1 = M \cap HEAD(P_1)$, $N_1 = N \cap \mathcal{L}(P_1)$, $M_2 = M \cap HEAD(P_2)$, and $N_2 = N \cap \mathcal{L}(P_2)$.

Corollary 1. *Let P_1, P_2 two normal programs. Let $\mathcal{L}$ be the set of atoms that occur in $P_1 \cup P_2$. Suppose that for every atom $x \in Head(P_2)$ then x does not occur in P_1. Let M be a set of atoms such that $M \subseteq \mathcal{L}$. Let $N = \mathcal{L} \setminus M$. For every atom z that occurs negated in P then $z \in M$. Let $I_{\mathcal{L}}$ be the interpretation defined as $I(x) = 1$ iff $x \in M$. Assume that I models $P_1 \cup P_2$. Then $P_1 \cup P_2 \models M$ iff $P_1 \models M_1$ and $REDU(P_2, M_1, N_1) \models M_2$.*

4 Relevant Modules of a Program

A program P induces a notion of *dependency* between atoms from $\mathcal{L}_P$ as we now define.

Definition 4. *Let r be a normal clauses, $a \in H(r)$ and $b \in B(r)$. We say that a depends immediately on b.*

For our example EXA, *e depends immediately on c, c depends immediately on e* and so on.

The two place relation *depends on* is the transitive closure of *depends immediately on*. For instance, in example EXA, we have that *d depends on c*.

We define an equivalence relation $\equiv$ between atoms of $\mathcal{L}_P$ as follow: $x \equiv y$ iff $x = y$ or (x *depends-on* y and y *depends-on* x). We write $[x]$ to denote the equivalent class induced by the atom x. We use *rel* to the the partial order induced by $\equiv$ on its equivalent classes. So, $[x]$ *rel* $[y]$ iff y *depends-on* x. For every program, we can always define a linear order among our equivalence classes by applying a topological sort. We use $\leq$ to denote such total order. Call an atom to *of order* 0 , if $[a]$ is minimal in $\leq$. If, for some atom a, the maximal order of the atoms of which a depends is n, then a is called to be of order $n + 1$. We say that a program P is of order n if n is the maximum order of its atoms. For i, $0 \leq i \leq n$, we define A_i as the set of atoms of order i. For our example EXA, $A_0 = \{e, c\}, A_1 = \{a, b\}, A_2 = \{d\}$.

We can break a program P (of order n) into the disjoint union of programs P_i ($0 \leq i \leq n$) such that P_i is the set of rules such that the head of each atom if of order i (with respect to P). We say that $P_0...P_n$ are the relevant modules of P. Consider again our program EXA:

Our linear order is $\{e, c\} \leq \{a, b\} \leq \{d\}$. Then, we have

$$EXA_0 : e \leftarrow c. \qquad EXA_1 : a \leftarrow \neg b, c. \qquad EXA_2 : d \leftarrow b$$
$$c \leftarrow e. \qquad\qquad b \leftarrow \neg a, \neg e.$$

Definition 5. *Let $P_1, ..., P_n$ be a partition of relevant modules of a program P. Then M_i is defined as $M \cap HEAD(P_i)$. N_i is defined as $HEAD(P_i) \setminus M_i$. MM_i is defined as $\bigcup_{1 \leq j \leq i} M_j$. NN_i is defined as $\bigcup_{1 \leq j \leq i} N_j$.*

We can now formalize an algorithm for the test of logical consequence using the result of Corollary 1 and the partition of a program in relevant modules. In Algorithm 1 we assume already done all the polynomial preprocessing phase of finding the relevant modules and building a topological order [4].

Theorem 2. *Let P be a normal logic program and $M \subseteq \mathcal{L}(P)$ a set of atoms such that for every atom z that occurs negated in P then $z \in M$. Let $\{P_1, \ldots, P_n\}$ be the partition of relevant modules of P, we define $k = |\mathcal{L}(P_{max})|$, where $P_{max} \in \{P_1, \ldots, P_n\}$ is such that $|\mathcal{L}(P_{max})| \geq |\mathcal{L}(P_i)|$ for $i = 1, \ldots, n$. Let $I_{\mathcal{L}}$ be the interpretation defined as $I(x) = 1$ iff $x \in M$. Let assume that I models P. Then, Algorithm 1 with P, M and $\{P_1, \ldots, P_n\}$ as input performs in time $O(n2^k + (nk)^2)$.*

---

input : A normal logic program P, a set of atoms $M \subseteq \mathcal{L}(P)$ such that for every atom z that occurs negated in P then $z \in M$ and such that there exist an interpretation $I_{\mathcal{L}}^M$ modeling P. A partition $\{P_1, \ldots, P_n\}$ of relevant modules of P in topological order.
output: *TRUE* if $P \models M$, *FALSE* if $P \nvDash M$.

for $i \leftarrow 1$ **to** n **do**
    $P_i \leftarrow$ REDU$(P_i, MM_{i-1}, NN_{i-1})$;
    if $P_i \nvDash M_i$ **then**
        return *FALSE*;
    end
end
return *TRUE*;

Algorithm 1. Test for logical consequence

Observe that if the given k is small the algorithm is efficient. For a given program P, we also say that the program has parameter k. If k is a given constant we let C_k the class of of programs such that the parameter of each program in C_k is less or equal to k. Then the decision problem of whether $P \models M$ (where M satisfies the restrictions of the last theorem) is polynomial time computable.

5   Deciding Whether M Is a P-Stable Model of a Normal Program

We study the problem *w.r.t.* the less general Corollary 1 because more interesting from the point of view of Definition 2. Let consider the following normal program:

$$
\begin{array}{llll}
P: & a \leftarrow not\ b. & (1) & d \leftarrow e. \qquad\qquad (5) \\
& b \leftarrow a. & (2) & d \leftarrow not\ e. \qquad (6) \\
& b \leftarrow c. & (3) & e \leftarrow d. \qquad\qquad (7) \\
& a \leftarrow not\ d. & (4) & c \leftarrow not\ e, c. \quad (8)
\end{array}
$$

We can individuate for P the following partition in relevant modules: $P_1 = \{(1),(2),(3),(4)\}$, $P_2 = \{(5),(6),(7)\}$, $P_3 = \{(8)\}$. We have $P_2 \leq P_1$, $P_2 \leq P_3$, $P_3 \leq P_1$. Consider the following P-Stable semantics decision problem. Let P be a program in C_k. Let M be a set of atoms, such that $M \subseteq \mathcal{L}$, where $\mathcal{L}$ is the vocabulary of P. Is M a P-Stable model of P?

We can state the following result:

Theorem 3. *The P-Stable semantics k-decision problem is polynomal time computable.*

Given the above program P and the set of atoms $M = \{e, d, b\}$ we can verify that $I_{\mathcal{L}}^M$ is a model of P. Given the topological order $[P_2, P_3, P_1]$ of the relevant modules of P, we can test if $RED(P, M) \models M$ by testing if: $P_2 \models \{e, d\}$, $REDU(P_3, \{e, d\}, \emptyset) \models \emptyset$ and $REDU(P_1, \{e, d\}, \{c\} \models \{b\}$.

6 Conclusions

We presented an algorithm for the test of logical consequence exploiting the subdivision of a program in relevant modules. Such an algorithm has a computational complexity bounded by a relation exponential in a parameter k and polynomial in the size of the input. We showed an example of application $w.r.t.$ P-Stable semantics.

References

1. Chen, J.-E.: Parameterized Computation and Complexity: A New Approach Dealing with NP-Hardness. J. Comput. Sci. Technol. 20(1), 18–37 (2005)
2. Osorio, M., Navarro, J.A., Arrazola, J.: Applications of intuitionistic logic in answer set programming. Theory and Practice of Logic Programming 4(3), 325–354 (2004)
3. Galindo, M.O., Pérez, J.A.N., Ramírez, J.A., Macías, V.B.: Logics with Common Weak Completions. Journal of Logic and Computation 16(6), 867–890 (2006)
4. Pascucci, S.: Normal Logic Programs and P-Stable Semantics: Optimization Techniques. M.Sc. Thesis, Universitá La Sapienza, Italy (2009) (in progress)
5. Pearce, D.: Stable inference as intuitionistic validity. Logic Programming 38, 79–91 (1999)
6. Gelfond, M., Lifschitz, V.: The stable model semantics for logic programming. In: Kowalski, R., Bowen, K. (eds.) Proceedings of the 5th Conference on Logic Programming, pp. 1070–1080. MIT Press, Cambridge (1988)
7. Gelfond, M., Lifschitz, V.: Classical negation in logic programs and disjunctive databases. New Generation Computing 9, 365–385 (1991)
8. Koch, C., Leone, N., Pfeifer, G.: Enhancing disjunctive logic programming systems by SAT checkers. Artif. Intell. 151(1-2), 177–212 (2003)
9. Pascucci, S., Fernandéz, A.L.: Syntactic transformation rules under P-Stable semantics: theory and implementation. Revista Iberoamericana de Inteligencia Artificial, Asociación Española para la Inteligencia Artificial, AEPIA (2009), http://dx.doi.org/10.4114/ia.v13i41.1028
10. Osorio, M., Ramirez, J.R.A., Carballido, J.L.: Logical weak completions of paraconsistent logics. Journal of Logic and Computation 18(6), 913–940 (2008)
11. Gottlob, G., Scarcello, F., Sideri, M.: Fixed-parameter complexity in AI and non-monotonic reasoning. Artif. Intell. 138(1-2), 55–86 (2002)
12. Baral, C.: Knowledge Representation, Reasoning and Declarative Problem Solving. Cambridge University Press, Cambridge (2003)

Social Default Theories

Chiaki Sakama

Department of Computer and Communication Sciences
Wakayama University, Sakaedani, Wakayama 640 8510, Japan
sakama@sys.wakayama-u.ac.jp

Abstract. This paper studies a default logic for social reasoning in multiagent systems. A *social default theory* is a collection of default theories with which each agent reasons and behaves by taking attitudes of other agents into account. The semantics of a social default theory is given as *social extensions* which represent the agreement of beliefs of individual agents in a society. We show the use of social default theories for representing social attitudes of agents and for reasoning in *cooperative planning* and *negotiation* among multiple agents.

1 Introduction

In a multiagent society, individual agents are requested to act interactively with other agents. Any problem, which is not solved by a single agent, could be solved cooperatively by exchanging information or sharing resources. It is usually the case, however, that an agent does not always have exact information of other agents. In this case, an agent performs *default reasoning* on belief states of other agents. For instance, suppose that two agents 1 and 2 live in the same apartment and share a car. The agent 1 usually uses the car for shopping if the agent 2 does not use it. The agent 2, on the other hand, normally uses the car to go to a school for picking up her child if the agent 1 does not use it. In this situation, the action of one agent depends on the action of another agent, but each agent does not know the exact plan of another agent. One day, the agent 1 plans to go shopping around noon. She knows that the agent 2 usually uses the car in the afternoon, then she goes to a parking lot. If she finds a car at the parking, she can use it; otherwise, she gives up shopping or waits for the agent 2 to come back. In this example, the agent 1 performs default reasoning on the behavior of another agent and takes an action.

A *default theory* [6] provides a logic for representing and reasoning with incomplete belief of an agent, so that it is natural to represent incomplete beliefs of multiple agents as a collection of default theories. In this case, an individual default theory contains default assumption on the beliefs of other agents as well as assumption on his/her own belief. A default theory has an extension which represents a collection of beliefs of an agent. In the presence of multiple default theories, the notion of extensions would be extended to those representing beliefs consented by every agent. To construct a logic for default reasoning in

E. Erdem, F. Lin, and T. Schaub (Eds.): LPNMR 2009, LNCS 5753, pp. 470–476, 2009.

multiagent systems, this paper introduces the framework of *social default theories*. We show the use of social default theories for representing a variety of social attitudes of individual agents, and for reasoning in *cooperative planning* and *negotiation* among multiple agents.

2   Social Default Theory

In this paper we consider theories represented by a propositional logic language. Every propositional atom has an *annotation* representing an *agent identifier*. For a propositional variable p in the language and an integer i, p^i is called an *annotated atom*. An *annotated formula* is inductively defined as follows: (i) An annotated atom p^i is an *annotated formula*. (ii) If α is an annotated formula, so is $\neg\alpha$. (iii) If α and β are annotated formulas, so are $\alpha \vee \beta$, $\alpha \wedge \beta$, $\alpha \supset \beta$, and $\alpha \equiv \beta$. The meanings of logical connectives are the same as those in classical logic. Annotation is introduced to distinguish beliefs among different agents, so the meaning of a formula, $p^i \vee q^i$ for instance, is equivalent to the formula $p \vee q$ that is a belief of the agent i. Note that two annotated atoms p^i and p^j represent different formulas, so $p^i \wedge \neg p^j$ is consistent as far as $i \neq j$. An annotated formula α in which every atom in α has the annotation i is called an *i-annotated formula*. An annotated formula is simply said a *formula* hereafter. $\mathcal{F}$ is the set of all formulas in the language, and $\mathcal{F}_i$ represents the set of all i-annotated formulas ($\mathcal{F}_i \subseteq \mathcal{F}$). A multiagent society is a finite set of *agents*. Formally, a society is represented by a *social default theory* defined as follows.

Definition 1. (social default theory) A *social default theory* (SDT, for short) $\mathcal{S}$ is a tuple of theories ($\Delta_1, \ldots, \Delta_m, \Gamma$) defined as follows.[1]

1. Each Δ_i ($1 \leq i \leq m$) is a *default theory* which is a finite set of *default rules* $\delta = \frac{\alpha : \beta_1, \ldots, \beta_n}{\gamma}$ where $\alpha \in \mathcal{F}$ and $\gamma \in \mathcal{F}_i$, and for each $1 \leq j \leq n$ there is some $1 \leq k \leq m$ such that $\beta_j \in \mathcal{F}_k$. α, $\beta_1, \ldots, \beta_n$, and γ are called a *prerequisite*, *justifications* and a *consequent*, respectively. A default rule δ is called a *social default rule* if its justifications contain at least one formula $\beta_j \in \mathcal{F}_k$ such that $k \neq i$. A default rule $\frac{\cdot}{\gamma}$ is identified with the formula γ.
2. Γ is a finite set of default rules such that each default rule in Γ has the consequent $\gamma = false$. Any default rule in Γ is called a *social constraint*.

Each default theory Δ_i in $\mathcal{S}$ represents the set of beliefs of an individual agent in the society. By contrast, Γ represents constraints that each agent must obey in the society. Informally, the default rule δ means that "if an agent i believes α, and each of $\beta_1, \ldots, \beta_n$ is consistently assumed with respect to the beliefs of agents k ($1 \leq k \leq m$), then the agent i believes γ". Note that in Δ_i the prerequisite and the justifications of any default rule may contain beliefs of other agents as well as beliefs of the agent i, while the consequent only contains beliefs of the agent i. Beliefs of other agents in the prerequisite represent strong conditions which affect the application of the default rule, while those in the justification represent weak conditions to reach the conclusion.

---

[1] We define a default theory as in [5].

Example 1. Compare $\Delta_1 = \{ \frac{\neg anti\text{-}smoker^3:}{smoke^1} \}$ and $\Delta_2 = \{ \frac{:\neg anti\text{-}smoker^3}{smoke^2} \}$. In Δ_1 the agent 1 smokes if it is known that the agent 3 is not an anti-smoker. In Δ_2, on the other hand, the agent 2 smokes unless the agent 3 is known to be an anti-smoker. In this sense, the social attitude of the agent 1 is more *considerate* than that of the agent 2.

Thus, SDTs can represent different social attitudes of agents. In this paper, a default theory Δ_i is identified with an agent i, and an SDT S is identified with a society. If the set of social constraints is empty, $(\Delta_1, \ldots, \Delta_m, \emptyset)$ is simply written as $(\Delta_1, \ldots, \Delta_m)$. In an SDT $S = (\Delta_1, \ldots, \Delta_m, \Gamma)$, the notion of *extension* for each Δ_i $(1 \leq i \leq m)$ is defined as usual [6]. A default theory Δ_i is *consistent* if it has a consistent extension; otherwise, it is *inconsistent*. An SDT S is *rational* if every Δ_i is consistent. In an SDT an individual agent would have its own extensions, while the society would also have extensions.

Definition 2. (social extension) A set E of formulas is a *social extension* of an SDT $S = (\Delta_1, \ldots, \Delta_m, \Gamma)$ if E is an extension of the default theory $\bigcup_{i=1}^{m} \Delta_i \cup \Gamma$.

A social extension represents a collection of beliefs of individual agents, which are consented by each agent and accord with constraints in the society.

Example 2. The car-sharing example in the introduction is represented by the SDT $S = (\Delta_1, \Delta_2)$ where $\Delta_1 = \{ shopping^1, \frac{shopping^1 : use\_car^1, \neg use\_car^2}{use\_car^1} \}$ and $\Delta_2 = \{ school^2, \frac{school^2 : use\_car^2, \neg use\_car^1}{use\_car^2} \}$. Then, S has two social extensions: $E_1 = Th(\{ shopping^1, school^2, use\_car^1 \})$ and $E_2 = Th(\{ shopping^1, school^2, use\_car^2 \})$, which represent two possibilities of using a shared car.

Note that in Example 2 if (Δ_1, Δ_2) is replaced with $(\Delta_1', \Delta_2', \Gamma)$ where $\Delta_1' = \{ shopping^1, \frac{shopping^1 : use\_car^1}{use\_car^1} \}$, $\Delta_2' = \{ school^2, \frac{school^2 : use\_car^2}{use\_car^2} \}$, and $\Gamma = \{ \frac{use\_car^1 \wedge use\_car^2:}{false} \}$, $(\Delta_1', \Delta_2', \Gamma)$ has no social extension. In fact, $Th(\{ shopping^1, school^2, use\_car^1, use\_car^2 \})$ is the extension of $\Delta_1' \cup \Delta_2'$, but is not the extension of $\Delta_1' \cup \Delta_2' \cup \Gamma$. Thus, social constraints are effective to eliminate useless extensions. Note also that if (Δ_1, Δ_2) in Example 2 is replaced with (Δ_1', Δ_2), (Δ_1', Δ_2) has the single social extension E_1. In this situation, Δ_1' does not take care of the usage of the car by the agent 2, and in this sense, the agent 1 is *self-interested*. A default theory Δ_i tends to be more self-interested if it contains less social default rules. When a society consists of self-interested agents, it is hard to reach an agreement. $(\Delta_1', \Delta_2', \Gamma)$ represents such a situation.

Proposition 1. *For any social extension E of a rational SDT S, there is an extension G of some Δ_i in S such that $(E \cap \mathcal{F}_i) \subseteq G$.*

Proposition 1 represents that a social extension includes a part of beliefs of some individual agents. This reflects the situation that belief or desire of individual agents are often suppressed in a society. The existence of a single agent which is self-interested and inconsistent, would eliminate social extensions.

Proposition 2. *Let S be an SDT such that some Δ_i in S is inconsistent and contains no social default rule. Then S has no social extension.*

To make an agreement, some agents can form a *party* by excluding those who have no interaction with them. This is also effective to isolate agents who are self-interested and inconsistent. For an SDT $\mathcal{S} = (\Delta_1, \ldots, \Delta_m, \Gamma)$, let $\mathcal{S}_P = (\Delta_1, \ldots, \Delta_l, \Gamma')$ $(l \leq m)$ where every default rule in Δ_i $(1 \leq i \leq l)$ and Γ' contains no formula from $\mathcal{F}_k$ $(l+1 \leq k \leq m)$.

Proposition 3. *If an SDT $\mathcal{S}$ has a social extension E, $\mathcal{S}_P$ has a social extension G such that $E \cap (\mathcal{F}_1 \cup \cdots \cup \mathcal{F}_l) = G$.*

The converse of Proposition 3 does not hold in general.

3 Social Reasoning by SDT

3.1 Cooperative Planning

In cooperative planning, multiple agents are supposed to have a common goal to be accomplished, and they build a joint plan by working cooperatively.

Definition 3. (cooperative planning framework) A *cooperative planning framework* is defined as a tuple $(\mathcal{S}, \mathcal{A}, \omega)$, where $\mathcal{S} = (\Delta_1, \ldots, \Delta_m, \Gamma)$ is an SDT, $\mathcal{A} \subseteq \mathcal{F}_1 \cup \cdots \cup \mathcal{F}_m$ is a set of *actions*, and $\omega \in \mathcal{F}$ is a *goal*. Given $(\mathcal{S}, \mathcal{A}, \omega)$, let $\Gamma^\omega = \Gamma \cup \{ \frac{:\neg\omega}{false} \}$. Then, a set $\Phi \subseteq \mathcal{A}$ is a *solution* of a cooperative planning framework $(\mathcal{S}, \mathcal{A}, \omega)$ if $\bigcup_{i=1}^m \Delta_i \cup \Gamma^\omega$ has an extension E such that $\Phi = E \cap \mathcal{A}$.

In cooperative planning, a common goal is given as a constraint to be satisfied in a society, and each agent computes their role of actions to achieve the goal.

Example 3. A robot 1 has a blue block and another robot 2 has a red block. There is a yellow block on the floor. The goal is to put the blue block on the yellow one, and to put the red block on the blue one. To achieve the goal, these two robots make a cooperative plan. Each robot can sense the change of the block world. The situation is represented by the cooperative planning framework $(\mathcal{S}, \mathcal{A}, \omega)$ with $\mathcal{S} = (\Delta_1, \Delta_2, \Gamma)$ where

$$\Delta_1 = \{ \; \frac{X\_on\_Y^1_{(T)} \wedge Z\_to\_X^1_{(T)} : \neg W\_to\_X^2_{(T)}, Z\_on\_X^1_{(T+1)}}{Z\_on\_X^1_{(T+1)}}, \quad \frac{has\_X^1_{(T)} : X\_to\_Y^1_{(T)}}{X\_to\_Y^1_{(T)}},$$

$$\frac{X\_on\_Y^2_{(T)} : X\_on\_Y^1_{(T)}}{X\_on\_Y^1_{(T)}}, \quad \frac{X\_on\_Y^1_{(T)} : X\_on\_Y^1_{(T+1)}}{X\_on\_Y^1_{(T+1)}}, \; has\_blue^1_{(0)}, \; yellow\_on\_floor^1_{(0)} \; \}.$$

Here, uppercase letters represent variables which are shorthand of their instances, and (T) means time steps. In Δ_1 the first rule represents that if a block X is on Y and the robot 1 moves another block Z to the location of X, and if it is assumed that the robot 2 does not move another block W to the location of X, the block Z is normally put on the block X at the next time step. The second rule says if a robot 1 has a block X then the robot can take an action of moving X to the location of Y. The third rule represents that if a block X is put on a block Y by the robot 2, the robot 1 can recognize the situation. The fourth rule represents the inertial rule: if a block X is on Y at time T and it is consistent to assume the existence of X on Y at time $T+1$, it is indeed at that

location. The fifth fact represents that the robot 1 has a blue block and the sixth
fact represents that the yellow block is on the floor at the time 0. Δ_2 has default
rules similar to Δ_1 such that agent identifiers 1 and 2 of Δ_1 are exchanged, and
Δ_2 has the fact $has\_red^2_{(0)}$ instead of $has\_blue^1_{(0)}$ in Δ_1. Γ is used for specifying
state constraints in planning, but here we put $\Gamma = \emptyset$ for simplicity.

The set of actions is put $\mathcal{A} = \{ X\_to\_Y^1_{(T)}, X\_to\_Y^2_{(T)} \}$ where variables represent their instances. If the goal ω is to be achieved at time 3, it is represented as
$$\Gamma^\omega = \{ \frac{: \neg\, (red\_on\_blue^1_{(3)} \wedge blue\_on\_yellow^1_{(3)} \wedge red\_on\_blue^2_{(3)} \wedge blue\_on\_yellow^2_{(3)})}{false} \}, \text{which}$$
states that two robots recognize the goal to be accomplished at time 3: red block
is on the blue one which is on the yellow one. A solution of a plan then becomes
$\Phi = \{ blue\_to\_yellow^1_{(1)}, red\_to\_blue^2_{(2)} \}$, which represents that the robot 1 moves
the blue block to the location of the yellow block at time 1, and the robot 2 moves
the red block to the location of the blue block at time 2.

3.2   Negotiation

In negotiation, individual agents have their own goals and make a deal to accomplish them. Here we consider negotiation between two agents.

Definition 4. (negotiation framework) A (one-to-one) *negotiation framework*
is defined as a tuple $(\mathcal{S}, \omega_1, \omega_2)$, where $\mathcal{S} = (\Delta_1, \Delta_2, \Gamma)$ is an SDT, $\omega_1 \in \mathcal{F}_1$ and
$\omega_2 \in \mathcal{F}_2$ are *goals* of agents 1 and 2, respectively. Put $\Delta_i^\omega = \Delta_i \cup \{\frac{:\neg\omega_i}{false}\}$.

In contrast to cooperative planning, each agent (or at least one of two agents) has
its own goal in a negotiation framework. If an agent i has no goal, put $\omega_i = true$.
Two agents negotiate with each other to achieve their own goals. If a *proposal*
is made by an agent, the opponent agent decides whether it is acceptable or
not. If it is unacceptable, the opponent tries to make a *counter-proposal*. A
negotiation proceeds by exchanging and evaluating mutual proposals until it
reaches a (dis)agreement. Suppose two agents Ag_1 and Ag_2, and its negotiation
framework $(\mathcal{S}, \omega_1, \omega_2)$. When Ag_1 has its goal, a negotiation proceeds according
to the following protocol.

1. If Δ_1^ω is inconsistent and $\Delta_1^\omega \cup \{\phi\}$ has a consistent extension for some $\phi \in \mathcal{F}_2$,
   Ag_1 makes a proposal ϕ to Ag_2.
2. If Δ_2 has an extension E such that $E \cup \{\phi\}$ is consistent, Ag_2 returns the
   proposition $accept_\phi$ to Ag_1. Else if for some $\psi \in \mathcal{F}_1$, $\Delta_2 \cup \{\psi\}$ has an extension G such that $G \cup \{\phi\}$ is consistent, Ag_2 returns a counter-proposal ψ to
   Ag_1. Otherwise, Ag_2 returns the proposition $reject_\phi$ to Ag_1.
3. If the response made by Ag_2 is $accept_\phi$, the negotiation ends in success.
   Else if the response is $reject_\phi$, Ag_1 seeks another condition $\phi' \in \mathcal{F}_2$ to
   satisfy Step 1. If any condition exists, repeat Step 2; if nothing exists, the
   negotiation ends in failure. Otherwise, repeat Step 2 for evaluating ψ in Δ_1.
4. Iterate Step 2 and Step 3 until one of the agents gets a response of $accept_\phi$
   for some ϕ (negotiation succeeds) or $reject_\phi$ for any ϕ (negotiation fails).

Example 4. A buyer agent 1 wants to buy a PC with a discount price. She has no cash and wants to pay by card. A seller agent 2 sells a PC with a normal price, but a discount price is applied if the buyer pays by cash or accepts a used PC. The situation is represented by a negotiation framework $(\mathcal{S}, \omega_1, \omega_2)$ where

$$\Delta_1 = \{\ \neg pay\_cash^1,\quad \frac{discount^2:}{buy^1}\ \},$$

$$\Delta_2 = \{\ \frac{:normal^2}{normal^2},\ \frac{:\neg discount^2}{\neg discount^2},\ \frac{discount^2:}{\neg normal^2},\ \frac{pay\_cash^1:discount^2}{discount^2},\ \frac{used\_pc^1:discount^2}{discount^2}\ \}.$$

The buyer has the goal $\omega_1 = buy^1$, then the agent 1 starts negotiation. As $\Delta_1^\omega \cup \{\ discount^2\ \}$ has a consistent extension, the agent 1 proposes $\phi = discount^2$ to the seller. As Δ_2 has no extension which is consistent with $\{\phi\}$, the seller cannot accept ϕ as it is. The agent 2 then seeks a condition to accept it. Since $\Delta_2 \cup \{pay\_cash^1\}$ has an extension which is consistent with $\{\phi\}$, the seller returns the counter-proposal $\psi = pay\_cash^1$ to the agent 1. The buyer does not accept the proposal ψ because $\Delta_1 \cup \{\psi\}$ is inconsistent. Then, the agent 1 returns $reject_\psi$ to the agent 2. In return to this, the agent 2 seeks another condition and finds $\psi' = used\_pc^1$ as a counter-proposal. As Δ_1 has an extension which is consistent with $\{\psi'\}$, the buyer accepts the proposal ψ' and sends $accept_{\psi'}$ to the agent 2. Then, negotiation succeeds.

4   Related Work

There are some work studying default logic in distributed environments. Baral et al. [1] consider the problem of combining multiple default theories. Given a set $\{\Delta_1, \ldots, \Delta_n\}$ of default theories and a set IC of integrity constraints as first-order formulas, they compute maximal subsets of the combination $\Delta_1 \cup \cdots \cup \Delta_n$ which are consistent with IC. Their goal is to resolve inconsistencies that may arise by combining different default theories. Ryzko and Rybinski [7] introduce *distributed default logic* to realize distributed problem solving in multiagent systems. The purpose is computing extensions of a default theory by its partitions using a distributed algorithm. Brewka et al. [2] introduce *contextual default systems* (CDS) as a tuple $(\Delta_1, \ldots, \Delta_n)$ of contextual default theories. Here, contextual default theories corresponds to default theories of our Definition 1. Compared with SDT, CDS has no notion of social constraints in its syntax. In semantics, *contextual extensions* of CDS are defined as tuples of extensions of individual theories, which is different from social extensions in SDT. Moreover, CDS does not exhibit its application to cooperative planning or negotiation. Sakama [8] uses default logic to represent weak belief of an agent, which could be abandoned during negotiation. He uses *super normal defaults* of the form $\frac{:\gamma}{\gamma}$ for this purpose, but does not use other types of default rules in negotiation. Buccafurri and Caminiti [3] introduce a *social logic program* (SOLP) which has rules of the form: $head \leftarrow [selection\_condition]\{body\}$, where $selection\_condition$ specifies social conditions concerning either the cardinality of communities or particular individuals satisfying the body. SDT is different from SOLP in both language and semantics. In particular, SOLP does not have social constraints which describe regulations in a society. Using the connection to logic programming [4], a subclass of SDT is represented by a collection of logic programs and social extensions are computed as *answer sets* of such programs. A number of studies provide logical frameworks for negotiation or cooperative planning. The purpose of this paper is not only formulating particular social reasoning, but providing a logic for representing and reasoning about social attitudes of multiple agents.

References

1. Baral, C., Kraus, S., Minker, J., Subrahmanian, V.S.: Combining default logic databases. J. Cooperative Information Systems 3, 319–348 (1994)
2. Brewka, G., Roelofsen, F., Serafini, L.: Contextual Default Reasoning. In: Proc. IJCAI 2007, pp. 268–273 (2007)
3. Buccafurri, F., Caminiti, G.: A social semantics for multi-agent systems. In: Baral, C., Greco, G., Leone, N., Terracina, G. (eds.) LPNMR 2005. LNCS (LNAI), vol. 3662, pp. 317–329. Springer, Heidelberg (2005)
4. Gelfond, M., Lifschitz, V.: Classical negation in logic programs and disjunctive databases. New Generation Computing 9, 365–385 (1991)
5. Gelfond, M., Lifschitz, V., Przymusinska, H., Truszczynski, M.: Disjunctive defaults. In: Proc. KR 1991, pp. 230–237 (1991)
6. Reiter, R.: A logic for default reasoning. Artificial Intelligence 13, 81–132 (1980)
7. Ryzko, D., Rybinski, H.: Distributed default logic for multi-agent system. In: Proc. Int'l Conf. Intelligent Agent Technology, pp. 204–210 (2006)
8. Sakama, C.: Inductive negotiation in answer set programming. In: Baldoni, M., Son, T.C., van Riemsdijk, M.B., Winikoff, M. (eds.) DALT 2008. LNCS (LNAI), vol. 5397, pp. 143–160. Springer, Heidelberg (2009)

nfn2dlp and nfnsolve: Normal Form Nested Programs Compiler and Solver

Annamaria Bria, Wolfgang Faber, and Nicola Leone

Department of Mathematics, University of Calabria, 87036 Rende (CS), Italy
{a.bria,faber,leone}@mat.unical.it

Abstract. Normal Form Nested (*NFN*) programs have recently been introduced in order to allow for enriching the syntax of disjunctive logic programs under the answer sets semantics. In particular, heads of rules can be disjunctions of conjunctions, while bodies can be conjunctions of disjunctions. Different to many other proposals of this kind, *NFN* programs may contain variables, and a notion of safety has been defined for guaranteeing domain independence. Moreover, *NFN* programs can be efficiently translated to standard disjunctive logic programs (*DLP*).

In this paper we present the tool nfn2dlp, a compiler for *NFN* programs, which implements an efficient translation from safe *NFN* programs to safe *DLP* programs. The answer sets of the original *NFN* program can be obtained from the answer sets of the transformed program (which in turn can be obtained by using a *DLP* system) by a simple transformation. The system has been implemented using the object-oriented programming language Ruby and Treetop, a language for Parsing Expression Grammars (PEGs). It currently produces *DLP* programs in the format of DLV. The separate script nfnsolve uses DLV as a back-end to compute answer sets for *NFN* programs. Thus, combining the two tools we obtain a system which supports the powerful *NFN* language, and is available for experiments.

1 Introduction

Disjunctive logic programming under the answer set semantics (*DLP*, *ASP*) has been acknowledged as a versatile formalism for knowledge representation and reasoning during the last decade. The heads (resp. the bodies) of *DLP* rules are disjunctions (resp. conjunctions) of simple constructs, viz. atoms and literals. In [1], we proposed Normal Form Nested programs that are an extension of Disjunctive Logic Programs with variables. In particular the head of an *NFN* rule is a formula in disjunctive normal form; while the body is a formula in conjunctive normal form. We provided also a polynomial translation from *NFN* programs to *DLP* programs. The main idea of the algorithm is to introduce new atoms in order to rewrite conjunctions appearing in the head of the rules and disjunctions appearing in the bodies. This result allows for evaluating *NFN* programs using DLP systems, such as DLV [2], GnT [3], or Cmodels3 [4].

In this paper we describe a tool implementing the efficient translation from safe *NFN* programs to safe *DLP* programs presented in [1], called nfn2dlp. The system

E. Erdem, F. Lin, and T. Schaub (Eds.): LPNMR 2009, LNCS 5753, pp. 477–482, 2009.

provides an *NFN* parser and safety checker, and an efficient translation to an equivalent *DLP* program. The output program is in the format of DLV, state-of-the-art implementation for disjunctive logic programs under the answer set semantics, and thus allows for effective answer set computation of *NFN* programs. A second tool, called `nfnsolve`, automates this procedure and directly computes answer sets for *NFN* programs by translating them into *DLP* programs (in the same way as `nfn2dlp`), and then invoking DLV on them, filtering out all symbols that have been introduced during the translation to produce the answer sets of the input *NFN* programs.

2   Normal Form Nested Programs

In this section, we briefly introduce syntax, semantics and safety of *NFN* programs. For a more detailed discussion, we refer to [1].

Syntax. We consider a first-order language without function symbols. *NFN* programs are finite sets of rules of the form

$$C_1 \vee \ldots \vee C_n :\text{-} D_1, \ldots, D_m. \qquad n, m \geq 0$$

where each of $C_1, \ldots, C_n$ is a positive basic conjunction $(a_1, \ldots, a_n)$ of atoms $a_1, \ldots, a_n$ and each of $D_1, \ldots, D_m$ is a basic disjunction $(l_1 \vee \ldots \vee l_n)$ of literals $l_1, \ldots, l_n$. The parentheses around basic conjunctions and disjunctions may be omitted. $C_1 \vee \ldots \vee C_n$ is the *head*, and $D_1, \ldots, D_m$ is the *body* of a rule. An *NFN* program is called *standard* if all basic conjunctions and disjunctions are singleton literals.

In our experience, the need for going beyond DLP arises relatively often in real world applications. As an example, we recall a consistent query answering setting from [1]: According to [5], a global relation $p(ID, name, surname, age)$ (for persons) with a key-constraint on the first attribute ID is "repaired" by intensionally deleting one of them whenever two tuples would share the same key. In DLP, this is done by the following rules ($\overline{p}$ denotes deleted tuples, p' the resulting consistent relation).

$$\overline{p}(I, N, S, A) \vee \overline{p}(I, M, T, B) :\text{-} p(I, N, S, A), p(I, M, T, B), N \neq M.$$
$$\overline{p}(I, N, S, A) \vee \overline{p}(I, M, T, B) :\text{-} p(I, N, S, A), p(I, M, T, B), S \neq T.$$
$$\overline{p}(I, N, S, A) \vee \overline{p}(I, M, T, B) :\text{-} p(I, N, S, A), p(I, M, T, B), A \neq B.$$
$$p'(I, N, S, A) :\text{-} p(I, N, S, A), \textbf{not } \overline{p}(I, N, S, A).$$

The first three DLP rules can be written as a single *NFN* rule.

$$\overline{p}(I, N, S, A) \vee \overline{p}(I, M, T, B) :\text{-} p(I, N, S, A), p(I, M, T, B), (N \neq M \vee S \neq T \vee A \neq B).$$

Safety. Let r be an *NFN* rule. A variable X in r is *restricted* if there exists a positive basic disjunction D in the body of r, such that, for each $a \in D$, X occurs in a; we also say that D saves X and X is made safe by D. A rule is safe if each variable appearing in the head and each variable that appears in a negative body literal are restricted. An *NFN* program is safe if each of its rules is safe.

Safe programs have the important property of domain independence, that is, their semantics is invariant with respect to the given universe (as long as it is large enough).

Semantics. We consider ground instantiations of *NFN* programs with respect to a given universe. When considering safe *NFN* programs, the Herbrand universe is sufficient. An *interpretation* for a safe *NFN* program P can therefore be denoted as a subset of the Herbrand base. The satisfaction of ground rules by interpretations is defined in the classical way, interpreting rules as implications. An interpretation that satisfies a program is called a *model*.

The *reduct* of a ground program P with respect to an interpretation I, denoted by P^I, is obtained by (1) deleting all false literals w.r.t. I from rule bodies, and (2) deleting all rules s.t. any basic disjunction becomes empty after (1). An interpretation I is an *answer set* for P iff I is a subset-minimal model for P^I. We denote the set of answer sets for P by $AS(P)$.

3   An Efficient Translation from *NFN* to *DLP*

In this section we will review the rewriting algorithm *rewriteNFN* from [1], to which we refer for a more detailed description. The basic structure of *rewriteNFN* is shown in Fig. 1. The input for *rewriteNFN* is a safe *NFN* program P and it builds and eventually returns a safe standard *DLP* program, P_{DLP}. The algorithm transforms one rule at a time. For each *NFN* rule, it constructs one *major rule*, which maintains the structure of the *NFN* rule, replacing complex head and body structures by appropriate labels. Head and body of the major rule are built independently by means of functions *buildHead* and *buildBody*, respectively, which will be described in the sequel of this section. These functions may also create a number of auxiliary rules, for defining labels and auxiliary predicates which are needed mostly for guaranteeing safety of the transformed program.

3.1   Head Transformation

Function *buildHead* is comparatively lightweight and replaces non-singular nested structures by fresh label atoms. For each head conjunction C of a rule r containing more than one atom, a label atom with the fresh predicate name $auxh_C^r$ and all variables in C is created in its place. In order to act as a substitute for C, the function also creates auxiliary rules $auxh_C^r(\ldots) :- C.$ and $a_i :- auxh_C^r(\ldots).$ for each $a_i \in C$. The safety of the auxiliary rules is straightforward, and the safety of the major rule is guaranteed by the safety of the original *NFN* program and the body transformation described next.

begin *rewriteNFN*
Input: *NFN* program P
Output: *DLP* program P_{DLP}.
var B: conjunction of literals; H: disjunction of atoms;
  $P_{DLP} := \emptyset$;
  for each rule $r \in P$ **do**
    $H := buildHead(H(r), P_{DLP})$;
    $B := buildBody(B(r), P_{DLP})$;
    $P_{DLP} := P_{DLP} \cup \{H :- B.\}$;
  return P_{DLP};

Fig. 1. Algorithm *rewriteNFN*

3.2   Body Transformation

More care has to be taken in function *buildBody*. Since not all variables in a safe *NFN* rule body have to be restricted, just replacing body disjunctions by labels as for *NFN* heads may result in an unsafe auxiliary rule because of an unrestricted variable. If the variable in question occurs only in its body disjunction, it can be safely dropped from the label atom, but if this variable occurs also elsewhere in the rule, the values it represents must match in each of its occurrences, while in some occurrences the variable may not be bound to any value. Therefore, *buildBody* focuses on *shared variables*, where a variable X is *shared* in a rule r, if it appears in two different body disjunctions of r, or if X appears in both head and body of the rule.

For creating the body of the major rule, *buildBody* replaces each body disjunction D of a rule r containing more than one literal by a label atom $aux_D^r(V_1, \ldots, V_n)$, where aux_D^r is a fresh symbol and $V_1, \ldots, V_n$ are the shared variables of r occurring in D. An auxiliary rule for defining $aux_D^r(V_1, \ldots, V_n)$ is added for each literal in D, where variables not occurring in a literal are replaced by the special constant $\#u$, representing that the respective variable is not bound in this occurrence. Moreover, if the literal is negative, some new *universe* atoms (see [1]) are added to the body defining the label atom, which in turn are defined by appropriate auxiliary rules. Since $\#u$ has to match with any other constant, matching has to be made explicit in the body of the major rule by adding dedicated atoms, which are also defined by auxiliary rules.

3.3   Properties of the Algorithm

Let P a safe *NFN* program, $P_{DLP} = rewriteNFN(P)$, and $\mathcal{A}_N$ and $\mathcal{A}_D$ be the sets of predicate symbols that appear in P and in P_{DLP}, respectively ($\mathcal{A}_N \subseteq \mathcal{A}_D$). Then, there is a bijection between $AS(P_{DLP})$ and $AS(P)$ such that $J \in AS(P_{DLP})$ iff $J \cap \mathcal{A}_N \in AS(P)$. As mentioned previously, all rules generated by *rewriteNFN* are safe. Moreover, the complexity of the algorithm is a small polynomial.

4   Systems nfn2dlp and nfnsolve

Algorithm *rewriteNFN*, along with an *NFN* parser and safety checker has been implemented as a front-end to *DLP* systems. Currently, the syntax of the system DLV is supported, but the implementation is decoupled from DLV and can easily be modified for supporting other *DLP* systems such as GnT or Cmodels3. The resulting tools, called nfn2dlp (for translating only) and nfnsolve (for additionally invoking a *DLP* backend), are publicly available at http://www.mat.unical.it/software/nfn2dlp/ . In the following we provide some information about issues in the implementation mainly of nfn2dlp. Moreover, we give a description of the usage of nfn2dlp and nfnsolve.

4.1   Implementation of nfn2dlp and nfnsolve

The tools nfn2dlp and nfnsolve have been implemented using the language Ruby [6], an object-oriented language rooted also in functional and scripting languages.

Both tools exploit an *NFN* parser, implemented using the tool `treetop` [7], which provides a parser generator for Parsing Expression Grammars (PEGs) [8] for Ruby. PEGs are a novel concept for parser specification, which look similar to classical grammars but differ in semantics; most importantly these grammars avoid ambiguity.

The tools rely on a code base which has been constructed using an object-oriented design: For all language constructs, such as atoms, literals, basic disjunctions, basic conjunctions and rules, appropriate Ruby classes exist, and the respective objects are created during parsing. The safety check has been implemented as a method of the rule class.

Moreover, two classes for handling rewriting have been defined, RewriteHead and RewriteBody, respectively. These classes contain as attributes the respective *NFN* structure (head and body, respectively), a corresponding *DLP* structure for constructing the major rule, and a set of auxiliary *DLP* rules. The methods of these classes effectively implement *buildHead* and *buildBody*.

For `nfnsolve`, all predicate symbols of the *NFN* program are collected during parsing, which are then used to filter the answer sets of the rewritten program computed by the external solver (exploiting the `-filter` option of DLV), which then represent precisely the answer sets of the *NFN* program.

Both `nfn2dlp` and `nfnsolve` provide a basic commandline interface, which we overview in Sections 4.2 and 4.3.

4.2 Using `nfn2dlp`

The interface of `nfn2dlp` is via the command-line. By default, `nfn2dlp` reads the files provided as arguments, treats their contents as one *NFN* program, analyzes its well-formedness and safety, and eventually translates it into a *DLP* program, which will be provided on standard output.

Example 1. Consider the program P represented in the text file `ex.nfn` as

$$a, b(X) :\text{-} c(X) \lor d(X, Y). \quad c(1). \quad d(2, 3).$$

In order to test for safety and to transform P into a *DLP* program, we issue

    $ nfn2dlp.rb ex.nfn

on the command line. Since the program is safe, the rewritten program is printed on standard output:

$a :\text{-} auxh1\_0(X). \quad b(X) :\text{-} auxh1\_0(X). \quad auxh1\_0(X) :\text{-} a, b(X). \quad c(1).$
$aux1\_0(X) :\text{-} c(X). \quad aux1\_0(X) :\text{-} d(X, Y). \quad auxh1\_0(X) :\text{-} aux1\_0(X). \quad d(2, 3).$

The answer sets of the *NFN* program can be computed by pipelining the output into DLV using the command

    $ nfn2dlp.rb ex.nfn | DLV --

yielding answer set $\{c(1), d(2, 3), a, auxh1\_0(1), auxh1\_0(2), b(1), b(2), aux1\_0(1), aux1\_0(2)\}$. The answer sets of the original *NFN* program P can be obtained by filtering on the original predicates in P:

    $ nfn2dlp.rb ex.nfn | DLV -- -filter=a,b,c,d

yielding the answer set $\{c(1), d(2, 3), a, b(1), b(2)\}$.

4.3 Using `nfnsolve`

Also `nfnsolve` possesses a command-line interface. As `nfn2dlp`, `nfnsolve` reads the files provided as arguments, and treats their contents as one *NFN* program, analyzes its well-formedness and safety, and eventually translates it into a *DLP* program. In addition, it invokes DLV as a backend. The location of the DLV executable can be specified following option `-d` or alternatively `--dlv`, the default being DLV in the path. Moreover, additional options can be passed on to DLV by means of the option `--dlvoptions`; care should be taken that those options should form one word for the shell, which means that usually those options should be quoted.

Example 2. Continuing Example 1 and program *P* represented in file `ex.nfn`, we can issue (provided that the default DLV is an executable in the path):

```
$ nfnsolve.rb ex.nfn
DLV [build BEN/Oct 11 2007 gcc 4.1.2]

{c(1), d(2,3), a, b(1), b(2)}
```

If the DLV executable is to be invoked as `./d` and if this executable is to be passed options `-silent` (suppressing the banner with version and compiler information) and `-nofacts` (not printing facts), we issue and obtain:

```
$       nfnsolve.rb -d ./d --dlvoptions '-silent -nofacts'
ex.nfn
{a, b(1), b(2)}
```

References

1. Bria, A., Faber, W., Leone, N.: Normal form nested programs. In: Hölldobler, S., Lutz, C., Wansing, H. (eds.) JELIA 2008. LNCS (LNAI), vol. 5293, pp. 76–88. Springer, Heidelberg (2008)
2. Leone, N., Pfeifer, G., Faber, W., Eiter, T., Gottlob, G., Perri, S., Scarcello, F.: The DLV System for Knowledge Representation and Reasoning. ACM TOCL 7(3), 499–562 (2006)
3. Janhunen, T., Niemelä, I.: GNT — A solver for disjunctive logic programs. In: Lifschitz, V., Niemelä, I. (eds.) LPNMR 2004. LNCS (LNAI), vol. 2923, pp. 331–335. Springer, Heidelberg (2004)
4. Lierler, Y.: Disjunctive Answer Set Programming via Satisfiability. In: Baral, C., Greco, G., Leone, N., Terracina, G. (eds.) LPNMR 2005. LNCS (LNAI), vol. 3662, pp. 447–451. Springer, Heidelberg (2005)
5. Bertossi, L.E.: Consistent query answering in databases. SIGMOD Record 35(2), 68–76 (2006)
6. Flanagan, D., Matsumoto, Y.: The Ruby Programming Language. O'Reilly, Sebastopol (2008)
7. Sobo, N.: Treetop homepage, http://treetop.rubyforge.org/
8. Ford, B.: Parsing expression grammars: a recognition-based syntactic foundation. In: POPL 2004, pp. 111–122 (2004)

An ASP System with Functions, Lists, and Sets

Francesco Calimeri, Susanna Cozza, Giovambattista Ianni, and Nicola Leone

Department of Mathematics, University of Calabria, I-87036 Rende (CS), Italy
{calimeri,cozza,ianni,leone}@mat.unical.it

Abstract. We present DLV-Complex, an extension of the DLV system that features the support for a powerful (possibly recursive) use of functions, list and set terms in the full ASP language with disjunction and negation.

Any computable function can be encoded in a rich and fully declarative KRR language, ensuring termination on all programs belonging to the recently introduced class of finitely-ground programs; furthermore, termination can be "a priori" guaranteed on demand by means of a syntactic restriction check that ensures a finite-domain property.

The system, which is already successfully used in many universities and research institutes, comes also equipped with a rich library of built-in functions and predicates for the manipulation of complex terms.

1 System Language

A *term* is either a *simple term* or a *complex term*. A simple term is either a constant or a variable. A complex term is either a *functional*, a *list* or a *set* term. If $t_1 \ldots t_n$ are terms, then complex terms are defined as follows.

- A *functional term* is defined as $f(t_1, \ldots, t_n)$, where f is a function symbol (*functor*) of arity n.
- A *list term* can be of the two forms:
  - $[t_1, \ldots, t_n]$, where t_1, ..., t_n are terms;
  - $[h|t]$, where h (the head of the list) is a term, and t (the tail of the list) is a list term.
- A *set term* is defined as: $\{t_1, \ldots, t_n\}$, where $\{t_1, \ldots, t_n\}$ do not contain any variable.

Some functional terms, referred to as *built-in* functions, are predefined, and have a fixed meaning. Syntactically, their functors are prefixed by $\#$. Such kind of functional term is supposed to be substituted by the value resulting from the application of the functor to its arguments, according to some predefined semantics. For this reason built-in functions are also referred to as *interpreted* functions.[1]

If $t_1, \ldots, t_k$ are terms, then $p(t_1, \ldots, t_k)$ is an *atom*, where p is a predicate of fixed arity $k \geq 0$; by $p[i]$ we denote its i-th argument. Atoms prefixed by $\#$ are instances

---

[1] The system herein presented provides a number of interpreted functions, including arithmetic operations and list and set manipulation functions.

E. Erdem, F. Lin, and T. Schaub (Eds.): LPNMR 2009, LNCS 5753, pp. 483–489, 2009.

of *built-in* predicates. Such kind of atoms are evaluated as true or false by means of operations performed on their arguments, according to some predefined semantics.[2]

A *literal* l is of the form a or not a, where a is an atom; in the former case l is *positive*, and in the latter case *negative*. A *rule* r is of the form $\alpha_1 \vee \cdots \vee \alpha_k :- \beta_1, \ldots, \beta_n, \text{not } \beta_{n+1}, \ldots, \text{not } \beta_m.$ where $m \geq 0$, $k \geq 0$; $\alpha_1, \ldots, \alpha_k$ and $\beta_1, \ldots, \beta_m$ are atoms. We define $H(r) = \{\alpha_1, \ldots, \alpha_k\}$ (the *head* of r) and $B(r) = B^+(r) \cup B^-(r)$ (the *body* of r), where $B^+(r) = \{\beta_1, \ldots, \beta_n\}$ (the *positive body* of r) and $B^-(r) = \{\text{not } \beta_{n+1}, \ldots, \text{not } \beta_m\}$ (the *negative body* of r). If $H(r) = \emptyset$ then r is a *constraint*; if $B(r) = \emptyset$ and $|H(r)| = 1$ then r is referred to as a *fact*.

A rule is safe if each variable in that rule also appears in at least one positive literal in the body of that rule. For instance, the rule $p(X, f(Y, Z)) :- q(Y), \text{not } s(X).$ is not safe, because of both X and Z. From now on we assume that all rules are safe and there is no constraint.[3] An ASP program is a finite set P of rules. As usual, a program (a rule, a literal) is said to be *ground* if it contains no variables.

A thorough discussion about the semantics, based on the notion of Answer Set proposed in [1], can be found in [2].

2   Knowledge Representation with DLV-Complex

The main strength point of DLV-Complex, besides all those shared with DLV [3], is the featured language, that includes a full usage of function symbols and a native support for list and set terms. Function symbols, and other complex terms, allow to aggregate atomic data, manipulate complex data structures and generate new symbols (*value invention*). This paves the way to a more natural knowledge representation.

Example 1. The term `notify(delete(F))` in the atom `request(S, I, notify (delete(F)), T_0)` is a functional term. Other examples of functional terms are `father(X)` and `mother(X)` in the atom `family_trio(X, father(X), mother (X))`, or `f(g(a))` in the atom `p(f(g(a)))`.

Lists can be profitably exploited in order to explicitly model collections of objects where position matters, and repetitions are allowed.

Example 2. The term `[a,d,a]` in the atom `palindromic([a,d,a])` is a list term. Other examples of list terms are `[jan,feb,mar]`, or `[jan,feb|[mar,apr,may, jun]]`, or `[[jan,31]| [[feb,28],[mar,31],[apr,30],[may,31],[jun, 30]]]`.

Set terms are used to model collections of data having the usual properties associated with the mathematical notion of set. They satisfy idempotence (i.e., sets have no duplicate elements) and commutativity (i.e., two collections having the same elements but with a different order represent the same set) properties.

---

[2] The system herein presented provides some simple built-in predicates, such as the comparative predicates equality, less-than, and greater-than ($=, <, >$), list and set predicates like #member, #subset.

[3] Under Answer Set semantics, a constraint :- B(r) can be simulated through the introduction of a standard rule fail :- B(r), not fail, where fail is a fresh predicate not occurring elsewhere in the program.

Example 3. The term $\{john, carl, ann, mary\}$ in the atom seatAt(1,{john,carl, ann, mary}) is a set term. Other examples of set terms are {red,green,blue}, or {[red,5], [blue,3], [green,4]}, or {{red,green},{red,blue},{green, blue}}. It is worth remembering that duplicated elements are ignored, thus the terms {red,green,blue} and {green, red,blue,green} actually represent the same set.

In the following we illustrate the usage of DLV-Complex as a tool for knowledge representation and reasoning, by means of some examples.[4]

Example 4. Suppose that a system administrator wants to model the following security policy about file deletion: 'A certain subject (a person, an agent,..) S is permitted to delete a file F if at a time T_0 she/it sends a request to a target institution I, notifying the intention of deleting F, and there are no explicit request from I to S to retain F in the next ten units of time'. The following rule can be used to naturally express this policy thanks to proper functional terms.

$$permitted(S, delete(F), T_1) :\text{-} \ request(S, I, notify(delete(F)), T_0),$$
$$\textbf{not} \ requestInBetween(I, S, retain(F), T_0, T_1),$$
$$T_1 = T_0 + 10.$$

Example 5. We show now how lists can be used to model strings of characters. Let us consider the following facts:

$$word([a, d, a]). \quad word([g, i, b, b, i]). \quad word([a, n, n, a]).$$

The following rule:

$$palindromic(X) :\text{-} \ word(X), \#reverse(X) = X.$$

allows to easily deduce words that are palindromic by using the interpreted function $\#reverse$ for inverting the order of the elements in a list, and the built-in predicate '=' for comparison. The program consisting of the two rules above would have the unique following answer set:

$$\{palindromic([a, d, a]), palindromic([a, n, n, a])\}.$$

Example 6. Let us consider the famous "Towers of Hanoi" puzzle [5]. Assume that a possible move is represented by the predicate possible_move, featuring three attributes: the first represents the move number, the second the state of the three stacks before applying the current move, and the last the state of the three stacks after the move has been applied. Then, we can encode a possible move from the first stack to the second stack by means of the following rule:

$$possible\_move(\#succ(I), towers([X|S1], S2, S3), towers(S1, [X|S2], S3)) :\text{-}$$
$$possible\_state(I, towers([X|S1], S2, S3)),$$
$$legalMoveNumber(I), legalStack([X|S2]).$$

Roughly, the top element of the first stack can be moved on top of the second stack if: (i) the current state for stacks is admissible, i.e. this state can be reached after applying

---

[4] Full encodings for problems presented in Example 6 and Example 7 can be found at [4].

a sequence of "I" moves(possible_state(I,towers([X|S1],S2,S3))); (ii) the number "I" is in the range of allowed move numbers (legalMoveNumber(I)); (iii) the new resulting configuration for the second stack is legal, i.e. there is no larger disc on top of a smaller one (legalStack([X|S2])).

It is worth noting the use "à la Prolog" ([Head|Tail]) for lists, and the role of the interpreted function #succ, whose meaning is straightforward.

Example 7. If facts like: $sons(\ someone,\ \{son_1, ..., son_n\})$ model the association between a parent and her/his sons, one can make the transitive closure and obtain the names of all descendants of someone by means of the following rules:

$$ancestor(A, Ss) :\!- sons(A, Ss).$$
$$ancestor(A, \#union(Ds, Ss)) :\!- ancestor(A, Ds), \#member(S, Ds),$$
$$sons(S, Ss).$$

where the first argument of the predicate "ancestor" represents the name of a person (ancestor) and the second argument represents the set of names of descendants of this person. The first rule says that all sons of A are descendants of A. The second rule says that if Ds are descendants of A, S belongs to the set of descendants Ds, and Ss is the set of all sons of S, then the set resulting from the union of the sets Ds and Ss is also a set of descendants for A. The second rule makes use of the interpreted function *#union* and the built-in predicate *#member*, which have intuitive meaning.

Example 8. Let us imagine that the administrator of a social network wants to increase the connections between users. In order to do that, (s)he decides to propose a connection to pairs of users that result, from their personal profile, to share more than two interests. If the data about users are given by means of EDB atoms of the form $user(id,\ \{interest_1, \dots, interest_n\})$, the following rule would compute the set of common interests between all pairs of users:

$$sharedInterests(U_1, U_2, \#intersection(S_1, S_2)) :\!- user(U_1, S_1), user(U_2, S_2), U_1 \neq U_2.$$

where the interpreted function *#intersection* takes as input two sets and returns their intersection. Then, the predicate selecting all pairs of users sharing more than two interests could be defined as follows:

$$proposeConnection(pair(U_1, U_2)) :\!- sharedInterests(U_1, U_2, S), \#card(S) > 2.$$

Here, the interpreted function *#card* returns the cardinality of a given set, which is compared to the constant 2 by means of the built-in predicate ">".

3   Implementation and Usage

As previously stated, the system herein presented has been built on top of the state-of-the-art ASP system DLV [3]. The support for complex terms has been achieved by means of a proper rewriting strategy.

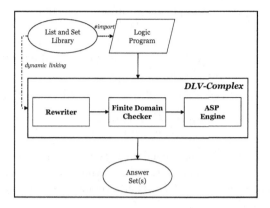

Fig. 1. *System Architecture*

Figure 1 shows an overview of the system architecture. At first, the program is processed by a rewriting module, which removes all complex terms and introduces a number of instances of predefined built-in predicates (described below). Since the evaluation of an ASP program with complex terms is not guaranteed to terminate in general [6], the rewritten program is passed to a module (that can be bypassed on demand) that checks the membership to a class that is "a priori" known to be computable, namely the class of *finite-domain* programs, recently introduced in [2]. Finally, the rewritten program is given as input to DLV, that deals with the built-in predicates introduced in the rewriting phase by means of the framework defined in [7].[5]

The DLV-Complex core (consisting of the aforementioned modules) can also link, at run-time, a library of list and set manipulation functions and predicates, if required by the original ASP program being evaluated.

We briefly illustrate in the following how the rewriting module acts in presence of functional terms. Firstly, any functional term $t = f(X_1, \ldots, X_n)$, appearing in some rule $r \in P$, is replaced by a fresh variable F; then, one of the following atom is added to $B(r)$:

- $\#function\_pack(f, X_1, \ldots, X_n, F)$ if t appears in $H(r)$;
- $\#function\_unpack(F, f, X_1, \ldots, X_n)$ if t appears in $B(r)$.

This transformation is applied to the rule r until no functional terms appear in it anymore. The arguments of the $\#function\_pack/unpack$ built-ins are:

1. the fresh variable F representing the whole functional term;
2. the function symbol f;
3. all of the arguments: $X_1, \ldots, X_n$ for the original functional term.

Therefore, an instance of the $\#function\_pack$ built-in predicate is in charge of building a functional term, starting from a functor and its arguments, while an instance of

---

[5] The DLV-EX system, discussed in [8,7], allows the definition of custom built-in atoms in the $\#p(t_1, \ldots, t_n)$ syntactic form.

the $\#function\_unpack$ built-in predicate unfolds a functional term in order to give values to its arguments. Hence, the former built-in predicate binds the F variable, provided that all other terms are already bound; the latter binds (or checks the values, in case they are already bound) the $X_1, \ldots, X_n$ variables according to the binding for the F variable (the whole functional term).

Example 9. The rule: $p(f(f(X))) :- q(X, g(X,Y)).$ will be rewritten as follow:

$$p(F_1) :- \#function\_pack(F_1, f, F_2), \#function\_pack(F_2, f, X),$$
$$q(X, F_3), \#function\_unpack(F_3, g, X, Y).$$

Note that rewriting the nested functional term $f(f(X))$ requires two $\#function\_pack$ atoms in the body: (i) for the inner f function having X as argument and (ii) for the outer f function having as argument the fresh variable F_2, representing the inner functional term.

List and set terms are treated by means of proper "pack/unpack" built-in predicates, that act analogously to what described above, with some minor adjustments, such as a flag used in order to identify the kind of list term to be managed (comma-separated enumeration or "à la prolog").

3.1   Usage

DLV-Complex per se is a command-line tool (as usual, for a brief list of command-line options, specify "-help"). In addition to the DLV standard ones (the reader may refer to the official DLV documentation [9]), some options are available in order to modify the behavior of the system in presence of complex terms.

A finite-domain syntactic checker is enabled by default. If the user is confident that the program can be grounded in finite time (that is it belongs to the class of finitely-ground programs), then she can skip the finite-domain check. This can be done by specifying the command-line option -nofdcheck. Another way that lead to the guarantee of termination is the choice of a maximum allowed nesting level N for functional terms, via the command-line option -MAXNL=<N>.

The system comes equipped with a rich library of list and set manipulation functions. This library actually contains the definition of many interpreted functions and predicates; in order to exploit them, the logic program must contain, in the preamble, a line that tells the system to include the library itself.

Example 10. Let's suppose that we have a file named palindromic.dl, containing the program reported in Example 5. Since the program exploits the #reverse interpreted function, we must also add a very first line with the instruction #include<ListAndSet>.

If the palindromic.dl file is placed in the same folder as the DLV-Complex executable (let it be dlv-complex), and the list and set library is placed in the /usr/mylib folder we can invoke the system by the following command line:[6]

```
$ dlv-complex palindromic.dl -silent -nofacts -libpath=/usr/mylibs/
```

---

[6] We suppose to work on a Linux-like platform. The system behaves analogously on other platforms.

4   Conclusions

We have presented the DLV-Complex system, an extension of the DLV system that features the support for functions, list and set terms in the full ASP language with disjunction and negation.

DLV-Complex is already successfully used in many universities and research institutes, and it is already exploited for some real-world applications; for instance, in [10] it is used for the computation of minimum cardinality diagnoses, and functional terms are here employed to replace existential quantification.

The system, which is free for non-commercial use, is available for download at [4]. The library for list and set terms manipulation is available for free download as well, together with a reference guide, a tutorial and a number of examples. For any further detail about system options and synopsis the reader is referred to such documentation.

References

1. Gelfond, M., Lifschitz, V.: Classical Negation in Logic Programs and Disjunctive Databases. NGC 9, 365–385 (1991)
2. Calimeri, F., Cozza, S., Ianni, G., Leone, N.: Computable Functions in ASP: Theory and Implementation. In: Garcia de la Banda, M., Pontelli, E. (eds.) ICLP 2008. LNCS, vol. 5366, pp. 407–424. Springer, Heidelberg (2008)
3. Leone, N., Pfeifer, G., Faber, W., Eiter, T., Gottlob, G., Perri, S., Scarcello, F.: The DLV System for Knowledge Representation and Reasoning. ACM TOCL 7(3), 499–562 (2006)
4. Calimeri, F., Cozza, S., Ianni, G., Leone, N.: DLV-Complex homepage (since 2008), http://www.mat.unical.it/dlv-complex
5. Buneman, P., Levy, L.S.: The Towers of Hanoi Problem. Information Processing Letters 10(4/5), 243–244 (1980)
6. Dantsin, E., Eiter, T., Gottlob, G., Voronkov, A.: Complexity and Expressive Power of Logic Programming. ACM Computing Surveys 33(3), 374–425 (2001)
7. Calimeri, F., Cozza, S., Ianni, G.: External sources of knowledge and value invention in logic programming. AMAI 50(3-4), 333–361 (2007)
8. Calimeri, F., Ianni, G.: External sources of computation for Answer Set Solvers. In: Baral, C., Greco, G., Leone, N., Terracina, G. (eds.) LPNMR 2005. LNCS (LNAI), vol. 3662, pp. 105–118. Springer, Heidelberg (2005)
9. Faber, W., Pfeifer, G.: DLV homepage (since 1996), http://www.dlvsystem.com/
10. Friedrich, G., Ivanchenko, V.: Diagnosis from first principles for workflow executions. Tech. Rep., http://proserver3-iwas.uni-klu.ac.at/download_area/Technical-Reports/technical_report_2008_02.pdf

A Simple Distributed Conflict-Driven Answer Set Solver

Enrico Ellguth[1], Martin Gebser[1], Markus Gusowski[1], Benjamin Kaufmann[1],
Roland Kaminski[1], Stefan Liske[1], Torsten Schaub[1,*], Lars Schneidenbach[2],
and Bettina Schnor[1]

[1] Institut für Informatik, Universität Potsdam, D-14482 Potsdam, Germany
[2] IBM Ireland, Dublin Software Lab, Mulhuddart, Dublin 15, Ireland

Abstract. We propose an approach to distributed Answer Set Solving based on
Message Passing. Our approach aims at taking advantage of modern ASP solvers
rather than proposing a genuine yet involved parallel ASP solver. To this end, we
rely upon a simple master-worker architecture in which each worker amounts to
an off-the-shelf ASP solver augmented with a separate communication module
being only lightly connected to the actual solver. The overall communication is
driven by the workers' communication modules, which asynchronously exchange
messages with the master. We have implemented our approach and report upon
an empirical study demonstrating its computational impact.

1 Introduction

Despite the progress of sequential Answer Set Solving technology, only little advance-
ment is observed in the parallel setting. This is deplorable in view of the rapidly growing
availability of clustered, multi-processor, and/or multi-core computing devices. We address this shortcoming and furnish a distributed approach to ASP solving by focusing
on the parallelization of search. Our approach builds upon the Message Passing Interface (MPI; [1]), realizing communication and data exchange between computing units
via message passing. Interestingly, MPI abstracts from the actual hardware and lets us
execute our system on clusters as well as multi-processor and/or multi-core machines.

We aim at a simple and transparent approach in order to be able to take advantage
of the high performance offered by modern off-the-shelf ASP solvers. To this end, we
have chosen a simple master-worker architecture, in which each worker consists of an
ASP solver along with an attached communication module. The solver is linked to its
communication module via an elementary interface requiring only marginal modifications to the solver. All major communication is initiated by the workers' communication
modules exchanging messages with the master in an asynchronous way.

2 Distributed Answer Set Solving

We have implemented our approach in C++ using MPI [1]. The resulting system is
called *claspar*, alluding to its underlying ASP solver *clasp* [2]. Although we tried to
keep our design generic, we took advantage of some design features of *clasp*, whose
basic search procedure can be outlined by means of the following loop [3]:

---

* Affiliated with Simon Fraser University, Canada, and Griffith University, Australia.

E. Erdem, F. Lin, and T. Schaub (Eds.): LPNMR 2009, LNCS 5753, pp. 490–495, 2009.

loop
    propagate   // compute deterministic consequences
    if no conflict **then**
        if all variables assigned **then return** variable assignment
        else *decide*   // choose a non-deterministic consequence
    else
        if top-level conflict **then return** unsatisfiable
        else
            analyze   // analyze conflict and add a conflict constraint
            backjump   // undo assignments until conflict constraint is unit

At first, the closure under deterministic consequence operations is computed. Then, four cases are distinguished. In the first one, a non-conflicting complete assignment is returned. In the second case, an unassigned variable is non-deterministically chosen and assigned. Or at last, a conflict is encountered. All assignments made before the first non-deterministic choice constitute the *top-level*. Hence, a top-level conflict indicates unsatisfiability. Otherwise, the conflict is analyzed and learned in form of a conflict constraint. Then, the algorithm backjumps by undoing a maximum number of successive assignments so that exactly one literal of the constraint is unassigned.

The *clasp* solver extends the static concept of a top-level by additionally providing a dynamic variant referred to as *root-level* [3]. As with the top-level, conflicts within the root-level cannot be resolved given that all of its variable assignments are precluded from backtracking. We build upon this feature for splitting the search space. Splitting is accomplished according to a so-called *guiding path* [4], the sequence of all non-deterministic choices. Given a root-level $i-1$, a guiding path $(v_1, \ldots, v_{i-1}, v_i, \ldots, v_n)$ can be divided into a prefix $(v_1, \ldots, v_{i-1})$ of non-splittable variables and a postfix $(v_i, \ldots, v_n)$ of splittable variables. We can split the search space at the first splittable variable by incrementing the root-level by one and dissociating a guiding path composed of the first $i-1$ variables and the complement of the ith variable, yielding $(v_1, \ldots, v_{i-1}, \overline{v_i})$. Note that the local assignment remains unchanged, and only the root-level is incremented to i. We have chosen to split at the first splittable variable because, first, this results in cutting off the largest part of the search space and, second, this way the backjumping is least restricted.

Upon enumerating answer sets, (locally) using the scheme in [5], the assignment can contain complements of non-deterministically assigned variables of previously enumerated answer sets. Such complements $\overline{u_1}, \ldots, \overline{u_j}$ indicate that the search spaces for answer sets containing $(v_1, \ldots, v_{i-1})$ and at least one of $u_1, \ldots, u_j$ have already been explored, while v_i or $\overline{v_i}$ may have belonged to already enumerated answer sets. In order to avoid repetitions, it is thus important to pass guiding path $(v_1, \ldots, v_{i-1}, \overline{u_1}, \ldots, \overline{u_j}, \overline{v_i})$ in response to a split request. This refinement for repetition-free answer set enumeration is implemented in *claspar*.

Finally, *clasp* incorporates constraint database simplifications wrt variables assigned at the top-level. In particular, conflict constraints can lead to top-level assignments, in which case the corresponding variables are eliminated from all resident constraints. The root-level plays a crucial role for whether such simplifications are applied, as variables assigned at or below the root-level but beyond the top-level are not subject to simplification. This feature is also inherited by *claspar* setting the root-level to the number of

variables in a guiding path. As a consequence, conflicts and resulting assignments due
to a nonempty guiding path (or subproblem, respectively) do not involve simplification,
while top-level assignments independent of the guiding path lead to simplifications.

3   Communication

Our approach to distribution builds upon message passing, accounting for communication and data exchange. For the sake of simplicity, we have adopted a classical master-
worker model. While the purpose of the single master is to handle the overall message
exchange, each worker amounts to an ASP solver enhanced by message handling capacities. The workers constitute the active components, initiating all requests, while the
master mainly reacts by processing the workers' requests.

Master. The main task of the master is the reception and transmission of search
(sub)problems. To accomplish this, the master divides its assigned workers into a set
of active and inactive workers. The active workers, i.e., workers assigned a not yet processed guiding path, are arranged in a queue ordered by a workload parameter. On the
other hand, the inactive workers have either finished processing their guiding paths or
have not yet been assigned any.

At the beginning, the search space has to be distributed among the workers. As initially all workers are inactive, the master receives a work request from each worker.
The first incoming work request obtains the empty guiding path, representing the entire
search space. It is then successively split and distributed among the other workers.

The overall routine of the master is driven by load balancing. A work request by a
worker normally results in a split request to another worker. The split request is sent
to a worker with putatively high workload, namely, to one with a short guiding path.
Notably, each worker determines whether and/or how often it is asked (and thus interrupted) for work. The master merely maintains its priority queue according to the
information supplied by the workers.[1] When a subproblem is returned to the master by
a split response, it is forwarded to the first worker in the request queue or put into a
cache to allow for immediate response to the next work request. Apart from the guiding
path, a split response also contains information on the workload of the sending worker.

Whenever all workers are inactive and the cache is empty, the given problem is found
to be unsatisfiable. As soon as the requested number of answer sets is computed or
unsatisfiability is established, the master asks all workers to gather runtime statistics
and to then terminate. Once all statistics are received, they are aggregated and printed
before the master terminates itself.

Worker. Our worker design is driven by the desire to minimize modifications to the
given ASP solver while keeping the overall approach as simple as possible. To this end,
we attach a module handling communications, providing an interface reacting to incoming messages during search. The interface is used at the end of the conflict analysis
in the solver loop given in Section 2. This is the only change done to the ASP solver at
hand (except for redirecting its output operations).

---

[1] Currently, this information consists of the length of the initial guiding path and the number of
choices made by the worker's ASP solver since the guiding path was received.

However, in general, the worker's communication module has two modes of operation. When out of work (no guiding path received yet or completed subproblem), the first mode cares about raising a work request and launching its ASP solver with the guiding path obtained from the master. Notably, upon such a response, *claspar* reuses the previously learnt clauses and heuristic information like variable activities. The second mode addresses split requests. For this purpose, the communication module is equipped with a heuristic function for deciding whether a guiding path is extracted from the ASP solver. If so, a guiding path is sent to the master (accompanied with some information on the workload). The current strategy is to return the shortest guiding path.

If the worker decides not to split its search space, it can signal to retry later or to send no further split requests. Once an answer set is found, the worker sends it to the master. In the case of unsatisfiability, conflict constraints accumulated over time may eventually yield a top-level conflict that is also signaled to the master, which then asks all workers to send their statistics and to terminate.

4   Experiments

Our experiments consider *claspar*[2] (0.1.0) based on *clasp* (1.0.5); they ran under MPI (mpich2-1.0.7) on the cluster described at http://www.cs.uni-potsdam.de/bs/research/labs/highland.html, each individual run restricted to 900s time and 2GB RAM per worker. Each solver instance of *claspar* was run with the default settings of *clasp* (except for the second group of *PigeonHole* benchmarks).

Table 1 summarizes benchmark results capturing the scaling capacities of *claspar*.[3] We consider a master run on a single machine plus increasing numbers of workers running on machines with double cores (and thus at most two MPI processes per machine). The single worker setting amounts to that of a serial run of *clasp*: the same number of choices and conflicts are obtained for each run, and the message passing overhead leads to an increase in execution time of less than one percent. We have selected popular benchmark classes for evaluating our approach. Among them, *BlockedQueens-sat* are satisfiable and terminated after an answer set was obtained. All remaining benchmarks are unsatisfiable and thus necessitate a complete traversal of the search space. Each setting is described by the sums of times,[4] and timeouts over all underlying instances.[5] Moreover, we give the relative speedup wrt the single worker setting and indicate the efficiency by normalizing the speedup with the number of workers in each setting. In general, we observe a steady increase in speedup although the efficiency goes down with more workers due to greater overhead. An unsteady speedup is observed on the class of satisfiable benchmarks. Even though the search for one answer set can be boosted by lucky strikes, it is surprising to see that 8 workers performed better than 16. This is different on the unsatisfiable instances in *BlockedQueens-unsat* that show a steady yet suboptimal speedup. The *GraphColoring* benchmarks are taken from this year's ASP

---

[2] Available at http://potassco.sourceforge.net

[3] The detailed table is available at http://www.cs.uni-potsdam.de/claspar

[4] Timeouts are taken as maximum time, viz., 900s.

[5] A tarball containing all benchmark instances is available at the URL given in Footnote 3.

Table 1. Scaling *claspar* from 1 to 16 workers

claspar 0.1.0	1 worker	2 workers	4 workers	8 workers	16 workers
Benchmark	*BlockedQueens-sat (7 instances)*				
Time (Timeouts)	678.82 (0)	444.06 (0)	248.47 (0)	99.74 (0)	116.23 (0)
Speedup (Efficiency)	1.00 (1.00)	1.53 (0.76)	2.73 (0.68)	6.81 (0.85)	5.84 (0.37)
Benchmark	*BlockedQueens-unsat (9 instances)*				
Time (Timeouts)	1528.72 (0)	1223.84 (0)	649.57 (0)	401.74 (0)	191.31 (0)
Speedup (Efficiency)	1.00 (1.00)	1.25 (0.62)	2.35 (0.59)	3.81 (0.48)	7.99 (0.50)
Benchmark	*GraphColoring (20 instances)*				
Time (Timeouts)	18000 (20)	16993.80 (18)	12841.70 (11)	9910.80 (6)	7063.68 (3)
Speedup (Efficiency)	1.00 (1.00)	1.06 (0.53)	1.40 (0.35)	1.82 (0.23)	2.55 (0.16)
Benchmark	*PigeonHole (2 instances)*				
Time (Timeouts)	975.66 (0)	988.66 (1)	469.45 (0)	190.48 (0)	153.20 (0)
Speedup (Efficiency)	1.00 (1.00)	0.99 (0.49)	2.08 (0.52)	5.12 (0.64)	6.37 (0.40)
Benchmark	*PigeonHole-norestarts (2 instances)*				
Time (Timeouts)	984.67 (1)	587.38 (0)	293.10 (0)	169.63 (0)	85.32 (0)
Speedup (Efficiency)	1.00 (1.00)	1.68 (0.84)	3.36 (0.84)	5.80 (0.73)	11.54 (0.72)

Table 2. Scaling *claspar*'s enumeration of answer sets within 900s from 1 to 16 workers

claspar 0.1.0	1 worker	2 workers	4 workers	8 workers	16 workers
Benchmark	*ClumpyGraphs (12 instances)*				
Models	83,866,664	153,764,698	312,272,614	610,673,252	1,247,804,500
Speedup (Efficiency)	1.00 (1.00)	1.83 (0.92)	3.72 (0.93)	7.28 (0.91)	14.88 (0.93)

competition. Here the speedup values are insignificant because one worker alone cannot even solve a single instance. However, increasing the number of workers leads to a steady decrease of timeouts, demonstrating that distribution can make a difference. Finally, we consider the well-known *PigeonHole* example, spanning a very uniform search tree. The default behavior of *claspar* allows solver instances to restart, which has a negative effect on the individual and thus global solver performance. Once restarts are inhibited, reasonable speedups are obtained.

Finally, we consider in Table 2 how *claspar* scales regarding the enumeration of answer sets. For this, we substitute runtimes by the number of answer sets obtained within 900s. We consider Hamiltonian cycles in *ClumpyGraphs*, a benchmark designed in [6] for showing the advantage of conflict-driven learning ASP solvers. We observe that the distribution of search incurs only minor overhead in *claspar*'s efficiency, and the speedup closely follows the number of workers. This nicely demonstrates the computational impact of distributed ASP solving.

5 Discussion

We proposed a simple approach to distributed ASP solving based on the well-known Message Passing Interface. To this end, we rely upon a master-worker architecture in which each worker amounts to an off-the-shelf ASP solver augmented with a separate communication module being only lightly connected to the actual solver. The simplicity is pragmatically best reflected by the fact that *claspar* changes less than a dozen lines of code in *clasp* while adding another ~350 lines for handling distribution.

Our approach differs from existing work in several respects. In fact, it provides the first distributed version of an ASP solver using conflict-driven learning. Unlike this, existing distributed ASP solvers rely on classical backtracking schemes that provide a much tighter control over the search space. Second, our approach endows each solver instance with great independence, principally allowing for a variety of different ASP solvers at the same time. Although the *platypus* framework [7,8] also aims at a certain genericity, this applies only to the deterministic part of the solvers, while distributed search is controlled by *platypus* itself. The approach of [9,10] goes even further in building a genuine parallel solver based on sequential ASP solver *smodels*.

Acknowledgments. This work was partially funded by DFG grant SCHA 550/8-1.

References

1. Gropp, W., Lusk, E., Thakur, R.: Using MPI-2: Advanced Features of the Message-Passing Interface. MIT Press, Cambridge (1999)
2. Gebser, M., Kaufmann, B., Neumann, A., Schaub, T.: Conflict-driven answer set solving. In: Proceedings IJCAI 2007, pp. 386–392. AAAI Press/MIT Press (2007)
3. Eén, N., Sörensson, N.: An extensible SAT-solver. In: Giunchiglia, E., Tacchella, A. (eds.) SAT 2003. LNCS, vol. 2919, pp. 502–518. Springer, Heidelberg (2004)
4. Zhang, H., Bonacina, M., Hsiang, J.: PSATO: a distributed propositional prover and its application to quasigroup problems. Journal of Symbolic Computation 21(4), 543–560 (1996)
5. Gebser, M., Kaufmann, B., Neumann, A., Schaub, T.: Conflict-driven answer set enumeration. In: Baral, C., Brewka, G., Schlipf, J. (eds.) LPNMR 2007. LNCS (LNAI), vol. 4483, pp. 136–148. Springer, Heidelberg (2007)
6. Ward, J., Schlipf, J.: Answer set programming with clause learning. In: Lifschitz, V., Niemelä, I. (eds.) LPNMR 2004. LNCS (LNAI), vol. 2923, pp. 302–313. Springer, Heidelberg (2004)
7. Gressmann, J., Janhunen, T., Mercer, R., Schaub, T., Thiele, S., Tichy, R.: Platypus: A platform for distributed answer set solving. In: Baral, C., Greco, G., Leone, N., Terracina, G. (eds.) LPNMR 2005. LNCS (LNAI), vol. 3662, pp. 227–239. Springer, Heidelberg (2005)
8. Gressmann, J., Janhunen, T., Mercer, R., Schaub, T., Thiele, S., Tichy, R.: On probing and multi-threading in platypus. In: Proceedings ECAI 2006, pp. 392–396. IOS Press, Amsterdam (2006)
9. Pontelli, E., Balduccini, M., Bermudez, F.: Non-monotonic reasoning on Beowulf platforms. In: Dahl, V., Wadler, P. (eds.) PADL 2003. LNCS, vol. 2562, pp. 37–57. Springer, Heidelberg (2003)
10. Balduccini, M., Pontelli, E., El-Khatib, O., Le, H.: Issues in parallel execution of non-monotonic reasoning systems. Parallel Computing 31(6), 608–647 (2005)

An Implementation of Belief Change Operations Based on Probabilistic Conditional Logic*

Marc Finthammer[1], Christoph Beierle[1], Benjamin Berger[1],
and Gabriele Kern- Isberner[2]

[1] Dept. of Computer Science, FernUniversität in Hagen, 58084 Hagen, Germany
[2] Dept. of Computer Science, TU Dortmund, 44221 Dortmund, Germany

Abstract. Probabilistic conditionals are a powerful means for expressing uncertain knowledge. In this paper, we describe a system implemented in Java performing probabilistic reasoning at optimum entropy.
It provides nonmonotonic belief change operations like revision and update and supports advanced querying facilities including diagnosis and
what-if-analysis.

1 Introduction

Conditionals of the form *"if A then B with probability x"* use the well-studied
mathematical foundation of probability theory for expressing uncertain knowledge. The MEcoRe system implements reasoning over probabilistic conditional
knowledge bases by employing the concept of optimum entropy [1,2,3]. Besides
providing the core functionalities needed for probabilistic reasoning at optimum
entropy, the main objective of MEcoRe is to support advanced belief management operations like revision, update, diagnosis, or what-if-analysis in a most
flexible and easily extendible way. MEcoRe can be looked upon as the realization
of an intelligent agent being able to accept knowledge from the environment and
to change her own epistemic state in the light of new information. In this paper, we briefly sketch the background of probabilistic conditional logic, provide
a system walkthrough by means of a small example, and provide an overview of
MEcoRe's implementation.

2 Probabilistic Logic and Optimum Entropy in a Nutshell

We start with a propositional language $\mathcal{L}$, generated by a finite set Σ of (binary)
atoms $a, b, c, \ldots$. The formulas of $\mathcal{L}$ will be denoted by uppercase Roman letters
$A, B, C, \ldots$. For conciseness of notation, we will omit the logical *and*-connector,
writing AB instead of $A \wedge B$, and overlining formulas will indicate negation, i.e.
$\overline{A}$ means $\neg A$. Let Ω denote the set of possible worlds over $\mathcal{L}$; Ω will be taken here
simply as the set of all propositional interpretations over $\mathcal{L}$ and can be identified

---

* The research reported here was supported by the Deutsche Forschungsgemeinschaft
(grants BE 1700/7-1 and KE 1413/2-1).

E. Erdem, F. Lin, and T. Schaub (Eds.): LPNMR 2009, LNCS 5753, pp. 496–501, 2009.

with the set of all complete conjunctions over Σ. For $\omega \in \Omega$, $\omega \models A$ means that the propositional formula $A \in \mathcal{L}$ holds in the possible world ω. By introducing a new binary operator $|$, we obtain the set $(\mathcal{L} \mid \mathcal{L}) = \{(B|A) \mid A, B \in \mathcal{L}\}$ of (unquantified) *conditionals* (or *rules*) over $\mathcal{L}$. $(B|A)$ formalizes "*if A then B*" and establishes a plausible, probable, possible etc connection between the *antecedent* A and the *consequent* B. The set Sen_C contains all *probabilistic conditionals* (or *probabilistic rules*) of the form $(B|A)[x]$ where x is a probability value $x \in [0, 1]$.

To give appropriate semantics to conditionals, they are usually considered within richer structures such as *epistemic states*. Besides certain (logical) knowledge, epistemic states also allow the representation of e.g. preferences, beliefs, assumptions of an intelligent agent. In a quantitative framework, most appreciated representations of epistemic states are provided by *probability functions* (or *probability distributions*) $P : \Omega \to [0, 1]$ with $\sum_{\omega \in \Omega} P(\omega) = 1$. Thus, in this setting, the set of *epistemic states* we will consider is $EpState = \{P \mid P : \Omega \to [0, 1]$ is a probability function$\}$. The probability of a formula $A \in \mathcal{L}$ is given by $P(A) = \sum_{\omega \models A} P(\omega)$, and the probability of a conditional $(B|A) \in (\mathcal{L} \mid \mathcal{L})$ with $P(A) > 0$ is defined as $P(B|A) = P(AB)/P(A)$, the corresponding conditional probability. Conditionals are interpreted via conditional probability. So the satisfaction relation $\models_C \subseteq EpState \times Sen_C$ of probabilistic conditional logic is defined by $P \models_C (B|A)[x]$ iff $P(B|A) = x$.

A central notion for the optimum entropy is the cross-entropy $H_{ce}(P', P)$ between two distributions P', P that can be taken as a directed (i.e. asymmetric) information distance [3]. Given a set R of probabilistic conditionals and a distribution P, $MinCEnt(P, R)$ denotes the *unique* distribution satisfying R and having minimum cross-entropy to P, i.e. $MinCEnt(P, R)$ solves the minimization problem [1,2,3]

$$\arg \min_{P' \models_C R} H_{ce}(P', P) = \sum_{\omega} P'(\omega) \log \frac{P'(\omega)}{P(\omega)} \tag{1}$$

3   Walkthrough by Example

We will demonstrate the purpose and the application of the MECoRe system by a small example. This example will show (in the actual MECoRe input syntax) how to build a knowledge base and how to perform some of the knowledge management operations, which are all be based on an epistemic state `currState`. For a larger application example from the medical domain see [4].

Building Up the Knowledge Base. In this example, we want to model some (uncertain) knowledge about a person being a student (s), young (y), and unmarried(u). We start with defining the corresponding propositional variables:

(1)   `var -> s, y, u;`

We can express that 70% of unmarried people are young and that 80% of students are young by defining a set of probabilistic conditionals, i.e. probabilistic rules:

(2)   `kbase := ((y|u)[0.7], (y|s)[0.8]);`

In this probabilistic knowledge base `kbase` e.g. the rule $(y|u)[0.7]$ can be read as "If *a person is unmarried*, then *this person is young with a probability of 0.7*".

Initializing an Epistemic State. The knowledge base `kbase` contains all the (uncertain) knowledge we have about the relations between students, young, and unmarried people. Of course, this knowledge is incomplete, e. g. we cannot directly answer queries like *"What is the probability of a student being unmarried?"* denoted by (u|s). To answer such queries, we need a full epistemic state (i. e. probability distribution) that represents our knowledge. In general, there is an unlimited number of epistemic states which represent some given (incomplete) knowledge. From these possible epistemic states we want to choose the one which is as unbiased as possible, i. e. the one that does not unnecessarily presuppose any additional information apart from the explicitly given knowledge `kbase`, and therefore represents this knowledge most faithfully. By making use of the concept of optimum entropy, such a most unbiased epistemic state can be uniquely determined. Since the entropy of a probability distribution delivers a measure for the "indeterminateness" of the distribution, the most unbiased epistemic state is exactly that one which has the maximum entropy of all epistemic states satisfying the given knowledge. It is well-known [1,2] that this maximum entropy distribution is identical to the distribution satisfying the given knowledge and having minimal cross-entropy to the uniform distribution $P_=$ that assigns the same probability to all worlds, thus representing complete ignorance.

The MECoRe system allows inductive probabilistic reasoning by computing a full probability distribution from the knowledge base via the maximum entropy method. This distribution serves as the initial epistemic state and its building is called by the command:

(3)   `currState := epstate.initialize(kbase);`

In our example, MECoRe computes $P^* = MinCEnt(P_=, \mathtt{kbase})$ as given in the following table and assigns it to `currState`:

ω	$P^*(\omega)$	ω	$P^*(\omega)$	ω	$P^*(\omega)$	ω	$P^*(\omega)$
syu	0.1950	$sy\bar{u}$	0.1758	$s\bar{y}u$	0.0408	$s\bar{y}\,\bar{u}$	0.0519
$\bar{s}yu$	0.1528	$\bar{s}y\bar{u}$	0.1378	$\bar{s}\,\bar{y}u$	0.1081	$\bar{s}\,\bar{y}\,\bar{u}$	0.1378

Querying an Epistemic State. Once an epistemic state has been calculated, we can ask any queries about the relations between students, young, and unmarried people, determining the corresponding beliefs holding in `currState`. The antecedent and the consequent of a conditional can be arbitrary formulas, e. g. (y|u & s), *"What is the probability that unmarried students are young?"* (where "&" denotes logical conjunction). MECoRe can also handle named sets of queries so that they can be reused later on:

(4)   `someQueries := ((u|s),(y|u & s));`
(5)   `currState.query(someQueries);`

MECoRe returns (u|s)[0.5087] and (y|u & s)[0.8270]. The answer to the second query shows that in the current epistemic state it is believed that the probability of an unmarried student to be young is 0.8270, expressing the joint influences of both rules in `kbase` in an information-theoretic optimal way.

Belief Change: Updating an Epistemic State. MECoRe provides the belief change operator *update* to processes new knowledge about an evolving world,

i. e. it enables us to adjust an epistemic state to new, (possibly) changed knowledge. To determine a unique updated epistemic state, the principle of minimum cross-entropy (see Sec. 2) is used. The updated epistemic state is determined by calculating the *unique* distribution $MinCEnt(P, R)$ which has the minimal information distance to the current distribution P *and* which satisfies the new knowledge R. Thus, the updated epistemic state results from the previous one by modifying it only as much as necessary (in terms of the entropy measure) to incorporate the new knowledge.

Continuing our example, suppose that some time later we learn that the relationship between students, unmarried, and young people has changed. Now, the probability that a student is young has increased to 0.9 and we get to know that young people are unmarried with a probability of 0.85. Therefore, we must update our current epistemic state with this new information to $MinCEnt(\texttt{currState}, (u|y)[0.85], (y|s)[0.9]))$:

(6)   `currState.update((u|y)[0.85], (y|s)[0.9]);`

(7)   `currState.query(someQueries);`

This gives us the following result:

ω	$P^*(\omega)$	ω	$P^*(\omega)$	ω	$P^*(\omega)$	ω	$P^*(\omega)$
syu	0.3304	$sy\overline{u}$	0.0583	$s\overline{y}u$	0.0190	$s\overline{y}\,\overline{u}$	0.0242
$\overline{s}yu$	0.2331	$\overline{s}y\overline{u}$	0.0411	$\overline{s}\,\overline{y}u$	0.1293	$\overline{s}\,\overline{y}\,\overline{u}$	0.1646

Asking `someQueries` again, the probabilities $(u|s)[0.8090]$ and $(y|u\,\&\,s)[0.9456]$ are higher than before. In particular, based on the new information, the belief that students are unmarried has increased, and it is now believed that the probability of an unmarried student to be young is 0.9456.

4   Components of the MEcore System

Figure 1 illustrates the essential functionalities of the MECoRe system. All belief change and reasoning operations make use of a probability distribution as representation of the current epistemic state. A new epistemic state is initialized by a set of probabilistic conditionals (`kbase` in our example). Certain operations can be performed on an epistemic state, as depicted in Fig. 1. The required parameters of each operation are also denoted there, where *knowledge*, *additional_knowledge*, *changed_knowledge*, and *assumptions* are sets of probabilistic conditionals, whereas *queries* are sets of unquantified conditionals, and *evidence* resp. *diagnoses* are sets of quantified resp. unquantified (simple) formulas.

Some of the operations not covered by our example are: `revise`, which incorporates additional knowledge by maintaining previous knowledge; `margin`, which outputs the marginal distribution with respect to some given variables; `whatif`, which facilitates a what-if-analysis in terms of *"What would be the probability of* queries *if the* assumptions *held?"*, and `diagnosis`, which determines the probability of diagnoses given some factual evidence.

One of MECoRe's advanced features is to handle multiple epistemic states. Therefore, every epistemic state is assigned to a unique identifier (`currState` in

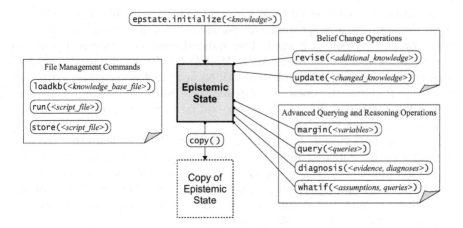

Fig. 1. Functionalities of the MECoRe system and operations on epistemic states

our example), i. e. an epistemic state is referenced via its identifier. A scenario with multiple epistemic states comes up if the copy operation is used to duplicate an epistemic state, i. e. an independent copy is assigned to another identifier. This can be very useful for experimenting, e. g. to compare a modified epistemic state with a previously created copy to analyze certain changes in detail.

Moreover, for practical applications, there are flexible file management operations e. g. for the execution of script files or for loading knowledge base files. For more details of MECoRe's operations and their background information, see [5,2]. MECoRe is implemented in Java [5]. In its current version, it uses a straight-forward, direct implementation of a well-known *MinCEnt* algorithm and provides a very powerful and flexible interface.

Computation of *MinCEnt*. MECoRe computes the distribution $MinCEnt(P, R)$ in an iterative way; a detailed description of the used algorithm can be found in [6]. In principle, the algorithm iterates over all rules in R in a cyclical order. In each iteration step, only one rule is considered and the current distribution (starting with P) is appropriately adjusted to satisfy the considered rule. It can be shown that this iterative process converges to $MinCEnt(P, R)$. Although MECoRe's current representation operating on the full probability distribution has to cope with the exponential size in the number of variables, it is still efficient enough to handle knowledge bases with 20 variables in about 30 seconds.

The User Interface. The present version of MECoRe can be controlled by a text command interface or by scripts, i. e. text files that allow the batch processing of command sequences. These scripts and the text interface use a programming language-like syntax that allows to define, manipulate and display variables, propositions, rule sets and epistemic states. Hence, one is able to use both previously defined rule sets and rules that are entered just when they are needed, and combinations of both. The ability to manipulate rule sets, to automate sequences of updates and revisions, and to output selected (intermediate) results for comparing, yields a very expressive command language. This command language is a powerful tool for experimenting and testing with different setups. All core functions of the MECoRe system are also accessible through a software interface (in terms of a Java API). So MECoRe can easily be extended by a GUI or be integrated into another software application.

5 Related Work and Conclusions

The aim of the MECoRe project is to implement subjective probabilistic reasoning, as it could be performed by agents, making various belief operations possible. In particular, it allows changing of beliefs in a very flexible way by taking new, complex information into account.

There are other systems performing inferences in probabilistic networks, especially in Bayesian networks. One system built upon such network techniques to implement reasoning at optimum entropy is the expert system shell SPIRIT [7]. Graph based methods are known to feature a very efficient representation of probability distributions via junction trees and hypergraphs, whereas MECoRe works on a model based representation of probabilities, which is not feasible if the number of variables gets very large. Since MECoRe's *MinCEnt* implementation is independent from the actual representation of the distribution, the system can orthogonally be extended by a more sophisticated kind of representation, e. g. by junction trees [8], which we plan to do in the future. However, efficient methods of restructuring probabilistic networks needed for complex belief changes as provided by MECoRe still have to be developed.

References

1. Paris, J., Vencovska, A.: In defence of the maximum entropy inference process. International Journal of Approximate Reasoning 17(1), 77–103 (1997)
2. Kern-Isberner, G.: Characterizing the principle of minimum cross-entropy within a conditional-logical framework. Artificial Intelligence 98, 169–208 (1998)
3. Shore, J.: Relative entropy, probabilistic inference and AI. In: Uncertainty in Artificial Intelligence, pp. 211–215. North-Holland, Amsterdam (1986)
4. Finthammer, M., Beierle, C., Berger, B., Kern-Isberner, G.: Probabilistic reasoning at optimum entropy with the MECORE system. In: Proceedings 22nd International FLAIRS Conference, FLAIRS 2009. AAAI Press, Menlo Park (2009)
5. Berger, B.: Realisierung einer Komponente für die Revision probabilistischen Wissens. BSc Thesis, Dep. of Computer Science, FernUniversität Hagen (2008)
6. Csiszár, I.: I-divergence geometry of probability distributions and minimization problems. Ann. Prob. 3, 146–158 (1975)
7. Rödder, W., Reucher, E., Kulmann, F.: Features of the expert-system-shell SPIRIT. Logic Journal of the IGPL 14(3), 483–500 (2006)
8. Teh, Y.W., Welling, M.: On improving the eciency of the iterative proportional fitting procedure. In: Proc. of the Ninth International Workshop on Artificial Intelligence and Statistics, Society for Artificial Intelligence and Statistics (2003)

On the Input Language of ASP Grounder *Gringo*

Martin Gebser, Roland Kaminski, Max Ostrowski, Torsten Schaub*, and Sven Thiele

Universität Potsdam, Institut für Informatik, August-Bebel-Str. 89, D-14482 Potsdam

Abstract. We report on recent advancements in the development of grounder *Gringo* for logic programs under answer set semantics. Like its relatives, *DLV* and *Lparse*, *Gringo* has in the meantime reached maturity and offers a rich modeling language to program developers. The attractiveness of *Gringo* is fostered by the fact that it significantly extends the input language of *Lparse* while supporting a compatible output format, recognized by many state-of-the-art ASP solvers.

1 Introduction

Answer Set Programming (ASP; [1]) is an attractive paradigm for knowledge representation and reasoning. On the one hand, its popularity is due to the availability of efficient off-the-shelf solvers (cf. [2]). But equally or even more important under the aspect of usability is its rich modeling language, including first-order variables, function symbols, aggregates, etc. In fact, search problems are in ASP usually modeled in a uniform way by means of a data part, called instance, and a general part, called encoding (cf. [3,4,5,6,7]). The computation of answer sets, corresponding to problem solutions, is then typically performed in two phases: first, *grounding* the encoding on the problem instance, and second, solving the resulting propositional program.

In contrast to the multitude of available solvers, the field of ASP grounders is still underrepresented. To the best of our knowledge, there are only three popular grounders, namely, (the grounding component of) *DLV* [8], *Lparse* [9], and *Gringo* [10]. While *DLV* processes the grounding result internally or prints it as text, the numerical output format of *Lparse* and *Gringo* is recognized by many state-of-the-art ASP solvers. In view of this transparency from the solver side and the progress made since the first description of *Gringo* [11], *Gringo* has become a real alternative to *Lparse*. In particular, the attractiveness of *Gringo* is fostered by the fact that it significantly extends the input language of *Lparse*, providing advanced modeling features to ASP programmers. This paper reports on such new features of *Gringo*, potentially useful for ASP applications.

2 Modeling Features

The input language of *Gringo* is designed to be in large parts compatible to the one of *Lparse* [9], so that the majority of *Lparse* programs can still be grounded with *Gringo*. Assuming basic familiarity with *Lparse*, we focus our description on extensions available in *Gringo* and also mainly take *Lparse* as the grounder to contrast with.

---

* Affiliated with Simon Fraser University, Canada, and Griffith University, Australia.

E. Erdem, F. Lin, and T. Schaub (Eds.): LPNMR 2009, LNCS 5753, pp. 502–508, 2009.

λ-Restricted Programs [11]. The class of programs processable with *Gringo* is a proper superclass of ω-restricted programs [12] accepted by *Lparse*. The underlying idea is that all relevant ground instances of a rule (that is, ground instances whose bodies can potentially be true wrt an answer set) are implicitly given if, for each variable in the rule, we find some atom in the positive rule body such that its predicate's relevant ground instances are known. As the basic grounding algorithm of *Gringo* works rule-wise, the latter is the case when all rules with the predicate occurring in the head have already been instantiated. In fact, before beginning with instantiation, *Gringo* computes an ordering such that all rules with a predicate in the head are completely processed before the predicate is used to restrict variable domains in other rules where it occurs in the positive body. Notably, *Gringo* imposes no additional restrictions, such as being definite or stratified, on the rules to be ordered. To see this, consider the following example [10]:

```
     zig(0)   :- not zag(0).           zig(1)   :- not zag(1).
     zag(0)   :- not zig(0).           zag(1)   :- not zig(1).
zigzag(X,Y)   :- zig(X), zag(Y).   zagzig(Y,X)  :- zigzag(X,Y).
```

Here, *Gringo* first looks at the (ground) rules with zig/1 and zag/1 in the head, and so it determines that 0 and 1 are all argument values for which the predicates can hold wrt answer sets. This is used to restrict ground instances of X and Y in the rule with zigzag/2 in the head, which in turn restricts X and Y in the rule with zagzig/2. Hence, *Gringo* grounds the above program without complaints, while *Lparse* rejects it because of not being ω-restricted. In order to use *Lparse*, we would have to add a domain predicate, saying that X and Y must be 0 or 1, to the bodies of the last two rules. Of course, such information would be redundant, and thus λ-restrictedness helps to write more focused programs, concentrating on the relevant information within rules.

Uninterpreted Functions. The input language of *Gringo* allows for using functions in the heads and bodies of rules, and unification is applied for instantiating variables in uninterpreted functions. For instance, this enables *Gringo* to ground the program:

```
parent(joan,mother(jane)).   female(Y)  :- parent(X,mother(Y)).
parent(joan,father(john)).     male(Y)  :- parent(X,father(Y)).
```

Though *Lparse* also tolerates uninterpreted functions, it internally handles them like interpreted (arithmetic) functions, and so it refuses to instantiate Y in the above program. For modeling, the full support of uninterpreted functions by *Gringo* can be beneficial.

Conditions. Conditions are indicated by ":" in the input languages of *Lparse* and *Gringo*. Their purpose is to instantiate "local" variables on the left-hand side with values for which (a set of) literals over domain predicates on the right-hand side holds (cf. [9,10]). For illustration, consider the following program:

```
od(1).             ne(1).
      ev(2). pr(2).
od(3).         pr(3).
and_1 :-       pr(X) : od(X).              % and_1 :- pr(1), pr(3).
and_2 :-       ne(X) : od(X) : not pr(X).  % and_2 :- ne(1).
and_3 :- not ev(X) : ev(X) : not pr(X).    % and_3.
or(X) : od(X).                             % or(1) | or(3).
```

For comparison, the ground rules qualified by the rules with conditions are provided in comments. First, observe that *Gringo* expands conditions in the bodies of rules into conjunctions of the required length, while disjunction is used for conditions in rule heads. Furthermore, default negation via not can be used on the right-hand side and, in rule bodies, also on the left-hand side of a condition. A particular case is illustrated by the rule with head and_3, where the left-hand side is the negation of an atom on the right-hand side. In this situation, the set of literals on the right-hand side must be unsatisfied by all of its ground instances, as it happens with the above rule, or the expansion of the condition is immediately unsatisfied. This phenomenon can be exploited for testing whether certain properties do not hold wrt all ground instances of a set of literals. Comparing with conditions in *Lparse* yields that it accepts only the rule with head and_1, while neither negative literals nor occurrences in rule heads are supported. To illustrate the usefulness of the latter, let us consider a disjunctive encoding of N-Coloring:

```
#const n=3.
col(X,C) : C = 1..n :- node(X).
% col(X,1) | ... | col(X,n) :- node(X).
:- col(X,C), col(Y,C), edge(X,Y).
```

To keep the encoding general, we make use of a constant n for defining the number of available colors. All values from 1 to n can successively be assigned to C in the condition of the first rule. Hence, we obtain a disjunction ranging over all colors for each node X. Without this opportunity, it is more involved to make use of disjunction for arbitrary N. In fact, as the length of the required disjunction is open (illustrated also by the uncommented rule), other ways of encoding it would have to be used instead.

Aggregates. Aggregates (and associated comparison operations), like the ones supported by *DLV* [13] or cardinality and weight constraints of *Lparse* [9], permit a compact representation of (numerical) constraints on sets of literals. The aggregates currently supported by *Gringo* are: #count, #sum, #times, #avg, #min, #max, #even, and #odd. Each aggregate applies to either a set of literals, enclosed in curly brackets, or a multiset of literals with associated weights, enclosed in square brackets, where 1 is used as a default for omitted weights. The result of applying an aggregate can be compared to a lower bound ($-\infty$ if omitted) and an upper bound (∞ if omitted) in order to obtain a truth value. The only exceptions to this are #even and #odd whose meanings are fixed independently of bounds. Before we illustrate individual aggregates, we note that [14] provides a general semantics for them. An objective of *Gringo* is to respect this semantics as far as possible, with the modification of applying "choice semantics" [15,16] instead of minimization to atoms occurring positively in an aggregate being the head of a rule. However, some compromises are needed for compatibility to the output format of *Lparse*, supporting only #count and #sum (all other aggregates are compiled into them), and only non-negative weights in #sum aggregates (negative weights are eliminated by translation [15]). As a consequence, compliance with the "choice version" of the semantics in [14] is only guaranteed if dependencies through #avg as well as #sum and #times aggregates with negative weights are not subject to (positive) recursion, i.e., an atom appearing in such an aggregate in a rule body should not be defined (directly or indirectly) by any atom occurring positively in the rule head. Now illustrating the available aggregates, we begin with the ones familiar from *Lparse*:

```
1 #count {a,    not b, c} 2.    % 1 {a,    not b,    c} 2.
1 #count {a,a, not b, c} 2.    % 1 {a,    not b,    c} 2.

2 #sum [a=1, not b=1, c=2] 3. % 2 [a=1, not b=1, c=2] 3.
2 #sum [a,    not b,    c,c] 3. % 2 [a=1, not b=1, c=2] 3.
```

The above (ground) facts specify #count and #sum aggregates. In comments, we provide their notations in terms of cardinality and weight constraints, also accepted by *Gringo* for compatibility to *Lparse*. Note that *Gringo* properly deals with the set semantics of #count and multisets of #sum, while *Lparse* turns a, a as in the second fact into a=2. However, given that the above facts contain negative literal not b (in the head), *Lparse* would not accept them either. Such restrictions do not apply to *Gringo*, capable of handling negative literals in aggregates occurring as rule heads. The next examples demonstrate the use of the further aggregates supported by *Gringo*:

```
2 #times [a=1,      not b=2, c=3] 3.
2 #times [a, not b,not b, c,c,c] 3.
    % 2 #times [a=1, not b=1, c=1] 3.

2 #avg [a=3, not b=1, c=0] 2.
2 #avg [a,a, not b=2].        % 2 #avg [a=1, a=1, not b=2].

2 #min [a=1,      not b=2, c=3] 2.
2 #min [a, not b,not b, c,c,c] 2.
    % 2 #min [a=1, not b=1, c=1] 2.

2 #max [a=1,      not b=2, c=3] 2.
2 #max [a, not b,not b, c,c,c] 2.
    % 2 #max [a=1, not b=1, c=1] 2.

    #even {a,    not b, c}.     #odd  {a,    not b, c}.
    #even {a,a, not b, c}.     #odd  {a,a, not b, c}.
    % #even {a, not b, c}.     % #odd  {a, not b, c}.
```

For aggregates over repeated literals and omitted weights, semantically equivalent counterparts are provided in comments. With multisets, repeated literals appear also repeatedly in the output of *Gringo*, while the effect of such repetitions depends on the aggregate at hand. As regards the #avg aggregate, a=2 contributes one and a, a two addends to the numerator and denominator, respectively, in the average calculation. Also note that the meanings of a=2 and a, a are different from one another in #times, #min, and #max aggregates. Finally, as #even and #odd determine the parity of the number of (true) literals in sets, repeated literals are collapsed into one. Though not demonstrated above, aggregates (and associated comparison operations) can also be used in rule bodies. In addition to comparing aggregate results to bounds, *Gringo* supports assigning the result to a variable, as exemplified in the following program:

```
q(X) :- X = #sum [p(Y) : p(Y) : Y #mod 2 != 0 = Y].    p(1..3).
```

Such assignments of aggregate results are also possible with *DLV* (cf. [17]), but not in *Lparse*. Their main application is to identify deterministic properties of instances that can be calculated from the stratified part of a program [10].

Runtime Options. The default output format of *Gringo* is the same as the one of *Lparse* [9]. Via (experimental) option `--aspils`, the output is printed in one of the normal forms of ASPils, an intermediate format proposed in [18]. As with *Lparse*, option `--text` (or `-t`) makes *Gringo* print ground rules in human-readable text format. In addition, option `--debug` can be provided to investigate internal representations of (non-ground) rules during grounding. Via `--const` (or `-c`), also available in *Lparse*, occurrences of a constant can be replaced with another term, e.g., beneficial with N-Coloring as encoded above. For disjunctive programs, in particular, "head-cycle-free" ones, option `--shift` replaces disjunction in rule heads with default negation in rule bodies, so that solvers for non-disjunctive ASP can be applied. To improve efficiency, if an input program is already ground, it can be signaled to *Gringo* via option `--ground`. This allows *Gringo* to avoid unnecessary yet non-negligible overhead, which is useful, e.g., for running in a mode similar to category SCore of the first ASP system competition [2]. Note that any occurrences of variables are considered as syntax errors if *Gringo* expects an input program to be ground. The binder-splitting technique [11], applied by default, can be switched off via option `--bindersplit`; this is mainly to admit experimental comparisons. Finally, options `--ifixed` and `--ibase` enable *Gringo* to ground incremental programs, written for *iClingo* [19] and containing meta-directives `#base`, `#cumulative`, and `#volatile` [10].

3   Discussion

We have presented relevant features of grounder *Gringo* (version 2.0.3), significantly extending the functionalities of *Lparse*. *Gringo* constitutes an integral part of *Potassco*, the Potsdam Answer Set Solving Collection bundling tools for ASP, for which sources (and binaries) are publicly available at `http://potassco.sourceforge.net`. In addition to its executable, *Gringo* comes as library. As such, it is used inside ASP systems *Clingo*, *iClingo*, and *Clingcon*, all belonging to the Potassco suite. However, due to supporting the output format of *Lparse*, *Gringo* is not limited to work only in integrated tools, but can be used as a front-end for many state-of-the-art ASP solvers. Recent applications modeled and grounded with *Gringo* include [20,21,22,23,24]. For the future, we plan to integrate grounding techniques beyond rule-wise working ones and, accordingly, to relax the required input program properties (currently λ-restrictedness).

Acknowledgments. This work was partially funded by DFG under Grant SCHA 550/8-1 and by the GoFORSYS project under Grant 0313924.

References

1. Baral, C.: Knowledge Representation, Reasoning and Declarative Problem Solving. Cambridge University Press, Cambridge (2003)
2. Gebser, M., Liu, L., Namasivayam, G., Neumann, A., Schaub, T., Truszczyński, M.: The first answer set programming system competition. In: [25], pp. 3–17
3. Schlipf, J.: The expressive powers of the logic programming semantics. Journal of Computer and System Sciences 51, 64–86 (1995)

4. Marek, V., Truszczyński, M.: Stable models and an alternative logic programming paradigm. In: The Logic Programming Paradigm: a 25-Year Perspective, pp. 375–398. Springer, Heidelberg (1999)

5. Niemelä, I.: Logic programs with stable model semantics as a constraint programming paradigm. Annals of Mathematics and Artificial Intelligence 25(3-4), 241–273 (1999)

6. Gelfond, M., Leone, N.: Logic programming and knowledge representation — the A-Prolog perspective. Artificial Intelligence 138(1-2), 3–38 (2002)

7. Lifschitz, V.: Answer set programming and plan generation. Artificial Intelligence 138(1-2), 39–54 (2002)

8. Leone, N., Pfeifer, G., Faber, W., Eiter, T., Gottlob, G., Perri, S., Scarcello, F.: The DLV system for knowledge representation and reasoning. ACM Transactions on Computational Logic 7(3), 499–562 (2006)

9. Syrjänen, T.: Lparse 1.0 user's manual,
   http://www.tcs.hut.fi/Software/smodels/lparse.ps.gz

10. Gebser, M., Kaminski, R., Kaufmann, B., Ostrowski, M., Schaub, T., Thiele, S.: A user's guide to gringo, clasp, clingo, and iclingo,
    http://potassco.sourceforge.net

11. Gebser, M., Schaub, T., Thiele, S.: GrinGo: A new grounder for answer set programming. In: [25], pp. 266–271

12. Syrjänen, T.: Omega-restricted logic programs. In: Eiter, T., Faber, W., Truszczyński, M. (eds.) LPNMR 2001. LNCS (LNAI), vol. 2173, pp. 267–279. Springer, Heidelberg (2001)

13. Faber, W., Pfeifer, G., Leone, N., Dell'Armi, T., Ielpa, G.: Design and implementation of aggregate functions in the DLV system. Theory and Practice of Logic Programming 8(5-6), 545–580 (2008)

14. Ferraris, P.: Answer sets for propositional theories. In: Baral, C., Greco, G., Leone, N., Terracina, G. (eds.) LPNMR 2005. LNCS (LNAI), vol. 3662, pp. 119–131. Springer, Heidelberg (2005)

15. Simons, P., Niemelä, I., Soininen, T.: Extending and implementing the stable model semantics. Artificial Intelligence 138(1-2), 181–234 (2002)

16. Ferraris, P., Lifschitz, V.: Weight constraints as nested expressions. Theory and Practice of Logic Programming 5(1-2), 45–74 (2005)

17. Terracina, G., De Francesco, E., Panetta, C., Leone, N.: Experiencing ASP with real world applications. In: Proceedings of the Fifteenth RCRA Workshop on Experimental Evaluation of Algorithms for Solving Problems with Combinatorial Explosion, RCRA 2008 (2008)

18. Gebser, M., Janhunen, T., Ostrowski, M., Schaub, T., Thiele, S.: A versatile intermediate language for answer set programming. In: Proceedings of the Twelfth International Workshop on Nonmonotonic Reasoning (NMR 2008), pp. 150–159 (2008)

19. Gebser, M., Kaminski, R., Kaufmann, B., Ostrowski, M., Schaub, T., Thiele, S.: Engineering an incremental ASP solver. In: [26], pp. 190–205

20. Mileo, A., Merico, D., Bisiani, R.: A logic programming approach to home monitoring for risk prevention in assisted living. In: [26], pp. 145–159

21. Boenn, G., Brain, M., de Vos, M., Fitch, J.: Automatic composition of melodic and harmonic music by answer set programming. In: [26], pp. 160–174

22. Gebser, M., Schaub, T., Thiele, S., Usadel, B., Veber, P.: Detecting inconsistencies in large biological networks with answer set programming. In: [26], pp. 130–144

23. Kim, T., Lee, J., Palla, R.: Circumscriptive event calculus as answer set programming. In: Proceedings of the Twenty-first International Joint Conference on Artificial Intelligence (IJCAI 2009). AAAI Press, Menlo Park (to appear, 2009)

24. Thielscher, M.: Answer set programming for single-player games in general game playing. In: Proceedings of the Twenty-fifth International Conference on Logic Programming (ICLP 2009). Springer, Heidelberg (to appear, 2009)
25. Baral, C., Brewka, G., Schlipf, J. (eds.): LPNMR 2007. LNCS (LNAI), vol. 4483. Springer, Heidelberg (2007)
26. Garcia de la Banda, M., Pontelli, E. (eds.): ICLP 2008. LNCS, vol. 5366. Springer, Heidelberg (2008)

The Conflict-Driven Answer Set Solver *clasp*: Progress Report

Martin Gebser, Benjamin Kaufmann, and Torsten Schaub*

Universität Potsdam, Institut für Informatik, August-Bebel-Str. 89, D-14482 Potsdam

Abstract. We summarize the salient features of the current version of the answer set solver *clasp*, focusing on the progress made since version RC4 of *clasp*. Apart from enhanced preprocessing and search-supporting techniques, a particular emphasis lies on advanced reasoning modes, such as cautious and brave reasoning, optimization, solution projection, and incremental solving.

1 Introduction

The solver *clasp* for Answer Set Programming (ASP; [1]) is based upon advanced Boolean constraint solving technology. The theoretical foundations and basic algorithms underlying *clasp* can be found in [2,3]. It is freely available as open source package at [4]. This paper reports on the progress made since the first system description of *clasp* [5] covering the features of version RC4: it mainly dealt with an empirical evaluation of *clasp*'s features related to conflict-driven nogood learning, comparing various strategies for restarts, nogood deletion, and decision heuristics. In the meantime, *clasp* won the solving categories *SCore* and *SLparse* at the first ASP system competition and is currently participating in the second one. Also, *clasp* qualified for this year's final round of the industrial Satisfiability checking (SAT) competition and competed in SAT-Race 2008 as well as in the 2007 Pseudo-Boolean (PB) evaluation[1].

2 Features

This section describes the major features of version 1.2.1 of *clasp* added since RC4.

Reasoning Modes. As almost all ASP solvers, *clasp* relies on a grounder providing a representation of a propositional logic program. Its major input format is *Lparse* output, provided by either *Lparse* [6] or *Gringo* [7]. Although *clasp*'s primary use case is the computation of a given number of answer sets, it also allows for computing the supported models of a logic program (via command line option `--supp-models`). As detailed below, in either case, options `--cautious` and `--brave` permit computing the intersection and union, respectively, of the respective types of models. Finally, the `--dimacs` option allows for using *clasp* as a SAT solver computing the classical models of a propositional formula supplied in DIMACS format.

---

* Affiliated with Simon Fraser University, Canada, and Griffith University, Australia.

[1] Thanks to Gayathri Namasivayam and Mirosław Truszczyński, University of Kentucky.

E. Erdem, F. Lin, and T. Schaub (Eds.): LPNMR 2009, LNCS 5753, pp. 509–514, 2009.

Preprocessing. At the beginning, a propositional logic program is subject to extensive preprocessing [8]. The idea is to simplify a logic program while identifying equivalences among its relevant constituents. These equivalences are then used for building a compact representation of the program (in terms of Boolean constraints). Notably, sometimes preprocessing is able to turn a non-tight program into a tight one (cf. [9]). Preprocessing is configured via option --eq, taking an integer value fixing the number of iterations. Once a program has been transformed into a set of Boolean constraints, it is subject to further preprocessing, mostly borrowed from the area of SAT [10]. SAT-based preprocessing is invoked with option --sat-prepro and further parameters. However, care must be taken when adapting such techniques from SAT because preprocessing must not eliminate variables that are relevant to the unfounded set checker or that occur in optimize statements or weight rules.

Dedicated Propagation. Not all parts of a logic program are turned into nogoods by *clasp* (in its default setting). Rather *clasp* employs specialized propagation algorithms and has a dedicated implementation for cardinality and weight constraints [11]. These are particular count and sum aggregates offered by *Lparse* and *Gringo*. As detailed in [11], their treatment involves a dedicated, source-pointer-based unfounded set algorithm that computes loop nogoods only on demand, while aiming at lazy unfounded set checking and backtrack-freeness. Although per default all cardinality and weight constraints are subject to dedicated propagation, their treatment can be configured through option --trans-ext. Propagation with loop nogoods is influenced by option --loops controlling their creation.

Model Enumeration. Different ways of enumerating models are supported by *clasp*. In fact, solution enumeration is non-trivial in the context of backjumping and nogood learning. A popular approach consists in recording solutions as nogoods and exempting them from nogood deletion. Although *clasp* supports this via option --solution-recording, it is prone to blow up in space in view of an exponential number of solutions in the worst case. Unlike this, the default enumeration algorithm of *clasp* runs in polynomial space [3]. Both approaches also allow for projecting solutions on a subset of atoms [12]; invoked with --project and configured via the well-known directives #hide and #show of *Lparse* and *Gringo*. For example, the program consisting of the choice rule {a,b,c}. has eight (obvious) answer sets. When augmented with directive #hide c., still eight solutions are obtained, yet including four duplicates. Unlike this, invoking *clasp* with --project yields only four duplicate-free solutions. This option is of great practical value whenever one faces overwhelmingly many answer sets, involving solution-irrelevant variables having a proper combinatorics. As regards implementation, it is interesting to note that *clasp* offers a dedicated interface for enumeration. This allows for abstracting from how to proceed once a model was found and thus makes the search algorithm independent of the concrete enumeration strategy. One further strategy implemented via the enumeration interface consists of computing the intersection or union of all answer sets of a program (via --cautious and --brave, respectively). Rather than computing a set of (possibly) exponentially many answer sets, the idea is to compute a first answer set, record a constraint eliminating it from further solutions, then compute a second answer set, strengthen the constraint to represent the intersection (or union) of the first two answer sets, and to continue in this way until no more answer sets are obtained. This process involves computing at most as many answer sets as there are atoms in the input program. Either the cautious or the brave consequences are then given by the atoms captured by the final constraint.

Optimization. Another application-oriented feature is optimization. As common in *Lparse*-like languages, an objective function is specified by a sequence of #minimize and #maximize statements. For finding optimal solutions, *clasp* offers several options. First, *clasp* allows for computing one or all (--opt-all) optimal solutions. Second, the objective function can be initialized via --opt-value. The latter turns out to be useful when one is interested in computing consequences belonging to all optimal solutions (in combination with --cautious). One starts with a search for an optimum and then re-launches *clasp* by bounding its search with the value of the optimum. Doing the latter with --cautious yields all consequences true in all optimum answer sets. On applications, it turned out to be very useful to optimize using the option --restart-on-model (making *clasp* restart after each (putative) minimum solution) in order to ameliorate the convergence to an optimum solution. Again, optimization is implemented via the aforementioned enumeration interface. When a solution is found, the optimization constraint is updated by the corresponding value. Then, the decision level invalidating the updated constraint is identified and backtracked; if the constraint is violated on the top-level, search terminates. Furthermore, it is worth mentioning that *clasp* also propagates over optimization statements. For this, optimization statements are themselves stored as Boolean constraints [5] in the solver. As such, they can derive (and provide reasons for) implications during unit propagation.

Restarts. The robustness of *clasp* is boosted by advanced restart strategies. Apart from the policies already discussed in [5], namely, geometric, fixed-interval, and Luby-style policies, a nested policy, first used in *picosat* [13], is meanwhile also offered by *clasp*. This policy takes three parameters x, y, z and makes restarts follow a two dimensional pattern that increases geometrically in both dimensions. The geometric restart sequence $x * y^i$ is repeated when it reaches an outer limit $z * y^j$, where i counts the number of restarts and j how often the outer limit was hit so far. Usually, restart strategies as listed above are based on a global number of conflicts. Moreover, *clasp* features local restarts [14]. Here, one counts the number of conflicts at each decision level in order to localize the measure of difficulty. For this, we maintain a counter $c(d)$ for each decision level d. When a new decision level d is created, $c(d)$ is set to the global number of conflicts. When backtracking to level d, a restart is only initiated if the difference between the global number of conflicts and $c(d)$ is now larger than the strategy-dependent threshold. It is worth noting that despite the fact that recent SAT solvers use rather aggressive restart strategies (cf. Section 3), *clasp* still defaults to a more conservative geometric policy because this performs better on our ASP-specific benchmarks.

Progress Saving. Another search-related feature of *clasp* is progress saving, as described in [15]. The idea is as follows. On backjumping (or restarting), the values of variables whose assignment is about to be erased are saved for all but those variables assigned on the last decision level. The saved values are then used during decision making.

That is, when a variable for which a value was saved is selected by the decision heuristic, it is assigned to that value. The intuition behind this strategy is that the assignments made prior to the last decision level did not lead to a conflict and may have satisfied some subproblem. Hence, repeating those assignments may help to avoid solving subproblems multiple times. Progress saving is invoked with option `--save-progress`; its computational impact depends heavily on the structure of the application at hand.

Application Programming Interface. A major yet internal feature of *clasp* is that it can be used in a stateful way. That is, *clasp* may keep its state, involving program representation, learned constraints, heuristic values, etc, and be invoked under additional (temporary) assumptions and/or by adding new atoms and rules. The corresponding interfaces are fundamental for supporting incremental ASP solving as realized in *iClingo* [16], a combination of *Gringo* and *clasp* for incremental grounding and solving. Furthermore, they allow for solving under assumptions [17]; an important feature that is, for example, used in our parallel ASP solver *claspar* [18].

3   Fine-Tuning

Advanced Boolean constraint solving technology adds a multitude of degrees of freedom to ASP solving. For instance, currently, *clasp* has roughly 40 options, half of which control the search strategy. Although considerable efforts were taken to find default parameters for optimizing robustness and speed, the default setting still leaves room for drastic improvements on specific benchmark classes by fine-tuning the parameters. The question arises how to deal with this vast "configuration space" and how to conciliate it with the idea of declarative problem solving. Currently, there seems to be no alternative to manual fine-tuning when addressing highly demanding application problems.

As rules of thumb, we usually start by investigating the following options:

`--heuristic`: Try *VSIDS* instead of *clasp*'s default *BerkMin*-style heuristic.

`--sat-prepro`: SAT-based preprocessing works best on tight programs with few cardinality and weight constraints. It should (almost) always be used if extended rules are transformed into nogoods (via `--trans-ext`).

`--restarts`: Try aggressive restart policies, like *Luby-256* or the *nested policy*, or try disabling restarts, whenever a problem is deemed to be unsatisfiable.

`--save-progress`: Progress saving typically works nicely if the average backjump length (or the #choices/#conflicts ratio) is high (≥ 10). It usually performs best if combined with aggressive restarts.

`--trans-ext`: Applicable if the program contains extended rules, that is, rules including cardinality and weight constraints. Try at least the *dynamic* transformation.

The impact of simple fine-tuning can be seen on the following examples. As shown in [19], *clasp* times out on satisfiable *4-Coloring* problems. However, with `--save-progress`, *clasp* solves all instances in less than 2 sec (the average backjump length is >60). For another example, consider the benchmark *WeightBounded-DominatingSet* from the second ASP competition. The default configuration of *clasp* results in six timeouts, all of which vanish once aggressive restarts are used. Similar effects are observed on application problems featuring yet different characteristics.

Although such fine-tuning may greatly improve the efficiency of *clasp*, it is hard to accomplish for an unpracticed user, and after all it takes us away from the ideals of declarative problem solving. To this end, we advocate an extension of *clasp*, called *claspfolio*, that maps benchmark features to solver configurations (via machine learning techniques). It is interesting future work to see whether this allows for an automatic selection of effective parameter settings.

4 Discussion

Since its inception in 2007, *clasp* has become an efficient, full-fledged ASP solver. Beyond its computational power, it meanwhile features various reasoning modes that make it an attractive tool for knowledge representation and reasoning. This is witnessed by an increasing number of applications relying on *clasp* or derivatives as reasoning engine, e.g., [20,21,22,23,24,25]. *clasp* constitutes a central component of *Potassco*, the Potsdam Answer Set Solving Collection bundling tools for Answer Set Programming developed at the University of Potsdam. An extension of *clasp*, called *claspD* [26], allows for dealing with disjunctive ASP programs. Meanwhile, the family has grown and two new systems, *Clingo* and *iClingo* [16], have emerged. *Clingo* is a monolithic combination of *clasp* and *Gringo*. *iClingo* is an ASP system that allows for dealing incrementally with parametrized problems, as encountered for instance in bioinformatics, planning, and model checking. The latest addition to the family is *Clingcon* [27], augmenting *Clingo* with (non-Boolean) constraint processing capacities. Also, there is a distributed version of *clasp*, called *claspar* [18], designed for running on clusters with MPI. Sources (and binaries) of our systems are publicly available at [4].

Acknowledgments. This work was partially funded by DFG under Grant SCHA 550/8-1 and by the GoFORSYS project under Grant 0313924.

References

1. Baral, C.: Knowledge Representation, Reasoning and Declarative Problem Solving. Cambridge University Press, Cambridge (2003)
2. Gebser, M., Kaufmann, B., Neumann, A., Schaub, T.: Conflict-driven answer set solving. In: Veloso, M. (ed.) Proc. of IJCAI 2007, pp. 386–392. AAAI Press, Menlo Park (2007)
3. Gebser, M., Kaufmann, B., Neumann, A., Schaub, T.: Conflict-driven answer set enumeration. In: [28], pp. 136–148
4. http://potassco.sourceforge.net/
5. Gebser, M., Kaufmann, B., Neumann, A., Schaub, T.: clasp: A conflict-driven answer set solver. In: [28], pp. 260–265
6. Syrjänen, T.: Lparse 1.0 user's manual
7. Gebser, M., Schaub, T., Thiele, S.: GrinGo: A new grounder for answer set programming. In: [28], pp. 266–271
8. Gebser, M., Kaufmann, B., Neumann, A., Schaub, T.: Advanced preprocessing for answer set solving. In: Ghallab, M., Spyropoulos, C., Fakotakis, N., Avouris, N. (eds.) Proc. of ECAI 2008, pp. 15–19. IOS Press, Amsterdam (2008)
9. Babovich, Y., Lifschitz, V.: Computing answer sets using program completion (2003)

10. Eén, N., Biere, A.: Effective preprocessing in SAT through variable and clause elimination. In: Bacchus, F., Walsh, T. (eds.) SAT 2005. LNCS, vol. 3569, pp. 61–75. Springer, Heidelberg (2005)
11. Gebser, M., Kaminski, R., Kaufmann, B., Schaub, T.: On the implementation of weight constraint rules in conflict-driven ASP solvers. In: [29], pp. 250–264
12. Gebser, M., Kaufmann, B., Schaub, T.: Solution enumeration for projected Boolean search problems. In: van Hoeve, W., Hooker, J. (eds.) Proc. of CPAIOR 2009, pp. 71–86. Springer, Heidelberg (2009)
13. Biere, A.: PicoSAT essentials. Journal on Satisfiability, Boolean Modeling and Computation 4, 75–97 (2008)
14. Ryvchin, V., Strichman, O.: Local restarts. In: Kleine Büning, H., Zhao, X. (eds.) SAT 2008. LNCS, vol. 4996, pp. 271–276. Springer, Heidelberg (2008)
15. Pipatsrisawat, K., Darwiche, A.: A lightweight component caching scheme for satisfiability solvers. In: Marques-Silva, J., Sakallah, K.A. (eds.) SAT 2007. LNCS, vol. 4501, pp. 294–299. Springer, Heidelberg (2007)
16. Gebser, M., Kaminski, R., Kaufmann, B., Ostrowski, M., Schaub, T., Thiele, S.: Engineering an incremental ASP solver. In: [30], pp. 190–205
17. Eén, N., Sörensson, N.: Temporal induction by incremental SAT solving. Electronic Notes in Theoretical Computer Science 89(4) (2003)
18. Ellguth, E., Gebser, M., Gusowski, M., Kaminski, R., Kaufmann, B., Liske, S., Schaub, T., Schneidenbach, L., Schnor, B.: A simple distributed conflict-driven answer set solver. In: Erdem, E., Lin, F., Schaub, T. (eds.) Proc. of LPNMR 2009. Springer, Heidelberg (to appear, 2009)
19. Janhunen, T., Niemelä, I., Sevalnev, M.: Computing stable models via reductions to Boolean circuits and difference logic. In: Denecker, M. (ed.) Proc. of LaSh 2008, pp. 16–30 (2008)
20. Mileo, A., Merico, D., Bisiani, R.: A logic programming approach to home monitoring for risk prevention in assisted living. In: [30], pp. 145–159
21. Boenn, G., Brain, M., de Vos, M., Fitch, J.: Automatic composition of melodic and harmonic music by answer set programming. In: [30], pp. 160–174
22. Gebser, M., Schaub, T., Thiele, S., Usadel, B., Veber, P.: Detecting inconsistencies in large biological networks with answer set programming. In: [30], pp. 130–144
23. Ishebabi, H., Mahr, P., Bobda, C., Gebser, M., Schaub, T.: Answer set vs integer linear programming for automatic synthesis of multiprocessor systems from real-time parallel programs. Journal of Reconfigurable Computing (to appear, 2009)
24. Kim, T., Lee, J., Palla, R.: Circumscriptive event calculus as answer set programming. In: Boutilier, C. (ed.) Proc. IJCA 2009. AAAI Press, Menlo Park (to appear, 2009)
25. Thielscher, M.: Answer set programming for single-player games in general game playing. In: [29] (to appear)
26. Drescher, C., Gebser, M., Grote, T., Kaufmann, B., König, A., Ostrowski, M., Schaub, T.: Conflict-driven disjunctive answer set solving. In: Brewka, G., Lang, J. (eds.) Proc. KR 2008, pp. 422–432. AAAI Press, Menlo Park (2008)
27. Gebser, M., Ostrowski, M., Schaub, T.: Constraint answer set solving. In: [29], pp. 235–249
28. Baral, C., Brewka, G., Schlipf, J. (eds.): LPNMR 2007. LNCS (LNAI), vol. 4483. Springer, Heidelberg (2007)
29. Hill, P., Warren, D. (eds.): Proc. of ICLP 2009. Springer, Heidelberg (2009)
30. Garcia de la Banda, M., Pontelli, E. (eds.): Proc. ICLP 2008. Springer, Heidelberg (2008)

System F2LP – Computing Answer Sets of First-Order Formulas

Joohyung Lee and Ravi Palla

Computer Science and Engineering
Arizona State University, Tempe, AZ, USA
{joolee,ravi.palla}@asu.edu

Abstract. We present an implementation of the general language of
stable models proposed by Ferraris, Lee and Lifschitz. Under certain conditions, system F2LP turns a first-order theory under the stable model
semantics into an answer set program, so that existing answer set solvers
can be used for computing the general language. Quantifiers are first
eliminated and then the resulting quantifier-free formulas are turned into
rules. Based on the relationship between stable models and circumscription, F2LP can also serve as a reasoning engine for general circumscriptive
theories. We illustrate how to use F2LP to compute the circumscriptive
event calculus.

1 Introduction

One advantage of classical logic over logic programs is that the former allows us
to encode knowledge in a complex formula, which is often more convenient than
encoding in conjunctive normal form only. While the input languages of answer
set solvers have evolved to allow various constructs for facilitating encoding
efforts, such as choice rules, cardinality constraints and aggregates, the syntax is
still limited to rule forms and does not allow quantifiers and connectives nested
arbitrarily as in classical logic.

Recently, there have been some efforts in lifting the syntactic restriction by
extending the stable model semantics to arbitrary first-order formulas, under
which an answer set program is viewed as the conjunction of the implications
corresponding to the rules [1,2]. The generality of the language allows to view
choice rules and cardinality constraints as abbreviations of first-order formulas
without involving grounding [3].

System F2LP[1] is a step towards implementing this general language. It translates an arbitrary first-order formula under the stable model semantics into an
answer set program. By calling existing answer set solvers on the resulting program, we can compute Herbrand stable models of a first-order formula. The
system extends the previous version described in [4], which computes stable
models of an arbitrary propositional formula. The translation implemented in
F2LP is based on the following recent theoretical results.

---

[1] http://reasoning.eas.asu.edu/f2lp

E. Erdem, F. Lin, and T. Schaub (Eds.): LPNMR 2009, LNCS 5753, pp. 515–521, 2009.
© Springer-Verlag Berlin Heidelberg 2009

- Every first-order formula is strongly equivalent to its prenex form [4, Theorem 2] and can be also rewritten as a universal formula under certain conditions at the price of introducing new predicate constants [5, Proposition 3].
- Every quantifier-free formula (including propositional formula) is strongly equivalent to a logic program [6,7,4].

We expect that F2LP will facilitate encoding efforts. It can also serve as a tool for computing general circumscriptive theories, in view of the relationship between the stable models and circumscription described in [5]. We illustrate how F2LP can be used for computing circumscriptive event calculus [8,9], whose syntax is not necessarily in the rule form. System CIRC2DLP [10] is another implementation of circumscription using answer set solvers, which can even handle prioritized circumscription and allows varied constants. On the other hand, F2LP allows more general syntax.

2   Review: Stable Models for First-Order Formulas

We follow the definition of a stable model from [2], a journal version of [1]. The definition is also reproduced in [11]. There stable models are defined using "stable model operator SM" with "intensional predicates," similar to circumscription.

Let $\mathbf{p}$ be a list of distinct predicate constants $p_1, \ldots, p_n$ other than equality. For any first-order sentence F, by $\mathrm{SM}[F; \mathbf{p}]$ we denote the second-order sentence

$$F \wedge \neg \exists \mathbf{u}((\mathbf{u} < \mathbf{p}) \wedge F^*(\mathbf{u})),$$

where $\mathbf{u}$ is a list of n distinct predicate variables $u_1, \ldots, u_n$. Expression $\mathbf{u} < \mathbf{p}$ stands for a formula expressing that $\mathbf{u}$ is "stronger than" $\mathbf{p}$, defined same as in circumscription. Formula $F^*(\mathbf{u})$ is defined recursively.

- $p_i(\mathbf{t})^* = u_i(\mathbf{t})$ for any tuple $\mathbf{t}$ of terms;
- $F^* = F$ for any atomic F that does not contain members of $\mathbf{p}$;
- $(F \odot G)^* = (F^* \odot G^*)$, $\odot \in \{\wedge, \vee\}$;
- $(F \to G)^* = (F^* \to G^*) \wedge (F \to G)$;
- $(QxF)^* = QxF^*$, $Q \in \{\forall, \exists\}$.

A model of F (in the sense of first-order logic) is *stable* (relative to the set $\mathbf{p}$ of *intensional* predicates) if it satisfies $\mathrm{SM}[F; \mathbf{p}]$. Let $\sigma(F)$ be the signature consisting of the object, function and predicate constants occurring in F. If F contains at least one object constant, an Herbrand interpretation of $\sigma(F)$ that satisfies $\mathrm{SM}[F; \mathbf{p}]$ where $\mathbf{p}$ is the list of all predicate constants occurring in F, is called an *answer set* of F. The answer sets of a logic program Π are defined as the answer sets of the FOL-representation of Π (i.e., the conjunction of the universal closures of implications corresponding to the rules). It turns out that this definition, applied to the syntax of logic programs, is equivalent to the traditional definition of answer sets based on grounding and fixpoint construction [1].

3   Quantifier Elimination

Given a set of formulas, F2LP first eliminates all quantifiers and then applies the transformation defined in [7] that turns the resulting quantifier-free formulas into logic program rules. In this section we describe how quantifier elimination is done in F2LP.

Obviously, if the domain is known and finite, quantifiers can be replaced with multiple disjunctions and conjunctions. For instance, consider the formula

$$r \wedge \neg \exists x (p(x) \wedge q(x)) \to s \tag{1}$$

occurring in a program that contains n object constants $\{a_1, \ldots, a_n\}$. Replacing $\exists x(p(x) \wedge q(x))$ with multiple disjunctions and then turning the result into a logic program yields 2^n rules. Also this translation is not modular as it depends on the underlying domain, so that the multiple disjunctions need to be updated when the domain changes. Alternatively, we can introduce a new predicate constant p', and turn (1) into

$$s \leftarrow r, not\ p'$$
$$p' \leftarrow p(x), q(x)$$

which does not involve grounding so that the translation is not dependent on the domain.

Under the general stable model semantics, maximal negative occurrences of $\exists$ and maximal positive occurrences of $\forall$ in the formula can be dropped in view of the fact that the standard prenex normal form conversion turns such occurrences into outermost $\forall$ while preserving strong equivalence [4]. As shown in the example above, positive occurrences of $\exists$ can be eliminated using new predicate constants if the quantified formula is in the scope of negation. This condition is further generalized in the proposition below. We say that an occurrence of a predicate constant in a formula F is *strictly positive* if that occurrence is not in the antecedent of any implication. (For instance, in $(p \to q) \to r$, only r has a strictly positive occurrence.) About a formula F, we say that it is *negative* on a tuple $\mathbf{p}$ of predicate constants if members of $\mathbf{p}$ have no strictly positive occurrences in F [11]. The following proposition is a slight generalization of [5, Proposition 3] in view of Theorem on Double Negations from [11].

Proposition 1. *Let F be a sentence, let $\mathbf{p}$ be a list of distinct predicate constants and let q be a predicate constant that does not belong to the signature of F. For any positive occurrence of a subformula $\exists x G(x, \mathbf{y})$ of F where $\mathbf{y}$ is the list of all free variables in $\exists x G(x, \mathbf{y})$, let F' be the formula obtained from F by replacing that occurrence with $\neg\neg q(\mathbf{y})$. If the occurrence of $G(x, \mathbf{y})$ is in a subformula of F that is negative on $\mathbf{p}$, then the models of*

$$\mathrm{SM}[F' \wedge \forall x \mathbf{y}(G(x, \mathbf{y}) \to q(\mathbf{y})); \mathbf{p}, q]$$

restricted to the signature of F are precisely the models of $\mathrm{SM}[F; \mathbf{p}]$.

Negative occurrences of $\forall$ can also be eliminated using the proposition by first rewriting $\forall x G$ as $\neg \exists x \neg G$.

For example, $\exists x(p(x) \wedge q(x))$ in formula (1) is contained in a negative formula (relative to any set of intensional predicates). According to Proposition 1 $SM[(1); \ p, q, r, s]$ has the same models as

$$SM[(r \wedge \neg\neg\neg p' \rightarrow s) \wedge \forall x(p(x) \wedge q(x) \rightarrow p'); \ p, q, r, s, p']$$

if we disregard p'.

These ideas lead to the following procedure for quantifier elimination, which is implemented in F2LP.

Definition 1. *Given a formula F, repeat the following until there are no occurrences of quantifiers remaining:*

Select a maximal occurrence of $QxG(x, \mathbf{y})$ in F where Q is $\forall$ or $\exists$ and $\mathbf{y}$ is the list of all free variables in $QxG(x, \mathbf{y})$.

(a) *If Q is $\exists$ and the occurrence of $QxG(x, \mathbf{y})$ in F is negative, or if Q is $\forall$ and the occurrence of $QxG(x, \mathbf{y})$ in F is positive, then set F to be the formula obtained from F by replacing the occurrence of $QxG(x, \mathbf{y})$ with $G(z, \mathbf{y})$ where z is a new variable.*

(b) *If Q is $\exists$ and the occurrence of $QxG(x, \mathbf{y})$ in F is positive, then set F to be*

$$F' \wedge (G(x, \mathbf{y}) \rightarrow p_G(\mathbf{y}))$$

where F' is the formula obtained from F by replacing the occurrence of $QxG(x, \mathbf{y})$ with $\neg\neg p_G(\mathbf{y})$ where p_G is a new predicate constant.

(c) *If Q is $\forall$ and the occurrence of $QxG(x, \mathbf{y})$ in F is negative, then set F to be the formula obtained from F by replacing the occurrence of $QxG(x, \mathbf{y})$ with $\neg\exists x \neg G(x, \mathbf{y})$.*

4   F2LP Implementation

Formulas can be encoded in the language of F2LP using the following ASCII characters.

Symbol	$\neg$	$\wedge$	$\vee$	$\rightarrow$	$\bot$	$\top$	$\forall xyz$	$\exists xyz$
ASCII	-	&	|	->	false	true	![X,Y,Z]:	?[X,Y,Z]:

F2LP turns a formula into the corresponding LPARSE program.[2] The usual LPARSE encoding is also allowed in F2LP: it is simply copied to the output. The LPARSE program returned by F2LP can be passed to ASP grounders and solvers that accept LPARSE language. While function symbols are allowed in the input language of F2LP, it is left to the grounder to handle them.

The current version of F2LP does not check if the condition to apply quantifier elimination (Proposition 1) is satisfied, which is left to the users. Also F2LP does not check if the given formula is safe (according to [3]), and may turn a safe formula into an unsafe program. For instance, F2LP turns the safe formula

---

[2] http://www.tcs.hut.fi/Software/smodels

```
p(X) -> ((q(Y)->r(Y)) | s(X)).
```

into an unsafe program

```
r(Y)|s(X) :- q(Y),p(X).
s(X) :- {not q(Y)}0,not r(Y),p(X).
```

However, this may not be a serious limitation since we usually declare variables using the #domain directive in LPARSE language, which is the same as appending domain predicates to the body of each rule.

5   Computing Circumscriptive Theories

Kim *et al.* [5] show that for a certain class of formulas called "canonical," circumscription and the general stable model semantics coincide. This allows F2LP to be used for computing circumscription of canonical formulas. For example, consider the formula

$$F = \exists x (p(x) \wedge r(x)) \rightarrow q(b)$$

and the intensional predicates $\{p, q\}$. According to [5], the formula is "canonical" relative to $\{p, q\}$ so that $\mathrm{CIRC}[F; p, q]$ is equivalent to $\mathrm{SM}[F; p, q]$, and furthermore to $\mathrm{SM}[F \wedge \forall x (r(x) \vee \neg r(x)); p, q, r]$. Formula $F \wedge \forall x (r(x) \vee \neg r(x))$ can be encoded in the language of F2LP (In addition, let us assume that the domain is $\{a, b, c\}$):

```
objects(a;b;c).
#domain objects(X).
?[X]:(p(X)&r(X)) -> q(b).
{r(X)}.
```

Canonical theories cover a wide range of action formalisms based on circumscription, such as circumscriptive event calculus. Here we illustrate how to use F2LP to compute an event calculus description.

A circumscriptive event calculus domain description is defined as

$$\mathrm{CIRC}[\Sigma \; ; \; \mathit{Initiates, Terminates, Releases}] \wedge \; \mathrm{CIRC}[\Delta \; ; \; \mathit{Happens}] \wedge \Xi \quad (2)$$

where Σ, Δ, Ξ are first-order sentences such that all positive occurrences of $\exists x G$ in these formulas are contained in subformulas that are negative on $\{\mathit{Initiates}, \mathit{Terminates, Releases, Happens}\}$. Theorem 1 from [5] shows that this theory can be turned into

$$\mathrm{SM}[\Sigma \wedge \Delta \wedge \Xi \wedge \mathit{Choice}(\mathbf{p} \setminus \{\mathit{Initiates, Terminates, Releases, Happens}\}); \mathbf{p}] \quad (3)$$

where $\mathbf{p}$ is the set of all predicates occurring in the description. (By $\mathit{Choice}(\mathbf{p})$ we denote the conjunction of "choice formulas" $\forall \mathbf{x}(p(\mathbf{x}) \vee \neg p(\mathbf{x}))$ for all predicate constants p in $\mathbf{p}$ where $\mathbf{x}$ is a list of distinct object variables whose length is the same as the arity of p.) Note that the condition on Σ, Δ, Ξ above satisfies the condition for eliminating existential quantifiers in Proposition 1.

In view of Theorem 1 from [5], F2LP can be used for computing the models of (2). To compute the models, a user can encode

$$\Sigma \wedge \Delta \wedge \Xi \wedge Choice(\mathbf{p} \setminus \{Initiates, Terminates, Releases, Happens\})$$

in (3) in the language of F2LP, and run F2LP to turn it into an answer set program. For instance, an action precondition axiom (in Ξ) for the Blocks World can be encoded in F2LP as

```
T < maxstep & happens(pickUp(X),T)
  -> holdsAt(clear(X),T) & X != table & -?[Y]:holdsAt(holding(Y),T).
```

("picking up X is possible only if X is clear, the agent is not already holding another object and the object being picked up is not the table.")

F2LP turns the axiom into the following rules.

```
holdsAt(clear(X),T) :- T<maxstep,happens(pickUp(X),T).
  :- {not holdsAt(holding(NV1),T)}0,T<maxstep,happens(pickUp(X),T).
  :- X=table,T<maxstep,happens(pickUp(X),T).
```

A full encoding of the Blocks World in the language of F2LP is available on the F2LP webpage (Footnote 1).

Acknowledgements

We are grateful to Yuliya Lierler, Vladimir Lifschitz, and anonymous referees for their useful comments on this paper. This work was partially supported by the National Science Foundation under Grant IIS-0839821.

References

1. Ferraris, P., Lee, J., Lifschitz, V.: A new perspective on stable models. In: Proceedings of International Joint Conference on Artificial Intelligence (IJCAI), pp. 372–379 (2007)
2. Ferraris, P., Lee, J., Lifschitz, V.: Stable models and circumscription. Artificial Intelligence (to appear, 2010)
3. Lee, J., Lifschitz, V., Palla, R.: A reductive semantics for counting and choice in answer set programming. In: Proceedings of the AAAI Conference on Artificial Intelligence (AAAI), pp. 472–479 (2008)
4. Lee, J., Palla, R.: Yet another proof of the strong equivalence between propositional theories and logic programs. In: Working Notes of the Workshop on Correspondence and Equivalence for Nonmonotonic Theories (2007)
5. Kim, T.W., Lee, J., Palla, R.: Circumscriptive event calculus as answer set programming. In: Proceedings of International Joint Conference on Artificial Intelligence, IJCAI (to appear, 2009)
6. Cabalar, P., Ferraris, P.: Propositional theories are strongly equivalent to logic programs. TPLP 7(6), 745–759 (2007)
7. Cabalar, P., Pearce, D., Valverde, A.: Reducing propositional theories in equilibrium logic to logic programs. In: Bento, C., Cardoso, A., Dias, G. (eds.) EPIA 2005. LNCS (LNAI), vol. 3808, pp. 4–17. Springer, Heidelberg (2005)

8. Shanahan, M.: A circumscriptive calculus of events. Artif. Intell. 77(2), 249–284 (1995)
9. Mueller, E.: Commonsense reasoning. Morgan Kaufmann, San Francisco (2006)
10. Oikarinen, E., Janhunen, T.: circ2dlp - translating circumscription into disjunctive logic programming. In: Baral, C., Greco, G., Leone, N., Terracina, G. (eds.) LPNMR 2005. LNCS (LNAI), vol. 3662, pp. 405–409. Springer, Heidelberg (2005)
11. Ferraris, P., Lee, J., Lifschitz, V., Palla, R.: Symmetric splitting in the general theory of stable models. In: Proceedings of International Joint Conference on Artificial Intelligence, IJCAI (to appear, 2009)

The First Version of a New ASP Solver : ASPeRiX

Claire Lefèvre and Pascal Nicolas

LERIA – University of Angers – France
2, bd Lavoisier – F-49045 Angers Cedex 01
claire.lefevre@univ-angers.fr, pascal.nicolas@univ-angers.fr

Abstract. We present the first version of our ASP solver ASPeRiX that implements a new approach of answer set computation. The main specifity of our system is to realize a forward chaining of first order rules that are grounded on the fly. So, unlike all others available ASP systems ASPeRiX does not need a pregrounding processing.

1 Introduction

When someone uses Answer Set Programming (ASP) to represent and solve a problem, (s)he writes a first order normal logic program such that its *stable models* [4], also called *answer sets*, represent the solutions of the problem. To compute the answer sets, traditional ASP systems proceed in two steps. The program with variables is first given to a *grounder* which builds a propositional version by instantiating all rules with all constants occurring in the program. Then, a *solver* takes this ground program and computes the answer sets.

Following our approach of answer set computing [5], we have implemented in C++ a new ASP solver called ASPeRiX[1]. Its innovative strategy escapes the pregrounding phase since it applies a forward chaining of first order rules that are grounded on the fly during the computation of the answer sets.

2 Description of ASPeRiX

ASPeRiX deals directly with any *normal logic program* containing rules with variables, function symbols and arithmetic calculus. The only syntactic restriction is the *safety* of rules, ie: all variables occurring in a rule must occur in its positive body. To avoid the possible problem of infinite Herbrand universe due to function symbols and arithmetic calculus, the user can pass to ASPeRiX the command line parameters -F k to limit to k the number of nestings in functional terms and -N k to limit the set of numbers to $[-k, \ldots, k]$.

The search procedure of ASPeRiX [5] follows a forward chaining that alternates two steps : a monotonic propagation phase applying the largest possible number of rules instances, and a choice point applying an instance of a non monotonic rule. These inferences build incrementally two atom sets IN and OUT representing atoms occurring or not in the solution.

---

[1] Available at http://www.info.univ-angers.fr/pub/claire/asperix

E. Erdem, F. Lin, and T. Schaub (Eds.): LPNMR 2009, LNCS 5753, pp. 522–527, 2009.
© Springer-Verlag Berlin Heidelberg 2009

At each step, some ground instances of rules are built by unifying atoms of the positive body with atoms occurring in IN and propagating this unifier in the head and the negative body. In order to limit the number of these unifications, ASPeRiX uses dependencies between predicates. If a is an atom, we note $pred(a)$ the predicate of a, the notation is extended to atom sets. The *dependency graph* of a program P is a graph whose nodes are predicates occurring in P and arcs are $\{(p, q) \mid \exists r \in P, \ p = pred(head(r)), \ q \in pred(body(r))\}$. We say that a predicate p *depends on* q if there is a path from p to q in the dependency graph. Predicates are grouped according to maximal strongly connected components (scc for short) of the graph. Components are themselves ordered as $\langle C_1, ..., C_n \rangle$ such that if $i < j$ then no predicate in C_i depends on some predicate in C_j. In the following we denote *rules from a scc* C the rules whose head predicate is in C. In ASPeRiX, the rules considered in order to be eventually applied come from the *current* scc, knowing that the first one is C_1.

In the propagation phase, to apply a monotonic rule consists in adding in IN its ground head provided that its ground body is included in IN. On its side, the choice point step adds in IN the ground head of a non monotonic rule such that its ground positive body is included in IN, and no atom in its ground negative body occurs in IN. It also adds to OUT all the ground negative body of the rule. All along the search process, the condition $IN \cap OUT = \emptyset$ is checked and if it is violated a backtrack is done. The last applied non monotonic rule is then retracted and its negative body is added as a new constraint in order to record that this rule has to be blocked.

When neither propagation nor choice point is possible in one scc, all predicates from it are said *solved* and the next scc in the order defined above becomes the current one. To be solved for a predicate p means that all rules concluding p are exhausted or, in other words, that the extension of p (i.e., the set of all atoms whose predicate is p belonging to IN) is entirely known. In this case, for every propositional atom a such that $pred(a) = p$, $a \notin IN \Rightarrow a \in OUT$. Note that if all predicates appearing in the negative body of a rule are solved, the rule can be considered as a monotonic one and be completely processed by propagation without leading to a choice point. By this way, only truly non monotonic first order rules can generate choice points. In particular, if the program is stratified, the propagation phase is enough to compute the only answer set.

Our approach implies that all atoms in IN are *supported*, in the sense that they result from the application of a sequence of rules, each of them having its positive body included in IN when it is fired. But in special case of rule instances r such that the head is in OUT, the positive body is included in IN and the negative body is restricted to a singleton $\{b_1\}$ with $b_1 \notin IN \cup OUT$, one can conclude that b_1 must be true otherwise contradiction occurs. ASPeRiX uses this information by keeping a set MBT of atoms that *must be true*. These atoms can be used for propagation, in order to prune the search space. But they are not supported yet and thus can not be used as such for generating choice points.

$$\text{Let us take } P_1 = \left\{ \begin{array}{ll} \texttt{n(1).} & (r_2) \text{ a(X) :- n(X), not b(X).} \\ (r_1) \text{ n(X+1) :- n(X).} & (r_3) \text{ b(X) :- n(X), not a(X).} \\ & (r_4) \text{ c(X) :- n(X), not b(X+1).} \end{array} \right\}$$

processed by ASPeRiX with option -N 2. For the dependency graph of P_1 the scc are: $\langle C_1 = \{n\}, C_2 = \{a, b\}, C_3 = \{c\}\rangle$. At the beginning, $IN = \{\texttt{n(1)}\}$ and $OUT = \emptyset$. A propagation step completely processes rule r_1: n(X) is unified with n(1), and n(2) is added in IN, then n(X) is unified with n(2) but $X + 1 = 3$ is out of range of accepted integers $[-2, \ldots, 2]$ for this example. C_1 is thus entirely computed, and n is solved. In scc C_2, there is nothing to propagate and we do the first choice point, for instance rule r_2 where n(X) is unified with n(1). b(1) is then added to OUT and a(1) is added to IN. It does not allow new propagation and a second choice point takes place: rule r_2 where n(X) is unified with n(2), b(2) is then added to OUT and a(2) is added to IN. Instances of r_3 can not be chosen because, for each instance of n(X) in IN, the corresponding a(X) is also in IN and thus r_3 is blocked. So, C_2 is entirely computed and a and b are solved. In scc C_3, a propagation step is enough to treat r_4: the rule is considered as a monotonic one because b is solved. It can be applied only for $X = 1$ (b(2) $\notin IN$) and c(1) is added to IN. Finally, the first answer set is $\{\texttt{n(1), n(2), a(1), a(2), c(1)}\}$. If an other answer set is asked, we backtrack on the last choice point, here a(2):- n(2), not b(2). by retracting a(2) and c(1) from IN and b(2) from OUT and adding :- not b(2). in the program. This leads to add b(2) in MBT. Then a new choice is done: r_3 with $X = 2$ that leads to move b(2) from MBT to IN and to obtain a second answer set.

3   Evaluation of ASPeRiX

In the following we give some results of evaluation of ASPeRiX highlighting its adequacy to some particular problems. In no case it is a deep evaluation, but the results reported below illustrate that our system is operational and very efficient on certain classes of problem[2]. Since we want to compare our system to others by taking into account the two steps of ASP computation (grounding and solving), we have formed the following couples of systems : Lparse 1.1.1+Smodels 2.32 [9,8], Gringo 2.0.3+Clasp 1.2.1 [3,2], DLV Oct 11 2007 [6] or its extension DLV-complex[1] supporting recursive functions, lists and sets, and GASP [7].

Let P_{birds} be a program encoding a taxonomy about flying (f) and non flying (nf) birds (b) such penguins (p), super penguins (sp) and ostriches (o).

$$P_{birds} = \left\{ \begin{array}{lll} \text{p(X) :- sp(X).} & \text{b(X) :- p(X).} & \text{b(X) :- o(X).} \\ \text{f(X) :- b(X), not p(X), not o(X).} & & \text{f(X) :- sp(X).} \\ \text{nf(X) :- p(X), not sp(X).} & & \text{nf(X) :- o(X).} \end{array} \right\}$$

In fact, P_{birds} contains also also the atoms encoding N birds with 10% of ostriches, 20% of penguins whose half of them are super penguins. And let $P_{locstrat}$ be the following locally stratified program.

---

[2] In addition, we have submitted ASPeRiX to the second ASP competition http://www.cs.kuleuven.be/~dtai/events/ASP-competition.

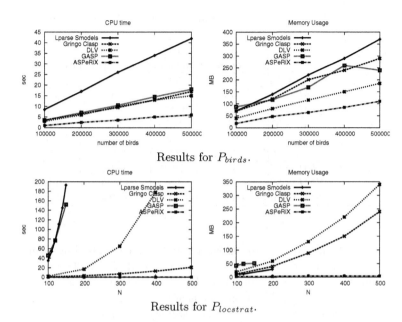

Results for P_{birds}.

Results for $P_{locstrat}$.

Fig. 1. Comparative evaluation of ASPeRiX

$$P_{locstrat} = \left\{ \begin{array}{ll} \texttt{p(1). ... p(N).} & \\ \texttt{a(X) :- p(X), not b(X).} & \texttt{aa(X,Y) :- a(X), p(Y), not a(Y).} \\ \texttt{b(X) :- p(X), not a(X).} & \texttt{bb(X,Y) :- b(X), p(Y), not b(Y).} \end{array} \right\}$$

The CPU time and the memory usage needed to compute the unique answer set of P_{birds} and one answer set of $P_{locstrat}$ are summarized in Fig. 1. For a stratified program, as P_{birds}, well-founded semantics coincides with the stable model semantics. So we have also used the system XSB [10], a logic programming and deductive database system that is able to compute the well-founded semantics of a program. But, for 100000 birds XSB needs more than 500 seconds to end.

The preceding examples illustrate the ability of ASPeRiX to manage very efficiently (locally) stratified programs. On their side, for a definite or stratified program, all traditional grounders do not generate all ground instances of rules but compute in fact the answer set of the program and the solver has nothing to do (except to read the input and it can be very time consuming). On the contrary, for $P_{locstrat}$ that is only locally stratified, a solver is necessary to deal with predicates a and b. But, once the choices are made, a and b are solved and the rest of the program becomes stratified and can be easily evaluated by ASPeRiX. But, seeing instantiation as a pretreatment forces the traditional grounders to generate all ground instances of rules for aa and bb and the used memory quickly becomes prohibitive. GASP, which builds a CSP at each local grounding step, is not either efficient for this last example.

Table 1. CPU time of computation for Hanoi tower problem

		original encoding		modified encoding	
		`DLV-complex`	`ASPeRiX`	`DLV-complex`	`ASPeRiX`
$NbD = 4$	$NbM = 15$	3.7	0.08	3.8	0.14
	$NbM = 30$	10.7	0.34	11	0.14
	$NbM = 60$	25.8	0.84	27	0.14
$NbD = 5$	$NbM = 31$	61	0.6	62	1.02
	$NbM = 40$	94	1.2	97	1.02
	$NbM = 50$	134	2.2	137	1.03
$NbD = 6$	$NbM = 63$	1377	12.4	1412	16
	$NbM = 80$	> 2000	28	> 2000	16.1
	$NbM = 100$	> 2000	51	> 2000	16.4

To illustrate the ability of `ASPeRiX` to manage function symbols we treat the Hanoi tower problem as it is encoded for `DLV-complex`[3]. For this system the problem is encoded by a disjunctive logic program but only one rule has a disjunction in head: `move(I,towers(S1,S2,S3))` v `nomove(I,towers(S1,S2,S3)):-....` For `ASPeRiX` we have translated this rule in a pair of non-disjunctive rules: `move(I,towers(S1,S2,S3)) :- ...,` not `nomove(I,towers(S1,S2,S3))` and `nomove(I,towers(S1,S2,S3)):-...,` not `move(I,towers(S1,S2,S3))..` Since `ASPeRiX` does not treat (yet) specifically lists we have substituted terms like `[4,3,2,1]` by `l(4,l(3,l(2,l(1,nil))))`, `l` being the functor for list. In Table 1 we report under the title *original encoding* the performances of `DLV-complex` and `ASPeRiX` to compute one answer set (and the only one if it exists) of this program. NbD is the number of disks in the problem and NbM is the maximum number of moves that are allowed to move all disks from the first rod to the third. The least value of NbM is $2^{NbD} - 1$ since it is the minimum number of moves required to achieve the goal. We can see that `ASPeRiX` outperforms `DLV-complex` but the performances of both systems are deteriorated by the presence of useless possible moves (beyond the $2^{NbD} - 1$ required).

Under the title *modified encoding* in Table 1 we report results for the same programs modified by the adding of a default negation in the body of rules generating all possible moves. The role of this default negation is to block the applicability of rules when the global goal is satisfied and then to not generate useless possible moves. `DLV-complex` does not take advantage of this new encoding because of the a priori generation of numerous rules that can not be discarded during the pregrounding since only the answer set semantics justifies their cancellation. On its side `ASPeRiX` takes advantage of this encoding[4] by not dealing with blocked rules when the goal is reached and then the computation time remains the same even if the number of possible moves is too large.

---

[3] All encodings used here are available at `ASPeRiX` home page.
[4] We also use this kind of schema to limit the computation of recursive functions in [5].

4 Conclusion

In this paper we have presented the first version of our new ASP solver ASPeRiX. It has been developed under a strategy radically different from that of traditional solvers. To summarize, ASPeRiX deals with first order rules with function symbols, does not require a grounder and is rule-oriented. It has proved its efficiency (in term of performances and power of knowledge representation) on particular classes of problems: (locally) stratified programs, programs with big numbers and arithmetic calculus (because we have not to ground rules with all existing numbers but only those useful, see [5]) and some programs with calculus that need to or can be dynamically bound during the search (like Hanoi example). In the future, we plan to improve our system in two directions. First, by better taking into account ASP semantics, we may integrate better propagation strategies, earlier detection of unsatisfiable constraints, intelligent backtracking in order to prune more efficiently the search space. Second, on the side of the software development, we may improve our grounding techniques, first order rule management and tuples handling. We may also envisage to allow some built-in predicates in the rules in order to help the debugging process in ASP.

References

1. Calimeri, F., Cozza, S., Ianni, G., Leone, N.: Computable functions in ASP: Theory and implementation. In: Garcia de la Banda, M., Pontelli, E. (eds.) ICLP 2008. LNCS, vol. 5366, pp. 407–424. Springer, Heidelberg (2008)
2. Gebser, M., Kaufmann, B., Neumann, A., Schaub, T.: Conflict-driven answer set solving. In: Proceedings of the 20th International Joint Conference on Artificial Intelligence (IJCAI 2007), pp. 386–392 (2007)
3. Gebser, M., Schaub, T., Thiele, S.: GrinGo: A new grounder for answer set programming. In: Baral, C., Brewka, G., Schlipf, J. (eds.) LPNMR 2007. LNCS (LNAI), vol. 4483, pp. 266–271. Springer, Heidelberg (2007)
4. Gelfond, M., Lifschitz, V.: The stable model semantics for logic programming. In: Kowalski, R.A., Bowen, K. (eds.) ICLP, pp. 1070–1080. MIT Press, Cambridge (1988)
5. Lefèvre, C., Nicolas, P.: A first order forward chaining approach for answer set computing. In: Erdem, E., Lin, F., Schaub, T. (eds.) LPNMR 2009. LNCS, vol. 5753. Springer, Heidelberg (2009)
6. Leone, N., Pfeifer, G., Faber, W., Eiter, T., Gottlob, G., Perri, S., Scarcello, F.: The DLV system for knowledge representation and reasoning. ACM Transactions on Computational Logic 7(3), 499–562 (2006)
7. Dal Palù, A., Dovier, A., Pontelli, E., Rossi, G.: Answer set programming with constraints using lazy grounding. In: Hill, P., Warren, D. (eds.) ICLP 2009. LNCS, vol. 5649. Springer, Heidelberg (2009)
8. Simons, P., Niemelä, I., Soininen, T.: Extending and implementing the stable model semantics. Artificial Intelligence 138(1-2), 181–234 (2002)
9. Syrjänen, T.: Implementation of local grounding for logic programs for stable model semantics. Technical report, Helsinki University of Technology (1998)
10. http://xsb.sourceforge.net

An ASP-Based Data Integration System

Nicola Leone, Francesco Ricca, and Giorgio Terracina

Dipartimento di Matematica, Università della Calabria, 87030 Rende, Italy
{leone,ricca,terracina}@mat.unical.it

Abstract. The task of an information integration system is to combine data residing at different sources, providing the user with a unified view of them, called *global schema*. Simple data integration scenarios have been widely studied and efficient systems are already available. However, when some constraints are imposed on the quality of the global data, the integration process becomes difficult and, often, it may provide ambiguous results. Important research efforts have been spent in this area, but no actual system efficiently implementing the corresponding techniques is available yet. This paper is intended to be a step forward in this direction; it proposes a new data integration system, based on Answer Set Programming (ASP) and many optimizations, allowing to carry out consistent query answering (CQA) over massive amounts of data.

1 Introduction

The task of an *information integration system* is to combine data residing at different sources, providing the user with a unified view of them, called *global schema*. Users formulate queries over the global schema, and the system suitably queries the sources, providing an answer. Users are not obliged to have any information about the sources.

Recent developments in IT such as the expansion of the Internet have made available to users a huge number of information sources, generally autonomous, heterogeneous and widely distributed: as a consequence, information integration has emerged as a crucial issue in many application domains, e.g., distributed databases, cooperative information systems, data warehousing, or on-demand computing.

However, information integration is, in general, an extremely complex task. Both state-of-the-art commercial software solutions (e.g., [1]) and academic systems (see e.g. [2,3] for a survey) fulfill only partially the ambitious goal of integrating information in complex application scenarios. Moreover, comprehensive, formal methodologies and coherent tools for designing information integration systems are still missing.

The main objectives a data integration system should address are the following:

1. **A comprehensive information model**, through which the knowledge about the integration domain can be easily specified. The possibility of defining expressive integrity constraints (ICs) over the global schema, the precise characterization of the relationship between global schema and the local data sources, the formal definition of the underlying semantics, as well as the use of a powerful query language, are mandatory for the specification of complex integration applications.

E. Erdem, F. Lin, and T. Schaub (Eds.): LPNMR 2009, LNCS 5753, pp. 528–534, 2009.

2. **Capability of dealing with data that may result inconsistent** with respect to global ICs. Even if some solutions to the problem of query answering for inconsistent data (e.g., [3,4,5]) have been already proposed, they did not produce so far effective and scalable system implementations, mainly due to the high computational complexity of the problem.
3. **Advanced information integration algorithms**; these should provide a formal correspondence between the data integration system and the expected query answers, especially in the handling of inconsistent data.
4. **Mass-memory-based evaluation strategies**. In fact, real world scenarios involve massive amounts of data; thus, techniques that require all-in-memory evaluations would not provide the needed scalability.

The system proposed in this paper is an attempt to address the above issues; it starts from the experience we gained in the INFOMIX [6] project to overcome some limitations experienced in real-world scenarios. In fact, it is based on Answer Set Programming (ASP) and exploits datalog-based methods for answering user queries, which are sound and complete with respect to the semantic of query answering. This guarantees meaningful data integration and solves issue 1. It incorporates a number of optimization techniques that "localize" and limit the inefficient computation, due to the handling of inconsistencies, to a very small fragment of the input. This allows obtaining fast query-answering, even in such a powerful data-integration framework, thus solving issue 2. The problem of consistent query answering (CQA) is reduced to cautious reasoning on disjunctive datalog programs, which allows to effectively compute the query results precisely, by using state-of-the-art disjunctive datalog systems. The formal query semantics is captured also in presence of inconsistent data. This solves issue 3. Finally, the system adopts DLV^{DB} [9,11] as internal query evaluation engine, which allows for mass-memory evaluations and distributed data management features, and solves issue 4.

In the following sections we introduce some details on the key components of the system and its overall architecture.

2   Key System Components

In our setting, a data integration system $\mathcal{I}$ is a triple $\langle \mathcal{G}, \mathcal{S}, \mathcal{M} \rangle$, where $\mathcal{G}$ is the global schema, which provides a uniform view of the information sources to be integrated, $\mathcal{S}$ is the source schema, which comprises the schemas of all the sources to be integrated, and $\mathcal{M}$ is the mapping establishing a relationship between $\mathcal{G}$ and $\mathcal{S}$. $\mathcal{G}$ may contain integrity constraints (ICs). $\mathcal{M}$ is a *Global-As-View* (GAV) mapping [5], i.e., $\mathcal{M}$ is a set of logical implications $\forall x_1 \cdots \forall x_n . \Phi_{\mathcal{S}}(x_1, \ldots, x_n) \supset g_n(x_1, \ldots, x_n)$, where g_n is a relation from $\mathcal{G}$, n is the arity of g_n, $\Phi_{\mathcal{S}}$ is a conjunction of atoms on $\mathcal{S}$ and $x_1, \ldots, x_n$ are the free variables of $\Phi_{\mathcal{S}}$. Each global relation is thus associated with a *union of conjunctive queries* (UCQs).Both $\mathcal{G}$ and $\mathcal{S}$ are assumed to be represented in the relational model, whereas $\mathcal{M}$ is represented as a set of datalog rules.

As an example consider a bank association that desires to unify the databases of two branches. The first database models managers by using a table $man(code, name)$ and employees by a table $emp(code, name)$, where code is a primary key for both tables.

The second database stores the same data in table $employees(code, name, role)$. Suppose that the data has to be integrated in a global schema with two tables: $m(code)$, and $e(code, name)$, having both code and name as keys and the inclusion dependency $m[code] \subseteq e[code]$, indicating that manager codes must be employee codes. GAV mappings are defined as follows:

$$e(C, N) :- emp(C, N). \quad e(C, N) :- employee(C, N, \_).$$

$$m(C) :- man(C, \_). \quad m(C) :- employee(C, \_, man).$$

If emp stores $(e1, john)$, $(e2, mary)$, $(e3, willy)$, man stores $(e1, john)$, and $employees$ stores $(e1, ann, man)$, $(e2, mary, man)$, $(e3, rose, emp)$, it is easy to verify that, while the source databases are consistent w.r.t. local constraints, the global database obtained by evaluating the mappings violates the key constraint on e (e.g. both $john$ and ann have the same code $e1$ in table e). Basically, when data are combined in a unified schema with its own integrity constraints the resulting global database might be inconsistent; any query posed on an inconsistent database would then produce an empty result.

In this context, user queries must be re-modelled according to the mappings and violated constraints, in order to compute *consistent* answers, i.e. answers which consider as much as possible of *correct* input data. This task is accomplished in our system by the CQA Rewriter component.

Another key component of the system is the Query Evaluator, which is in charge of actually evaluating, over source databases, the queries posed over the global schema. These two key components are described in some detail in the following subsections.

2.1 CQA Rewriter

In the field of data-integration several notions of consistent query answering have been proposed (see [3] for a survey), depending on whether the information in the database is assumed to be *correct* or *complete*. Basically, the incompleteness assumption coincides with the *open world assumption*, where facts missing from the database are not assumed to be false. In our system, we assume that sources are complete; as argued in [8], this choice strengthens the notion of minimal distance from the original information.[1] Moreover, there are two important consequences of this choice: integrity restoration can be obtained by only *deleting tuples*; and, computing CQA for conjunctive queries remains *decidable* even for arbitrary sets of denial constraints and inclusion dependencies [8].

More formally, given a global schema $\mathcal{G}$ and a set C of integrity constraints, let $\mathcal{DB}$ and $\mathcal{DB}^r$ be two global database instances. $\mathcal{DB}^r$ is a *repair* [8] of $\mathcal{DB}$ w.r.t. C, if $\mathcal{DB}^r$ satisfies all the constraints in C and the instances in $\mathcal{DB}^r$ are a maximal subset of the instances in $\mathcal{DB}$. Basically, given a conjunctive query Q, *consistent answers* are those query results that are not affected by axioms violations and are true in any possible repair [8]. Thus, given a database instance $\mathcal{DB}$ and a set of constraints C, a conjunctive query Q is consistently true in $\mathcal{DB}$ w.r.t. C if Q is true in every repair of

---

[1] It is worth noting that, in relevant cases like denial constraints, query results coincide for both correct and complete information assumptions.

$\mathcal{DB}$ w.r.t. C. Moreover, if Q is non-ground, the consistent answers to Q are all the tuples $\bar{t}$ such that the ground query $Q[\bar{t}]$ obtained by replacing the variables of Q by constants in $\bar{t}$ is consistently true in $\mathcal{DB}$ w.r.t. C. Note that, in this setting, the problem of computing consistent answers to queries in the case of denial constraints and inclusion dependencies (the most common schema constraints) belongs to the Π_2^P complexity class [8]. The CQA Rewriter, takes as input a conjunctive query Q, a set of integrity constraints C, and a global database $\mathcal{DB}$ and builds both an ASP program Π_{cqa} and a query Q_{cqa}, such that: Q is consistently true in $\mathcal{DB}$ w.r.t. C iff Q_{cqa} is true in every answer set of Π_{cqa}, in symbols: $\Pi_{cqa} \models_c Q_{cqa}$.

Just to show an example on how the rewriter works, consider the example introduced previously; the program obtained by rewriting the global constraints is:

$$\bar{e}(X,Y) \vee \bar{e}(X,Z) :- e(X,Y), e(X,Z), Y <> Z.$$
$$\bar{e}(X,Y) \vee \bar{e}(Z,Y) :- e(X,Y), e(Z,Y), X <> Z.$$

$$e^r(X,Y) :- e(X,Y), not\ \bar{e}(X,Y).$$
$$m^r(X) :- m(X), e^r(X, \_).$$

Here, the disjunctive rules guess atoms to be cancelled for satisfying key constraints, whereas the following rules remove atoms violating also referential integrity constraints, and build repaired relations. Note that the minimality of answer sets guarantees that deletions are minimized. This sub-program is then fed along with the mappings and the user query to the query evaluator (see next Section).

Suppose now that we ask for the list of manager codes; since both $m^r(e1)$ and $m^r(e2)$ are in all the answer sets of the resulting program, both $m(e1)$ and $m(e2)$ are derived as the consistent answers.

It is important to point out that our rewriting procedure has been devised in such a way that the evaluation of produced ASP programs is complexity-wise optimal according to the complexity classification of constraints and queries of [8]. Indeed, polynomial, co-NP and Π_2^P queries are dealt with by exploiting normal (stratified) programs, head-cycle-free, and non-head-cycle-free programs, respectively.

2.2 Query Evalutator

The core query evaluation engine of our integration system is DLV^{DB} [9]. It is a DLP evaluator born as a database oriented extension of the well known DLV system [10]. It has been recently extended [11] for dealing with unstratified negation, disjunction and external function calls.

The main peculiarities of DLV^{DB} related to the data integration system are:

- It allows the handling of (possibly distributed) massive amounts of data; in fact, it exploits ODBC connections to link the ASP program to database relations and implements an evaluation strategy working mostly onto the database, where input data reside. More precisely, it carries out the grounding of the logic program completely on the DBMS, and then loads in main memory only the minimal amount of data necessary for the generation of stable models, thus handling disjunction and unstratified negation. With respect to the data integration setting, this strategy perfectly fits the system's needs; in fact, if input data is "clean" and no global constraints are violated, the query answering process can be completed during the grounding, and no data must be loaded in main memory. On the contrary, if some constraint is violated, only conflicting data must be loaded in main memory; this amount of data is usually much smaller than the overall input size.

- GAV mappings defining the integration system can be directly evaluated without further elaboration; this provides a direct correspondence between what the designer specifies and what the system should evaluate. This simplifies the application of optimization and rewriting techniques.

- DLV^{DB} extends ASP with external function calls; this is, in general, particularly suited for solving inherently procedural sub-tasks. External functions must be defined as stored functions in the database coupled with DLV^{DB}. In the context of data integration, the possibility of calling external functions provides rich capabilities in cleaning input data. In fact, it allows to embed, in a declarative setting, purely procedural tasks such as string manipulation. As an example, consider the integration of two databases representing dates in different formats: one as dd/mm/yy, and the other one as mm/dd/yy. In order to guarantee consistency in the integrated database, one of these representations must be transformed and this can be easily carried out by string manipulation. This procedurally simple task would be tedious in a purely declarative setting.

- DLV^{DB} embodies some query-oriented optimization strategies, like magic-sets, query unfolding, and static filtering (see [12] for a survey), capable of significantly improving query evaluation performances. In the context of data integration, such optimizations find their perfect application when queries contain constants, since they allow to "localize" the computation and drastically reduce the amount of data to reason about.

3 The Integration System

The general architecture of the proposed system is shown in Figure 1. It is intended to simplify both the integration system design and the querying activities by exploiting a user-friendly GUI. Specifically, at design time, the user can:

- Graphically design the global schema and the mappings (which we recall are expressed by UCQs) between global relations and source schemas.
- Specify data transformation rules on source data; these can be implemented by suitable functions defined in the working database as stored functions.
- Specify global constraints, in order to define quality parameters that global integrated data must satisfy.

At query time, the user can exploit a QBE-like interface to express queries over the global schema; these are internally expressed in datalog as UCQs. The "plain" query is then elaborated by the CQA Rewriter which takes into account both mappings and global constraints to express the query over the sources and to handle inconsistencies

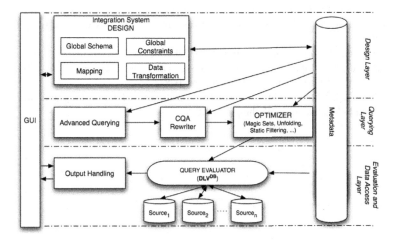

Fig. 1. System Architecture

possibly involving the query answers; the output of the CQA Rewriter is then a (possibly disjunctive) datalog program which is fed to the Optimizer for further elaboration. The Optimizer applies rewriting strategies which aim at pushing down selections directly onto the sources and at "localizing" over conflicting data as much as possible of the needed reasoning. Finally, the optimized program is fed to the Query Evaluator (DLV^{DB}) which executes the grounding phase totally on the DBMS and loads in main-memory only data strictly necessary to resolve conflicts. The output of this evaluation is then the query answer, which is proposed graphically back to the user.

References

1. Hayes, H., Mattos, N.: Information on demand. DB2 Magazine 8(3) (2003)
2. Halevy, A.Y.: Data integration: A status report. In: 10th Conference on Database Systems for Business, Technology and Web (BTW 2003), pp. 24–29 (2003)
3. Bertossi, L., Hunter, A., Schaub, T. (eds.): Inconsistency Tolerance. LNCS, vol. 3300. Springer, Heidelberg (2005)
4. Bravo, L., Bertossi, L.: Logic programming for consistently querying data integration systems. In: Int. Joint Conference on Artificial Intelligence (IJCAI 2003), pp. 10–15 (2003)
5. Lenzerini, M.: Data integration: A theoretical perspective. In: Proc. PODS 2002, pp. 233–246 (2002)
6. Leone, N., et al.: The INFOMIX System for Advanced Integration of Incomplete and Inconsistent Data. In: Proc. ACM SIGMOD 2005, pp. 915–917 (2005)
7. Arenas, M., Bertossi, L.E., Chomicki, J.: Consistent Query Answers in Inconsistent Databases. In: Proc. PODS 1999, pp. 68–79. ACM Press, New York (1999)
8. Chomicki, J., Marcinkowski, J.: Minimal-change integrity maintenance using tuple deletions. Information and Computation 197(1-2), 90–121 (2005)
9. Terracina, G., Leone, N., Lio, V., Panetta, C.: Experimenting with recursive queries in database and logic programming systems. Theory and Practice of Logic Programming (TPLP) 8(2), 129–165 (2008)

534   N. Leone, F. Ricca, and G. Terracina

10. Leone, N., Pfeifer, G., Faber, W., Eiter, T., Gottlob, G., Perri, S., Scarcello, F.: The DLV
    System for Knowledge Representation and Reasoning. ACM Trans. Comput. Log. 7(3), 499–
    562 (2006)
11. Terracina, G., De Francesco, E., Panetta, C., Leone, N.: Enhancing a DLP system for
    advanced database applications. In: Calvanese, D., Lausen, G. (eds.) RR 2008. LNCS,
    vol. 5341, pp. 119–134. Springer, Heidelberg (2008)
12. Minker, J. (ed.): Foundations of Deductive Databases and Logic Programming. Morgan
    Kaufmann Publishers, Inc., Washington (1988)

GORGIAS-C: Extending Argumentation with Constraint Solving

Victor Noël[1] and Antonis Kakas[2]

[1] Université de Toulouse, Institut de Recherche en Informatique de Toulouse,
118, route de Narbonne, 31 062 Toulouse Cedex, France
victor.noel@irit.fr
[2] Department of Computer Science, University of Cyprus,
75 Kallipoleos St. Nicosia, Cyprus
antonis@cs.ucy.ac.cy

1 General Information

GORGIAS-C is a system implementing a logic programming framework of argumentation that integrates together preference reasoning and constraint solving.

The framework of argumentation with preference reasoning [1,2] was first implemented in the GORGIAS system[1] which has mainly been used in the multi-agent domain [2,3,4], medical informatics [5] and network security [6].

We have extended this framework and the GORGIAS system to integrate constraint solving at the representation and argumentation level of the framework. GORGIAS-C is the first framework and system that we are aware of that links argumentation with domain constraints.

In GORGIAS-C problems are expressed in the combined language of Logic Programming with Priorities and Constraint Logic Programming (CLP) [7] thus allowing us to address problems with large computational domains while at the same time exploiting the high expressive power of the framework to capture complex requirements and preferences.

Technically, GORGIAS-C is implemented as a modular meta-interpreter on top of the object-oriented Logtalk preprocessor[2] [8] with SWI-Prolog[3] and its CLP(FD) library [9] but can be used with other Prolog interpreters with CLP supported by Logtalk.

Furthermore, thanks to this modularity, GORGIAS-C is easy to extend, allowing researchers of the domain to prototype new extensions of the argumentation framework.

2 Description of the System

GORGIAS-C computes answers to queries asked on a logic program with priorities on rules and domain constraints on variables. It also allows, as GORGIAS, the use of abductive predicates as a special kind of argumentation. The general format of such a program is as follows.

---

[1] http://www2.cs.ucy.ac.cy/~nkd/gorgias/
[2] http://logtalk.org/
[3] http://www.swi-prolog.org/

E. Erdem, F. Lin, and T. Schaub (Eds.): LPNMR 2009, LNCS 5753, pp. 535–541, 2009.

Rules. Rules are labeled logic programming rules of the form:

```
rule(Label, Head, Body).
```

where, `Label` is a term for referencing the rule and `Head` is a positive or negative logic program literal (with negation applied using the operator `neg/1` on positive atoms). `Body` is a list containing positive or negative logic program literals, or domain constraints over variables in the rule.

A special case of these rules, where the `Head` is an atom on the special predicate `prefer/2`, is used to define the priorities between the rules of the program. Both arguments of `prefer(label1,label2)` are labels of rules, denoting that the the first is preferred over the second. For the implementation of GORGIAS-C, these two rules need to have contradictory conclusions.

Abducibles. Abducible predicates are declared as such with the following construct:

```
abducible(Predicate).
```

where `Predicate` is a user-defined predicate name.

Abducible predicates can be partially defined in the program with rules whose head refers to them. Integrity constraints (ICs) [10] on an abducible predicate can be specified with a rule whose head is the negation of an abducible and whose body is the condition under which the abducible cannot be assumed.

Domain Constraints. Domain constraints usable within GORGIAS-C are the ones available in the underlying constraint solver. In the SWI-Prolog's CLP(FD) library, they are finite domain constraints like `#>=`, `#=`, `#\=` or `#\`, `#\/`, `#==>`. In other Prolog interpreters with CLP, other types of constraints can be used (e.g. over reals). As mentioned above these are used in the body of the logic program rules.

Easing Development. GORGIAS-C proposes some facilities to ease the development of programs:

- `::/1` and `::/2` to call predicates directly using the underlying Logtalk interpreter preventing argumentation on them: they are explained in the tutorial on the website
- `complement/2` to specify complementarity between logic program literals, extending the notion of negation

For example, `complement(give(Object, Time), keep(Object, Time))` means that `give/2` and `keep/2` are contradictory and thus we cannot build arguments with both and arguments for one can potentially attack arguments for the other.

2.1 Computing Answers

Given a program as defined above and a query $\exists Q(X)$, GORGIAS-C computes an answer of the form $\langle \Delta, A, C \rangle$ where:

1. Δ is a set of program rules used to conclude an argument for Q.
2. A is a (possibly empty) set of abducible hypotheses (literals on abducible predicates) needed for Δ to conclude Q.
3. C is a set of domain constraints on X and variables in $\Delta \cup A$.

An answer $\langle \Delta, A, C \rangle$ fulfils the following conditions:

1. C is satisfiable.
2. For every valuation σ of C:
   (a) Δ_σ is admissible: consistent and attacking all its attacking arguments
   (b) $\Delta_\sigma \cup A_\sigma \models Q_\sigma$.

The precise definition of attacks is given in [11,2]. Informally, a set of rules (an argument) attacks another argument if they have a complementary conclusion and the attacking argument renders its rules of higher priority than the rules of the attacked argument or at least not of lower priority.

The rules in an answer form an argument for the query, giving not only the rules that derive it but also the rule priorities that make the answer preferred over contrary ones. The set A contains information (that is missing from the program due to the incompleteness of the abducible predicates) needed for the arguments in Δ to be enabled. The constraints C impose necessary restrictions on the cases (variables) of the arguments needed in order for the argument to admissibly conclude the query, as described above.

The set of domain constraints C are computed by integrating within the argumentation reasoning a new (symmetric) attacking relation where any two domain constraints c_1 and c_2, which together are non-solvable, attack each other. This works together with the underlying domain constraint solving to ensure that the computed constraints C are solvable.

The classical Tweety example (extended to allow for incomplete knowledge on the relation bird/1) can be represented in GORGIAS-C (and GORGIAS since this example does not involve domain constraints) as follows:

```
rule(r1(X), fly(X), [bird(X)]).
rule(r2(X), neg(fly(X)), [::penguin(X)]).     abducible(bird(_)).
rule(r3(X), bird(X), [::penguin(X)]).         penguin(tweety).
rule(pr1(X), prefer(r2(X), r1(X)), []).
```

GORGIAS-C answers the query fly(leon) and neg(fly(tweety)) respectively with <{r1(leon)},{bird(leon)},{}> and <{r2(tweety), f1},{},{}>. An answer for fly(tweety) cannot be built as the argument given by the second rule, r2, for the contrary conclusion, is stronger thanks to pr1 and thus cannot be counter-attacked.

A second example where we use the new constraint handling facility of GORGIAS-C is the following:

```
rule(r1(P), buy(P), [::offer(OP), P #> OP]).                  offer(27).
rule(r2(P), neg(buy(P)), [::offer(OP), P #>= OP + 5]).
rule(pr1(P), prefer(r2(P),r1(P)), []).              top-price(30).
```

The answer to the query buy(P) is <{r1(P)},{},{P in 28..31}>, because to prevent r2 to be preferred over r1, P must be greater than the last offer of 27 and less than the last offer plus 5: 31. The resulting domain of 28..31 for the variable P was computed by applying the constraint P #> 27 together with the application of negating the constraint P #>= 27 + 5.

If we then add the rules rule(r3(P), neg(buy(P)), []). and
rule(pr2(P), prefer(r3(P),r1(P)), [::top-price(Top), P #> Top]).

The answer to the query `buy(P)` is `<{r1(P)},{},{P in 28..30}>`, the resulting domain was computed by applying the constraint `P #> 27` together with the application of negating each of the constraints `P #> 30` and `P #>= 27 + 5`.

Hence, GORGIAS-C uses, like GORGIAS, rules and abduction to conclude goals, and in addition finds domains of finite domain variables where these are applicable. Furthermore, to ensure that there are no stronger opposing arguments, GORGIAS-C will find variables domains that make potential opposing arguments inapplicable or prevent them from being stronger.

2.2   Implementation

GORGIAS-C is implemented as a modular meta-interpreter for its logic programs on top of Logtalk using SWI-Prolog and its CLP(FD) library and has successfully been used with ECLIPSe[4] with CLP over reals.

Using the features of Logtalk for object and component-based programming [12], GORGIAS-C is a modular meta-interpreter, consisting of:

- A Logtalk's category that implements:
  - The core resolution algorithm consisting of interleaving phases of attacks and defences.
  - The reduction of the constraint domains by using the underlying constraint solver at key points of the computation.
- Categories for the 'modules' implementing different aspects of this framework, called by the core algorithm when needed:
  - 'LPwNF' for attacks based on priority rules.
  - 'Abducible' for attacks and resolution based on abduction.
  - 'CLP' for attacks based on domain constraints.

Core. The core algorithm alternates two phases of attack and defence on a set of rules. For every attack possible on the set, a defence against it (a counter-attack) is found and added to the set. This is the recursively applied to the original extended set until it is admissible, i.e. it defends against every attack and does not attack itself.

More specifically, when querying the system this will first resolve the query, resulting on an initial set Δ_0 that concludes the query. Then it will extend Δ_0 by finding all attacks on it, and for each of them, find a counter-attack to add to Δ_0. If it can not counter-attack an attack, then the computation fails and backtracks to the last choice-point.

Domain Constraints. When constraints are present in an attack on Δ_0, to defend it GORGIAS-C will construct counter-attacks by negating the constraints in the attacks to make them inapplicable.

To be able to negate them, these constraints should be reifiable (which means that we can reflect their truth value into boolean values represented by 0 and 1).

We can also note that, because GORGIAS-C only relies on the fact that constraints must be reifiable:

---

[4] htt://www.eclipse-clp.org

- It can use all reifiable constraints added to the constraint solver by users.
- It can use any constraint solver (other than finite domain ones) that uses reifiable constraints (such as the IC library of ECLIPSe).

3   Applying the System

Argumentative programs are embedded in Logtalk's objects importing the core category and other needed 'modules'. Depending on what is enabled, the Logtalk preprocessor will generate the smallest Prolog code, with the possibility to enable a debugging trace of the argumentation process.

Querying Programs. Once the user's program is loaded we can call the predicate `prove(Query, Delta)` to compute an answer to the query `Query`, by unifying `Delta` with a set of rules labels and assumed abducibles, and an associated constraint store maintained by the underlying constraint solver. Variables in this answer can then be labeled with the different algorithms provided by the constraint solver.

Methodology. Writing an argumentative program is a task requiring different skills. Apart from the (constraint) logic programming background, the following will help.

Typically, programs written for GORGIAS-C (and GORGIAS) can be separated into 2 or more layers. The first one would be composed of rules for and against some conclusion of interest to the user, using negation or 'complement' predicates in heads. The second one would be composed of preference rules that the user wants to apply when the rules of the first layer come into conflict. And so on, if the second layer contains conflicting preferences.

Abduction is used to model facts for which there is incomplete knowledge, but, above all, can be used to model the solution of a problem: rules are used to construct the solution (as assumed abducibles) and ICs enforce properties on the solution.

4   Evaluating the System: Performance and Expressiveness

In term of performance, compared to Prolog, GORGIAS-C adds two overheads: the monotonic resolution (meta-interpretation) and the argumentative process of constructing attacks and counter-attacks while maintaining the constraint store. But at the same time, GORGIAS-C provides a declarative expressiveness available through abduction and preferences combined with CLP.

Classical constraint problems (e.g. N-Queens, Graph Colouring) can be resolved using GORGIAS-C, not exploiting its argumentative aspects but only the underlying constraint solver. Thus they will only suffer from the monotonic resolution overhead which is negligible according to benchmarks made with the N-Queens example. Compared to Prolog and Logtalk, times for one solution are unchanged with 10, 15, 20 or 25 queens

Abduction was used in planning and scheduling problems: advantages of the approach is well covered in [13]. Applications presented in this paper were ported to GORGIAS-C and presented equivalent performances for job-shop 20x5, 10x10 and 20x10 when using ECLIPSe.

Preference reasoning has been used with GORGIAS in autonomous agents to specify interaction rules or protocols, and policies for negotiation processes, but also in network security to specify firewall policies. This kind of applications can be extended to handle large computational domains with the help of domain constraints. Pricing and firewall policies are currently being worked on: constraints should simplify the programs and improve performances in some problems by moving complex computations from the Prolog interpreter to the constraint solver.

After adding domain constraints to preferences-enabled problems, we will investigate the addition of preferences to constraint solving problems and study the benefits and performance that GORGIAS-C will have. For example, planning and scheduling problems could benefit from priorities to guide the search, by reducing the domains of the solutions based on criteria such as action costs or job importance.

5   Obtaining the System

GORGIAS-C can be found on its website at `http://dev.crazydwarves.org/trac/Gorgias`, together with sample examples of its use, applications and benchmarks.

References

1. Kakas, A.C., Mancarella, P., Dung, P.M.: The acceptability semantics for logic programs. In: ICLP, pp. 504–519 (1994)
2. Kakas, A., Moraitis, P.: Argumentation Based Decision Making for Autonomous Agents. In: AAMAS 2003, pp. 883–890 (2003)
3. Kakas, A., Maudet, N., Moraitis, P.: Modular Representation of Agent Interaction Rules through Argumentation. Autonomous Agents and Multi-Agent Systems 11(2), 189–206 (2005)
4. Kakas, A., Mancarella, P., Sadri, F., Stathis, K., Toni, F.: Computational Logic Foundations of KGP Agents. Journal of Artificial Intelligence Research 33, 285–348 (2008)
5. Letia, I.A., Acalovschi, M.: Achieving Competence by Argumentation on Rules for Roles. In: Gleizes, M.-P., Omicini, A., Zambonelli, F. (eds.) ESAW 2004. LNCS (LNAI), vol. 3451, pp. 45–59. Springer, Heidelberg (2005)
6. Bandara, A.K., Kakas, A.C., Lupu, E.C., Russo, A.: Using Argumentation Logic for Firewall Policy Specification and Analysis. In: State, R., van der Meer, S., O'Sullivan, D., Pfeifer, T. (eds.) DSOM 2006. LNCS, vol. 4269, pp. 185–196. Springer, Heidelberg (2006)
7. Jaffar, J., Lassez, J.L.: Constraint Logic Programming. In: POPL 1987, pp. 111–119 (1987)
8. Moura, P.: Logtalk - Design of an Object-Oriented Logic Programming Language. PhD thesis, University of Beira Interior, Portugal (September 2003)
9. Triska, M., Neumerkel, U., Wielemaker, J.: A Generalised Finite Domain Constraint Solver for SWI-Prolog. In: WLP 2008, pp. 89–91 (2008)
10. Denecker, M., Kakas, A.: Abduction in logic programming. In: Kakas, A.C., Sadri, F. (eds.) Computational Logic: Logic Programming and Beyond. LNCS (LNAI), vol. 2407, pp. 402–436. Springer, Heidelberg (2002)
11. Noël, V., Kakas, A.: Extending argumentation with constraint solving. Technical report, University of Cyprus (2009)

12. Moura, P.: Logtalk 2.6 Documentation. Technical Report DMI 2000/1, University of Beira Interior, Portugal (July 2000)
13. Kakas, A., Mourlas, C.: ACLP: Flexible Solutions to Complex Problems. In: Fuhrbach, U., Dix, J., Nerode, A. (eds.) LPNMR 1997. LNCS, vol. 1265, pp. 388–399. Springer, Heidelberg (1997)

ANTON: Composing Logic and Logic Composing

Georg Boenn[1], Martin Brain[2], Marina De Vos[2], and John ffitch[2]

[1] Cardiff School of Creative & Cultural Industries
University of Glamorgan
Pontypridd, CF37 1DL, UK
gboenn@glam.ac.uk
[2] Department of Computer Science
University of Bath
Bath, BA2 7AY, UK
{mjb,mdv,jpff}@cs.bath.ac.uk

Abstract. In most styles of music, composition is governed by a set of rules. We demonstrate that approaching the automation and analysis of composition declaratively, by expressing these rules in a suitable logical language, powerful and expressive intelligent composition tools can easily be built. This paper describes the use of answer set programming to construct an automated system, named ANTON, that can compose both melodic and harmonic music, diagnose errors in human compositions and serve as a computer-aided composition tool.

1 Introduction

In this paper we investigate the use of declarative logic programming in the automatic composition of music. We show that it is possible to use Answer Set Programming (ASP) [2] to create *ab initio* short musical pieces that are both melodic and harmonic. Our system, ANTON, named in honour of our favourite composer of the second Viennese School, is presented as both a design and as a practical working system, showing that rule-based declarative systems can be effectively used. We report on our experience in using ASP for this system, and indicate a number of potentially exciting directions in which this system could develop, both musically and computationally. This paper provides a summary ANTON 1.0. More detail can be found in [4]. The full system is publicable available from http://www.cs.bath.ac.uk/~mjb/anton/ and a selection of compositions can be found at http://dream.cs.bath.ac.uk/Anton

2 Algorithmic Composition

One can distinguish between *improvisation* systems and *composition* systems. In the former the note selection progresses through time, without detailed knowledge of what is to come. In practice this is informed either by knowing the chord progression or similar musical structures [7], or using some machine listening. However in this paper we are concerned with *composition*, so the process takes place out of time, and we can make decisions in any order.

E. Erdem, F. Lin, and T. Schaub (Eds.): LPNMR 2009, LNCS 5753, pp. 542–547, 2009.

A common problem in musical composition can be summarised in the question "where is the next note coming from?". For many composers over the years the answer has been to use some process to generate notes. It is clear that in many pieces from the Baroque period simple note sequences are being elaborated in a fashion we would now call algorithmic. It is usual to credit Mozart's Musikalisches Würfelspiel (Musical Dice Game) [8] as the oldest classical algorithmic composition[1]. This form of composition is in essence stochastic, a system of algorithmic composition that leads to Xenakis's *oeuvre*.

There is a variety of different approaches to algorithmic composition, including chaotic processes, and Markov chains [9]. We also note the variation of the accompanist, where either the chord structure and style is known in advance, or using machine-listening techniques follows a melody.

A more recent trend is to cast the problem as one of constraint satisfaction. For example, PWConstraints [12] is an extension for IRCAM's Patchwork, a Common-Lisp-based graphical programming system for composition. It uses a custom constraint solver employing backtracking over finite integer domains. OMSituation and OMClouds are similar and were more recently developed for Patchwork's successor OpenMusic. A detailed evaluation of them can be found in [1], where the author gives an example of a 1st-species counterpoint (two voices, note against note) after [10] developed with Strasheela, a constraint system for music built on the multi-paradigm language Oz. Our musical rules however implement the melody and counterpoint rules described by [16], which we believe give better musical results.

It should also be noted that these algorithmic systems compose pieces of music in either a melodic or a harmonic fashion, and are frequently associated with computer-based synthesis. The system we propose is different as it deals with both simultaneously in an integrated fashion.

Systems that for instance harmonise Bach chorales start with a melody for which at least one valid harmonisation exists[2], and the program attempts to find one; this problem is clearly soluble. This differs significantly from our system: as we generate the melody and harmonisation together, the requirement for harmonisation affects the melody. The system is also capable of other modes of operation including computer aided composition and diagnosis of existing pieces.

3 ANTON

ANTON applies ASP techniques to compositional rules to produce an algorithmic composition system. *AnsProlog* is used to write a description of the rules that govern the melodic and harmonic properties of a correct piece of music; in this way the program works as a model for music composition that can be used to assist the composer by suggesting, completing and verifying short pieces.

The composition rules are modelled so that the *AnsProlog* program defines the requirements for a piece to be musically valid, and thus every answer set corresponds to a different valid piece. To generate a new piece the composition system simply has to

---

[1] Although there is some doubt if the game form is really his.

[2] If we may so characterise Bach.

```
% Every choosen note must be in the key
#const err_nik="Choosen note not in key".
reason(err_nik).
error(P,T,err_nik) :- choosenNote(P,T,N), chromatic(N,C), not key(C).

% The last two notes of a minor scale are dependant on
% direction upwards the last two are 10,12
error(P,T + 1,err_ism) :- choosenNote(P,T + 1,N), chromatic(N,9), upAt(P,T), keyMode(minor).

% Parts can only ever meet at a single point, and this
% can only happen once.
haveMet(P,T+1) :- choosenNote(P,T,N), choosenNote(P+1,T,N), not haveMet(P,T), part(P+1).
haveMet(P,T+1) :- haveMet(P,T).

% At every time step the note must change.
% It changes by stepping (moving one note in the scale) or leaping (moving more than one note)
% Upwards or downwards
1 { stepAt(P,T), leapAt(P,T) } 1 :- T != t.
1 { downAt(P,T), upAt(P,T) } 1 :- T != t.

% Leaps can only use consonant intervals)
1 { leapBy(P,T,LS) : leapSize(LS) : LS > 0 } 1 :- leapUp(P,T).

% When a part leaps up by I, the note at time T+1
% is I steps higher than the current note
choosenNote(P,T + 1,N + L) :- choosenNote(P,T,N), leapAt(P,T), leapBy(P,T,L), note(N + L).
```

Fig. 1. An ANTON fragment

generate an (arbitrary) answer set. Fig. 1 presents a simplified fragment of the *AnsProlog* program used in ANTON. The model is defined over a number of time steps, given by the variable T. The key proposition is chosenNote(P,T,N) which represents the concept "At time T, part P plays note N". To encode the options for melodic progress ("the tune either steps up or down one note in the key, or it leaps more than one note"), choice rules are used. To encode the melodic limits on the pattern of notes and the harmonic limits on which combinations of notes may be played at once, constraints are included.

To allow for verification and diagnosis, each constraint is given an error message, as shown in Fig. 1. These error-atoms are in a way used as constraints. Depending on how the system is used, composition or diagnosis, you will either be interested in those pieces that do not result in errors at all, or in an answer set that mentions the error messages. For the former we simply specify the constraint :- error(P,T,R). effectively making any error rule into a constraint. For the latter we include the rules: errorFound :- error(P,T,R). and :- not errorFound. requiring that an error is found (i.e. returning no answers if the diagnosed piece is error free).

By adding constraints on which notes can be included, it is possible to specify part or all of a melody, harmony or complete piece. This allows ANTON to be used for a number of other tasks beyond automatic composition. By fixing the melody it is possible to use it as an automatic harmonisation tool. By fixing part of a piece, it can be used as computer aided composition tool. By fixing a complete piece, it is possible to check its conformity with the rules, for marking student compositions or harmonisations. Alternatively we could request the system to complete part of a piece. In order to do so, we provide the system with a set of *AnsProlog* facts expressing the mode (major, minor, etc.), the notes which are already fixed, the number of notes in the piece, the configuration and the number of parts.

We provide a number of output formats: CSOUND [5] with a suitable selection of sounds, text, *AnsProlog* facts or the LILYPOND score language. Other output formats can easily be created and plugged into the system.

4   Evaluation

Musical Evaluation. Musical evaluation is difficult to encapsulate. Pieces by ANTON have been played to a number of musicians, who apart from the rhythmic deficiency considered later have agreed that it is valid music. The interested reader can find examples on the web[3], but here in figure 2 we present a few fragments in the Dorian mode that ANTON composed especially for this paper; the audio and score can be found in the same location as the other works. This piece was chosen to show that the musical rules are for more than just major and minor scales, and can create in various modes.

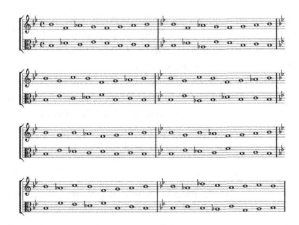

Fig. 2. Fragments in the Dorian Mode by ANTON

Technical Evaluation. The *AnsProlog* programs used in ANTON contains less than 200 lines (not including comments, empty lines and user defined pieces) and encodes 28 melodic and harmonic rules. It should be noted that ANTON's 200 lines of code (800 including all support scripts and translators) contrast with the 8000 lines in Strasheela [1] and 88000 in Bol [3], making it in our opinion easier to understand.

We performed a number of tests to assess the practicality of using answer set programming as the underlying technology of a composition system. Table 1 contains the times taken by a number of answer set solvers (SMODELS [14], SMODELS-IE [6], SMODELSCC [17], CMODELS [13] and CLASP [11]) to compose a solo or duet of a given length. LPARSE [15] and GRINGO were used to ground the programs. Grounding with the former typically took around 30-60 seconds while the later only needed 0.5-12 seconds. Grounding time is omitted from the results reported in the tables.

---
[3] http://dream.cs.bath.ac.uk/Anton

Table 1. Time taken (in seconds) for a number of solvers generating a solo and duet pieces

	smodels 2.32		smodels-ie 1.0.0		smodelscc 1.08	cmodels 3.75		clasp 1.0.5
Length	Default	Restarts	Default	Restarts	No lookahead	w/ zchaff	w/ MiniSAT	Default
Solo								
4	1.02	1.03	0.09	0.09	1.17	0.33	0.39	0.22
6	2.43	2.43	0.38	0.38	2.58	0.64	0.85	0.46
8	5.16	5.16	1.03	1.04	4.94	1.06	1.62	1.01
10	12.25	11.72	2.58	2.59	8.55	1.54	2.63	1.33
12	28.25	46.13	8.08	15.14	11.36	2.42	4.04	2.27
14	40.62	140.00	10.50	43.54	18.78	3.14	6.05	3.48
16	101.05	207.25	29.40	69.53	27.94	4.01	9.40	4.62
Duet								
4	3.77	3.77	0.31	0.32	4.08	1.18	1.26	0.77
6	10.36	11.24	1.89	1.89	13.90	2.17	2.81	1.60
8	54.64	77.10	14.71	21.84	26.07	3.88	5.93	3.73
10	Time out	Time out	Time out	500.26	78.72	9.51	11.12	9.34
12	Time out	Time out	Time out	Time out	103.81	14.50	18.14	16.84
14	Time out	Time out	Time out	Time out	253.92	32.41	32.34	25.59
16	Time out	Time out	Time out	Time out	452.38	82.64	49.29	29.63

All times where recorded using a 2.4GHz AMD Athlon X2 4600+ processor, running a 64 bit version of OpenSuSE 10.3. All solvers were built in 32 bit mode. Each run was limited to 20 minutes of CPU time and 2Gb of RAM. The *AnsProlog* programs used are available from the same location as ANTON.

The results show that ANTON, when using the more powerful solvers, is sufficiently fast to be used as a component in an interactive composition tool, although further work would be needed to support real time generation of music. It is also interesting to note that the only solvers able to generate longer sequences using two parts all implement clause learning strategies, suggesting that the problem is particularly susceptible to this kind of technique.

Why ASP? While music appreciation is matter of personal taste, musicologists use sets of rules which determine to which style a musical composition belongs or whether a piece breaks. These sets of rules also govern the composition. So an intuitive and obvious way for automatic composition is to encode these rules and use a rule based algorithm to produce valid music compositions. This natural and simple way of encoding things is show in terms of speed of development, roughly 2 man-months, sophistication of the results, the amount of code (about 200 lines of code) and flexibility; we can not only easily encode different styles but the same application not only for automated composition but also diagnosis and human assisted composition. Furthermore, we automatically gain from any improvements in the underlying solver.

5 Future Work

An obvious extension to the composition of duets is to expand this to three and four parts, by adding inner voices. It should perhaps be noted that inner voices obey different rules, and these await implementation.

So far we have only considered a particular style of Western music, Renaissance Counterpoint. However the framework should be applicable to other styles, especially formal ones. Given the musical knowledge of the team and collaborators, we are considering applying ANTON to Twelve Tone and Hindustani musics.

In real life pieces some of the rules are sometimes broken. This could be simulated by one of a number of extensions to answer set semantics (preferences, consistency restoring rules, defensible rules, etc.). However how to systematise the knowledge of when it is acceptable to break the rules and in which contexts it is 'better' to break them is an open problem.

The major deficiency in ANTON version 1.0 is the lack of rhythm, as all parts play all the time (with no rests), with notes of equal duration, which, while usual in some styles, stands in the way of a whole range of interesting variety. We have plans and some theory in this area.

References

1. Anders, T.: Composing Music by Composing Rules: Design and Usage of a Generic Music Constraint System. PhD thesis, Queen's University, Belfast, Department of Music (2007)
2. Baral, C.: Knowledge Representation, Reasoning and Declarative Problem Solving, 1st edn. Cambridge University Press, Cambridge (2003)
3. Bel, B.: Migrating Musical Concepts: An Overview of the Bol Processor. Computer Music Journal 22(2), 56–64 (1998)
4. Boenn, G., Brain, M., De Vos, M., ffitch, J.: Automatic composition of melodic and harmonic music by answer set programming. In: Garcia de la Banda, M., Pontelli, E. (eds.) ICLP 2008. LNCS, vol. 5366, pp. 160–174. Springer, Heidelberg (2008)
5. Boulanger, R. (ed.): The Csound Book: Tutorials in Software Synthesis and Sound Design. MIT Press, Cambridge (2000)
6. Brain, M., De Vos, M., Satoh, K.: The significance of memory costs in answer set solver implementation. Joural of Logic and Computation (2008); Advance Access Online
7. Brothwell, A., Ffitch, J.: An Automatic Blues Band. In: Proceedings of 6th International Linux Audio Conference, Kunsthochscule für Medien Köln, March 2008, pp. 12–17 (2008)
8. Chuang, J.: Mozart's Musikalisches Würfelspiel (1995),
   http://sunsite.univie.ac.at/Mozart/dice/
9. Cope, D.: A Musical Learning Algorithm. Computer Music Journal 28(3), 12–27 (Fall 2006)
10. Fux, J.J.: The Study of Counterpoint from Johann Joseph Fux's Gradus ad Parnassum. W.W. Norton (1965); orig 1725
11. Gebser, M., Kaufmann, B., Neumann, A., Schaub, T.: Conflict-Driven Answer Set Solving. In: Proceeding of IJCAI 2007, pp. 386–392 (2007)
12. Laurson, M.: PWConstraints Reference Manual. IRCAM, Paris (1996)
13. Lierler, Y., Maratea, M.: Cmodels-2: SAT-based Answer Set Solver Enhanced to Non-tight Programs. In: Lifschitz, V., Niemelä, I. (eds.) LPNMR 2004. LNCS (LNAI), vol. 2923, pp. 346–350. Springer, Heidelberg (2004)
14. Syrjänen, T., Niemelä, I.: The Smodels System. In: Eiter, T., Faber, W., Truszczyński, M. (eds.) LPNMR 2001. LNCS (LNAI), vol. 2173, p. 434. Springer, Heidelberg (2001)
15. Syrjänen, T.: Lparse 1.0 User's Manual. Helsinki University of Technology (2000)
16. Thakar, M.: Counterpoint. New Haven (1990)
17. Ward, J., Schlipf, S.: Answer Set Programming with Clause Learning. In: Lifschitz, V., Niemelä, I. (eds.) LPNMR 2004. LNCS, vol. 2923, pp. 302–313. Springer, Heidelberg (2004)

Modelling Normative Frameworks Using Answer Set Programing

Owen Cliffe, Marina De Vos, and Julian Padget

Department of Computer Science
University of Bath, BATH BA2 7AY, UK
{occ,mdv,jap}@cs.bath.ac.uk

Abstract. Norms and regulations play an important role in the governance of human society. Social rules such as laws, conventions and contracts prescribe and regulate our behaviour, however it is possible for us to break these rules at our discretion and face the consequences. By providing the means to describe and reason about norms in a computational context, normative frameworks may be applied to software systems allowing for automated reasoning about the consequences of socially acceptable and unacceptable behaviour. In this paper, we outline our mathematical formulation for normative frameworks and describe how its semantics can be represented in ASP, thus enabling the construction of models of normative systems that can be subjected to formal verification and that can act as functional repositories of normative knowledge for the software components that participate in them.

1 Introduction

Normative frameworks are provide a mechanism to capture and reason about "correct" and "incorrect" behaviour within a certain context. The participants of a normative framework are governed by social norms and regulations. The framework monitors the permissions, empowerment and obligations of their participants and generate violations when norms are not adhered to. Information of the norms and the effects of participants actions is stored in the state of the framework. The constant change of the state over time as a result of these actions provides participants information about each others behaviour. The information can also be used by the designer to verify the normative structures of the framework. In this paper we will only look mainly at the latter.

We do not include an extensive and detailed case for the purpose and value of normative frameworks here—this can be found in [2] and [1]— as this paper focuses on applications of answer set programming and because of the need for brevity. What we will present is: (i) an outline of a formal event-based model of the specification of normative frameworks that captures all the essential properties, namely empowerment, permission, obligation and violation (ii) a summary of its formal translation to ASP that is sound and complete (essentially a long mechanical case analysis which not included here), resulting in a decidable and executable model of an normative frameworks that can be used for querying and verification.

E. Erdem, F. Lin, and T. Schaub (Eds.): LPNMR 2009, LNCS 5753, pp. 548–553, 2009.

2   The Normative Model

The components of a normative framework are events, which change the states and fluents, that characterize the state. The framework defines the interplay between these concepts over time and we here give the syntax and semantics.

2.1   Syntax

We define a normative framework as a 5-tuple $I := \langle \mathcal{E}, \mathcal{F}, \mathcal{C}, \mathcal{G}, \Delta \rangle$ consisting of a set of events $\mathcal{E}$, a set of fluents $\mathcal{F}$, a set of causal rules $\mathcal{C}$, a set of generation rules $\mathcal{G}$ and an initial state Δ.

Events $\mathcal{E}$. Each normative framework defines a set of event symbols $e \in \mathcal{E}$, each of which denotes a type of event that may occur. We identify two disjoint subsets, $\mathcal{E}_{ex}$ consisting of *exogenous events* and $\mathcal{E}_{inst}$ consisting of *normative events*.

- **Exogenous events $\mathcal{E}_{ex}$:** consists of events that fall outside the scope of the normative framework. These may include agent communication events such as $tell(a, b, yes)$ (a tells b "yes") and other externally-defined events, such as time-outs. The exogenous events are in effect the interface to the normative framework. It is the responsibility of the designer of the specification to define this set carefully.
- **Normative events $\mathcal{E}_{inst}$:** that in contrast to exogenous events are those events that are generated *within* the normative framework as a, possibly indirect, consequence of an exogenous event.

  We identify two disjoint subsets of normative events: *normative actions*, $\mathcal{E}_{act}$ that capture significant changes in normative state, and *violation events*, $\mathcal{E}_{viol}$ denoting points at which violations have occurred. The set of violation events is defined such that it contains at least one violation event corresponding to each normative action and each exogenous event: $\forall e \in \mathcal{E}_{act} \cup \mathcal{E}_{ex} : viol(e) \in \mathcal{E}_{viol}$. Specifications may also define additional violation events not directly associated with particular exogenous or normative actions. Violation and sanction play an important role in the specification of normative frameworks. Violations may arise either from explicit generation, from the occurrence of a non-permitted event, or from the failure to fulfil an obligation. In these cases sanctions that may include obligations on violating agents or other agents and/or changes in agents' permission to do certain actions, may then be expressed as consequences of the occurrence of the associated violation event in the subsequent normative state.

Normative Fluents $\mathcal{F}$. We now turn to the definition of the normative state which we model through the definition of a set of fluent properties. We make a distinction between *normative fluents*, which express normative properties of the state such as permissions, powers and obligations, and *domain fluents* which correspond to properties which are specific to the normative framework itself. In both cases we model fluents as propositions which may be true or false in a given normative state. We define the set $\mathcal{D}$ to include all normative framework-specific fluents which may be true in the normative state. The set of normative fluents is broken down into sets of fluents for powers, permissions and obligations as follows:

- W: A set of normative power fluents of the form $\mathrm{pow}(e) : e \in \mathcal{E}_{act}$ where each power proposition denotes the capability of some action e to be brought about in the normative framework.
- $\mathcal{P}$: A set of event permission fluents: $\mathrm{perm}(e) : e \in \mathcal{E}_{act} \cup \mathcal{E}_{ex}$, where each permission proposition denotes that it is permitted for action e to be brought about. We do not define a proposition for "forbidden" but just treat it as the absence of permission for that event to be brought about.
- $\mathcal{O}$: A set of obligations, of the form $\mathrm{obl}(e, d, v)$ with $e \in \mathcal{E}, d \in \mathcal{E}, v \in \mathcal{E}_{viol}$. The presence of an obligation fluent in the normative state denotes that event e should occur before the (deadline) event d or violation v will be generated.

The set of states, Σ, of an normative framework is defined as: $\Sigma = 2^{\mathcal{F}^*}$ but not all of these states may be reachable. Rules in the normative framework may have an effect in multiple normative states, and this effect may be conditional on the normative state. In order to qualify the effect of normative rules over a specific set of states, we define a language $\mathcal{X}$ of *state formulae* that allows us to specify to which normative states the rule may apply. The set of state formulae is defined as: $\mathcal{X} = 2^{\mathcal{F} \cup \neg \mathcal{F}}$

Causal Rules $\mathcal{C}$. Each normative framework has a relation $\mathcal{C}$ that describes which fluents are initiated and terminated by the performance of a given action in a state matching some expression. The function is expressed as $\mathcal{C} : \mathcal{X} \times \mathcal{E} \rightarrow 2^{\mathcal{F}} \times 2^{\mathcal{F}}$, where the first set in the range of the function describes which fluents are initiated by the given event and the second those fluents terminated by the event. We use the notation $\mathcal{C}^{\uparrow}(\phi, e)$ to denote the fluents which are initiated by the event e in a state matching ϕ and $\mathcal{C}^{\downarrow}(\phi, e)$ to denote those fluents which are terminated by event e in a state matching ϕ.

Generation Rules $\mathcal{G}$. Each normative framework defines an event generation function $\mathcal{G}$ which describes when the performance of one event counts-as, or generates, another: $\mathcal{G} : \mathcal{X} \times \mathcal{E} \rightarrow 2^{\mathcal{E}_{inst}}$. The generation of events may be conditional on some properties in the normative state, and one event may generate multiple (consequent) events. It should be noted that this relation describes explicitly specified relationships between events in the normative framework. There are cases when events may be generated which are not in this relation, for instance in the case of unsatisfied obligations. Additionally this function represents cases where event generation *may* occur, however as normative events require empowerment, they will only be generated when this property holds.

Initial State Δ. Each normative framework defines the set $\Delta \subseteq \mathcal{F}$ that denotes the set proposition that shall hold when the normative framework is created.

2.2 Semantics

Due to space restrictions, we are only able to provide an informal overview of our semantics for normative frameworks. A full discussion with definitions can be found in [2]. The semantics of this framework is defined over a sequence, called traces, of exogenous events. Starting from the initial state, each exogenous event will be responsible for a state change, i.e. a set fluent of additions and deletions. Using the event generation function, the transitive closure of all events that count as this exogenous event are being obtained. Add to this set of events, all violations of events that were not permitted and obligations that were not fulfilled and we have the set of all events whose consequences will result in the new state. Using the consequence relation on this set of events, we obtain all fluents that need to be initiated and deleted from the current state in order to obtain the new state. So with each trace, we can associate a sequence of states. The combination of the two is referred to as a model or history. When verifying normative frameworks we are interested in finding out whether these models and traces have certain properties. To provide designer support for normative frameworks, we need a computation tools that can compute all traces/models or those that have specific features. In the next section, we demonstrate that ASP can be such a tool.

3   ASP Implementation

We summarize the representation of a normative framework in ASP through reference to the framework components identified earlier and a case-by-case translation of the mathematical model into an ASP representation. A crucial aspect of the model is how we have chosen to represent time and its relationship with the predicate holdsat/2 to represent the truth of a particular fluent at a particular time. Time in this representation is linked to state transitions. Time zero is part of the initial state.

If the causal relation initiates a fluent, we want it to *hold at* the next time instant. A consequence of this representation of time is that unless explicit action is taken, a fluent does *not* hold at the *next* time instant, that is, the maintenance of inertia requires action. These two issues are addressed in this way:

$$f \in \mathcal{F}^* \; \mathtt{holdsat(f, T1)} \leftarrow \mathtt{initiated(f, T), instant(T),}$$
$$\mathtt{instant(T1), next(T, T1).}$$

$$f \in \mathcal{F}^* \; \mathtt{holdsat(f, T1)} \leftarrow \mathtt{holdsat(f, T),}$$
$$\mathbf{not} \; \mathtt{terminated(f, T), instant(T),}$$
$$\mathtt{instant(T1), next(T, T1).}$$

Events cause the evolution of the normative state through the generation relation, thus an exogenous event is observed and its occurrence is added to the normative state:

$$e_{ex} \in \mathcal{E}_{ex} \; \mathtt{occurred(e_{ex}, T)} \leftarrow \mathtt{observed(e_{ex}, T), instant(T),}$$
$$\mathtt{instant(T1), next(T, T1).}$$

To model conditions on the state we use the auxiliary function EX:

$$EX(x_1 \wedge x_2 \wedge \ldots x_n, t_i) \overset{\mathrm{def}}{\equiv} EX(x_1, t_i), EX(x_2, t_i), \ldots EX(x_n, t_i)$$
$$EX(\neg f, t_i) \overset{\mathrm{def}}{\equiv} \mathbf{not} \; EX(f, t_i)$$
$$EX(f, t_i) \overset{\mathrm{def}}{\equiv} \mathtt{holdsat(f, t_i)}$$

Thus, in the first of the following rules, the empowered event e_2 is added to the normative state through conventional generation as a result of the occurrence of e_1 and in the second, the event e_1 is added as a result of the occurrence of the violation event e_2, in both cases subject to a condition ϕ^1:

$$\left.\begin{array}{l} e_1 \in \mathcal{E}, \phi \in \mathcal{X}, \\ e_2 \in \mathcal{G}(\phi, e_1), \\ e_2 \in \mathcal{E}_{act} \end{array}\right\} \begin{array}{l} \texttt{occurred}(e_2, T) \leftarrow \texttt{occurred}(e_1, T), \texttt{EX}(\phi, T), \\ \qquad\qquad\qquad\qquad \texttt{holdsat}(\texttt{pow}(e_2), T), \texttt{instant}(T). \end{array}$$

$$\left.\begin{array}{l} e_1 \in \mathcal{E}, \phi \in \mathcal{X}, \\ e_2 \in \mathcal{G}(\phi, e_1), \\ e_2 \in \mathcal{E}_{viol} \end{array}\right\} \begin{array}{l} \texttt{occurred}(e_1, T) \leftarrow \texttt{occurred}(e_2, T), \texttt{EX}(\phi, T), \\ \qquad\qquad\qquad\qquad \texttt{instant}(T). \end{array}$$

Violations arise when there is either an event that is not permitted (the first rule below) or the deadline event of an obligation (second rule):

$$e \in \mathcal{E}_{act} \cup \mathcal{E}_{ex} \quad \begin{array}{l} \texttt{occurred}(\texttt{viol}(e), T) \leftarrow \texttt{occurred}(e, T), \\ \qquad\qquad\qquad \texttt{not holdsat}(\texttt{perm}(e), T), \texttt{instant}(T). \end{array}$$

$$\texttt{obl}(e, d, v) \in \mathcal{O} \quad \begin{array}{l} \texttt{occurred}(v, T) \leftarrow \quad \texttt{holdsat}(\texttt{obl}(e, d, v), T), \\ \qquad\qquad\qquad \texttt{occurred}(d, T), \texttt{instant}(T). \end{array}$$

The causal relation, covering the initiation and termination of fluents arising from some normative state, is captured in the following two rules:

$$\left.\begin{array}{l} e \in \mathcal{E}, \phi \in \mathcal{X}, \\ f \in \mathcal{C}^{\uparrow}(\phi, e) \end{array}\right\} \quad \texttt{initiated}(f, T) \leftarrow \texttt{occurred}(e, T), \texttt{EX}(\phi, T).$$

$$\left.\begin{array}{l} e \in \mathcal{E}, \phi \in \mathcal{X}, \\ f \in \mathcal{C}^{\downarrow}(\phi, e) \end{array}\right\} \quad \begin{array}{l} \texttt{terminated}(f, T) \leftarrow \texttt{occurred}(e, T), \texttt{EX}(\phi, T), \\ \qquad\qquad\qquad\qquad \texttt{instant}(T). \end{array}$$

Obligations are removed (terminated) subject to the occurrence of either the event necessary to satisfy the obligation or the deadline event associated with the obligation[2]:

$$\texttt{obl}(e, d, v) \in \mathcal{O} \quad \texttt{terminated}(\texttt{obl}(e, d, v), T) \leftarrow \texttt{occurred}(e, T), \texttt{instant}(T).$$

$$\texttt{obl}(e, d, v) \in \mathcal{O} \quad \texttt{terminated}(\texttt{obl}(e, d, v), T) \leftarrow \texttt{occurred}(d, T), \texttt{instant}(T).$$

3.1 Traces and Queries

The purpose of constructing a (event-driven) model of a normative framework is to be able to test it in various ways through querying and this in turn depends on the construction of all the traces of events that could arise between one state and another. Thus query formulation and trace construction are intimately tied up. However, before traces can be generated, the program must be grounded, which means being explicit about the meaning of time instants and precisely how many there are, which in turn determines the length of the trace. For time instants $t_i : 0 \leq i \leq n$, we define the following three rules: $\texttt{instant}(t_i).$, $\texttt{next}(t_i, t_{i+1}).$ and $\texttt{final}(t_n).$ denoting each ground instant of time, relative order and final state, respectively.

---

[1] Note that for violations, we do not need to verify that they are empowered.

[2] See violations above for how the deadline event is then translated into a violation.

The general trace program generates answer sets containing all possible combinations of n exogenous events, but by the addition of further constraints, the answer sets can be limited to those containing changes, being either the initiation or termination of fluents. We define an effective transition as:

$$\text{changed}(T) \leftarrow \text{initiated}(F, T), \text{not holdsat}(F, T),$$
$$\text{ifluent}(F), \text{instant}(T).$$
$$\text{changed}(T) \leftarrow \text{terminated}(F, T), \text{holdsat}(F, T),$$
$$\text{ifluent}(F), \text{instant}(T).$$

so that something changes at time instant T (indicated by changed(T)) because there is either a fluent F which does (not) hold at time instant T which is initiated/terminated at time instant T. Of these, the *valid* states are those in which in which something has changed and the rest are discarded. This can further be enhanced to select traces of *up to* length n, but for the sake of space, those details are also not included here.

4 Extensions and Conclusions

So far we have written of a single normative framework and its model. However, there are two motivations to address the concept of several interacting normative frameworks: (i) one is that for design modularity, it is attractive to focus on small self-contained frameworks as well as in so doing enabling re-use, while (ii) another is that in the real world actors are typically simultaneously governed by many such frameworks. The technical details of this formalization and the notion of power and permission of one framework over another appear in [3]. The important point from the perspective of this paper is the ease with which such models can be combined by virtue of the ASP representation, so that given the necessary linkages between frameworks, their combination can be processed in just the same way as a single framework definition and subsequently a model for the combination is constructed that can answer queries about events and states in their combination.

The normative framework we have presented is designed to capture the essence of regulated environments and through the use of ASP enables its expression in a form that allows the construction of a computable model that can then be subjected to queries through the traces it admits. Subsequently, we have developed a high level action language to specify normative frameworks and a complementary query language.

References

1. Cliffe, O.: Specifying and Analysing Institutions in Multi-Agent Systems using Answer Set Programming. PhD thesis, University of Bath (2007)
2. Cliffe, O., De Vos, M., Padget, J.A.: Answer set programming for representing and reasoning about virtual institutions. In: Inoue, K., Satoh, K., Toni, F. (eds.) CLIMA 2006. LNCS (LNAI), vol. 4371, pp. 60–79. Springer, Heidelberg (2007)
3. Cliffe, O., De Vos, M., Padget, J.A.: Specifying and reasoning about multiple institutions. In: Noriega, P., Vázquez-Salceda, J., Boella, G., Boissier, O., Dignum, V., Fornara, N., Matson, E. (eds.) COIN 2006. LNCS (LNAI), vol. 4386, pp. 67–85. Springer, Heidelberg (2007)

Generating Optimal Code Using Answer Set Programming

Tom Crick, Martin Brain, Marina De Vos, and John Fitch

Department of Computer Science
University of Bath
Bath BA2 7AY, UK
{tc,mjb,mdv,jpff}@cs.bath.ac.uk

Abstract. This paper presents the *Total Optimisation using Answer Set Technology* (TOAST) system, which can be used to generate optimal code sequences for machine architectures via a technique known as superoptimisation. Answer set programming (ASP) is utilised as the modelling and computational framework for searching over the large, complex search spaces and for proving the functional equivalence of two code sequences. Experimental results are given showing the progress made in solver performance over the previous few years, along with an outline of future developments to the system and applications within compiler toolchains.

1   Introduction

Within the field of compiler development the term *optimisation* is something of a misnomer. Compilers typically use a series of templates to generate machine level instructions from the parse tree of the target program. An optimisation phase [1] then attempts to improve this code (with respect to both size and performance) by applying a set of transforms, reductions and equivalences. In many modern compilers, this results in significant improvements but it is very unlikely to produce optimal sequences of instructions; and if it does, it will not be able to determine that they are indeed optimal. To further complicate matters, it is often not clear which order these improvement techniques should be applied as they may enable or inhibit further improvements. The current order of application in most compilers is a result of experience and trial and error rather than design.

In a relatively narrow, but significant range of applications, this approach to code generation is not sufficient. In the inner loops of high-performance computing tasks, performance critical system libraries, many embedded applications [6] and even the templates used for code generation within compilers [5] (both conventional and Just In Time (JIT) compilers within virtual machines), if it is possible to generate optimal code, then it would be desirable to do so.

Superoptimisation [8] is an approach that views code generation for loop-free segments of code as a combinatorial search problem. Thus by utilising appropriate search techniques it is possible to generate genuinely optimal instruction sequences. The *Total Optimisation using Answer Set Technology* (TOAST) system uses answer set programming (ASP) [3] as a computational framework to solve the superoptimisation search

E. Erdem, F. Lin, and T. Schaub (Eds.): LPNMR 2009, LNCS 5753, pp. 554–559, 2009.

problem. A model of the machine architecture is created in *AnsProlog* and answer set solvers are used to generate and verify candidate optimal instruction sequences. In this way, developments in solver technology can thus directly improve the performance of the superoptimiser. The flexibility of *AnsProlog* also allows arbitrary constraints to be added to the search with minimal effort, something that is very difficult in the case of procedural superoptimisers, but of huge importance, as it allows a superoptimiser to be used to augment its own set of constraints. For reasons of compactness, this paper does not include a description of the answer set semantics or ASP; an in-depth description can be found in [3].

2   Superoptimisation

Massalin [8] coined the term superoptimisation to refer to an alternative approach to code generation for short, loop-free sections of machine code. Rather than starting with crude, template-generated code and running multiple improvement passes, a superoptimiser starts with the specification of a function and performs a directed search for a sequence of instructions that meets this specification.

Superoptimisation naturally decomposes into two tasks: *searching* for candidate sequences that meet a reduced set of conditions and then *verifying* that they meet the required specification. The raw search space of possible sequences of a given length is at least exponential in the number of instructions; potentially factorial if the order of inputs to the sequence is considered. However, a number of constraints and heuristics exist that can considerably reduce the space that has to be searched. For example, if an instruction computes a commutative function (such as addition) then only one ordering of inputs needs to be considered; likewise, if instructions can be reordered then only one ordering need by searched. Handling the size and complexity of this space is the current limit on superoptimiser performance.

Despite significant potential, superoptimisation has received relatively little research within the field of code generation and optimisation. Recent work [5] has utilised a range of techniques to handle the large search spaces involved in superoptimisation, including automatic theorem proving [7] and satisfiability testing [2], showing the viability of the approach for specific application areas. However, a major deficiency of the existing superoptimising implementations is that there is no guarantee of optimality. Due to the significant computational burden of proving the functional equivalence of two non-trivial sequences of code, most of the existing implementations use a representative test to shortcut the verification, or timeout and discard sequences that take too long to verify.

3   The TOAST System

The TOAST provably optimal code generation system consists of modular interacting components that generate answer set programs and parse answer sets, with a controlling interface that utilises these components to generate a shorter, superoptimised version of the original input sequence. A preliminary version of the TOAST system was first presented in [4].

```
in: v32
in: v32
inst: land i1 i2
inst: add i1 1
inst: add i1 2
inst: sub i0 3
out: v32
```

Listing 1. A program in TOAST input format

3.1 Architecture

The TOAST system supports multiple processor types, with processor specific information stored in a description file which provides meta-information about the processor, as well as which instructions are available. The TOAST system currently supports the following architectures: MIPS R2000, SPARC V7 and SPARC V8, with more proposed. Porting to a new architecture is simple and takes between a few hours and a week, depending on how many of the instructions used have already been modelled.

TOAST accepts programs in an assembly language-like format as input. These are used as the target of the search, to find the shortest sequence of instructions that has the same output. The example given in Listing 1 defines a program of four instructions, with two 32 bit inputs and one 32 bit output.

We assume that the cost of each instruction in the input program is the same; for RISC-like processors where there are no cache or memory issues, and no pipeline breaks this is a fair simplifying assumption. In the case of minimising memory taken by the instruction stream, as might be used in mobile devices, this is the correct measure.

A set of vectors, binary values for each bit of each input, are generated in ASP. This give a set of vectors for each possible path through the input code. The input program is then 'run' with these vectors to generate constraints, giving the 'correct' values of the outputs for each set of vectors.

By using the input vectors and output constraints (essentially start and end values), we search for candidate sequences of length one, two, and so on, up to one less than the length of the input sequence. The set of instruction sequences given by this necessarily contains any optimal sequences, but may contain extra sequences that only give the correct output for one particular set of inputs. Thus once a candidate set has been found, TOAST searches within this candidate set, picking new vectors each time, until either the the set of candidates is empty, in which case the search moves on to the next length, or until the set of candidates stabilises.

When one or more candidate is found, they have to be verified for equivalence to the original sequence, over all inputs. Searching can generate a large amount of candidate sequences, so two verification steps are performed: an initial representative heuristic test and a full equivalence test. *pre-verify* is a fast heuristic that uses a directed set of vectors to perform a fast representative verify on the two programs. If *pre-verify* returns false, then the candidate is discarded (i.e. it is definitely not equivalent); if true, then a full verify must be performed to prove full equivalence. Empirical evidence suggests that the *pre-verify* heuristic is important in quickly discarding invalid sequences, but it is still necessary to validate a sequence with a full verify, as it is possible to generate cases which can pass the heuristic, but will fail the full verification.

```
value(C,T,B)  :- istream(C,P,land,R1,R2,none), pc(C,P,T), value(C,R1,B),
                 value(C,R2,B), register(R1), register(R2), colour(C),
                 position(C,P), time(C,T), bit(B).
-value(C,T,B) :- istream(C,P,land,R1,R2,none), pc(C,P,T), not value(C,T,B),
                 register(R1), register(R2), colour(C), position(C,P),
                 time(C,T), bit(B).
symmetricInstruction(land).
```

Listing 2. Logical AND (land) instruction encoded in *AnsProlog*

As noted in Section 2, the number of combinations covered by search is factorial in the length of the sequences. In practise this is handled by a series of calls to a solver with progressively increasing program sizes. Likewise verify searches for a refutation in a space of combinations that is exponential in the number of input bits; effectively a co-NP task and handled by a single invocation of a solver.

One key recent development is the *buildMultiple* tool which uses TOAST to build and refine a series of additional constraints which augment the search component. It is based on the observation that an optimal sequence of instructions will not contain a sub-optimal instruction sequence. The search component of TOAST is used to generate a set of all possible instructions sequences of a given length using a fixed number of inputs. These are then superoptimised using TOAST; if they are sub-optimal or equivalent to another sequence then they are abstracted to form additional constraints. Although this procedure is time consuming, it produces very strong sets of constraints and only ever needs to be run once for a given architecture. Critically, it shows a key advantage of using ASP; the flexibility to add extra constraints without changing the search algorithm. With a procedural system, *buildMultiple* would simply not be possible.

3.2 *AnsProlog* Encodings

TOAST uses *AnsProlog* to model the integer processing unit of the target processors. The majority of the model is at bit level, with *AnsProlog* rules relating input bits of an instruction to the output bits.

The *instruction sequence* is represented as a series of facts, or in the case of searching, a set of choice rules. These literals are then used by the instruction definitions to control the value literals that give the value of various registers within the processor. If the literal is in the answer set, the given bit is taken to be a 1, if the classically-negated version of the literal is in the answer set then it is a 0. An example instruction definition for a logical AND (land) is given in Listing 2. Note the use of negation as failure to reduce the number of rules needed and the declaration that AND is symmetric, which is used to reduce the search space.

Flow control rules define which instruction will be 'executed' at a given time step by controlling the program counter (pc) literal. As ASP programs may need to simultaneously model multiple independent code streams (for example, when trying to verify their equivalence), all literals are tagged with the abstract property 'colour'. The inclusion of the colour(C) literal in each rule then allows copies to be created for each code stream during instantiation. In most cases, when only one code stream is used, only one value of colour is defined and only one copy of each set of rules is produced; the overhead involved is negligible. An example encoding is shown in Listing 3.

```
haveJumped(C,T)   :- jump(C,T,J), colour(C), time(C,T), jumpSize(C,J).
pc(C,PCV+J,T+1)   :- pc(C,PCV,T), jump(C,T,J), colour(C), position(C,PCV),
                     time(C,T), jumpSize(C,J).
pc(C,PCV+1,T+1)   :- pc(C,PCV,T), not haveJumped(C,T), colour(C),
                     position(C,PCV), time(C,T).
pc(C,1,1).
```

Listing 3. Flow control rules encoded in *AnsProlog*

Flag control rules model the setting and checking of processor flags such as carry, overflow, zero and negative; although generally only used for controlling conditional branches and multi-word arithmetic, these flags are a source of many superoptimised sequences [8].

4 Benchmarks

Benchmarks for the two main tasks of the TOAST system are given: searching for candidate sequences and then verification of two sequences to show full equivalence for all inputs.

The tests[1] are for the SPARC V8 [9], a 32 bit RISC architecture family. All tests were run on quad-core Intel 2.8GHz Xeon E5462 processors with 32GB RAM, running a variant of Scientific Linux. Programs were ground with GRINGO (2.0.0) and tested with the following four solvers: CLASP, SMODELS, SMODELS-IE and SUP; all tools were built in 32 bit mode. None of the *AnsProlog* programs generated within the TOAST system require disjunction, aggregates or any other non-syntactic extensions. All programs generated by the TOAST system are tight.

The search test (*sequence4*) attempts to find shorter optimal sequences for a four instruction program, with two 32 bit inputs, as given in Listing 1. This sequence was selected as an example of an optimal sequence that cannot be improved via superoptimisation, giving an approximate ceiling on the performance of the system. Programs *ss1* to *ss4* are searches over the spaces of 1 to 4 instructions respectively.

We performed two types of verification tests: one is which the two programs are (non-trivially) the same, returning zero answer sets (*verifytest1*); and the second in which the two programs differ on only one set of inputs, hence returning one answer set (*verifytest2*). In these tests, we amended the input programs to demonstrate that the TOAST system is able to verify sequences for 8 bit, 16 bit and 32 bit architectures.

Table 1 presents timings for the search and verify tests, with solver time outs occurring after 100 hours. These results demonstrate that we are able to superoptimise sequences for 32 bit architectures, while the projected growth figures suggest that a fully verified build-once architecture library is feasible as is done in *buildMultiple*.

5 Future Work

The results in Table 1 show a significant improvement from the initial benchmarks provided in [4]. Some of this is due to improvements in hardware, although the majority of the improvement is due to the progress made in answer set solvers, particularly the

---

[1] Available online from: http://www.cs.bath.ac.uk/tom/toast/

Table 1. Timings (in sec) for TOAST search and verify tests for SPARC V8

	sequence4				verifytest1			verifytest2		
Solver	ss1	ss2	ss3	ss4	8 bit	16 bit	32 bit	8 bit	16 bit	32 bit
clasp-1.1.1	123.20	105.72	578.37	12355.16	0.46	0.48	15.81	0.31	0.37	8.67
smodels-2.32	123.25	266.17	6880.87	-	0.18	11.33	-	0.20	4.75	-
smodels-ie-1.0.0	123.21	281.60	1983.94	-	0.20	11.08	-	0.21	4.79	-
sup-0.2	123.22	103.46	768.36	-	0.40	3.38	-	0.15	0.14	8.70
Atoms	853	1411	2098	2941	904	2212	6940	1030	1526	2518
Rules	42740	118779	238212	410902	1622	4870	17122	3591	6591	12583

inclusion of techniques from SAT solvers, notably clause learning. This demonstrates one of the advantages of ASP; that improvements in solver performance directly benefit applications using them, and that more advanced solvers can be 'plugged-in' with minimal integration needed. Our approach incorporates the concept of a full verification rather than a plausibility test as is done in other systems, taking the need for a human out of the operation.

Making use of these advances in solver technology and the flexibility of ASP (especially with *buildMultiple*), it is hoped that TOAST can be built into a competitive superoptimising system. Key application areas are seen in improving the quality of templates and peephole optimisers used in both conventional and JIT compilers.

References

1. Aho, A., Lam, M., Sethi, R., Ullman, J.: Compilers: Principles, Techniques and Tools, 2nd edn. Addison Wesley, Reading (2006)
2. Bansal, S., Aiken, A.: Automatic Generation of Peephole Superoptimizers. In: Proceedings of the 12th International Conference on Architectural Support for Programming Languages and Operating Systems (ASPLOS XII), pp. 394–403. ACM Press, New York (2006)
3. Baral, C.: Knowledge Representation, Reasoning and Declarative Problem Solving. Cambridge University Press, Cambridge (2003)
4. Brain, M., Crick, T., De Vos, M., Fitch, J.: TOAST: Applying Answer Set Programming to Superoptimisation. In: Etalle, S., Truszczyński, M. (eds.) ICLP 2006. LNCS, vol. 4079, pp. 270–284. Springer, Heidelberg (2006)
5. Granlund, T., Kenner, R.: Eliminating Branches using a Superoptimizer and the GNU C Compiler. In: Proceedings of the ACM SIGPLAN 1992 Conference on Programming Language Design and Implementation (PLDI 1992), pp. 341–352. ACM Press, New York (1992)
6. Hall, M., Padua, D., Pingali, K.: Compiler Research: the Next 50 Years. Communications of the ACM 52(2), 60–67 (2009)
7. Joshi, R., Nelson, G., Randall, K.: Denali: A Goal-Directed Superoptimizer. In: Proceedings of the ACM SIGPLAN 2002 Conference on Programming Language Design and Implementation (PLDI 2002), pp. 304–314. ACM Press, New York (2002)
8. Massalin, H.: Superoptimizer: A Look at the Smallest Program. In: Proceedings of the 2nd International Conference on Architectural Support for Programming Languages and Operating Systems (ASPLOS II), pp. 122–126. IEEE Computer Society Press, Los Alamitos (1987)
9. SPARC International, Inc. The SPARC Architecture Manual, Version 8, Revision SAV080SI9308 (1992)

Logic Programming Techniques in Protein Structure Determination: Methodologies and Results*

Alessandro Dal Palù[1], Agostino Dovier[2], and Enrico Pontelli[3]

[1] Dept. of Mathematics, University of Parma
alessandro.dalpalu@unipr.it
[2] Dept. of Maths and Computer Science, University of Udine
dovier@dimi.uniud.it
[3] Dept. of Computer Science, New Mexico State University
epontell@cs.nmsu.edu

Abstract. The purpose of this paper is to provide a brief overview of how logic programming technology has been used by our team in addressing the problem of tertiary protein structure determination. The proposed approach tackles the problem from the perspective of viewing protein structure as a folding of protein sequences in a discrete representation of space (a crystal lattice structure). Logic programming and constraint programming technologies can be effectively used to provide an elegant and effective solution.

1 Introduction

In recent years, we have witnessed the rapid growth of a new research area, whose results have already made a profound impact [22] on traditional disciplines such as biology, chemistry, medicine, and agriculture—here denoted globally as *"Bio."* This area, known as *Bioinformatics*, uses algorithms and methodologies developed by computer scientists to solve challenging problems in "Bio" areas. In turn, the emerging problems in these areas have produced stimuli for computer scientists to develop new algorithms and methods. Bioinformatics, in broad terms, deals with the use of computational techniques to organize and extract knowledge from biological data; bioinformatics has successfully addressed problems in areas like recognition and analysis of DNA sequences, biological systems simulations, prediction of the spatial conformation of biological polymers, and ontological analysis of biomedical knowledge.

Logic programming has already asserted itself as a strong technology for bioinformatics. Logic programming environments have been developed, either to serve the needs of specific applications or to provide general frameworks for bioinformatics development (e.g. [8,24]). In particular, the search capabilities provided by logic programming provide a natural platform for addressing combinatorial problems—which occur frequently in biomedical research areas.

In this paper, we overview our work in the area of logic programming applications to the problem of protein structure prediction. We address the problem of tertiary structure prediction using ab initio techniques [28], from the perspective of folding a protein

---

* The research has been partially supported by the FIRB Project RBNE03B8KK.

E. Erdem, F. Lin, and T. Schaub (Eds.): LPNMR 2009, LNCS 5753, pp. 560–566, 2009.

sequence in a discretized representation of the three-dimensional space (viewed as a crystal lattice structure), optimizing an objective function which is related to the potential energy function of the resulting configuration. In the rest of this paper we briefly introduce this challenging problem and give an overview of our results.

2   The Protein Structure Prediction Problem

The *Primary structure* of a protein is a linked sequence of amino acids. There are 20 types of amino acids—each commonly identified by a distinct letter. For the scope of this paper, the primary structure of a protein can be simply viewed as a string $s_1 \cdots s_n$ with $s_i \in \{A, \ldots, Z\} \setminus \{B, J, O, U, X, Z\}$.

The *Tertiary Structure* (or *native state*) of a protein is a 3D conformation associated to the primary structure. The *protein structure prediction problem* is the problem of predicting the tertiary structure of a protein, given its primary structure. *Local* 3D conformations, called α-helices and β-sheets, are often present in a protein native state. The set of these local structures is typically referred to as *secondary structures*. We do not discuss in this paper the *quaternary structure*, which is the arrangement of several folded proteins in a more general complex.

Figure 1 shows an abstract representation of the conformation of protein 1FVS: on the left a full atom view and on the right a cartoon representation obtained by linking the C_α atoms (intuitively, the central atoms of each amino acid). This picture highlights a 4-strands β-sheet, plotted with cyan (light gray), and two α-helices in red (dark gray). At the bottom, the primary sequence, viewed as a sequence of amino acids, is reported.

When modeling the protein structure prediction problem, we impose physical constraints to the conformations. Let us denote with $\mathcal{D}$ the set of admissible points in the 3D space where an amino acid can be placed. Let c, d be two fixed distances. For two points $p, q \in \mathcal{D}$, we say that $\mathsf{next}(p, q)$ holds if and only if $|p - q| = d$. d is related to 3.8Å, namely the distance between two consecutive C_α in the primary sequence. We define the Boolean function contact as follows: $\mathsf{contact}(p, q) = 1$ if and only if $|p - q| \leq c$. A *folding* of a n-element primary sequence is a function $\omega_n : \{1, \ldots, n\} \longrightarrow \mathcal{D}$ satisfying the following properties: (1) for all $1 \leq i, j \leq n$, if $i \neq j$ then $\omega_n(i) \neq \omega_n(j)$, and (2) for all $1 \leq i < n$, it holds that $\mathsf{next}(\omega_n(i), \omega_n(i + 1))$.

arevilavhgmtcsactntintqlralkgvtkcdislvtnecqvtydnevtadsikeiiedcgfdceilrd

Fig. 1. Primary and Tertiary structures (all-atoms and C_α–C_α structure) of Protein 1FVS

Let Pot be a function that maps pairs of amino acids to integer numbers; the function estimates the energy contribution of a pair of amino acids that are in contact. The *free energy of a folding* ω_n, denoted by $E(\omega_n)$, is computed as follows:

$$E(\omega_n) = \sum_{\substack{1 \le i < n \\ i + 2 \le j \le n}} \text{contact}(\omega_n(i), \omega_n(j)) \cdot \text{Pot}(s_i, s_j)$$

The *protein structure prediction problem (PSP)* is the problem of determining the folding(s) ω_n that minimize $E(\omega_n)$. The problem contains some symmetries that can be avoided by symmetry breaking search (see, e.g., [2]). The simplest way to remove symmetries is to fix the positions $\omega_n(1)$ and $\omega_n(2)$ of the first two amino acids.

Two main approximations can be made: (1) *space*: the set of admissible points, and (2) *energy*: the details of the Pot function used. Lattice-based models are realistic approximations of the set of the admissible points for the C_α atoms of a protein [27]. Lattices are 3D graphs with repeated patterns; e.g., the *face centered cube (FCC)* lattice is defined as: $\mathcal{D} = \{(x, y, z) \in \mathbb{N}^3 : x + y + z \text{ is even}\}$, $E = \{(p, q) \in \mathcal{D}^2 : |p - q| = \sqrt{2}\}$. In this case, $d = \sqrt{2}$ (that corresponds to 3.8Å), while c is set to 2 ($\sim$ 5.4Å, a reasonable distance for considering in *contact* two non consecutive amino acids).

The contact energy contribution Pot assigns a (positive or negative) value to a pair of aminoacids in contact. In literature, there are proposals with aminoacids clusterized into 2 classes: the HP model [18], into 4 classes: the HPNX model [4], or into 20 classes (one for each aminoacid): the 20×20 model [6].

3 Related Work

The HP model [18] divides amino acids in two categories: *hydrophobic (H)* and *polar (P)*. Two hydrophobic amino acids in contact contribute -1 to the energy. The other contacts are not relevant. Even with these simplifications, the problem of determining the optimal folding in the simplest lattice structure ($\mathcal{D} = \mathbb{N}^2$, $E = \{(p, q) \in \mathcal{D}^2 : |p - q| = 1\}$, and $c = d = 1$) is NP-complete, as originally proved in [9]. In particular, given a sequence made of symbols P and H, determining the existence of a folding with at least k contacts between occurrences of the symbol H is a NP-complete problem. In these simple settings, this problem has been encoded in *Answer Set Programming* [20].

Backofen and Will solved this problem using constraint programming techniques, for sequences of length 160 and more on the FCC lattice [3,1,2]. Efficiency is obtained using symmetry breaking and introducing the notion of *core*. Basically, the patterns of optimal configurations of H amino acids in contact are pre-computed. This kind of approach is not suitable to versions of the problem where a more detailed energy model is used or where additional structural constraints (e.g., known α-helices and β-sheets) are introduced. In [4], they consider an energy model in which amino acids are partitioned in four categories. Other researchers (e.g., [26]) provide approximated solutions to the same problem using local search and refined meta-heuristics.

Barahona and Kripphal provided a constraint based solution to the problem on a more general space model (off-lattice cubes). They also considered proteins docking and developed the tool *Chemera*, which is extensively used by by biochemists [23,5].

4 Our Contribution

In our research, we focused on the use of the FCC model to discretize the 3D space, and the 20×20 energy model [6], more precise than the HP model.

CLP(FD) encoding. In [19] we encoded the problem using the library clpfd of SICStus Prolog. Since contact energy is not suitable to predict α-helices and β-sheets in the FCC lattice, we pre-computed secondary structure elements using other well-known tools. The results of these pre-computations have been introduced as constraints within the main code. In this first encoding the number of admissible angles for secondary structure elements was too limited. We relaxed this limitation in [10], where a more general and precise handling of secondary structure constraints has been implemented. However, the exponential growth of the search space w.r.t. the protein length made impossible to explore the whole search space for proteins of length greater than 30/40. Therefore, we proposed an ad-hoc labeling search strategy with biologically motivated heuristics and we introduced a data structure (the potential matrix) that allowed us to reduce calculations during this phase. This approach was later extended by relaxing some constraints and developing alternative search heuristics [11].

In all these approaches, we used a dual representation of the tertiary structure: a cartesian one, based on the set of points, and a polar one, based on the torsional angles generated by the protein during the folding. The cartesian representation is useful for defining the notion of self-avoiding walk and the notion of constraint-based energy function. The polar representation simplifies the encoding of secondary structure constraints. However, a significant number of additional constraints needs to be introduced to manage the conversion between the two representations, and this has a significant impact on the scalability—for large proteins (e.g., length 60 or higher), the constraint solver quickly exhausts its memory. The polar representation was abandoned in later stages of the research [13], and secondary structure constraints were encoded directly using cartesian constraints. The consequence of this choice is the loss of the chirality property of helices, but the overall definition becomes simpler and more scalable.

In the same paper, we also developed a novel search heuristics (Bounded Block Fail—BBF). The list of variables is dynamically split into blocks of k variables, that are labeled at the same time. When the variables in the block B_i are instantiated to a tentative solution, the search moves to the successive block B_{i+1}, if any. If the labeling of the block B_{i+1} fails, the search backtracks to the block B_i. If the number of times that B_{i+1} has failed is below a certain threshold, then the process continues, by generating one new solution for B_i and re-entering the labeling of the block B_{i+1}. Otherwise, the heuristics generates a failure for B_i and backtracks to B_{i-1}. The key idea is that small local changes do not significantly change the protein shape. When the search has attempted a sufficiently large number of similar conformations without success, it is better to abandon that search branch.

Ad-hoc constraint solver. In [12] we developed an ad-hoc constraint solver written in C, named COLA (COnstraint solving on LAttices). Previously, each 3D point was viewed as a triple of FD variables $\langle X, Y, Z \rangle$. In COLA, instead, the lattice point is an elementary element, associated with a 3D domain (a box). We developed and

Table 1. Running time of the various approaches on some small proteins

ID–n	[19]	[10]	[11]	[13]	[16]
1LE3–16	1m 43s	1.6 s	25.3 s	1.2 s	0.2 s
1ZDD–34	13m 16 s	7m 26 s	1m 14 s	2m 18	0.4 s
2GP8–40	30m 36 s	9.9 s (*)	7h 42 m	1m 12s	0.16 s
1ENH–54	40m 45 s	40m 24s (*)	>10h	9h 46m	1 m 20 s

implemented ad-hoc constraint propagation techniques and the BBF heuristics. This approach, which also was parallelized to enhance scalability, was presented in [16].

Just to give an idea of the evolution of our proposals, we report the running times of the systems on the prediction of some small proteins in Table 1. Timings have been recomputed on an AMD Opteron 2.2GHz Linux machine, using SICStus Prolog 4.0.4 for the first 4 columns and COLA 2.1 for the last one. The solutions found with different techniques are not always the same, but (save for the first column related to a too strict encoding) they have comparable energy and shape. More importantly, the solutions are very close to the real tertiary structure (see original papers for details). The times marked by a (*) are lower than [11] and [13], since some extra strong structural constraints among secondary structure were added. Observe that [16] improves those results even without extra information. The protein 1FVS of Figure 1 is predicted by COLA 2.1 with BBF in less than one hour. All codes are available from www.dimi.uniud.it/dovier/PF.

Towards generalization and integration. The *ab-initio* approach used by COLA is still computationally infeasible when applied to proteins with more than hundred amino acids. Only the presence of other types of partial information—e.g., known folds for sub-blocks extracted from the Protein Data Bank (PDB)—can significantly speed-up the search. This is indeed what is done by other predictors (e.g., ROSETTA [25]), where partial information is extracted from PDB from similar structures/substructures, and only small subsequences need to be arranged in 3D space.

We started a systematic study about *global constraints* needed in a solver for structure predictions on lattice models. We studied the definition and the complexity of testing satisfiability and applying propagation for the constraints alldifferent, contiguous, self avoiding walk, alldistant, chain, and rigid block in [14]. These global constraints are currently incorporated in COLA, which processes additional information coming from known proteins and from partial predictions. We also investigated a global constraint that accounts for partial information coming from protein density maps [15].

We have also studied how to use model checking results for analyzing the folding process [17], and how to model the protein folding problem as a *planning problem* using a variant of the well-known action description language $\mathcal{B}$ [21]. An approach to the protein folding problem using Agent-Based simulation has been proposed in [7].

5 Conclusions

This work represents a typical use of logic programming paradigm for problem solving. The problem can be encoded easily and solutions (for small inputs) can be computed by built-in mechanisms of (constraint) logic programming. Heuristics and alternative encodings can be easily programmed and tested. When the encoding becomes stable, enhanced performance can be obtained using less declarative methods.

References

1. Backofen, R.: The protein structure prediction problem: A constraint optimization approach using a new lower bound. Constraints 6, 223–255 (2001)
2. Backofen, R., Will, S.: Excluding Symmetries in Constraint-Based Search. Constraints 7(3-4), 333–349 (2002)
3. Backofen, R., Will, S.: A Constraint-Based Approach to Fast and Exact Structure Prediction in 3-Dimensional Protein Models. Constraints 11(1), 5–30 (2006)
4. Backofen, R., et al.: Application of constraint programming techniques for structure prediction of lattice proteins with extended alphabets. Bioinformatics 15(3), 234–242 (1999)
5. Barahona, P., Krippahl, L.: Constraint Programming in Structural Bioinformatics. Constraints 13(1-2), 3–20 (2008)
6. Berrera, M., Molinari, H., Fogolari, F.: Amino acid empirical contact energy definitions for fold recognition in the space of contact maps. BMC Bioinformatics 4(8) (2003)
7. Bortolussi, L., Dovier, A., Fogolari, F.: Agent-based Protein Structure Prediction. Multiagent and Grid Systems 3(2), 183–197 (2007)
8. Calzone, L., Fages, F., Soliman, S.: BIOCHAM: an environment for modeling biological systems and formalizing experimental knowledge. Bioinformatics 22(14), 1805–1807 (2006)
9. Crescenzi, P., Goldman, D., Papadimitriou, C., Piccolboni, A., Yannakakis, M.: On the Complexity of Protein Folding. Journal of Computational Biology 5(3), 423–466 (1998)
10. Dal Palù, A., Dovier, A., Fogolari, F.: Protein Folding in CLP(FD) with Empirical Contact Energies. In: Apt, K.R., Fages, F., Rossi, F., Szeredi, P., Váncza, J. (eds.) CSCLP 2003. LNCS (LNAI), vol. 3010, pp. 250–265. Springer, Heidelberg (2004)
11. Dal Palù, A., Dovier, A., Fogolari, F.: Constraint logic programming approach to protein structure prediction. BMC Bioinformatics 5(186) (2004)
12. Dal Palù, A., Dovier, A., Pontelli, E.: A Constraint Logic Programming Approach to 3D Structure Determination of Large Protein Complexes. In: Sutcliffe, G., Voronkov, A. (eds.) LPAR 2005. LNCS (LNAI), vol. 3835, pp. 48–63. Springer, Heidelberg (2005)
13. Dal Palù, A., Dovier, A., Pontelli, E.: Heuristics, Optimizations, and Parallelism for Protein Structure Prediction in CLP(FD). In: Proc. of PPDP, pp. 230–241. ACM Press, New York (2005)
14. Dal Palù, A., Dovier, A., Pontelli, E.: Global constraints for Discrete Lattices. In: Proc. of WCB (2006)
15. Dal Palù, A., Dovier, A., Pontelli, E.: The density constraint. In: Proc. of WCB (2007)
16. Dal Palù, A., Dovier, A., Pontelli, E.: A constraint solver for discrete lattices, its parallelization, and application to protein structure prediction. Software Practice and Experience 37(16), 1405–1449 (2007)
17. De Maria, E., Dovier, A., Montanari, A., Piazza, C.: Exploiting Model Checking in Constraint-based Approaches to the Protein Folding. In: Proc. of WCB (2006)
18. Dill, K.A.: Dominant forces in protein folding. Biochemistry 29, 7133–7155 (1990)
19. Dovier, A., Burato, M., Fogolari, F.: Using Secondary Structure Information for Protein Folding in CLP(FD). In: Proc. of WFLP. ENTCS, vol. 76 (2002)
20. Dovier, A., Formisano, A., Pontelli, E.: A Comparison of CLP(FD) and ASP Solutions to NP-Complete Problems. In: Gabbrielli, M., Gupta, G. (eds.) ICLP 2005. LNCS, vol. 3668, pp. 67–82. Springer, Heidelberg (2005)

21. Dovier, A., Formisano, A., Pontelli, E.: Multivalued Action Languages with Constraints in CLP(FD). In: Dahl, V., Niemelä, I. (eds.) ICLP 2007. LNCS, vol. 4670, pp. 255–270. Springer, Heidelberg (2007)
22. Johnson, G.: All Science is Computer Science. The New York Times (3/25/2001)
23. Krippahl, L., Barahona, P.: PSICO: Solving Protein Structures with Constraint Programming and Optimisation. Constraints 7, 317–331 (2002)
24. Mungall, C.: Biomedical Logic Programming Integration Toolkit,
    http://www.blipkit.org
25. Simons, K., Bonneau, R., Ruczinski, I., Baker, D.: Ab initio protein structure prediction of CASP III targets using ROSETTA. Proteins: Struct. Fund. Genet. 3, 171–176 (1999)
26. Shmygelska, A., Hoos, H.H.: An ant colony optimisation algorithm for the 2D and 3D hydrophobic polar protein folding problem. BMC Bioinformatics 6(30) (2005)
27. Skolnick, J., Kolinski, A.: Reduced models of proteins and their applications. Polymer 45, 511–524 (2004)
28. Zhang, Y.: Progress and Challenges in Protein Structure Prediction. Curr. Opin. Struct. Biol. 18(3) (2008)

PHYLO-ASP: Phylogenetic Systematics with Answer Set Programming

Esra Erdem

Faculty of Engineering and Natural Sciences, Sabancı University, Istanbul 34956, Turkey

Abstract. This note summarizes the use of Answer Set Programming to solve various computational problems to infer phylogenetic trees and phylogenetic networks, and discusses its applicability and effectiveness on some real taxa.

1 Introduction

Cladistics (or phylogenetic systematics), developed by Willi Hennig [1], is the study of evolutionary relations between species based on their shared traits. Represented diagrammatically, these relations can form a tree whose leaves represent the species, internal vertices represent their ancestors, and edges represent the genetic relationships between them. Such a tree is called a "phylogenetic tree" (or a "phylogeny"). We consider reconstruction of phylogenies as the first step of reconstructing the evolutionary history of a set of taxa (taxonomic units). The idea is then to reconstruct (temporal) phylogenetic networks, which also explain the contacts (or borrowings) between taxonomic units, from the reconstructed phylogenies.

We studied both steps using Answer Set Programming: the first step is studied in [2,3,4], and the second step is studied in [5,6]. We call our ASP-based approach to phylogenetic tree and phylogenetic network reconstruction as PHYLO-ASP. We illustrated the applicability and effectiveness of PHYLO-ASP for the historical analysis of languages, and to the historical analysis of parasite-host systems.

Histories of individual languages give us information from which we can infer principles of language change. This information is not only of interest to historical linguists but also of interest to archaeologists, human geneticists, physical anthropologists as well. For instance, an accurate reconstruction of the evolutionary history of certain languages can help us answer questions about human migrations, the time that certain artifacts were developed, when ancient people began to use horses in agriculture [7,8,9,10].

Parasites occur worldwide, causing malnutrition, sickness, and even sometimes the death of their hosts. Historical analysis of parasites gives us information on where they come from and when they first started infecting their hosts. The phylogenies of parasites, with the phylogenies of their hosts, and with the geographical distribution of their hosts, can be used to understand the changing dietary habits of a host species, to understand the structure and the history of ecosystems, and to identify the history of animal and human diseases. This information allows predictions about the age and duration of specific groups of animals of a particular region or period, identification of regions of evolutionary "hot spots" [11], and thus can be useful to assess the importance of specific habitats, geographic regions, and biotas—all the plant and animal life of a

E. Erdem, F. Lin, and T. Schaub (Eds.): LPNMR 2009, LNCS 5753, pp. 567–572, 2009.
© Springer-Verlag Berlin Heidelberg 2009

particular region—and areas of critical genealogical and ecological diversity [12,11]. Identification of the most vulnerable members of a community by this way allows us to make more reliable predictions about the impacts of perturbations (natural or caused by humans) on ecosystem structure and stability [12].

With PHYLO-ASP, we studied evolutionary history of 7 Chinese dialects based on 15 lexical characters, and 24 Indo-European languages based on 248 lexical, 22 phonological and 12 morphological characters. Some of the phylogenetic trees and networks computed by PHYLO-ASP are plausible from the point of view of historical linguistics. We also studied evolutionary history of 9 species of *Alcataenia* (a tapeworm genus) based on their 15 morphological characters. Some of the phylogenetic trees and networks computed by PHYLO-ASP are plausible from the point of view of coevolution— the evolution of two or more interdependent species each adapting to changes in the other, and from the point of view of historical biogeography—the study of the geographic distribution of organisms.

This note summarizes the use of PHYLO-ASP to solve various computational problems related to the inference of phylogenetic trees and phylogenetic networks, and discusses its applicability and effectiveness on some real taxa.

2   Phylogeny Reconstruction

A *phylogenetic tree* (or *phylogeny*) for a set of taxa is a finite rooted binary tree $\langle V, E \rangle$ along with two finite sets I and S and a function f from $L \times I$ to S, where L is the set of leaves of the tree. The set L represents the given taxonomic units whereas the set V describes their ancestral units and the set E describes the genetic relationships between them. The elements of I are usually positive integers ("indices") that represent, intuitively, qualitative characters, and elements of S are possible states of these characters. The function f "labels" every leaf v by mapping every index i to the state $f(v, i)$ of the corresponding character in that taxonomic unit.

A character $i \in I$ is *compatible* with a phylogeny (V, E, L, I, S, f) if there exists a function $g : V \times \{i\} \to S$ such that

(C1) for every leaf v of the phylogeny, $g(v, i) = f(v, i)$;
(C2) for every $s \in S$, if the set $V_{is} = \{x \in V : g(x, i) = s\}$ is nonempty then the digraph $\langle V, E \rangle$ has a subgraph with the set V_{is} of vertices that is a rooted tree.

A character is *incompatible* with a phylogeny if it is not compatible with that phylogeny.

The computational problem we are interested in is, given the sets L, I, S, and the function f, to build a phylogeny (V, E, L, I, S, f) with the maximum number of compatible characters. This problem is called the *maximum compatibility problem*. It is NP-hard even when the characters are binary [13]. We solve the maximum compatibility problem, by means of the following decision problem: given sets L, I, S, a function f from $L \times I$ to S, and a nonnegative integer n, decide the existence of a phylogeny (V, E, L, I, S, f) with at most n incompatible characters. In [2,4], we describe this decision problem as an ASP program whose answer sets correspond to such phylogenies.

3 Phylogenetic Network Reconstruction

A contact between two taxonomic units can be represented by a horizontal edge added to a pictorial representation of a "temporal phylogeny"—a phylogeny along with a function τ from vertices of the phylogeny to real numbers denoting the times when these taxonomic units emerged (Fig. 1). The two endpoints of the edge are simultaneous "events" in the histories of these communities. An event can be represented by a pair $v{\uparrow}t$, where v is a vertex of the phylogeny and t is a real number.

A finite set C of contacts defines a *(temporal) phylogenetic network* $\langle V \cup V_C, E_C \rangle$— a digraph obtained from $T = \langle V, E \rangle$ by inserting the elements $v{\uparrow}t$ of the contacts from C as intermediate vertices and then adding every contact in C as a bidirectional edge. We say that a set C of contacts is *simple* if the endpoints of all lateral edges are different from the vertices of T, and each lateral edge subdivides an edge of T into exactly two edges.

About a simple set C of contacts (and about the corresponding phylogenetic network $\langle V \cup V_C, E_C \rangle$) we say that it is *perfect* if there exists a function $g : (V \cup V_C) \times I \to S$ such that the function g extends f from leaves to all internal nodes of the phylogenetic network, and that every state s of every character i could evolve from its original occurrence in some "root" (i.e., every character i is compatible with the phylogenetic network).

We are interested in the problem of turning a temporal phylogeny into a perfect phylogenetic network by adding a small number of simple contacts. For instance, given the phylogeny in Fig. 1(a), the single contact $\{B{\uparrow}1750, D{\uparrow}1750\}$ is a possible answer.

It is clear that the information included in a temporal phylogeny is not sufficient for determining the exact dates of the contacts that turn it into a perfect phylogenetic network. To make this idea precise, let us select for each $v \in V \setminus \{R\}$ a new symbol $v{\uparrow}$, and define the *summary* of a simple set C of contacts to be the result of replacing

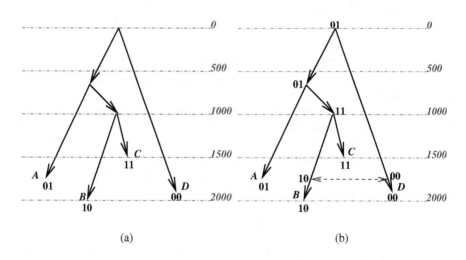

(a)   (b)

Fig. 1. A temporal phylogeny (a), and a perfect temporal network (b) with a lateral edge connecting $B{\uparrow}1750$ with $D{\uparrow}1750$

each element $v{\uparrow}t$ of every contact in C with $v{\uparrow}$. Thus summaries consist of 2-element subsets of the set $V{\uparrow}= \{v{\uparrow} : \ v \in V \setminus \{R\}\}$. For instance, the summary of the set of contacts of Fig. 1(b) is $\{\{B{\uparrow}, D{\uparrow}\}\}$.

An *IPSTN problem* (for "Increment to Perfect Simple Temporal Network") is defined by a phylogeny $\langle V, E, I, S, f \rangle$ and a function $v \mapsto (\tau_{min}(v), \tau_{max}(v))$ from the vertices of the phylogeny to open intervals. A *solution* to the problem is a set of 2-element subsets of $V{\uparrow}$ that is the summary of a perfect simple set of contacts for a temporal phylogeny $\langle V, E, I, S, f, \tau \rangle$ such that, for all $v \in V$, $\tau_{min}(v) < \tau(v) < \tau_{max}(v)$.

In [6], we describe IPSTN problem as an ASP program. We solve IPSTN problems in two steps: use an ASP system to compute summaries so that every character is compatible with the phylogenetic network, and then use a constraint programming system to check, for each of summary, whether the corresponding contact occurs within the given time intervals.

4   Experimental Results

We applied PHYLO-ASP to three sets of taxa: Chinese dialects, Indo-European languages, and Alcataenia (a tapeworm genus) species. For each taxa and a given integer n, first we computed all phylogenies with at most n incompatible characters, iteratively with a script as follows: at iteration i, compute the i'th phylogeny with the input program, and then add to the input program a constraint that prevents generation of the answer sets that describe the i'th phylogeny. After that, we identified the phylogenies that are plausible. For the Chinese dialects and Indo-European languages, the plausibility of phylogenies depends on the linguistics and archaeological evidence; for *Alcataenia*, the plausibility of the phylogeny we compute is dependent on the knowledge of host phylogeny (e.g., phylogeny of the seabird family *Alcidae*), chronology of the fossil record, and biogeographical evidence. Then, for each plausible phylogeny, we computed phylogenetic networks that require minimum number of lateral edges, and identified the plausible ones.

Experiments with Chinese dialects. We considered the Chinese dialects Xiang, Gan, Wu, Mandarin, Hakka, Min, and Yue. We used the dataset, originally gathered by Xu Tongqiang and processed by Wang Feng, described in [14]. In this dataset, there are 15 lexical characters; each character has 2–5 states.

After preprocessing this dataset, we computed 33 phylogenies with 6 incompatible characters, and we found out that there is no phylogeny with less than 6 incompatible characters. These phylogenies are presented in [4]. The sub-grouping of the Chinese dialects is not yet established. However, many specialists agree that there is a Northern group and a Southern group. That is, for the dialects we chose in our study, we would expect a (Wu, Mandarin, Gan, Xiang) Northern grouping and a (Hakka, Min) Southern grouping. (It is not clear which group Yue belongs to.) We identified 5 plausible phylogenies with respect to this hypothesis.

For each plausible phylogeny, we reconstructed phylogenetic networks. We observed that, among these phylogenies, two of them require at least 2 lateral edges (representing borrowings between Gan and Wu, and between (Mandarin, Wu) and Min) to turn into a plausible perfect phylogenetic network.

Experiments with Indo-European languages. We applied PHYLO-ASP to reconstruct the evolutionary history of the Indo-European language groups Balto-Slavic (BS), Italo-Celtic (IC), Greco-Armenian (GA), Anatolian (AN), Tocharian (TO), Indo-Iranian (IIR), Germanic (GE), and the language Albanian (AL). We used the dataset assembled by Don Ringe and Ann Taylor [15], with the advice of other specialist colleagues. There are 282 informative characters in this dataset, each with 2–22 states.

After preprocessing this dataset, we computed 45 phylogenies with at most 20 incompatible characters, taking into account the given domain-specific information (e.g., Anatolian is the outgroup for all the other subgroups, Albanian cannot be a sister of Indo-Iranian or Balto-Slavic). Out of these 45 phylogenies, 34 are identified by Don Ringe as plausible from the point of view of historical linguistics. These phylogenies are presented in [4]. The most plausible one with 16 incompatible characters is (AN, (TO, (IC, ((GE, AL), (GA, (IIR, BS))))))).

Based on this phylogeny, and given some time intervals for each node in the phylogeny, we reconstructed 3 plausible temporal phylogenetic networks with 3 lateral edges, taking also into account some domain-specific information (e.g., a contact between IC and BA is unlikely because the former was spoken in western Europe, while the Balts were probably confined to a fairly small area in northeastern Europe).

Experiments with Alcatenia species. We used PHYLO-ASP to reconstruct the evolutionary history of 9 species of *Alcataenia*—a tapeworm genus whose species live in alcid birds (puffins and their relatives): *A. Larina, A. Fraterculae, A. Atlantiensis, A. Cerorhincae, A. Pygmaeus, A. Armillaris, A. Longicervica, A. Meinertzhageni, A. Campylacantha.* We used the dataset described in [16]. In this dataset, there are 15 characters, each with 2–3 states.

After preprocessing this dataset, we computed 18 phylogenies, with 5 incompatible characters, for *Alcataenia*, taking into account some domain-specific information (e.g., the outgroup for all *Alcataenia* species is *A. Larina*). For the plausibility of the phylogenies for *Alcataenia*, we consider the phylogenies of its host *Alcidae* (a seabird family) and the geographical distributions of *Alcidae*. For instance, according to host and geographic distributions over the time, diversification of *Alcataenia* is associated with sequential colonization of puffins (parasitized by *A. Fraterculae* and *A. Cerorhincae*), razorbills (parasitized by *A. Atlantiensis*), auklets (parasitized by *A. Pygmaeus*), and murres (parasitized by *A. Armillaris, A. Longicervica,* and *A. Meinertzhageni*). This pattern of sequential colonization is supported by the phylogeny of *Alcidae* in [17]. Out of the 18 trees we computed, only two are consistent with this pattern. Each plausible tree needs 3 lateral edges to turn into a perfect phylogenetic network.

5   Conclusion

We have briefly described the use of ASP to reconstruct the evolutionary history of a set of taxonomic units (as in [5,3,6,2,4]), calling this ASP-based approach to phylogenetic systematics as PHYLO-ASP. We have discussed the applicability and effectiveness of PHYLO-ASP with three sets of taxa: Indo-European languages, Chinese dialects, and *Alcatenia* species. Our ongoing work involves extending PHYLO-ASP to analyze and compare phylogenetic trees and networks [18,19].

Acknowledgments. This work has involved, at various stages, close collaborations with Dan Brooks, Selim Erdoğan, Vladimir Lifschitz, Luay Nakhleh, James Minett, Don Ringe, Feng Wang. It has been partially supported by TUBITAK Grant 107E229.

References

1. Hennig, W.: Phylogenetic Systematics. University of Illinois Press (1966); by Davis, D.D., Zangerl, R.: Translated from Grundzuege einer Theorie der phylogenetischen Systematik (1950)
2. Brooks, D.R., Erdem, E., Minett, J.W., Ringe, D.: Character-based cladistics and answer set programming. In: Hermenegildo, M.V., Cabeza, D. (eds.) PADL 2004. LNCS, vol. 3350, pp. 37–51. Springer, Heidelberg (2005)
3. Erdem, E., Wang, F.: Reconstructing the evolutionary history of Chinese dialects. Accepted for presentation at the 39th International Conference on Sino-Tibetan Languages and Linguistics, ICSTLL 2006(2006)
4. Brooks, D.R., Erdem, E., Erdoğan, S.T., Minett, J.W., Ringe, D.: Inferring phylogenetic trees using answer set programming. Journal of Automated Reasoning 39(4), 471–511 (2007)
5. Erdem, E., Lifschitz, V., Nakhleh, L., Ringe, D.: Reconstructing the evolutionary history of Indo-European languages using answer set programming. In: Dahl, V., Wadler, P. (eds.) PADL 2003. LNCS, vol. 2562, pp. 160–176. Springer, Heidelberg (2003)
6. Erdem, E., Lifschitz, V., Ringe, D.: Temporal phylogenetic networks and logic programming. Theory and Practice of Logic Programming 6(5), 539–558 (2006)
7. Mair, V.H. (ed.): The Bronze Age and Early Iron Age Peoples of Eastern Central Asia. Institute for the Study of Man, Washington (1998)
8. Mallory, J.P.: In Search of the Indo-Europeans. Thames and Hudson, London (1989)
9. Roberts, R.G., Jones, R.M., Smith, M.A.: Thermoluminescence dating of a 50,000-year-old human occupation site in Northern Australia. Science 345, 153–156 (1990)
10. White, J.P., O'Connell, J.F.: A Prehistory of Australia, New Guinea, and Sahul. Academic Press, New York (1982)
11. Brooks, D.R., McLennan, D.A.: Phylogeny, Ecology, and Behavior: A Research Program in Comparative Biology. Univ. Chicago Press, Chicago (1991)
12. Brooks, D.R., Mayden, R.L., McLennan, D.A.: Phylogeny and biodiversity: Conserving our evolutionary legacy. Trends in Ecology and Evolution 7, 55–59 (1992)
13. Day, W.H.E., Sankoff, D.: Computational complexity of inferring phylogenies by compatibility. Systematic Zoology 35(2), 224–229 (1986)
14. Minett, J.W., Wang, W.S.Y.: On detecting borrowing: distance-based and character-based approaches. Diachronica 20(2), 289–330 (2003)
15. Ringe, D., Warnow, T., Taylor, A.: Indo-European and computational cladistics. Transactions of the Philological Society 100(1), 59–129 (2002)
16. Hoberg, E.P.: Congruent and synchronic patterns in biogeography and speciation among seabirds, pinnipeds, ans cestodes. J. Parasitology 78(4), 601–615 (1992)
17. Chandler, R.M.: Phylogenetic analysis of the alcids. PhD thesis, University of Kansas (1990)
18. Cakmak, D., Erdem, E., Erdogan, H.: Computing weighted solutions in answer set programming. In: Proc. of LPNMR (to appear, 2009)
19. Eiter, T., Erdem, E., Erdogan, H., Fink, M.: Finding similar or diverse solutions in answer set programming. In: Proc. of ICLP (to appear, 2009)

HAPLO-ASP: Haplotype Inference Using Answer Set Programming

Esra Erdem[1], Ozan Erdem[1], and Ferhan Türe[2]

[1] Faculty of Engineering and Natural Sciences, Sabancı University, Istanbul 34956, Turkey
[2] Department of Computer Science, University of Maryland, College Park, MD 20742, USA

Abstract. Identifying maternal and paternal inheritance is essential to be able to find the set of genes responsible for a particular disease. However, due to technological limitations, we have access to genotype data (genetic makeup of an individual), and determining haplotypes (genetic makeup of the parents) experimentally is a costly and time consuming procedure. With these biological motivations, we study haplotype inference—determining the haplotypes that form a given set of genotypes—using Answer Set Programming; we call our approach HAPLO-ASP. This note summarizes the range of problems that can be handled by HAPLO-ASP, and its applicability and effectiveness on real data in comparison with the other existing approaches.

1 Introduction

Each genotype (the specific genetic makeup of an individual) in a diploid organism has two copies, one from the mother and one from the father. These two copies are called haplotypes, and they combine to form the genotype. The genetic information contained in haplotypes can be used for early diagnosis of diseases, detection of transplant rejection, and creation of evolutionary trees. However, although it is easier to access to the genotype data, due to technological limitations, determining haplotypes experimentally is a costly and time consuming procedure. With these biological motivations, researchers have been studying haplotype inference—determining the haplotypes that form a given set of genotypes—by means of some computational methods.

One haplotype inference problem that has been extensively studied is Haplotype Inference by Pure Parsimony (**HIPP**) [1]. This problem asks for a minimal set of haplotypes that form the given genotypes; the decision version of **HIPP** is NP-hard [1,2]. **HIPP** has been studied with various approaches, such as, HYBRIDIP (based on integer linear programming) [3], HAPAR (based on a branch and bound algorithm) [4], SHIPs (based on a SAT-based algorithm) [5], RPOLY (based on pseudo-boolean optimization methods) [6].

Another haplotype inference problem that has been studied so far is Haplotype Inference from Present-Absent Genotype (**HIPAG**) [7,8]; it is a variation of **HIPP** that takes into account biallelic genotypes only, possibly with missing information, and domain-specific information, like haplotype patterns observed for some specific gene family. **HIPAG** has been studied by [7] with a greedy algorithm (HAPLO-IHP); [9] studies a variation of **HIPAG** for polyallelic and polyploid genotypes but does not take into

E. Erdem, F. Lin, and T. Schaub (Eds.): LPNMR 2009, LNCS 5753, pp. 573–578, 2009.

account domain-specific information, and computes a solution using a SAT-based algorithm (SATLOTYPER).

Recently, we have presented a novel approach to solving both sorts of haplotype inference problems, **HIPP** and **HIPAG**, and their variations, using Answer Set Programming (ASP) [8]; this ASP-based approach to haplotype inference is called HAPLO-ASP. We have extended the applicability of HAPLO-ASP, thanks to its expressive representation language, to solve also variations of **HIPP** and **HIPAG** where we can specify preferences over parts of haplotypes by assigning weights to their sites/alleles in accordance with their importance, and/or where we can consider polyallelic and polyploid genotypes. This note summarizes the range of problems handled by HAPLO-ASP, and its applicability and effectiveness on real data in comparison with the other approaches.

2   Haplotype Inference by Pure Parsimony

Haplotype Inference by Pure Parsimony (**HIPP**) [1] asks for a minimal set of haplotypes that explain the given genotypes. The decision version of **HIPP** (i.e., deciding that a set of k haplotypes that explain the given genotypes exists) is NP-hard [1,2].

A standard definition of the concept of two haplotypes "explaining" a genotype appears in [1]. According to this definition, we view a genotype as a vector of sites, each site having a value 0, 1, or 2; and a haplotype as a vector of sites, each site having a value 0 or 1. The values 0 and 1 (called *alleles*) correspond to complementary bases, like c and G. The sites correspond to single nucleotide polymorphisms (SNPs). A site of a genotype is *ambiguous* if its value is 2; and *resolved* otherwise. Two haplotypes h_1 and h_2 *form (explain)* a genotype g if for every site j the following hold: if $g[j] = 2$ then $h_1[j] = 0$ and $h_2[j] = 1$ or $h_1[j] = 1$ and $h_2[j] = 0$; if $g[j] = 1$ then $h_1[j] = 1$ and $h_2[j] = 1$; and if $g[j] = 0$ then $h_1[j] = 0$ and $h_2[j] = 0$. For instance, the genotype 20110 can be explained by the haplotypes 10110 and 00110.

We consider the following decision version of **HIPP**:

HIPP-DEC Given a set G of n genotypes each with m sites, and a positive integer k, decide whether there is a set H of at most k unique haplotypes such that each genotype in G is explained by two haplotypes in H.

Erdem and Türe have presented in [8] an ASP program that describes **HIPP-DEC**. An instance of **HIPP** can be solved with that ASP program, by trying various values for k (the number of unique haplotypes explaining the given genotypes). HAPLO-ASP computes an approximate lower bound l for k and an upper bound u for k, using ASP, and tries to find the optimal value for k by a binary search between l and u. The computation of such a lower bound and upper-bound is discussed in [8].

3   Haplotype Inference from Present-Absent Genotype

Haplotype Inference from Present-Absent Genotype (**HIPAG**) is another haplotype inference problem, which asks for the minimal set of haplotypes "compatible" with the given genotypes. Both haplotypes and genotypes can be viewed as vectors, as in **HIPP**.

In this problem, each site of a haplotype takes one of the two values 0 or 1 specifying the presence/absence of a particular gene. Sites of genotypes are biallelic; the value of each site is a pair of numbers from $\{0, 1, ?\}$. For instance, $(1, ?)(1, ?)(1, 1)$ is a genotype.

For a genotype g of the form $(g_{11}, g_{12})...(g_{m1}, g_{m2})$, let us denote by g^1 the vector $g_{11}g_{21}...g_{m1}$ and by g^2 the vector $g_{12}g_{22}...g_{m2}$. We say that two alleles i and j are *compatible* if they are identical or if one of them is ?. Two haplotypes h_1 and h_2 are *compatible* with a genotype g, all with m sites, if, for every site j, $h_1[j]$ is compatible with one of the two alleles, $g^1[j]$ or $g^2[j]$, and $h_2[j]$ is compatible with the other. For instance, the haplotypes 011 and 111 are compatible with $(1, 0)(1, 1)(1, 1)$ and $(1, ?)(1, ?)(1, 1)$ but not with $(1, ?)(1, 0)(1, 1)$. Note that, we can discard the sites with missing information while computing a solution for **HIPAG**.

We consider the following decision version of **HIPAG**:

HIPAG-DEC Given a set G of n genotypes each with m biallelic sites, and a positive integer k, decide that there is a set H of at most k unique haplotypes such that each genotype in G is compatible with two haplotypes in H.

We have shown that **HIPAG-DEC** is also NP-complete.

[8] discusses how to describe **HIPAG-DEC** as an ASP program. Then an instance of **HIPAG** can be solved with that ASP program, by trying various values for k, as in **HIPP** instances.

4 Other Variations of Haplotype Inference

HAPLO-ASP can solve also variations of **HIPP** and **HIPAG** where we can specify domain-specific information (like haplotype patterns observed for some gene family), preferences over parts of haplotypes by assigning weights to their sites/alleles in accordance with their importance in detecting the cause of a disease, and/or where we can consider polyallelic and polyploid genotypes. In the following, we will briefly discuss these problems.

Observed Haplotype Patterns. In haplotype studies of some gene families, sometimes some patterns may be observed. For instance, [7] derived three patterns of haplotypes for the family of KIR genes of Caucasian population, from the observations in KIR haplotype studies like [10]. These patterns may help generating more accurate haplotypes, if included in the computation of haplotypes. Such domain-specific information can be described by an ASP program, as described in [8], and, during computation of haplotypes, it can be given to the answer set solver as an input in a separate program file.

Weighted Haplotype Inference. In some populations, some sites/alleles may have a more significant role in identifying, for instance, the cause of a disease, and thus have a larger weight. With this motivation, we have studied modified versions of **HIPP** and **HIPAG**: *Weighted Haplotype Inference by Pure Parsimony* (**WHIPP**) and *Weighted Haplotype Inference from Present-Absent Genotype* (**WHIPAG**). We have studied the decision versions of **WHIPP** and **WHIPAG** and presented them in ASP. We have been testing HAPLO-ASP on some real data.

Haplotype Inference with Polyallelic and Polyploid Genotypes. In the haplotype inference problems above, every genotype is explained by two haplotypes (since the individuals are *diploid*) and every genotype has at most two kinds of alleles (e.g., in **HIPAG**, the genes are *biallelic*). However, there are many species, like some varieties of the potato (e.g., Solanum tuberosum), with polyploid and polyallelic genes in nature, where every genotype is explained by more than two haplotypes, and each genotype consists of more than two kinds of alleles. *Haplotype Inference from Polyallelic and Polyploid Genotype* (**HIPPG**) extends **HIPAG** to such polyallelic and polyploid genotypes. We have studied the decision version of **HIPPG** and presented it in ASP. We have tested HAPLO-ASP on some real data for cultivated potato genes (described in [11]), and obtained promising results as in [9].

5   Experimental Results for HIPP and HIPAG

We have implemented a haplotype inference system, also called HAPLO-ASP, based on the ASP-based approach above; it is a PERL script including system calls to answer set solvers. HAPLO-ASP is available at http://people.sabanciuniv.edu/~esraerdem/haplo-asp.html. We have performed two groups of experiments using this system [8].

In these experiments, the executable for SHIPs is obtained from their authors. RPOLY (executables) and HAPLO-IHP (source files) are available at their web pages. We use the versions of these systems available on January 28, 2008. In the experiments, as their search engines, RPOLY uses MINISAT+ (Version 1.0), SHIPs uses MINISAT (Version 2.0), and HAPLO-ASP uses CMODELS (Version 3.74) with LPARSE (Version 1.0.17) and MINISAT (Version 2.0). In our experiments, we have used a workstation with 1.5GHz Xeon processor and 4x512MB RAM, running Red Hat Linux(Version 4.3).

Experimenting with **HIPP** *Problems.* In these experiments, we have compared HAPLO-ASP with the other state-of-the-art haplotype inference systems, RPOLY [6] (based on pseudo-boolean optimization methods) and SHIPs [5] (based on a SAT-based algorithm); these systems can solve **HIPP** problems only. We have excluded from our experiments the systems based on integer linear programming (ILP), such as HYBRIDIP [3], and HAPAR [4] (based on a branch and bound algorithm) since RPOLY and SHIPs perform much better than these systems [5,6].

We have experimented with 334 instances of **HIPP**, used also in the experiments of [3,5,6]: 40 instances generated using MS of [12] (20 *uniform*, 20 *nonuniform*), 294 real instances (24 *hapmap*, 90 *abcd*, 90 *ace*, 90 *ibd*). (These datasets are explained in detail in the cited articles above.) All problem instances are simplified by eliminating duplicates (of genotypes and haplotypes) as described in [5,6] before our experiments. For each haplotype inference system we have assigned 1000 sec.s of CPU time to solve each problem. Table 1 shows the number of problem instances solved by each system. According to this table, HAPLO-ASP solves the most number of problems (332 out of 334 problems). RPOLY aborts for only one problem that HAPLO-ASP solves, but in most of the problems for which it computes a solution in 1000 sec.s it is faster than HAPLO-ASP by a magnitude of up to 300.

Table 1. Number of problems solved, with a timeout of 1000 sec.s for each problem

Group of problems	# of problems	# of problems solved		
		SHIPS	HAPLO-ASP	RPOLY
abcd	90	90	90	90
ace	90	90	90	90
hapmap	24	24	23	23
ibd	90	78	89	88
unif	20	20	20	20
nonunif	20	20	20	20

Both SHIPs and HAPLO-ASP use MINISAT as their search engine. Usually the propositional theory prepared for MINISAT by SHIPs is smaller than the one prepared by HAPLO-ASP.

Experimenting with **HIPAG** *Problems.* We have compared HAPLO-ASP with the haplotype inference system HAPLO-IHP [7] with respect to the computation time and the accuracy of generated haplotypes. We have considered the accuracy measure of [8] to check how much the inferred haplotypes match the original ones. HAPLO-IHP is based on statistical methods and it can compute approximate solutions to instances of **HIPAG** only. The other haplotype inference systems can not solve **HIPAG**.

We have experimented with one of the data sets generated by Yoo et. al for 17 KIR genes of Caucasian population. This data set contains 200 genotypes with 14 biallelic sites. Yoo et al. derived three patterns of haplotypes for this family of genes from the observations in KIR haplotype studies like [10]. As in our experiments with **HIPP** problems, we have first simplified this data set by eliminating the duplicates, and modified the haplotype patterns accordingly. After eliminating the two genotypes that do not match any patterns, the simplified data set contains 28 genotypes with 11 sites. Without the given haplotype patterns, no solution can be found in 30 minutes by HAPLO-IHP; whereas HAPLO-ASP finds 11 haplotypes in 57.08 CPU sec.s, with the accuracy rate 0.702 (by performing a binary search between 1 and 56.) With the given haplotype patterns, HAPLO-IHP finds 19 haplotypes compatible with the given genotypes in 9.1 CPU sec.s, with the accuracy rate 0.732; whereas HAPLO-ASP finds 11 haplotypes in 640 CPU sec.s, with the accuracy rate 0.768. Here 640 sec.s include 1.4 sec.s to infer the set of haplotypes (for $k = 11$), and 631 sec.s to verify its minimality (for $k = 10$). HAPLO-ASP computes an exact solution to **HIPAG**, whereas HAPLO-IHP computes an approximation with a greedy algorithm; this explains the difference between the computation times as well the higher accuracy rate of HAPLO-ASP's solution.

6 Conclusion

We have briefly described HAPLO-ASP, an ASP-based approach to solving various haplotype inference problems, such as **HIPP** and **HIPAG**, possibly by taking into account the given domain-specific information. We have also discussed the range of problems that can be handled by HAPLO-ASP: some of these problems (**WHIPP** and **WHIPAG**) allow us to specify preferences over parts of haplotypes by assigning weights to their sites/alleles in accordance with their importance; and some problems (**HIPPG**) allow us to handle polyallelic and polyploid genotypes. HAPLO-ASP is the first haplotype inference approach/system that can solve all these variations of haplotype inference. For some of these problems (**HIPP, HIPAG, HIPPG**), we have illustrated the applicability and effectiveness of HAPLO-ASP on some real data, in comparison with the other existing approaches.

References

1. Gusfield, D.: Haplotype inference by pure parsimony. In: Baeza-Yates, R., Chávez, E., Crochemore, M. (eds.) CPM 2003. LNCS, vol. 2676, pp. 144–155. Springer, Heidelberg (2003)
2. Lancia, G., Pinotti, M.C., Rizzi, R.: Haplotyping populations by pure parsimony: Complexity of exact and approximation algorithms. INFORMS Journal on Computing 16(4), 348–359 (2004)
3. Brown, D., Harrower, I.: Integer programming approaches to haplotype inference by pure parsimony. IEEE/ACM Transactions on Bioinformatics and Computational Biology 3, 348–359 (2006)
4. Wang, L., Xu, Y.: Haplotype inference by maximum parsimony. Bioinformatics 19(14), 1773–1780 (2003)
5. Lynce, I., Marques-Silva, J.: Efficient haplotype inference with boolean satisfiability. In: AAAI (2006)
6. Graça, A., Marques-Silva, J.P., Lynce, I., Oliveira, A.: Efficient haplotype inference with pseudo-boolean optimization. In: Anai, H., Horimoto, K., Kutsia, T. (eds.) AB 2007. LNCS, vol. 4545, pp. 125–139. Springer, Heidelberg (2007)
7. Yoo, Y.J., Tang, J., Kaslow, R.A., Zhang, K.: Haplotype inference for present absent genotype data using previously identified haplotypes and haplotype patterns. Bioinformatics 23(18), 2399–2406 (2007)
8. Erdem, E., Ture, F.: Efficient haplotype inference with answer set programming. In: Proc. of AAAI (2008)
9. Neigenfind, J., Gyetvai, G., Basekow, R., Diehl, S., Achenbach, U., Gebhardt, C., Selbig, J., Kersten, B.: Haplotype inference from unphased snp data in heterozygous polyploids based on sat. BMC Genomics 9(1), 356 (2008)
10. Hsu, K.C., Chida, S., Geraghty, D.E., Dupont, B.: The killer cell immunoglobulin-like receptor (kir) genomic region: gene-order, haplotypes and allelic polymorphism. Immunological Reviews 190(1), 40–52 (2002)
11. Pajerowska-Mukhtar, K., Stich, B., Achenbach, U., Ballvora, A., Lübeck, J., Strahwald, J., Tacke, E., Hofferbert, H.R., Ilarionova, E., Bellin, D., Walkemeier, B., Basekow, R., Kersten, B., Gebhardt, C.: Single nucleotide polymorphisms in the allene oxide synthase 2 gene are associated with field resistance to late blight in populations of tetraploid potato cultivars. Genetics 181, 1115–1127 (2009)
12. Hudson, R.: Generating samples under a wrightfisher neutral model of genetic variation. Bioinformatics 18, 337–338 (2002)

Using Answer Set Programming to Enhance Operating System Discovery

François Gagnon and Babak Esfandiari

Carleton University, Canada
{fgagnon,babak}@sce.carleton.ca

Abstract. Although knowing the operating systems running in a network is becoming more and more important (mainly for security reasons), current operating system discovery tools are not sufficiently accurate to acquire the information in a fully automated way. Many design choices explain this lack of accuracy, but they all come down to a poor knowledge representation scheme. In this paper, we study how answer set programming can be used to guide the design of a knowledge-oriented operating system discovery tool. The result is significantly more accurate than today's state of the art tools.

1   Introduction

Knowing which Operating Systems (OSes) are running in a network is increasingly important from a security point of view [2]. As networks are growing larger and more dynamic, it becomes essential to develop tools to automatically gather such knowledge. Operating System Discovery (OSD) tools rely on idiosyncrasies in the communication behavior of the computers to identify specific OSes. These idiosyncrasies exist because communication protocols are often under-specified, leaving OS vendors the freedom to implement the behavior of their choice, see Example 1.

Example 1 (OS Identification Based on a TCP Syn Packet). *When two computers want to communicate together using the TCP protocol they must first open a TCP connection (TCP handshake). Opening a TCP connection requires sending a TCP Syn packet. One of the fields in a TCP Syn packet is the Time To Live (TTL). The TCP protocol specification RFC 791 suggests an initial TTL value of at least 40. As a result, not all OS vendors use the same initial TTL value: Linux uses 64, MacOS uses 255, most Windows versions use 128, but Windows NT 3.51 uses 32.*

Several tools exist to identify operating systems. Unfortunately, they are highly inaccurate [2]. This inaccuracy can be explained by three design choices:

- No memory: this prevents the use of past information to narrow down the current deduction.
- Packet-by-packet approach: this does not allow modeling complex communication patterns that require more than one packet (e.g., stimulus-response).
- No reasoning: this results in unfocussed search.

E. Erdem, F. Lin, and T. Schaub (Eds.): LPNMR 2009, LNCS 5753, pp. 579–584, 2009.
© Springer-Verlag Berlin Heidelberg 2009

Basically, these limitations are the result of an inadequate knowledge representation scheme. In this paper, we discuss a new OSD tool with a strong knowledge representation module which is implemented using answer set programming. This results in a significant improvement over today's state of the art tools.

2   Background on Operating System Discovery

Passive OSD tools simply listen to the network and deduce information from the recorded packets. In particular, they do not probe a machine to check how it reacts in a specific situation. An example of a passive OSD test follows from Example 1; since TCP Syn packets are used as part of normal communications, a passive tool can wait for such a packet and analyze it.

The main problem with this approach is that the information may not be available when needed. It also seems obvious that passive tools should monitor the network and update their knowledge base on a continuous basis. However, existing tools simply analyze packets one by one and try to guess the OS based solely on the current packet, regardless of any other information that could have been known beforehand.

Active OSD tools, on the other hand, can directly probe a machine to identify its OS, depending on the reaction of the target to the synthesized stimuli. For instance, they can initiate a TCP handshake (i.e., with a SYN) and analyze the way the target responds (e.g., what value is used as a TTL in the SYN/ACK). Another possibility is to stimulate the target with a semantically malformed packet (e.g., with a SYN/FIN packet[1]) and analyze how it behaves. Since the stimulus is malformed, there is no standard way to respond and each OS may respond differently.

The main problem with active OS discovery is the large amount of traffic generated to discover the OS. As no single test can determine the OS with certainty, it is necessary to come up with a sequence of tests. Most active tools simply execute all available tests. To prevent generating too much traffic, they limit the number of tests available. Active tools do not take advantage of the information freely available on the network (i.e. normal network communications), instead, they rely entirely on the execution of their tests.

Every limitation of current OSD tools is related to the absence of a solid knowledge representation scheme. This results in a significant lack of accuracy.

3   Using ASP for OSD

Building a knowledge-oriented OSD tool will provide three enhancements:

- Unify the passive and active techniques in a single hybrid approach.
- Enhance the passive module with a memory.

---

[1] Such a packet is malformed since it simultaneously requests both the initialization and the termination of a TCP connection.

– Enable reasoning for test selection in the active module, to reduce the number of tests executed.

The hybrid approach works as follows: Initially, we consider that each computer can run any OS. Then, as traffic is seen for a specific computer, we update its set of possible OSes by eliminating OSes that cannot generate the traffic observed (passive module). Finally, when a user needs information about a particular computer and the knowledge base does not contain enough information, active tests are executed in order to obtain the missing information (active module).

To achieve our objectives, we needed a knowledge representation language that:

– Is declarative (an order independent), to allow automatically generating the program from a database of OS fingerprints.
– Supports disjunction, as more than one OS can generate a specific event.
– Supports non-monotonic reasoning, when new events are obtained, we need to eliminate some previously possible OSes.
– Supports reasoning about action and change (active module).

Below we discuss the ASP description of both the passive and active modules. We use the DLV engine in our implementation. See [1] for more information.

3.1   Passive Module

The passive module has two components. An Intensional Database (IDB) which is a set of ASP rules linking communication patterns with their corresponding OSes. And an Extensional Database (EDB) consisting of a set of ASP facts representing the observed network events.

3.1.1   Passive IDB

Figure 1 presents a fragment of our IDB. The weak constraint in line 2 states that, unless necessary, one machine should not be assigned two different operating systems in a single answer set (each answer set will correspond to a *possible* assignment of OSes for the computers).

The set of rules in the **TCP Syn** group represents different behaviors for the sender of the first packet of a TCP handshake, see Example 1. The predicate $tcp(X, Y, Xport, Yport, DF, Flags, TTL)$ represents a TCP packet sent from X through port $Xport$ to Y on port $Yport$ where DF indicates whether or not the "Don't Fragment" bit is set, $Flags$ lists the TCP flags that are set (SYN, ACK, RST, ...), and TTL contains the time to live.

3.1.2   Passive EDB

Every time a network packet is captured, a corresponding fact is added to the EDB. That way, the information provided by a packet is combined to the one provided by other packets, allowing to narrow down the set of possible OSes.

Example 2 (passive module). *Suppose the only packet captured so far is a TCP SYN packet from 10.1.1.6 to 10.1.1.1 with the DF bit set and a TTL of 64;*

%One IP should not correspond to two different OSes
:~ os(X,Y), os(X,Z), Z != Y.
%TCP Syn
os(X,linuxRH5_2) :- tcp(X,_,_,_,no,syn,64).
os(X,win2k) ∨ os(X,winXP) :- tcp(X,_,_,_,yes,syn,128).
os(X,sunOS) ∨ os(X,linuxRH7_1) ∨ os(X,macOS) ∨ os(X,linuxRH8_0) ∨
os(X,freeBSD5_0) :- tcp(X,_,_,_,yes,syn,64).

Fig. 1. Some Rules for Passive OS discovery

this is represented by the fact $tcp(ip10\_1\_1\_6, ip10\_1\_1\_1, 3952, 80, yes, syn, 64)$. *The program has two answer sets and the OS for 10.1.1.6 is either win2k or winXP.*

The weak constraint is flexible enough to differentiate between OS ambiguity and multiple OSes hidden behind a single IP address (e.g. due to NAT).

To avoid a combinatorial explosion of answer sets, we maintain one EDB file for each monitored IP address. Moreover, we evaluate the program for every 100 packets gathered. After each evaluation, we drop those packets and replace them with a logical formula, in the EDB, summarizing the information acquired so far. After Example 2, we would replace the packet with the following summary:

$$os(ip10\_1\_1\_6, win2k) \vee os(ip10\_1\_1\_6, winXP).$$

3.2 Active Module

When a query is made to the system and the current knowledge is not sufficient to answer it, the job of the active module is to select relevant tests to be executed. After the execution of those tests, the knowledge base will have enough information to answer the query. Here, we use ASP's planning abilities to select tests based on the query and the current knowledge state.

The active module has two components. IDB is a set of ASP rules describing the effect of executing the tests and EDB is a set of ASP facts describing the current knowledge state as well as the query (i.e., the goal).

3.2.1 Active IDB

A fragment of our actual intensional database for the active module is presented in Figure 2. The first four rules define the test outcome interpretation module; their meaning is as follows[2]:

1) if an association of IP address to OS $\langle I, OS \rangle$ explicitly does not hold before the execution of a test, it will not hold after its execution;
2) if an association $\langle I, OS \rangle$ holds before the execution of a test and the outcome of this test confirms that possibility, $\langle I, OS \rangle$ will still hold after the execution of the test;

---

[2] The predicate *holds/3* expresses the possible OSes at each state, while *possible/3* denotes what is possible with respect to the outcome of a test.

```
%Test Outcome Interpretation
1) -holds(I,OS,τ₁) :- -holds(I,OS,τ), τ < τ₁.
2) holds(I,OS,τ₁) :- holds(I,OS,τ), next(τ,τ₁), possible(I,OS,τ₁).
3) holds(I,OS,τ₁) :- holds(I,OS,τ), next(τ,τ₁), not actionExecuted(I,τ).
4) -possible(I,OS,τ) :- not possible(I,OS,τ).
```

%T8 - The TCP SynAck test
```
oT8A(I,τ₁) v oT8B(I,τ₁) v oT8C(I,τ₁) v oT8D(I,τ₁) :-
    next(τ,τ₁), execute(testT8,I,τ).
```

%oT8A (OS is mac or sun)
```
possible(I,macOS,τ₁):- oT8C(I,τ₁).
possible(I,sunOS,τ₁):- oT8C(I,τ₁).
```
%oT8B (OS is linux Red Hat5.2)
```
possible(I,linuxRH5_2,τ₁):- oT8A(I,τ₁).
```
...

Fig. 2. Some Rules for Active OS Discovery

3) all associations $\langle I, OS \rangle$ that are true remain true when no test is executed;
4) what is not explicitly said to be possible by the outcome of a test should be considered impossible.

The last part of Figure 2 describes the possible outcomes of the active test $T8$. If $T8$ is executed against I at time T, this will cause at least (and exactly) one of the predicates $oT8A(I, \tau_1)$, $oT8B(I, \tau_1)$, $oT8C(I, \tau_1)$, or $oT8D(I, \tau_1)$ to be true ($oT8A$ denotes outcome A of test $T8$). Each outcome then provides the OSes that are possible after the execution of $T8$. For instance, MacOS and SunOS are the only possible explanations of outcome A for $T8$.

3.2.2 Active EDB

The EDB contains a description of the current[3] state of knowledge which is a set of facts of the form $holds(I, X, 0)$ where X ranges over the set of currently possible OSes for machine I. The EDB also contains a representation of the query. For instance, to make sure that after the execution of at most 5 tests we know the actual OS of a given machine, i.e., the set of possible OSes is a singleton, we could use this constraint

$$: -\#count\{X : holds(I, X, 5)\} > 1.$$

4   Experimental Results

To compare our new OSD tool, called HOSD, with other existing tools, we measure their ability to answer the following query: "Does the OS of a computer belong to a given set Θ of OSes?" This query is extremely useful from a security

---

[3] At the time the query is made.

F. Gagnon and B. Esfandiari

Table 1. Experimental Results

OSD Tools	SinFP	Siphon	Nmap	p0f	ettercap	Xprobe	HOSD
%Good Answers	4.3%	6.0%	12.1%	20.6%	21.2%	33.2%	83.3%

point of view: to test if a target is vulnerable to a specific attack. See [2] for more details on this experiment[4].

Our experiment was done using 95 different OSes installed in virtual machines. We used 5,761 queries, each with a different initial state of knowledge (i.e. a traffic trace from which the tool can deduce the OS) or a different goal (i.e. the given set Θ). Table 1 gives the percentage of queries correctly answered by each tool. HOSD clearly provides a significant improvement over current OSD tools. The good results of HOSD are mainly explained by the enhancements related to its knowledge-oriented approach. These enhancements specifically address the limitations of current OSD tools as mentioned in Section 2.

4.1 Time Benchmark

One of the main concerns with our ASP implementation is the running time. On the 5,761 runs mentioned above, the passive module runs fast enough on average (386ms), but the worst case (16s) is problematic. The active module runs quite slowly both on average (9s) and in the worst case (25s), but this was expected as we are doing planning to avoid executing useless tests.

5 Discussion

ASP was instrumental in the quick development of our tool as it provided a very intuitive way to describe the reasoning related to OS discovery.

We are currently working on optimizing our implementation, basically by replacing full planning with heuristics in the active module. We are also working to encode more complex OSD phenomena, such as DHCP leases, using ASP.

References

[1] Gagnon, F., Esfandiari, B., Bertossi, L.: A Hybrid Approach to Operating System Discovery Using Answer Set Programming. In: Proceedings of the 10th IFIP/IEEE Symposium on Integrated Management (IM 2007), pp. 391–400 (2007)
[2] Gagnon, F., Massicotte, F., Esfandiari, B.: Using Contextual Information for IDS Alarm Classification. In: Flegel, U., Bruschi, D. (eds.) DIMVA 2009. LNCS, vol. 5587, pp. 147–156. Springer, Heidelberg (2009)

---

[4] See also http://hosd.sourceforge.net/experiments/2009-04.html

Non-monotonic Reasoning Supporting Wireless Sensor Networks for Intelligent Monitoring: The SINDI System

Alessandra Mileo, Davide Merico, and Roberto Bisiani

Department of Informatics, Systems and Communication, University of
Milan-Bicocca, viale Sarca 336/14, I–20126 Milan

Abstract. In recent years there has been growing interest in solutions
for the delivery of clinical care for the elderly, due to the large increase
in aging population. Monitoring a patient in his home environment is
necessary to ensure continuity of care in home settings, but this activity
must not be too invasive and a burden for clinicians. We prototyped
a system called SINDI (Secure and INDependent lIving), focused on i)
collecting data about the person and the environment through Wireless
Sensor Networks (WSN), and ii) reasoning about these data both to
contextualize them and to support clinicians in understanding patients'
well being as well as in predicting possible evolutions of their health.

1 Overview

The life-expectancy in several countries continues to grow, but people who live
longer tend to be in a state in which chronic conditions substantially cripple
their quality of life and their autonomy.

Future Independent Living systems should go beyond data collection to focus on assisting caregivers in understanding health evolution and enhancing autonomy of monitored patients. We refer to an *intelligent monitoring system* as
a monitoring system that is able to reason about gathered data and support
decisions. Most of the pervasive systems for healthcare proposed so far use a
probabilistic approach to behaviour analysis and activity recognition aimed at
enhancing autonomy [1]. These approaches are sometimes coupled with logic-
based planning techniques. When we talk about intelligent monitoring, though,
we refer to a different (potentially complementary) view of artificial intelligence
applied to home healthcare. In our view, expressive knowledge representation and
reasoning techniques are needed to analyse the context and to understand health
evolution. In the SINDI system we address this issue by using non-monotonic
logical reasoning in a two-step inference process: i) summarizing and correlating
sensor data in a consistent interpretation of the context in which the person
lives in terms of clinical profile, environment, movements, and ii) predicting possible evolutions of the person's health in order to devise effective preventive
strategies.

E. Erdem, F. Lin, and T. Schaub (Eds.): LPNMR 2009, LNCS 5753, pp. 585–590, 2009.
© Springer-Verlag Berlin Heidelberg 2009

The logical framework of Answer Set Programming (ASP) [2] is well suited to deal with similar complex knowledge representation and reasoning tasks, in that it overcomes most of the limitation of previous logic programming systems such as Prolog. Compared to pure statistical approaches, logic inference based on ASP is highly expressive and computationally more performant because it can deal with first-order representations, which are much richer than the propositional ones characterizing probabilistic inference. Furthermore, ASP can deal with incomplete information and commonsense reasoning using defaults. Cardinality and weight constraints together with program optimization techniques can also be used to model different degrees of uncertainty [3,4,5].

In the remainder of this paper we will describe how ASP-based Knowledge Representation and Reasoning supports Wireless Sensor Networks technologies in the SINDI Intelligent Monitoring system.

2   Context Data: Acquisition and Interpretation

Wireless Sensor Networks (WSNs) [6] consist of nodes that are capable of interacting with the environment by sensing or controlling physical parameters. These networks are the enabling technology for acquiring all possible information about the context in which the user lives.

Data gathered by the sensors are processed and aggregated according to specific algorithms for feature analysis, features are then processed by the ASP program and made available at the upper levels. The SINDI's context model aims at being generic and computationally rich at the same time, being based on a high level description of the home environment in terms of rooms, areas, objects, properties, relations and observations mapped into a set of logic predicates to be manipulated by the inference engine for health assessment [7]. The information flow is illustrated in Figure 1.

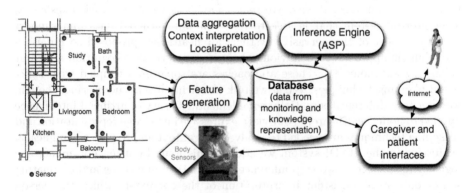

Fig. 1. Flow of Data in the SINDI System

In the context of home monitoring, the more intuitively relevant aspects of a context are where you are, who you are (clinical profile and personal information), which resources you are using, what you are doing and when [8,9].

In order to represent this information, SINDI's context model considers four entities on the first level: the *Person* entity, the *Room* entity, the *Area* entity and the *Object* entity. A small subset of generic[1] spatial relations among entities are also defined, such as *in, under, on, near*.

All the other pieces of information are at a second level and can be indexed by the primary context because they are attributes of the entities at the primary level.

Values of both attributes and spatial relations (except the inclusion of an area in a room which is static) are dynamic and need to be associated to an interval of time. In this way, the reasoning system can take into account their dynamic evolution in context interpretation.

Reasoning about the collected data is used to better characterize movements and to localize the person in the areas of the house. If we consider SINDI's localization component , given proximity values with a certain accuracy P and defined over (possibly overlapping) time intervals T_i, T_j, the ASP program takes all available sensor data as input and identifies all possible consistent sequences of moves across rooms and areas. Logical rules for disambiguation state that, by default, proximity to an area A of a room R in a temporal segment T1,T2 is given by the fact that a signal has been received from the corresponding node in that temporal segment. This holds unless there is a more reliable signal received in the same interval from another node. This other signal determines proximity unless additional contextual data make it invalid (e.g. a mat sensor indicating pressure in a different area $A1$ of room $R1$) and in case several of these data are available, a measure of reliability can be used to identify the best solution.

The high level representation of the collected data is automatically mapped into logic predicates resulting in a set of facts which is combined with the ASP logic program. Besides disambiguation, context interpretation is also a crucial phase in understanding basic behaviours that may be important for health assessment. Results are used to evaluate meaningful aspects of the elderly care as detailed in Section 3.

3   Reasoning Support for Intelligent Monitoring

In order to support caregivers in understanding and controlling the patient's health-evolution in his home environment, the reasoning component of the SINDI system uses the results of context interpretation in two reasoning steps: static and dynamic evaluations of significant aspects of the patient's quality of life (referred to as *indicators*) and prediction of possible evolutions of the patient's health state to plan appropriate preventive strategies. To do this, SINDI uses a knowledge representation model of health described in Section 3.1.

---

[1] The term "generic" here refers to the fact that they refer to relative spatial position of two entities, no matter where they are in the physical space. This results in greater generality in that we do not need an absolute physical description of the environment.

3.1   Knowledge Representation: SINDI's Model of Health

SINDI's Model of Health formalizes medical knowledge and correlations between indicators and health-related factors (referred to as *items*)[2]. A careful analysis of health care in home settings suggests that *items* can be classified into three classes: i) *functionalities*, evaluated in terms of functional disabilities; ii) *daily activities*, evaluated in terms of level of dependence in performing daily activities and iii) *risks*, representing complex aspects of the elderly care.

The relation between items and indicators is the following: each indicator can contribute to the evaluation of one or more items; items are correlated by dependency relations[3] indicating how the value of an item may impact values of other items and how. The reasoning tasks detailed in Section 3.2 are performed by analyzing the graph of dependencies connecting an item I with related indicators as well as with other items.

3.2   Reasoning Tasks: Evaluation, Prediction and Prevention

Evaluation. At each reasoning cycle, results of the context interpretation process are used to infer consistent *absolute* and *differential* evaluations of indicators and items. In general, absolute evaluation of items is available only as inputs from caregivers according to results of specific tests. As for indicators, their absolute evaluations can be based on i) results of specific evaluation by clinicians (e.g. hearing functionalities), results of data aggregation (e.g. quality of movement), results of ad-hoc logic rules (e.g. quality of sleep). Differential evaluations are obtained, when possible, as a measure of the value increase or decrease derived by comparing values at the current inference cycle with values at the previous inference cycle and it has four possible outcomes: worsening, improvement, no substantial change, undefined. Given that different kinds of potentially contradictory dependencies are allowed in SINDI's knowledge model, the combination of such influences of several indicators Ind_i on item I so as to provide a coherent differential evaluation for I is not always possible and the evaluation process returns a partially labelled graph as output. The following reasoning step of SINDI starts from this incomplete information and uses the computational power of the ASP framework in order to predict possible evolutions in terms of differential evaluations of items that have not been labeled as a result of the evaluation task and provide a qualitative analysis of results.

Prediction. is identified as the identification of plausible effects of certain changes in items' values on values of unlabeled items; intuitively, this is done by considering all possible consistent values for the missing information according to SINDI's logic model of dependencies between items; prediction makes it possible to act *before* major symptoms and to plan appropriate short- and long-term interventions, thus reducing risks. Prediction involves

---

[2] Indicators and Items have been identified according to the medical practice in health assessment of the elderly [10] and encoded in our declarative framework [11].

[3] Dependency relations are specified by knowledge engineers and medical experts.

several subtasks: i) contextualization, referred to as the correlation of context evaluation results with the clinical profile of the person; ii) local prediction, combining results of contextualization with medical dependencies to generate all total consistent labeling of the dependency graph resulting from evaluation and iii) local explanation referred to as the identification of all possible minimal set of dependency arcs that may justify the worsening of a person's health (provided as a result of the prediction task). These tasks are *local* processes rather than a case-based ones, in that they take into account results of reasoning under particular clinical and environmental conditions.

Prevention. *Prevention* [12] is referred to as all those interventions (feedback) that may keep health changes within safe boundaries. The combination of inference results (prediction) and context-related knowledge about the person and the environment is used to determine i) *what* should be provided as feedback, ii) in *which form* and iii) *when*. The content of the feedback is determined according to the medical literature (evidence-based studies) and encoded in the system. When the system determines a set of feedback actions that should be performed at a given time, they are qualitatively analyzed in order to infer which action is more urgent. A reaction to a feedback, when detected, is logged to be used at a later time. Exploring this history, caregivers can improve the way feedback actions are performed and identify the most effective communication patterns.

4   Preliminary Evaluation

So far we have run the full system for short periods of time (days) in a mock-up environment without real users. Our preliminary evaluation took into account how SINDI addresses most of the expected requirements [11].

The combination, in the reasoning process, of different sources of information (sensors, medical knowledge, clinical profile, user defined constraints) that change over time makes the system more reliable (i.e. much better able to disambiguate situations, thus reducing false positives) and adaptive (e.g. easily extended on the face of new available information). The modularity, espressivity and computational efficiency of the ASP framework make us strongly believe it is a good logic programming paradigm for automated reasoning in knowledge-intensive domains. In the first implementation of our system we evaluated ASP programs by using *Gringo* as grounder [13] and the *Clasp* solver [5] as inference engine[4]. Security and privacy are guaranteed by the use of security standards and techniques and the use of off-the-shelf components in SINDI considerably reduces overall costs.

Research-wise we are considering the problems of the encoding and integration of the appropriate medical knowledge and the use of apparently inconsistent prediction results to discover missing dependencies in the initial medical knowledge.

---

[4] http://potassco.sourceforge.net/

Application-wise, given the high variability among trials and studies addressing prediction and prevention issues, it is still difficult to extract a coherent picture of what leads to disability and to develop coherent prevention strategies. In this respect, our system has the potential to automatically collect a massive amount of data in oder to evaluate context-related prediction patterns and effective communication strategies for prevention.

These aspects are being concretely taken into account in the context of a real deployment of SINDI in a geriatrics hospital.

References

1. Haigh, K.Z., Yanco, H.: Automation as caregiver: A survey of issues and technologies. In: Proc. of AAAI 2002 Workshop on Automation as Caregiver, pp. 39–53 (2002)
2. Gelfond, M., Lifschitz, V.: The stable model semantics for logic programming. In: Proc. of ICLP 1988, pp. 1070–1080. MIT Press, Massachussets (1988)
3. Simons, P., Niemelä, I., Soininen, T.: Extending and implementing the stable model semantics. Artificial Intelligence Journal 138(1-2), 181–234 (2002)
4. Leone, N., Pfeifer, G., Faber, W., Eiter, T., Gottlob, G., Perri, S., Scarcello, F.: The dlv system for knowledge representation and reasoning. ACM Transactions on Computational Logic 7(3), 499–562 (2006)
5. Gebser, M., Kaufmann, B., Neumann, A., Schaub, T.: Conflict-driven answer set solving. In: Proc. of IJCAI 2007, pp. 386–392. AAAI Press, Massachussets (2007)
6. Akyildiz, I.F., Su, W., Sankarasubramaniam, Y., Cayirci, E.: A survey on sensor networks. IEEE Communications Magazine 40(8), 102–114 (2002)
7. Mileo, A., Merico, D., Bisiani, R.: Support for context-aware monitoring in home healthcare. In: Proc. of IE 2009 (to appear, 2009)
8. Ryan, N.S., Pascoe, J., Morse, D.R.: Enhanced reality fieldwork: the context-aware archaeological assistant. In: Gaffney, V., van Leusen, M., Exxon, S. (eds.) Computer Applications in Archaeology 1997. British Archaeological Reports, Tempus Reparatum (1998)
9. Schilit, B., Adams, N., Want, R.: Context-aware computing applications. In: Proc. of the Workshop on Mobile Computing Systems and Applications, pp. 85–90. IEEE Computer Society, Los Alamitos (1994)
10. Fleming, K.C., Evans, J.M., Weber, D.C., Chutka, D.S.: Practical functional assessment of elderly persons: A primary-care approach. Mayo Clinic 70, 890–910 (1995)
11. Merico, D., Mileo, A., Pinardi, S., Bisiani, R.: A Logical Approach to Home Healthcare with Intelligent Sensor-Network Support. The Computer Journal (2009) bxn074
12. Mileo, A., Bisiani, R.: Context-aware prediction and prevention to extend healthy life years:preventing falls. In: Proc. of the IJCAI Workshop on Intelligent Systems for Assisted Cognition (to appear, 2009)
13. Gebser, M., Schaub, T., Thiele, S.: Gringo: A new grounder for answer set programming. In: Baral, C., Brewka, G., Schlipf, J. (eds.) LPNMR 2007. LNCS (LNAI), vol. 4483, pp. 266–271. Springer, Heidelberg (2007)

Some DLV Applications for Knowledge Management

Giovanni Grasso[1], Salvatore Iiritano[2], Nicola Leone[1], and Francesco Ricca[1]

[1] Dipartimento di Matematica, Università della Calabria, 87030 Rende, Italy
{leone,ricca}@mat.unical.it
[2] Exeura Srl, Via Pedro Alvares Cabrai - C.da Lecco 87036 Rende (CS), Italy
salvatore.iiritano@exeura.com

Abstract. Even if the industrial exploitation of the DLV system has started very recently, DLV already has a history of applications on the industrial level. The most valuable applications from a commercial viewpoint are those in the area of Knowledge Management. They have been realized by the company EXEURA s.r.l. - a spin-off company of the University of Calabria having a branch also in Chicago - with the support of the DLVSYSTEM s.r.l.. DLV applications in this area have not been realized directly, but through some specializations of DLV into Knowledge Management (KM) products for Text Classification, Information Extraction, and Ontology Representation and Reasoning. After briefly describing these KM products, we report on their recently-released successful applications.

1 Introduction

Answer Set Programming (ASP) [2,3] is a powerful logic programming language. In its general form, allowing for disjunction in rule heads [1] and nonmonotonic negation in rule bodies, ASP can represent *every* problem in the complexity class Σ_2^P and Π_2^P (under brave and cautious reasoning, respectively) [4]. The high knowledge modeling power of ASP, and the availability of efficient ASP systems, has implied a renewed interest in this formalism in recent years, due to the need for representing and manipulating complex knowledge, arising in Artificial Intelligence as well as in other emerging areas, like Knowledge Management and Information Integration.

One of the most relevant ASP systems is DLV [13]. The DLV system is the product of more than twelve years of research and development and is the state-of-the-art implementation of disjunctive ASP. DLV is widely used by researchers all over the world, and it is competitive, also from the viewpoint of efficiency, with the most advanced ASP systems. Indeed, at the First Answer Set Programming System Competition [5] DLV won in the Disjunctive Logic Programming category; and DLV finished first also in the general category MGS (Modeling, Grounding, Solving — also called royal competition, which is open to all ASP systems). Importantly, DLV is profitably employed in many real-word applications, and has stimulated quite some interest also in industry. In the following, we report on the most valuable applications of DLV, which are industrially developed by the company EXEURA s.r.l., a spin-off company of the University of Calabria, with the support of the DLVSYSTEM s.r.l., another spin-off company maintaining the system. Actually, many DLV applications have not been realized directly, but through some specializations of the system in products for Text Classification, Information Extraction, and Ontology Representation and Reasoning.

E. Erdem, F. Lin, and T. Schaub (Eds.): LPNMR 2009, LNCS 5753, pp. 591–597, 2009.

In this paper, we first describe the DLV-based Knowledge Management products, we then overview some successful applications of these products, and we briefly report also on some further applications exploiting DLV directly. The described applications fall in many different domains, including Team Building in a Seaport, E-Tourism, E-Government, etc..

2   DLV-Based Commercial Systems

The three main industrial products of Exeura s.r.l. that are strongly based on the DLV system are, namely: OntoDLV [6,7], OLEX [8,9], H𝑖L𝜀X [10,11]. OntoDLV is an ontology management and reasoning system; OLEX is a document classification system; and, H𝑖L𝜀X is an information extraction system. In the following, we provide a brief description of the main features of those systems.

OntoDLV. Traditional ASP in not well-suited for ontology specifications, since it does not directly support features like classes, taxonomies, individuals, etc. Moreover, ASP systems are a long way from comfortably enabling the development of industry-level applications, mainly because they lack important tools for supporting programmers. Both the above-mentioned issues were addressed in OntoDLV [6,7] a system for ontologies specification and reasoning. Indeed, OntoDLV implements a powerful logic-based ontology representation language, called OntoDLP, which is an extension of (disjunctive) ASP with all the main ontology constructs including classes, inheritance, relations, and axioms. OntoDLP is strongly typed, and includes also complex type constructors, like lists and sets. Importantly, OntoDLV supports a powerful interoperability mechanism with OWL, allowing the user to retrieve information from external OWL Ontologies and to exploit this data in OntoDLP ontologies and queries.

Using OntoDLV, domain experts can create, modify, store, navigate, and query ontologies thanks to a user-friendly visual environment; at the same time, application developers can easily implement knowledge-intensive applications embedding OntoDLP specifications using a complete Application Programming Interface (API) [12]. Moreover, OntoDLV facilitates the development of complex applications in a user-friendly visual environment; it is endowed with a robust persistency-layer for saving information transparently on a DBMS, and it seamlessly integrates the DLV system [13].

OLEX. The OntoLog Enterprise Categorizer System (OLEX) [8,9] is a corporate classification system supporting the entire content classification life-cycle, including document storage and organization, ontology construction, pre-processing and classification. OLEX exploits a reasoning-based approach to text classification which synergically combines: (i) ontologies for the formal representation of the domain knowledge; (ii) pre-processing technologies for a symbolic representation of texts and (iii) ASP as categorization rule language. Logic rules, indeed, provides a natural and powerful way to encode how document contents may relate to ontology concepts.

More in detail, the main task of OLEX is text categorization, which is is the task of assigning documents to predefined categories on the basis of their content. To this end, in the system, ontologies are exploited for modeling the domain knowledge; and, with each concept of a given ontology is associated a specific ASP program (containing the *classification rules*) that is used to recognize concepts in a text. Classification rules can be either manually specified or automatically determined [9]. Clearly, the system has to pre-process the input documents in order to produce a logic representation of their content. The OLEX pre-processor performs the following tasks: Pre-Analysis and Linguistic Analysis. The former consists of document normalization, structural analysis and tokenization; whereas the latter includes lexical analysis, which determines the Part of Speech (PoS) of each token, reduction (elimination of the stop words), and frequency analysis. The output of the pre-processing phase for a document is a set of facts modeling its content. The obtained facts are then fed into the DLV system together with the classification rules to compute an association between the processed document and ontology concepts.

HᵢLεX. [10,11] is an advanced system for ontology-based information extraction from semi-structured and unstructured documents. In practice, HᵢLεX implements a semantic approach to the information extraction problem by exploiting: (i) ontologies as knowledge representation formalism; (ii) a general document representation model able to unify different document formats (html, pdf, doc, ...); and, (iii) the definition of a formal attribute grammar able to describe, by means of declarative rules, objects/classes w.r.t. a given ontology.

HᵢLεX is based on OntoDLP for describing ontologies, since this language perfectly fits the definition of semantic extraction rules. Regarding the unified document representation, the idea is that a document (unstructured or semi-structured) can be seen as a suitable arrangement of objects in a two-dimensional space. Each object has its own semantics, is characterized by some attributes and is located in a two-dimensional area of the document called *portion*. A portion is defined as a rectangular area univocally identified by four cartesian coordinates of two opposite vertices. Each portion "contains" one or more objects and an object can be recognized in different portions.

The language of HᵢLεX is founded on the concept of *ontology descriptor*. A "descriptor" looks like a production rule in a formal attribute grammar, where syntactic items are replaced by ontology elements, and where extensions for managing two-dimensional objects are added. Each descriptor allows us to describe: (i) an ontology object in order to recognize it in a document; or (ii) how to "generate" a new object that, in turn, may be added in the original ontology. Note that an object may also have more than one descriptor, thus allowing one to recognize the same kind of information when it is presented in different ways. It is worth noting that, most of the existing information extraction approaches do not work in a semantical way and they are not independent of the specific type of document they process. On the contrary, the approach implemented in HᵢLεX allows for recognizing, extracting and structuring relevant information form heterogeneous sources.

3 Some Commercial Applications

In this Section, we report a brief description of a number of applications developed by Exeura s.r.l. which employ the commercial products based on DLV.

Team Building in the Gioia-Tauro Seaport. The port authority of Gioia Tauro is employing a system, based on OntoDLV, for the automatic generation of the teams of employees. The problem here is to produce an optimal allocation of the available personnel of the international seaport of Gioia Tauro in such a way that the right processing of the shoring cargo boats is guaranteed at the minimum cost. To this end several constraints have to be satisfied, concerning the size and the slot occupied by cargo boats, the allocation of each employee (e.g. each employee might be employed in several roles of different responsibility, roles have to be played by the available units by possibly applying a round-robin policy, etc.), etc.. The system can build new teams or complete the allocation automatically when the roles of some key employees are fixed manually.

In this application, the domain is modeled by exploiting OntoDLV, and a set of suitably defined reasoning modules is exploited for finding the desired allocation. In this application, the pure declarative nature of the language allowed for refining and tuning both problem specifications and encodings together while interacting with the stakeholders of the seaport. It is worth noting that, the possibility of modifying (by editing text files) in a few minutes a complex reasoning task (e.g. by adding new constraints), and testing it "on-site" together with the customer is a great advantage of our approach.

E-Tourism. IDUM is an e-tourism system developed in the context of the project "IDUM: Internet Diventa Umana" funded by the administration of the Calabria Region. The IDUM system helps both employees and customers of a travel agency in finding the best possible travel solution in a short time. It can be seen as a "mediator' system finding the best match between the offers of the tour operators and the requests of the turists. More in detail, in the IDUM system, behind the web-based user interface, there is an intelligent core that exploits an OntoDLV ontology for both modeling the domain of discourse (i.e., geographic information, user preferences, and touristic offers, etc.) and storing the available data. The ontology is automatically populated by extracting the information contained in the touristic leaflets produced by tour operators and received by the travel agency attached to email messages. It is worth noting that, the received e-mails are human-readable, and the details are often contained in email-attachments of different format (plain text, pdf, gif, or jpeg files) and structure that might contain a mix of text and images. The H$\imath$LεX system allows for automatically processing the received contents, and to populate the ontology with the data extracted from touristic leaflets. Once the information is loaded on the ontology, the user can perform an intelligent search for selecting the holiday packages that best fit his needs. Basically, IDUM tries to mimic the behavior of the typical employee of a travel agency by running a set of specifically devised logic programs that reason on the information contained in the ontology. The result is a system that is able to search in a huge database of automatically classified offers. IDUM combines the speed of computers with the knowledge of a travel agent.

Automatic Itinerary Search. In this application, a web portal conceived for better exploiting the whole transportation system, including both public and private companies, of the Italian region Calabria. The user can ask for the automatic construction of a complete itinerary from a given place to another in the region, and the system provides it with several possible solutions depending on both the available resources and user-selected options (e.g., preferred mean, preferred transportation company, minimization of travel distances and/or travel times etc.) The system is very precise, it tells you where and what time to catch your bus/train, where to get off and transfer, how long your trip will take, walking directions etc. This service was implemented by exploiting an OntoDLV ontology that models all the available transportation means, their timetables, and a map with all the streets, bus stops, railways and train stations etc. A set of specifically devises ASP programs are used to build the required itineraries.

e-Government. In this field, an application of the OLEX system was developed, in which legal acts and decrees issued by public authorities are classified. The system employes an ontology based on both TE.SE.O (Thesaurus of the Senato della Repubblica Italiana), an archive that contains a complete listing of words arranged in groups of synonyms and related concepts regarding juridical terminology employed by the Italian Parliament, and a set of categories identified by analyzing a corpus of 16,000 documents of the public administration. The system was validated with the help of the employees of the Calabrian Region administration, and it performed very well by obtaining an f-measure of 92% and a mean precision of 96% in real-world documents.

e-Medicine. OLEX was employed for developing a system able to classify automatically case histories and documents containing clinical diagnoses. The system was commissioned, with the goal of conducting epidemiological analyses, by the ULSS n.8 (which is, a local authority for health services) of the area of Asolo, in the Italian region Veneto. Basically, available case histories are classified by the system in order to help the analysts of the ULSS while browsing and searching documents regarding specific pathologies, supplied services, or patients living in a given place etc. The application exploits an ontology of clinical case histories based on both the MESH (Medical Subject Headings) ontology and ICD9-CM a system employed by the Italian Ministry of the Heath for handling data regarding medical services (e.g. X-Rays analyses, plaster casts, etc.). The analyzed documents are stored in PDF documents and contain medical reports, hospital discharge forms, clinical analysis results etc. Classification rules were manually devised and taken into account, beside the extracted linguistic information, also the metadata contained in the case history forms. The system has been deployed and is currently employed by the personnel of the ULSS of Asolo.

4   Other Applications

The European Commission funded a project on Information Integration, which produced a sophisticated and efficient data integration system, called INFOMIX, which uses DLV at its computational core [14]. The powerful mechanisms for database interoperability, together with magic sets [15,16] and other database optimization techniques, which are implemented in DLV, make DLV very well-suited for handling information integration tasks. And DLV (in INFOMIX) was succesfully employed to develop in a real-life integration system for the information system of the University of Rome "La Sapienza" The DLV system has been experimented also with an application for Census Data Repair [17], in which errors in census data are identified and eventually repaired. DLV has been employed at CERN, the European Laboratory for Particle Physics, for an advanced deductive database application that involves complex knowledge manipulation on large-sized databases. The Polish company Rodan Systems S.A.

has exploited DLV in a tool for the detection of price manipulations and unauthorized use of confidential information, which is used by the Polish Securities and Exchange Commission. In the area of self-healing Web Services, moreover, DLV is exploited for implementing the computation of minimum cardinality diagnoses [18].

References

1. Minker, J.: On Indefinite Data Bases and the Closed World Assumption. In: Loveland, D.W. (ed.) CADE 1982. LNCS, vol. 138, pp. 292–308. Springer, Heidelberg (1982)
2. Gelfond, M., Lifschitz, V.: The Stable Model Semantics for Logic Programming. In: Proc. of ICLP, pp. 1070–1080. MIT Press, Cambridge (1988)
3. Gelfond, M., Lifschitz, V.: Classical Negation in Logic Programs and Disjunctive Databases. New Generation Computing 9, 365–385 (1991)
4. Eiter, T., Gottlob, G., Mannila, H.: Disjunctive Datalog. ACM Transactions on Database Systems 22(3), 364–418 (1997)
5. Gebser, M., Liu, L., Namasivayam, G., Neumann, A., Schaub, T., Truszczyński, M.: The first answer set programming system competition. In: Baral, C., Brewka, G., Schlipf, J. (eds.) LPNMR 2007. LNCS (LNAI), vol. 4483, pp. 3–17. Springer, Heidelberg (2007)
6. Ricca, F., Gallucci, L., Schindlauer, R., Dell'Armi, T., Grasso, G., Leone, N.: OntoDLV: an ASP-based system for enterprise ontologies. Journal of Logic and Computation (2009)
7. Ricca, F., Leone, N.: Disjunctive Logic Programming with types and objects: The DLV$^+$ System. Journal of Applied Logics 5(3), 545–573 (2007)
8. Cumbo, C., Iiritano, S., Rullo, P.: OLEX – A Reasoning-Based Text Classifier. In: Alferes, J.J., Leite, J. (eds.) JELIA 2004. LNCS (LNAI), vol. 3229, pp. 722–725. Springer, Heidelberg (2004)
9. Rullo, P., Policicchio, V.L., Cumbo, C., Iiritano, S.: Effective Rule Learning for Text Categorization. IEEE Transactions on Knowledge and Data Engineering - TKDE-2007-07-0386.R3
10. Ruffolo, M., Manna, M.: HiLeX: A System for Semantic Information Extraction from Web Documents. In: ICEIS (Selected Papers). LNCS(LNBIP), vol. 3, pp. 194–209. Springer, Heidelberg (2008)
11. Ruffolo, M., Leone, N., Manna, M., Saccà, D., Zavatto, A.: Exploiting ASP for Semantic Information Extraction. In: Proc. of ASP 2005, Bath, UK, July 2005, pp. 248–262 (2005)
12. Gallucci, L., Ricca, F.: Visual Querying and Application Programming Interface for an ASP-based Ontology Language. In: Vos, M.D., Schaub, T. (eds.) Proc. of SEA 2007, pp. 56–70 (2007)
13. Leone, N., Pfeifer, G., Faber, W., Eiter, T., Gottlob, G., Perri, S., Scarcello, F.: The DLV System for Knowledge Representation and Reasoning. ACM TOCL 7(3), 499–562 (2006)
14. Leone, N., Gottlob, G., Rosati, R., Eiter, T., Faber, W., Fink, M., Greco, G., Ianni, G., Kałka, E., Lembo, D., Lenzerini, M., Lio, V., Nowicki, B., Ruzzi, M., Staniszkis, W., Terracina, G.: The INFOMIX System for Advanced Integration of Incomplete and Inconsistent Data. In: Proc. of ACM SIGMOD (SIGMOD 2005), Baltimore, USA, pp. 915–917. ACM Press, New York (2005)
15. Cumbo, C., Faber, W., Greco, G., Leone, N.: Enhancing the magic-set method for disjunctive datalog programs. In: Demoen, B., Lifschitz, V. (eds.) ICLP 2004. LNCS, vol. 3132, pp. 371–385. Springer, Heidelberg (2004)
16. Faber, W., Greco, G., Leone, N.: Magic Sets and their Application to Data Integration. Journal of Computer and System Sciences 73(4), 584–609 (2007)

17. Franconi, E., Palma, A.L., Leone, N., Perri, S., Scarcello, F.: Census Data Repair: A Challenging Application of Disjunctive Logic Programming. In: Nieuwenhuis, R., Voronkov, A. (eds.) LPAR 2001. LNCS (LNAI), vol. 2250, pp. 561–578. Springer, Heidelberg (2001)
18. Friedrich, G., Ivanchenko, V.: Diagnosis from first principles for workflow executions. TR, Alpen Adria University, Klagenfurt, Austria (2008),
http://proserver3-iwas.uni-klu.ac.at/download_area/Technical-Reports/technical_report_2008_02.pdf

Application of ASP for Automatic Synthesis of Flexible Multiprocessor Systems from Parallel Programs*

Harold Ishebabi, Philipp Mahr, Christophe Bobda, Martin Gebser, and Torsten Schaub**

University of Potsdam, Institute for Informatics, August-Bebel-Str. 89, D-14482 Potsdam
{ishebabi,pmahr,bobda,gebser,torsten}@cs.uni-potsdam.de

Abstract. Configurable on chip multiprocessor systems combine advantages of task-level parallelism and the flexibility of field-programmable devices to customize architectures for parallel programs, thereby alleviating technological limitations due to memory bandwidth and power consumption. Given the huge size of the design space of such systems, it is important to automatically optimize design parameters in order to facilitate wide and disciplined explorations. Being a combinatorial problem, system design can be modeled and solved as such, but the amount of parameters renders the problem difficult to solve for large instances. However, as the synthesis problem usually exhibits structure, Answer Set Programming (ASP), for which solvers utilizing techniques from the propositional satisfiability domain are available, can be effectively employed. This paper presents a design flow based on ASP that uses the solver clasp as back-end engine. Synthesis experiments demonstrate the effectiveness of the approach.

1 Design Flow

The input to the flow in Figure 1 is a parallel program, and optionally information on task periods. The application is simulated and analyzed to obtain inter-task data traffic and task precedence information. This information is used to specify an instance of an Integer Linear Programming (ILP) problem or an ASP program. Similar to related work in this area, the other input to the design flow is information on available processing elements and communication networks, as well as their costs and constraints. In our approach, the design space is not pre-constrained, and the problem dimensions are not ranked, which ensures the optimality of solutions. For realtime systems, it is often sufficient to meet timing constraints so that the interest is not to find the fastest solution. In such situations, the flow can be used to find the smallest system instead.

The solution obtained from an ILP/ASP solver is used to generate an abstract description of the system, which is passed to further tool chains described in [1] to generate the configuration bit-stream. Because post-synthesis results could deviate from initial cost models used, new cost models can optionally be extracted after placement and routing to start a new iteration.

---

* Long version of this paper will appear in International Journal of Reconfigurable Computing.
** Affiliated with Simon Fraser University, Canada, and Griffith University, Australia.

E. Erdem, F. Lin, and T. Schaub (Eds.): LPNMR 2009, LNCS 5753, pp. 598–603, 2009.
© Springer-Verlag Berlin Heidelberg 2009

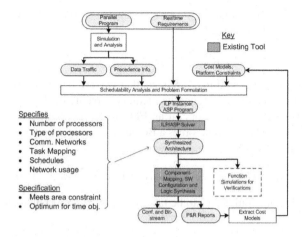

Fig. 1. Architecture Synthesis Flow

2   ASP Approach

For specifying synthesis in ASP, we build upon the ILP model proposed in [2, 3] and convert existing ILP instances into ASP programs. The general problem is to map a number of concurrent tasks on processors and communication resources such that hard space constraints are satisfied, while the throughput or the overall execution time of a parallel program is subject to optimization. We use the following notations: $I_i \in \{I_0, \ldots, I_n\}$ is a task in a parallel program, $J_j \in \{J_0, \ldots, J_m\}$ is a processor in an Intellectual Property (IP) library, and $C_k \in \{C_0, \ldots, C_p\}$ is a communication resource in an IP library. Boolean variables x_{ij} are used to indicate whether a task I_i is mapped on processor J_j in a synthesized multiprocessor system.

To represent linear constraints of 0-1 ILP instances in ASP, we use weight constraints [4] having the general form:

$$l\,[\,\ell_0 = w_0,\ \ell_1 = w_1,\ \ldots,\ \ell_n = w_n\,]\,u. \tag{1}$$

In (1), literals $\ell_0, \ell_1, \ldots, \ell_n$ are associated with weights $w_0, w_1, \ldots, w_n$. The lower bound l and upper bound u can be omitted, in which case they are identified with $-\infty$ and ∞, respectively. Given this, a weight constraint (1) represents linear (in)equality $l \leq \ell_0 * w_0 + \ell_1 * w_1 + \cdots + \ell_n * w_n \leq u$. Notably, weights and bounds of weight constraints are integers, while ILP usually admits rational numbers. If not mentioned otherwise, we deal with this by rounding l as well as $w_0, w_1, \ldots, w_n$ up and u down when translating ILP to weight constraints. In principal, rounding can change the semantics of a constraint, but it was unproblematic in our application.

In what follows, we describe how ASP programs are derived from ILP instances. To begin with, for each task I_i, the associated task mapping constraint [2] is given by

$$1\,[\,x_{i0} = 1,\ x_{i1} = 1,\ \ldots,\ x_{im} = 1\,]\,1. \tag{2}$$

Such constraints stipulate that each task is mapped on exactly one processor J_j.

Processor sharing constraints [2] on program sizes s_{ij} of tasks and program memory s_j of a processor J_j are specified as

$$[\, x_{0j} = s_{0j}, \ x_{1j} = s_{1j}, \ \ldots, \ x_{nj} = s_{nj} \,]\, s_j. \tag{3}$$

The omission of a lower bound in (3) reflects that we admit J_j to remain unallocated, while the upper bound makes sure that the memory capacity of J_j is not exceeded.

The area constraint for processors on an FPGA can be specified in terms of three linear constraints [2]. The first one forces a Boolean variable v_j to be true if at least one task is mapped on processor J_j:

$$[\, x_{0j} = 1, \ x_{1j} = 1, \ \ldots, \ x_{nj} = 1, \ v_j = -(n+1) \,]\, 0. \tag{4}$$

The second constraint stipulates v_j to be false when J_j remains unallocated:

$$[\, x_{0j} = -1, \ x_{1j} = -1, \ \ldots, \ x_{nj} = -1, \ v_j = 1 \,]\, 0. \tag{5}$$

Finally, the third constraint connects areas (coefficients a_j) required by processors J_j on an FPGA with the available area A_J on the FPGA:

$$[\, v_0 = a_0, \ v_1 = a_1, \ \ldots, \ v_m = a_m \,]\, A_J. \tag{6}$$

Note that, if $a_0 + a_1 + \cdots + a_m > A_J$, it is not permissible to allocate all of the available processors.

We next consider network resources, needed whenever two communicating tasks I_{i_1}, I_{i_2} are mapped on different processors J_{j_1}, J_{j_2} for $i_1 < i_2$ and $j_1 \neq j_2$. Such a situation is indicated by an auxiliary atom $\alpha_{i_1 i_2 j_1 j_2}$, defined by the following rule:

$$\alpha_{i_1 i_2 j_1 j_2} \leftarrow x_{i_1 j_1}, \ x_{i_2 j_2}. \tag{7}$$

Given this, the next constraint stipulates another auxiliary atom $\lambda_{i_1 i_2}$ to be true precisely when communicating tasks I_{i_1}, I_{i_2} require a communication resource:

$$0\,[\, \alpha_{i_1 i_2 j_0 j_1} = 1, \alpha_{i_1 i_2 j_0 j_2} = 1, \ldots, \alpha_{i_1 i_2 j_m j_{m-2}} = 1, \alpha_{i_1 i_2 j_m j_{m-1}} = 1, \lambda_{i_1 i_2} = -1 \,]\, 0. \tag{8}$$

The following constraints then stipulate exactly one communication resource C_k to be allocated for distributed communicating tasks I_{i_1}, I_{i_2}, and y_k indicates allocation of C_k:

$$0\,[\, z_{0 i_1 i_2} = 1, \ z_{1 i_1 i_2} = 1, \ \ldots, \ z_{p i_1 i_2} = 1, \ \lambda_{i_1 i_2} = -1 \,]\, 0. \tag{9}$$

$$[\, z_{k i_1 i_2} = 1, \ y_k = -1 \,]\, 0. \tag{10}$$

For a communication resource allocation, the next constraints finally check that the capacity M_k of C_k as well as the available FPGA area A_C for communication infrastructure, where a coefficient a_k gives the area required by C_k, are not exceeded:

$$[\, z_{k i_1 i_2} = 1, \ z_{k i_1 i_3} = 1, \ \ldots, \ z_{k i_{n-2} i_n} = 1, \ z_{k i_{n-1} i_n} = 1 \,]\, M_k. \tag{11}$$

$$[\, y_0 = a_0, \ y_1 = a_1, \ \ldots, \ y_p = a_p \,]\, A_C. \tag{12}$$

Furthermore, we consider scheduling feasibility constraints [3], where a parameter $F_{gj} \in \{0,1\}$ indicates whether any group $G = \{I_{g_0}, I_{g_1}, \ldots, I_{g_l}\}$ of tasks can be mapped on a processor J_j without violating realtime constraints. The fact that all tasks in G are mapped on the same processor and the feasibility requirement are combined in the following rule:

$$1\,[\,\mathcal{M}_{gj} = 1\,]\,F_{gj} \;\leftarrow\; x_{g_0j},\, x_{g_1j},\, \ldots,\, x_{g_lj}. \tag{13}$$

Note that an atom $\mathcal{M}_{gj}$ is derived when all tasks in the group are mapped on J_j, and when each task in the group can meet its deadline. In the worst case, all nonempty groups of tasks in the power set of I are considered in (13), so that the cardinality of I is critical for the problem size. Avoiding such space blow-up is a subject to the future (cf. Section 4).

Finally, we turn our attention to the objective function to be minimized, dealing with scheduling costs of largest groups mapped on processors [3]. For a group G mapped on processor J_j and associated super-groups indicated by $\mathcal{M}_{s_1j}, \ldots, \mathcal{M}_{s_hj}$, we use the next rule to derive γ_{gj} precisely when G is the largest group mapped on J_j:

$$\gamma_{gj} \;\leftarrow\; \mathcal{M}_{gj},\, \text{not } \mathcal{M}_{s_0j},\, \text{not } \mathcal{M}_{s_1j},\, \ldots,\, \text{not } \mathcal{M}_{s_hj}. \tag{14}$$

The objective function can now be stated in terms of a minimize statement [4]:

$$\begin{aligned}
\text{minimize } [\, \ldots,\, & x_{ij} = T_{ij},\, \ldots,\, \gamma_{gj} = (T'_{gj} * t_j + O_j),\\
\ldots,\, & z_{ki_1i_2} = (L_k * D_{i_1i_2} + \tau_k * p_k * B_{i_1i_2}),\, \ldots\,].
\end{aligned} \tag{15}$$

Note that, when optimizing in makespan mode, the weights T_{ij} expressing task execution times are set to zero for tasks I_i not on any critical path. Similarly, weights $T'_{gj} * t_j$ and O_j representing scheduling and operating system overhead are set to zero when the corresponding group G contains no critical task. Weights $L_k * D_{i_1i_2}$ and $\tau_k * p_k * B_{i_1i_2}$ represent data transfer latency and network arbitration overhead [2], respectively.

Notably, it is important to recognize that the weights in (15) cannot simply be rounded up (as with constraints) because doing so disrupts the cost structure of a problem unless small time units are used. However, using such small units can result in huge numbers, which can easily overflow the computation of the value of the objective function. Instead of time units, we thus use processor cycles, normalized by the slowest processor and the smallest weight occurring in the objective function.

3   Comparison of Synthesis Results

We applied conflict-driven learning ASP solver clasp [5], embedded into ASP system clingo [6] (version 2.0.2), to synthesis problems translated from ILP. Previous empirical investigations [5] have shown that learning solvers perform well, at least for structured problems like the ones on synthesis. For comparison, we used ILP solver lp_solve [7] (version 5.5.0.14). We conducted experiments with five applications described in [2]: filtering (FIR), Derivation, Simpson's method, N-Body problem, and matrix Inversion. Figure 2 summarizes runtimes for synthesis scenarios using 16 processors and 5 communication resources for increasing number of tasks from 4 to 22. All benchmarks were run on a machine equipped with a 1.66GHz T5500 processor and 2048Mb of main memory, using a timeout of 28,800 seconds.

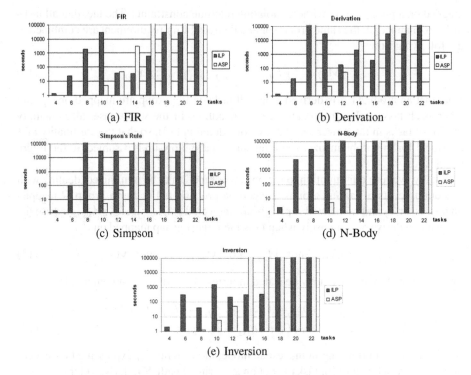

Fig. 2. ASP-ILP Comparison: Increasing the number of tasks

All columns that terminate at the boundary of 28,800s in Figure 2 indicate that a suboptimum solution was found. Columns that exceed the boundary, i.e., those which touch the 100,000s line, indicate that no solution was found by timeout. Otherwise, optimum solutions were obtained. The results show that ASP-based synthesis outperforms ILP for 4 to 12 tasks. Beyond that region, ILP-based synthesis is either faster or at least finds a suboptimum solution by timeout. A closer look however showed that, when ASP-based synthesis performs badly, then most of the runtime is spent on reading ASP programs. In fact, in almost all cases where clingo timed out, clasp did not get any chance to start searching. The reason was that, when reading large ASP programs generated for greater number of tasks, clingo was running out of memory and swapped. As scheduling feasibility (13) is mainly responsible for the observed file size explosion, a more compact representation of it would likely enable scaling up ASP-based synthesis.

4 Summary and Conclusion

We have presented a method for automated architecture synthesis of FPGA multiprocessor systems using ASP. We showed that ASP-based synthesis has a great potential for solving difficult system design problems. Our continuing work addresses the compact representation of scheduling feasibility constraints to avoid file size explosion. Moreover, as our ASP programs were obtained from existing ILP instances, we made use of available ASP solving technology, but not (yet) of knowledge representation capacities of ASP. Future work includes the development of direct ASP solutions for automated synthesis, building on a uniform encoding and grounding.

Acknowledgments. This work was partially funded by DFG under Grant BO 2480/5-1 and under Grant SCHA 550/8-1.

References

1. Bobda, C., Haller, T., Mühlbauer, F., Rech, D., Jung, S.: Design of adaptive multiprocessor on chip systems. In: Petraglia, A., Pedroni, V., Cauwenberghs, G. (eds.) Proceedings of the Twentieth Annual Symposium on Integrated Circuits and Systems Design (SBCCI 2007), pp. 177–183. ACM Press, New York (2007)
2. Ishebabi, H., Bobda, C.: Automated architecture synthesis for parallel programs on FPGA multiprocessor systems. Microprocessors and Microsystems 33(1), 63–71 (2009)
3. Ishebabi, H., Mahr, P., Bobda, C.: Automatic synthesis of multiprocessor systems from parallel programs under preemptive scheduling. In: Torres, C. (ed.) Proceedings of the International Conference on Reconfigurable Computing and FPGAs (ReConFig 2008), pp. 19–24. IEEE Press, Los Alamitos (2008)
4. Simons, P., Niemelä, I., Soininen, T.: Extending and implementing the stable model semantics. Artificial Intelligence 138(1-2), 181–234 (2002)
5. Gebser, M., Kaufmann, B., Neumann, A., Schaub, T.: Conflict-driven answer set solving. In: Veloso, M. (ed.) Proceedings of the Twentieth International Joint Conference on Artificial Intelligence (IJCAI 2007), pp. 386–392. AAAI Press/MIT Press (2007)
6. Gebser, M., Kaminski, R., Kaufmann, B., Ostrowski, M., Schaub, T., Thiele, S.: A user's guide to `gringo`, `clasp`, `clingo`, and `iclingo`, `http://potassco.sourceforge.net`
7. `http://lpsolve.sourceforge.net/5.5/`

Optimal Multicore Scheduling:
An Application of ASP Techniques

Viren Kumar and James Delgrande

Simon Fraser University
8888 University Drive
Burnaby BC, Canada
{vka4,jim}@sfu.ca
http://www.cs.sfu.ca

Abstract. Optimal scheduling policies for multicore platforms have been
the holy grail of systems programming for almost a decade now. In this
paper, we take two instances of the optimal multicore scheduling problem, the dual-core and the quad-core instances, and attempt to solve
them using Answer Set Programming. We represent the problem using
a graph theoretic formulation, whereby finding the minimum edge cover
of a weighted graph aids us in finding an optimal multicore schedule. We
then use an answer set solver to obtain answer sets of varying sizes for
our formulation. While a polynomial time algorithm for finding the minimum edge cover of a weighted graph exists in imperative languages and
is suitable for dual-core processors, our approach uses ASP techniques
to solve both the dual-core and quad-core versions of this problem. We
discuss some optimizations to reduce execution runtime, and conclude
by discussing potential uses of our application.

Keywords: ASP, clingo, multicore, scheduling, edge cover.

1 Introduction

Multicore processors have established themselves as the *de facto* standard in
computing today. The problem of optimally scheduling applications on multicore
platforms to maximize performance remains wide open even today, with many
possible solutions being proposed over the last decade.

Background: Moore's Law posits that the maximum possible number of tran-
sistors on an integrated circuit doubles every eighteen months. This law ground
to a halt for uniprocessors in the last few years, due to coming up against several physical barriers, such as the speed of light and heat constraints. Increasing
the number of cores on a processor is one way to maintain Moore's law and
uphold the performance curve. Performance gains were subsequently realized at
the substantial price of requiring additional parallelism from programs. This, in
turn, has created new challenges in the field of multicore computing, the most
prominent of which is the optimal scheduling problem.

E. Erdem, F. Lin, and T. Schaub (Eds.): LPNMR 2009, LNCS 5753, pp. 604–609, 2009.

Processor Composition: Modern processors are comprised of several components. Uniprocessors consist of a single core that runs on these components and uses them exclusively. Multicore processors, by definition have more than one core on a physical processor. Having multiple cores leads to contention for on-chip resources such as caches. When an application suffers a cache miss, some data present in the cache is evicted. If an application evicts data belonging to another co-scheduled application running on the same processor, it negatively affects the other application, because the latter application now has to re-fetch its data back into the cache. In this fashion, two co-scheduled applications can negatively impact each other.

In this paper, we attempt to find an optimal schedule for two types of multicore processors, dual-core and quad-core. Our test applications are taken from the SPEC CPU 2006 suite of benchmarks[4], both CINT and CFP. For the dual-core case, each application is run offline against all others to measure performance effects. This gives all possible pairs of co-schedules and performance impacts. Since running all possible sets of four applications was not feasible, the quad-core problem uses simulated performance impact data. It should be noted that our application generates an optimal schedule based on this *a priori* data but cannot itself be directly used as a real scheduler in an operating system.

2   Optimal Scheduling

What makes scheduling applications on a multicore platform difficult is the impact co-scheduled applications have on each other. Performance is usually measured with the Instructions Per Cycle (IPC) metric, which measures how many instructions a processor can execute per clock cycle. The goal of multicore scheduling is to optimally schedule simultaneously executing applications on all cores of a multicore processor, such that the performance degradation is minimized, usually attained by minimizing contention for shared resources[5]. Unfortunately, even if all possible combinations of co-scheduled applications and their deleterious effects were known beforehand, optimal scheduling for more than 2 cores on a processor is an NP-complete problem.

In this paper, we focus on dual-core processors and quad-core processors. For the dual-core case, each processor in our test system has two cores, with the overall system composed of a varying number of processors. Finding an optimal co-schedule thus involves finding pairs of applications that can be concurrently run together with minimal IPC degradation. The quad-core case is similar, with varying numbers of processors consisting of four cores. In this case, the optimal solution consists of quartets of applications that can be run together with minimal performance impact.

A polynomial time algorithm to find an optimal schedule exists for the dual-core case, but like our approach, it requires complete *a priori* knowledge of the applications and all possible co-schedules and performance impacts. It too requires that the problem be formulated using graph theory, which is the next step in our solution.

2.1   Problem Formulation

First, we consider the dual-core case. If we consider applications as nodes and the co-scheduled IPC degradation as the cost of an edge between two nodes, then it is straightforward to create a graph that embodies the optimal scheduling problem. In Figure 1(a), we see that there are four applications: A, B, C and D. An edge between two nodes has a cost equal to the IPC degradation suffered when the two applications represented by the nodes are co-scheduled. In the figure, running applications A and B together results in a performance degradation of 13, while running applications A and C instead will result in a performance degradation of 17.

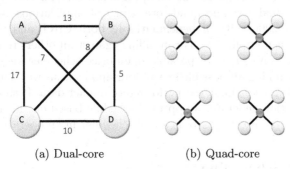

(a) Dual-core   (b) Quad-core

Fig. 1. Optimal groupings

Figure 1(a) gives us an insight into the dual-core optimal scheduling problem: finding an optimal schedule with a minimal degradation is equivalent to finding a minimum cost edge cover for the given graph. Since an edge cover includes all nodes, we can guarantee that all applications will be scheduled. Finding the minimum cost edge cover will give us a set of edges that cover all vertices and have the lowest possible cost, i.e., the least IPC degradation possible. Thus, for the dual-core processor case, finding a minimal edge cover is tantamount to finding an optimal schedule. From Figure 1(a), we see that a minimum edge cover is {AD, BC} with a total cost of 15.

In the quad-core case, since there are no edges between applications, but rather quadruplets of applications, as seen in Figure 1(b), the edge cover is less useful. The one concept we do borrow from the edge cover theory is to select combinations of four vertices, such that all the vertices in the graph are covered. Additionally, no two combinations can have common vertices. From there, we find the combinations with minimal performance impacts, thereby resulting in a minimal cost optimal schedule.

Having expressed the optimal scheduling problem graph theoretically, the next step is to encode it using Answer Set Programming and solve it.

3   Expressing the Problem in ASP

Answer Set Programming (ASP) is a form of logic programming, which uses declarative syntax to solve problems. We follow the usual ASP technique of writing the problem description in a declarative language such as lparse[3], which is then fed to an ASP solver such as *smodels*[3].

The Paths, Trees and Flowers[2] algorithm by Edmonds can be used to find minimal edge covers in polynomial time, but its iterative specification precludes it from easily being used by ASP solvers. We used *clingo*[1] as our ASP solver.

3.1 Clingo

clingo, like other ASP solvers uses the generate and test strategy, which requires that the solver generate all possible answer sets to the problem, then discard those answer sets that do not meet the constraints set by the original problem. For the dual-core case, the performance degradation data was presented to *clingo* in the form `corun(A,B,n)` where A and B are benchmarks from the suite, and n is the performance degradation resulting from running A and B together. Similarly,the quad-core case presented *clingo* with data in the form `corun(A,B,C,D,n)` where A, B, C and D are benchmarks from the suite, and n is the performance degradation resulting from simultaneously running A along with B,C and D. Our formulation accepts these co-runs as the input of the problem and then builds a solution from them. For the dual-core case with n cores, our implementation generates all possible edge covers by selecting all $\binom{n}{2}$ edges and rejecting those combinations of edges that do not touch all vertices, i.e. are not edge covers. The minimize keyword is then used to select the minimal edge cover from these edge covers. The implementation and test data for this paper can be found at `http://www.sfu.ca/~vka4/multicore_asp.zip`.

Our implementation works for varying numbers of dual-core and quad-core processors, i.e. it can be used to find optimal schedules for simultaneously running applications on these processors. For both the dual-core and quad-core case, we are able to find optimal schedules for up to 16 applications co-scheduled simultaneously. The key difference is that the dual-core schedule will consist of pairs of applications, with each pair co-scheduled on a dual-core processor, while the quad-core schedule will consist of quartets of applications, with each quartet co-scheduled on a quad-core processor.

3.2 Optimizations

Our brute-force implementation takes exponential time in the number of nodes, as we shall see in the next section. *Clingo* has several optimizations related to pruning and searching which can result in a significant speedup. To determine if any of these optimizations would result in a quicker execution time, we ran *clingo* on an edge cover of size 6 for the dual-core case, with several different optimizations and measured the runtime of each. These optimizations did not affect the output of a minimal edge cover, except for the − −cautious flag, which resulted in a non-minimal edge cover.

Of all the different optimizations we tried, the one with the smallest execution time that was also correct was the − −save−progress flag. It dropped the execution time for the 6-edge case from 513 seconds to 83 seconds. We tried several

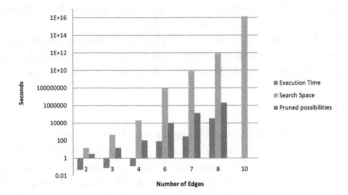

Fig. 2. Execution Times of Edge Covers

combinations of the most promising optimization switches, hoping for reduced execution times. However, none of the combinations resulted in a smaller time footprint, leaving us with the −−save−progress flag as the best optimization to use in the rest of the project.

3.3 Results

All tests were run on a machine with eight Pentium IV Xeon X5365 cores and 8 GB of RAM. The system ran Linux 2.6 as its operating system, with the prebuilt binary of *clingo* 2.03 for the x86-Linux platform being used in our experiments.

Figure 2 shows the time taken to find the minimum edge cover for the dual-core case, the size of the search space and the pruning done by *clingo*, for minimum edge covers of varying sizes. The vertical axis is logarithmic in scale and shows us that finding the optimal schedules for up to 14 applications can be done in a reasonable period of time, not exceeding five minutes. However, when we move up to 16 applications (eight dual-core processors), finding an optimal schedule takes around ten hours. While we were not able to complete the run to find a minimum edge cover for 10 edges, we can estimate that it would take a long time, approximately ten years.

Similarly, the quad-core approach also showed analogous time increases. Finding an optimal schedule for 8 nodes took under a second, while finding optimal quartets for 12 nodes took around a minute. However, the 16-node case is expected to take around 4 days, based on our estimations.

The observed exponentially increasing time isn't surprising, given the worst-case exponential time behaviour of the underlying ASP solver, along with the fact that our approach is brute force.

4   Conclusion

In this paper, we have shown how two instances of the optimal multicore scheduling problem can be represented in a form amenable to being solved by Answer

Set Programming. We used ASP tools to solve as large an instance of the problem as currently feasible. ASP solvers use elegant formalizations and powerful pruning mechanisms to narrow a large search space, but have their limits in the face of exponential, combinatorial blowup.

An optimal schedule is beneficial in many ways, besides the obvious. If an optimal schedule can be generated in a relatively short period of time, it can actually be used for scheduling long-term jobs. While this window of usage remains narrow, an optimal schedule has other theoretical uses. Many scheduler implementations often compare their performance to current system schedulers in operating systems such as Solaris and Linux. The system schedulers in these operating systems are agnostic of optimal schedules themselves. Having an optimal schedule provides a reliable measuring-point, by which to gauge all other scheduler implementations.

Our implementation finds optimal schedules based on real performance data. While gathering this data for the quad-core case or higher is prohibitive in terms of time, it is certainly possible for the dual-core case. Despite the exponential blowup for larger versions of the dual-core and quad-core scheduling problem, our implementation is efficient enough to be used for current, realistic scenarios involving fewer processors.

With some work, our current implementation could be modified to find optimal schedules for n-core processors as well. The next step in the hierarchy would be a six-core or eight-core processor, even though the former is only just appearing in the mainstream processor market today. Finding optimal schedules involving sets of six or eight applications co-scheduled together could be done with our implementation, provided some parameters and constraints were changed.

Acknowledgments. We thank Dr. Alexandra Fedorova for her help and the gracious use of the SPEC CPU 2006 benchmarks and systems equipment.

References

1. clingo, http://potassco.sourceforge.net/
2. Edmonds, J.: Paths, Trees and Flowers. Canada J. Math. 17, 449–467 (1965)
3. smodels and lparse, http://www.tcs.hut.fi/Software/smodels/
4. SPEC CPU (2006), http://www.spec.org/cpu2006/
5. Tam, D., Azimi, R., Soares, L., Stumm, M.: Managing Shared L2 Caches on Multicore Systems in Software. In: WIOSCA (June 2007); held in conjunction with the (ISCA) (2007)

From Data Integration towards Knowledge Mediation*

Gerhard Brewka[1] and Thomas Eiter[2]

[1] Universität Leipzig, Augustusplatz 10-11, D-04109 Leipzig, Germany
brewka@informatik.uni-leipzig.de
[2] Vienna University of Technology, Favoritenstraße 9-11, A-1040 Vienna, Austria
eiter@kr.tuwien.ac.at

Abstract. A major goal of AI are powerful reasoning capabilities which empower the development of advanced knowledge processing systems. To this end, current research is focused on information and knowledge integration, addressing the needs that emerge in the realm of the World Wide Web and future information systems. Following Wiederhold's mediator concept, powerful knowledge mediation is envisioned that fruitfully exploits LPNMR technology, as already done in the data integration context. Realizing this vision raises in turn challenging issues for research.

1   The Knowledge Mediation Vision

The IT developments of the last decade have rapidly changed the possibilities for data and knowledge access. The World Wide Web and the underlying Internet provide a backbone for the information systems of the 21st century, which will possess powerful reasoning capabilities that enable one to combine various pieces of information, stored in heterogeneous formats and with different semantics.

To fully exploit the wealth of knowledge, information from plain sources and software packages with plain semantics will have to be mixed with semantically rich sources like domain ontologies, expert knowledge bases, temporal reasoners etc. in a suitable manner, bridging the gap between different sources. However, more than simple *integration* of data and knowledge, as it is targeted in current research (and has been targeted in the KR area since decades, cf. [1]), is desirable. In fact, *mediation* of data and knowledge, which provides services at an abstract level that go far beyond technical aspects of integration, and take aspects such as situatedness, social context, and goals of a user into account. The vision of powerful knowledge mediation is not new, and goes back at least to Wiederhold's classic paper [2], and ramifying efforts like the Knowledge Interchange Format (KIF). Yet, like many visions of AI, it has not yet materialized satisfactorily.

We imagine that tools and techniques from the logic programming and nonmonotonic reasoning area are very helpful to help realizing this vision, and that in fact nonmonotonic features, working with defaults and implicit assumptions are inherent to intelligent mediation.

---

* Work supported by the Austrian Science Fund (FWF) projects P20841 and the Vienna Science and Technology Fund (WWTF) project ICT08-020.

E. Erdem, F. Lin, and T. Schaub (Eds.): LPNMR 2009, LNCS 5753, pp. 610–612, 2009.

2   Where We Are

Driven by the need to combine (possibly heterogenous) knowledge bases, numerous proposals were made to solve this problem, and extensions of declarative KR formalisms were conceived that allow to access external data and knowledge sources; LPNMR technology was fruitfully deployed to data integration cf. [3].

A recent framework are *nonmonotonic multi-context systems (MCSs)* [4], which allow for a principled integration of different logic-based formalisms. An MCS $(C_1, \ldots, C_n)$ comprises information in contexts $C_i = (L_i, kb_i, br_i)$ of an abstract "logic" L_i, a knowledge base in L_i, and a set br_i of *bridge rules*

$$p \leftarrow (r_1 : p_1), \ldots, (r_j : p_j), \textbf{not } (r_{j+1} : p_{j+1}), \ldots, \textbf{not } (r_m : p_m)$$

where p and p_i are from the language of L_i resp. L_{r_i}, which specify the information flow between the contexts depending on presence or absence of the p_i in the local belief sets of the contexts. Such MCS have a global-state semantics based on LP-like rules, and generalize previous work on monotonic context systems by the Trento School around Giunchiglia and Serafini.

MCS empower the combination of "typical" monotonic KR logics like Description Logics, and nonmonotonic formalisms like ASP or Default Logic, but also formal argumentation systems; given the very abstract notion of "logic", a wide range of other formalisms can be accommodated as well. However, while a formal tool at the technical level, MCSs are far from solving integration, yet the mediation problem at a semantic level, simply because high level integration and mediation featured are not reflected in the framework. Extensions are possible, and even bootstrapping might help to arrive at richer systems.

This is exemplified by *Argumentation Context Systems (ACSs)* presented at this conference, which specialize and generalize MCSs at the same time. In contrast to a general MCS, an ACS is homogeneous as all reasoning components are of the same type, viz. Dung-style argumentation frameworks [5]. On the other hand, an ACS context (called *module M_1*) may influence another M_2 much stronger than in an MCS; M_1 may directly affect M_2's KB and reasoning mode, by invalidating arguments or attack relationships in M_2's argumentation framework, and by determining the semantics to be used by M_2. This is achieved by a designated language to "configure" the argumentation framework dynamically.

In addition, ACSs feature an integration-oriented mediation service, by special *mediator* components in the modules which manage inconsistencies. A mediator collects the information coming in from connected modules and turns it into a consistent configuration for its module, using a pre-specified consistency handling method. Different such methods allow to model a range of scenarios, from strictly hierarchical to more "democratic" forms of decision making.

3   Where to Go

In order to materialize the long term vision of powerful knowledge mediation systems on the basis of LPNMR technology, clearly a lot of efforts must be devoted to solve the many problems that arise in this context. We think that work

on ensembles of (possibly non-monotonic) knowledge sources (be it knowledge
bases, modules, contexts etc), that are connected through well-defined interfaces,
is an important direction. Among the issues that need attention are:

- Distributed execution platforms. Currently, we lack platforms that can efficiently run an ensemble of knowledge sources. Here, aspects like communication media and cost play a role.
- Inconsistency management. Inconsistency in an MCS may arise through the
  interaction at a global level, perhaps by mediation itself. Novel mechanisms
  to handle this are needed (e.g., avoidance, or recognition and resolution).
- Interpretation/transformation of vocabularies. The ubiquitous problem of
  aligning different terminology and models should be solved at the mediation
  level. Especially, bridge rules may take advantage of such services.
- Communication and interaction between entities. Beyond mere information
  flow, modules may engage in more involved communication according to well-
  defined protocols. Mediators in particular may maintain data connections to
  various modules, relaying and controlling the knowledge exchange.
- Text analysis and understanding. Knowledge sources may be unstructured,
  and to elicit logical structure and meaning from text so that it can be rendered in an MCS context is desirable.
- Combining quantitative and qualitative uncertainty. While current MCS are
  geared to qualitative uncertainty, means for a quantitative account of uncertainty are needed, as well as for the combination of the two.
- Preference and goal handling. Knowledge integration and mediation does not
  end in itself but serves a user. Situatedness, preferences and goals of a user
  must be adequately respected; current MCSs provide no support for this.

Many of these issues are not new, and existing results and technology may be
taken up and advanced. Text analysis and understanding using LPNMR technology is a particular challenge that received attention recently.

We are investigating how to extend MCSs (not necessarily based on argumentation), with (limited) mediation components to a generalization of both MCS
and ACS. This will be an important step forward, yet many more steps remain.

References

1. Baral, C., Kraus, S., Minker, J., Subrahmanian, V.: Combining knowledge bases
   consisting of first-order analysis. Comp. Intell. 8, 45–71 (1992)
2. Wiederhold, G.: Mediators in the architecture of future information systems. IEEE
   Computer 25, 38–49 (1992)
3. Eiter, T.: Data integration and answer set programming. In: Baral, C., Greco, G.,
   Leone, N., Terracina, G. (eds.) LPNMR 2005. LNCS (LNAI), vol. 3662, pp. 13–25.
   Springer, Heidelberg (2005)
4. Brewka, G., Eiter, T.: Equilibria in Heterogeneous Nonmonotonic Multi-Context
   Systems. In: AAAI 2007, pp. 385–390. AAAI Press, Menlo Park (2007)
5. Dung, P.M.: On the acceptability of arguments and its fundamental role in nonmonotonic reasoning, logic programming and n-person games. Artif. Intell. 77, 321–358
   (1995)

Integrating Answer Set Modules into Agent Programs

Stefania Costantini

Università degli Studi di L'Aquila
Dipartimento di Informatica
Via Vetoio, Loc. Coppito, I-67010 L'Aquila - Italy
Stefania.Costantini@univaq.it

The use of ASP in agents has been advocated since long, with ASP mainly taking the form of Action Description Languages. These kind of ASP-based languages were first introduced in [1] and [2] and have been since then extended and refined in many subsequent papers by several authors. Action Description Languages are formal models used to describe dynamic domains, by focusing on the representation of effects of actions. In particular, an action specification represents the direct effects of each action on the state of the world, while the semantics of the language takes care of all the other aspects concerning the evolution of the world (e.g., the ramification problem).

The first approaches have been extended in many ways, recently also in order to cope with, interpret, and recover from, exogenous events and unexpected observations. In this direction we mention [2], [3], and the very recent work presented in [4]. In this work, an architecture (called AAA) is described where both the description of the domain's behavior and the reasoning components are written in Answer Set Programming, selected because of its ability to represent various forms of knowledge including defaults, causal relations, statements referring to incompleteness of knowledge, etc. An AAA agent behaves according to the *Observe-Think-Act* model proposed in the seminal paper [5], which works as follows. An agent:

(1) observes the world, explains the observations, and updates its knowledge base;
(2) selects an appropriate goal, G;
(3) finds a plan (sequence of actions) to achieve G;
(4) executes part of the plan, updates the knowledge base, and goes back to step 1.

Unexpected observations are coped with by hypothesizing the undetected occurrence of exogenous actions. In [6], this notion of an agent is extended to enable communication between agents through the introduction of special named sets of fluents known as "requests".

In other directions, we mention a different line of work, focusing upon modeling agent decisions in an extended ASP by means of game theory [7]. In [8,9] and other papers by the same group, ASP is exploited to model dynamic updates of an agent's knowledge base. Also, we are aware of ongoing work about modeling properties of multi-agent systems in ASP.

Despite this corpus of work is technically and conceptually very well-developed, the view of an agent based upon having an ASP program as its "core" does not appear to be fully convincing. One reason is that the basic feature of ASP, which is that a program may have several answer sets that correspond to alternative coherent views of the

E. Erdem, F. Lin, and T. Schaub (Eds.): LPNMR 2009, LNCS 5753, pp. 613–615, 2009.
© Springer-Verlag Berlin Heidelberg 2009

world, is in our opinion not fully exploited. Another reason is that the architecture outlined above appears to be too rigid with respect to the other approaches to defining agent architectures in computational logic, among which one has to mention at least MetateM, 3APL, AgentSpeak, Impact, KGP and DALI [10,11,12,13,14,15,16,17,18] (for a recent survey the reader may refer to [19]). All these architectures, and their operational models, are in practice or at least in principle more dynamic and flexible. If we consider for instance the KGP [17,18] architecture, we find many modules ("capabilities") and many knowledge bases, integrated by *control theories* that can be interchanged according to the agent's present context and tasks. In KGP, capabilities are supposed to be based upon abductive logic programming [20] but the architecture might in principle accommodate modules defined in different ways.

We believe that an "ideal" agent architecture should exploit the potential of integrating several modules/components representing different behaviors/forms of reasoning, with these modules possibly based upon different formalisms. The "overall agent" should emerge from dynamic, non-deterministic combination of these behaviors that should occur also in consequence of the evolution of the agent's environment. Therefore, in our view an important present and future direction of ASP is that of being able to encapsulate ASP programs into modules suitable to be integrated into an overall agent program, the latter expressed in whatever language/formalism. There is a growing corpus of literature about modules in ASP. However, the existing approaches mainly refer to traditional programming techniques and to software engineering methodologies (for a review of the state of the art the reader may refer to [21] and to the references therein). To the best of our knowledge, except for the approach of [22] in the context of action theories, there is no existing approach which is specifically aimed at the agent realm.

We are convinced that modularization of ASP is possible under various perspective. A first perspective is that of "Reactive ASP modules", aimed at defining complex reaction strategies to cope with external events and decide what to do, independently of overall planning strategies. A second particularly relevant perspective is that of "Modal ASP modules", that should exploit the multi-model nature of answer set semantics to allow for reasoning about possibility and necessity in agents, at a comparatively low complexity. From the implementation point of view, we are confident upon the integration of such modules into logic-based agent architectures being easily feasible in conceptually clear ways, e.g., by adopting blackboards that should act as input/output structures. The forms of reasoning which would be made possible are interesting, and allow for real applications at a comparatively low complexity, which implies satisfactory efficiency assuming a reasonable size of modules.

References

1. Gelfond, M., Lifschitz, V.: Action languages. ETAI, Electronic Transactions on Artificial Intelligence (6) (1998)
2. Baral, C., Gelfond, M.: Reasoning agents in dynamic domains. In: Minker, J. (ed.) Workshop on Logic-Based Artificial Intelligence, pp. 257–279. Kluwer Academic Publishers, Dordrecht (2001)
3. Balduccini, M.: Answer Set Based Design of Highly Autonomous, Rational Agents. PhD thesis (2005)

4. Balduccini, M., Gelfond, M.: The AAA architecture: An overview. In: AAAI Spring Symposium 2008 on Architectures for Intelligent Theory-Based Agents, AITA 2008 (2008)
5. Kowalski, R.A., Sadri, F.: From logic programming towards multi-agent systems. Annals of Mathematics and Artificial Intelligence 25(3-4), 391–419 (1999)
6. Gelfond, G., Watson, R.: Modeling cooperative multi-agent systems. In: Costantini, S., Watson, R. (eds.) Proc. of ASP 2007, 4th International Workshop on Answer Set Programming at ICLP 2007 (2007)
7. Vos, M.D., Vermeir, D.: Extending answer sets for logic programming agents. Annals of Mathematics and Artifical Intelligence, Special Issue on Computational Logic in Multi-Agent Systems 42(1-3), 103–139 (2004)
8. Alferes, J.J., Brogi, A., Leite, J.A., Pereira, L.M.: Evolving logic programs. In: Flesca, S., Greco, S., Leone, N., Ianni, G. (eds.) JELIA 2002. LNCS (LNAI), vol. 2424, pp. 50–61. Springer, Heidelberg (2002)
9. Alferes, J.J., Dell'Acqua, P., Pereira, L.M.: A compilation of updates plus preferences. In: Flesca, S., Greco, S., Leone, N., Ianni, G. (eds.) JELIA 2002. LNCS (LNAI), vol. 2424, pp. 62–74. Springer, Heidelberg (2002)
10. Rao, A.S., Georgeff, M.: Modeling rational agents within a bdi-architecture. In: Proc. of the Second Intl. Conf. on Principles of Knowledge Representation and Reasoning (KR 1991), pp. 473–484. Morgan Kaufmann, San Francisco (1991)
11. Rao, A.S.: Agentspeak(l): BDI agents speak out in a logical computable language. In: Perram, J., Van de Velde, W. (eds.) MAAMAW 1996. LNCS (LNAI), vol. 1038. Springer, Heidelberg (1996)
12. Hindriks, K.V., de Boer, F., van der Hoek, W., Meyer, J.C.: Agent programming in 3APL. Autonomous Agents and Multi-Agent Systems 2(4) (1999)
13. Fisher, M.: Metatem: The story so far. In: Bordini, R.H., Dastani, M.M., Dix, J., El Fallah Seghrouchni, A. (eds.) PROMAS 2005. LNCS (LNAI), vol. 3862, pp. 3–22. Springer, Heidelberg (2006)
14. Subrahmanian, V.S., Bonatti, P., Dix, J., Eiter, T., Kraus, S., Ozcan, F., Ross, R.: Heterogeneous Agent Systems. MIT Press/AAAI Press, Cambridge (2000)
15. Costantini, S., Tocchio, A.: A logic programming language for multi-agent systems. In: Flesca, S., Greco, S., Leone, N., Ianni, G. (eds.) JELIA 2002. LNCS (LNAI), vol. 2424, p. 1. Springer, Heidelberg (2002)
16. Costantini, S., Tocchio, A.: The DALI logic programming agent-oriented language. In: Alferes, J.J., Leite, J. (eds.) JELIA 2004. LNCS (LNAI), vol. 3229, pp. 685–688. Springer, Heidelberg (2004)
17. Kakas, A.C., Mancarella, P., Sadri, F., Stathis, K., Toni, F.: The KGP model of agency. In: Proc. ECAI 2004 (2004)
18. Bracciali, A., Demetriou, N., Endriss, U., Kakas, A., Lu, W., Mancarella, P., Sadri, F., Stathis, K., Terreni, G., Toni, F.: The KGP model of agency: Computational model and prototype implementation. In: Priami, C., Quaglia, P. (eds.) GC 2004. LNCS (LNAI), vol. 3267, pp. 340–367. Springer, Heidelberg (2005)
19. Fisher, M., Bordini, R.H., Hirsch, B., Torroni, P.: Computational logics and agents: a road map of current technologies and future trends. Computational Intelligence Journal 23(1), 61–91 (2007)
20. Kakas, A.C., Kowalski, R.A., Toni, F.: The role of abduction in logic programming. In: Gabbay, D., Hogger, C., Robinson, A. (eds.) Handbook of Logic in Artificial Intelligence and Logic Programming, vol. 5, pp. 235–324. Oxford University Press, Oxford (1998)
21. Oikarinen, E.: Modularity in Answer Set Programs. PhD thesis, Faculty of Information and Natural Sciences, Helsinki University of Technology (2008) ISBN 978-951-22-9581-4
22. Baral, C., Son, T.: Relating theories of actions and reactive control. ETAI (Electronic transactions of AI) 2(3-4), 211–271 (1998)

What Next for ASP?
(A Not-Entirely-Well-Informed Opinion)

James Delgrande

School of Computing Science
Simon Fraser University
Burnaby BC, V5A 1S6, Canada
jim@cs.sfu.ca

Preamble. The inception of Answer Set Programming (ASP) can be marked by the appearance of the stable model semantics [GL88], something over 20 years ago. The roots of ASP in turn can be traced to work in nonmonotonic reasoning, notably Default Logic [Rei80]. With the advent of efficient ASP solvers, as exemplified by *smodels* [NS97] and *dlv* [ELM+97], there was a great deal of interest and excitement over the application of ASP (broadly taken) to various problems, along with its use as a modelling tool. Indeed, applications have been proposed in a wide variety of fields, including bioinformatics, configuration, database integration, diagnosis, hardware design, insurance industry applications, model checking, phylogenesis, planning, security protocols, and high-level control of the US space shuttle [Sch08]. In concert with these applications, there has been a widespread flowering of ASP solvers built on various technologies [DVB+09].

With these successes, attention has turned to the application of ASP to other interesting and challenging problems. Since it can be argued (or at least taken as a position for debate) than *any* research in CS should be with an eye to eventual practical application, the question of the application of ASP can be seen as obliquely asking, *What is the potential role of ASP in AI/CS/the world at large?*

Applying ASP. Broadly, and for purposes of discussion,[1] we can consider three non-exclusive and non-exhaustive areas of application: to other areas of AI, to areas of interest in mathematics, and to "real world" problems. Consider each in turn:

ASP and AI: To consider the role of ASP in AI is to essentially ask about the suitability of ASP for providing general KR languages. Certainly, earlier surveys such as [BG94, DB96] regarded Knowledge Representation and Reasoning as the principal focus of ASP, as implicitly does [Bar03].

Two points can be made in this regard. First, the trend with respect to applications is *away* from KR. Witness, for example, the problem suite in [DVB+09], which would seem to not have a great deal to do with KR.[2] As well, as implementations have developed and evolve, enhancements have been added to the

---

[1] which is to say this classification shouldn't be taken too seriously.

[2] For a perhaps unfair comparison, consider in contrast the Challenge Problems for Commonsense Reasoning at www-formal.stanford.edu/leora/commonsense

E. Erdem, F. Lin, and T. Schaub (Eds.): LPNMR 2009, LNCS 5753, pp. 616–618, 2009.
© Springer-Verlag Berlin Heidelberg 2009

language; these can be declarative (e.g. aggregates and cardinality constraints) or procedural (e.g. adding options to control search). In the latter case, representational force is lost, in favour of procedural gain. Second, and counter to the first point, research in ASP on KR can nonetheless be regarded as alive and well, given work on strongly related topics such as action languages and causal representations, and representations of interesting domains using such formalisms (e.g. [AEL+04]). This suggests of course the potential role of ASP as a target language for higher-level encodings of problems.[3]

ASP and mathematics: Most of the challenge problems for ASP are from graph theory, combinatorics, or number theory. This arguably reflects the shift in emphasis toward solving constraint problems (or maybe just the prevalence of toy problems from these areas). Regardless, most proposed applications seem to lie in these areas, one way or another. So one possibility is to attempt to solve specific open problems in mathematics via ASP. One example is determining the fifth Schur number;[4] a second is suggested at the conclusion of this article.

ASP and "real world" problems: This of course is a highly interesting, difficult, and potentially important long-term use of ASP. There are various impediments that need to be addressed; in particular, current implementations would need to scale up in various ways. Grounding is clearly a bottleneck. As well, there is a need for a general programming methodology, and perhaps a better understanding of the relation between problem type and search strategy. Similarly there would seem to be a need for programming environments and tools for the construction of large programs. Work on locality (perhaps involving conditional independence structures) and structuring blocks of rules would be useful. Certainly if similar work in KR (e.g. [Mor98]) is anything to go by, such applications promise to be messy.

A Modest Proposal. A specific problem area that falls into the second category above, yet may have practical application while skirting issues concerning scalability and software engineering, is that of *balanced incomplete block designs* (BIBD) [CD06]. Roughly a BIBD is a set X and a collection of subsets of X, called *blocks*, such that each block is the same size and each element of X appears in the same number of blocks. Hence BIBDs come with a compact, austere, theoretical specification. They also have significant real world application, as they are fundamental in experimental design, and have applications in software testing and cryptography. They encompass a large class of problems, and include as subareas Steiner triple systems, finite projective planes, and Latin squares.

---

[3] This also suggests an interesting project, comparing the two leading logic programming paradigms via their respective ease (or lack thereof) for representing action formalisms: The University of Toronto action group translates the situation calculus into PROLOG typically, while action languages may be translated into extended logic programs.

[4] I've shamelessly cribbed this example from Torsten Schaub's position paper, who in turn credits it to Mirek Truszczyński.

Moreover, they would seem ideally suitable for ASP encodings, as the general problem for BIBDs is to find a solution for a given set of parameters or show that no solution exists.

References

[AEL+04]   Akman, V., Erdoğan, S., Lee, J., Lifschitz, V., Turner, H.: Representing the zoo world and the traffic world in the language of the causal calculator. Artificial Intelligence 153(1-2), 105–140 (2004)

[Bar03]   Baral, C.: Knowledge Representation, Reasoning and Declarative Problem Solving. Cambridge University Press, Cambridge (2003)

[BG94]   Baral, C., Gelfond, M.: Logic programming and knowledge representation. Journal of Logic Programming 19, 73–148 (1994)

[CD06]   Colbourn, C.J., Dinitz, J.H.: Handbook of Combinatorial Designs, 2nd edn. Chapman & Hall/CRC, Boca Raton (2006)

[DB96]   Dix, J., Brewka, G.: Knowledge representation with logic programs. Fachberichte Informatik 15–96, Universität Koblenz-Landau (1996)

[DVB+09]   Denecker, M., Vennekens, J., Bond, S., Gebser, M., Truszczyński, M.: The second answer set programming competition. In: Erdem, E., Lin, F., Schaub, T. (eds.) LPNMR 2009. LNCS (LNAI), vol. 5753. Springer, Heidelberg (2009)

[ELM+97]   Eiter, T., Leone, N., Mateis, C., Pfeifer, G., Scarcello, F.: A deductive system for nonmonotonic reasoning. In: Dix, J., Furbach, U., Nerode, A. (eds.) LPNMR 1997. LNCS (LNAI), vol. 1265, pp. 363–374. Springer, Heidelberg (1997)

[GL88]   Gelfond, M., Lifschitz, V.: The stable model semantics for logic programming. In: Kowalski, R., Bowen, K. (eds.) Proc. ICLP, pp. 1070–1080. MIT Press, Cambridge (1988)

[Mor98]   Morgenstern, L.: Inheritance comes of age: Applying nonmonotonic techniques to problems in industry. Artificial Intelligence 103, 1–34 (1998)

[NS97]   Niemelä, I., Simons, P.: Smodels: An implementation of the stable model and well-founded semantics for normal logic programs. In: Dix, J., Furbach, U., Nerode, A. (eds.) LPNMR 1997. LNCS, vol. 1265, pp. 420–429. Springer, Heidelberg (1997)

[Rei80]   Reiter, R.: A logic for default reasoning. Artificial Intelligence 13(1-2), 81–132 (1980)

[Sch08]   Schaub, T.: Here's the beef: Answer set programming! In: Dovier, A., Garcia de la Banda, M., Pontelli, E. (eds.) ICLP 2008. LNCS, vol. 5366, pp. 93–98. Springer, Heidelberg (2008)

Using Lightweight Inference to Solve Lightweight Problems

Marc Denecker and Joost Vennekens*

Dept. Computerscience, K.U. Leuven
Celestijnenlaan 200A
B-3001 Heverlee, Belgium
{marc.denecker,joost.vennekens}@cs.kuleuven.be

Traditionally, Logic Programming and related Non-monotonic Reasoning formalisms have mainly been applied to "hard" AI problems, such as planning, scheduling, constraint solving, belief revision, etc. In the real world of software engineering, however, these hard problems are vastly outnumbered by more mundane tasks. A significant part of software that is written today consists of applications that any reasonably experienced programmer could write in a couple of weeks, using whatever imperative language happens to be the industry standard *du jour*.

A typical example of this kind of applications is so-called *configuration software*. The idea is here that there a number of allowable configurations—say of a computer system, or of a server network, or a study program, or a life insurance, or a bicycle, or so on—and that the user must be guided through a number of choices that will eventually result in the configuration that is best suited to his desires. Programming such an application poses no great challenges, neither conceptually nor computationally. The main source of complexity here lies in the domain knowledge concerning valid configurations: this might imply complex dependencies, which could be tedious to encode in an imperative language, especially if the application is also supposed to proactively narrow down the options still available to the user, as more and more choices are made.

While such applications might seem trivial in comparison to the great open problems of AI, we believe that computational logic might also have an interesting role to play here. The goal is in this context not to accomplish the previously impossible, but rather to do existing things *better*. By leveraging the expressive knowledge representation capabilities of logic, such applications could be written more easily, in less lines of "code", and hopefully considerably quicker—maybe even in days rather than weeks. An additional advantage is that once domain knowledge has been declaratively expressed, it can also be reused for different applications in the same domain. For instance, a specification of the requirements for a schedule of courses at some university might be used at the start of the semester to produce a full schedule, but it might also be used later—by a different UI front-end—to help a secretary adapt the schedule to unforseen

---

* Joost Vennekens is a post-doctoral researcher of FWO-Vlaanderen.

E. Erdem, F. Lin, and T. Schaub (Eds.): LPNMR 2009, LNCS 5753, pp. 619–621, 2009.

circumstances, such as an absent teaching assistant or urgent maintenance in some class room.

The setting of configuration software is somewhat different from the setting traditionally considered in LP, though, and therefore it comes with its own set of requirements. For instance, solving a scheduling problem is a task that is typically performed off-line and on a reasonably good machine. Configuration software, on the other hand, is interactive, and might be running on a web-server, together with who knows how many other threads. So, the computational requirements are much more stringent in this case. There is also good news, however, and that is that configuration software does not need to be complete. Because it runs interactively, it is not absolutely necessary that *all* consequences of a user's choice are immediately shown on-screen; if one is missed at first, it will still eventually be detected after the user has filled in more of the form.

This kind of lightweight interactive applications, therefore, seems to call for anytime algorithms, that may only be approximative and offer no guarantees of completeness, but do run fast in limited memory and will provide useful output even when interrupted before completion. These requirements are rather incompatible with the ground-and-solve approach that most Answer Set Programming systems currently take. An integrated "ground on demand" approach would be much more suited.

In [6], we have presented an algorithm that could work well in such cases. The roots of this algorithm lie in the IDP-system [3]. This is a model expansion system [4] for the language FO($\cdot$), an extension of first-order logic with, among others, a LP-based representation of inductive definitions. IDP is a quite competitive system, which ranked fourth in the global ranking of the second ASP competition [2]. It uses a standard ground-and-solve architecture, in which the grounder already performs a number of logical propagation steps that attempt to minimize the size of the grounding. This propagation algorithm has been extended to a polytime approximate reasoning algorithms for full FO($\cdot$) [6]. In [5], we have used this algorithm to develop an FO($\cdot$) framework for the implementation of configuration software.

Approximation is not the only way of providing inference algorithms that are able to deal with the strict computational requirements of interactive software. In certain contexts, it might also be sometimes possible to exploit computations that have previously been carried out off-line. A prime example is railroad scheduling: the regular timetables are computed far in advance, but if some technical failure occurs, the schedule has to be quickly adapted online, possibly in interaction with an engineer. This can only be accomplished if the existing schedule is leveraged as much as possible and only minimal changes are applied—instead of a "from skratch" model generation system, we therefore need an incremental model *revision* approach. [1] outlines one such algorithm.

In summary, we believe that, in addition to the traditional "hard" AI problems that have been tackled, there are many "easy" problems in software engineering that could benefit from a declarative approach. However, this calls for different inference algorithms, which satisfy the requirements of interactiveness by means of anytime approximation, or by leveraging as much as possible the result of computations that were previously performed off-line. The development and further refinement of such algorithms is a useful step towards increasing the impact that logic formalisms have on the everyday life of software engineers world-wide.

References

1. De Cat, B.: Developments of algorithms for model revision with applications for train rescheduling and netwerk reconfiguration. Master thesis, Department of Computer Science, K.U.Leuven (in Dutch) (2009)
2. Denecker, M., Vennekens, J., Bond, S., Gebser, M., Truszczyński, M.: The second answer set programming competition. In: LPNMR (2009)
3. Mariën, M., Wittocx, J., Denecker, M.: The IDP framework for declarative problem solving. In: LASH (2006)
4. Mitchell, D.G., Ternovska, E.: A framework for representing and solving NP search problems. In: AAAI, pp. 430–435 (2005)
5. Vlaeminck, H., Vennekens, J., Denecker, M.: A logical framework for configuration software. In: PPDP (2009)
6. Wittocx, J., Denecker, M.: Approximate reasoning in first-order logic theories. In: KR (2008)

Present and Future Challenges for ASP Systems
(Extended Abstract)

Agostino Dovier[1] and Enrico Pontelli[2]

[1] Dip. di Matematica e Informatica, Università of Udine
dovier@dimi.uniud.it
[2] Dept. Computer Science, New Mexico State University
epontell@cs.nmsu.edu

1 Introduction

The advent of ASP has reinvigorated the field of logic programming. ASP has a simple syntax, which from the beginning appealed logic programmers—thanks also to its similarity to PROLOG's syntax. ASP has a well-understood declarative semantics, which can adequately capture non-monotonicity. ASP benefits from good implementations, which guarantee an adequate level of efficiency and reflect the declarative nature of the paradigm—differently from the case of PROLOG, where there is a disconnect between declarative semantics and the implementation of its operational semantics.

In this presentation, we would like to draw on our personal experiences in comparative development of applications using both ASP and other declarative paradigms, and use these to indicate what we believe are strong directions of development for future generations of ASP languages and systems.

2 Applications

Applications of ASP range over a wide spectrum. Since its inception, ASP has found a natural application in the domain of knowledge representation and reasoning (e.g., planning, legal reasoning, abduction, expert systems) [1].

The continuous improvements of efficiency of ASP solvers enabled this paradigm to become an effective alternative to solve constraint satisfaction problems, such as timetabling, scheduling, configuration problems, and NP-complete problems on graphs [4].

Recently, other emerging applications that require both representation expressiveness and good execution performance have emerged, in areas like bioinformatics, semantic web, and music composition [6,2,9]. Whenever the problem has limited requirements in terms of numerical computations (e.g., Boolean problems), the behavior of ASP systems is excellent. Instead, if problems require extensive use of arithmetic operations with large integers or with floating point numbers, other techniques are currently more promising.

E. Erdem, F. Lin, and T. Schaub (Eds.): LPNMR 2009, LNCS 5753, pp. 622–624, 2009.

3   Language Extensions

We have witnessed several proposals for extensions of the basic ASP language (i.e., standard normal logic programs) with various features, often dictated by the needs of specific applications. Two relevant directions being pursued are:

- The introduction of forms of aggregations and set/multiset-based reasoning.
- The introduction of capabilities for numerical reasoning.

Recently, several interesting proposals have been put forward describing syntactic and semantic extensions addressing these problems (e.g., new semantics for aggregates, numerical constraints, abstract constraints), yet there is still a lack of general agreement and a relatively limited exploration of implementation issues [10,8].

4   Solvers Integration

ASP is good for certain classes of applications (e.g., dealing with inferential reasoning, abduction, non-monotonicity, Boolean constraint satisfaction problem) but inadequate for other types of reasoning or other levels of expressiveness. For example, even the simple problem of determining the minimal length plan for a planning problem is beyond the reach of standard ASP systems, while a simple loop around the execution of a planner would be sufficient to solve the problem. This suggests the need for:

- Embedding ASP within other popular declarative paradigms (e.g., PROLOG).
- Embedding other declarative paradigms within ASP (e.g., constraint programming).

Limited forms of these types of extensions have been proposed, e.g., DLV with external predicates, iClingo [5]. Interesting solutions in this direction include also the recent proposals on modular ASP [7].

This perspective would also contribute to enable ASP to join the ongoing large scale integration efforts in the domain of constraint satisfaction/optimization. In this realm, a new community has been emerging with the aim of integrating the knowledge and the solving capabilities of various paradigms, such as integer linear programming, constraint programming, and of other techniques coming from artificial intelligence (such as local search).

ASP technology should be part of this overall unifying project.

5   Execution Models

Although successful for a series of applications, the traditional execution model for ASP is inadequate for others:

- ASP computations are bottom-up, and this makes it hard to employ the user's intention in guiding the search for solutions.

- ASP computations are mostly viewed as black box computations, with limited opportunities for interaction (e.g., user-defined heuristics) and communication.
- ASP computations requires a grounding stage.

Other applications require different execution models; in particular, we have often encountered the need for:

- Avoiding grounding, or at least not performing the whole grounding step at once [3];
- Top down computations, or at least computation models that are capable of focusing on specific goal(s);
- Introducing user-defined heuristics and programmable search strategies.

These alternative execution models will, in turn, enable a wide range of other extensions, such as integration with finite domain constraints (or constraints over reals) and parallel execution models.

All these alternatives should co-exist within a unique ASP framework, allowing the programmer to select the most appropriate options for each specific application.

References

1. Baral, C.: Knowledge representation, reasoning and declarative problem solving. Cambridge University Press, Cambridge (2003)
2. Boenn, G., Brain, M., De Vos, M., Fitch, J.: Automatic Composition of Melodic and Harmonic Music by Answer Set Programming. In: Garcia de la Banda, M., Pontelli, E. (eds.) ICLP 2008. LNCS, vol. 5366, pp. 160–174. Springer, Heidelberg (2008)
3. Dal Palù, A., Dovier, A., Pontelli, E., Rossi, G.: Answer Set Programming with Constraints using Lazy Grounding. In: ICLP, pp. 115–129. Springer, Heidelberg (2009)
4. Dovier, A., Formisano, A., Pontelli, E.: An empirical study of CLP and ASP solutions of combinatorial problems. Journal of Experimental & Theoretical Artificial Intelligence 21(2), 79–121 (2009)
5. Eiter, T., Ianni, G., Schindlauer, R., Tompits, H.: A Uniform Integration of Higher-Order Reasoning and External Evaluations in Answer-Set Programming. In: IJCAI, pp. 90–96 (2005)
6. Erdem, E., Türe, F.: Efficient Haplotype Inference with Answer Set Programming. In: AAAI, pp. 436–441. AAAI/MIT Press (2008)
7. Jarvisalo, M., Oikarinen, E., Janhunen, T., Niemela, I.: A Module-Based Framework for Multi-Language Constraint Modeling. In: LPNMR. Springer, Heidelberg (2009)
8. Liu, L., Pontelli, E., Son, T., Truszczynski, M.: Logic Programs with Abstract Constraint Atoms: The Role of Computations. In: Dahl, V., Niemelä, I. (eds.) ICLP 2007. LNCS, vol. 4670, pp. 286–301. Springer, Heidelberg (2007)
9. McIlraith, S., Son, T.: Adapting Golog for Composition of Semantic Web Services. In: KRR, pp. 482–496. AAAI/MIT Press (2002)
10. Mellarkod, V., Gelfond, M.: Integrating Answer Set Reasoning with Constraint Solving Techniques. In: Garrigue, J., Hermenegildo, M.V. (eds.) FLOPS 2008. LNCS, vol. 4989, pp. 15–31. Springer, Heidelberg (2008)

ASP: The Future Is Bright

A Position Paper

Marina De Vos

Department of Computer Science
University of Bath,
BATH BA2 7AY, UK
mdv@cs.bath.ac.uk

In the twenty years since the creation of the stable model semantics[4] for logic programs and inclusion of both classical and negation as failure, answer set programming (ASP)[1] has grown from a small fledgeling field within logic programming to a maturing field of its own. In this extended abstract we discuss some future application areas for ASP and theoretic and implementational problems that need to be addressed in order to make them feasible. With the increase in efficiency of the answer set solvers and a better understanding of the formalism, its advantages and disadvantages, more applications areas for ASP are found. While traditional application domains were mainly in the knowledge representation and reasoning area, we now also see more and more applications in areas were competitors like SAT or CSP were considered better alternatives. We believe, these current applications are just the tip of the iceberg. In years to come, we will see more and more successful applications plus further expansion and improvements on existing applications.

In recent years we have seen a lot of ASP activity in the fields of the semantic web and web-services with specific workshops dedicated to the topic[1]. We believe that these fields hold more potential for years to come. For example, ASP could also be used for match-making and service coordination and orchestration.

Another application area of ASP is multi-agent systems. In most cases this was always on the more theoretical side of applications. To our knowledge, nobody has used ASP to construct a running multi-agent system where the agents use ASP for reasoning purposes. With our current solvers, we should be able to provide such a system.

ASP has also be applied in wide variety of other domains like planning, diagnosis, language-evolution, policy-design, bio-informatics, compiler optimisation, music composition, security, cryptography and game design. Most of these application are in a prototype phase, demonstrating that ASP could be of value in this domain and are slowly making their way to conferences and journals in the domain they are addressing. By doing some domain experts will get to know ASP and its capabilities. In time this will lead to them pointing ASP researchers to similar problems in the same domain. So in the medium term, we believe we can look forward to more ASP applications in bio-informatics, e-health, medicine, e-government and engineering.

Normally we refer to ASP as a problem solving paradigm. We model the problem as an ASP program such that the answer sets correspond to the solutions of this problem.

---

[1] http://events.deri.at/alpsws2006/

E. Erdem, F. Lin, and T. Schaub (Eds.): LPNMR 2009, LNCS 5753, pp. 625–627, 2009.
© Springer-Verlag Berlin Heidelberg 2009

But some applications also perform use what could be referred to as content generation. In this case we are doing something subtly, but importantly, different; we are using a solver to generate representative objects given a specification. This is more knowledge presentation, since the language is not only used to describe the objects but also to come up with a computational description. This opens a new range of possibilities for ASP. What is needed is consistent, sensible background content, which is currently hard to generate in large amounts as at the moment it is done by humans, which is expensive and slow. Music, games, virtual worlds and puzzle magazines would just a few examples.

With the current state-of-the-art solvers, our theoretical understanding and a wide variety of ASP applications, we believe that ASP technology should become marketable. As a community we should move beyond the proof of concepts and simple prototypes we have and make people use them. While making existing and new applications marketable will push the boundaries of our solver implementations and the underlying theory in ways we cannot foresee at the moment, a number of novel research avenues already clear.

In order to move to a marketable technology, we need solvers to be a single, coherent, push button tool. The implementation needs to be robust, documented, maintained and integrated. We need to go from talking about things starting with "this is answer set semantics" to "this is the tool that can solve your problems".

Linked in with this are questions of methodology and education. How do we use these tools, how do we teach others to use these tools? While other successful applications can bring ASP into the spotlight of a certain domain, it is also important for domain experts to use ASP for themselves. At presents, ASP provides little programming support tools for both novices and expert programmers alike. To make writing applications more easy for both domain experts and ASP programmers alike, we believe that we need a software methodology and development environments specifically tailored for ASP. The last few years work has started in this direction with modules, debuggers and prototypes of IDEs. Another approach to making ASP technology more accessibly outside the community is to provide domain specific front-ends, in much the same way as has been done for action domains. At present, writing answer set programs relies on experience and trial and error. It is often not clear why a certain encoding works better or more efficient than another. To provide more consistent implementations, we need a better understanding of encodings and how solvers react to them. When understood, we could move to a more encompassing methodology for answer set programming.

There is also the question of standardisation. Standardisation is the commons of the scientific world, everyone benefits from having standardised tools but no one wants to compromise and put time towards doing so. Efforts are being made to standardise. The draft intermediate format presented at NMR'08[3] was a good step forward and the work should be continued. Language extensions could be a serious problem when it comes to language extensions. It is hard for a whole community to agree on special purpose constructs. An interesting theoretical question is when a construct should be implemented natively and when to emulate/translate it.

While solvers have become faster, grounding is still a bottleneck. While grounding on the fly would be very welcome, we believe it will be some time before a theoretical solution will be found for this.

In ASP, variables are used for two distinct, and largely separate processes. Some variables are templates; creating multiple copies of rules, the values given to these variables are effectively just labels and some variables carry actual semantic value. We should be able to identify, automatically (to a fairly high degree of accuracy) which variables are which. Semantic variables can be handled in an unground fashion such as CSP style finite domains (if the domains are big enough, if they are small instantiation is probably a better strategy). This would vastly reduce the size of the instantiation and will give corresponding speed ups due to caching effects, etc.

To improve the usability of ASP, we need move away from implementation tricks. When there is a trade off between clarity of expression and speed of implementation, we should add to the tools (probably via unground preprocessing) to convert the clear version to the faster version automatically. A good example is the automatic removal of instantiation symmetries.

While our solvers have become much more efficient over the years, especially with the introduction of clause learning[2, 6], we believe that further increases are possible and needed for applications, especially for ASP applications that require real-time performance. With more machines becoming multi-core with shared memory, solvers need to be able to take advantage of these architectures. Having transparent portability to parallel architectures would be a big selling point. Platypus[5] has already shown that linear or super linear speed ups are not only theoretically possible but also in practice.

In this paper, we argued that ASP has a lot of potential when it comes to marketable applications but that there is still some work ahead of us to make that happen. As researchers we should rise to the challenge.

Acknowledgements. The author would like to thank Martin Brain and Owen Cliffe for their input and comments on the earlier versions of this document.

References

[1] Baral, C.: Knowledge Representation, Reasoning and Declarative Problem Solving. Cambridge Press, Cambridge (2003)
[2] Gebser, M., Kaufmann, B., Neumann, A., Schaub, T.: Conflict-driven answer set solving. In: Veloso, M. (ed.) Proceedings of the Twentieth International Joint Conference on Artificial Intelligence (IJCAI 2007), pp. 386–392. AAAI Press/The MIT Press (2007), http://www.ijcai.org/papers07/contents.php
[3] Gebser, M., Janhunen, T., Ostrowski, M., Schaub, T., Thiele, S.: A versatile intermediate language for answer set programming. In: Proceedings of the 12th International Workshop on Nonmonotonic Reasoning, Sydney, Australia, pp. 150–159 (2008)
[4] Gelfond, M., Lifschitz, V.: The stable model semantics for logic programming. In: Kowalski, R.A., Bowen, K.A. (eds.) Logic Programming, Proceedings of the Fifth International Conference and Symposium, Seattle, Washington, August 1988, pp. 1070–1080. MIT Press, Cambridge (1988)
[5] Gressmann, J., Janhunen, T., Mercer, R., Schaub, T., Thiele, S., Tichy, R.: Platypus: A Platform for Distributed Answer Set Solving. In: Baral, C., Greco, G., Leone, N., Terracina, G. (eds.) LPNMR 2005. LNCS (LNAI), vol. 3662, pp. 227–239. Springer, Heidelberg (2005)
[6] Lierler, Y.: Abstract Answer Set Solvers. In: Garcia de la Banda, M., Pontelli, E. (eds.) ICLP 2008. LNCS, vol. 5366, pp. 377–391. Springer, Heidelberg (2008)

Exploiting ASP in Real-World Applications: Main Strengths and Challenges

Nicola Leone

Department of Mathematics, University of Calabria, 87030 Rende, Italy
leone@mat.unical.it

Abstract. We have recently started a couple of spin-off companies which aim at exploiting ASP, and the DLV system in particular, in real-world applications. We briefly report on the first experiences, evidentiating the positive aspects of ASP which allowed us to develop some successful applications, and some obstacles which should be overcome to improve ASP systems, and make them best suited for industry-level applications.

1 Principal Classes of DLV Applications

The industrial exploitation of the ASP system DLV [1] is explored by two spin-off companies of University of Calabria, namely, EXEURA s.r.l. and DLVSYSTEM s.r.l.. EXEURA developes products and applications in the area of Knowledge Management, also exploiting DLV; while DLVSYSTEM developes and maintains the DLV system.

Looking at the way how the ASP system is used, we can group our applications in three main classes:

1. Knowledge Management products incorporating DLV as the computational core;
2. Artificial intelligence applications, characterized by
3. the high complexity of the underlying computational problem; Advanced deductive database applications, characterized by a large amount of data.

The first class contains the three industrial products of Exeura s.r.l., namely, OntoDLV (ontology management) [2,3], OLEX (document classification) [5,4], and HιLεX (information extraction) [6,7], incorporating the DLV system as the computational core. OntoDLV is an ontology management and reasoning system; OLEX is a document classification system; and, HιLεX is an information extraction system. The large number of real-world applications that have been developed on top of these systems can be seen as "indirect" DLV applications.

The third class contains, for instance, the DLV application developed at CERN (the European Laboratory for Particle Physics) involving knowledge manipulation on large-sized databases, together with the Automatic Itinerary Search, developed for the Government of the Calabria Region.

The most interesting Team Building application, developed for the port authority of the Gioia-Tauro Seaport, belongs to the second class.

Information Integration applications, like INFOMIX [8], are somehow across the second and the third class, since they need to solve a computationally hard task (CONP-complete) on a large amount of input data.

E. Erdem, F. Lin, and T. Schaub (Eds.): LPNMR 2009, LNCS 5753, pp. 628–630, 2009.

2   ASP Strengths for Applications

In the "Society of Knowledge" there is an increasing need of methods and tools for knowledge representation and management. The high knowledge-modeling power and the inference capabilities of ASP can be profitably used to satisfy this need.

Killer ASP applications exploit the distinguishing features of ASP, like Variables, Default Negation, Disjunction, and Reasoning (rules and inference). Importantly, the high expressiveness of ASP allows us to often encode the applications "directly" in ASP, with a very limited usage of an host (imperative) language (if any).

The availability of an executable specification-language allows us to speed-up the specification phase, and obtain a very fast prototyping. For instance, in the Team Building application for the Seaport of Gioia Tauro, a key problem was the clarification of very complex specifications that the user was unable to formalize. We went to the Seaport of Gioa Tauro with DLV on the laptop, and, thanks to the high-level language of DLV, we could immediately encode the complex constraints with the users, let the user check the results of the execution, and refine the constraints. In this way, we could quickly formalize the specifications, have them cross-checked and approved by the user, and rapidly obtain a prototype. Also the easy way to access the databases was important here, since we could immediately retrieve the data and simulate the real executions.

It is worth remarking that the key factor for the success of this application was not the DLV efficiency, rather it was the expressiveness of the ASP language and the ease of use of the ASP system DLV.

In our opinion, it is very unlikely that ASP will be exploited in industry for solving combinatorial problems because of its major efficiency. Even if we convince people to use a logic-based approach, then lower-level formalisms, like SAT or CSP, will likely be more efficient and preferred to ASP on this basis. Obviously, the ASP systems should be efficient enough to support the applications, but ASP should exploit the major expressiveness of its knowledge representation language, its advanced reasoning features, and the ease of use.

3   Lessons Learned and Challenges

While develping applications, we have learned many useful lessons.

Average engineers are unable to write (correct and efficient) ASP Programs. We need, above all, suitable ASP programming methodologies, to drive the design of ASP programs, together with debugging techniques, and methodologies to optimize ASP programs. We need also some tools for ASP programmers, ASP programming environments, debuggers, and friendly interfaces.

Applications programs are frequently easy (stratified or nearly such), but have often to deal with **huge** amount of data. We need to further improve the ASP instantiators, bringing in more from database technologies, designing partial evaluation techniques which avoid a full instantiation (like Magic Sets [9,10]) or perform grounding "on demand".

Input data often reside on databases or on the web. We need to empower ASP systems with suitable mechanisms for the interoperability with both DBMSs and with OWL/RDF systems.

ASP misses some "practical" linguistic features: (1) Data Types (both data and methods) for floating-point number, strings, dates, etc. (2) Other aggregate functions (e.g., Average); (3) Complex terms like Lists and Sets. Moreover, adding some application specific functions is often needed and should be allowed.

Finally, ASP systems are often used as the "intelligent" kernel of complex systems. Application Programming Interfaces (API) together with mechanisms and tools for interoperability are definitely needed.

References

1. Leone, N., Pfeifer, G., Faber, W., Eiter, T., Gottlob, G., Perri, S., Scarcello, F.: The DLV System for Knowledge Representation and Reasoning. ACM TOCL 7(3), 499–562 (2006)
2. Ricca, F., Gallucci, L., Schindlauer, R., Dell'Armi, T., Grasso, G., Leone, N.: OntoDLV: an ASP-based system for enterprise ontologies. Journal of Logic and Computation (2009)
3. Ricca, F., Leone, N.: Disjunctive Logic Programming with types and objects: The DLV$^+$ System. Journal of Applied Logics 5(3), 545–573 (2007)
4. Rullo, P., Policicchio, V.L., Cumbo, C., Iiritano, S.: Effective Rule Learning for Text Categorization. IEEE Transactions on Knowledge and Data Engineering - TKDE-2007-07-0386.R3
5. Cumbo, C., Iiritano, S., Rullo, P.: OLEX - A Reasoning-Based Text Classifier. In: Alferes, J.J., Leite, J. (eds.) JELIA 2004. LNCS (LNAI), vol. 3229, pp. 722–725. Springer, Heidelberg (2004)
6. Ruffolo, M., Manna, M.: HiLeX: A System for Semantic Information Extraction from Web Documents. In: ICEIS (Selected Papers). LNBIP, vol. 3, pp. 194–209. Springer, Heidelberg (2008)
7. Ruffolo, M., Leone, N., Manna, M., Saccà, D., Zavatto, A.: Exploiting ASP for Semantic Information Extraction. In: Proc. of ASP 2005, Bath, UK, July 2005, pp. 248–262 (2005)
8. Leone, N., Gottlob, G., Rosati, R., Eiter, T., Faber, W., Fink, M., Greco, G., Ianni, G., Kałka, E., Lembo, D., Lenzerini, M., Lio, V., Nowicki, B., Ruzzi, M., Staniszkis, W., Terracina, G.: The INFOMIX System for Advanced Integration of Incomplete and Inconsistent Data. In: Proc. of ACM SIGMOD (SIGMOD 2005), Baltimore, USA, pp. 915–917. ACM Press, New York (2005)
9. Cumbo, C., Faber, W., Greco, G., Leone, N.: Enhancing the magic-set method for disjunctive datalog programs. In: Demoen, B., Lifschitz, V. (eds.) ICLP 2004. LNCS, vol. 3132, pp. 371–385. Springer, Heidelberg (2004)
10. Faber, W., Greco, G., Leone, N.: Magic Sets and their Application to Data Integration. Journal of Computer and System Sciences 73(4), 584–609 (2007)

Making Your Hands Dirty Inspires Your Brain!
Or How to Switch ASP into Production Mode

Torsten Schaub[*]

Universität Potsdam, Institut für Informatik, August-Bebel-Str. 89,
D-14482 Potsdam
torsten@cs.uni-potsdam.de

Rising from strong theoretical foundations in Logic Programming and Nonmonotonic Reasoning, Answer Set Programming (ASP) came to life as a declarative problem solving paradigm [1,2,3] in the late nineties. The further development of ASP was greatly inspired by the early availability of efficient and robust ASP solvers, like *smodels* [4] and *dlv* [5]. The community started modeling with ASP and a first milestone was the conception of TheoryBase [6] providing a systematic and scalable source of benchmarks stemming from combinatorial problems. Although the scalability of such benchmarks is of great value for empirically evaluating systems, the need for application-oriented benchmarks was early perceived. The demand for systematic benchmarking led to the Dagstuhl initiative and with it the creation of the web-based benchmark archive *asparagus* [7]. This repository has in the meantime grown significantly, mainly due to the two past ASP competitions [8,9], and contains nowadays a whole variety of different types of benchmarks, although it is still far from being comprehensive.

Meanwhile, the prospect of ASP has been demonstrated in numerous application scenarios.[1] A highlight among them is arguably the usage of ASP for the high-level control of the space shuttle [10]. What makes this application so special is the fact that it was solving an application problem in a real-world environment. Although we still need many more elaborated proofs of concept, showing how ASP addresses different application scenarios, solving such real(-world) problems is yet another issue. Let me approach this by answering some preliminary questions.

What is a real problem? Such a problem could be an unsolved combinatorial or mathematical problem. Also, it could stem from an application that is traditionally solved with different methods. In either case, the problem is not academic anymore, but rather about producing an effective solution. To accomplish this, we have to switch to a *production mode*,[2] that is, the process of organizing the production of a solution to truly challenging problems. This mode of operation goes (currently) quite beyond conceptual modeling and benchmarking that most of us are used to so far.

Why should we solve real problems? Apart from the fact that the prospect of doing so partly nourishes our right of existence, real problems are a tremendously fruitful source of new research questions. For instance, concepts like cardinality and weight constraints [11], magic set transformations [12], constraint additions to ASP [13,14],

---

[*] Affiliated with Simon Fraser University, Canada, and Griffith University, Australia.

[1] See http://www.kr.tuwien.ac.at/research/projects/WASP and/or
http://www.cs.uni-potsdam.de/~torsten/asp for an overview.

[2] This term goes back to Marxist theory and has been adopted in informatics in various ways.

E. Erdem, F. Lin, and T. Schaub (Eds.): LPNMR 2009, LNCS 5753, pp. 631–633, 2009.

or projection [15], had never been pushed so hard without a real problem driving their development. In other words, making our hands dirty inspires our brain!

Where do we find real problems? One source are open problems known to academia. For instance, deciding the fifth Schur number.[3] In fact, the benchmarking repository of the Automated Theorem Proving community used to contain unsolved challenge problems that were a driving force for the development of theorem provers. And after all, they made it into the New York Times by proving Robbins' conjecture.[4] Actually, interesting challenge problems are often simply down the hall, sitting on your colleagues' desks. Getting interested in their problems, makes them discover ASP and exhibits us to non-artificial problems. The least expected result is a proof of concept or a benchmark suite, the greatest is to make a difference with ASP!

How do we solve real problems? To begin with, one still starts with a proof of concept addressing a toy version of the actual problem. If this scales and the solution is satisfactory, one just hit the jackpot. Unfortunately, this never happens (to me) and brings us back to the aforementioned production mode of ASP. The dilemma is that we must address real problems in order to further develop ASP as a tool for addressing real applications. This vicious cycle makes current production processes far from ideal and dominated by pragmatics, and often not even addressable by means of ASP only.

The first bottleneck in the ASP production mode is the encoding. This has become a true art and often the initial, rather declarative problem specification bears little resemblance with the final stream-lined encoding reducing combinatorics. This also applies to automatically generated encodings. This is not to say that the final encoding is not indicative but it needs quite some experience to be produced. Moreover, the optimization of encodings is also tightly connected to the target ASP system, and in particular, its grounding component. It makes quite a difference whether the grounder relies on domain predicates or not, and whether it provides special-purpose methods, like constraint handling techniques or unification for avoiding grounding large domains. Clearly, this track leaves the idea of declarative problem solving behind and the burden of optimizing encodings has to be partly taken off the user and handled (semi-)automatically in the long run.

The second bottleneck is the configuration of the actual ASP solver. Modern ASP solvers relying on Boolean constraint technology offer a manifold arsenal of parameters for controlling the search for answer sets. For instance, *clasp* has roughly forty options [16], half of which control the search strategy. Choosing the right parameters often makes the difference between being able to solve a problem or not. But again this takes us away from the idea of declarative problem solving and automatic methods must be conceived for partially relieving this second burden.[5]

As a matter of fact, the two aforementioned bottlenecks are not regarded as problematic in the SAT community. Rather encodings are often presented to a solver at its convenience and industrial problems are solved with particular parameter settings (eg. aggressive restart strategies). This marks a true difference in the philosophy of both communities.

---

[3] Thanks to Mirosław Truszczyński for pointing this out.

[4] http://www.nytimes.com/library/cyber/week/1210math.html

[5] For instance, a first such prototype is *claspfolio* using machine learning techniques for mapping problem features to solver parameters.

Finally, it is surprising that the lack of software engineering tools is yet not even an issue in ASP's production mode. The reason is simply that the production mode in ASP is up to now accomplished by experts in ASP and no end users. This lack will become a true bottleneck once ASP would principally be ready for real applications.

So, let's make our hands dirty and get inspired!

References

1. Niemelä, I.: Logic programs with stable model semantics as a constraint programming paradigm. Annals of Mathematics and Artificial Intelligence 25(3-4), 241–273 (1999)
2. Marek, V., Truszczyński, M.: Stable models and an alternative logic programming paradigm. In: Apt, K., Marek, W., Truszczyński, M., Warren, D. (eds.) The Logic Programming Paradigm: a 25-Year Perspective, pp. 375–398. Springer, Heidelberg (1999)
3. Lifschitz, V.: Answer set planning. In: de Schreye, D. (ed.) Proceedings of ICLP 1999, pp. 23–37. MIT Press, Cambridge (1999)
4. Niemelä, I., Simons, P.: Evaluating an algorithm for default reasoning. In: Working Notes of the IJCAI 1995 Workshop on Applications and Implementations of Nonmonotonic Reasoning Systems, pp. 66–72 (1995)
5. Eiter, T., Leone, N., Mateis, C., Pfeifer, G., Scarcello, F.: A deductive system for nonmonotonic reasoning. In: Dix, J., Furbach, U., Nerode, A. (eds.) LPNMR 1997. LNCS (LNAI), vol. 1265, pp. 363–374. Springer, Heidelberg (1997)
6. Cholewiński, P., Marek, V., Mikitiuk, A., Truszczyński, M.: Experimenting with nonmonotonic reasoning. In: Sterling, L. (ed.) Proceedings of ICLP 1995, pp. 267–281. MIT Press, Cambridge (1995)
7. Borchert, P., Anger, C., Schaub, T., Truszczyński, M.: Towards systematic benchmarking in answer set programming: The Dagstuhl initiative. In: Lifschitz, V., Niemelä, I. (eds.) LPNMR 2004. LNCS (LNAI), vol. 2923, pp. 3–7. Springer, Heidelberg (2003)
8. Gebser, M., Liu, L., Namasivayam, G., Neumann, A., Schaub, T., Truszczyński, M.: The first answer set programming system competition. In: Baral, C., Brewka, G., Schlipf, J. (eds.) LPNMR 2007. LNCS (LNAI), vol. 4483, pp. 3–17. Springer, Heidelberg (2007)
9. Denecker, M., Vennekens, J., Bond, S., Gebser, M., Truszczyński, M.: The second answer set programming competition. In: [17] (to appear)
10. Nogueira, M., Balduccini, M., Gelfond, M., Watson, R., Barry, M.: An A-prolog decision support system for the space shuttle. In: Ramakrishnan, I.V. (ed.) PADL 2001. LNCS, vol. 1990, pp. 169–183. Springer, Heidelberg (2001)
11. Simons, P., Niemelä, I., Soininen, T.: Extending and implementing the stable model semantics. Artificial Intelligence 138(1-2), 181–234 (2002)
12. Faber, W., Greco, G., Leone, N.: Magic sets and their application to data integration. Journal of Computer and System Sciences 73(4), 584–609 (2007)
13. Mellarkod, V., Gelfond, M.: Integrating answer set reasoning with constraint solving techniques. In: Garrigue, J., Hermenegildo, M.V. (eds.) FLOPS 2008. LNCS, vol. 4989, pp. 15–31. Springer, Heidelberg (2008)
14. Gebser, M., Ostrowski, M., Schaub, T.: Constraint answer set solving. In: Hill, P., Warren, D. (eds.) Proceedings of ICLP 2009, pp. 235–249. Springer, Heidelberg (2009)
15. Gebser, M., Kaufmann, B., Schaub, T.: Solution enumeration for projected Boolean search problems. In: van Hoeve, W., Hooker, J. (eds.) Proceedings of CPAIOR 2009, pp. 71–86. Springer, Heidelberg (2009)
16. Gebser, M., Kaufmann, B., Schaub, T.: The conflict-driven answer set solver clasp: Progress report. In: [17] (to appear)
17. Erdem, E., Lin, F., Schaub, T. (eds.): LPNMR 2009. Springer, Heidelberg (2009)

Towards an Embedded Approach to Declarative Problem Solving in ASP

Jia-Huai You

Department of Computing Science
University of Alberta
Edmonton, Alberta, Canada

Abstract. The strength of answer set programming (ASP) lies in solving computationally challenging problems declaratively, and hopefully efficiently. A similar goal is shared by two other approaches, SAT and Constraint Programming (CP). As future applications of ASP hinge on its underlying solving techniques, in this note, I will briefly comment on the related techniques, and argue for the need of ASP systems to integrate with state-of-the-art techniques for constraint solving, and in general to serve as the core reasoning engine to glue other logics and reasoning mechanisms together.

1 Constraint Solving Paradigms

Started in the field of Operations Research, an optimization/search problem can be specified in forms of mathematical expressions, and an implemented system is responsible for efficient computation of solutions.

A similar goal is shared by SAT and CP. In this note, by CP, I mean the systems that implement the Constraint Satisfaction Problem (CSP). This includes various CSP packages and CP languages developed by the CSP/CP community, and Constraint Logic Programming over finite domains (CLP(FD), or just CLP).

SAT in a simple language, and SAT solvers are generally regarded as black-boxes. A standard language plus the state-of-the-art DPLL architecture allows the SAT research to focus on a small number of issues, and (it is often said) a competitive SAT solver can be implemented under a thousand lines of code. Revolutionized by Chaff around 2000, there are now standard techniques for complete SAT solvers, including activity-based heuristics, the (right) conflict analysis mechanism, two-watched literals, and restarts.

As the underlying solving techniques have a direct impact on potential applications, let's recall the parallels between SAT and ASP. ASP may be said as a *variant* of SAT, though a very distinct one. While SAT deals with clauses, ASP expresses Clark's predicate completion satisfying loop formulas [6]. This parallel is so striking that the loop formulas are rarely needed in practical applications. In fact, a huge majority of normal programs written for benchmarks and applications are just *2-literal programs* [4], where a (ground) rule consists of at most two literals. In this case, completion can be translated to clauses rather directly. It is therefore not a surprise that the key techniques in more recent improvements to ASP solvers are largely adopted from those of SAT and related techniques. This trend is expected to continue.

E. Erdem, F. Lin, and T. Schaub (Eds.): LPNMR 2009, LNCS 5753, pp. 634–636, 2009.

The expectations of ASP are somewhat different from those of SAT. The goal of declarative knowledge representation and problem solving calls for a wider scope of investigation, from grounding to incorporation of disjunction, functions, aggregates, abstract constraints, probabilities, etc. While SAT mainly focuses on a uniform but low level language, ASP attempts to address declarative knowledge representation from all angles. As a result, ASP languages are more expressive and easier to use.

As a variant of SAT, ASP is also weak on the handling of numeric constraints and some of the computational tasks that seem to have a natural mapping to constraints in the sense of CP, e.g., for scheduling tasks. Let me cite two examples. One is the *Channel Routing Problem* proposed by Neng-Fa Zhou for the 2nd ASP Competition; for a typical weight program encoding, ASP solvers are competitive for unsatisfiable instances, but for satisfiable ones the best ASP solvers are orders of magnitude slower than a B-Prolog solution. Although grounding in this case is a significant factor, the statement still holds even if the grounding time is discounted.[1] Another representative example is the so-called *Newspapers Problem*, which schedules a number of readers to finish reading a number of papers at an earliest possible time, where the key constraints are that no reader can read more than one paper at a time and no paper can be read by more than one reader at a time. For a weight program encoding, grounding can be done fairly efficiently. But for the computation of answer sets, ASP solvers can be orders of magnitude slower than a typical CLP implementation (e.g., Sicstus Prolog[2]).

In comparison, CP possesses powerful modeling features, typically in the form of *global constraints* and a rich set of constraint handling techniques. However, besides the tuning issues associated with various domain reduction techniques, a problem for CP is the lack of a standard language, or the lack of consensus of what such a language should be. A constraint is a relation, but how such a relation is defined in a language may differ in different implementations. Although some uniformity is agreed upon for CLP(FD), there are some difficulties for how such (possibly recursively) defined relations may be decomposed into a *constraint store*. In my experience, this had made it harder to program in CLP than in ASP, at least for ordinary users.[3] Furthermore, some problems do not seem to have a natural encoding in CP. One example is the planning problem, where the current practice (cf. [2]) is to specify a skeleton of a plan, which is a list of states each of which consists of fluents that either hold or do not hold in the state. One may argue that reasoning with constraints over $\{0, 1\}$-domains is not the strength of a CP's search engine.

For a recent comparative study on SAT vs. CP, see [1].

2   An Embedded Approach

To summarize the discussion so far, we conclude that (1) For declarative problem solving, ASP stands out as a very promising approach, but the underlying solver techniques may continue to draw lessons from SAT and related techniques; and (2) ASP can benefit from integrating with other techniques especially from constraint solving.

---

[1] The findings reported here were discovered experimentally by Rei Thiessen.

[2] The experiments were due to Guohua Liu.

[3] The author has taught CLP in a senior undergraduate course and ASP in a graduate course.

Thanks to the formulation of logic programs with abstract constraints [7] and the recent extension to handle arbitrary abstract constraints [10], one framework for such an integration may already be laid out. Essentially, constraints of all sorts, whenever they can be specified by a domain and a set of admissible solutions, can be embedded into a logic program. Such a constraint may be pre-defined/built-in, or a global constraint that comes with a special dedicated propagator or an efficient solver.

The problem of integrating user-defined constraints into ASP is trickier. One approach is to design a new ASP language, where interface predicates can be explicitly defined by the user, or identified by the system [8]. In the approach of [5], the user writes a logic program with functions which is translated to an instance of CSP so that the answer sets can be computed by a CSP solver. One immediate benefit is that CSP facilities, such as global constraints, become accessible to ASP. It remains to be seen whether the approach can be lifted to a model of integration.

In theory, the embedded approach is powerful enough to embed anything that can be expressed as constraints. For example, it is recently discovered that the approach of [3] to embed description logics into ASP can be seen as a case of logic programs with constraint atoms under the default approach of [9]. The idea of the embedded approach is closely related to that of *CLP Scheme* and the more recent development on *Satisfiability Modulo Theories*. I fully anticipate that future ASP systems will be capable of embedding various logics and reasoning facilities, thus serving as the core reasoning engine to glue various reasoning mechanisms together.

References

1. Bordeaux, L., Hamadi, Y., Zhang, L.: Propositional satisfiability and constraint programming: a comparative survey. ACM Computing Surveys 38(4), Article No 12 (2006)
2. Dovier, A., Formisano, A., Pontelli, E.: An experimental comparison of constraint logic programming and answer set programming. In: Proc. AAAI 2007, pp. 1622–1625 (2007)
3. Eiter, T., Ianni, G., Lukasiewicz, T., Schindlauer, R., Tompits, H.: Combining answer set programming with description logics for the semantic web. Artif. Intell. 172(12-13), 1495–1539 (2008)
4. Huang, G., Jia, X., Liau, C., You, J.: Two-literal logic programs and satisfiability representation of stable models: a comparison. In: Cohen, R., Spencer, B. (eds.) Canadian AI 2002. LNCS (LNAI), vol. 2338, pp. 119–131. Springer, Heidelberg (2002)
5. Lin, F., Wang, Y.: Answer set programming with functions. In: Proc. KR 2008, pp. 454–464 (2008)
6. Lin, F., Zhao, Y.: ASSAT: Computing answer sets of a logic program by SAT solvers. Artificial Intelligence 157(1-2), 115–137 (2004)
7. Marek, V., Truszczyński, M.: Logic programs with abstract constraint atoms. In: Proc. AAAI 2004, pp. 86–91 (2004)
8. Mellarkod, V.S., Gelfond, M., Zhang, Y.: Integrating answer set programming and constraint logic programming. Annals of Mathematics and Artificial Intelligence 53(1-4), 251–287 (2008)
9. Shen, Y., You, J.: A default approach to semantics of logic programs with constraint atoms. In: These proceedings (2009)
10. Son, T.C., Pontelli, E., Tu, P.H.: Answer sets for logic programs with arbitrary abstract constraint atoms. Journal of Artificial Intelligence Research 29, 353–389 (2007)

The Second Answer Set Programming Competition

Marc Denecker[1], Joost Vennekens[1], Stephen Bond[1], Martin Gebser[2],
and Mirosław Truszczyński[3]

[1] Department of Computer Science, Katholieke Universiteit Leuven Celestijnenlaan 200A,
B-3001-Heverlee
{marcd,joost,stephen}@cs.kuleuven.be
[2] Institut für Informatik, Universität Potsdam, August-Bebel-Str. 89,
D-14482 Potsdam, Germany
gebser@cs.uni-potsdam.de
[3] Department of Computer Science, University of Kentucky, Lexington, KY 40506-0046, USA
mirek@cs.uky.edu

Abstract. This paper reports on the *Second Answer Set Programming Competition*. The competitions in areas of Satisfiability checking, Pseudo-Boolean constraint solving and Quantified Boolean Formula evaluation have proven to be a strong driving force for a community to develop better performing systems. Following this experience, the Answer Set Programming competition series was set up in 2007, and ran as part of the International Conference on Logic Programming and Nonmonotonic Reasoning (LPNMR). This second competition, held in conjunction with LPNMR 2009, differed from the first one in two important ways. First, while the original competition was restricted to systems designed for the *answer set programming language*, the sequel was open to systems designed for other modeling languages, as well. Consequently, among the contestants of the second competition were a CLP(FD) team and three model generation systems for (extensions of) classical logic. Second, this latest competition covered not only satisfiability problems but also optimization ones. We present and discuss the set-up and the results of the competition.

1   Introduction

In many real-life problems, we search for objects of complex nature — plans, schedules, assignments. Several research areas within computer science, operations research and mathematics are concerned with the development of systems that compute such objects from their specifications. Researchers in these areas design and study languages to describe objects of interest, as well as algorithms to extract them from these descriptions. Depending on the language and the area, such objects are called "answer sets", "valuations", "structures", "interpretations", "functions" or "arrays". Answer Set Programming (ASP), propositional Satisfiability (SAT) and Constraint Programming (CP) are arguably the three most prominent areas developing such languages and techniques.

In the context of logic, it is often the case that once constraints on objects to search for are given as formulas (rules), models of the resulting theory determine answers to the search problem — objects satisfying the constraints. For instance, each model of a theory specifying a scheduling domain typically defines a correct schedule. Thus,

E. Erdem, F. Lin, and T. Schaub (Eds.): LPNMR 2009, LNCS 5753, pp. 637–654, 2009.

a model generator applied to such a theory will solve the corresponding scheduling problem. This idea of model generation as a declarative problem solving paradigm has been pioneered in the area of ASP [1,2,3].

ASP has three fundamental characteristics: a *modeling language* based on the syntax of logic programs, the use of the *answer set semantics* [4] to interpret programs in that language, and a *problem-solving methodology* in which a program is written so that its answer sets provide solutions. ASP has its origins in Logic Programming (LP) [5,6], in particular in the attempts in the 1980s to develop a declarative semantics for logic programs with negation and to turn this logic into a formalism suitable for knowledge representation. Gelfond and Lifschitz sought inspiration in nonmonotonic reasoning [7,8] and proposed to interpret logic programs as special default theories under the semantics of Reiter [8]. Based on this view, they developed the *stable model semantics* for logic programs [9], and extended it later to the *answer set semantics* for disjunctive logic programs with classical negation [4], which is the *core* ASP language today.

The area made a major leap in 1997, when the first two systems to compute answer sets of logic programs were developed: *dlv* [10] and *smodels* [11]. These systems demonstrated that effective tools for processing answer set programs are possible. Following that, in 1999, Marek and Truszczyński [1] and Niemelä [2] proposed answer set computation as a new declarative problem solving paradigm, and Lifschitz dubbed the area answer set programming [12]. It turned out that a rich class of problems could be modeled elegantly as answer set programs according to this paradigm. That became a strong driving force for the development of fast computational techniques in ASP, and for studies of practical applications where the ASP tools could be used effectively.

Experience in areas concerned with checking propositional or Pseudo-Boolean (PB) satisfiability and evaluating Quantified Boolean Formulas (QBF) shows that a programming competition gives an effective incentive to the research community to work on developing better performing systems. ASP sought to emulate that experience. The first preliminary competitions for ASP systems were held in 2002 and 2005 at two Dagstuhl meetings [13]. In 2007, the *First Answer Set Programming System Competition* [14] was organized as part of LPNMR. That competition consisted of four tracks. In three tracks, solvers were tested on prespecified answer set programs. In the SCore-v and SCore tracks, input consisted of a ground logic program, respectively with and without disjunction; in the SLparse track, the programs used *lparse*'s output language (including aggregates). In the fourth track, called *Model, Ground, Solve* (MGS), contestants encoded problems in a language of their choice. Ten teams competed. The *clasp* solver won the SCore and SLparse tracks, and the competition version of *dlv* won in SCore-v and MGS.

In the past two years, some important developments have occurred in ASP. First, as it is clear from the results of the second ASP competition (see Section 5), existing systems have improved considerably, both in available language features and in the speed of the solvers. In addition, new systems have been built, and more teams competed. Clearly, the first ASP competition has had its desired effect! Second, the ASP community has been gradually opening up to other domains. The fields of SAT [15], SAT Modulo Theories (SMT) [16], CP [17] and, in some way, also Abductive Logic

Programming (ALP) [18] are in one key respect very close to ASP. Namely, the ASP declarative problem solving paradigm does not depend on the answer set programming language but applies to and, in fact, has been used with other declarative languages too. For example, Mitchell and Ternovska [19] proposed to use model expansion (a form of model generation) for (extensions of) first-order logic as a declarative problem solving paradigm for NP search problems. (Ground) abduction in ALP is similar to model generation [18], and integrations of ALP and CLP have been used for planning, scheduling and constraint solving problems [20,21]. Moreover, workshops organized by the ASP community such as Answer Set Programming and Other Computing Paradigms (AS-POCP), held in conjunction with the International Conferences on Logic Programming in Udine in 2008 and Pasadena in 2009, and Logic and Search (LaSh), held in Seattle in 2006 and Leuven in 2008, explicitly aimed to bring together researchers from all fields that share the problem solving methodology based on model generation.

The *Second Answer Set Programming Competition*, organized in conjunction with LPNMR 2009, further fortified this trend. Having a competition that would be open not only to the ASP community but also to the communities of SAT, LP, CP, etc. was an important objective of the program committee, and a key condition for accepting the charge of organizing it put forth by the members of the Knowledge Representation and Reasoning (KRR) group at the K.U. Leuven. The underlying idea was that only such an open competition can lead to progress and result in new insights into the strengths of different technologies in the context of diverse applications. An important ramification was that the competition had to be restricted to the *Model and Solve* track, as only that mode allows teams using tools based on different languages and logics to compete. On the other hand, the scope of the competition was expanded to include not only decision problems but also optimization ones.

The first step in organizing the competition was to collect *benchmarks* — specific problems together with sets of instances. Many researchers contributed, and we gratefully acknowledge their efforts. In total, we received 38 benchmarks, nine of which were optimization problems. Most benchmarks came from ASP researchers, some came from the Constraint Logic Programming (CLP) [22] community. We split these benchmarks into four categories: the *P decision class* consisting of polynomially solvable decision problems, the *NP decision class* consisting of decision problems in NP not known to be polynomially solvable, the *global decision class* consisting of all problems in the first two groups and of one Σ_2^p-complete problem and, finally, the *optimization class*.

The competition was a success. Sixteen teams competed, with nine of them for the first time! Twelve teams used the ASP language and tools. The call for participation to other communities had limited success. Three teams, Enfragmo, IDP and amsolver, used model generators/expanders for first-order logic extended with aggregates, arithmetic and, in the case of IDP and amsolver, also with inductive definitions. However, the amsolver team competed only on eight benchmarks. One team, BPSolver-CLP(FD), led by Neng-Fa Zhou, used the CLP(FD) solver B-Prolog. There were no SAT, PB nor SMT participants, but SAT, PB or SMT solvers were used in many systems.

It is important to note that the second ASP competition had a policy of openness about benchmark solutions (programs encoding problem specifications). Several teams made their codes publicly available, allowing other teams to profit from their efforts.

For example, at least eight teams used the benchmark solutions by the Potassco team available at [23].

The paper is organized as follows. In the next section, we give an outline of the competition and specify its format. In Section 3, we discuss the collection of benchmarks, the ranking system and the competition platform. In Section 4, we introduce the teams that competed. We present the results in Section 5. Finally, we close by summarizing our experience and outlining potential future improvements. For more detailed information on the results of this competition, we refer to [24].

2   Outline of the Competition

The second ASP competition had one track, referred to as *Model and Solve*, and included both decision and optimization problems.

The first step of the competition was a call for submitting benchmark problems. This call was sent to research communities of ASP, LP, SAT and CP. Benchmark authors were invited to provide an (informal) problem description, input and output predicates, a set of instances and a *checker* program to verify the correctness of solutions (see Section 3 for details).

Contributed benchmarks were then evaluated by the organizers and the program committee (from now on referred to simply as organizers). The overall benchmark pool was a mixture of many diverse decision problems and several optimization problems. There were fundamental differences between decision benchmarks, and some seemed to favor particular types of systems quite strongly. To understand better which features were useful for particular types of problems, the organizers decided to split decision benchmarks into the following sub-categories:

Decision-in-P class: Problems that can be solved in polynomial time. While they are simple, the challenge is the sheer size of the instances requiring highly optimized grounding techniques.

Decision-in-NP class: Problems that are in NP but are not known to be in P. They are the problems that require highly effective search algorithms.

Decision-global class: Problems from the previous two groups and one Σ_2^p-complete problem, the well-known Strategic Companies problem. The goal of this category is to evaluate solving systems with respect to both their grounding and search effectiveness on a broad range of problems of varying complexity.

In the competition, each team was to solve each benchmark problem for several instances. To this end, each team had to submit a script for each benchmark on which the team wanted to compete. This script was to call the solver and apply it to the benchmark encoding and a problem instance given through standard input. During the competition, this script was called repeatedly for each instance of the benchmark.

For both decision and optimization problems, a problem instance was represented as a sequence of atomic clauses (an atom followed by a period) over predicates from the input vocabulary. The output depended on the type of problem:

Decision problems: The script should write the following to standard output:
– **UNSATISFIABLE**, if the instance has no solution.

- A sequence of atomic clauses in the output vocabulary, if the instance is satisfiable. Such a sequence should represent a "witness" to the satisfiability of the instance, that is, it should be the set of atoms over the output predicates in an answer set or model (for first-order logic formalisms) determining a solution for the instance. The format is the same as with the input except that it must not contain line breaks.
- **UNKNOWN**, if the solver decides to give up before timeout.

Optimization problems: The script should output the following:
- **UNSATISFIABLE**, in case of an unsatisfiable instance.
- A series of witnesses of the search problem, if the instance is satisfiable. The format of a witness is the same as for decision problems. Witnesses are separated by line breaks. The last (and hopefully best) witness is considered as the generated solution.
- **OPTIMUM FOUND**, if the instance is satisfiable and the solver can ascertain the optimality of the last produced witness.

Remark. The use of separate scripts for different benchmarks gives teams the freedom to fine-tune parameter settings to specific benchmarks, or even to use different solvers. During the installation phase of the competition, this raised some controversy among the participants. Our point of view is that for an open competition, there is no option but to offer this freedom. For example, a SAT team should have the freedom to develop for each benchmark a program to compute the CNF theory corresponding to an instance. However, only two teams, Potassco and BPSolver-CLP(FD) made extensive use of the facility to tune systems towards benchmarks. Potassco used various grounders and solvers, with different parameter settings. The BPSolver-CLP(FD) team used B-Prolog's control structures to implement various labeling strategies. Other teams used the same combination of systems and parameters for all decision problems and for all optimization problems. This distinction is relevant for the interpretation of the competition results and will be taken into account in Section 5.

3 Benchmarks

The collected benchmarks constitute the result of efforts by many researchers, mostly from the ASP community; some were provided by the BPSolver-CLP(FD) team. The author(s) of a benchmark problem provided us with the following information:

- a clear non-ambiguous (informal) problem description,
- a specification of the input and output vocabulary,
- for optimization problems, an integer-valued cost function to be minimized,
- a set of instances,
- a checker program to test the correctness of witnesses and, for optimization problems, to additionally compute the cost.

Verifying unsatisfiability of an NP-hard decision problem or optimality of a witness of an NP-hard optimization problem is intractable in general (assuming the polynomial

Table 1. Benchmarks of the Second Answer Set Programming Competition

Benchmark	Class	Contributors	#Instances
HydraulicPlanning	Decision,P	M. Gelfond, R. Morales and Y. Zhang	15
HydraulicLeaking	Decision,P	M. Gelfond, R. Morales and Y. Zhang	15
CompanyControls	Decision,P	Mario Alviano	15
GrammarBasedInformationExtraction	Decision,P	Marco Manna	29
Reachability	Decision,P	Giorgio Terracina	15
BlockedNQueens	Decision,NP	G. Namasivayam and M. Truszczyński	29
Sokoban	Decision,NP	Wolfgang Faber	29
15Puzzle	Decision,NP	L. Liu, M. Truszczyński and M. Gebser	16
HamiltonianPath	Decision,NP	L. Liu, M. Truszczyński and M. Gebser	29
SchurNumbers	Decision,NP	L. Liu, M. Truszczyński and M. Gebser	29
TravellingSalesperson	Decision,NP	L. Liu, M. Truszczyński and M. Gebser	29
WeightBoundedDominatingSet	Decision,NP	L. Liu, M. Truszczyński and M. Gebser	29
Labyrinth	Decision,NP	Martin Gebser	29
GeneralizedSlitherlink	Decision,NP	Wolfgang Faber	29
HierarchicalClustering	Decision,NP	G. Namasivayam and M. Truszczyński	12
ConnectedDominatingSet	Decision,NP	G. Namasivayam and M. Truszczyński	21
GraphPartitioning	Decision,NP	G. Namasivayam and M. Truszczyński	13
Hanoi	Decision,NP	G. Namasivayam, M. Truszczyński and G. Terracina	15
Fastfood	Decision,NP	Wolfgang Faber	29
WireRouting	Decision,NP	G. Namasivayam and M. Truszczyński	23
Sudoku	Decision,NP	Neng-Fa Zhou	10
DisjunctiveScheduling	Decision,NP	Neng-Fa Zhou	10
KnightTour	Decision,NP	Neng-Fa Zhou	10
ChannelRouting	Decision,NP	Neng-Fa Zhou	11
EdgeMatching	Decision,NP	Martin Brain	29
GraphColouring	Decision,NP	Martin Brain	29
MazeGeneration	Decision,NP	Martin Brain	29
Solitaire	Decision,NP	Martin Brain	27
StrategicCompanies	Decision,Σ_2^P	M. Alviano, M. Maratea and F. Ricca	17
GolombRuler	Optimization	Martin Brain	24
MaximalClique	Optimization	Johan Wittocx	29
15PuzzleOptimize	Optimization	L. Liu, M. Truszczyński and M. Gebser	16
TravellingSalespersonOptimize	Optimization	L. Liu, M. Truszczyński and M. Gebser	29
WeightBoundedDominatingSetOptimize	Optimization	L. Liu, M. Truszczyński and M. Gebser	29
LabyrinthOptimize	Optimization	Martin Gebser	28
SokobanOptimize	Optimization	Wolfgang Faber	29
FastfoodOptimize	Optimization	Wolfgang Faber	29
CompanyControlsOptimize	Optimization	Mario Alviano	15

hierarchy does not collapse). For this reason, checkers only had to test the correctness of witnesses, and not of the answers **UNSATISFIABLE** or **OPTIMUM FOUND**.

Table 1 gives an overview of all benchmarks of the competition. For each problem, the entry specifies the author(s), the category and the number of instances used in the competition.

3.1   Detecting Errors

A team was disqualified for a benchmark as soon as an error was detected for one of the benchmark instances. As we wrote before, checker programs did not test correctness of **UNSATISFIABLE** and **OPTIMUM FOUND**. To check the correctness of those answers, we used the following incomplete strategy. An erroneous **UNSATISFIABLE** answer for an instance is reported if another solver finds a correct witness. An erroneous **OPTIMUM FOUND** answer for an instance is reported if another solver finds a strictly better solution. This method can only detect an erroneous **UNSATISFIABLE** answer if at least one solver finds a correct witness. A similar comment holds for erroneous **OPTIMUM FOUND** answers.

3.2   Ranking System

Ranking on decision problems. The ranking system for the three decision problem tracks is very similar to the one used in the first ASP competition. The primary criterion is the number of instances solved. For each problem instance of each benchmark, the teams are given a time bound of 600 seconds to solve the instance. Each participant is assigned points according to the number of instances that can be solved within this time bound. For each benchmark, a team is awarded points as follows:

- No points, if the solver makes an error on one of the instances of the benchmark.
- Otherwise, $20(S/N)$ points, where S is the number of instances solved by the team and N is the number of instances of this benchmark that were solved by at least one team.

This boils down to the weighted sum over solved instances, the weight of a solved instance from a particular benchmark being $1/N$, where N is the number of instances of this benchmark that were solved by at least one team. By defining weights in this way (normalizing with respect to the total number of instances solved within a benchmark), the method prevents benchmarks with (very) many instances from dominating the score.

Ties are resolved by comparing actual running time where, for **UNKNOWN** answers, the timeout time is taken as the running time. During the competition, system failures frequently occurred, most often due to systems running out of memory. When a system failure occurred on an instance, it was treated as an **UNKNOWN** answer and assigned the timeout time as the running time.

Ranking on optimization problems. The ranking of teams on optimization problems depends on the quality of the solution they find. We recall that each problem has a cost function mapping a solution to an integer. Solutions with minimal cost are desirable for each instance. The checker programs of the optimization problems not only check the correctness of a witness for an instance, but also compute its cost. Teams are awarded points proportionally to the distance between the cost of the returned solution and the lowest cost of any solution found by any team. Ties are resolved according to the running time. Extra points are given to systems that correctly (or not incorrectly) return the keyword **OPTIMUM FOUND**.

A team could earn 100 points per instance of a benchmark. These points were distributed as follows:

- No points, if the solver makes an error on one of the instances of the benchmark.
- 100 points, if the solver correctly outputs **UNSATISFIABLE**.
- If the solver produces a correct witness:
  - it is awarded an initial 25 points;
  - in addition, it can receive up to 50 points for the quality of the solution. Let M be the lowest cost of a solution that was produced by any solver for this instance. If the cost of the solution produced by the team is Q, then the team is awarded $50(M/Q)$ points;

- in addition, 25 points are awarded if the solver correctly outputs **OPTIMUM FOUND**.

The ranking score for this category is a weighted sum of the earned points. The weight for an instance in a benchmark is $1/N$, where N is the number of solved instances from that benchmark (by any team). This is done to prevent a disproportionate effect benchmarks with many instances might otherwise have on the scores.

Global rankings. We also used two global rankings: one for all decision problems, and another one for all decision and optimization problems. The weighting of individual decision benchmarks within the global decision category is calculated in the same fashion as described above. The global ranking awards each team a score that is the average of its score for the global decision raking and its score for the optimization ranking (divided by 100). This gives equal importance to both categories of the competition.

3.3   Competition Software and Platform

The competition was run on a pool of computers of the DTAI research group of the K.U. Leuven. The system consisted of one network server through which participants could login via ssh, one database server for storing state and results, and a pool of five identical Linux machines reserved for the competition and only accessible through the network server. One of these machines was reserved for participants to install and test solutions, while the other four served for the actual testing of solvers on benchmark instances. The system was installed and maintained by the system group of the Computer Science Department of the K.U. Leuven, in particular Bart Swennen and Kris Vangeneugden.

The competition software was derived from software that had been developed for *De Vlaamse Programmeerwedstrijd*[1] (the Flemish Programming competition) by Pieter Wuille (PhD student of DTAI). Pieter helped us greatly by adapting, maintaining and running the benchmarking software for this competition.

The operation mode of the competition system can be sketched as follows. When a team submits a benchmark solution to the competition server, the latter registers it in its database. The competition software maintains identical copies of all submitted solvers and benchmarks on the four competition machines. The tests of the instances are controlled by the database server and four client processes that run on the four test machines. Each client process requests a task from the database server. The task consists of executing a submitted benchmark script on an instance. The database server distributes the tasks and maintains their status and the obtained results. When a client process is assigned a task, it executes it with limited time and memory. When the benchmark script returns a witness, the checker script is called on it. For optimization problems, the checker program is called on the last generated witness. The client process collects all necessary data (time, correctness of witness, system error, timeout), sends the data to the database server and requests a new task.

The five competition machines had identical hardware and software. The installed Linux version was Kubuntu Hardy (8.04). The details of the hardware are: *Dell OptiPlex 745 (1 CPU with 2 cores: GenuineIntel Intel(R) Core(TM)2 CPU 6600 2.40GHz), 4096 MB*

---

[1] http://www.vlaamseprogrammeerwedstrijd.be/?page=main

RAM (4x1024 MB 667 MHz 1.5 ns), Disk capacity 160 GB (model ATA ST3160815AS 3.AD). Although these machines have two cores, the choice was made to use only one core per task and per computer. Thus, effective parallelism was impossible. Benchmark solutions were executed with a timeout of 600 seconds and a memory limit of 2.79 GB RAM.

4 Competitors

Sixteen teams registered with participants from more than fifteen universities. Each team was assigned a user account on a competition machine. Solvers and benchmark solutions were installed from 1/4/2009 till 15/5/2009 (and later). Most participants did not submit solutions for all benchmarks, often because of limitations of their systems. For instance, only four teams proposed a solution for the Strategic Companies problem (a Σ_2^p-complete problem). Some groups submitted multiple systems: Potsdam joined with two teams each using multiple systems, Helsinki (TKK) with three systems, and a team uniting the forces of researchers at the universities of Kentucky and of Texas at Tyler and at Microsoft also submitted three different systems.

The first ASP systems, *dlv* and (a direct descendant of) *smodels*, were still in the competition and scored very well. The Smodels-IE solver is an updated version of *smodels* developed at the University of Bath. A variety of languages and of solver techniques were present in the competition. As for the languages, twelve teams used different dialects of ASP, three used extensions of first-order logic, and one team used B-Prolog with CLP(FD) and a planning preprocessor.

Five teams (Potassco, DLV, Claspfolio, Smodels-IE, ASPeRiX) participated with "native" ASP solvers, the one of ASPeRiX performing grounding on the fly, without a separate grounder. Other teams used a variety of back-ends: existing or modified SAT solvers (IDP, CMODELS, SUP, Enfragmo, LP2SAT+MINISAT, sabe), SMT solvers (LP2DIFF+BCLT, LP2DIFF+YICES), a PB solver (pbmodels) and a new solver for propositional logic with weight constraints (amsolver). Eight teams used the grounder *gringo* and the benchmark solutions available at the Asparagus system [23].

Table 2. Participating teams and systems

Team	Affiliation	Language	Systems
IDP	K.U. Leuven, KRR	FO(·)	*idp* (*gidl* + *minisatid*)
Potassco	U. of Potsdam	ASP	*clasp, claspd, gringo, clingo, iclingo, clingcon, bingo*
DLV	U. of Calabria	ASP	*dlv*
Claspfolio	U. of Potsdam	ASP	*gringo* + *clasp*
Smodels-IE	U. of Bath	ASP	*gringo* + *smodelsie*
ASPeRiX	U. of Angers	ASP	*asperix*
CMODELS	U. of Texas at Austin	ASP	*gringo* + *cmodels*
SUP	U. of Texas at Austin	ASP	*gringo* + *sup*
BPSolver-CLP(FD)	International B-Prolog team	CLP(FD)	*bprolog* (tabling, CLP(FD), B$_{mv}^{fd}$)
Enfragmo	Simon Fraser U., Computational Logic Laboratory	FO(·)	*enfragmo* (grounder + SAT solver)
LP2DIFF+BCLT	Helsinki U. of Technology (TKK)	ASP	*gringo* + *smodels* + *lp2diff* + *bclt*
LP2SAT+MINISAT	Helsinki U. of Technology (TKK)	ASP	*gringo* + *smodels* + *lp2exp* + *minisat*
LP2DIFF+YICES	Helsinki U. of Technology (TKK)	ASP	*gringo* + *smodels* + *lp2diff* + *yices*
pbmodels	U. of Kentucky, U. of Texas at Tyler, Microsoft	ASP	*pbmodels* (uses *minisat+*)
sabe	U. of Kentucky, U. of Texas at Tyler, Microsoft	ASP	*sabe* (uses *minisat*)
amsolver	U. of Kentucky, U. of Texas at Tyler, Microsoft	FO(·)	*amsolver*

Table 3. Submitted benchmark solutions per team

Team	HydraulicPlanning	HydraulicLeaking	CompanyControls	GrammarBasedInformationExtraction	Reachability	BlockedNQueens	Sokoban	15Puzzle	HamiltonianPath	SchurNumbers	TravellingSalesperson	WeightBoundedDominatingSet	Labyrinth	GeneralizedSlitherlink	HierarchicalClustering	ConnectedDominatingSet	GraphPartitioning	Hanoi	Fastfood	WireRouting	Sudoku	DisjunctiveScheduling	KnightTour	ChannelRouting	EdgeMatching	GraphColouring	MazeGeneration	Solitaire	StrategicCompanies	GolombRuler	MaximalClique	15PuzzleOptimize	TravellingSalespersonOptimize	WeightBoundedDomSetOptimize	LabyrinthOptimize	SokobanOptimize	FastfoodOptimize	CompanyControlsOptimize
	Decision in P					Decision in NP																							Σ_2^p	Optimization								
IDP	y	y	n	y	y	y	y	y	y	y	y	y	y	y	y	y	y	y	y	y	y	y	y	y	y	y	y	y	n	y	y	y	y	y	y	y	y	n
Potassco	y	y	y	y	y	y	y	y	y	y	y	y	y	y	y	y	y	y	y	y	y	y	y	y	y	y	y	y	y	y	y	y	y	y	y	y	y	y
DLV	y	y	y	y	y	y	y	y	y	y	y	y	y	y	y	y	y	y	y	y	y	y	y	y	y	y	y	y	y	y	y	y	y	y	y	y	y	y
Claspfolio	y	y	y	y	y	y	y	y	y	y	y	y	y	y	y	y	y	y	y	y	y	y	y	y	y	y	y	y	n	y	y	y	y	y	y	y	y	y
Smodels-IE	y	y	y	n	y	y	y	y	y	y	y	y	y	y	y	y	y	y	y	y	y	y	y	y	y	y	y	y	n	y	y	y	y	y	y	y	y	y
ASPeRiX	y	y	n	y	y	y	y	y	y	n	y	n	n	n	n	n	n	n	n	y	n	n	y	y	n	y	n	n	n	n	n	n	n	n	n	n	n	n
CMODELS	y	y	y	y	y	y	y	y	y	y	y	y	y	y	y	y	y	y	y	y	y	y	y	y	y	y	y	y	y	n	n	n	n	n	n	n	n	n
SUP	y	y	y	y	y	y	y	y	y	y	y	y	y	y	y	y	y	y	y	y	y	y	y	y	y	y	y	y	n	n	n	n	n	n	n	n	n	n
BPSolver-CLP(FD)	y	y	y	y	y	y	n	y	y	y	y	y	n	y	y	y	y	y	y	y	y	y	y	y	y	y	y	y	y	y	y	y	y	y	n	n	y	y
Enfragmo	y	y	y	y	y	y	n	y	y	y	y	y	n	y	y	y	y	y	y	n	y	y	y	n	y	y	n	y	n	n	y	n	n	n	n	n	n	n
LP2DIFF+BCLT	y	y	y	n	y	y	y	y	y	y	y	y	y	y	y	y	y	y	y	y	y	y	y	y	y	y	y	y	n	n	n	n	n	n	n	n	n	n
LP2SAT+MINISAT	y	y	y	n	y	y	y	y	y	y	y	y	y	y	y	y	y	y	y	y	y	y	y	y	y	y	y	y	n	n	n	n	n	n	n	n	n	n
LP2DIFF+YICES	y	y	y	n	y	y	y	y	y	y	y	y	y	y	y	y	y	y	y	y	y	y	y	y	y	y	y	y	n	n	n	n	n	n	n	n	n	n
pbmodels	y	y	n	n	y	y	y	y	y	y	y	n	y	y	y	y	y	y	y	y	y	y	y	n	y	y	y	y	n	y	n	n	n	n	n	y	n	n
sabe	y	y	n	n	y	y	y	y	y	y	y	n	y	y	y	y	y	y	y	y	y	n	y	y	y	y	y	y	n	y	n	y	n	n	n	y	n	n
amsolver	n	n	n	n	n	y	n	n	y	y	y	n	n	n	y	y	n	n	n	n	y	n	n	n	n	y	n	n	n	n	n	n	n	n	n	n	n	n

The teams and their systems are summarized in Table 2. Many teams did not participate on all benchmarks. Table 3 specifies the benchmarks on which teams competed.

5   Results

This section presents the results of the *Second Answer Set Programming System Competition*. For each category of decision problems, we report the score of each team, the number of solved instances, and the total time. For the optimization category, we report the score and total time per participating team. In the rankings, distinction is made between single-system and multi-system teams. The latter used multiple systems/parameter settings for different benchmarks and are marked by *. More statistics and details are available at [24].

Before actually giving the results, we would like to warn the reader to be careful in interpreting them. Below, we point out the most important issues:

– The score is a **weighted sum** of numbers of solved instances per benchmark. Solving an instance of a benchmark with a large number of instances has a smaller contribution than solving one of a benchmark with fewer instances. Thus, it is possible that one team solves more instances but has a smaller score than another team. The weights were introduced to prevent benchmarks with many instances from dominating the competition.

- Most teams did not participate on all benchmarks. The only teams that participated on all benchmarks are Potassco and DLV. When a benchmark solution was missing, a team was assigned score 0 and 600 seconds time per instance. Consequently, the rankings of teams for which benchmark solutions were missing may not give an accurate account of the quality of their systems. We refer to [24] for detailed data per benchmark.
- Fine-tuning a benchmark solution (for instance, by adding certain redundant constraints) may have a major impact on speed. Not all teams were in a position to spend the same amount of time and care on this, which complicates an objective comparison between different solvers. This factor is not significant among the teams that used the encodings available at [23]: Potassco, Claspfolio, CMODELS, SUP, Smodels-IE, LP2DIFF+BCLT, LP2SAT+MINISAT and LP2DIFF+YICES.

5.1 Decision Problems: NP

Benchmarks in this track belong to NP and are not known to be in P. These problems require that solvers search quickly through large search spaces.

The winners in this category are:

FIRST PLACE WINNER	Potassco*	
SECOND PLACE WINNER	Claspfolio	FIRST SINGLE-SYSTEM TEAM
THIRD PLACE WINNER	CMODELS	SECOND SINGLE-SYSTEM TEAM
	IDP	THIRD SINGLE-SYSTEM TEAM

The ranking of all teams is provided in Figure 1. Figure 2 gives a comprehensive graphical overview of the results of all systems. The x-axis represents the number of (solved) benchmark instances, and the y-axis represents the maximum time needed for solving one of these. To compute this plot, the instances solved by each team were ordered according to running times, and a point (x, y) in the chart expresses that the xth instance was solved in y seconds. The more to the right the curve of a team ends, the more benchmark instances were solved within the allocated time and space.

Place	Team	Score	#Solved	Time
1	Potassco*	0.97	491 / 516 = 95%	021253
2	Claspfolio	0.89	451 / 516 = 87%	049513
3	CMODELS	0.85	434 / 516 = 84%	072283
4	IDP	0.83	409 / 516 = 79%	077428
5	LP2SAT+MINISAT	0.82	430 / 516 = 83%	075883
6	SUP	0.80	405 / 516 = 78%	083248
7	DLV	0.76	391 / 516 = 75%	100496
8	LP2DIFF+BCLT	0.73	378 / 516 = 73%	108715
9	LP2DIFF+YICES	0.72	373 / 516 = 72%	096989
10	Smodels-IE	0.61	309 / 516 = 59%	137300
11	Enfragmo	0.59	291 / 516 = 56%	156298
12	BPSolver-CLP(FD)*	0.57	274 / 516 = 53%	155559
13	pbmodels	0.44	214 / 516 = 41%	201563
14	sabe	0.40	203 / 516 = 39%	215250
15	amsolver	0.12	83 / 516 = 16%	265833
16	ASPeRiX	0.12	32 / 516 = 06%	293363

Fig. 1. Decision problems in NP: Ranking   **Fig. 2.** Decision problems in NP: Plot

Place	Team	Score	#Solved	Time
1	Potassco*	1.00	89 / 89 = 100%	00735
2	BPSolver-CLP(FD)*	1.00	89 / 89 = 100%	01342
3	DLV	1.00	89 / 89 = 100%	04861
4	Claspfolio	0.80	60 / 89 = 67%	17982
5	Smodels-IE	0.80	60 / 89 = 67%	18021
6	LP2SAT+MINISAT	0.80	60 / 89 = 67%	18270
7	SUP	0.80	60 / 89 = 67%	18606
8	LP2DIFF+BCLT	0.80	60 / 89 = 67%	18713
9	CMODELS	0.80	60 / 89 = 67%	19072
10	LP2DIFF+YICES	0.78	59 / 89 = 66%	18864
11	Enfragmo	0.76	57 / 89 = 64%	24157
12	ASPeRiX	0.69	66 / 89 = 74%	18051
13	IDP	0.54	41 / 89 = 46%	29594
14	sabe	0.41	31 / 89 = 34%	36426
15	pbmodels	0.38	29 / 89 = 32%	36656
16	amsolver	0.00	00 / 89 = 0%	53845

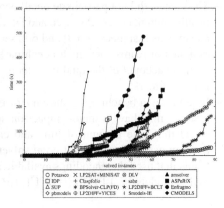

Fig. 3. Decision problems in P: Ranking   **Fig. 4.** Decision problems in P: Plot

5.2 Decision Problems: P

Problems in this class are polynomially solvable. Difficulty in solving them stems from the sheer size of the instances, which grounders may not be able to handle.

The winners in this category are:

FIRST PLACE WINNER   Potassco*
SECOND PLACE WINNER   BPSolver-CLP(FD)*
THIRD PLACE WINNER   DLV   FIRST SINGLE-SYSTEM TEAM
                      Claspfolio   SECOND SINGLE-SYST. TEAM
                      Smodels-IE   THIRD SINGLE-SYSTEM TEAM

The results for all teams are presented in Figure 3 and in Figure 4.

5.3 Decision Problems: Global

This track consists of all previous decision problems and one Σ_2^p problem, the well-known Strategic Companies problem.

The winners in this category are:

FIRST PLACE WINNER   Potassco*
SECOND PLACE WINNER   Claspfolio   FIRST SINGLE-SYSTEM TEAM
THIRD PLACE WINNER   CMODELS   SECOND SINGLE-SYSTEM TEAM
                      DLV   THIRD SINGLE-SYSTEM TEAM

The global decision problem results are provided in Figure 5 and in Figure 6.

5.4 Optimization Problems

The competition included nine optimization problems. Most are optimization versions of decision benchmarks, with the exception of Golomb Ruler and Maximal Clique. Only nine of the sixteen teams submitted solutions to optimization problems.

The winners in this category are:

Place	Team	Score	#Solved	Time
1	Potassco*	0.95	585 / 622 = 94%	029607
2	Claspfolio	0.84	511 / 622 = 82%	077780
3	CMODELS	0.82	498 / 622 = 80%	099721
4	DLV	0.81	497 / 622 = 79%	108448
5	LP2SAT+MINISAT	0.79	490 / 622 = 78%	104438
6	SUP	0.77	465 / 622 = 74%	112641
7	IDP	0.75	450 / 622 = 72%	117223
8	LP2DIFF+BCLT	0.72	438 / 622 = 70%	137713
9	LP2DIFF+YICES	0.70	432 / 622 = 69%	126138
10	BPSolver-CLP(FD)*	0.63	365 / 622 = 58%	165902
11	Smodels-IE	0.62	369 / 622 = 59%	165607
12	Enfragmo	0.60	348 / 622 = 55%	190741
13	pbmodels	0.42	243 / 622 = 39%	248505
14	sabe	0.39	234 / 622 = 37%	261961
15	ASPeRiX	0.21	98 / 622 = 15%	321700
16	amsolver	0.10	83 / 622 = 13%	329963

Fig. 5. Decision problems globally: Ranking   **Fig. 6.** Decision problems globally: Plot

FIRST PLACE WINNER	Potassco*
SECOND PLACE WINNER	Claspfolio
THIRD PLACE WINNER	DLV
	IDP

FIRST SINGLE-SYSTEM TEAM
SECOND SINGLE-SYSTEM TEAM
THIRD SINGLE-SYSTEM TEAM

The results for the participating teams are presented in Figure 7 and in Figure 8.

5.5   Decision and Optimization Problems: Global

The goal of this track is to select the systems with widest applicability. Scores are obtained as the average of the scores in the global decision and optimization categories to give decision and optimization problems the same importance.

The winners in this category are:

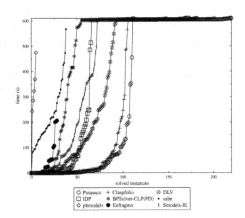

Place	Team	Score	Time
1	Potassco*	81.12	74317
2	Claspfolio	69.61	78333
3	DLV	61.04	92889
4	IDP	50.88	101081
5	Smodels-IE	49.88	103176
6	BPSolver-CLP(FD)*	35.8	113551
7	sabe	6.74	122848
8	Enfragmo	5.07	121598
9	pbmodels	1.19	135883

Fig. 7. Optimization problems: Ranking   **Fig. 8.** Optimization problems: Plot

FIRST PLACE WINNER	Potassco*	
SECOND PLACE WINNER	Claspfolio	FIRST SINGLE-SYSTEM TEAM
THIRD PLACE WINNER	DLV	SECOND SINGLE-SYSTEM TEAM
	IDP	THIRD SINGLE-SYSTEM TEAM

The combined results for decision and optimization problems are shown in Figure 9 and in Figure 10.

Place	Team	Score	Time
1	Potassco*	0.88	103925
2	Claspfolio	0.77	156113
3	DLV	0.71	201338
4	IDP	0.63	218304
5	Smodels-IE	0.56	268783
6	BPSolver-CLP(FD)*	0.49	279453
7	CMODELS	0.41	237661
8	LP2SAT+MINISAT	0.39	242378
9	SUP	0.38	250581
10	LP2DIFF+BCLT	0.36	275653
11	LP2DIFF+YICES	0.35	264078
12	Enfragmo	0.32	312339
13	sabe	0.23	384810
14	pbmodels	0.21	384388
15	ASPeRiX	0.10	459640
16	amsolver	0.05	467903

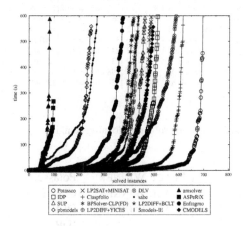

Fig. 9. Global category: Ranking   **Fig. 10.** Global category: Plot

5.6 Summary of Results

The Potassco team of the University of Potsdam is the clear winner of the competition. The team won in every category, and in all but the P track with a margin of around 10%. Potassco won twenty of the thirty-eight benchmarks. This is the result of a large effort for developing an excellent library of systems (*clasp, claspd, gringo, clingo, iclingo, clingcon, bingo*) and intensive work of an experienced team on benchmark solutions and parameter tuning. We congratulate the team on this success!

Potassco spent by far the most effort in fine-tuning their systems to each benchmark, and this paid off. Given that the goal of declarative problem solving is to minimize the effort of programmers, it is of equal interest to investigate the performance of teams that used a single collection of systems with uniform parameter settings.

In all rankings except for the P decision track, the best single-system was Claspfolio[2]. Claspfolio ran *gringo* and the *clasp* solver whose settings were chosen from instance features. On decision problems in P, the best team using a single setting for all benchmarks is DLV. BPSolver-CLP(FD) also performed excellently on P problems.

As for the winners of individual benchmarks, we already mentioned that Potassco won twenty. Five benchmarks were won by BPSolver-CLP(FD), and four by DLV. Claspfolio and Smodels-IE each won two, and IDP, Enfragmo and amsolver one each[3].

---

[2] In fact, Claspfolio, just like CMODELS and SUP, used a different grounder for one benchmark. We can ignore this here because they had a zero score for this benchmark.

[3] This list does not include the multiple ex aequo winners of HydraulicLeaking and Hydraulic-Planning.

6 Discussion

The principal goals of this competition, namely, taking a snapshot of the state of the art of declarative programming paradigms and fostering future improvements, are similar to related competitions on SAT, PB, QBF, SMT, CP, etc. However, in contrast to those and the first ASP competition held in 2007, the form of this competition, aiming at openness towards alternative paradigms, was quite different. This manifests itself in the fact that participants were allowed (and actually required) to provide their individual modelings for the benchmarks used in the competition. In contrast to the other competitions mentioned, the inputs to systems run in this competition were not fixed by the organizers, except for the (arbitrarily chosen) format of problem instances. As a consequence, the results of this competition may indicate trends on the simplicity or difficulty of developing effective problem solutions using particular systems, but they cannot provide a perfect picture of the efficiency of the systems themselves.

Several SAT, PB and SMT systems were involved in the competition, as back-ends of ASP or FO($\cdot$) systems. Techniques from these areas are also applied in "native" ASP solvers like those used by Potassco and Claspfolio. That there were no teams from these areas in the competition, and only one team from CLP, may have different explanations. These fields have their own, well-established competitions, while the ASP competition is relatively recent and open to them only for the first time. Another explanation is that the difficulty of modeling the benchmark problems was very high for them. We know of one SAT team that considered to participate in the competition, but gave up because of this reason. Despite the flexibility of CLP(FD) in modeling constraint problems, BPSolver-CLP(FD) had difficulties in modeling certain benchmarks and did not submit solutions for all of the planning problems. ASP and FO($\cdot$) appear to offer superior modeling facilities, which is hardly surprising given that these languages were developed for knowledge representation. On the other hand, BPSolver-CLP(FD) came in second in the P track and won on three benchmarks of the NP track. This shows that tabling as well as the constraint programming techniques featured in B-Prolog can be very useful for some kinds of problems, in particular, those in which large domains would make exhaustive grounding blow up in space. Other new entrants in the competition were the FO-based systems IDP, Enfragmo and amsolver. IDP ended fourth in the NP and global track but was less successful for P decision problems. Enfragmo and amsolver performed very well on certain benchmarks but did not compete in enough benchmarks to obtain a good ranking.

In their invited talks at LPNMR 2007, both Nicola Leone[4] and Jack Minker[5] appealed for using real-world application problems in the ASP competition. Although the benchmarks of the current contest covered a variety of different modeling or computational aspects, only a few benchmarks came from such applications. This issue was discussed with several contributors of benchmarks. The problem is not that there are no real-world applications. In fact, the contrary is shown in the application summary track of LPNMR 2009. But such real-world problems tend to be very complex, making it harder for a contributor to describe the problem in an informal yet unambiguous way

---

[4] http://lpnmr2007.googlepages.com/nicola-lpnmr07.pdf
[5] http://lpnmr2007.googlepages.com/LPNMR-07.ppt

and to create suitable problem instances. Moreover, the effort of modeling such problems may become too high for some contestants. This problem is inherent to a Model and Solve competition and does not occur in competitions where the formal theory is given, as in the SAT competition or the categories SCore, SCore-v and SLparse of the first ASP competition.

As motivated in Section 2, teams were free to fine-tune their solving systems towards particular benchmarks. To this end, they could use a number of instances that were available during the installation phase. Two teams effectively fine-tuned their systems in this way. BPSolver-CLP(FD) may have no option than to do so, since the programmer needs to specify the search strategy in B-Prolog. Potassco took the opportunity to test its library of systems and system parameters for controling preprocessing, heuristics and restarts[6]. In the previous section, we therefore distinguished these teams from the other single-system teams. Given that Claspfolio and Potassco used the same benchmark solutions and mostly the same technology, the competition gives a fairly accurate account of the impact of Potassco's effort on fine-tuning. Globally, Claspfolio lost 10% on Potassco and it was outperformed by Potassco in a few benchmarks. On the one hand, this shows that in the current state of the art, fine-tuning pays off and may be imperative to build hard real-world applications. On the other hand, the long term goal of declarative problem solving is to allow a programmer to focus on the declarative properties of the problem and to relieve him or her of tedious control issues. The example of Potassco and Claspfolio allows us to evaluate our current progress towards this goal. In this respect, it is encouraging to see that globally, Claspfolio lost by *only* 10%.

We would like to end with some reflections and recommendations on the competition format. We believe that an open model and solve competition like this one fosters cross-fertilization between different areas of declarative programming and gives valuable global information on the quality of modeling and solving technologies. On the other hand, it does not allow for a precise and unbiased comparison of system performance, due to the use of different encodings. To allow for more detailed and objective comparisons, separate competitions are needed in which problem encodings are given and fixed. This is the format used in the SAT competition and also in the SCore, SCore-v and SLparse tracks of the previous ASP competition. In future competitions, it would be of interest to have such tracks for both grounders and solvers. Such tracks will not be accessible for areas that do not rely on grounding and solving. A further prerequisite is the availability of common input languages. As regards a grounder competition, a common high-level fragment of ASP and FO($\cdot$) is currently lacking and should be strived for. For an unbiased comparison of solvers, some low-level language similar to DIMACS for SAT, Lparse for ASP or MNF for FO($\cdot$) would need to be selected.

A final recommendation concerns the participation of SAT teams or systems in the competition. SAT technology is heavily used in ASP and FO($\cdot$). The model and solve track would be a good occasion to compare different SAT systems in the context of typical ASP benchmarks. Participation of a SAT team can be made very easy by providing a grounder to DIMACS, benchmark solutions in the language of the grounder and a script translating SAT output into the competition format. The grounders of Enfragmo and LP2SAT+MINISAT could be used for this purpose.

---

[6] For details, see Potassco's team webpage [24].

Acknowledgments

A great number of people have contributed to the success of this competition. We thank the members of the program committee who helped to set up the competition format. We are indebted to all members of the ASP and CLP community that submitted benchmarks. We are extremely grateful to everyone that participated in the contest. Thank you for your enthusiasm and your encouragements.

We wish to thank the DTAI group of the K.U. Leuven for providing the pool of computers on which the competition was run. Special thanks go to Pieter Wuille for developing and managing the competition software, and to Bart Swennen and Kris Vangeneugden for setting up the computer infrastructure. We are grateful also to Hanne Vlaeminck and Maarten Mariën, who together with the first three authors formed the local organization team. The KRR group participated in the competition with the IDP team, mostly through the efforts of Johan Wittocx who did 90% of the work. The group thanks him for this.

On behalf of the program committee and the participants, Martin Gebser and Mirosław Truszczyński would like to thank the KRR group of the K.U. Leuven for the huge efforts taken to prepare and to conduct this competition. Their excellent work was invaluable to make this competition, first, possible and, second, a great success for the ASP community in the most general sense, in view of presenting and representing a broad spectrum of alternative yet tightly bonded modeling and solving approaches.

Martin Gebser acknowledges partial funding by DFG under Grant SCHA 550/8-1.

References

1. Marek, V., Truszczyński, M.: Stable models and an alternative logic programming paradigm. In: Apt, K., Marek, W., Truszczyński, M., Warren, D. (eds.) The Logic Programming Paradigm: a 25-Year Perspective, pp. 375–398. Springer, Heidelberg (1999)
2. Niemelä, I.: Logic programming with stable model semantics as a constraint programming paradigm. Annals of Mathematics and Artificial Intelligence 25(3-4), 241–273 (1999)
3. Baral, C.: Knowledge Representation, Reasoning and Declarative Problem Solving. Cambridge University Press, Cambridge (2003)
4. Gelfond, M., Lifschitz, V.: Classical negation in logic programs and disjunctive databases. New Generation Computing 9(3-4), 365–385 (1991)
5. Colmerauer, A., Kanoui, H., Pasero, R., Roussel, P.: Un systeme de communication homme-machine en Francais. Technical report, University of Marseille (1973)
6. Kowalski, R.: Predicate logic as a programming language. In: Rosenfeld, J. (ed.) Proceedings of the Congress of the International Federation for Information Processing, pp. 569–574. North Holland, Amsterdam (1974)
7. McCarthy, J.: Circumscription — a form of nonmonotonic reasoning. Artificial Intelligence 13(1-2), 27–39 (1980)
8. Reiter, R.: A logic for default reasoning. Artificial Intelligence 13(1-2), 81–132 (1980)
9. Gelfond, M., Lifschitz, V.: The stable model semantics for logic programming. In: Kowalski, R., Bowen, K. (eds.) Proceedings of the International Conference on Logic Programming, pp. 1070–1080. MIT Press, Cambridge (1988)
10. Eiter, T., Leone, N., Mateis, C., Pfeifer, G., Scarcello, F.: A deductive system for nonmonotonic reasoning. In: Dix, J., Furbach, U., Nerode, A. (eds.) Proceedings of the International Conference on Logic Programming and Nonmonotonic Reasoning, pp. 364–375. Springer, Heidelberg (1997)

11. Niemelä, I., Simons, P.: Smodels — an implementation of the stable model and well-founded semantics for normal logic programs. In: Dix, J., Furbach, U., Nerode, A. (eds.) Proceedings of the International Conference on Logic Programming and Nonmonotonic Reasoning, pp. 420–429. Springer, Heidelberg (1997)

12. Lifschitz, V.: Answer Set Planning. In: De Schreye, D. (ed.) Proceedings of the International Conference on Logic Programming, pp. 23–37. MIT Press, Cambridge (1999)

13. Borchert, P., Anger, C., Schaub, T., Truszczyński, M.: Towards Systematic Benchmarking in Answer Set Programming: The Dagstuhl Initiative. In: Lifschitz, V., Niemelä, I. (eds.) LPNMR 2004. LNCS (LNAI), vol. 2923, pp. 3–7. Springer, Heidelberg (2004)

14. Gebser, M., Liu, L., Namasivayam, G., Neumann, A., Schaub, T., Truszczyński, M.: The first answer set programming system competition. In: Baral, C., Brewka, G., Schlipf, J. (eds.) LPNMR 2007. LNCS (LNAI), vol. 4483, pp. 3–17. Springer, Heidelberg (2007)

15. Biere, A., Heule, M., van Maaren, H., Walsh, T. (eds.): Handbook of Satisfiability. IOS Press, Amsterdam (2009)

16. Nieuwenhuis, R., Oliveras, A., Tinelli, C.: Solving SAT and SAT modulo theories: From an abstract Davis-Putnam-Logemann-Loveland procedure to DPLL(T). Journal of the ACM 53(6), 937–977 (2006)

17. Rossi, F., van Beek, P., Walsh, T. (eds.): Handbook of Constraint Programming. Elsevier, Amsterdam (2006)

18. Denecker, M., Kakas, A.C.: Abduction in Logic Programming. In: Kakas, A.C., Sadri, F. (eds.) Computational Logic: Logic Programming and Beyond. LNCS (LNAI), vol. 2407, pp. 402–436. Springer, Heidelberg (2002)

19. Mitchell, D., Ternovska, E.: A framework for representing and solving NP search problems. In: Veloso, M., Kambhampati, S. (eds.) Proceedings of the National Conference on Artificial Intelligence, pp. 430–435. AAAI Press / MIT Press (2005)

20. Kakas, A., Michael, A.: Air-Crew scheduling through abduction. In: Imam, I., Kodratoff, Y., El-Dessouki, A., Ali, M. (eds.) Proceedings of the International Conference on Industrial and Engineering Applications of Artificial Intelligence and Expert Systems, pp. 600–611. Springer, Heidelberg (1999)

21. Pelov, N., De Mot, E., Denecker, M.: Logic programming approaches for representing and solving constraint satisfaction problems: A comparison. In: Parigot, M., Voronkov, A. (eds.) LPAR 2000. LNCS (LNAI), vol. 1955, pp. 225–239. Springer, Heidelberg (2000)

22. Van Hentenryck, P.: Constraint Satisfaction in Logic Programming. MIT Press, Cambridge (1989)

23. http://asparagus.cs.uni-potsdam.de

24. http://www.cs.kuleuven.be/~dtai/events/ASP-competition

Author Index